Science of Olfaction

Michael J. Serby
Karen L. Chobor
Editors

Science of Olfaction

With 99 Illustrations in 119 Parts

Springer-Verlag
New York Berlin Heidelberg London Paris
Tokyo Hong Kong Barcelona Budapest

Michael J. Serby
Department of Psychiatry
The Mount Sinai Medical Center
New York, New York 10029
USA

Karen L. Chobor
Department of Neurology
New York University Medical Center
New York, New York 10016
USA

Library of Congress Cataloging-in-Publication Data

Science of olfaction/Michael J. Serby, Karen L. Chobor, editors.
p. cm.
Includes bibliographical references and index.
ISBN-13: 978-1-4612-7690-6 e-ISBN-13: 978-1-4612-2836-3
DOI: 10.1007/978-1-4612-2836-3
1. Smell. I. Serby, Michael J. II. Chobor, Karen L.
QP458.S42 1992
152.1'66—dc20 91-33812

Printed on acid-free paper.

Production coordinated by Chernow Editorial Services, Inc. and managed by Linda H. Hwang; manufacturing supervised by Jacqui Ashri.
Typeset by Asco Trade Typesetting Limited, Hong Kong.

9 8 7 6 5 4 3 2 1

For Susan, Adam, Jed, and Benjamin, with love.

M.J.S.

For my parents, John and Mary, with loving gratitude.

K.L.C.

"I should think we might fairly gauge the future of biological science, centuries ahead, by estimating the time it will take to reach a complete, comprehensive understanding of odor. It may not seem a profound enough problem to dominate all the life sciences, but it contains, piece by piece, all the mysteries."

Lewis Thomas (1980)

Preface

A book such as this one necessarily entails the effort of many individuals. Its first seed was a workshop in 1988 that one of us (K.L.C.) designed and directed at the City University of New York Graduate Center as part of a series of conferences and workshops on topics in neuropsychology sponsored by the Institute for Research in Behavioral Neuroscience (IRBN). At that time, studies in the efficacy of olfaction as an alternative modality of stimulation in brain-injured adults were taking place at IRBN. The workshop helped to spark many new ideas among IRBN's staff, leading to various projects in clinical intervention, some of which are discussed in this volume. We soon discovered that clinical interest in olfaction was steadily increasing, and that this modality was in many ways on a par with the rest. Along with progress in the clinical investigation of olfaction, an abundance of work has been published in the basic science of the field. Ideas such as the regeneration of olfactory neurons and the use of these neurons in transplantation took hold and were proved to be viable ones. These developments along with a profusion of techniques that allow a closer look at the anatomy and physiology of olfaction, have contributed to the accelerated pace with which this field is now advancing. Dr. Jason Brown encouraged us to seek out investigators who could contribute to a comprehensive volume that would bridge the gap between basic science and clinical work in this area. We offer sincere thanks to him for getting us started.

Olfaction now holds the interest of a wide range of disciplines, many of which are represented in this book. Each of the contributors has added ideas for solutions to the puzzle of brain function, and we congratulate them for their eagerness to share their ideas in such a unique forum. We thank them for their hard work and patience in seeing this through to fruition.

M.J.S. wishes to thank the patients and families who have been so cooperative; a number of essential research assistants, including Bart Rypma, Nicole Etienne, Davina Kalkstein, and Pamela Larson; Dr. Michael Shipley, who was instrumental in suggesting and recruiting a number of authors; and Dr. Susan Katz for love and encouragement.

K.L.C. would like to thank Chuck Snyder, Dr. Victor Mark and Dr. Judith Jaeger, all of whom encouraged a step off the beaten track. She would also like to thank Bartholomew Lawson for his inspiring curiosity and loving support.

Both of us would like to thank Frank Zinn, Beth Levine, and Rosemarie Basile for compiling the subject index, and those at Springer-Verlag involved in the production of this book.

Michael J. Serby
Karen L. Chobor

Contents

Contributors

Robert R.H. Anholt, PhD, Department of Neurobiology, Box 3209, Duke University Medical Center, Durham, North Carolina 27710, USA

Linda Bastone, PhD, Department of Psychology, CUNY Graduate Center, 33 West 42nd Street, The City University of New York, New York, New York 10036, USA

Gary K. Beauchamp, PhD, Monell Chemical Senses Center, 3500 Market Street, Philadelphia, Pennsylvania 19104, USA

Irving Bieber, MD*, 132 East 72nd Street, New York, New York, 10021, USA

Tony B. Bieber, PhD, 132 East 72nd Street, New York, New York, 10021, USA

William S. Cain, PhD, John B. Peirce Laboratory, 290 Congress Avenue, Yale University, New Haven, Connecticut 06519, USA

Karen L. Chobor, MS, Department of Neurology, New York University Medical Center, New York, New York 10016, USA

J. Enrique Cometto-Muñiz, PhD, Member of the Carrera del Investigador Cientifico, Consejo Nacional de Investigaciones Cientificas y Técnicas, República Argentina

June Corwin, PhD, The Johns Hopkins School of Medicine, Division of Psychology, Mayor 218, 600 North Wolfe Street, Baltimore, Maryland 21205, USA

Richard M. Costanzo, PhD, Department of Physiology, Medical College of Virginia, Virginia Commonwealth University, Richmond, Virginia 23298-0551, USA

---

*Deceased.

Kathleen M. Dorries, BA, Department of Psychology, Uris Hall, Cornell University, Ithaca, New York 14853, USA

Heather J. Duncan, PhD, University of Cincinnati Taste and Smell Center, Deparment of Otolaryngology and Maxillofacial Surgery, 231 Bethesda Avenue, University of Cincinnati College of Medicine, Cincinnati, Ohio 45267-0528, USA

Howard Ehrlichman, PhD, Department of Psychology, CUNY Graduate Center, 33 West 42nd Street, The City University of New York, New York, New York 10036, USA

Howard Eichenbaum, PhD, Department of Biological Sciences, Wellesley College, Wellesley, Massachusetts 02181, USA

D.L. Flint, PhD, Department of Psychology, University of New Hampshire, Durham, New Hampshire 03824, USA

Richard C. Friedman, MD, 225 Central Park West, New York, New York 10024, USA

Avery N. Gilbert, PhD, Director of Olfactory Science, Roure, Inc., 1775 Windsor Road, Teaneck, New Jersey 07666, USA

Mark S. Greenberg, PhD, Department of Psychiatry, Cambridge Hospital and Harvard Medical School, Cambridge, Massachusetts 02138, USA

Anne Harrington, PhD, Department of The History of Science, Science Center 235, Harvard University, Cambridge, Massachusetts 02138, USA

Davina Kalkstein, BA, Aging and Dementia Research Program, New York University, New York, New York 10016, USA

Pamela M. Larson, MA, Department of Psychiatry, The Mount Sinai Medical Center, One Gustave L. Levy Place, New York, New York 10029, USA

R.G. Mair, PhD, Department of Psychology, University of New Hampshire, Durham, New Hampshire 03824, USA

John H. McLean, PhD, Division of Basic Medical Sciences, Memorial University Faculty of Medicine, St John's Newfoundland A1B 3V6, Canada

Edward E. Morrison, PhD, Department of Physiology, Medical College of Virginia, Virginia Commonwealth University, Richmond, Virginia 23298-0551, USA. Present address: Department of Anatomy and Histology, School of Veterinary Medicine, 109 Greene Hall, Auburn University, Auburn, Alabama 36849-5518, USA

W.T. Nickell, PhD, Department of Anatomy and Cell Biology, Division of Neurobiology, University of Cincinnati College of Medicine, Cincinnati, Ohio 45267-0521, USA

Tim Otto, PhD, Department of Biological Sciences, Wellesley College, Wellesley, Massachusetts 02181, USA. Present address: Department of Psychology, University of North Carolina, Chapel Hill, North Carolina 27599, USA

Vernon Rosario, PhD, Department of The History of Science, Science Center 235, Harvard University, Cambridge, Massachusetts 02138, USA

Susan S. Schiffman, PhD, Medical Psychology, Duke University Medical Center, Durham, North Carolina 27706, USA

Hilary J. Schmidt, PhD, 185 E. Orange Avenue, University of Medicine and Dentistry of New Jersey, Newark, New Jersey 07103, USA

James E. Schwob, PhD, Department of Anatomy and Cell Biology and the Clinical Olfactory Research Center, SUNY Health Science Center, Syracuse, New York 13210, USA

Michael J. Serby, MD, Department of Psychiatry, The Mount Sinai Medical Center, One Gustave L. Levy Place, Box 1230, New York, New York 10029, USA

Michael T. Shipley, PhD, Department of Anatomy and Cell Biology, Division of Neurobiology, University of Cincinnati College of Medicine, Cincinnati, Ohio 45267-0521, USA

David V. Smith, PhD, Department of Otolaryngology and Maxillofacial Surgery, University of Cincinnati Taste and Smell Center, 231 Bethesda Avenue, University of Cincinnati College of Medicine, Cincinnati, Ohio 45267-0528, USA

John D. Ward, PhD, Division of Neurosurgery, Medical College of Virginia, Virginia Commonwealth University, Richmond, Virginia 23298-0551, USA

René A. de Wijk, PhD, John B. Pierce Laboratory, 290 Congress Avenue, Yale University, New Haven, Connecticut 06519, USA

Harold F. Young, PhD, Division of Neurosurgery, Medical College of Virginia, Virginia Commonwealth University, Richmond, Virginia 23298-0551, USA

Introduction

We have edited this book in an effort to present the current state of knowledge about a particular sensation, the sense of smell, and its relationship to the brain. The study of olfactory physiology and clinical function has lagged behind research in the auditory and visual domains, but this regrettable state of affairs has been changing in recent years. The sense of smell is increasingly becoming an easily measured phenomenon that provides a window to the central nervous system. It is unique in several ways, including the fact that olfactory sensory neurons are the only regenerating neurons in the adult mammal. The fairly direct access of external stimuli to the brain via olfactory circuits is another important feature. These facts, along with its idiosyncratic observable behavior, set olfaction apart from the commonly explored modalities in a way that will be understood only through the efforts of multidisciplinary investigations. This book details advances in basic science and clinical research that, taken together, accentuate the vital interaction of olfactory and brain function.

Section One presents an historical perspective. Chapter 1 by Harrington and Rosario focuses on nineteenth-century scientific, pseudo-scientific, and literary thought. Efforts to understand olfactory phenomena were plagued by ambivalence and dogma, and the "primitive" sense of smell was both trivialized and romanticized. Viewpoints frequently reflected purely prejudicial judgments. At the same time, however, the earliest trends of our current research can be traced. For example, there was interest in the structure of the nasal mucosa, the coexistence of cognitive dysfunction and anosmia, the smell–sexuality link, and the brain areas subserving olfaction.

Section Two comprises seven chapters that detail aspects of basic science research. As Morrison and Costanzo note in Chapter 2, the olfactory system is a unique part of the nervous system. The fact that this is the sole site of persistent neurogenesis and regeneration has obvious implications for future efforts to treat central nervous system disorders. The puzzle fragments of neural development are just beginning to suggest clear outlines that could enhance the technology of transplantation. In fact, olfactory neurons demonstrate exceptional plasticity in that they can grow and organize morphological environments in various brain regions. The olfactory system may give us insight into general central nervous system development and regeneration.

In Chapter 3, Anholt discusses odor recognition and chemosensory transduction. The molecular mechanisms of olfactory receptor cells are a topic of great interest. Anholt focuses on the recognition of odors at the sensory membrane, including the nature of olfactory binding and odorant receptor proteins. There are several mechanisms of signal transduction, including electric currents, gated channels, adenyl cyclase systems, and calcium/calmodulin. The chapter finishes with a review of olfactory neuron-specific proteins. Schwob (Chapter 4) summarizes our knowledge of the biochemical phenotypes of olfactory neurons. Chemical markers that identify cell stages can help us understand neuronal differentiation and regulation at various levels. These markers should prove useful in terms of neuron function (e.g., odorant specificity) and growth capacity. For example, the presence of juvenile phenotypic traits in mature olfactory neurons might indicate persistent growth potential. It may also be likely that chemically distinct neuronal subsets have specific odor sensitivities.

Chapters 5 and 6 explore the central role of the olfactory bulb in this sensory system. McLean and Shipley (Chapter 5) discuss anatomical substrates, dealing with the complexities of the circuitry and the neurochemistry involved. The interaction of olfactory pathways with vital parts of the central nervous system takes on added importance as we learn more about clinical deficits in the sense of smell. For example, "nonolfactory modularity" tracts connecting the olfactory bulb with the diagonal band, the locus coeruleus, and the dorsal raphe nucleus suggest that olfaction may be affected in a number of neuropsychiatric disorders in which these nuclei are known or suspected to be involved. These anatomical foundations serve as the starting point for Nickell and Shipley (Chapter 6) to discuss the physiology of olfaction. They explain the possible significance of anatomical organization in odor coding. This chapter also reviews neurotransmitters and their putative roles in olfactory bulb physiology.

Chapters 7 and 8 are a link to the clinical section that follows. Otto and Eichenbaum (chapter 7) summarize studies of rodent olfaction and suggest that this may be a model system for understanding learning and memory in general. Their anatomical focus is on higher brain areas than the bulb, namely, the amygdala, dorsal medial nucleus of the thalamus, prefrontal cortex, and hippocampus. They discuss these locations as the sites of links between olfactory and memory functions and point out that the development of "electrical odors" and their effect on long-term potentiation provides a useful research mechanism for the study of cognition. Dorries concludes this section with Chapter 8 on gender differences in olfaction in mammals. Although we are certain of anatomical dimorphism in olfactory systems, we are relatively ignorant of the functional significance of these differences. Dorries reviews the evidence for the usefulness of specific animal models in efforts to expand our knowledge in this fascinating area. In the pig and hamster, there are sex differences in significant olfactory responses. These animals seem to be excellent models for approaching a series of questions relating to dimorphism of olfactory systems, the correlation of hormonal milieu with olfactory sensitivity, and the underpinnings of odor-mediated behaviors.

Section Three deals with clinical issues in olfaction and is headed by Cain, Cometto-Muñez, and de Wijk (Chapter 9), who give a comprehensive review of techniques and procedures relevant to the study of the functional properties of human olfaction. They point out the difficulties of controlling and delivering the olfactory stimulus and how this inhibits many investigators from attempting to carry out their ideas. They describe the use of static and dynamic olfactometry as the means by which vapors may be delivered to the nose and emphasize the benefits of utilizing environmental chambers, as they provide a way to perform odor and pungency research with environmental realism. The differences in measurement of threshold are discussed, including choice of presentation in ascending, descending, or random orders and the impact of time for recovery from the effects of stimulation. These and other issues remind us that only through precise and accurate measurement techniques will the corpus of data on olfactory performance be considered valid.

Gilbert and Greenberg (Chapter 10) discuss how individual differences in odor hedonics, the ontogeny of odor preferences, the accessibility of odor memory, and the waning of odor detection in normal and abnormal aging are central to our conception of olfaction. They identify as a striking feature of olfactory investigations the predominant use of food-related odors as experimental stimuli, and point out that this may limit the generalizability of conclusions drawn from the experiments. Their chapter is punctuated by lists of items used in stimulus selection by a broad range of investigators. This information, they claim, reveals a trend that is both unsettling and deserving of the attention of experimentalists.

Given that olfactory stimuli have qualitatively different properties from visual and auditory stimuli, the issues raised by Corwin (Chapter 11) are important to address. On pragmatic and theoretical grounds, she suggests using the simplest tasks that will give an interpretable result or, if possible, more than one task to reveal different aspects of a performance. Corwin applies an information processing approach to the examination of standard and novel olfactory assessment paradigms, with particular attention to the assessment of olfactory identification in cognitively impaired subjects and in normals. She points out the necessity of reducing the memory and processing requirements to an absolute minimum, as is achievable through a simple binary yes/no response.

Chobor (Chapter 12) presents a review of work involving the linguistic limitations on olfactory performance, especially the tip-of-the-nose state in odor identification through naming. She proposes that clinicians and investigators use probes to elicit knowledge about odors beyond a simple label. She reviews some of her own work with aphasic and demented individuals who show intact semantic knowledge, such as category membership, item associations, or experiential relations. Given the idiosyncratic nature of olfaction, such as the tendency for all-or-none coding, this modality offers an opportunity for the exploration of atypically organized linguistic systems. Further, creativity in stimulus presentation and response mode allows greater insight into the mechanisms of olfaction.

Schmidt and Beauchamp (Chapter 13) reiterate the potential superiority of the olfactory system relative to other sensory modalities in newborns, raising the possibility that olfaction is of considerable importance during development, although they note that much of the available evidence on this topic is speculative. They list three important functions for olfaction in infancy, including the modulation of feeding behavior, avoidance behavior, and the recognition of individuals. Based on analogy to animal research, they suggest that lifelong preferences and aversions for odors, as well as other strong emotional associations with odors, are formed during infancy and early childhood. The authors wisely encourage the use of infants and young children in the study of olfactory learning, as they are less burdened by the effects of individual experience.

Bieber, Bieber, and Friedman (Chapter 14) point out that from both an historical and a contemporary point of view, the linkage between olfaction and sexuality has been mostly ignored. Ellis, Kraft-Ebing, and Freud, the three major sexologists among the psychiatric pioneers of their age, were ambivalent about the role of olfaction, as were the major theory builders around Freud, both disciples and dissidents. This included such major theorists as Horney, Sullivan, Thompson, Rado, Kardiner, and others. In 1959, however, Bieber delineated the importance of early olfactory impressions in psychosocial development and heterosexual organization. His rich history of observation and the interpretation of dream material of patients in analysis is shared with us in this volume. According to his theory, olfaction is linked to human sexual development from its earliest manifestations. Bieber and his colleagues suggest that it is not penis envy, but rather sensitivity to the male odor that initiates female heterosexual reactivity.

With the goal of better understanding olfaction and emotion, Ehrlichman and Bastone (Chapter 15) outline several propositions connecting olfaction to affect. They review numerous studies that provide support for the claim that the hedonic quality of odors can influence mood. As it is quite possible that odors can affect mood without their having any special ability to evoke emotional memories, they argue that at present there is no basis for claims that odors are more emotional than other types of stimuli. Until very recently, olfactory research has largely ignored the affective consequences of the odor experience, yet these authors point out that in a number of psychological domains, the use of odors can serve as an important complement to studies using more cognitively mediated affective stimuli.

Smith and Duncan's goal in Chapter 16 is to discuss the characteristics and common etiologies of primary olfactory disturbances and to offer a comprehensive "dictionary" of terms commonly used to describe them. Disturbances of smell may result from a wide variety of underlying etiologies, making the diagnosis and treatment of these disorders a difficult and challenging problem. Most patients, however, fall into one of a few etiologic categories, including nasal or sinus disease, prior upper respiratory infection, head trauma, or idiopathic causes. With a goal of easy management, it is especially necessary to separate the conductive losses of smell that result from interference with odor access to the receptors from the sensorineural losses, as the former are often amenable to treatment.

Greenberg (Chapter 17) provides us with a broad range of clinical information relating to olfactory hallucinations, including the differences between this and the other modalities. He notes that olfactory hallucinations (or false perceptions) are unique in that their descriptions not only rely heavily on the subjective report of the experiencer, but that they are influenced by the inherently tenuous relationship between cognition and olfaction. The focus of Greenberg's chapter is on neurological entities in which olfactory hallucinations have been described. Diagnostic dilemmas in the psychiatry of this phenomenon are considered, and case reports are used in support of his search for a pattern that might help to tease the verbal report of the hallucinatory experience from the experience itself.

Schiffman describes chemosensory losses in the elderly, resulting from a combination of causes including normal aging, diseases (especially Alzheimer's disease), and drug therapy. She discusses the nature of the aged individual's decreased ability to recognize and identify odors, and offers a theoretical model based on the "across-fiber pattern" theory of Erickson (1963), which proposes that the pattern of neural activity across many neurons codes the sensation of a particular stimulus. The amount of neural mass determines olfactory perception in the elderly, and the consequences of decreased amounts of evoked neural mass (resulting from neural loss or histopathological changes) are elevated thresholds, reduced intensity of sensation, and small distinctions between stimuli. Along with her descriptions of anatomic and physiologic changes that give rise to olfactory loss, she offers an innovative means of treatment through the addition of commercial "flavors" to food before and after preparation.

In Chapter 19, Mair and Flint raise numerous possibilities about the causes of impaired olfactory performance in patients with Korsakoff's disease. They point out that the deficits in odor perception in this population are as consistent a sign of this disease as are the better-known memory impairments. There is evidence of a global mental impairment that affects attentional and endogenous cognitive processes. Taken with evidence of multimodal perceptual deficits, these results raise the possibility that olfactory discrimination deficits of this disease may result from a more general, nonolfactory disorder. It has not been established that these patients have a selective impairment of odor perception, although this remains a possibility.

Costanzo, Ward, and Young (Chapter 20) review the mechanisms of head injury, its frequency of occurrence, clinical evaluation and testing, patient management, and prognosis. They encourage the use of olfactory testing as part of all neurological evaluations for head injury. This should include detection and identification of real odors and testing of trigeminal nasal sensation. They point out that the danger of a cursory examination is that there are cases in which simple odor detection is normal, yet olfactory dysfunction of central origin is present. Treatment for anosmia is not often feasible, though they stress the importance of counseling for depression and related social factors. The likelihood of spontaneous recovery is much higher in the first six months following injury, and although animal studies show that damaged olfactory cells can replace themselves and functional connections with the central nervous system can be reestablished, this has not been demonstrated in humans.

The final chapter, by Serby, Larson, and Kalkstein (Chapter 21), discusses impaired olfactory performance in a variety of neurological and psychiatric disorders, namely, Alzheimer's disease, depression, schizophrenia, and the dementia of the acquired immune deficiency syndrome (AIDS). A focus of the chapter is on the course of impairment in Alzheimer's disease, a condition that warrants much definition. The authors encourage the use of olfactory testing in demented patients, as there is some evidence that it correlates well with stage of deterioration. The delineation of such patterns and the comparison of these patterns between groups of demented patients will serve as a step toward resolving the specificity of the hypsomia.

Considering the sum of these chapters, it is clear that the sense of smell is more than a primitive phenomenon with little to tell us about higher brain function. The range of material in this book, from the molecular architecture of olfaction to its cognitive-sensory links and dissolution in disease, reminds us that the Proustian connection of odor, memory, and affect is not poetic license but the inevitable consequence of experience. There is a richness of meaning in olfaction that links us to the primitive and to the complexities of human behavior. We hope that this book contributes to the organization of knowledge in our field and serves as a stimulus for further study of olfaction and the central nervous system.

Part I
Historical Perspectives

1
Olfaction and the Primitive: Nineteenth-Century Medical Thinking on Olfaction

ANNE HARRINGTON and VERNON ROSARIO

Introduction

In his social history of odor, *The Foul and the Fragrant* (1986), Alain Corbin notes the extent to which, historically, science and scholarship have been ambivalent in their attitudes towards the phenomena of olfaction. Subtle and not so subtle prejudices against this sensory organ can be traced, among other things, to its association with lust, desire, and impulsive sensuality, its links to the undignified sniffing behavior of animals, the poverty of human language to categorize and analyze the phenomena with which it is concerned, and its apparent relative uselessness in civilized society (Corbin, 1986: 6–7). At the same time, a counterpoint to this main song of disapproval has also left its trace on history: early students of animal behavior saluted olfaction as the primary organ of self-preservation; nineteenth-century men of letters elevated it to the privileged organ of feelings, recollection, and intimacy; it was the organ "capable of shaking man's inner life more profoundly than were the senses of hearing or of sight. It seemed to reach to the roots of life" (Corbin, 1986: 7–8).

A variety of taboos and mysterious attractions thus form the cultural framework for the present investigation into the nature of biomedical thinking on the problem of olfaction. The chapter is divided, for the sake of clarity, into five parts, but we will see that connections across themes and between authors are abundant. We begin (Section 1) with an overview of nineteenth century clinical models of olfactory pathology, focusing on so-called nasal reflex neuroses and olfactory disorders of organic cerebral origin. We then (Section 2) set these medical debates against the backdrop of broader biomedical interest in the evolutionary significance of olfaction, focusing on the problem of sexual selection, the nose of the "savage," and the meaning of the declining place of human olfaction in the evolution of modern civilization. In Section 3, we examine the way in which evolutionary thinking (and the social values associated with it) would be translated into the language of neuroanatomy, with Paul Broca's concept of the "great limbic lobe," the original "smell brain." Section 4 turns to the role of olfaction in medical thinking about sexuality and sexual deviance. In the final section (Section 5), we attempt to bring the story of olfaction and the nervous system more or less up to the modern context—and, in the process, to reveal twentieth-century links to key nineteenth-century themes of earlier sections—by briefly tracing the transformation of the nineteenth-century "smell brain" into the limbic system of James Papez and Paul MacLean.

Pathologies of the Olfactory Sense

Conceptual Orientation

Since at least the eighteenth-century, it had been known that exposure to certain odors could stimulate a range of apparently involuntary behavioral responses in nonhuman animals, particularly salivation and gustatory displays. This fact led the materialist philosopher Pierre Cabanis (1757–1808) to propose a "sympathy"—an unspecified functional linkage—between the nose and (among other things) the digestive tract (Cabanis, 1805: 526–528). By 1812, the British physician William Prout had reinforced this early idea of a nasal-gastric connection by demonstrating that odor was largely the basis for distinguishing among flavors (Corbin, 1986: 112). It was puzzling that the functioning of gustation should turn out to be dependent on one of the other senses (we should never expect audition to turn out to be a necessary basis for vision), and the discovery reinforced the sense of a peculiar sympathy of the olfactory organs with other internal organs.

Reflection at this stage, however, remained fairly inchoate and undeveloped. In 1821, Hippolyte Cloquet, a student of Cabanis, sifted through a scattered range of scientific, philosophical, and folk observations and theorizings to produce history's first reference work on the problem of olfaction: *Osphrésiologie, ou Traité des odeurs, du sens et des organs de l'olfaction, avec l'histoire détaillée des maladies du nez et des fosses nasales et des operations qui leur conviennent* (1821). This enormous volume was to remain the classic work on olfaction throughout the century; Havelock Ellis would still cite it extensively in his *Sexual Selection in Man* as late as 1905.

From the outset, the monograph established Cloquet's commitment to the mechanistic, sensationist orientation of his teacher, who (himself influenced by Albrecht von Haller and Julien O. de la Mettrie) had sought to reduce the study of man to physics and physiology. We know, Cloquet wrote, that all passions reduce to desire and nondesire, and that these two passions in turn can be understood according to mechanical laws of attraction and repulsion. Given this, we see that the olfactory sense functions by reacting either positively or negatively to odorous particles impinging on the olfactory organs, a process which results in approach (desire) or withdrawal (repugnance) by the animal.[1] All the sympathetic bodily responses associated with these initial nasal responses serve the constant struggle (*lutte continuel*) for existence (Cloquet, 1821: 20). Indeed, in subhuman species, Cloquet identifed olfaction as the central perceptual organ of survival. After all, was it not intimately bound up in a sympathetic relationship with the digestive and (as was increasingly also clear) reproductive systems? While he went on to admit that sight and hearing had taken pride of place in the sensory life of humans, he noted that odors continued to produce sneezing, lacrymation, sleep and wakefulness, strong affective reactions, strong vegetative reactions, nausea, and even death (Cloquet, 1821: 79).

Humans were also still vulnerable to the effects of odors in the realm of sexual activity. The secretion by animals of odorous substances to facilitate male-female recognition and coitus found its human counterpart in the exploitation by civilized women of the seductive power of perfume. In addition, Cloquet noted that the symptoms of a "genital neurosis," such as hysteria, could be excited or relieved by certain odors (Cloquet, 1821: 126).[2]

Notwithstanding the interest of the numerous anecdotes, remarks, and conjectures that Cloquet presents, his basic premise is of particular importance to us, since it helped set the direction of a very important subset of the later clinical literature on olfaction. This was his adoption of Cabanis's

---

[1]Debates about the physics of odor transmission (atomistic versus humoral theories) had been (at least temporarily) resolved by 1798 when Foucroy, and later Bertholet, apparently established that smell depended on odorous particles of a body disengaging and dissolving in air or liquid (Corbin, 1986: 112).

[2]The term "neurosis" began its history with a meaning almost diametrically opposed to its present one. Far from referring to a disorder of nonorganic origin, the term "neurosis" was used by late eighteenth- and early nineteenth-century authorities to refer to a physiological disorder of the nervous system (either of the motor or of the sensory type). In the cage of hysteria, it was still possible to consider the genitals or uterus to be the source of nervous troubles, because of the "extensive sympathies of the uterus" with the nerves (Richard Bright, *Reports of Medical Cases*, 1827, cited in Spillane, 1981: 187).

By the middle of the nineteenth-century, the uterus argument would increasingly fall out of favor, and more and more physicians would begin to accept the idea of a direct nervous origin for the "neurosis" of hysteria. When pioneering medical neurologists such as Jean-Pierre Charcot in France failed to identify any postmortem lesions associated with this disorder, the definition of "neurosis" was further refined to mean a "functional" disorder of the nervous system without morbid structural change. Matters became more confusing as it became increasingly clear that "suggestion" and other psychological means could often influence the disorder (cf. in this paper the "rose cold" case of John MacKenzie). The final transformation of "neurosis" from a puzzling and protean organic disease of the nerves (the great "imitator") to a behavioral phenomenon that could only be explained in psychological language would have to await the *fin-de-siècle* work of men like Pierre Janet in Paris and Freud and Breuer in Vienna.

idea of a basic sympathy, or functional relationship, between the olfactory system and other bodily systems. Without making the physiological links clear, Cloquet opened the door to the possibility that the perceiving, sensing organ of the human soul—the brain—was closely bound to the animal world of vegetative functions served by the internal organs. The effect of such an integrative approach to physiology was to suggest a closer link between the human intellectual faculties and animal instinctual functioning than most people were probably comfortable with (Richards, 1987: 109). Indeed, the long silence in France on the subject of olfaction in the years after Cloquet may be related to the general political disfavor into which sensationalist psychology fell—for making just these sorts of radical claims—during the conservative middle years of the nineteenth-century.

Nasal Reflex Neuroses

When olfaction did reemerge in the later nineteenth century as a subject of medical interest, attention returned to the sympathetic association. In discussions of this period, we find that "sympathy" was now understood to function in both afferent and efferent directions: i.e., it was felt that olfactory disease or irritation could produce a range of psychological and physiological sympathetic reactions ("nose to body" type disorders), *and* that pathologies having their source in other body organs could lead to sympathetic olfactory dysfunctioning ("body to nose" type disorders).[3] Both of these pathological processes were, somewhat confusedly, sorted under the single term "nasal reflex neurosis" (Hack, 1882).

Sneezing served as an excellent model of the "nose to body" form of nasal reflex neurosis. One of the more intriguing dimensions of this reflex response was that it engaged the whole body in an involuntary spasmodic act. As W. Spencer Watson, British author of *The Diseases of the Nose and Accessory Cavities*, explained in 1875:

> During this expiration no part of the body is quiet, and hence perhaps some authors have regarded sneezing as a sort of temporary epilepsy. . . . Immediately after the cessation of movements the frame sinks into a general but agreeable lassitude; a copious effusion of serum moistens the cavities of the nose; the eyes are wet with tears (Watson, 1875: 344).

For Watson, the importance of these observations lay in the intriguing similarities they revealed between the reflex act of sneezing and two other paroxysmatic acts of great physiological and medical interest: sexual orgasm and epileptic convulsion. Some physicians had already concluded that the relationship went beyond the phenomenological level. Watson cited a certain Wedelius who felt that "sneezing often ushers in the access of epilepsy or marks the termination of the fit" (in Watson, 1875: 345). In

[3] The nosology here is ours!

his *Observations rares de médecine* (also cited by Watson), Stalpat der Wiel had spoken of "individuals in whom the act of coitus was often preceded by sneezing." Watson noted that other emotional states may trigger sneezing, and he felt that this pointed to an analogy between sneezing and asthma (1875: 343f; cf. Fraenkel, 1881 on nasal irritation and asthma).

Lest it be thought that sneezing was a relatively benign clinical condition, Watson reported that sneezing fits had been known to cause pulmonary hemorrhages, amaurosis, anosmia, and sudden death (Watson, 1875: 345). One woman suffered a miscarriage after a "fit of sneezing." Her doctor noted, though, that "this lady was of slightly hysterical temperament" (Gray, 1875). A link between hysteria and (at least some) sneezing paroxysms was more firmly asserted in that same year by a Dutch physician, who went on to recommend various "anti-hysterical medicines" designed to cure such attacks (Mayer, 1875).

In the United States, John MacKenzie (cf. below) also argued in the direction of classifying at least some nasal paroxysms as hysterical phenomena. In 1886, he reported on a lady of "nervous temperament" who was cured of her violent reaction to pollen-producing flowers when MacKenzie induced an attack of coryza in her through exposure to an artificial rose, then revealed the nature of his deception to her. "When told that the rose was an artificial one, her amazement was great. . . . A few days later, she called to see me again, and on that occasion she buried her nostrils in a large fragrant specimen of the genuine article and inhaled its pollen without the slightest tendency to the production of reflex acts" (MacKenzie, 1886: 49).

Meanwhile, the Freiburg otolaryngologist Wilhelm Hack (1882) had expanded the range of "nose to body" type reflex neuroses to include neuralgias, headaches, migraines, and nausea [cf. also the classification system offered by Wilhelm Fliess (1893)]. A leading figure in the nasal neuroses field, Hack believed the immediate cause of all these reflex disorders to be a reactive inflammation of the inferior concha of the nose; not a septic inflammatory reaction, but an engorgement of the erectile tissues of the muscosa (Schneider's membrane). He advocated surgical electrocautery excision of the membrane as a general remedy for sufferers. Others cautioned against such radicalism; thus Moldenhauer (1888) prescribed cocaine as an alternative means of reducing nasal inflammation effective in most cases.

Nasal Reflex Neuroses: The Nasogenital Link

Cases of "body to nose" style nasal reflex neuroses were also a subject of lively medical interest in this time. Much medical attention came to focus on one critical body region: the genitals. Although the histological similarities between the erectile tissues of the nose and the genitals had long elicited coy speculation, systematic investigation into the possibility of a link between nasal neuroses and genital activity seems to have begun with the work of the leading Baltimore laryngologist and rhinologist John Noland MacKenzie (1853–1925).

In his 1884 report, MacKenzie began by recalling the anatomical relationship between the erectile tissue of the nose and that of the penis. Stimulation of the former organ, like that of the latter, leads to erection and 'orgasm' (the term is MacKenzie's) in the nasal tissue. Overindulgence in "venery," then,

> constitutes the anatomical explanation of the stoppage of the nose in acute coryza, and their permanent enlargement is the distinctive feature of hypertrophic nasal catarrh. . . Indeed, the changes which [this erectile tissue] undergoes seems to lie at the foundation of nasal pathology, and furnish the key not only to the correct interpretation of nasal disease, but also to many obscure affections in other and remote organs of the body (MacKenzie, 1884: 362; cf. also MacKenzie, 1898).

That there was an intimate physiological relationship between the genitals and the nose was "sufficiently evident" from:

1. Clinical observations that a certain proportion of women regularly suffer from engorgement of the nasal cavernous tissue during menstruation, but at no other times
2. Clinical evidence of vicarious nasal menstruation in women unable to menstruate normally and of nasal hemorrhages in a proportion of boys upon achieving puberty
3. The "well-known sympathy" between erectile tissue in the sexual organs and other erectile tissues of the body ("There is no reason why the sexual excitement which leads to congestion and erection of these organs may not. . . cause engorgement of the nasal erectile spaces")
4. Cases of spontaneous irritation of the nasal passages during copulation or other forms of sexual activity, especially expressed through sneezing
5. Cases of genito-urinary irritation caused by nasal dysfunctioning
6. The importance of the sense of smell in sexual arousal and selection in lower animals and certain humans—MacKenzie thought it "quite possible that irritation and congestion of the nasal erectile tissue" could be a facilitator or producer of the "olfactory [erotic] impressions" in question

The "well-known law of economy" explained how it came that a normal physiological relationship could also be a source of pathology. "If. . . excitation be carried beyond its physiological limits, there comes a time, sooner or later, when that which is a normal process becomes translated into a pathological state" (MacKenzie, 1884: 364). The natural therapeutic conclusion would be that cases of nasal reflex disease of genital origin could be helped only by sexual moderation or abstinence.[4]

The years following MacKenzie's presentation saw a rapidly growing interest in the important problem of a nasogenital connection.[5] Of the new workers, probably the most important historically was Wilhelm Fliess, the Berlin physician who would have such an important influence on Sigmund Freud (Fliess, 1897 and 1910). Although Fliess largely followed MacKenzie (of whose work he seemed unaware) in the arguments he set out in defense of the physiological relationship between the nose and the genitals, he did go beyond him in actually attempting to localize the nasogenital meeting ground in certain special "genital zones" (*Genitalstellen*) in the nose itself (Fliess, 1897).[6] Fliess himself claimed to be the first to have established the importance of cocaine both as a diagnostic tool and as a therapeutic agent in olfactory disorders (Fliess, 1893), though we have seen how cocaine's effectiveness in reducing nasal inflammation had been claimed by Moldenhauer in 1886.

Fliess also attempted to set his concept of a nasogenital connection in a larger dynamic framework by linking it to a model of biological periodicity, based on 28- and 23-day physiological cycles (for details, see Sulloway, 1979). This preoccupation with periodicity led him to focus particular attention on the often-asserted association between menses and nasal disorders, especially nasal congestion (Fliess, 1897). Since the usefulness of cocaine in treating olfactory disorders had already been established, it was a small step now for Fliess to experiment with its efficacy as "a new treatment for dysmenorrhea." This therapy, in which cocaine would be applied to the nasal passages of afflicted women, with expected resulting sympathetic working on the uterus and other afflicted organs, was considered eccentric but (in the eyes of some at least) surprisingly effective (cf. Ries, 1903).

Sensitized to the importance of the nasogenital link through Fliess, Freud granted olfaction a major role in the biophysical theory of psychosexual behavior that he attempted to develop in his 1895 (ultimately aborted) "Project for a Scientific Psychology" (Freud, 1895). For him, the sexual processes associated with olfaction were to be understood as chemical. Like all chemicals, in some cases they could also have toxic effects;

[4]Thus MacKenzie reported a case in 1898 of a young married woman who suffered from asthmatic breathing "associated with stoppage of the nostrils." It turned out she had been involved in "intemperate venery" with her husband every night for five years. "Moderation in their sexual relations caused rapid disappearance of the symptoms" (MacKenzie, 1898: 120).

[5]A Wurzburg University Inaugural Dissertation reviewing the literature on the relationship between the nasal passages and the sexual organs was produced in 1892 by Gustav Endriss, "Die bisherigen Beobachtungen von physiologischen und pathologischen Beziehungen der oberen Luftwege zu den Sexualorganen" (cited in Hagen, 1901: 280).

[6]In his 1898 appreciative review of Fliess's work (originally held as a talk at a medical congress in Montreal), MacKenzie noted how "curiously enough, the genital zones of Fliess correspond exactly with the most sensitive portions of the sensitive [nasal] refex mapped out by me in 1883." Many masturbators, MacKenzie believed, suffered from nasal disease and nosebleeding from these areas.

Freud suggested implicating them, for example, in the migraines associated with nasal reflex neuroses of genital origin. In an 1896 letter to Fliess on all these issues, Freud wrote:

The question of the source of the states of stimulation in the nasal organs now arises. The idea suggests itself that the qualitative organ for olfactory stimuli may be Schneider's membrance [in the nasal septum] and the quantitative organ (distinct from this) may be the corpora cavernosa [of the genitals]. Olfactory substances . . . are the breakdown products of the sexual metabolism; they would act on both these organs. During menstruation and other sexual processes the body produces an increased Q [quantity] of these substances and therefore of these stimuli. . . . Thus the nose would, as it were, receive information about internal olfactory stimuli by means of the corpora cavernosa, just as it does about external stimuli by Schneider's membrane: one would come to grief from one's own body [through the development of a nasal reflex neurosis such as a migraine]. The two ways of acquiring migraine—spontaneously and through smells, or human toxic emanations—would thus be equivalent, and their effects could at any time be brought about by summation (Freud, 1985: 161–62).[7]

Olfactory Disorders of Neurological Origin

Although it is true that "nasal reflex neuroses" were associated in some of their variations with diffuse neurological symptoms (such as migraines), this nineteenth-century syndrome must be clearly distinguished from clinical work and theory on olfactory sensory and perceptual disorders either of cerebral origin or resulting from damage to the olfactory nerve. In his 1934 master work on brain pathologies resulting from war injuries, the German neurologist Karl Kleist (1934: 1267–1278) would identify five classes of such disorders:

1. Anosmia, or loss of sense of smell of a high or lower grade, unilateral or bilateral
2. Parosmia, or perverted olfactory perception
3. Olfactory hallucinations (i.e., subjective perception of smells in the absence of stimuli)
4. Osmo-motor disorders (i.e., compulsive movements of the nostrils, nose, and often lips and tongue, etc., associated with temporal epilepsy)
5. Olfactory agnosia (osmagnosia)

[7] Although we are limiting this brief intellectual history of nasal-genital relations to the nineteenth century, readers might want to be aware of an important twentieth century development in the nasogenital story: the identification by the German psychiatrist Franz J. Kallmann of a genetically based (probably X-chromosome) disorder that combines hypogonadotropic hypogonadism with anosmia. Victims of Kallman syndrome both fail to attain sexual maturity and are impaired olfactorily (*Roche Lexikon Medizin*, 1987: 916). One suspects that our nineteenth-century theorists would hardly have been surprised.

Kleist lamented the fact that these various disturbances had been treated so inadequately by his colleagues.[8] The complaint was not without substance: there is no question that, for various theoretical and subtle social reasons, the emerging field of clinical neurology in the late nineteenth century would devote considerably less attention to brain pathologies affecting olfaction than it did to other sensory and cognitive brain dysfunctions. As one of the esoteric voices in a much larger cultural dialogue about what, in a post-Darwinian world, still made the human animal special, clinical neurologists were interested above all in those functions that most profoundly defined the human psychological economy, that took pride of place in the phylogenetic hierarchy of human powers. For many, olfaction appeared a primitive function, crucial to the life of lower animals, but largely displaced in the evolution of human civilization by vision and audition. Thus the French surgeon Notta reported in 1870 on some cases of anosmia he linked to atrophy of the olfactory nerve, but admitted that olfaction (with gustation) was the least important of the senses, and that its dysfunction hardly represented a serious, life-threatening infliction (Notta, 1870). J. Wickham Legg of London's St. Bartholomew's Hospital announced to *The Lancet* readership in 1873 that loss of smell "barely amounts to a discomfort" (Legg, 1873). And Julius Althaus in England echoed the common sentiment eleven years later when he remarked that "if Prince Bismarck and M. Gambetta were to become suddenly blind and deaf, the destinies of Europe would no doubt be changed; while if these two men were to lose their smell and taste, things would probably go on much in the same manner as they do now" (Althaus, 1881: 771).

In spite of such condescension, it turns out that a few scattered papers of interest were published in the nineteenth century. Olfactory hallucinations came in for some attention, with recognized causes including cerebral tumors, tabes dorsalis, neurasthenia ("exhaustion" of the nerves), insanity, hysteria, and especially epilepsy of temporal lobe origin (Collet, 1904: 84–85). As regards this latter disorder, the British neurologist John Hughlings Jackson was one of the first to call attention to the frequent tendency for "disagreeable smells" to precede what he called "uncinate attacks," or temporal epilepsy (Jackson, 1864). Jackson's observations would be widely cited by the olfactory researchers, who recognized that the subjective smells associated with temporal epilepsy were almost invariably unpleasant (Watson, 1875: 333). These smells were also very difficult to characterize,

[8]Two years after the publication of Kleist's book, this situation would improve notably with the publication of Walther Riese's extensive "Symptomatologie der Erkrankungen des Riechnerven" and "Symptomatologie der Erkrankungen der Riechrinde" (Riese, 1936), works which deserve to be better known. Riese covered a wide clinical range and supplemented quantitative methods and phenomenological description with theoretical perspectives drawn from gestalt psychology and the various holistic psychologies.

possibly because of some ill-defined synaesthetic quality. One of Jackson's close colleagues, David Ferrier, described a temporal epilepsy patient whose attacks were regularly preceded by a smell she described as reminiscent of "green thunder" (cited in Critchley, 1979: 106).

The widening research into the phenomenology of temporal lobe epilepsy also led to recognition of at least a temporal correlation between epileptically induced hallucinations of smell on the one hand and various peculiar sensations of forced memory or inexplicable familiarity on the other (*déjá vu, déjá vecu*)—elaborate conditions that Jackson called "dreamy mental states" (Jackson, 1880–1881). At the same time, it is important to stress (especially in light of present-day interest in this question) that no nineteenth-century neurologist ever went so far as to suggest a direct relationship between brain structures serving memory and those serving olfaction.[9]

Relatively straightforward clinico-anatomical reports of anosmia (of varying etiologies) were not uncommon in the nineteenth century, and a standard discussion was provided in all the rhinology textbooks of the time. Recognized key figures in the literature included Notta (1870), Knight (1877), Watson (1875), Cohen (1884), Moldenhauer (1888), Moure (1893), and Collet (1904). In 1870, William Ogle, well known in the history of aphasiology for his work on agraphia, called attention to what he believed to be a "frequent coincidence of anosmia [of central origin] and aphasia," and suggested some diagnostic conclusions implied by this observation—"the presence or absence of anosmia in cases of aphasia will for the future be perhaps of some use in enabling us to fix with greater precision the site and extent of the brain lesion" (Ogle, 1870: 275). Ogle also reflected extensively on the relationship between keenness of smell and extent of nasal pigmentation, especially in different human races; we will return to these sociologically revealing contributions by him later.

Parosmia and hyperosmia came in for some attention; both were recog-

---

[9]This is *not* to say that nineteenth-century medicine failed to recognize a connection between olfaction and memory in a general way: in 1819, the *Dictionnaire des sciences médicales* called the sense of smell the sense of "tender memories" (cited in Corbin, 1986: 202). some fifty years later, the alienist Henry Maudsley would remark that the sense of smell in humans "is singularly effective in recalling vividly the ideas and images of forgotten scenes and places" (Maudsley, 1868: 134).

Nevertheless, nineteenth century understanding of the relationship between olfaction and memory was not essentially a medical understanding. Such understanding, rather, developed in the wider culture of *belles lettres* under the influence of the sensual romanticism of the early nineteenth century. The nostalgic pleasures of olfactory reminiscences were explored by such figures as Rousseau, Tennyson, George Sand, and of course Marcel Proust (in whose book the subtle aroma of a madeline soaked in tea evokes thousands of pages of reminiscence). Insofar as medicine also recognized links between olfaction and remembering, it seems to have reflected the wider cultural consensus, without appreciably modifying or expanding on it in any original way (cf. Corbin, 1986: 200–204).

nized as invariably resulting from nervous dysfunction, including cerebral dysfunction associated with insanity. This was in contrast to anosmia, where the etiology could also sometimes be an internal stoppage of the nasal passages, etc. Thus a patient of the French physician Castex (cited in Collet, 1904: 84). smelled onions when presented with a sweet, agreeable odor, and vanilla when presented with an disagreeable odor. Cohen (1884) cites the case of hyperosmia presented by the patient Caspar Hauser who

> perceived odors where others could detect nothing of the kind. He was overpowered by the smell of a rose, and could distinguish fruit-trees from each other, at a considerable distance, by the odor of the leaves. The smell of old cheese produced nausea and vomiting; and that from a churchyard occasioned a paroxysm of fever. . . . In fact, every odor, excepting those of bread, fennel, anise, and caraway, was more or less disagreeable to him, so much so that the only food he would partake of was bread and water (Cohen, 1884, citing "Feuerbach's [1834] Account of Caspar Hauser," 404).

The Place of Olfaction in Nineteenth-Century Evolutionary Theory

Introduction

The clinic was not the only arena where discussions of olfaction would develop in fruitful directions. Natural historians and their later cousins, evolutionary biologists, also found increasing reason to ponder the significance of the nose in animal economy.

To begin, the differences in olfactory acuity among various animal species, and especially the decreasing importance of olfaction in primates and humans, had been a source of discussion since at least the beginning of the nineteenth century. Anatomists had long been aware of the dramatic differences in the size of the olfactory bulb and nerve (relative to the whole brain) in lower and higher vertebrates, especially primates and humans. Cloquet had further noted (1821: 130) that, although the olfactory nerve was found in all vertebrates, the extent of nerve innervation of the pituitary membrane (i.e., the area of the olfactory membrane) was proportional to the keenness of smell in the animal in question. By mid-century, such anatomical and physiological gradations would integrate themselves nicely into the new hierarchically oriented evolutionary theories.

One branch of speculation to result from the recasting of the problem of olfaction in evolutionary terms came to focus on the role of smell as a primary motor of evolution; another branch pondered on the significance of olfaction's apparently paradoxical diminution in the evolutionary development of higher life forms, including humans. A final branch of thought made creative points of contact with the other two in the course of finding the secret of human evolution in mankind's liberation from its animal "smell brain".

The Primal Olfactory Attraction of Gametes

Ernst Haeckel (1834–1919) was one of the most vocal advocates of Darwin on the Continent and is probably best known for his "biogenetic law" that ontogeny recapitulates phylogeny (for a critical, historical discussion, see Gould, 1977). In his *Anthropogenie, oder Entwickelungsgeschichte des Menschen* (1874), he pictured olfaction as a primordial attractive force in the copulatory union of male and female gonadal cells. He envisioned these cells as possessing a primitive consciousness (*Seelenthätigkeit*), and seeking each other out through a type of primitive smelling. The fact that this union between egg and sperm might also be explicable in physical-chemical terms (Haeckel desrcibed it as an *Erotischer Chemotropismus*) in no sense represented an argument against the likelihood of a primitive psychic component in the process as well; psychic and physic functions represented two faces of a monistic process.

Haeckel went on to identify this primitive gonadal instinct for union with one's sexual opposite as the source and motor of that complex psychic and chemical phenomenon we call "love." Although he did not discuss the role of smell in higher animals, the conclusion could easily be drawn that his theory of "erotic chemotropism" could allow one to see the erotic effects of certain odorous stimuli on sexually aroused animals (and humans) as an unconscious reaction of the organism to the promptings of its gonads (cf. Section 2.3 below).

Olfaction in Sexual Selection

We need only briefly mention the well-accepted recognition by nineteenth-century evolutionists of olfaction's critical role in the selection of sexual partners among animals. It is perhaps revealing of the lack of dissent on this issue that Charles Darwin did not find it necessary to spell out the details of the matter in his (1871) *The Descent of Man and Selection in Relation to Sex*. Noting the importance of odor in deterring enemies and attracting sexual partners, he contented himself with simply declaring that "the most odoriferous males are the most successful in winning females" (Darwin, 1871: 214–215, 605). In his early M (metaphysical) notebooks (written between 1838 and 1839, and not intended for publication), he had spoken more frankly on the importance of genital and anal odors in sexual attraction and social relationships among animals.

> What an animal likes taste of, likes smell of. Hyaena likes smell of that fatty substance it scrapes off its bottom.—it is relic of same thing that makes one dog smell posterior of the other.—

> Why do bulls and horses, animals of different orders turn up their nostrils when excited by love? . . .
> We need not feel so much surprise at male animals smelling vaginae of females.—when it is recollected that smell of ones own pud. not disagree.—Ourang outang at Zoolog Gardens touched pud. of young male and smelt its fingers (Darwin, 1836–1844: 536, 540).

However impolite it might have seemed to speak out publicly about such matters, the fact of odor as a sexual attractant at least posed no theoretical dilemmas for a Victorian naturalist like Darwin. It was quite a different story, however, with another indubitable fact associated with olfaction: its apparent *diminished* role in survival at the upper end of the evolutionary ladder. Most mammals have a relatively large percentage of forebrain material associated with olfaction. In primates, however, one finds a decreasing cephalic representation of this function as one proceeds through the monkeys to the great apes and finally to humans. This fact (along with its apparent duplication in the ontogenesis of the primates) had been known since the early nineteenth-century; it received extensive treatment, for example, in 1826 by a student of Cuvier, E. R. A. Serres, in a text that won the *Grand Prix* of the *Institut Royal de France* (Serres, 1826). There was thus an apparent paradox in the phylogeny of olfaction. On the one hand, evolutionists were dealing with a force that had driven the differentiation of species through sexual reproduction. On the other hand, this same force had apparently been emasculated in evolution's proudest products: the higher primates and human beings.

Olfaction and the "Savage": The Anthropological Reply to the Paradox

One way of dealing with this paradox was to subtly minimize it by stressing evidence that seemed to speak against it. Here, anthropological concepts of the "savage," the "primitive" man, were a Godsend. Pushed into fulfilling an intellectual need for a "missing link" between the animal kingdom and the civilized world of the white European, the savage's legendary acute sense of smell also served to downplay the problem of olfaction's diminution in evolution by suggesting that reports of its imminent demise in the higher animals had been exaggerated. This roundabout approach to the problem (which of course never addressed the neuroanatomical issues) then allowed evolutionists to identify the novel evolutionary pressures of modern civilization as the cause of the undeniable diminished role of olfaction in "advanced" white European humans.

Thus Cloquet (1821: 131), citing the popular example of the American Indian, whose sense of smell was said to be so acute that he could follow the trail of his prey without the aid of hunting dogs, remarked:

The habit of living in society and the assurance of finding amongst one's equals the intelligence (*lumiéres*) which has dispensed with the need to have recourse to smell, have left civilized man less sensitive to the impressions that act on that sense, which has therefore lost some of its sensitivity. Amongst individuals who have not risen in social state, there is a much superior acuity (Cloquet, 1821: 130).

Lest it seem too unfair that civilization should have left European man at any sort of disadvantage compared to his primitive fellows, some writers further suggested that acuity went hand in hand with crudity, and that civilized man's nose, while perhaps less powerful in absolute terms, was also significantly more sophisticated than that of the primitive. "The inhabitants of Kamchatka," one early writer noted, "could hardly smell spiritous melissa cordial or *eau de cologne* at all, whereas they could catch the scent of rotten fish or a grounded whale at a distance" (cited in Corbin, 1986: 140–141).

A telling variation on the argument for a link between savagery and olfactory acuity was made by William Ogle in 1870, in his analysis of a case first reported in 1852 (but not recognized, according to Ogle, for its theoretical import):

A boy in Kentucky, son of two black slaves, had up to his twelfth year a skin of the same dark colour as that of his parents. At this period a white patch appeared near the inner canthus of the left eye. This white patch spread gradually, until in ten years' time n extended over the whole external surface of the body: so that, but for his woolly hair, the body might have been taken for a very fair European. . . . At the same time that the boy began to change his colour, he also began to lose the sense of smell, and by the time he had become white, his smell was so seriously impaired that Dr. Hutchison, who records the case, states it to have been completely lost (Ogle, 1970: 276).

Ogle proposed a relationship between this pigmentation loss and the boy's anosmia, and found the connecting link in the fact that the mucous membrane of the olfactory region is also pigmented—and presumably was also destroyed by the boy's mysterious pigment disease. This suggested that olfactory keenness was related to pigmentation of this membrane, which in turn could be proportional in its intensity to the general level of skin pigmentation. If all this were true, it was only to be expected that the darker races should have more powerful senses of smell. Moreover, since pigmentation level was obviously an inherited trait, European civilizing influences were less likely to be in a position to modify the primitive noses of blacks and other nonwhite races than had been often believed.[10]

One other important theme in the story of olfaction and the "savage" still needs to be identified. This was the idea that the olfactory acuity of the

[10] Ogle also explored the possibility—based on data gathered by Darwin—that albino (and other kinds of white) animals possess less acute olfaction than darker animals, and are thus at an survival disadvantage when it comes to avoiding accidental poisoning, etc. (1870: 280–286).

nonwhite races could in turn be related to the more powerful (and usually unpleasant) body odors allegedly produced by these peoples, odors which they used to identify gender, tribal affiliation, etc. Certainly, there was something strained about this argument, since greater olfactory acuity in a people would rather logically incline one to suppose a capacity to carry out social and survival goals using relatively subtle olfactory cues (remember the American Indian?). However, social and ideological goals, not logical ones, were at stake here, and the idea that "primitives," especially Negroes, exuded a strong, unpleasant odor comes up again and again in the literature; even Darwin echoed the common consensus (1871: 225). In France, the psychiatrist Edgar Bérillon considered that the odor of the black races was responsible for keeping racial hatred alive in the United States, and an important motivation behind segregation (Bérillon, 1908). Havelock Ellis noted in 1905 that "all men and women are odorous," but Negroes especially so. On the other hand, Ellis was also one of the few to admit that Europeans were "considerably more odorous than are many other races—for instance the Japanese" because the latter are less hairy and therefore have fewer sweat glands than Europeans.

Olfaction and Brain Evolution: Broca and the "Great Limbic Lobe"

There was a more complex and subtle alternative explanation for the diminution of olfaction in the evolution of the highest mammals than the inchoate and contradictory arguments outlined above. It was championed in the 1870s by an individual better known for his contributions to the history of neuroanatomy and physical anthropology: the French physician Paul Broca. In Broca's explanation of the de-evolution of olfaction, evolutionary, neuroanatomical, and anthropological concepts on this theme found their first (however imperfect) synthesis. This is thus a story that deserves to be better known.

It begins in the late 1830s, when the French neuroanatomists Foville and Gratiolet pointed out that, despite the great variation in the various mammalian brains, one feature was common to all: a circular threshold (Latin *limen*) at the level of the third ventricle between the diencephalon and the cerebrum. In 1838, Gerdy and Foville noted further that this ring-like area was connected anteriorly to the olfactory lobe, interrupted only by the "temporal fissure" (the median aspect of the Sylvia fissure). This suggested that it played a critical role in the cerebral mediation of olfaction. Some forty years later, Paul Broca did not question this functional designation when he likened the area to the "circular border at the opening of a purse" (1878: 386). At an 1877 meeting of the *Société d'Anthropologie*, he baptized it the "limbic convolution" (Latin *limbus*, meaning "hem" or "border") (Broca, 1877); the following year he renamed it the "great limbic lobe" (the change of term being a result of his desire to stay true to Gratiolet's convention that true gyri or convolutions are limited to one lobe; the area Broca was interested in passed through all the lobes on the medial surface of the hemisphere).

In his 1878 discussion of the likely functional significance of this structure, Broca proceeded to examine an abundance of anatomical specimens (including fish, ducks, horses, otters, dolphins, chimpanzees, tapirs, foxes, and humans). Like past researchers, he concluded that olfaction was of highest survival value to submammalian vertebrates. He also noticed that as one progressed "up" the phylogenetic tree, the olfactory lobe was increasingly displaced rostrally, bringing it into closer contact with the forebrain. This fact had important behavioral implications. In lower animals, Broca observed that olfactory stimuli elicit only reflex responses; in fact, the olfactory apparatus "can even function independently of the brain proper, because in inferior vertebrates it has no connection with the encephalon that represents the cerebral hemispheres in mammals" (1878: 394). However, in "higher vertebrates" (reptiles, certain birds), the olfactory tubercle has direct connections to the cerebral peduncle, which allows for more controlled, reasoned reactions to olfactory stimuli. In most subprimate mammals, the olfactory tubercle attaches to the cerebral peduncle and the cerebrum, which allows the olfactory lobe "to function alternately or simultaneously as an autonomous center and as the vassal of the hemispheres" (1878: 396). Finally, in primates (including humans), the olfactory tubercle makes contact exclusively with the cerebrum, suggesting a high level of intellectual inhibitory control and mediation.

If one then follows the changes in the limbic lobe through Broca's four classes of primates (*cébiens, pithéciens, anthropoides, hominiens*), one sees that the limbic lobe progressively "regresses and atrophies" in proportion to the simultanous expansion of the supra-limbic cortex, particularly the frontal lobe, "seat of the highest intellectual faculties." This devolution of the limbic lobe was made all the more dramatic by its apparent stability of form up the phylogenetic scale, a lack of change which contrasted markedly with the visible increase of complication and folding in the brain areas surrounding it (1878: 396). In other words, the limbic lobe seemed to represent a part of the brain that had ceased to evolve functionally; it had, instead, simply decreased in raw bulk as other brain structures evolved and pushed their way into greater prominence. Broca was thus prompted to call the limbic lobe the "brute" cortex, noting that it predominated in brutes dominated by olfactory reflexes. The remaining mantle he named the "intellectual" cortex, since within it, all sensation was subordinated to the intellect (1878: 393–394). The moral of this neuroanatomical speculative reconstruction of the phylogeny of olfaction was that the increasing connection between the olfactory apparatus and the forebrain had led olfaction to become more and more "intelligent," until it finally became totally subordinate to intellect and effectively lost its instinctual, reflex power.[11]

Broca did not, however, simply want to argue that the frontal lobe came to dominate over the olfactory lobe in the evolution of higher mammalian organisms. He went so far as to propose that the forebrain had achieved its dominance through *exploitation* of its connections to that once all-powerful "brute cortex."

One can but wonder if it is not that reason which explains the intellectual superiority of the frontal lobe, called upon by its connections to the olfactory lobe to interpret, to incessantly discuss the sensations that have the greatest utility to the existence of the animal. Having thus become the seat of the most important determinations, and kept constantly alert, this lobe acquires a sort of hegemony in the hemisphere, and its intellectual predominance—once established as the law of the mammalian cerebral type—would be maintained and further developed on its own (Broca, 1879: 421).

The larger vision behind Broca's concept of "frontal predominance"—a vision of the human animal transformed from a ground-sniffing beast to a forebrain-dominated, intelligent biped—would be lyrically evoked by a colleague of Broca, Samuel Pozzi, in a posthumous elegy of Broca's scientific contributions:

Animals were quadrupeds, a position essentially favorable to the exercise of smell. The primate rose up, man lifted his head forever more away from the earth and directed his sight parallel to the horizon. The hegemony of the visual sense substituted itself for that of olfaction. . . . [T]he anterior extremities, now free, were modified, became hands, and touch took its place alongside of sight on the first rung of the senses.

What a radical revolution! Is it not curious that the anatomist can legitimately link to this first fact [bipedalism], so apparently simple, the development of the frontal lobe?

It is no less curious, from the same point of view, to discover in man's brain the vestiges of that dethroned organ, the limbic lobe. Instead of forming an imposing unit, it is reduced to the state of fragments, barely linked to each other; in other words, a collection of debris (Pozzi, in Broca, 1888).

If "predominance" of the forebrain over the limbic lobe was the hallmark of advanced evolution, then it is perhaps not surprising that Broca—

---

[11] As an aside, it is worth noting the extent to which this model of Broca's recalls the influential theory of "cortical dominance" that would be developed by the American neurologist John Fulton in the 1930s. Much like Broca (if more comprehensively), Fulton would argue for a progressive "encephalization" of function through the evolutionary series, so that a relatively greater proportion of functions would be mediated by the neocortex the higher one ascended up the primate scale (Fulton, 1938). While Broca's role in the history of neurology as the founder of the doctrine of "cerebral [left hemisphere] dominance" is generally accepted, it may be appropriate at this juncture to put in a modest claim for his priority in the development of the concept of "cortical dominance" as well.

a physical anthropologist at a time when France was preoccupied with solidifying its colonial empire—should have expected to find domination to be incomplete in the nonwhite, "primitive" human races. He found his evidence in the alleged presence of a "limbic fissure"[12] in the brains of the nonwhite human groups he looked at—an anatomical peculiarity allegedly not present in white Europeans, where cortical absorption of the limbic structures showed itself to be more complete. The "limbic fissure," then, represented a "trait of inferiority," a "sign of incomplete evolution, arrested at the simian level" (1878: 449; cf. 1877: 655).[13] Drawing on the traditions laid out by the authors mentioned in Section 2.4, Broca backed up his anatomical argument with pointed references to the extra-acute, if crudely instinctual, sense of smell possessed by the "man of nature." He contrasted this "survival" sense of hunting and war with the more refined, intellectualized "luxury" experience enjoyed by the "civilized man" with his perfumes, fine wines, and other fragrant pleasures (Broca, 1879: 450–451, 454). Thus we see how an esoteric theory on the evolutionary significance of olfaction in humans mapped elegantly onto social "knowledge" about the relative superiority of the white European, adult male over all other human species variations.[14]

Sexology and "Perversions" of Olfaction

A bit of recapitulation is now in order: we have seen that the idea that odor plays a role in sexual arousal and selection in humans had been widely accepted in the literature since at least the beginning of the nineteenth century. It was an uncomfortable idea for many, and early writers thus often qualified this observation by suggesting that "uncivilized" races were particularly sensitive to the erotic effects of odor. Yet, as we have seen,

---

[12] A small sulcus separating the hippocampal gyrus from the surrounding temporal lobe.

[13] For other examples of Broca's reliance on comparative neuroanatomical research to demonstrate nonwhite (and nonmale) evolutionary inferiority, see Gould (1981) and Harrington (1987).

[14] Insofar as white European women were widely believed in the nineteenth century to stand at an evolutionary level similar to that of the male nonwhite "primitive," one would expect to find remarks in the literature suggesting that white women were also more sensitive to olfactory stimuli than white men. The following remark from Havelock Ellis is suggestive in this direction: after reviewing a wide range of reports, he concluded that, while men and women were both susceptible to the "sexual influence of olfactory impressions, it is probable indeed that [women] are more affected, and more frequently affected, in this way, than are men" [Ellis, 1905: 86). At the same time, their relative evolutionary inferiority would help make women (along with their savage male counterparts) *less* susceptible to olfactory perversions than the highly civilized, dandified white male (see the next section).

they could not help but recognize that European civilization had cultivated an elaborate culture of sensual scents as well, perfumes being only the most obvious example (see Corbin, 1986: 176–199).

All this would probably have been still tolerable, so far as it went, but that was not the end of the matter: medical men had also begun to find reason to think that an *unnatural* preoccupation with olfactory stimuli might be most common *not* among "primitives," but among "civilized" peoples. Thus Cloquet related the embarrassing story of Cardinal Richelieu's olfactory addiction: "Richelieu had made such an abuse of perfumes in all their forms, that he could no longer sense their action, and he regularly lived in such an embalmed atmosphere that it made most of his visitors feel ill" (Cloquet, 1821: 137).

When Richard von Krafft-Ebing retold Cloquet's anecdote in his massive 1886 work *Psychopathia sexualis*, he described Richelieu as a sexual "pervert," a fetishist who needed ever-stronger olfactory stimuli to stir his overindulged sexual passions (1886: 119). At the time he was writing, Krafft-Ebing was only one of a small army of alienists who had been working to catalogue the so-called perversions and fetishes for several decades. Auguste Ambroise Tardieu used Latin in 1867 to describe the shocking practices of "sniffers" who got a sexual thrill from the smell of women defecating; in the 1880s, the French psychiatrists Charles Féré and Alfred Binet wrote of an olfactory fetishist in whom odors unleashed apparently irresistible impulses, compelling him to follow women in the streets whose scent fascinated him (Corbin, 1986: 208–209, 287).

Collective nightmares about some inevitable process of degeneration in European society—a society made "unnatural" as a result of its high level of civilization—gave an added emotional intensity to the growing morbid fascination with the prevalence of olfactory perversions in the white races (cf. Nye, 1984). Thus in his 1897 *Libido sexualis*, Albert Moll suggested that the decline in olfactory acuity in civilized man had had the disadvantage of depriving him of a powerful, natural spur to normal sexual choice; left to his own disoriented devices, he thus became prone to sexual perversion (cited in Sulloway, 1979: 304). Edgar Bérillon was similarly convinced that the dulling of the olfactory sense associated with the rise of civilization had led to a loss of instinctual orientation that carried in it the seeds of a degenerative process; at the same time, this French physician knew that a return to an animallike reliance on smell could also have regressive consequences (Bérillon, 1908). In a variation on the degeneration tune, John MacKenzie in the United States blamed olfactory perversions on the decadence and indulgent overstimulation of the upper-class life-style: "[I]n view of the relations of civilization to the sense of smell, we can readily understand why it is that this faculty is found more frequently deranged among the superior orders than in those lower down the social scale and in the savage state" (MacKenzie, 1898: 118n).

One prominent writer who argued, on the contrary, that the diminution of olfaction in modern life was a necessary and progressive process was Sigmund Freud. We already know that, through his contact with Fliess, Freud had been alerted to the problems associated with olfaction in sexuality. Yet, rather than seeing the decline of olfaction in humans as opening the door to increased sexual perversity, Freud identified repression of the olfactory function as essential to normal psychosexual development.

Following somewhat in the tradition of Broca, Freud proposed that, in human prehistory, upright carriage had made certain previously attractive olfactory stimuli (urine, feces, etc.) repulsive (Freud, 1895: 269). In a similar way, the anus and mouth, primitive sources for the "release of sexuality," no longer functioned as such in normally developed mature individuals, where sexual interest was firmly oriented towards the genitals. "Normal repression" (the extinction of inappropriate primitive sexual zones and their associated odors) was associated with the development of morality, modesty, etc.; the failure of repression could lead to a developmental inhibition that could in turn find expression in a sexual perversion, including an olfactory fetish.

Indeed, the olfactory fetishes were probably the most common—and most dangerous—of all. Like many other sexologists, Freud believed that perversion was a symptom of psychosexual regression, a descent into a prehuman brutish attitude. He thus remarked in an 1897 letter to Fliess: "The perversions regularly lead into zoophilia and have an animal character" [Freud, 1985 (letter to Fliess, Jan. 11, 1897)]. Smell, recognized by evolutionists as the dominant medium of sexual attraction in animals, would naturally present the largest field for perverted expression by regressed humans. It was therefore essential that the free working of this erotogenic zone be roundly suppressed. Considering the alternative scenario makes the urgency of the matter clear: if smell were permitted to remain a powerful source of sexual arousal in humans, the whole surface of the notoriously odorous human body would be at risk of acting as a sexual stimulant; an embarrassing situation impossible to reconcile with morality and civilized life.

From "Smell Brain" to Limbic System: The Work of Papez and MacLean

With Freud, we effectively conclude our survey of nineteenth century medical thinking on olfaction and the nervous system. However, our story is not quite finished. To conclude, we would now turn our attention briefly back to one of the earlier major players in our story—the "smell brain" of Paul Broca—and follow its fate into the first half of the twentieth century. Doing this will also help show the extent to which certain evolutionary and sexual themes explored in other sections of this paper have found new voices and guises in our century.

To begin, let us pass over several decades of relative quiescence in our story to 1937, the year the American anatomist James Papez published what was at the time a little-noticed paper on functional brain anatomy. Here he noted that the "general view of most morphologists, e.g. Ramon y Cajal, Déjerine, Turner and G.E. Smith, has been that the hippocampus, and even the entire limbic lobe, mediate in some obscure way the olfactory sector of functions" (Papez, 1937: 732). It was Papez's goal to suggest that matters were considerably more complex. In fact, his goal was to convince his colleagues that the brain areas associated for so long with olfaction were intimately involved with emotional experience and expression.

Certainly, there was a sense in which this implicit link between olfaction and affect was hardly novel: in their discussion of emotionally colored memories, the older writers on olfaction had long been conscious of it. Indeed, the Dutch physiologist H. Zwaardemaker was so impressed with the relationship between odors and strong sentiment that he called for a quantitative psychology of olfaction that would get to the bottom of these issues (Zwaardemaker, 1899: 203–204). Zwaardemaker's extensive work with "olfactometers" (which measured subjects' speed and acuity of response to selected olfactory stimuli) would later be praised by Havelock Ellis (1905) as the first investigations to put olfactory research squarely on the road of science[15] [for more on the history of olfactory research techniques, see William Cain (this volume)].

Papez, whose whole view of brain anatomy was deeply permeated by evolutionary principles, was functioning very much in the tradition of many of these nineteenth-century writers. He did not particularly question the olfactory function of at least some of the structures classically associated with the rhinecephalon, but he was convinced that emotion had evolved through the practices of feeding and sexual expression, in which olfaction and gustation play a critical role. Closing his long, detailed 1937 paper, he wrote challengingly: "Is emotion a magic product, or is it a physiologic process which depends on an anatomic mechanism?" It hardly needs to be said that he wished to affirm the latter. However, a final synthesis of his insights would still be needed before the "smell brain" could emerge out of its relative position of neglect at that time, and be reinvested with something of its earlier fascination as that slightly shameful organ "capable of

---

[15] Ellis himself had this to say on the phenomenological links between emotion, memory, and olfaction:

No sense has so strong a power of suggestion, the power of calling up ancient memories with a wider and deeper emotional reverberation, while at the same time no sense furnishes impressions which so easily change emotional color and tone, in harmony with the recipient's general attitude (Ellis, 1905: 55).

shaking man's inner life more profoundly" than any other (Corbin, 1986:8).

The job of synthesizer and visionary would fall to another man, Paul MacLean. MacLean saw that the evolution of the cerebrum out of the olfactory brain was "obtrusive evidence" of the central role of smell in the evolutionary development of the vertebrates. It seemed likely, though, that evolutionists had understood the functional meaning of this sense too narrowly. Olfaction's central relationship to feeding, fleeing, fighting, and copulation (the "four F's" of animal behavior) permitted the suggestion that "smell . . . might be thought of as an oral sense, or more broadly as a visceral sense, [and] . . . the rhinecephalon might be justifiably considered a visceral brain . . . [to be distinguished] from the neocortex which . . . subserves the functions of the intellect" (MacLean, 1949).

The multiplicity of needs served by the visceral brain areas—and the ever-present danger of crossed wires—might, MacLean felt, provide an explanation for a range of gustatory and olfactory perversions associated with sexual behavior, as well as cases of sexual sadism, sex murder, and so on:

> To use a crude analogy, it is as though the various elements of the sensorium were gathered together in the hippocampal gyrus and placed on a party line. The ringing of the bell for one part, particularly if persistent and intense, might bring one or more of the other parties to the phone. Much of the gossip going back and forth in this area is what we have commonly come to associate with the id, the beast, or sin in man (e.g. gluttony, lechery, etc.). In light of this it is interesting that the large uncinate fasciculus, the frontal lobes "stand guard" over this region. Could it be that feelings of guilt are fomented here? (MacLean, 1949: 198–199. Emphasis added).

But that was not all. MacLean also suggested in this article that the rhinecephalon—here renamed the "visceral brain" and soon to be christened the "limbic system"[16]—stood in the same relationship to the human cerebral cortex as that posited by Freud between the rational ego and the passion-driven id. Communication between the two levels was extremely problematic, largely because the "visceral" brain functioned nonverbally while the cortex functioned verbally. A great many psychosomatic disorders (MacLean published his initial 1949 article in the journal *Psychosomatic Medicine*) had their source in such faulty communication between brain levels.

---

[16]MacLean renamed his "visceral brain" in 1952, changing it to "limbic system," apparently because there had been some confusion about the scope of the term "visceral" in his concept (Durant, 1985: 7). Although he acknowledged Broca's role in creating the neologism "limbic" for the areas of the brain associated with the rhinecephalon, there is no indication that he was aware of Broca's evolutionary theories about these brain areas.

Considered in the light of Freudian ,psychology, the visceral brain would have many of the attributes of the unconscious id. One might argue, however, that the visceral brain is not at all unconscious (possibly not even in certain stages of sleep), but rather eludes the grasp of the intellect because its animalistic and primitive structure makes it impossible to communicate in verbal terms. Perhaps it were more proper to say, therefore, it was an animalistic and illiterate brain (MacLean, 1949: 200–201).

With MacLean, the nineteenth century's beast in the "smell brain" had returned, and this time it was clear that the creature was in every one of us.[17]

References

Althaus, J. (1881) A lecture on the physiology and pathology of the olfactory nerve. *Lancet, 1*: 771–773; 813–815.

Bérillon, E. (1908) Psychologie de l'olfaction: la fascination olfactive chez les animaux et chez l'homme. *Revue de l'Hypnotisme*, October 1908, 98ff.

Broca, P. (1877) Sur la circonvolution limbique et la scissure limbique. *Bulletins de la Société d'Anthropologie. 12*(2), 646–657.

Broca, P. (1878) Le grand lobe limbique et la scissure limbique. *Revue d'Anthropologie, 1*(2), 385–498.

Broca, P. (1879) Anatomie du lobe olfactif. *Bulletins de la Société d'Anthropologie*, 4th ser., *4*, 596–598.

Broca, P. (1888) *Mémoires sur le cerveau de l'homme et des primates*, intro. by S. Pozzi. Paris: C Reinwald.

Cabanis, G. (1805) *Rapports du physique et du morale de l'homme*. Paris.

Cloquet, H. (1821) *Osphrésiologie, ou Traité des odeurs, du sens et des organs de l'olfaction, avec l'histoire détaillée des maladies du nez et des fosses nasales et des opérations qui leur conviennent.* 2d ed. Paris: Méquignon-Marvis.

Cohen, J.S. (1884) *Diseases of the Thorax and Nasal Passages*. 2d ed. New York: William Wood & Co.

Collet [no first name] (1904) *L'Odorat et ses troubles*. Paris: J.-B. Ballière.

Corbin, A. (1986) *The Foul and the Fragrant: Odor and the French Social Imagination*. Cambridge, MA: Harvard University Press [originally published as *Le miasme et la jonquille*, Aubier Montaigne, 1982].

Critchley, M. (1979) Man's attitude to his nose. In *The Divine Banquet of the Brain, and Other Essays*. New York: Raven Press, pp. 106–114.

Darwin, C. (1836–1844) *Charles Darwin's Notebooks, 1836–1844*, P.H. Barrett, P.J. Gautrey, S. Herbert, D. Kohn, and S. Smith (Transcribers and eds.), Cambridge: Cambridge University Press, 1987.

Darwin, C. (1871) *The Descent of Man and Selection in Relation to Sex*. New York: A.L. Burt.

---

[17]For a more complete history of MacLean's "limbic system," its intellectual origins, and its wider gocial reverberations, see Durant (1985). For a discussion of current research directions on the limbic system (with an historical retrospective on Papez), see Doanne (1986).

Durant, J. (1985) The science of sentiment: the problem of the cerebral localization of emotion. In P.P.G. Bateson and P.H. Klopfer (Eds.), *Perspectives in Ethology*, Vol. 6: *Mechanisms*. New York: Plenum Press, pp. 1–31.

Ellis, H. (1905) Sexual selection. In *Studies in the Psychology of Sex*. New York: Random House, 1936.

Fliess, W. (1893a) *Neue Beiträge zur Klinik und Therapie der nasalen Reflexneurose*. Wien: F. Deuticke.

Fliess, W. (1893b) Les réflexes d'origine nasale. *Archives Internationales de Laryngologie*, *6*, 266–269.

Fliess, W. (1897) *Die Beziehungen zwischen Nase und weiblichen Geschlechtsorganen in ihrer biologischen Bedeutung dargestellt*. Leipzig: F. Deuticke.

Fliess, W. (1910) *Über den ursächlichen Zusammenhang von Nase und Geschlechtsorgan. Zugleich ein Beitrag zur Nervenphysiologie*. Halle: C. Marhold.

Fraenkel, B. (1881) Uber den Zusammenhang von Asthma nervosum und Krankheit der Nase. *Berliner Klinische Wochenschrift*, *16*, 217–220, 238–240.

Freud, S. (1895) Project for a scientific psychology. In J. Strachey (Ed.), *Standard Edition of the Complete Psychological Works of Sigmund Freud*, 24 vols. London: Hogarth Press, Vol. 1.

Freud, S. (1985) *The Complete Letters of Sigmund Freud to Wilhelm Fliess, 1887–1904*, J. Masson (Ed.). Cambridge, MA: Harvard University Press.

Fulton, J.F. (1938) *Physiology of the Nervous System*. London: Oxford University Press.

Gould, S.J. (1977) *Ontogeny and Phylogeny*. Cambridge, MA: Harvard University Press.

Gould, S.J. (1981) *The Mismeasure of Man*. New York: W.W. Norton.

Gray, H.A.C. (1875) [Letter to the Editor] Sneezing. *Lancet*, Jan. 16, 1875, p. 110.

Hack, W. (1882) Reflexneurosen und Nasenleiden. *Berliner klinische Wochenschrift*, *25*, 379–384.

Haeckel, E. (1874) *Anthropogenie, oder Entwickelungsgeschichte des Menschen*. Leipzig: W. Englemann.

Hagen, A. (1901) *Die sexuelle Osphresiologie. Oie Beziehungen des Geruchssinnes und der Gerüche zur menschlichen Geschlechtsthätigkeit* (Supplementary volume to: *Studien zur Geschichte des menschlichen Geschlechtslebens*). Charlottenburg: H. Barsdorf.

Harrington, A. (1987) *Medicine, Mind and the Double Brain: A Study in Nineteenth-Century Thought*. Princeton: Princeton University Press.

Jackson, J.H. (1864) Unilateral epileptiform seizures beginning by a disagreeable smell. *Medical Times and Gazette*, Aug. 13, 1864, p. 168.

Jackson, J.H. (1880–1881) On right or left-sided spasm at the onset of epileptic paroxysms, and on crude sensation warnings, and elaborate mental states. *Brain*, *3*, 192–205.

Kleist, K. (1934) *Kriegsverletzungen des Gehirns in ihrer Bedevtung für die Hirnlokalisation und Hirnpathologie*. In: *Handbuch der ärztlichen Erfahrung im Weltkrieg*, vol. 4, series O. von Schjerning (Ed.). 37 Leipzig: Barth.

Knight, F.I. (1877) Case of anosmia following a blow on the occiput. *Boston Med. Surg. J.*, *97*, 293–301.

Krafft-Ebing, R. von (1886) *Psychopathia sexualis*, Trans. H.E. Wedeck. New York: G.P. Putnam & Sons, 1965.

Legg, J.W. (1873) A Case of Anosmia Following a Blow, *Lancet*, *2*, 659–660.
MacKenzie, J. (1884) Irritation of the sexual apparatus as an etiological factor in the production of nasal disease. *Am. J. Med. Sci.*, April 1884, pp. 360–365.
MacKenzie, J. (1886) The production of the so-called "rose cold" by means of an artiflcial rose, with remarks and historical notes. *Am. J. Med. Sci.* (January 1886) pp. 45–57.
MacKenzie, J. (1898) The physiological and pathological relations between the nose and the sexual apparatus of man. *J. Laryngol., Phonol. Otol.*, *13*, 109–123.
MacLean, P.D. (1949) Psychosomatic disease and the "visceral brain": Recent developments bearing on the Papez theory of emotion. *Psychosomatic Med.*, *11*, 338–353.
Maudsley, H. (1868) *The Physiology and Pathology of Mind*, 2d ed. 1878.
Mayer, A. (1875) [Letter to the Editor] Ritual circumcision—Sneezing. *Lancet*, Jan. 9, 1875, p. 73.
Moldenhauer (1888) *Traité des maladies des fosses nasales des sinus et du pharynx nasal.* Trans. from the Gemman 1886 edition by Potiquet. Paris: Asselin et Houzeau.
Moure, E.J. (1893) *Manuel pratique des maladies des fosses nasales de la cavité naso-pharyngienne et des sinus de la face*, 2d ed. Paris: Octave Doin.
Notta. (1870) Recherches sur la perte de l'odorat. *Archives Générales de Médecine*, 6th series, *15*, 385–407.
Nye, R. (1984) *Crime, Madness and Politics in Modern France: The Medical Concept of National Decline*. Princeton: Princeton University Press.
Ogle, W. (1870) Anosmia, or, cases illustrating the physiology and pathology of the sense of smell. *Medico-Chirurgical Trans*, *53*, 263–290.
Papez, J.W. (1937) A proposed mechanism of emotion. *Arch. Neurolo. Psychi.*, *38*, 725–743.
Richards, R. (1987) *Darwin and the Emergence of Evolutionary Theories of Mind and Behavior*. Chicago: Universrty of Chicago Press.
Ries, E. (1903) A new treatment for dysmennorrhea. *Am. Gynecol.*, *3*, 375–379.
Riese, W. (1936) Symptomatologie der Erkrankungen des Riechnerven; Symptomatologie der Erkrankungen der Riechrinde [companion articles]. In O. Bumke and O. Foerster (Eds.), *Handbuch der Neurologie*, vol. 4: *Hirnnerven*. Berlin: Julius Springer, pp. 1–37.
Serres, E.R.A (1826) *Anatomie comparée du cerveau*. Paris: Gaben & Co.
Spillane, J.D. (1981) *The Doctrine of the Nerves: Chapters in the History of Neurology*. Oxford: Oxford University Press.
Sulloway, F.A. (1979) *Freud: Biologist of the Mind*. Bungay, Suffolk: Fontana Paperback, 1980.
Watson, W.S. (1875) *Diseases of the Nose and Accessory Cavities*. London: H.K Lewis.
Zwaardemaker, H. (1899) Les sensations olfactives, leurs combinaisons et leurs compensations. *L'Année psychologique*, *5*, 202–225.

Part II
The Basic Science of Olfaction

2 Morphology and Plasticity of the Vertebrate Olfactory Epithelium

EDWARD E. MORRISON and RICHARD M. COSTANZO

Introduction

The olfactory system detects and processes chemicals in the environment providing long-range information that is necessary for survival. Many vertebrates rely chiefly on olfactory input for social interactions, food locating, migration, reproduction, and alerting to the presence of predators.

In most vertebrates the receptor cells are in the olfactory epithelium, a pseudostratified structure that lines a relatively small portion of the nasal cavity (Graziadei, 1973a). The olfactory epithelium differs in several respects from other regions of the central nervous system. First, the olfactory receptor cell body has a peripheral location in direct contact with the external environment. Other CNS sensory neurons are generally protected by being more centrally located. Second, olfactory neurons have a single unbranched axon that projects directly to the telencephalon without synapsing on intermediate relay neurons. Third, the olfactory system replaces neural elements normally and when injured. The replacement neurons develop, mature, and grow axons that form new connections with second-order neurons of the olfactory bulb.

The evolution of continued olfactory neurogenesis is probably related to the receptor neurons' exposure to the environment, subject to insult and injury. Indeed, no other vertebrate CNS neuron has the capacity for replacement in adults. Postnatal neurogenesis and the ability to continuously form new anatomical and functional connections within the CNS makes the olfactory system an important model for studies of development and regeneration (Moulton, 1974; Graziadei and Monti Graziadei, 1978, 1979; Cancalon, 1982; Costanzo, 1985; Morrison and Costanzo, 1989).

Morphology

The nasal cavity is predominantly lined with a pseudostratified ciliated respiratory epithelium. This respiratory epithelium is continuous with a sensory (olfactory) neuroepithelium lining a small portion of the posterior

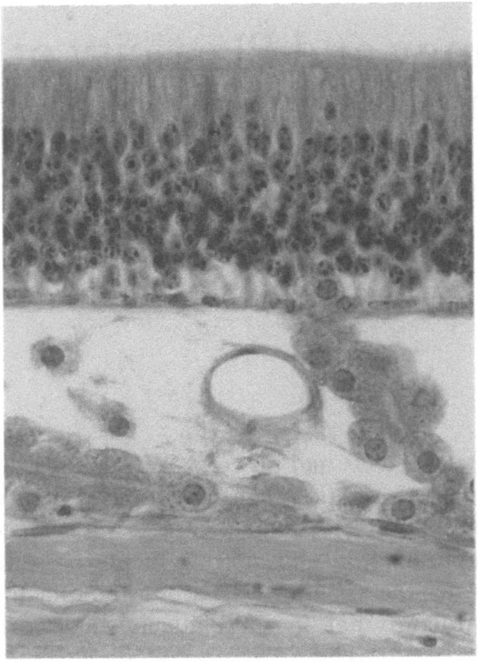

FIGURE 2.1. Light micrograph of olfactory neuroepithelium. The epithelium contains supporting cells, olfactory neurons, and basal cells. The lamina propria contains blood vessels, axon bundles, and Bowman's gland.

nasal cavity in most land vertebrates. The neuroepithelium has a pseudostratified columnar organization of three cell types: olfactory neurons, supporting cells, and basal cells (Graziadei, 1973b; Moulton, 1974). The nuclei of the three cellular elements are arranged in discrete layers (Figure 2.1). A fourth cell type, the microvillar cell, has occasionally been seen in the upper region of the olfactory neuroepithelium of some vertebrates (e.g., dog, Okano et al., 1967; humans, Moran et al., 1982; Morrison and Costanzo, 1990).

Olfactory Neurons

Olfactory neurons originate from a pair of head neural ectoderm swellings, the olfactory placodes (Carpenter, 1937; Cuschieri and Bannister, 1975). The placodes invaginate, forming the olfactory pits and, subsequently, the nasal cavity. The olfactory placode is composed of head nervous ectoderm, from which the olfactory receptor cells will develop, and a superficial nonnervous ectoderm layer that gives rise to nonneuronal components such as supporting cells and Bowman's gland. These two embryonic ectodermal tissue types combine and form the developing olfactory mucosa (Klein and Graziadei, 1983).

The olfactory receptor cell is a slender bipolar neuron generally located in the basal two-thirds of the epithelium. The mature neuronal cell body diameter is 5–8 μm; the nucleus typically contains 2–3 nucleoli. Olfactory neurons are in direct contact with others and with adjacent supporting cells (Graziadei, 1971; Breipohl et al., 1974; Rafols and Getchell, 1983; Costanzo and Morrison, 1989).

It has been shown that olfactory neurons arise from neuroblast cells (basal cells) at the base of the epithelium. These are the stem cells that mitotically divide, giving rise to new cells that develop into olfactory receptor neurons (Graziadei and Monti Graziadei, 1979). Three basic elements can be ultrastructurally identified during differentiation: basal cells, globose cells, and receptor neurons. Intermediate morphologies suggest that there are transitional forms between the basal cells proper and the maturing receptor neurons (Graziadei and Monti Graziadei, 1979).

Olfactory neurons, like other CNS neurons, develop an axon process that gradually tapers to a 0.1–0.3-μm process from a conical region at the base of the cell body (Ramon Y Cajal, 1916). Olfactory axons are always unbranched, unmyelinated, and grouped within the epithelium in small fascicles that pass through the basal lamina into the underlying lamina propria. Within the lamina propria, olfactory axons are ensheathed by Schwann or sheath cells. These cells are immunoreactive for the astrocytic protein glial fibrillary acidic protein (GFAP), and enclose olfactory axons in an unusual manner in which several mesaxon processes may surround up to 50 axons (Rafols and Getchell, 1983; Farbman and Squinto, 1985; Barber and Lindsay, 1982b). Olfactory axons then form larger fascicles that pass centrally through the foramina of the ethmoid bone's cribriform plate. Olfactory axons form their characteristic terminal structure, the glomerulus, in the olfactory bulb, where they make their first synapse with dendrites of second-order neurons.

The olfactory dendrite, with a diameter of 1–2 μm, originates from the apical part of the cell body. Dendrites are unbranched and extend to the epithelial surface in a straight or tortuous manner, passing along supporting cells. Fine cellular processes make contact with other dendrites and supporting cells (Graziadei, 1971; Breipohl et al., 1974). Olfactory dendrites terminate in expanded vesicles, the olfactory knobs at or near the epithelial surface. Olfactory knobs may be elongated or oval and are attached to adjacent cells by typical junctional complexes. A variable number (10–60) of sensory cilia arise from the olfactory knob and form a characteristic dense ciliary blanket over the mucosal surface (Figure 2.2).

Olfactory cilia arise from a basal body and have the "9 + 2" pattern of filaments typical of vertebrate motile cilia. As the cilia extend from the dendritic knob, the shaft tapers to where they may have only a few microtubules (Moran et al., 1982; Menco and Farbman, 1985). Olfactory cilia

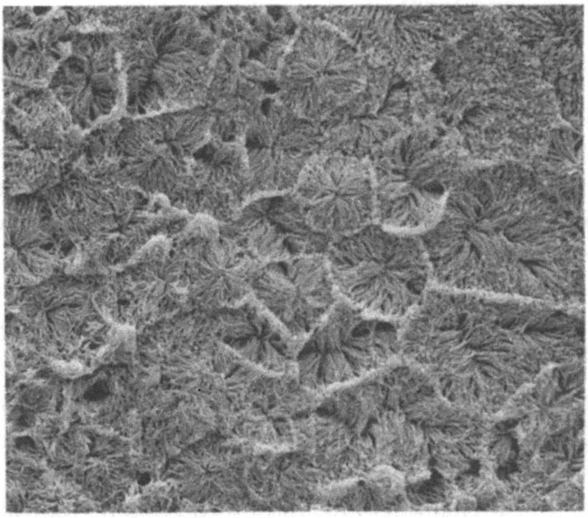

FIGURE 2.2. Scanning micrograph of olfactory epithelium surface showing the dense blanket of sensory cilia. (Reprinted, by permission, from Morrison and Costanzo, 1989a.)

have long been thought of as the site for initial sensory transduction (Vinnikov and Titova, 1957). Odorants enter the nasal cavity, are absorbed into the mucus layer, and are hypothesized to initiate olfactory transduction by binding to receptors on the cilia. There is evidence that binding odorant molecules to receptors activates a second messenger (cAMP) that acts on a channel protein, producing a receptor potential. Recent studies on isolated olfactory cilia show that they are enriched with odorant-sensitive adenylate cyclase and contain transductory enzymes such as cyclic nucleotide phosphatidase, signal-coupling guanine triphosphate (GTP) binding protein, and cAMP-dependent protein kinase (Anholt et al., 1987; Lancet, 1986; Pace and Lancet, 1986).

Sensory cilia are present in almost all vertebrate olfactory systems. In some species of fish, reptiles, and birds, the dendritic knob is covered with microvilli (Graziadei, 1973a). Recently, a population of cells in the human olfactory epithelium has been identified as having a microvillar rather then a ciliated surface (Moran et al., 1982; Morrison and Costanzo, 1990). These cells are typically near the epithelial surface and also have morphological characteristics (having an axonlike process) of neurons (Figure 2.3). Microvillar cells have also been observed near the epithelial surface in other species. Okano et al. (1967) observed a cell type in the dog olfactory epithelium with short microvilli and clear cytoplasm. Similar cells have also been observed in the turtle, rat, and cat. Although the microvillar cell has neuronal characteristics and may serve as a secondary chemoreceptor, its function is unknown. Sensory neurons of the vomeronasal organ, which

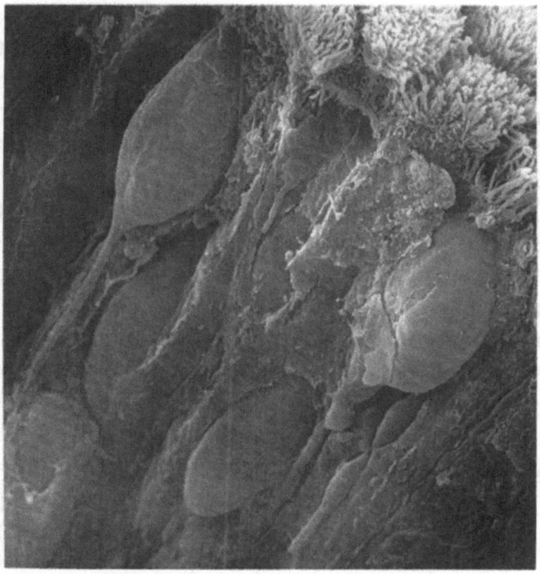

FIGURE 2.3. Scanning micrograph of human olfactory microvillar cell near the epithelial surface. (Reprinted, by permission, from Morrison and Costanzo, 1989a.)

also develop from the olfactory placode, have a microvillar surface. These cells have functional properties similar to those of olfactory receptors, although they lack cilia (Halpern, 1987).

Supporting Cells

The columnar supporting cells are major secretory components of the olfactory mucosa. They have apical microvillar surfaces, taper basally, and attach at the basal lamina by footlike processes. Supporting cells have fine cellular processes that form close contacts with each other and with olfactory receptors throughout the full depth of the epithelium. At the epithelial surface, the supporting cells attach to each other and to olfactory dendrites by typical junctional complexes. Supporting cells partially or completely surround dendrites in the apical epithelial region. In the basal region, a sleevelike process often surrounds small olfactory axon fascicles as they penetrate the basal lamina and enter the underlying lamina propria (Breipohl et al., 1974; Costanzo and Morrison, 1989). Supporting cells have a variety of basal footlike processes that attach at (and sometimes even below) the basal lamina directly onto capillaries and Bowman's glands (Wang and Halpern, 1980; Rafols and Getchell, 1983; Getchell et al., 1984; Costanzo and Morrison, 1989).

Until recently, supporting cells were thought to be a homogeneous population. Rafols and Getchell (1983), using Golgi impregnation, described two types (I and II) of columnar supporting cells in the tiger salamander. Type I cells are columnar and resemble ependyma tanycytes. Type II cells are similar to protoplasmic astrocytes, with lateral expansions surrounding several olfactory receptor cell bodies. Masukawa et al. (1985) also distinguished two types of supporting cells following olfactory nerve transection. They reported altered electrical activity in supporting cells during the degeneration and early recovery periods, and suggested that this is due to the breakdown of connections between supporting cells and degenerating olfactory neurons.

Recent scanning electron microscopy (SEM) studies of olfactory epithelium have provided detailed three-dimensional observations of supporting-cell morphology (Wang and Halpern, 1980; Breipohl and Ohyama, 1981; Costanzo and Morrison, 1989; Morrison and Costanzo, 1989). Hamsters have a heterogeneous population of supporting cells varying from tall, columnar, and cylinderlike to some with a tapering cellular process at the basal lamina. With the SEM we observed the fine cellular processes that make contact with adjacent cells. These processes form close contacts throughout the full extent of the epithelium. Such contacts have also been seen in human olfactory epithelium. The intimate connections between supporting cells and olfactory receptors have also been described in several vertebrates using transmission electron microscopy (TEM) (Graziadei, 1971; Breipohl et al., 1974). The importance of these contacts, which are observed at different levels of the epithelium, is unknown.

Supporting cells are derived from the nonnervous ectoderm layer of the olfactory placode and have been hypothesized to be functionally glial and ependymallike (Klein and Graziadei, 1983; Rafols and Getchell, 1983). Their functions may include: (a) insulation of olfactory receptor cells, (b) a glialike function within the neuroepithelium, (c) secretion (of acidic, sulfated, and/or neutral mucopolysaccharides from the apical pole of the cell surface), (d) transepithelial transport of molecules, and (e) guides for developing neurons and their neural processes (Graziadei, 1971; Getchell, 1977; Price, 1984; Rafols and Getchell, 1983; Trotier and MacLeod, 1986; Costanzo and Morrison, 1989; Morrison and Costanzo, 1989). Studies of the relations between supporting cells and olfactory receptor neurons are just beginning.

Basal Cells

Basal cells are irregularly shaped and located in the lower epithelium, near the lamina propria. Mitosis is often observed among basal cells. These cells are the stem cells or neuroblasts that give rise to replacement olfactory neurons normally and following injury (Graziadei and Monti Graziadei, 1979). Basal cells appear to be capable of neurogenesis for the life of the individual, and even well into old age (Costanzo and Morrison, 1987). The

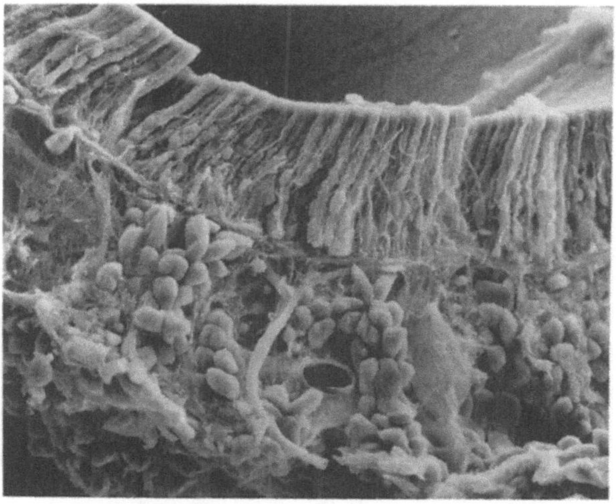

FIGURE 2.4. Low-power scanning micrograph of the olfactory mucosa. The olfactory epithelium and the underlying lamina propria. The lamina propria contains blood vessels and Bowman's glands. (Reprinted, by permission, from Costanzo and Morrison, 1989a.)

lifespan of olfactory neurons is thought to be around 30 to 45 days; they may survive for up to a year in animals kept in a sterile environment (Hinds et al., 1984).

Lamina Propria

The lamina propria, located directly below the epithelium, contains the tubuloalveolar Bowman's glands that are found throughout the olfactory region in most vertebrates (Figure 2.4) (Graziadei, 1973a). Bowman's glands and supporting cells are the major secretory units of the olfactory mucosa. Histochemically, Bowman's gland cells contain primarily neutral mucopolysaccharides (Getchell et al., 1984). A single duct projects from each gland to the epithelial surface. The lamina propria also contains olfactory axon fascicles, blood vessels, and loose connective tissue.

Biochemical Characteristics

Histochemical examination of the olfactory epithelium reveals a number of properties that appear to be unique to the olfactory system. The discovery and characterization of olfactory marker protein (OMP) has led to its use immunohistochemically in studies of the olfactory system (Margolis, 1972; Monti Graziadei et al., 1977; Carr et al., 1989). Olfactory marker protein is

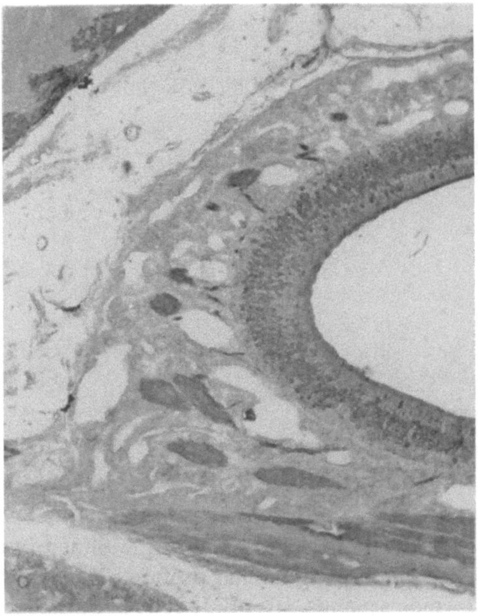

FIGURE 2.5. Olfactory epithelium stained to show the presence of olfactory marker protein (OMP). The OMP is seen in olfactory receptor neurons located in the lower two-thirds of the epithelium, and in axon fascicles within the lamina propria.

a low-molecular-weight soluble protein found almost exclusively in the mature olfactory neuron cell body, dendrites, and axon processes. It is absent from all other cell types of the olfactory neuroepithelium and is a convenient cell marker for mature olfactory neurons and their processes (Figure 2.5).

Developmentally, OMP's appearance coincides with synapse formation by olfactory axons with second-order neurons of the olfactory bulb. Some studies indicate that synapse formation or contact with the olfactory bulb is not prerequisite for OMP synthesis (Monti Graziadei, 1983; Morrison and Graziadei, 1983). However, *in vitro* studies of olfactory mucosa cultured alone and with olfactory bulb show that OMP synthesis is enhanced when olfactory receptors make contact with the target tissue (Chuah and Farbman, 1983). Since its discovery, OMP has been shown to be present in many species, including man. Olfactory marker protein appears to be strongly phylogenetically conserved and developmentally regulated. Its function remains unknown.

Olfactory neurons have some biochemical characteristics typical of immature CNS neurons. For example, in the adult rat, the early forms of brain microtubule associated protein (MAP) are expressed in the olfactory system (Viereck et al., 1989). The early forms of MAPs (MAP5,

MAP2, tau) are abundantly expressed in differentiating neurons throughout the developing brain, but its synthesis ceases when the mature neuronal morphology becomes established. The expression of these early forms of MAPs by the olfactory system suggests that they play a role in neurite outgrowth and neuronal plasticity.

In addition to expressing and maintaining early forms of MAPs, olfactory neurons have other characteristics that are generally present only in developing neurons. For example, Schwob et al. (1986) demonstrated that olfactory receptors of adult rats continue to synthesize vimentin (a protein usually found only in developing neurons) rather than neurofilament proteins. In addition, cell surface matrix antigens associated with neurite outgrowth (NILE, Stallcup et al., 1985), growth cone specific glycoprotein (Wallis et al., 1985), and embryonic forms of the neural cell adhesion molecule (NCAM) (Miragall et al., 1989) have been associated with olfactory neurons. Thus, olfactory receptor neurons maintain juvenile neural characteristics into adulthood. This is probably related to their persistent turnover, continued neurogenesis, neural morphogenesis, and plasticity.

Neural Plasticity

Olfactory neurons are replaced normally and following injury, apparently for the life of the individual. No other neurons of the adult central nervous system are replaced. The capacity for olfactory neurogenesis is assured by a persistent neuroblast population, the basal cells (Graziadei and Monti Graziadei, 1979). The neuroblasts generate replacement cells that develop, mature, and grow axon processes centrally. They not only form anatomical connections but reestablish function (Graziadei and Monti Graziadei, 1978; Costanzo, 1985). The olfactory system is an ideal model system for studies of neurogenesis, development, and regeneration in the vertebrate central nervous system.

Experimental Degeneration

Zinc Sulfate

In the early 1900s zinc sulfate solutions were used as a protective supplemental treatment against disease. Zinc sulfate solutions were used to irrigate the nasal cavity to prevent viruses (poliomyelitis, encephalitis) from entering the brain. Following treatment, patients complained of anosmia, which was due to destruction of the olfactory neuroepithelium (Smith, 1938). Since that time, zinc sulfate solutions have been commonly used by investigators to induce olfactory receptor cell degeneration (Smith, 1938; Harding et al., 1978; Matulionis et al., 1979; Cancalon, 1982). However, it is not selective for neurons and can affect the entire neuroepithelial cell population, basal lamina, and into the lamina propria.

Sensory Deprivation

The effect of sensory deprivation on the development of the olfactory system has been examined in neonatal and young adult rats (Meisami, 1976; Brunjes and Borror, 1983). When one naris in a neonatal animal is experimentally closed, the postnatal increase in size and weight of the ipsilateral olfactory bulb is decreased. This procedure is a functional, not an anatomical deprivation. Sensory deprivation also affects the olfactory receptor neurons. Following nares closure, the total number of olfactory neurons is decreased (Benson et al., 1984; Farbman et al., 1984). Farbman and coworkers (1988) have shown that the number of mature receptor neurons remains the same as on the contralateral side following sensory deprivation, but the rate of neurogenesis and the number of immature neurons is markedly decreased.

Olfactory receptors are exposed and vulnerable. This probably results in their short life span and the evolution of their neurogenesis and replacement. It has been suggested that olfactory neurons live longer if protected from the environment. Hinds et al. (1984) raised rats in a sterile environment, free from rhinitis. Autoradiography showed that olfactory receptor neurons had longer survival, up to one year. Studies on sensory deprivation also support these findings; a protected environment (nares closure) delays the degeneration of mature olfactory receptor cells (Farbman et al., 1988).

Experimental Axotomy

Olfactory nerve axotomy and bulbectomy also induce retrograde olfactory neuron degeneration. Unlike chemical exposure, these procedures are selective; degeneration occurs only in the olfactory neurons whose axons have been severed.

Degeneration Period

The olfactory neuroepithelium is profoundly affected by transecting olfactory axons. There is immediate retrograde degeneration of the mature olfactory neurons with transected axons. Degeneration is maximal by day 4 following injury in rodents. At that time the epithelium consists of supporting cells, a few developing neurons whose axons were spared at the time of axotomy, basal cells, and macrophages (Figure 2.6). Scanning electron microscopy during degeneration reveals that the epithelial surface lacks olfactory cilia; the underlying microvilli of supporting cells and the Bowman's gland duct openings are exposed. The absence of olfactory neurons during degeneration exposes the full length of the supporting cells (normally hidden by the receptor cells), revealing the intimate contact between adjacent supporting cells and the different types of basal foot processes that attach at the basal lamina (Figure 2.7).

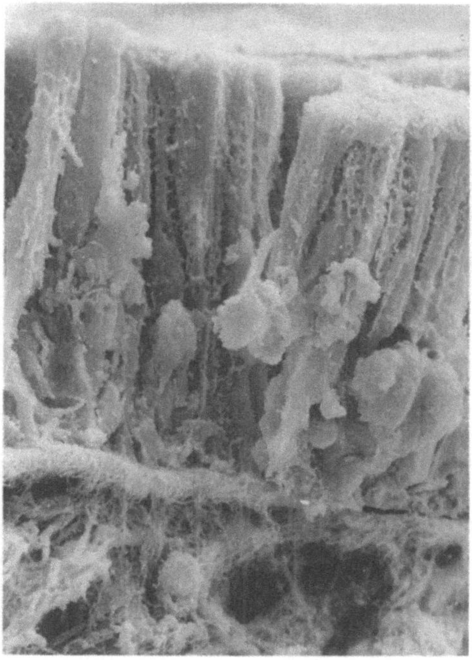

FIGURE 2.6. Scanning micrograph of olfactory epithelium 4 days after olfactory nerve transection. The epithelium contains supporting cells, degenerating neurons, and macrophage. A dividing basal cell can also be seen in the lower epithelium.

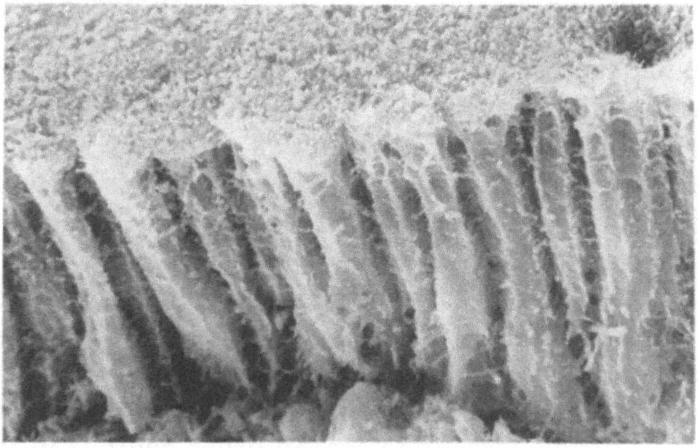

FIGURE 2.7. Scanning micrograph of olfactory epithelium 4 days following olfactory nerve transection. The intracellular connections between supporting cells can be seen. (Reprinted, by permission, from Morrison and Costanzo, 1989a.)

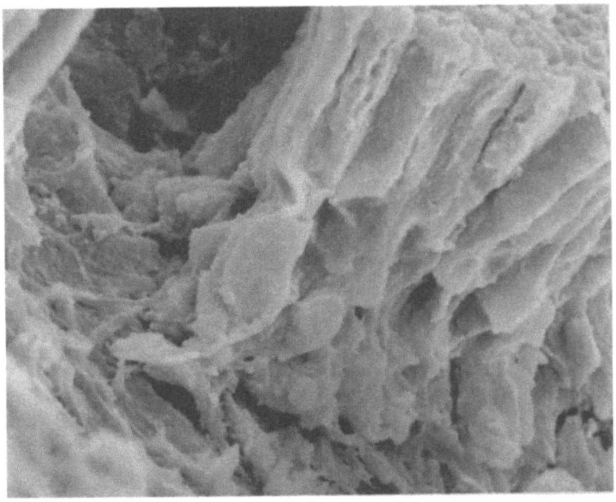

FIGURE 2.8. Scanning micrograph of the lower epithelium during recovery period. Note the olfactory neuron and clublike growth cone structure at the tip of the growing axon process. (Reprinted, by permission, from Morrison and Costanzo, 1989a.)

Recovery Period

The early recovery period (5–25 days) is marked by an increase in basal cell mitotic activity. The neuroepithelium is not uniformly active; regions or clusters of activity are often observed. Developing olfactory neurons move towards the epithelial center, possibly guided by adjacent supporting cells. Newly developed replacement neurons grow an axon process from the basal region of the cell body, then an apical dendritic process that extends to the epithelial surface (Morrison and Costanzo, 1989a). Olfactory neurons, like other developing neurons, have a specialized growth cone structure at the tip of growing processes (Figure 2.8). Olfactory growth cones can be flattened or club-shaped, having several fine filopodia extensions contacting adjacent cells. As the axons grow centrally, they form small fascicles surrounded by Schwann cell processes within the lamina propria. Newly developed axons pass through the cribriform plate and form characteristic spherical glomeruli, where they reestablish synaptic contacts with second-order cells of the olfactory bulb.

Costanzo (1984) noted that in hamsters the olfactory neuroepithelium recovers to approximately 80% of control thickness following complete nerve transection. Following neuroepithelial recovery, Costanzo (1985) found evidence that newly replaced olfactory neurons respond to odor stimuli and transmit signals directly to second-order neurons in the olfactory bulb. Therefore, following injury the olfactory neuron has the ability to re-establish anatomical and more importantly make functional connections with cells in the olfactory bulb.

The persistent stem cell population (basal cells) assures the recovery and reconstitution of the sensory epithelium, providing the basis for anatomical connection and reestablishing behavioral responses.

Bulbectomy

Surgically removing the olfactory bulb (total bulbectomy) in a variety of vertebrate species results in retrograde degeneration of mature olfactory neurons with severed axons. Neuroepithelial degeneration is followed by a wave of basal cell mitotic activity that reconstitutes the epithelium to approximately 60% of control thickness in hamsters (Costanzo and Graziadei, 1983). The new neurons mature, growing dendrite and axon processes. Newly developed neurons are OMP positive, but the amount is reduced relative to control. It is apparent that the neuroepithelium is capable of at least partial anatomical and biochemical recovery. If the olfactory bulb is removed in neonates, the forebrain projects forward towards the cribriform plate. The newly developed olfactory axons then grow centrally and penetrate the forebrain, forming ectopic glomeruli that contain synapses with local neurons (Graziadei et al., 1979). These experiments show the remarkable regenerative capacity of the olfactory system even in the absence of its normal target tissue (the olfactory bulb) and the ability of the newly developed olfactory axons to alter foreign neural environments.

Most studies on vertebrate olfactory regeneration have been on rodents. Following axotomy in adult squirrel monkeys, there is a pattern of degeneration followed by reconstitution of the neuroepithelium and axon growth, as in rodents. This suggests that neurogenesis and new synaptic connections can be formed in nonhuman primates and possibly in humans as well (Graziadei et al., 1980).

Studies on Aged Animals

Most studies on olfactory nerve regeneration have been on neonatal or young animals. Does the "old" olfactory system maintain the capacity for neurogenesis, replacement of neurons, and formation of central connections following injury?

Costanzo and Morrison have begun studying the olfactory system's capacity for recovery following axotomy in old mammals (hamster, 1–2 years of age). Following unilateral olfactory nerve transection there is rapid degeneration of mature olfactory neurons, similar to that in neonatal and young adult animals (Costanzo and Graziadei, 1983; Costanzo and Morrison, 1987). Degeneration is followed by an increase in mitotic activity in the basal epithelial region, with numerous mitotic figures in the basal cell layer. Later (25–120 days), the neuroepithelium becomes reconstituted with newly developed olfactory neurons. These replacement neurons are OMP positive, grow axons centrally to the olfactory bulb, forming characteristic glomeruli, and form synaptic connections with second-order olfactory bulb neurons. These preliminary results indicate that the neurogenesis and plasticity of the olfactory system, and its ability to replace neurons and form new synaptic connections, is maintained well into old age in mammals.

Recent Advances

The unique ability of the olfactory neurogenic matrix to continue neurogenesis has permitted examination of the potential for transplantation. Several studies have shown that olfactory axons can grow and invade different regions of the CNS (Graziadei et al., 1978, 1979). These newly developed axons also have the ability to reorganize morphological environments in nonolfactory brain regions. This exceptional neural plasticity has been further investigated by transplanting olfactory mucosa into the ante-

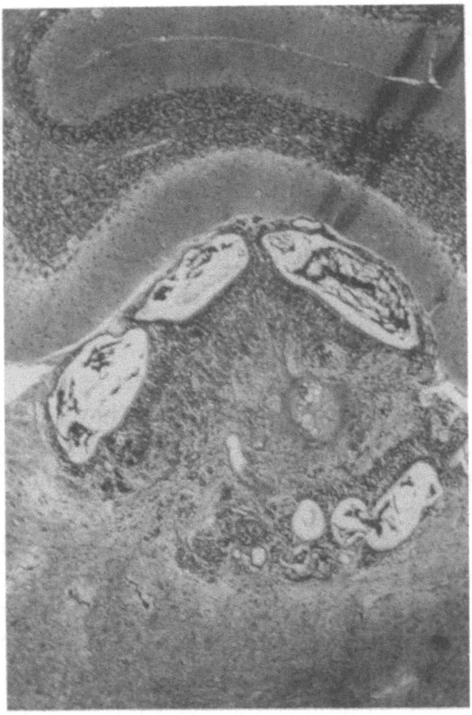

FIGURE 2.9. Light micrograph of olfactory tissue transplant placed within the cerebral cortex near the hippocampus. The transplant forms several vesicles lined by an olfactory epithelium.

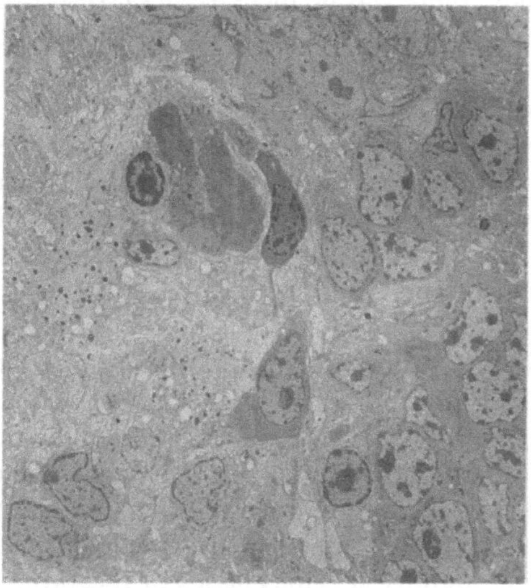

FIGURE 2.10. Transmission micrograph showing olfactory neurons that have migrated into the host brain tissue.

rior eye chamber and various regions of the CNS (Barber et al., 1982a; Heckroth et al., 1983; Morrison and Graziadei, 1983; Morrison and Costanzo, 1989b).

When fragments of olfactory mucosa are transplanted, they generally organize into a series of interconnecting vesicles, lined with a respiratory or olfactory epithelium (Figure 2.9). Following transplantation the mature olfactory neurons of the transplant degenerate because their axons were severed. The basal cells survive and produce new neurons. Some regions of the transplant have a well-developed epithelium consisting of supporting cells, olfactory neurons, and basal cells. In other regions the epithelium is thin and nearly simple cuboidal, and the transplant neurons are in direct contact with the host brain tissue. Transplant neurons migrate extraepithelially, passing into the lamina propria or directly into the host brain tissue along transplant axon bundles or blood vessels (Figure 2.10). This newly acquired property of migration brings the transplant neurons into direct contact with host neurons and glia.

Olfactory axons originate from the transplant and pass into the lamina propria. The new axons then grow into the host brain for quite some distance but do not organize into glomerular structures, even when fragments of olfactory mucosa are transplanted into the olfactory bulb (Graziadei and Monti Graziadei, 1986; Monti Graziadei and Graziadei, 1989).

Immunohistochemistry of transplant tissue reveals that not all transplant neurons contain OMP. Olfactory neurons that appear mature are OMP negative (Monti Graziadei and Morrison, 1988; Morrison and Costanzo,

1989b). Interestingly, there was no increase in OMP-positive neurons even when olfactory mucosa were transplanted into the olfactory bulb (Monti Graziadei and Graziadei, 1989).

Transplant experiments are still in their early phases. The results indicate that olfactory neurons can survive transplantation, develop, and enter the host brain. It is premature to offer the olfactory neuron as a "spare or replacement" element in the damaged or diseased CNS. However, having a neuroblast cell that survives transplantation and produces new neurons that develop processes and enter the host brain suggests the potential for such a development.

Organ and cell culture provide an opportunity to examine the importance of target tissue and epithelial organization (Chuah and Farbman, 1983). The advantage of *in vitro* studies is that one can determine the effect of manipulating the external environment on receptor neurons. Monoclonal antibodies also have great potential for examining the development and regenerative capacity of the olfactory system. Cell markers that can be used to explore and identify specific classes of receptors and their projections have been generated (Allen and Akeson, 1985; Fujita et al., 1985; Hempstead and Morgan, 1985). These methods, combined with other techniques, should allow rapid progress in our understanding of physiological mechanisms and development and regeneration in the olfactory system.

Summary

Although the vertebrate olfactory system has a seemingly simple arrangement, it has several unique characteristics. The olfactory system's capacity for continued neurogenesis and the ability to replace neurons normally and when injured makes it an important model for examining neural development and regeneration. It offers a unique opportunity for studies of neurogenesis, axon proliferation, and synaptogenesis in developing and aging animals. The discovery of specific cell markers identifying cellular components should provide new insights into neural development. The ability of the olfactory neuron to survive transplantation and continue to develop provides the investigator with a remarkable system that maintains its plasticity, even when removed from its natural environment. Future studies of the olfactory system should provide new insights into the development and regeneration of the central nervous system.

Acknowledgements. We wish to thank Dr. Steve Price for his comments during the preparation of this manuscript and Dr. Frank L. Margolis for generously donating olfactory marker protein antibody. We also thank Jeanette Acree and Chris LeGrand for their invaluable assistance. This project was supported in part by a grant from the Alzheimer's Association,

Neil Bluhm Pilot Research Grant (EEM), and by a grant from the National Institutes Deafness and Other Communication Disorders #DC00169 (RMC).

References

Allen, W.K., and R. Akeson (1985) Identification of a cell surface glycoprotein family of olfactory receptor neurons with a monoclonal antibody. *J. Neurosci.*, *5*, 284–296.

Anholt, R.R., S.M. Mumby, D.A. Stoffers, P.R. Girard, J.F. Kuo, and S.H. Snyder (1987) Transduction proteins of olfactory receptor cells: identification of guanine nucleotide binding proteins and protein kinase C. *Biochemistry*, *26*, 788–795.

Barber, P.C., S. Jensen, and J. Zimmer (1982a) Differentiation of neurons containing olfactory marker protein in adult rat olfactory epithelium transplanted to the anterior chamber of the eye. *Neuroscience*, *7*, 2687–2695.

Barber, P.C., and R.M. Lindsay (1982b) Schwann cells of the olfactory nerves contain glial fibrillary acidic protein and resemble astrocytes. *Neuroscience*, *7*, 3077–3090.

Benson, T.E., K. Ryugo, and J.W. Hinds (1984) Effects of sensory deprivation on the developing mouse olfactory system: A light and electron microscopic, morphometric analysis. *J. Neurosci.*, *4*, 638–653.

Breipohl, W., H.J. Laugwitz, and N. Bornfeld (1974) Topographical relations between the dendrites of olfactory sensory cells and sustentacular cells in different vertebrates. *J. Anat.*, *117*, 89–94.

Breipohl, W., and M. Ohyama (1981) Comparative and developmental SEM studies on olfactory epithelia in vertebrates (Biomedical aspects and speculations). *Biomedical Research*, *2* (Suppl), 437–448.

Brunjes, P.C., and M.J. Borror (1983) Unilateral odor deprivation: differential effects due to time of treatment. *Brain Res. Bull.*, *11*, 501–503.

Cancalon, P. (1982) Degeneration and regeneration of olfactory cells induced by $ZnSO_4$ and other chemicals. *Tissue Cell*, *14*, 717–733.

Carpenter, E. (1937) The head pattern in *amblystoma* studied by vital staining and transplantation methods. *J. Exp. Zool.*, *75*, 103–129.

Carr, V.M., A.I. Farbman, M.S. Lidow, L.M. Colletti, J.L. Hempstead, and J.I. Morgan (1989) Developmental expression of reactivity to monoclonal antibodies generated against olfactory epithelia. *J. Neurosci.*, *9*, 1179–1198.

Chuah, M.I., and A.I. Farbman (1983) Olfactory bulb increases marker protein in olfactory receptor cells. *J. Neurosci.*, *3*, 2197–2205.

Costanzo, R.M. (1984) Comparison of neurogenesis and cell replacement in the hamster olfactory system with and without a target (olfactory bulb). *Brain Res.*, *307*, 295–301.

Costanzo, R.M. (1985) Neural regeneration and functional reconnection following olfactory nerve transection in hamster. *Brain Res.*, *361*, 258–266.

Costanzo, R.M., and P.P.C. Graziadei (1983) A quantitative analysis of changes in the olfactory epithelium following bulbectomy in hamster. *J. Comp. Neurol.*, *215*, 370–381.

Costanzo, R.M., and E.E. Morrison (1987) The capacity for neurogenesis and axon reconnection persists in the old adult hamster. *Neurosci. Abstr.*, *17*, 462.16

Costanzo, R.M., and E.E. Morrison (1989) Three dimensional scanning electron microscopic study of the normal hamster olfactory epithelium. *J. Neurocytol.*, *18*, 381–391.

Cuschieri, A., and L.H. Bannister (1975) The development of the olfactory mucosa in the mouse: light microscopy. *J. Anat.*, *119*, 277–286.

Farbman, A.I., S.M. Ritz, and P.C. Brunjes (1984) The effect of odor deprivation on olfactory epithelium in developing rats. *Soc. Neurosci. Abstr.*, *10*, 530.

Farbman, A.I., and L.M. Squinto (1985) Early development of olfactory receptor axons. *Dev. Brain Res.*, *19*, 205–213.

Farbman, A.I., P.C. Brunjes, L. Rentfro, J. Michas, and S. Ritz (1988) The effect of unilateral naris occlusion on cell dynamics in the developing rat olfactory epithelium. *J. Neurosci.*, *8*, 3290–3295.

Fujita, S.C., K. Mori, K. Imamura, and K. Obata (1985) Subclasses of olfactory receptor cells and their segregated central projections demonstrated by a monoclonal antibody. *Brain Res.*, *326*, 192–196.

Getchell, T.V. (1977) Analysis of intracellular recordings from salamander olfactory epithelium. *Brain Res.*, *123*, 275–286.

Getchell, M.L., J.A. Rafols, and T.V. Getchell (1984) Histological and histochemical studies of the secretory components of the salamander olfactory mucosa: effects of isoproterenol and olfactory nerve section. *Anat. Rec.*, *208*, 553–565.

Getchell, T.V., F.L. Margolis, and M.L. Getchell (1984) Perireceptor and receptor events in vertebrate olfaction. *Prog. Neurobiol.*, *23*, 317–345.

Graziadei, P.P.C. (1971) Topographical relations between olfactory neurons. *Z. Zellforsck. Mikrosh., Anat.*, *118*, 449–466.

Graziadei, P.P.C. (1973a) Cell dynamics in the olfactory mucosa. *Tissue Cell.*, *5*, 113–131.

Graziadei, P.P.C. (1973b) The ultrastructure of vertebrates olfactory mucosa. In I. Friedman (Ed.), *The Ultrastructure of Sensory Organs*. Oxford: Elsevier, pp. 267–305.

Graziadei, P.P.C. and G.A. Monti Graziadei, (1978) The olfactory system: A model for the study of neurogenesis and axon regeneration in mammals. In C.W. Cotman (Ed.), *Neuronal Plasticity*, New York: Raven Press, pp. 113–153.

Graziadei, P.P.C., and G.A. Monti Graziadei (1979) Neurogenesis and neuron regeneration in the olfactory system of mammals. I. Morphological aspects of differentiation and structural organization of the olfactory sensory neurons. *J. Neurocytol.*, *8*, 1–18.

Graziadei, P.P.C., R.R. Levine, and G.A. Monti Graziadei (1979) Plasticity of connections of the olfactory sensory neurons: Regeneration into the forebrain following bulbectomy in neonatal mice. *Neurosci.*, *4*, 713–727.

Graziadei, P.P.C., M.S. Karlan, G.A. Mont. Graziadei, and J.J. Bernstein (1980) Neurogenesis of sensory neurons in the primate olfactory system after section of the fila olfactoria. *Brain Res.*, *186*, 289–300.

Graziadei, P.P.C., and G.A. Monti Graziadei (1986) Principles of organization of the vertebrate olfactory glomerulus: An hypothesis. *Neurosci.*, *19*, 1025–1035.

Halpern, M. (1987) The organization and function of the vomeronasal system. *Ann. Rev. Neurosci.*, *10*, 325–362.

Harding, J.W., T.V. Getchell, and F.L. Margolis (1978) Denervation of the primary olfactory pathway in mice. V. Long term effect of intranasal $ZnSO_4$ irrigation on behavior, biochemistry and morphology. *Brain Res.*, *140*, 271–285.

Heckroth, J.A., G.A. Monti-Graziadei, and P.P.C. Graziadei (1983) Intraocular transplants of olfactory neuroepithelium in rat. *Int. J. Dev. Neurosci.*, *1*, 273–287.

Hempstead, J.L., and J.I. Morgan (1985) A panel of monoclonal antibodies to the rat olfactory epithelium. *J. Neurosci.*, *5*, 438–449.

Hinds, J.W., P.L. Hinds, and N.A. McNelly (1984) An autoradiographic study of the mouse olfactory epithelium: evidence for long-lived receptors. *Anat. Rec.*, *210*, 375–383.

Klein, S.L., and P.P.C. Graziadei (1983) The differentiation of the olfactory placode in *Xenopus laevis*: a light and electron microscope study. *J. Comp. Neurol.*, *217*, 17–30.

Lancet, D. (1986) Vertebrate olfactory reception. *Annu. Rev. Neurosci.*, *9*, 329–355.

Margolis, F.L. (1972) A brain protein unique to the olfactory bulb. *Proc. Natl. Acad. Sci. USA*, *69*, 1221–1224.

Masukawa, L.M., B. Hedlund, and G.M. Shepherd (1985) Changes in the electrical properties of olfactory epithelial cells in the tiger salamander after olfactory nerve transection. *J. Neurosci.*, *5*, 136–141.

Matulionis, D.H., W. Breipohl, and R.P. Bhatangar (1982) Degeneration and regeneration of olfactory epithelium in the mouse. A scanning electron microscopic study. *Annals of Otology Rhinology Laryngology* (Suppl), *89*, 1–12.

Meisami, E. (1976) Effects of olfactory deprivation on postnatal growth of the rat olfactory bulb utilizing a new method for production of neonatal unilateral anosmia. *Brain Res.*, *107*, 437–444.

Menco, B.P.M. (1988) Tight junctional strands first appear in regions where three cells meet in differentiating olfactory epithelium: a freeze-fracture study. *J. Cell Sci.*, *89*, 495–505.

Menco, B.P.M., and A.I. Farbman (1985) Genesis of cilia and microvilli of rat nasal epithelia during pre-natal development. I. Olfactory epithelium, qualitative studies. *J. Cell Sci.*, *78*, 283–310.

Miragall, F., G. Kadmon, and M. Schachner (1989) Expression of L1 and N-CAM cell adhesion molecules during development of the mouse olfactory system. *Dev. Biol.*, *135*, 272–286.

Monti Graziadei, G.A. (1983) Experimental studies on the olfactory marker protein. III. The olfactory marker protein in the olfactory neuroepithelium lacking connections with the forebrain. *Brain Res.*, *262*, 303–308.

Monti Graziadei, G.A., F.L. Margolis, J.W. Harding, and P.P.C. Graziadei (1977) Immunocytochemistry of the olfactory marker protein. *J. Histochem. Cytochem.*, *25*, 1311–1316.

Monti Graziadei, G.A., and E.E. Morrison, (1988) Experimental studies on the olfactory marker protein. IV. Olfactory marker protein in the olfactory neurons transplanted within the brain. *Brain Res.*, *455*, 401–406.

Monti Graziadei, G.A., and P.P.C. Graziadei (1989) Experimental studies on the olfactory marker protein. V. Olfactory marker protein in the olfactory neurons transplanted within the olfactory bulb. *Brain Res.*, *484*, 157–167.

Moran, D.T., J.C. Rowley, B.W. Jafek, and M.A. Lovell (1982) The fine structure of the olfactory mucosa in man. *J. Neurocytol.*, *11*, 721–746.

Morrison, E.E., and P.P.C. Graziadei (1983) Transplants of olfactory mucosa in the rat brain. A light microscopic study of transplant organization. *Brain Res.*, *279*, 241–245.

Morrison, E.E., and R.M. Costanzo (1989a) Scanning electron microscopic study

of degeneration and regeneration in the olfactory epithelium after axotomy. *J. Neurocytol.*, *18*, 393–405.
Morrison, E.E., and R.M. Costanzo (1989b) Transplantation and survival of neurons in the brain of old hamsters. *Soc. Neurosci. Abstr.*, 542.2.
Morrison, E.E., and R.M. Costanzo (1990) Morphology of the human olfactory epithelium. *J. Comp. Neurol.*, *297*, 1–14.
Moulton, D.G. (1974) Dynamics of cell populations in the olfactory epithelium. *Ann. N.Y. Acad. Sci.*, *237*, 52–61.
Okano, M., A.F. Weber, and S.P. Frommes (1967) Electron microscopic studies of the distal border of the canine olfactory epithelium. *J. Ultrastruct. Res.*, *17*, 487–502.
Pace, U., and D. Lancet (1986) Olfactory GTP-binding protein: signal-transducing polypeptide of vertebrate chemosensory neurons. *Proc. Natl. Acad. Sci. USA*, *83*, 4947–4951.
Price, S. (1984) Mechanisms of stimulation of olfactory neurons: An essay. *Chem. Senses.*, *8*, 341–354.
Rafols, J.A., and T.V. Getchell (1983) Morphological relations between the receptor neurons, sustentacular cells and Schwann cells in the olfactory mucosa of the salamander. *Anat. Rec.*, *206*, 87–101.
Ramon Y Cajal (1909–1911) *Histologie du Systeme Nerveux de l'Homme et des Vertebres,* 2 vols. Translated by L. Azoulay, reprinted by Instituto Ramon Y Cajal del C.S.T.S., Madrid, 1952–1955.
Schwob, J.E., N.B. Farber, and D.I. Gottlieb (1986) Neurons of the olfactory epithelium in adult rats contain vimentin. *J. Neurosci.*, *6*, 208–217.
Smith, C.G. (1938) Changes in the olfactory mucosa and the olfactory nerves following intranasal treatment with one per cent zinc sulfate. *Can. Med. J.*, *39*, 138–140.
Stallcup, W.B., L.L. Beasley, and J.M. Levine (1985) Antibody against nerve growth factor-inducible large external (NILE) glycoprotein labels nerve fiber tracts in the developing rat nervous system. *J. Neurosci.*, *5*, 1090–1101.
Trotier, D., and P. MacLeod (1986) Intracellular recordings from salamander olfactory supporting cells. *Brain Res.*, *374*, 205–211.
Viereck, C., R. Tucker, and A. Matus (1989) The rat olfactory system expresses microtubule-associated proteins found in the developing brain. *J. Neurosci.*, *9*, 3547–3557.
Vinnikov, Y.A., and L.K. Titovia (1957) *Morphology of Olfactory Organs*. Moscow: Medgiz.
Wallis, I., L., Ellis, K. Suh, and K.H. Pfenninger (1985) Immunolocalization of a neuronal growth dependent membrane glycoprotein. *J. Cell Biol.*, *101*, 1990–1998.
Wang, R.T., and M. Halpern (1980) Scanning electron microscopic studies of the surface morphology of the vomeronasal epithelium and olfactory epithelium of garter snakes. *Am. J. Anat.*, *157*, 399–428.

3
Molecular Aspects of Olfaction

ROBERT R.H. ANHOLT

Introduction

Olfaction, the ability to recognize and discriminate with great accuracy and sensitivity myriads of airborne molecules either by themselves or as complex mixtures, is one of the most remarkable and least understood senses of virtually all vertebrates. The sense of smell, like the immune system, specializes in molecular recognition and is crucial for the survival of most animal species. This chapter will discuss biochemical aspects of odorant recognition and chemosensory transduction.

Functional Anatomy of the Olfactory Neuroepithelium

The olfactory neuroepithelium is a pseudostratified epithelium that contains mainly three cell types (Figure 3.1A): (1) olfactory receptor neurons; (2) sustentacular cells, which are glia-like columnar cells that traverse the width of the epithelium; and (3) basal cells, which lie adjacent to the basement membrane and represent a reservoir of neurogenic stem cells that can undergo cell division and differentiate into olfactory neurons throughout the life of the adult animal to replace damaged or senescent olfactory receptor cells (Harding et al., 1977 and 1978; Graziadei and Monti-Graziadei, 1978; Cancalon, 1982; Costanzo and Graziadei, 1983 and 1987; Camara and Harding, 1984; Hinds et al., 1984).

Olfactory receptor neurons are bipolar neurons that send dendrites toward the nasal lumen and axons to the olfactory bulb of the brain. These axons are thin and unmyelinated and form fascicles that are surrounded by a single Schwann cell. They do not synapse until after they cross the cribriform plate of the ethmoid bone and enter the olfactory bulb, where 100–1000 afferent axons contact the dendritic arbor of a single output neuron, forming complex synaptic structures known as "glomeruli." The dendrites of the olfactory receptor cells form dilatations as they reach the surface of the epithelium, the "olfactory knobs" or "olfactory vesicles," from which

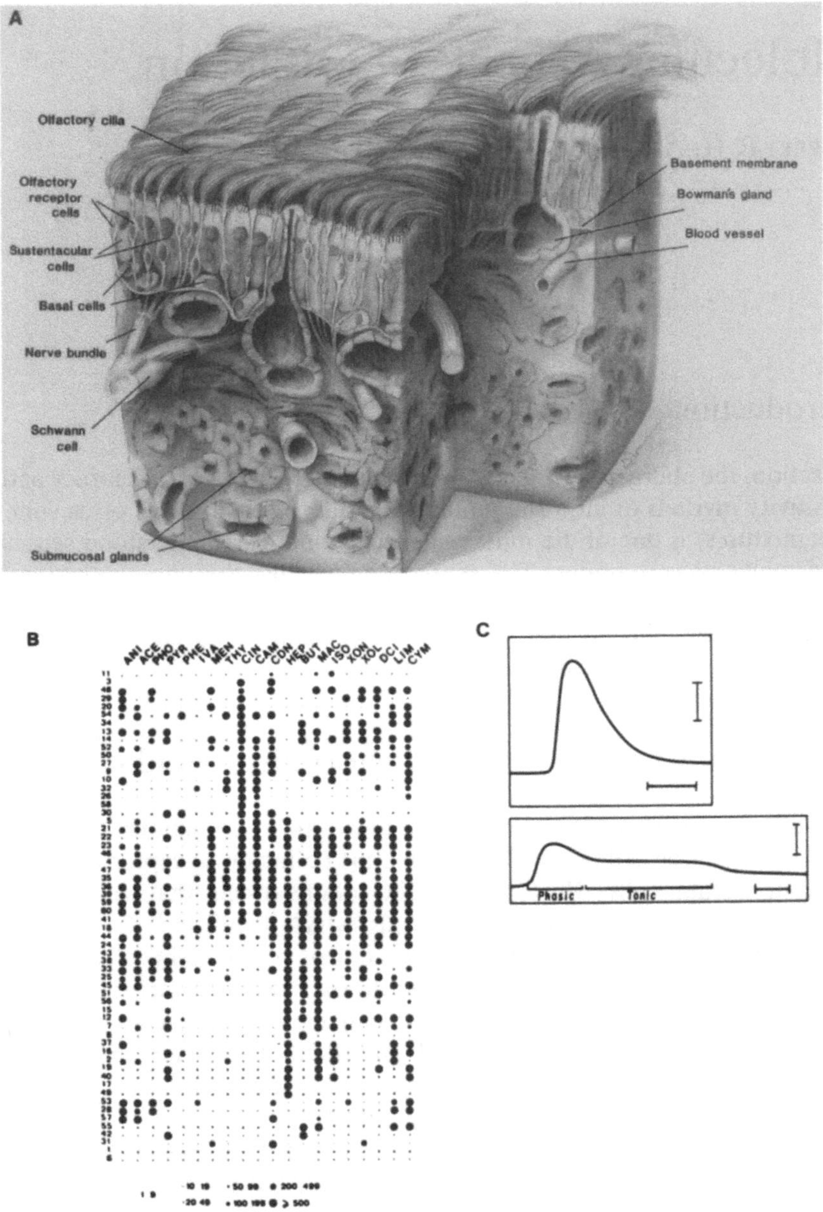

FIGURE 3.1. Functional anatomy of the olfactory neuroepithelium.

A. Histology of the olfactory epithelium.
Bipolar olfactory receptor neurons are surrounded by sustentacular cells, which span the width (~150 μm) of the epithelium and carry microvilli at their apical ends. The axons of the olfactory receptor neurons come together as fascicles that are ensheathed by a single Schwann cell and form the olfactory nerve that projects

groups of chemosensory cilia protrude into the layer of mucus that covers the epithelial surface. It is on this vastly expanded surface area of ciliary plasma membrane that odorants dissolved in the mucus interact with the olfactory neurons (Figure 3.1A; Getchell et al., 1984; Lancet, 1986; Anholt, 1987 and 1989).

Olfactory receptor cells differ in their responsiveness to different odorants, with each cell responding to more than one, but not to all odorants (Figure 3.1B; Duchamp et al., 1974; Sicard and Holley, 1984). Thus, different odorants activate different sets of cells and generate unique patterns of neural activity that encode odor quality and concentration. The macroscopic response of the olfactory epithelium to odorants can be monitored

---

to the olfactory bulb. Cuboidal cells on top of the basement membrane represent the basal cells, a population of neurogenic stem cells that retain the ability to differentiate into mature olfactory neurons after neuronal cell death occurs in the epithelium. Note that the epithelium is covered by a layer of cilia protruding from the dendritic tips of the olfactory receptor cells. Secretions from sustentacular cells, Bowman's glands, and numerous submucosal glands in the underlying lamina propria form an ~30-μm-thick layer of mucus (Reese, 1965) in which the cilia are embedded and in which interactions between odorants and the chemosensory membrane take place. Heteromorphous pigment cells (melanocytes) depicted in this scheme are typical for the olfactory epithelium of Amphibia. (Reprinted, by permission, from Anholt, 1987.)

B. Diagrammatic representation of olfactory receptor cell activity following stimulation by odorants.
The spot size is proportional to the spike frequency (spikes/min) measured by extracellular recording from single receptor cells in frog olfactory epithelium after odorant application. Receptor cells are identified by serial numbers on the left. The odorants tested are anisole (ANI), acetophenone (ACE), thiophenol (PHO), pyridine (PYR), phenol (PHE), isovaleric acid (IVA), L-menthol (MEN), thymol (THY), 1,8-cineole (CIN), (±)-camphor (CAM), cyclodecanone (CDN), *n*-heptanol (HEP), *n*-butanol (BUT), methyl amylketone (MAC), isoamyl acetate (ISO), cyclohexanone (XON), cyclohexanol (XOL), (+)-citronellol (DCI), (+)-limonene (LIM), and *p*-cymene (CYM). Note that different cells respond selectively to different sets of odorants. (Reprinted, by permission, Sicard and Holley, 1984.)

C. The electro-olfactogram.
The top panel shows the typical shape of a monophasic electro-olfactogram recorded from frog olfactory epithelium after application of 0.5 ml of odorized air. The stimulus is butanol. Vertical bar: 1 mV; time bar: 2 s. The lower panel shows the typical shape of the electro-olfactogram upon continuous stimulation of the olfactory epithelium with odorized air equilibrated with 0.01 M butanol for 15 s. Note the biphasic appearance of the electro-olfactogram. An initial transient "phasic" response decays into a prolonged "tonic" response of lower amplitude, probably reflecting adaptation of the epithelium to the stimulus (vertical bar: 1 mV; time bar: 3 s). (Modified, by permission, from Ottoson, 1956.)

electrically and appears as a negative voltage transient, referred to as the "electro-olfactogram" (Figure 3.1C; Ottoson, 1956; Tagaki et al., 1968 and 1969; Minor, 1971; Getchell, 1974; Getchell and Shepherd, 1978a and b). The response to odorants is terminated upon their removal from the olfactory epithelium via mucociliary clearance and absorbtion into the circulation (Getchell et al., 1984; Hornung and Mozell, 1977 and 1980). In addition, the olfactory epithelium contains a high activity of cytochrome P-450 (Dahl et al., 1982; Nef et al., 1989 and 1990), which not only may inactivate odorants, but conceivably also could transform odorants into compounds of altered odorous potency and quality and, thus, contribute to the complexity of the stimulus and its perception (Dahl, 1988).

The rate-limiting step in olfaction is the time it takes for the odorant to partition into the mucus and to diffuse to the ciliary membrane (Getchell et al., 1984). The physical parameters that govern this process, such as viscosity, the precise dimensions and composition of the mucus layer, and the effects of local air flow on partitioning, remain largely unknown. The mucus is produced by secretions from serous glands, Bowman's glands, and sustentacular cells and consists of a watery superficial layer and a more viscous lower layer (Reese, 1965). Although theoretically partitioning of odorants from the air into the mucus results in a 10–1000-fold increase in odorant concentration (Senf et al., 1980; Hornung and Mozell, 1981), estimates of odorant concentrations in the mucus should be considered speculative, since the effects of airflow, mucociliary movement, and exact composition of the mucus are unknown, and actual measurements of odorant concentrations in the mucus are not available. In addition, the precise role of odorant binding proteins in the mucus, described below, in transporting odorants remains to be determined.

Odorant Recognition

The Search for Odorant Receptors: Identification of Odorant Binding Proteins

Until recently, little was known about the nature and diversity of odorant receptors. Some investigators argued that differences in the physical properties of odorants and the olfactory receptor cells with which they interact could account for the discriminatory capability of the olfactory system without the need for specific odorant receptor proteins (Price, 1984; Kashiwayanagi and Kurihara, 1984 and 1985; Nomura and Kurihara, 1987a and b; Mozell et al., 1987). Others argued in favor of specific odorant receptor proteins, based on: (1) the common occurrence of specific anosmias, i.e. the inability to perceive a specific odorant or class of odorants while perception of other odorants is unaffected (Amoore, 1971 and 1982; Price, 1977; Wysocki and Beauchamp, 1984); (2) the stereospecificity of perceived odor quality in the case of some odorants, e.g. carvone, which smells like caraway in the D-form and like spearmint in the L-form; and (3) the observation that lectins (Wood et al., 1983; Shirley et al., 1983a; Kalinoski et al., 1987) and protein modifying reagents (Getchell and Gesteland, 1972; Menevse et al., 1978; Criswell et al., 1980; Delaleu and Holley, 1980 and 1983; Shirley et al., 1983b; Mason and Morton, 1984; Mason et al., 1984; Schafer et al., 1984a and b) can inhibit the response to odorants. It should, however, be noted that specific anosmias do not necessarily reflect the absence of an odorant receptor, but may be central rather than peripheral in origin. Similarly, lectins or protein modifying reagents may affect other transduction components rather than putative odorant receptors. However, addition of ethyl *n*-butyrate during treatment of the olfactory epithelium with *N*-ethyl-maleimide preserved responsiveness of the epithelium to ethyl *n*-butyrate and related odorants, while the response to unrelated odorants was abolished (Getchell and Gesteland, 1972). This observation and the demonstration that isolated olfactory receptor cells display odorant responses with some degree of selectivity and respond optimally when odorants are delivered near the cilia (Firestein and Shepherd, 1989; Frings and Lindemann, 1990) provided support for the notion that olfactory receptor cells possess odorant receptors that are located on the chemosensory cilia and apical dendritic region of the cell.

Initial attempts to identify odorant receptor proteins relied mostly on affinity labeling and ligand binding techniques. A major problem for affinity labeling studies was the unavailability of a functional *in vitro* assay for olfaction that would unambiguously identify labeled proteins as odorant receptors. The use of radioactive odorants led to the identification of several odorant binding proteins (Gennings et al., 1977; Price, 1978; Fesenko et al., 1978 and 1979; Goldberg et al., 1979; Pelosi et al., 1982; Bignetti et al., 1985; Pevsner et al., 1985), but none of those were identified as *bona fide* odorant receptors linked to cellular signal transduction systems. The best characterized of these proteins is an ~40 kD pyrazine binding protein, which is a homodimer of subunits of ~18kD (Pevsner et al., 1985). It is a soluble protein that is secreted in nasal mucus and lacrimal secretions and in the rat is present in high concentrations in the lateral nasal glands (Pevsner et al., 1986 and 1988a). Pyrazines as well as other odorants bind to this protein with low affinity (Pevsner et al., 1985, 1986, and 1990; Topazzini et al., 1985), and it has been suggested that this protein may function as a carrier for odorants in the mucus (Pevsner et al., 1986 and 1988a and b). Although the thermodynamic aspects of this hypothetical function remain to be clarified, analysis of the amino acid sequence of this odorant binding protein derived from the nucleotide sequence of cloned cDNA indicates that it belongs to a family of carrier proteins that includes plasma cholesterol binding proteins, α_2-macroglobulin, and retinal binding protein (Pevsner et al, 1988b), as well as a salivary protein secreted by von Ebner's glands that may bind hydrophobic tastants and play a role in gustation

(Schmale et al., 1990). A protein homologous to the mammalian odorant binding protein is also synthesized by the Bowman's glands of frog olfactory tissue, but this protein has not been shown to bind odorants (Lee et al., 1987). Since responses of olfactory receptor cells to odorants can be measured in the absence of mucus (Frings and Lindemann, 1988 and 1990; Firestein and Werblin, 1989; Firestein and Shepherd, 1989; Schild, 1989), it is clear that these odorant binding proteins are not *a priori* essential for olfactory transduction. Whether they affect *in vivo* the thresholds at which odorants are detected and/or the dose-response behavior of the olfactory receptor cells remains to be determined.

Another approach aimed at identifying odorant receptors consisted of the identification of gene products that are expressed in olfactory versus nonolfactory tissues or during regeneration of olfactory receptor cells after nerve transection or bulbectomy (Margolis, 1988). This approach has in some instances been frustrated by the abundance of glandular mRNA and led to the inadvertent identification of glandular proteins albeit specific to olfactory tissue (Lee et al., 1987; Margolis, 1988). However, small odorant responses could be elicited from Xenopus oocytes after injection of mRNA from olfactory tissue, indicating synthesis and assembly of olfactory transduction components (Getchell, 1988).

Identification of Putative Odorant Receptors: Members of the G Protein-Linked Superfamily of Receptors

Many receptor-mediated transduction pathways are controlled by regulatory GTP-binding proteins (G proteins; Gilman, 1987; Stryer and Bourne, 1986; Casey and Gilman, 1988). G proteins are membrane-associated proteins that are composed of three subunits, a hydrophilic α subunit that contains GDP, which is anchored to the membrane by a hydrophobic complex of the β and γ subunits (Sternweis, 1986). When this G protein interacts with a liganded membrane receptor, it dissociates, releasing the α subunit and the $\beta\gamma$ complex. The free α subunit exchanges GDP for GTP, and the GTP-α complex activates second messenger systems, such as the catalytic component of adenylate cyclase. Endogenous hydrolytic activity of the α subunit converts the bound GTP into GDP and is followed by reassociation of the three subunits and termination of the activation cycle. Activation of second messenger pathways, such as adenylate cyclase, via an intermediary regulatory component allows additional control over the transduction process and provides amplification, since one liganded receptor can activate multiple G proteins, each of which in turn can cause the generation of many intracellular second messenger molecules.

The G protein that links activation of membrane receptors to stimulation of adenylate cyclase is designated "G_s." A family of homologous G proteins has been identified that either mediate inhibition of adenylate cyclase (G_i; Gilman, 1987) or link activation of membrane receptors to the activation of phospholipase C (Cockcroft and Gomperts, 1985; Brass et al., 1986; Ashkenazi et al., 1989). The latter causes the release of inositol triphosphate from phosphatidylinositol, generating inositol triphosphate and diacylglycerol, intermediates of metabolic pathways that control the release of sequestered intracellular calcium (Downes, 1983; Majerus et al., 1986). Other members of the family of GTP-binding proteins include transducin, which mediates visual transduction by linking excitation of rhodopsin to the hydrolysis of cyclic GMP via activation of a phosphodiesterase (Stryer and Bourne, 1986), and G_o, a G protein specifically localized to nervous tissue (Huff et al., 1985; Asano et al., 1988), which has been implicated in the control of potassium and calcium channels (Hescheler et al., 1987; Nakajima et al., 1988; Ewald et al., 1988; Van Dongen et al., 1988).

Receptors that interact with G proteins are members of a superfamily and share structural homologies, including seven hydrophobic sequences that are thought to represent transmembrane domains (O'Dowd et al., 1989a). Based on the notion that olfactory transduction involves G-protein-mediated activation of second messenger systems, described in detail below, and on the hypotheses that odorant receptors would be expressed only in olfactory epithelium and would be members of a diverse family, Buck and Axel set out to identify new, olfactory tissue-specific, members of the G protein-linked superfamily of receptors (Buck and Axel, 1991). Using oligonucleotide probes based on conserved sequences of known G protein-linked receptors, they identified in a cDNA library derived from rat olfactory tissue a group of proteins that fulfilled these criteria. They reported the amino acid sequences of 18 putative odorant receptor proteins. These proteins showed characteristics that placed them within the superfamily of G protein-linked receptors. They contain seven hydrophobic sequences, proposed to form transmembrane domains with three alternating intracellular and extracellular loops, the N terminus being located extracellularly and the C terminus intracellularly (Figure 3.2). The fact that these newly discovered proteins are members of a multigene family which is expressed only in olfactory epithelium strongly suggests that they represent the long-sought-after odorant receptors. However, rigorous proof that these proteins are indeed expressed in olfactory neurons has not yet been provided. Binding of odorants or interactions between these receptors and second messenger systems have also not yet been demonstrated.

In addition to sequence motifs that indicate that these newly discovered receptors are members of the G protein-linked receptor superfamily, they contain sequence characteristics that clearly define them as a distinct family. The most striking feature of these proteins is the high degree of sequence variability in the third, fourth, and especially fifth transmembrane domains (Figure 3.2). These central transmembrane domains are thought to contribute to the odorant binding site, and their great variability may reflect a diverse array of odorant binding sites with different specificities. Furthermore, the third cytoplasmic loop is relatively short, consisting of

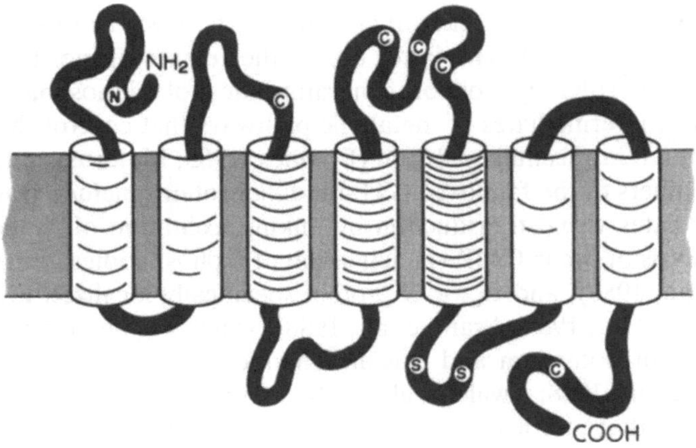

FIGURE 3.2. Schematic representation of a putative odorant receptor. Putative odorant receptors are thought to traverse the membrane seven times with the N terminus located extracellularly, the C terminus intracellularly, and three alternating cytoplasmic and extracellular loops. The N-terminal region contains a potential N-glycosylation site (N). Characteristic cysteines (C), which may form disulfide bonds, are present at fixed positions within the first and second extracellular loops. A cysteine (C) can be identified near the C terminus that may undergo palmitoylation to anchor this region of the molecule to the membrane (O'Dowd et al., 1989b). In addition, potential phosphorylation sites are present in the third cytoplasmic loop (S). Ligand-dependent phosphorylation could conceivably lead to odorant-specific desensitization. The shading of the transmembrane domains in the diagram is proportional to their degree of sequence variability. (Based on clone I15 described by Buck and Axel, 1991.)

only 17 amino acids. Studies on the β-adrenergic receptor have shown that this loop participates in interactions with the G protein (Kobilka et al., 1988; Hamm et al., 1988). The putative odorant receptors contain conserved sequence motifs at the N-terminal region of this loop. The conserved sequence is either (K/R)IVSSI or HIT(C/W)AV. It has been suggested that these two different conserved motifs may reflect sites of interactions with different G proteins, regulating different transduction pathways (Buck and Axel, 1991).

The 18 clones sequenced by Buck and Axel (1991) could be grouped into three subfamilies with higher degrees of sequence homology. These subfamilies may correspond to major structurally distinct categories of odorants, and individual members of each subfamily may detect different, but structurally related odorants. The degree of ligand specificity of these putative odorant receptors remains to be determined.

Genomic analysis showed that the number of genes encoding these putative odorant receptors may range into several hundreds. Expression of only a fraction of these genes in individual receptor cells could generate the diversity in response profiles characteristic of olfactory receptor neurons (Buck and Axel, 1991).

Signal Transduction in Olfactory Receptor Cells

Substantial progress in studies on the molecular mechanisms of olfaction has come from the development of preparations of isolated olfactory cilia. Procedures for detaching chemosensory cilia from olfactory epithelium initially were based on deciliation procedures for protozoa and were first successfully applied to the olfactory rosette of the rainbow trout (Cagan and Zeiger, 1978; Rhein and Cagan, 1980 and 1983; Brown and Hara, 1981). Subsequently, such procedures were extended and modified for amphibian and mammalian olfactory tissues, and protocols were developed that allow the rapid preparation of isolated olfactory cilia from frog, rat, and bovine olfactory epithelium with high yields (Chen and Lancet, 1984; Anholt et al., 1986; Chen et al., 1986; Shirley et al., 1986; Lazard et al., 1989). Olfactory cilia can readily be detached from frog olfactory tissue via a calcium shock followed by sucrose gradient centrifugation. The preparation that is obtained in this manner contains fragmented cilia, axonemal structures, and ciliary plasma membrane vesicles (Anholt et al., 1986). These membranes are expected to contain most, if not all, of the membrane-associated cellular machinery required for odorant recognition and olfactory transduction. Biochemical studies on these preparations along with patch-clamp studies on isolated olfactory receptor cells have provided a wealth of information on chemosensory transduction in olfactory receptor cells.

Olfactory adenylate cyclase

Early studies reported the effects of cyclic AMP and phosphodiesterase inhibitors on the electro-olfactogram (Minor and Sakina, 1973; Menevse et al., 1977). These observations suggested that cyclic AMP could play a role in the olfactory transduction process. Indeed, olfactory cilia were found to be highly enriched in adenylate cyclase, suggesting a role for cyclic AMP as second messenger in olfaction (Pace et al., 1985; Sklar et al., 1986; Shirley et al., 1986).

Adenylate cyclase is a 120-kD transmembrane glycoprotein, consisting of two large hydrophobic domains, which contain 6 putative transmembrane helices each, and two large intracellular hydrophilic domains, which show homologies to guanylate cyclases and are thought to contain nucleotide binding sites (Krupinski et al., 1989). Olfactory adenylate cyclase appears to be a structurally distinct isoform, designated adenylate cyclase type III (Bakalyar and Reed, 1990).

Stimulation of olfactory adenylate cyclase, like other cyclases, occurs via

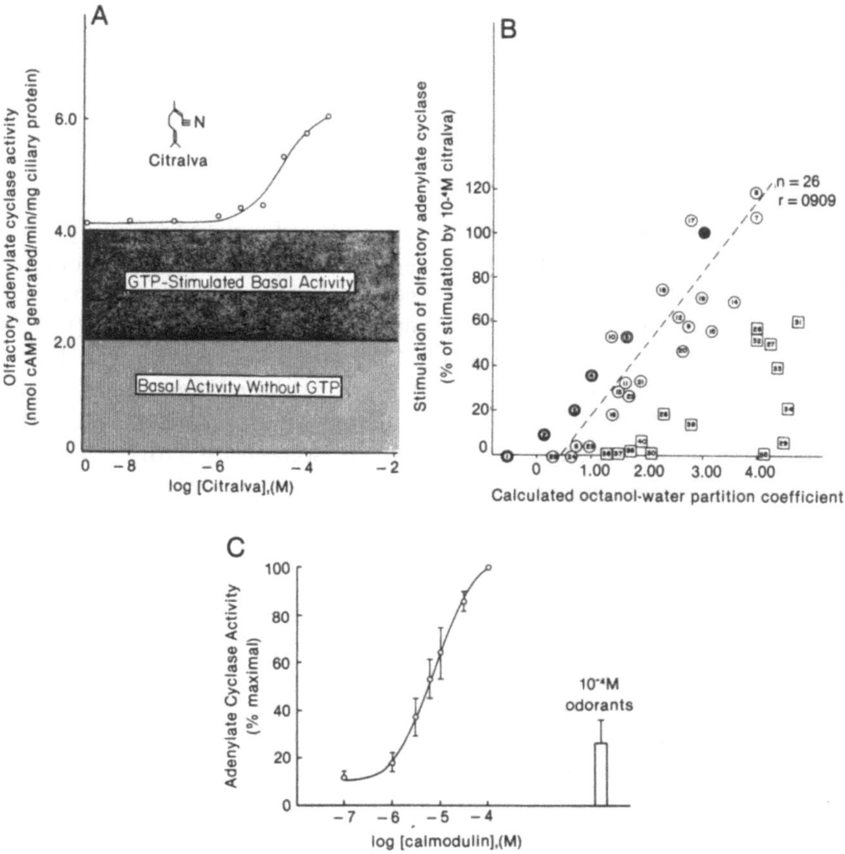

FIGURE 3.3. Stimulation of the olfactory adenylate cyclase by odorants and calmodulin.

A. Stimulation of the olfactory adenylate cyclase in chemosensory cilia from the olfactory epithelium of the bullfrog by the citrus odorant citralva (3,7-dimethyl-2,6-octadienenitrile).
Stimulation by citralva is shown relative to the basal activity in the absence and presence of 10 μM GTP. Stimulation by the odorant is GTP-dependent. (Based on data from Sklar et al., 1986.)

B. Stimulation of the olfactory adenylate cyclase by various odorants correlated with their octanol-water partition coefficient as index of their apparent hydrophobicity.
All odorants were tested at 100 μM and their potencies in stimulating the enzyme were standardized to citralva, indicated by the solid circle. The double circles represent a homologous series of methoxypyrazines. Odorants represented by circles (1–25, plus citralva) show a close correlation between stimulatory potency and apparent hydrophobicity. Odorants represented by squares (26–40) are less potent stimulators of the enzyme than would be expected from their octanol-water partition coefficients. The odorants indicated are methoxypyrazine (1), 2-methyl,3-methoxypyrazine (2), 2-ethyl,3-methoxypyrazine (3), 2-isopropyl,3-

a regulatory component, the G protein, described above. The olfactory epithelium contains several G proteins. cDNA clones from olfactory tissue have been identified that encode the α subunits of G_s, G_o, and three variants of G_i, designated G_{i1}, G_{i2}, and G_{i3}, (Jones and Reed, 1987). ADP-ribosylation studies (Pace et al., 1985) and the use of specific antisera (Anholt et al., 1987) have shown that olfactory cilia are enriched in G_s and contain small amounts of G_o. A cDNA clone has been identified that encodes the α subunit of the G_s protein expressed by olfactory receptor cells and localized to the chemosensory cilia. This G protein appears to be a variant of G_s, expressed only in olfactory receptor cells. It shows 88% homology with conventional G_s and has been named "G_{olf}" (Jones and Reed, 1989). Expression of G_{olf} in retrovirally transfected S49 cyc^- lymphocytes that contain adenylate cyclase, but are deficient in the expression of G_s, has provided evidence that this protein mediates activation of adenylate cyclase (Jones and Reed, 1989; Jones et al., 1990). The functional significance of why olfactory receptor neurons express tissue-specific variants of adenylate cyclase and G_s remains to be determined.

The olfactory adenylate cyclase can be stimulated by odorants (Figure 3.3; Pace et al., 1985; Sklar et al., 1986; Shirley et al., 1986). This activation occurs only in the presence of GTP and, thus, proceeds via the G protein. Quench-flow experiments using rat olfactory cilia have shown a rapid rise in cyclic AMP levels upon application of the odorant, isomenthone, at concentrations between 0.1 and 100 μM (Breer et al., 1990). This increase in cyclic AMP concentration occurs within milli-

---

methoxypyrazine (4), 2-isobutyl,3-methoxypyrazine (5), pyridine (6), 2-hexylpyridine (7), 3-hexylpyridine (8), geraniol (9), helional (10), acetophenone (11), hedione (12), furfuryl mercaptan (13), citral dimethyl acetal (14), citronellal (15), phenyl ethyl alcohol (16), isomenthone (17), L-and D-carvone (18), menthone (19), eugenol (20), cinnamic aldehyde (21), benzaldehyde (22), *n*-butanol (23), isobutyric acid (24), pyrollidine (25), β-ionone (26), citronellyl acetate (27), isoamyl acetate (28), limonene (29), lyral (30), coniferan (31), decanal (32), amylsalicylate (33), α-pinene (34), lilial (35), triethylamine (36), phenylethylamine (37), isovaleric acid (38), toluene (39), and chloroform (40). (Based on data from Sklar et al., 1986; reprinted, by permission, from Anholt, 1987.)

C. Stimulation of the olfactory adenylate cyclase by calmodulin.
Calmodulin stimulates the olfactory adenylate cyclase with an EC_{50} of 5.6 ± 0.9 μM (n = 6). Note that maximal stimulation observed with calmodulin is about 3 to 4 times greater than that observed with 10^{-4} M of a cocktail of odorants (menthone, 2-isobutyl,3-methoxypyrazine, and citralva, each at 10^{-4} M). These experiments were performed in the presence of 100 μM GTP. Maximal activation at 10^{-4} M calmodulin amounted to 4.85 ± 0.30 nmol/min/mg ciliary protein. Activation by calmodulin was dependent on the presence of micromolar concentrations of endogenous calcium and was enhanced by GTP. (Reprinted, by permission, from Anholt and Rivers, 1990.)

seconds after application of the odorant, reaches a maximal level after 25–50 ms, and then decays to the basal level over a period of 250–500 ms, most likely due to phosphodiesterase activity. However, when odorant-sensitive accumulation of cyclic AMP is measured over a longer time scale, concentrations of odorants in the high micromolar to millimolar range are required to elicit detectable stimulation. Dose-response curves generally do not reach saturation because of limitations imposed by the aqueous solubility of the odorants (Figure 3.3A). Furthermore, not all odorants increase the activity of the olfactory adenylate cyclase, and stimulation of the enzyme is mostly restricted to apolar compounds (Sklar et al., 1986). When homologous series of odorants are compared, their potencies for activating the olfactory adenylate cyclase closely parallel their lipid solubility (Figure 3.3B; Sklar et al., 1986; Anholt, 1987). Studies on the rapid activation of olfactory adenylate cyclase by the citrus odorant, citralva, revealed a biphasic dose-response curve with a saturable component at nanomolar concentrations and a second nonsaturating component at micromolar to millimolar concentrations (Boekhoff et al., 1990). It is likely that odorants at low concentrations activate adenylate cyclase via specific odorant receptor proteins, whereas at higher odorant concentrations, partitioning of odorants in the lipid membrane could cause some dissociation of the G protein into its component subunits and, consequently, activation of adenylate cyclase (Anholt, 1987 and 1988). Support for this possibility comes from the observation that odorants also stimulate adenylate cyclase when applied to cultured melanocytes from *Xenopus laevis*. These cells do not have an olfactory function, but, like olfactory cilia, contain a high activity of adenylate cyclase. Addition of odorants to these cells resulted in increased levels of intracellular cyclic AMP followed by spreading of their pigment granules (Lerner et al., 1988).

The olfactory adenylate cyclase, like other adenylate cyclases, can be persistently activated by the diterpene, forskolin, which acts directly on its catalytic moiety, and by nonhydrolyzable analogs of GTP, such as GppNHp and GTPγS. These GTP analogs stimulate the olfactory adenylate cyclase with a particularly high potency ($EC_{50} = \sim 4.0$ nM; Pace and Lancet, 1986; Anholt, 1988). This may reflect a property inherent in G_{olf}, which may represent a variant of G_s with increased affinity for GTP, or it may reflect the high concentration of adenylate cyclase and its associated G protein in the membrane and the resulting increased probability of collisional interactions between the two transduction components. Support for the latter notion comes from the observation that adenylate cyclase in membranes of the olfactory nerve, where it is present at ~20-fold lower specific activity, displays a much lower affinity for GTPγS than the ciliary enzyme (Anholt et al., 1989). It is of interest to note that adenylate cyclase in membranes from olfactory nerve is not sensitive to odorants. Reconstitution of olfactory adenylate cyclase from olfactory cilia in liposomes under conditions in which GTPγS- and forskolin-stimulated activities remain intact also results in a lowered affinity of the reconstituted enzyme complex for GTPγS (Anholt, 1988). Although it cannot be excluded that this reduced affinity is due to conformational alterations in G_{olf} as a result of the solubilization/reconstitution procedure, a lowered packing density of the adenylate cyclase and its G protein in the proteoliposomes can adequately explain these observations. Lowered concentrations of these membrane proteins in the liposomes could also account for the fact that the reconstituted enzyme is no longer sensitive to odorants, although denaturation of a putative odorant receptor protein during the reconstitution process cannot be excluded.

The consequences of increases in the intraciliary concentration of cyclic AMP appear to be twofold: cyclic AMP-dependent phosphorylation of cellular proteins and direct opening by cyclic AMP of ion channels. Phosphorylation by cyclic AMP-dependent protein kinase results in phosphorylation of ciliary proteins (Heldman and Lancet, 1986). Phosphorylation of ion-channel proteins is likely to alter their activation and inactivation behavior (Costa et al., 1982; Levitan, 1985; Reuter, 1987; Choquet et al., 1987; Rossie et al., 1987; Catterall, 1988) and could contribute to regulation of the membrane potential of olfactory receptor cells both at rest and during stimulation by odorants. Evidence for direct gating by cyclic AMP of ion channels has come from patch-clamp studies on excised membrane patches from olfactory cilia (Nakamura and Gold, 1987). These patches displayed a conductance that was gated directly by micromolar concentrations of both cyclic AMP and cyclic GMP. Cyclic nucleotide-activated conductances were not restricted to the ciliary membrane, but were also observed in membrane patches obtained from the olfactory knob or soma of olfactory receptor cells. Cloning of the cyclic nucleotide-activated channel from rat olfactory tissue showed it to be homologous to the cyclic GMP-activated channel of photoreceptor cells (Dhallan et al., 1990). It is not clear whether gating of the channel by cyclic GMP is merely a serendipitous experimental coincidence or whether cyclic GMP itself plays an as yet undiscovered role in olfaction. Since cyclic nucleotide-gated channels couple chemosensory transduction to excitation of the olfactory receptor cell, they play a central role in olfaction, and molecular studies on their mechanism of action are likely to be a major focus of future research.

Further evidence that supports the notion that cyclic AMP acts as a second messenger in olfaction comes from electrophysiological studies that showed that application of forskolin or analogues of cyclic AMP leads to depolarization of olfactory receptor cells (Firestein and Shepherd, 1989; Firestein et al., 1991; Frings and Lindemann, 1990 and 1991). Thus, one pathway for olfactory transduction consists of odorant-stimulated generation of cyclic AMP via G_{olf}. The resulting cyclic AMP opens cyclic nucleotide-activated channels to evoke the generator potential.

Alternative Transduction Pathways: Inositol Triphosphate and Calcium

Based on the early observations that some odorants fail to activate adenylate cyclase (Sklar et al., 1986; Shirley et al., 1986; Anholt, 1987), the notion that additional transduction pathways may participate in olfaction was pursued. Several investigators reported activation by odorants of ion channels in planar lipid bilayers to which olfactory membranes had been fused (Vodyanoy and Murphy, 1983; Vodyanoy and Vodyanoy, 1987; Labarca et al., 1988). These ion channels were cation specific and appeared upon application of nanomolar concentrations of odorants. It is difficult to extrapolate these observations made in planar lipid bilayers to the native olfactory membrane, and the precise nature of these channels and their possible involvement in olfactory transduction have, therefore, not been clarified.

There is, however, mounting evidence that inositol triphosphate may be generated during the response to certain odorants and that this second messenger may mediate calcium mobilization. Calcium is ideally suited as second messenger, since intracellular calcium concentrations in olfactory receptor cells at rest are estimated to be below 100 nM (Maue and Dionne, 1987), whereas the extracellular calcium concentration in the mucus surrounding the olfactory cilia is approximately 5 mM (Joshi et al., 1987). Thus, an ~50,000-fold chemical gradient for calcium exists across the ciliary membrane, several orders of magnitude higher than for any other ion. Consequently, only small increases in calcium permeability across the ciliary membrane can be expected to cause significant and rapid increases in the intraciliary calcium concentration.

Initial observations in the channel catfish, *Ictalurus punctatus*, indicated that odorous amino acids may cause breakdown of phosphatidyl inositol, liberating inositol triphosphate (Huque and Bruch, 1986). Furthermore, experiments using rapid kinetics in rat olfactory cilia indicated that some odorants that fail to stimulate adenylate cyclase, such as pyrazine, may cause transient increases in the concentration of inositol triphosphate (Boekhoff et al., 1990). In catfish olfactory receptor cells, an odorant-activated influx of calcium has been documented through the use of Fura-2, a calcium-dependent fluorescent dye (Restrepo et al., 1990). In this system, generation of inositol triphosphate opens calcium channels that are located in the ciliary plasma membrane. Although alternative transduction pathways have not yet been as extensively documented as the adenylate cyclase pathway, there is mounting evidence that olfactory transduction involves several second messenger systems. It is clear that regulatory mechanisms that allow communication between different transduction pathways must exist. The next section describes the role of calcium and calmodulin in mediating one aspect of crosstalk between second messenger systems.

Calcium/Calmodulin: Crosstalk between Second Messenger Systems

Several lines of evidence indicate that calcium is an important regulator of the response of olfactory receptor cells to odorants. Calcium at submicromolar concentrations modulates the activity of calcium-activated potassium channels (Maue and Dionne, 1987) and odorant-activated currents recorded from isolated olfactory receptor cells (Kurahashi, 1989). Patch-clamp studies on isolated olfactory receptor cells have identified an inward calcium current (Trotier, 1986; Schild, 1989; Trombley and Westbrook, 1991), and calcium channel blockers inhibit the electro-olfactogram (Winegar et al., 1988). Moreover, Kurahashi and Shibuya (1990) showed that isolated olfactory receptor cells of the newt fail to undergo inactivation when exposed to high concentrations of odorants in the absence of external calcium or in the presence of internal EGTA, suggesting that desensitization to odorants may be calcium-dependent. In addition, calcium inhibits the olfactory adenylate cyclase (Sklar et al., 1986; Shirley et al., 1986). This inhibition, however, occurs only at very high calcium concentrations ($IC_{50} = 200\ \mu M$) and may not be physiologically relevant.

Some of the most important regulatory functions of calcium may be mediated via calmodulin. Olfactory cilia, like other tissues, contain the ubiquitous calcium binding protein calmodulin (Anholt and Rivers, 1990), which, when complexed with calcium, performs a variety of regulatory functions (Stocklet et al., 1987; Means, 1988). Calcium/calmodulin activates neurotransmitter-regulated adenylate cyclases in regions of the brain, such as the striatum and the hippocampus (Ahlijanian and Cooper, 1988; Ojeda et al., 1988; Mollner and Pfeuffer, 1988; Cooper et al., 1988; Harrison et al., 1988; Uhlen and Wikberg, 1988). Studies on the effect of calmodulin on the olfactory adenylate cyclase revealed it to be a potent activator of this enzyme as well (Figure 3.3C; Anholt and Rivers, 1990). Indeed, calmodulin is more effective than odorants in stimulating the olfactory adenylate cyclase. The concentration of calmodulin that elicits half-maximal activation is approximately 6 μM, and maximal stimulation of the enzyme by calmodulin is about 3 times higher than that observed with odorants. The effects of calmodulin and odorants are additive and occur through independent mechanisms. Stimulation of the olfactory adenylate cyclase by calmodulin is dependent on calcium ($\sim 1\ \mu M$) and is enhanced by GTP, and the GTP-dependent component of this activation can be abolished by GDPβS, an analog of GDP that competes with GDP for its binding site on the α subunit of the G protein. Thus, calcium/calmodulin activates the olfactory adenylate cyclase, and this activation involves, at least in part, the G protein (Anholt and Rivers, 1990).

It is possible that, following the initial, rapid, transient increase in cyclic AMP concentration (Breer et al., 1990; Boekhoff et al., 1990), calcium influx during the response to odorants may cause a secondary, sustained rise in cyclic AMP concentration and that both cyclic AMP-dependent and calcium-dependent protein phosphorylation may contribute to desensitization. It is also likely that crosstalk between olfactory transduction pathways is reciprocal and that calcium influx may in turn be controlled by cyclic AMP. Precedence for this type of control, in which cyclic AMP-dependent protein phosphorylation inhibits calcium release, exists for the case of the inositol triphosphate-regulated calcium channel associated with endoplasmic reticulum (Suppatapone et al., 1988). Thus, olfactory transduction may involve a complex interplay of transduction pathways that together determine the activity of the olfactory receptor neuron.

Patch-Clamp Studies

Electrophysiological studies using the patch-clamp technique have provided a wealth of detailed information about the electrical properties of olfactory receptor neurons. Patch-clamp studies have shown the membrane resistance of olfactory receptor cells to be in the gigaohm range (Firestein and Werblin, 1987; Maue and Dionne, 1987; Schild, 1989), higher than estimated on the basis of intracellular recordings (Hedlund et al., 1987). This high resistance indicates that current introduced into these cells can be propagated electrotonically over long distances before dissipating. Thus, electrical events occurring at the chemosensory membranes of dendritic cilia can elicit action potentials at the axon hillock, and picoamperes of injected current are sufficient to cause excitation of the cell (Lynch and Barry, 1989). This high sensitivity of the olfactory receptor cell to injected current necessitates efficient control mechanisms to monitor the membrane potential at rest and during activation by odorants. Patch-clamp studies have identified in these cells several types of ion channels, including an array of calcium- and voltage-sensitive potassium channels that may regulate spike frequency and repolarization (Trotier, 1986; Maue and Dionne, 1987; Schild, 1989; Trombley and Westbrook, 1991).

Initial attempts to record odorant-induced currents from isolated olfactory receptor cells encountered difficulties, most likely because of proteolytic and/or mechanical damage to the cells during dissociation of the olfactory neuroepithelium. Subsequently, procedures were developed for the preparation of dissociated olfactory neurons without the use of proteolytic enzymes and with minimal mechanical damage. Odor responses could be obtained from cells prepared under such conditions, provided they possessed intact olfactory cilia, had a low resting membrane potential, and displayed a pattern of spontaneous activity consisting of action potentials and small irregular voltage bumps (Frings and Lindemann, 1988; Firestein and Werblin, 1989). Such cells responded reproducibly to 500 nM of the odorants cineole, amyl acetate, and 3-isobutyl,2-methoxypyrazine with periodic bursts of spike activity. This odorant-induced activity was potently inhibited by amiloride, indicating the involvement of an amiloride-sensitive sodium pathway in the response to odorants (Frings and Lindemann, 1988). Application of odorants in a solution of elevated potassium to isolated olfactory receptor cells or to cells in thin slices of the olfactory epithelium showed a small inward potassium current followed by a larger, slower odorant-induced generator current (Firestein and Werblin, 1989). The size of the potassium current was proportional to the potassium concentration in the delivery pipette and could be used to calculate the precise concentration of odorant that arrived at the cell surface. Dose-response curves obtained in this manner spanned a narrow range of concentrations between 10 and 100 μM. This narrow range of responsiveness to high concentrations of odorants is surprising in view of the wide dynamic range of the olfactory system and needs to be explained by further experimentation. Odor pulses directed at specific regions of isolated olfactory receptor cells showed the response to odorants to be confined to the ciliary and distal dendritic regions (Firestein and Shepherd, 1989; Frings and Lindemann, 1990). This observation, together with the fact that some cells that appear electrically and morphologically intact do not respond to a mixture of specific odorants (Firestein and Shepherd, 1989), provided further support for the notion that odorant recognition proceeds via specific receptor sites localized to the dendritic cilia. Furthermore, the long latency of several hundred milliseconds between the arrival of the odorant at the cell surface and the appearance of the generator potential could be accounted for by the involvement of intermediary transduction steps in the response to odorants, described above (Firestein and Werblin, 1989).

Olfactory Tissue-Specific Proteins of Undetermined Function

It was recognized early on that olfactory receptor neurons express several unique proteins. These proteins may play important, as yet not fully understood, roles in olfaction or in neurogenesis and are, therefore, of considerable interest. The first olfactory neuron specific protein to be identified was the "olfactory marker protein," a 19 kD cytoplasmic protein expressed only in mature olfactory receptor cells and found in many mammalian species (Margolis, 1980, 1985, and 1988). Later studies showed that olfactory marker protein is expressed also in some other cells of the central nervous system, which in the rat are confined to the preoptic region and hypothalamus (Baker et al., 1989). The amino acid sequence of olfactory marker protein was deduced from the nucleotide sequence of cDNA and shows no homology with other proteins (Margolis, 1985 and 1988; Sydor et al., 1986; Rogers et al., 1987). Characterization of the gene that encodes olfactory marker protein indicates the absence of introns and of the usual upstream CAAT or TATA boxes. However, a region was identified near the 5' end of the gene that regulates its tissue-specific expression (Danciger et al.,

1989). Although the developmental regulation and biochemistry of the olfactory marker protein have been studied in great detail (Margolis, 1985 and 1988), its function remains enigmatic.

Another cytoplasmic component that is enriched in olfactory receptor cells is the dipeptide carnosine, β-alanyl-L-histidine (Margolis and Grillo, 1977; Burd et al., 1982). Like the olfactory marker protein, the function of carnosine is unknown. It has been proposed that carnosine could serve as neurotransmitter released by primary afferent neurons at the glomerular synapses in the olfactory bulb (Hirsch et al., 1978; Rochel and Margolis, 1982), but conclusive evidence to substantiate this hypothesis is lacking. Furthermore, carnosinase, the enzyme that breaks down carnosine into its amino acid components, is found mostly in glands of the olfactory epithelium rather than at postsynaptic locations in the olfactory bulb (Margolis et al., 1983). In contrast to carnosinase, carnosine synthetase, the biosynthetic enzyme that links β-alanine to L-histidine to form carnosine, is expressed in olfactory neurons. Monoclonal antibodies have been obtained against this enzyme and have allowed its purification (Margolis et al., 1987; Margolis, 1988). In addition to its putative role as neurotransmitter, carnosine could serve as a reservoir for L-histidine. L-histidine can be decarboxylated to histamine, which may play a role in the regulation of neural regeneration following damage to the olfactory epithelium (Greene et al., 1984).

Olfactory receptor cells also contain vimentin, a neurofilament protein that is usually present only in developing and not in mature neurons (Schwob et al., 1986; Ophir and Lancet, 1988). The expression of vimentin in olfactory neurons may reflect the great degree of neural plasticity retained by these cells.

Several laboratories have produced monoclonal antibodies that react specifically with olfactory neurons (Allen and Akeson, 1985; Hempstead and Morgan, 1985; Fujita et al., 1985; Mori et al., 1985; Schwob and Gottlieb, 1986; Morgan, 1988; Onoda, 1988a and b; Onoda and Fujita, 1988; Carr et al., 1989; Anholt et al., 1990). Antibodies against developmentally regulated antigens (Onoda, 1988a and b; Carr et al., 1989) or antigens expressed only by subclasses of olfactory neurons (Fujita et al., 1985; Mori et al., 1985; Schwob and Gottlieb, 1986) have been described. In addition, monoclonal antibodies have been identified that react with olfactory-tissue-specific proteins that are associated with olfactory cilia (Anholt et al., 1990). These antibodies have identified a 57 kD glycoprotein that contains unique carbohydrate moieties and forms a 120 kD homo-dimer via intermolecular disulfide bonds. This protein, named "olfactomedin," is produced by sustentacular cells and submucosal glands and deposited in the mucous matrix which lines the epithelial surface, where it can adhere to olfactory cilia (Snyder et al., 1991). Although the massive production of olfactomedin and its striking deposition at the chemosensory surface of the olfactory neuroepithelium suggest a role for this protein in chemoreception, its function remains to be determined.

The functions of these olfactory-tissue-specific proteins, the specificity, regulation, and expression of putative odorant receptors, and the determination of the activity of the olfactory receptor neuron via the interplay of several second messenger systems will continue to be intriguing and challenging questions in this rapidly developing field.

Acknowledgments. Research in the author's laboratory is supported by grants from the National Institutes of Health (DC-00394), the U.S. Army Research Office (DAAL03-89-K-0178), and the Office of Naval Research (N00014-90-J-1056 to Dr. Sidney A. Simon).

References

Ahlijanian, M.K., and D.M. Cooper (1988) Distinct interactions between Ca^{2+}/calmodulin and neurotransmitter stimulation of adenylate cyclase in striatum and hippocampus. *Cell. Mol. Neurobiol.*, *8*, 459–469.

Allen, W.K., and R. Akeson (1985) Identification of a cell surface glycoprotein family of olfactory receptor neurons with a monoclonal antibody. *J. Neurosci.*, *5*, 284–296.

Amoore, J.E. (1971) Olfactory genetics and anosmia. In L.M. Beidler (Ed.), *Chemical Senses: Olfaction. Handbook of Sensory Physiology*, Vol. 4 (1). pp. 245–256.

Amoore, J.E. (1982) Odor theory and odor classification. In E.T. Theimer (Ed.), *Fragrance Chemistry: The Science of the Sense of Smell.* New York: Academic Press, pp. 27–76.

Anholt, R.R.H. (1987) Primary events in olfactory reception. *Trends Biochem. Sci.*, *12*, 58–62.

Anholt, R.R.H. (1988) Functional reconstitution of the olfactory membrane: Incorporation of the olfactory adenylate cyclase in liposomes. *Biochemistry*, *27*, 6464–6468.

Anholt, R.R.H. (1989) The molecular physiology of olfaction. *Am. J. Physiol.*, *257* (*Cell Physiol.*, *26*), C1043–C1054.

Anholt, R.R.H., U. Aebi, and S.H. Snyder (1986) A partially purified preparation of isolated chemosensory cilia from the olfactory epithelium of the bullfrog, *Rana catesbeiana. J. Neurosci.*, *6*, 1962–1969.

Anholt, R.R.H., S.M. Mumby, D.A. Stoffers, P.R. Girard, J.F. Kuo, and S.H. Snyder, (1987) Transduction proteins of olfactory receptor cells: Identification of guanine nucleotide binding proteins and protein kinase C. *Biochemistry*, *26*, 788–795.

Anholt, R.R.H., R.W. Farmer, and C.A. Karavanich (1989) Excitation by odorants of olfactory receptor cells: Molecular interactions at the ciliary membrane. In J.G. Brand, J.H. Teeter, M.R. Kare, and R.H. Cagan (Eds)., *Chemical Senses: Molecular Aspects of Taste and Odor Reception*. New York: Marcel Dekker Inc., Chapter 17, pp. 347–361.

Anholt, R.R.H., and A.M. Rivers (1990) Olfactory transduction: Crosstalk between second messenger systems. *Biochemistry 29*, 4049–4054.

Anholt, R.R.H., A.E. Petro, and A.M. Rivers (1990) Identification of a group of novel membrane proteins unique to chemosensory cilia of olfactory receptor cells. *Biochemistry*, *29*, 3366–3373.

Asano, T., R. Semba, N. Kamija, N. Ogaswara, and K. Kato (1988) G_o, a GTP-binding protein: Immunochemical and immunohistochemical localization in the rat. *J. Neurochem.*, *50*, 1164–1169.

Ashkenazi, A., E.G. Peralta, J.W. Winslow, J. Ramachandran, and D.J. Capon, (1989) Functionally distinct G proteins selectively couple different receptors to PI hydrolysis in the same cell. *Cell*, *56*, 487–493.

Bakalyar, H.A., and R.R. Reed (1990) Identification of a specialized adenylyl cyclase that may mediate odorant detection. *Science*, *250*, 1403–1406.

Baker, H., M. Grillo, and F.L. Margolis (1989) Biochemical and immunocytochemical characterization of olfactory marker protein in the rodent central nervous system. *J. Comp. Neurol.*, *285*, 246–261.

Bignetti, E., A. Cavaggioni, P. Pelosi, K.C. Persaud, R.T. Sorbi, and R. Tirindelli (1985) Purification and characterization of an odorant-binding protein from cow nasal tissue. *Eur. J. Biochem.*, *149*, 227–231.

Boekhoff, I., E. Tareilus, J. Strottman, and H. Breer (1990) Rapid activation of alternative second messenger pathways in olfactory cilia from rats by different odorants. *EMBO J.*, *9*, 2453–2458.

Brass, L.F., M. Laposata, H.S. Banga, and S.E. Rittenhouse (1986) Regulation of the phosphoinositide hydrolysis pathway in thrombin-stimulated platelets by a pertussis toxin-sensitive guanine nucleotide-binding protein: Evaluation of its contribution to platelet activation and comparisons with the adenylate cyclase inhibitory protein, G_i. *J. Biol. Chem.*, *261*, 16838–16847.

Breer, H., I. Boekhoff, and E. Tareilus (1990) Rapid kinetics of second messenger formation in olfactory transduction. *Nature*, *345*, 65–68.

Brown, B., and T.J. Hara (1981) Accumulation of chemostimulatory amino acids by a sedimentable fraction isolated from olfactory rosettes of rainbow trout (*Salmo gairdneri*). *Biochim. Biophys. Acta*, *675*, 149–162.

Bruch, R.C., and D.L. Kalinoski (1987) Interaction of GTP-binding regulatory proteins with chemosensory receptors. *J. Biol. Chem.*, *262*, 2401–2404.

Buck, L., and R. Axel, (1991) A novel multigene family may encode odorant receptors: A molecular basis for odor recognition. *Cell*, *65*, 175–187.

Burd, G.D., B.J. Davis, F. Macrides, M. Grillo, and F.L. Margolis (1982) Carnosine in primary afferents of the olfactory system: An autoradiographic and biochemical study. *J. Neurosci.*, *2*, 244–255.

Cagan, R.H., and W.N. Zeiger (1978) Biochemical studies of olfaction: Binding specificity of radioactively labeled stimuli to an isolated olfactory preparation from rainbow trout (Salmo *gairdneri*). *Proc. Natl. Acad. Sci. USA*, *75*, 4679–4683.

Camara, C.G., and J.W. Harding (1984) Thymidine incorporation in the olfactory epithelium of mice: early exponential response induced by olfactory neurectomy. *Brain Res.*, *308*, 63–68.

Cancalon, P. (1982) Degeneration and regeneration of olfactory cells induced by $ZnSO_4$ and other chemicals. *Tissue and Cell*, *14*, 717–733.

Carr, V.M., A.I. Farbman, M.S. Lidow, L. M. Coletti, J.L. Hempstead, and J.I. Morgan (1989) Developmental expression of reactivity to monoclonal antibodies generated against olfactory epithelia. *J. Neurosci.*, *9*, 1179–1198.

Casey, P.J., and A.G. Gilman, (1988) G-protein involvement in receptor-effector coupling. *J. Biol. Chem.*, *263*, 2577–2580.

Catterall, W.A. (1988) Structure and function of voltage-sensitive ion channels. *Science*, *242*, 50–61.

Chen, Z., and D. Lancet (1984) Membrane proteins unique to vertebrate olfactory cilia: Candidates for sensory receptor molecules. *Proc. Natl. Acad. Sci. USA*, *81*, 1859–1863.

Chen, Z., U. Pace, J. Heldman, A. Shapira, and D. Lancet (1986) Isolated frog olfactory cilia: A preparation of dendritic membranes from chemosensory neurons. *J. Neurosci.*, *6*, 2146–2154.

Choquet, D., P. Sarthou, D. Primi, P.A. Cazenave, and H. Kom (1987) Cyclic AMP-modulated potassium channels in murine B cells and their precursors. *Science*, *235*, 1211–1214.

Cockcroft, S., and B.D. Gomperts (1985) Role of guanine nucleotide binding protein in the activation of polyphosphoinositide phosphodiesterase. *Nature*, *314*, 534–536.

Cooper, D.M., M.K. Ahlijanian, and E. Perez-Reyes (1988) Calmodulin plays a dominant role in determining neurotransmitter regulation of neuronal adenylate cyclase. *J. Cell. Biochem.*, *36*, 417–427.

Costa, R.C., J.E. Casnellie, and W.A. Catterall (1982) Selective phosphorylation of the α subunit of the sodium channel by cAMP-dependent protein kinase. *J. Biol. Chem.*, *257*, 7918–1921.

Costanzo, R.M., and P.P.C. Graziadei (1983) A quantitative analysis of changes in the olfactory epithelium following bulbectomy in hamster. *J. Comp. Neurol.*, *215*, 370–381.

Costanzo, R.M., and P.P.C. Graziadei (1987) Development and plasticity of the olfactory system. In T.E. Finger and W.L. Silver (Eds), *Neurobiology of Taste and Smell* New York: John Wiley and Sons, pp. 233–250.

Criswell, D.W., F.L. McClure, R. Schafer, and K.R. Brower (1980) War gases as olfactory probes. *Science*, *210*, 425–426.

Dahl, A.R. (1988) The effect of cytochrome P450-dependent metabolism and other enzyme activities on olfaction. In F.L. Margolis and T.V. Getchell (Eds.), *Molecular Neurobiology of the Olfactory System: Molecular, Membranous, and Cytological Studies* New York: Plenum Press, pp. 51–70.

Dahl, A.R., W.M. Hadley, F.F. Hahn, J.M. Benson, and R.O. McClellan (1982) Cytochrome P-450-dependent monooxygenases in olfactory epithelium of dogs: Possible role in tumorigenicity. *Science*, *216*, 57–59.

Danciger, E., C. Mettling, M. Vidal, R. Morris, F. and Margolis (1989) Olfactory marker protein gene: Its structure and olfactory neuron-specific expression in transgenic mice. *Proc. Natl. Acad. Sci. USA*, *86*, 8565–8569.

Delaleu, J.C., and A. Holley (1980) Modification of transduction mechanisms in the frog's olfactory mucosa using a thiol reagent as olfactory stimulus. *Chemical Senses and Flavour*, *5*, 205–218.

Delaleu, J.C., and A. Holley (1983) Investigations of the discriminative properties of the frog's olfactory mucosa using a photoactivatable odorant. *Neurosci. Lett.*, *37*, 251–256.

Dhallan, R.S., K.W. Yau, K.A. Schrader, and R.R. Reed (1990) Primary structure and functional expression of a cyclic nucleotide-activated channel from olfactory neurons. *Nature*, *347*, 184–187.

Downes, C.P. (1983) Inositol phospholipids and neurotransmitter-receptor signalling mechanisms. *Trends Neurosci.*, *6*, 313–316.

Duchamp, A., M.F. Revial, A. Holley, and P. MacLeod (1974) Odor discrimination by frog olfactory receptors. *Chemical Senses and Flavour*, *1*, 213–233.

Ewald, D.A., P.C. Sternweis, and R.J. Miller (1988) Guanine nucleotide-binding protein G_o-induced coupling of neuropeptide Y receptors to Ca^{2+} channels in sensory neurons. *Proc. Natl. Acad. Sci. USA*, *85*, 3633–3637.

Fesenko, E.E., V.I. Novoselov, N.F. Mjasoedov, and G.V. Sidorov (1978) Molecular mechanisms of olfactory reception: III. Binding of camphor by rat and dog olfactory mucosa components. *Studia Biophysica*, *73*, 71–84.

Fesenko, E.E., V.I. Novoselov, and L.D. Krapivinskaya (1979) Molecular mechanisms of olfactory reception: IV. Some biochemical characteristics of the camphor receptor from rat olfactory epithelium. *Biochim. Biophys. Acta*, *587*, 424–433.

Firestein, S., and F.S. Werblin (1987) Gated currents in isolated olfactory receptor neurons of the larval tiger salamander. *Proc. Natl. Acad. Sci. USA*, *84*, 6292–6296.

Firestein, S., and F. Werblin (1989) Odor-induced membrane currents in vertebrate olfactory receptor neurons. *Science*, *244*, 79–82.

Firestein, S., and G.M. Shepherd (1989) Constraints on transduction mechanisms from analysis of olfactory receptor currents. Eleventh Annual Meeting of the Association for Chemoreception Sciences, Abst #152.

Firestein, S., B. Darrow, and G.M. Shepherd (1991) Activation of the sensory current in salamander olfactory receptor neurons depends on a G protein-mediated cAMP second messenger system. *Neuron*, *6*, 825–835.

Frings, S., and B. Lindemann (1988) Odorant response of isolated olfactory receptor cells is blocked by amiloride. *J. Membrane Biol.*, *105*, 233–243.

Frings, S., and B. Lindemann (1990) Single unit recording from olfactory cilia. *Biophys. J.*, *57*, 1091–1094.

Frings, S., and B. Lindemann (1991) Current recording from sensory cilia of olfactory receptor cells in situ. I. The neural response to cyclic nucleotides. *J. Gen. Physiol.*, *97*, 1–16.

Fujita, S.C., K. Mori, K. Imamura, and K. Obata (1985) Subclasses of olfactory receptor cells and their segregated central projections demonstrated by a monoclonal antibody. *Brain Res.*, *326*, 192–196.

Gennings, J.N., D.B. Gower, and L.H. Bannister (1977) Studies on the receptors to 5α-androst-16-en-3-one and 5α-androst-16-en-3α-ol in sow nasal mucosa. *Biochim. Biophys. Acta*, *496*, 547–556.

Getchell, M.L., and R.C. Gesteland (1972) The chemistry of olfactory reception: Stimulus-specific protection from sulfhydryl reagent inhibition. *Proc. Natl. Acad. Sci. USA*, *69*, 1494–1498.

Getchell, T.V. (1974) Electrogenic sources of slow voltage transients recorded from frog olfactory epithelium. *J. Neurophysiol.*, *37*, 1115–1130.

Getchell, T.V. (1988) Induction of odorant-evoked current transients in *Xenopus* oocytes injected with mRNA isolated from the olfactory mucosa of *Rana pipiens*. *Neurosci. Lett.*, *91*, 217–221.

Getchell, T.V., and G.M. Shepherd (1978a) Responses of olfactory receptor cells to step pulses of odour at different concentrations in the salamander. *J. Physiol. (Lond.)*, *282*, 521–540.

Getchell, T.V., and G.M. Shepherd (1978b) Adaptive properties of olfactory receptors analysed with odour pulses of varying durations. *J. Physiol. (Lond.)*, *282*, 541–560.

Getchell, T.V., F.L. Margolis, and M.L. Getchell (1984) Perireceptor and receptor events in vertebrate olfaction. *Prog. Neurobiol.*, *23*, 317–345.

Gilman, A.G. (1987) G-proteins: Transducers of receptor-generated signals. *Annu. Rev. Biochem.*, *56*, 615–649.

Goldberg, S.J., J. Turpin, and S. Price (1979) Anisole binding protein from olfactory epithelium: Evidence for a role in transduction. *Chemical Senses and Flavour*, *4*, 207–214.

Graziadei, P.P.C., and G.A. Monti-Graziadei (1978) Continuous nerve cell renewal in the olfactory system. In M. Jacobson (Ed.), *Handbook of Sensory Physiology. Development of Sensory Systems*. Berlin: Springer-Verlag, vol. 9, pp. 55–83.

Greene, S.M., F.L. Margolis, M. Grillo, and H. Fisher (1984) Enhanced carnosine (β-alanyl-L-histidine) breakdown and histamine metabolism following treatment with compound 48/80. *Eur. J. Pharmacol.*, *99*, 79–84.

Hamm, H.E., D. Deretic, A. Arendt, P.A. Hargrove, B. Koenig, and K.P. Hoffmann (1988) Site of G protein binding to rhodopsin mapping with synthetic peptides to the α subunit. *Science*, *241*, 832–835.

Harding, J.W., P.P.C. Graziadei, G.A. Monti-Graziadei, and F.L. Margolis (1977) Denervation in the primary olfactory pathway of mice. IV. Biochemical and morphological evidence for neuronal replacement following nerve section. *Brain Res.*, *132*, 11–28.

Harding, J.W., T.V. Getchell, and F.L. Margolis (1978) Denervation of the primary olfactory pathway in mice. V. Long-term effect of intranasal $ZnSO_4$ irrigation on behavior, biochemistry and morphology. *Brain Res.*, *140*, 271–285.

Harrison, J.K., C.K. Mickevicius, and M.E. Gnegy, (1988) Differential regulation by calmodulin of basal, GTP-, and dopamine-stimulated adenylate cyclase activities in bovine striatum. *J. Neurochem.*, *51*, 345–352.

Hedlund, B., L.M. Masukawa, and G.M. Shepherd (1987) Excitable properties of olfactory receptor neurons. *J. Neurosci.*, *7*, 2338–2343.

Heldman, J., and D. Lancet (1986) Cyclic AMP-dependent protein phosphorylation in chemosensory neurons: Identification of cyclic nucleotide-regulated phosphoproteins in olfactory cilia. *J. Neurochem.*, *47*, 1527–1533.

Hempstead, J.L., and J.I. Morgan (1985) A panel of monoclonal antibodies to the rat olfactory epithelium. *J. Neurosci.*, *5*, 438–449.

Hescheler, J., W. Rosenthal, W. Trautwein, and G. Schultz (1987) The GTP-binding protein, G_o, regulates neuronal calcium channels. *Nature*, *325*, 445–447.

Hinds, J.W., P.L. Hinds, and N.A. McNelly (1984) An autoradiographic study of the mouse olfactory epithelium: Evidence for long-lived receptors. *Anat. Rec.*, *210*, 375–383.

Hirsch, J.D., M. Grillo, and F.L. Margolis (1978) Ligand binding studies in the mouse olfactory bulb: Identification and characterization of a L-(^{3}H)carnosine binding site. *Brain Res.*, *158*, 407–422.

Hornung, D.E., and M.M. Mozell (1977) Odorant removal from the frog olfactory mucosa. *Brain Res.*, *128*, 158–163.

Hornung, D.E., and M.M. Mozell (1980) Tritiated odorants to monitor retention in the olfactory and vomeronasal organs. *Brain Res.*, *181*, 488–492.

Hornung, D.E., and M.M. Mozell (1981) Accessiblity of odorant molecules to the receptors. In R.H. Cagan and M.K. Kare (Eds.), *Biochemistry of Taste and Olfaction*. New York, Academic Press, pp. 33–45.

Huff, R.M., J.M. Axton, and E.J. Neer (1985) Physical and immunological characterization of a guanine nucleotide-binding protein purified from bovine cerebral cortex. *J. Biol. Chem.*, *260*, 10864–10871.

Huque, T., and R.C. Bruch, (1986) Odorant- and guanine nucleotide-stimulated phosphoinositide turnover in olfactory cilia. *Biochem. Biophys. Res. Commun.*, *137*, 36–43.

Jones, D.T., and R.R. Reed (1987) Molecular cloning of five GTP-binding protein cDNA species from rat olfactory neuro-epithelium. *J. Biol. Chem.*, *26*, 14241–14249.

Jones, D.T., and R.R. Reed (1989) G_{olf}: An olfactory neuron specific-G protein involved in odorant signal transduction. *Science*, *244*, 790–795.

Jones, D.T., S.B. Masters, H.R. Bourne, and R.R. Reed (1990) Biochemical characterization of three stimulatory GTP-binding proteins: The large and small forms of G_s and the olfactory-specific G-protein, G_{olf}. *J. Biol. Chem.*, *265*, 2671–2676.

Joshi, H., M.L. Getchell, B. Zielinski, and T.V. Getchell (1987) Spectrophotometric determination of cation concentrations in olfactory mucus. *Neurosci. Lett.*, *82*, 321–326.

Kalinoski, D.L., R.C. Bruch, and J.G. Brand (1987) Differential interaction of lectins with chemosensory receptors. *Brain Res.*, *418*, 34–40.

Kashiwayanagi, M., and K. Kurihara (1984) Neuroblastoma cell as model for olfactory cell: Mechanism of depolarization in response to various odorants. *Brain Res.*, *293*, 251–258.

Kashiwayanagi, M., and K. Kurihara (1985) Evidence for non-receptor odor discrimination using neuroblastoma cells as a model for olfactory cells. *Brain Res.*, *359*, 97–103.

Kobilka, B.K., T.S. Kobilka, K. Daniel, J.W. Regan, M.G. Caron, and R.J. Lefkowitz (1988) Chimeric α_2-β_2-adrenergic receptors: delineation of domains involved in effector coupling and ligand binding specificity. *Science*, *240*, 1310–1316.

Krupinski, J., F. Coussen, H.A. Bakalyar, W.J. Tang, P.G. Feinstein, K. Orth, C. Slaughter, R.R. Reed, and A.G. Gilman (1989) Adenylyl cyclase amino acid sequence: Possible channel- or transporter-like structure. *Science*, *244*, 1558–1564.

Kurahashi, T. (1989) Activation by odorants of cation-selective conductance in the olfactory receptor cell isolated from the newt. *J. Physiol.*, *419*, 177–192.

Kurahashi, T., and T. Shibuya (1990) Ca^{2+}-dependent adaptive properties in the solitary olfactory receptor cell of the newt. *Brain Res.*, *515*, 261–268.

Labarca, P., S.A. Simon, and R.R.H. Anholt (1988) Activation by odorants of a multistate cation channel from olfactory cilia. *Proc. Natl. Acad. Sci. USA*, *85*, 944–947.

Lancet, D. (1986) Vertebrate olfactory reception. *Ann. Rev. Neurosci.*, *9*, 329–355.

Lazard, D., Y. Barak, and D. Lancet (1989) Bovine olfactory cilia preparation: Thiol-modulated odorant-sensitive adenylyl cyclase. *Biochim. Biophys. Acta*, *1013*, 68–72.

Lee, K.H., R.G. Wells, and R.R. Reed (1987) Isolation of an olfactory cDNA: Similarity to retinol-binding protein suggests a role in olfaction. *Science*, *235*, 1053–1056.

Lerner, M.R., J. Reagan, T. Gyorgyi, and A. Roby (1988) Olfaction by melanophores: What does it mean? *Proc. Natl. Acad. Sci. USA*, *85*, 261–264.
Levitan, I.B. (1985) Phosphorylation of ion channels. *J. Membrane Biol.*, *87*, 177–190.
Lynch, J.W. and P.H. Barry (1989) Action potentials initiated by single channels opening in a small neuron (rat olfactory receptor). *Biophys. J.*, *55*, 755–768.
Majerus, P.W., T.M. Connolly, H. Deckmyn, T.S. Ross, T.E. Bross, H. Ishii, V.S. Bansal, and D.B. Wilson (1986) The metabolism of phosphoinositide-derived messenger molecules. *Science*, *234*, 1519–1526.
Margolis, F.L. (1980) A marker protein for the olfactory chemoreceptor neuron. In R.A. Bradshaw and D.M. Schneider (Eds.), *Proteins of the Nervous System*. New York: Raven Press, pp. 59–84.
Margolis, F.L. (1985) Olfactory marker protein: from PAGE band to cDNA clone. *Trends Neurosci.*, *8*, 542–546.
Margolis, F.L. (1988) Molecular cloning of olfactory-specific gene products. In F.L. Margolis and T.V. Getchell (Eds.), *Molecular Neurobiology of the Olfactory System: Molecular, Membranous and Cytological Studies*. New York: Plenum Press, pp. 237–265.
Margolis, F.L., and M. Grillo (1977) Axoplasmic transport of carnosine (β-alanyl-L-histidine) in the mouse olfactory pathway. *Neurochem. Res.*, *2*, 507–519.
Margolis, F.L., M. Grillo, N. Grannot-Reisfeld, and A.I. Farbman (1983) Purification, characterization and immunocytochemical localization of mouse kidney carnosinase. *Biochim. Biophys. Acta*, *744*, 237–248.
Margolis, F.L., M. Grillo, J. Hempstead, and J.L. Morgan (1987) Monoclonal antibodies to mammalian carnosine synthetase. *J. Neurochem.*, *48*, 593–600.
Mason, J.R., and T.H. Morton (1984) Fast and loose covalent binding of ketones as a molecular mechanism in vertebrate olfactory receptors. *Tetrahedron*, *40*, 483–492.
Mason, J.R., L. Clark, and T.H. Morton (1984) Selective deficits in the sense of smell caused by chemical modification of the olfactory epithelium. *Science*, *226*, 1092–1094.
Maue, R.A., and V.E. Dionne (1987) Patch-clamp studies of isolated mouse olfactory receptor neurons. *J. Gen. Physiol.*, *90*, 95–125.
Means, A.R. (1988) Molecular mechanisms of action of calmodulin. *Recent Prog. Horm. Res.*, *44*, 223–262.
Menevse, A., G. Dodd, and T.M. Poynder (1977) Evidence for the specific involvement of cyclic AMP in the olfactory transduction mechanism. *Biochem. Biophys. Res. Commun.*, *77*, 671–677.
Menevse, A., G. Dodd, and T.M. Poynder (1978) A chemical-modification approach to the olfactory code. *Biochem. J.*, *176*, 845–854.
Minor, A.V. (1971) Electro-olfactogram: Its origin and mechanisms of generation. *Fiziol. Zh. SSSR*, *57*, 1115–1122.
Minor, A.V., and N.L. Sakina (1973) The role of cyclic adenosine-3′,5′-monophosphate in olfactory reception. *Neurofiziologiya*, *5*, 415–422.
Mollner, S., and T. Pfeuffer (1988) Two different adenylyl cyclases in brain distinguished by monoclonal antibodies. *Eur. J. Biochem.*, *171*, 265–271.
Morgan, J.I. (1988) Monoclonal antibody mapping of the rat olfactory tract. In F.L. Margolis and T.V. Getchell (Eds.) *Molecular Neurobiology of the Olfactory*

System: Molecular, Membranous, and Cytological Studies New York: Plenum Press, pp. 269–296.

Mori, K., S.C. Fujita, K. Imamura, and K. Obata (1985) Immunohistochemical study of subclasses of olfactory nerve fibers and their projections to the olfactory bulb in the rabbit. *J. Comp. Neurol.*, *242*, 214–229.

Mozell, M.M., P.R. Sheehe, D.E. Hornung, P.F. Kent, S.L. Youngentob, and S.J. Murphy (1987) "Imposed" and "inherent" mucosal activity patterns: Their composite representation of olfactory stimuli. *J. Gen. Physiol.*, *90*, 625–650.

Nakajima, Y., S. Nakajima, and M. Inoue (1988) Pertussis toxin-insensitive G protein mediates substance P-induced inhibition of potassium channels in brain neurons. *Proc. Natl. Acad. Sci. USA*, *85*, 3643–3647.

Nakamura, T., and G.H. Gold (1987) A cyclic nucleotide-gated conductance in olfactory receptor cilia. *Nature*, *325*, 442–444.

Nef, P., J. Heldman, D. Lazard, T. Margalit, M. Jaye, I. Hanukoglu, and D. Lancet (1989) Olfactory-specific cytochrome P-450: cDNA cloning of a novel neuroepithelial enzyme possibly involved in chemoreception. *J. Biol. Chem.*, *264*, 6780–6785.

Nef, P., T.M. Larabee, K. Kagimoto, and U.A. Meyer (1990) Olfactory-specific cytochrome P-450 (P-450Olfl;IIG1): Gene structure and developmental regulation. *J. Biol. Chem.*, *265*, 2903–2907.

Nomura, T., and K. Kurihara (1987a) Liposomes as a model for olfactory receptor cells: Changes in membrane potential in response to various odorants. *Biochemistry*, *26*, 6135–6140.

Nomura, T., and K. Kurihara (1987b) Effects of changed lipid composition on responses of liposomes to various odorants: Possible mechanism of odor discrimination. *Biochemistry*, *26*, 6141–6145.

O'Dowd, B.F., R.J. Lefkowitz, and M.G. Caron (1989a) Structure of the adrenergic and related receptors. *Annu. Rev. Neurosci.*, *12*, 67–83.

O'Dowd, B.F., M. Hnatowich, M.G. Caron, R.J. Lefkowitz, and M. Bouvier (1989b) Palmitoylation of the human β_2-adrenergic receptor. Mutation of CYS^{341} in the carboxyl tail leads to an uncoupled non-palmitoylated form of the receptor. *J. Biol. Chem.*, *264*, 7564–7569.

Ojeda, S.R., H.F. Urbanski, K.H. Katz, and M.E. Costa (1988) Prostaglandin E2 releases luteinizing hormone-releasing hormone from the female juvenile hypothalamus through a Ca^{2+} dependent, calmodulin-independent mechanism. *Brain Res.*, *441*, 339–351.

Onoda, N. (1988a) Monoclonal antibody immunohistochemistry of rabbit olfactory receptor neurons during development. *Neuroscience*, *26*, 1003–1012.

Onoda, N. (1988b) Monoclonal antibody immunohistochemistry of degenerative and renewal patterns in rabbit olfactory receptor neurons following unilateral olfactory bulbectomy. *Neuroscience*, *26*, 1013–1022.

Onoda, N., and S.C. Fujita (1988) Monoclonal antibody immunohistochemistry of adult rabbit olfactory structures. *Neuroscience*, *26*, 993–1002.

Ophir, D., and D. Lancet (1988) Expression of intermediate filaments and desmoplakin in vertebrate olfactory mucosa. *Anat. Rec.*, *221*, 754–760.

Ottoson, D. (1956) Analysis of the electrical activity of the olfactory epithelium. *Acta Physiol. Scand.*, *35*, *Suppl.* 122, 1–83.

Pace, U., E. Hanski, Y. Salomon, and D. Lancet (1985) Odorant-sensitive adenylate cyclase may mediate olfactory reception. *Nature*, *316*, 255–258.

Pace, U., and D. Lancet (1986) Olfactory GTP-binding protein: Signal-transducing polypeptide of vertebrate chemosensory neurons. *Proc. Natl. Acad. Sci. USA*, *83*, 4947–4951.

Pelosi, P., N.E. Baldaccini, and A.M. Pisanelli (1982) Identification of a specific olfactory receptor for 2-isobutyl-3-methoxypyrazine. *Biochem. J.*, *201*, 245–248.

Pevsner, J., R.R. Trifiletti, S.M. Strittmatter, and S.H. Snyder (1985) Isolation and characterization of an olfactory receptor protein for odorant pyrazines. *Proc. Natl. Acad. Sci. USA*, *82*, 3050–3054.

Pevsner, J., P.B. Sklar, and S.H. Snyder (1986) Odorant-binding protein: Localization to nasal glands and secretions. *Proc. Natl. Acad. Sci. USA*, *83*, 4942–4946.

Pevsner, J., P.M. Hwang, P.B. Sklar, J.C. Venable, and S.H. Snyder (1988a) Odorant-binding protein and its mRNA are localized to lateral nasal gland implying a carrier function. *Proc. Natl. Acad. Sci. USA*, *85*, 2383–2387.

Pevsner, J., R.R. Reed, P.G. Feinstein, and S.H. Snyder (1988b) Molecular cloning of odorant-binding protein: Member of a ligand carrier family. *Science*, *241*, 336–339.

Pevsner, J., V. Hou, A.M. Snowman, and S.H. Snyder (1990) Odorant-binding protein: Characterization of ligand binding. *J. Biol. Chem.*, *265*, 6118–6125.

Price, S. (1977) Specific anosmia to geraniol in mice. *Neurosci. Lett.*, *4*, 49–50.

Price, S. (1978) Anisole binding protein from dog olfactory epithelium. *Chemical Senses and Flavour*, *3*, 51–55.

Price, S. (1984) Mechanisms of stimulation of olfactory neurons: An essay. *Chemical Senses*, *8*, 341–354.

Reese, T.S. (1965) Olfactory cilia in the frog. *J. Cell Biol.*, *25*, 209–230.

Reuter, H. (1987) Modulation of ion channels by phosphorylation and second messengers. *News Physiol. Sci.*, *2*, 168–171.

Restrepo, D., T. Miyamoto, B.P. Bryant, and J.H. Teeter (1990) Odor stimuli trigger influx of calcium into olfactory neurons of the channel catfish. *Science*, *249*, 1166–1168.

Rhein, L.D., and R.H. Cagan (1980) Biochemical studies of olfaction: Isolation, characterization, and odorant binding activity of cilia from rainbow trout olfactory rosettes. *Proc. Natl. Acad. Sci. USA*, *77*, 4412–4416.

Rhein, L.D., and R.H. Cagan (1983) Biochemical studies of olfaction: Binding specificity of odorants to a cilia preparation from rainbow trout olfactory rosettes. *J. Neurochem.*, *41*, 569–577.

Rochel, S., and F.L. Margolis (1982) Carnosine release from olfactory bulb synaptosornes is calcium-dependent and depolarization-stimulated. *J. Neurochem.*, *38*, 1505–1514.

Rogers, K.E., P. Dasgupta, U. Gubler, M. Grillo, Y.S. Khew-Goodal, and F.L. Margolis (1987) Molecular cloning and sequencing of a cDNA for olfactory marker protein. *Proc. Natl. Acad. Sci. USA*, *84*, 1704–1708.

Rossie, S., D. Gordon, and W.A. Catterall (1987) Identification of an intracellular domain of the sodium channel having multiple cAMP-dependent phosphorylation sites. *J. Biol. Chem.*, *262*, 17530–17535.

Schafer, R., D.W. Criswell, S.P. Frasek Jr., and K.R. Brower (1984a) Olfactory studies using ethyl bromoacetate and other chemically active odorants. *Chemical Senses*, *9*, 31–52.

Schafer, R., S.P. Frasek Jr., D.W. Criswell, and K.R. Brower (1984b) Protection of olfactory responses from inhibition by ethyl bromoacetate, diethylamine, and other chemically active odorants by certain esters and other compounds. *Chemical Senses*, *9*, 55–72.

Schild, D. (1989) Whole-cell currents in olfactory receptor cells of *Xenopus laevis*. *Exp. Brain Res.*, *78*, 223–232.

Schmale, H., H. Holtgreve-Grez, and H. Christiansen (1990) Possible role for salivary gland protein in taste reception indicated by homology to lipophilic-ligand carrier proteins. *Nature*, *343*, 366–369.

Schwob, J.E., and D.I. Gottlieb (1986) The primary olfactory projection has two chemically distinct zones. *J. Neurosci.*, *6*, 3393–3404.

Schwob, J.E., N.B. Farber, and D.I. Gottlieb (1986) Neurons of the olfactory epithelium in adult rats contain vimentin. *J. Neurosci.*, *6*, 208–217.

Senf, W., B.Ph.M. Menco, P.H. Punter, and P. Duyvensteyn (1980) Determination of odour affinities on the dose-response relationships of the frog's electro-olfactogram. *Experientia*, *36*, 213–215.

Shirley, S., E. Polak, and G.H. Dodd (1983a) Selective inhibition of rat olfactory receptors by concanavalin A. *Biochem. Soc. Trans.*, *11*, 780–781.

Shirley, S., E. Polak, and G.H. Dodd (1983b) Chemical modification studies on rat olfactory mucosa using a thiol-specific reagent and enzymatic iodination. *Eur. J. Biochem.*, *132*, 485–494.

Shirley, S.G., C.J. Robinson, K. Dickinson, R. Aujla and G.H. Dodd (1986) Olfactory adenylate cyclase of the rat: Stimulation by odorants and inhibition by Ca^{2+}. *Biochem. J.*, *240*, 605–607.

Sicard, G., and A. Holley (1984) Receptor cell responses to odorants: Similarities and differences among odorants. *Brain Res.*, *292*, 283–296.

Sklar, P.B., R.R.H. Anholt, and S.H. Snyder (1986) The odorant-sensitive adenylate cyclase of olfactory receptor cells: Differential stimulation by distinct classes of odorants. *J. Biol. Chem.*, *261*, 15538–15543.

Snyder, D.A., A.M. Rivers, H. Yokoe, B.Ph.M. Menco, and R.R.H. Anholt, (1991) Olfactomedin: Purification, characterization and localization of a novel olfactory glycoprotein. *Biochemistry*, *30*, 9143–9153.

Sternweis, P.C. (1986) The purified α subunits of G_o and G_i from bovine brain require $\beta\gamma$ for association with phospholipid vesicles. *J. Biol. Chem.*, *261*, 631–637.

Stocklet, J.C., D. Gerard, M.C. Kilhoffer, C. Lugnier, R. Miller, and P. Schaeffer (1987) Calmodulin and its role in intracellular calcium regulation. *Prog. Neurobiol.*, *29*, 321–364.

Stryer, L., and H.R. Bourne (1986) G-proteins: A family of signal transducers. *Ann. Rev. Cell. Biol.*, *2*, 391–419.

Supattapone, S., S.K. Danoff, A. Theibert, S.K. Joseph, J. Steiner, and S.H. Snyder (1988) Cyclic AMP-dependent phosphorylation of a brain inositol triphosphate receptor decreases its release of calcium. *Proc. Natl. Acad. Sci. USA*, *85*, 8747–8750.

Sydor, W., Z. Teitelbaum, R. Blacher, S. Sun, W. Benz, and F.L. Margolis (1986) Amino acid sequence of a unique neuronal protein: Rat olfactory marker protein. *Arch. Biochem. Biophys.*, *249*, 351–362.

Tagaki, S.F., G.A. Wyse, H. Kitamura, and K. Ito (1968) The roles of sodium and potassium ions in the generation of the electro-olfactogram. *J. Gen. Physiol.*, *51*, 522–578.

Tagaki, S.F., H. Kitamura, K. Imai, and H. Takeuchi (1969) Further studies on the roles of sodium and potassium in the generation of the electro-olfactogram. *J. Gen. Physiol.*, *53*, 115–130.

Topazzini, A., P. Pelosi, P.L. Pasqualetto, and N.E. Baldacinni (1985) Specificity of a pyrazine binding protein from cow olfactory mucosa. *Chemical Senses*, *10*, 45–49.

Trombley, P.Q., and G.L. Westbrook (1991) Voltage-gated currents in identified rat olfactory receptor neurons. *J. Neurosci.*, *11*, 435–444.

Trotier, D. (1986) A patch-clamp analysis of membrane currents in salamander olfactory receptor cells. *Pfluegers Arch.*, *407*, 589–595.

Uhlen, S., and J.E. Wikberg (1988) Relationship between forskolin and calcium-calmodulin stimulation of rat cerebral cortex adenylate cyclase: Enzyme activation modulates substrate (MgATP) affinity. *Pharmacol. Toxicol.*, *63*, 90–95.

Van Dongen, A.M.J., J. Codina, J. Olate, R. Mattera, R. Joho, L. Birnbaumer, and A.M. Brown (1988) Newly identified brain potassium channels gated by the guanine nucleotide binding protein G_o. *Science*, *242*, 1433–1437.

Vodyanoy, V., and R.B. Murphy (1983) Single-channel fluctuations in bimolecular lipid membranes induced by rat olfactory epithelial homogenates. *Science*, *220*, 717–719.

Vodyanoy, V., and I. Vodyanoy (1987) ATP and GTP are essential for olfactory response. *Neurosci. Lett.*, *73*, 253–258.

Winegar, B.D., E.R. Rosick, and R. Schafer (1988) Calcium and olfactory transduction. *Comp. Biochem. Physiol.*, *91A*, 309–315.

Wood, P., S. Shirley, and G. Dodd (1983) Effect of concanavalin A on frog olfactory mucosa. *Biochem. Soc. Trans.*, *11*, 781–782.

Wysocki, C.J., and G.K. Beauchamp (1984) Ability to smell androstenone is genetically determined. *Proc. Natl. Acad. Sci. USA*, *81*, 4899–4902.

4
The Biochemistry of Olfactory Neurons: Stages of Differentiation and Neuronal Subsets

JAMES E. SCHWOB

Introduction

The sensory neurons of the olfactory epithelium demonstrate two remarkable characteristics. First, although each individual neuron has a restricted (yet broadly tuned) sensitivity to odorants, as a population they are capable of transducing literally millions of odorants and, in response to these odorants, generating differential patterns of neuronal activity that permit the discrimination of olfactory stimuli by higher centers. The selective responsivity of individual neurons, in combination with the breadth of sensitivity provided by the population of neurons in aggregate, is highly reminiscent of the immune system. However, odorant receptor molecules have not yet been unambiguously identified, and the generation of odorant response diversity during neuronal development is poorly understood as a consequence (see the chapter by Anholt in the present volume). Second, olfactory neurons are the only neurons in the mature mammalian nervous system that can be replaced when they die. A population of stem cells is present in the olfactory epithelium from which new neurons can differentiate to substitute for those sensory cells that die as a consequence of damage, either mechanical or chemical, or at the end of their "natural" lifespan. The neurogenetic capacity of the epithelium is maintained throughout the life of the animal and makes the olfactory epithelium an attractive and intriguing model for studying many of the major events in the development of the nervous system as a whole, such as the birth of neurons, their differentiation (including the formation of specific axonal connections), and their eventual death. The intent of this chapter is to summarize our knowledge of the biochemical phenotype(s) of olfactory neurons, excluding those features that are known or suggested to be directly involved in stimulus reception or transduction. There are at least three reasons for reviewing this information. First, *all* newly born olfactory neurons must proceed through a series of stages during their differentiation into mature, functioning sensory neurons. To some extent these stages have been morphologically categorized (see below). However, to approach the more fundamental issue—namely, to understand in detail the events of olfactory neuronal differentiation and their regulation at a cell biological, biochemical, and, ultimately, molecular genetic level—requires the identification of various sets of biochemical markers that correlate with the stages of morphological and physiological differentiation. Moreover, the study of olfactory neuron differentiation is likely to provide insight into how newly born neurons begin to function as sensory cells and how they become selective for a particular subset of odorants, since differentiation and the expression of the odorant reception and transduction apparatus will occur *pari passu*. Second, olfactory neurons have the capability of reinnervating the mature CNS, which, in higher vertebrates, is unique to them. Do olfactory neurons have molecular phenotypic characteristics that correlate with this *growth capacity*? Third, the most obvious feature that differentiates one olfactory neuron from the next is odorant selectivity. Are there phenotypic characteristics that identify *subclasses* of olfactory neurons, and is there evidence that these subclasses are important for olfactory function? I will review these three aspects of the molecular phenotype of olfactory neurons. For the most part, I will focus on the vertebrate main olfactory epithelium, but I will include information obtained from studies on the vertebrate accessory olfactory system as well.

Biochemical Correlates of the Differentiation of Olfactory Neurons

Molecular Phenotype of Neuronal Progenitors

The olfactory epithelium is stratified, and the level at which soma and nucleus are situated (along the axis from the basal lamina to the apical surface of the epithelium) is an important clue in identifying the type of cell, whether basal cell, olfactory neuron, or supporting cell (from deep to superficial, respectively). Morphologically, the following stages in the differentiation of olfactory neurons in the mammalian olfactory epithelium have been identified by Graziadei and Monti Graziadei (1979), who also describe the existence of transitional forms between these stages. (1) Dark basal cells or basal cells proper, located directly against the basal lamina. Their cytoplasm is electron-dense and contains abundant tonofilaments and free ribosomes. Their other distinguishing feature is the formation of desmosomal junctions with the basal lamina and other cells in the epithelium and their cytoplasm. (2) Globose basal cells, located somewhat more superficially in the epithelium. These cells are more electron-lucent and contain basal bodies. (3) Olfactory neurons proper. The population of neurons is described as heterogenous on the basis of nuclear and cytoplasmic features, but can easily be distinguished from basal cells by the elaboration of processes and their more superficial position in the epithelium.

The proliferating stem cell population can be identified on the basis of incorporation of thymidine analogs into newly synthesized DNA or on the basis of mitotic figures at the EM level. Most labeled nuclei and mitotic figures are found in the basal cell zone of the epithelium, with a minor fraction observed in supporting cell nuclei. With increasing survival times after injection of ^{3}H-thymidine, the cohort of labeled cells can be completely "chased" into the population of neuronal cells.

These morphologically defined cell types in the epithelium provide a framework for understanding the stages in the differentiation of olfactory neurons from stem cells in the epithelium. However, these terms should not necessarily be considered functional descriptions, since there remain significant gaps in our knowledge of this process. For example, with regard to morphologically identified "basal cells," the question of whether both the dark and globose varieties are mitotically active remains open. Morphological observations made by Graziadei and Monti Graziadei (1979) and Vollrath et al. (1985) *suggest* that both cell types are capable of dividing, and they have shown that dividing cells can resemble either one or the other of these two cell types or an intermediate form. However, the identification and isolation of proliferating cells would be facilitated if there was a biochemical marker that was characteristic of these cells. Similarly, the issue of whether others of the globose basal cell population are immature olfactory neurons could be more easily resolved if the stages in neuronal differentiation were identifiable on the basis of selective expression of biochemical markers.

Recently, there have been attempts to characterize the functionally defined stem cell population in the olfactory epithelium more fully at an anatomical or biochemical level. As a first step, several laboratories have used an immunohistochemical approach to characterize the basal cell population *in situ*. For example, the monoclonal antibodies BCL-1, BCL-2 (Hempstead and Morgan, 1985a), and 6B7 (Akeson and Haines, 1989) and antibodies to the intermediate filament protein keratin (Vollrath et al., 1985) and the neural cell adhesion molecule (NCAM; Edelman et al., 1983) stain cells near the basal lamina of the olfactory epithelium. However, it is not clear whether any of these correspond to cells that are actively proliferating. Indeed, it is uncertain which of the two morphological classes of basal cells these markers recognize. It has been suggested that dark basal cells are NCAM (−) (Miragall et al., 1988), and that NCAM (+) cells located just superficially are globose basal cells and are likely to be proliferating, but these conclusions must be considered tentative, as they are based largely on position, rather than on morphology or evidence of proliferative potential (Miragall et al., 1988).

In order to better define the precursor population and thereby improve our understanding of the nature of the basal cell population(s), an immunohistochemical approach has been used in combination with agents that will functionally identify proliferating cells. Both *in vitro* and *in vivo* work indicates that the majority of proliferating cells in the olfactory epithelium express *neither* keratin *nor* NCAM. Calof and Chikaraishi (1989) used an *in vitro* approach to characterize the neuronal precursor population in the olfactory epithelium. In this case, the embryonic epithelium was dissociated and placed into relatively short-term tissue culture. Under these conditions, the differentiation of process-bearing neuronal cells was observed. Their data would suggest that the immediate neuronal precursor is not a keratin (+) epithelioid cell, but rather a keratin (−)/NCAM (−) small round cell. These cells incorporate ^{3}H-thymidine, and they were considered to be transformed into neurons based on the progressive decline in numbers of keratin (−)/NCAM (−) cells concomitant with the increase in thymidine-labeled NCAM (+) neurons as a function of time in culture. The authors suggest that these small round cells observed *in vitro* may correspond to the globose basal cells described *in vivo*. In addition, there is preliminary data *in vivo* from Schwartz et al. (1989) and from our laboratory to indicate that the *in vitro* results are substantially correct, and that the proliferating cells in the olfacatory epithelium by and large express neither NCAM nor keratin (Figure 4.1).

The role of the keratin (+) basal cells remains unclear. They may correspond to the population of dark basal cells described *in situ*, since this variety of basal cell contains tonofilaments, which are composed of keratin filaments in other epithelia. It does not appear that these keratin (+) cells are recruited into the pool of proliferating cells under conditions when the pool has been markedly expanded (for example, during the recovery of the neuronal population in the acute phase after olfactory bulb ablation; Schwartz et al., 1989). One intriguing hypothesis would postulate that these keratin (+) cells are recruited after widespread epithelial damage and are responsible for the replacement of the olfactory epithelium with respiratory epithelium, as has been reported in both man (Nakashima et al., 1984; Naessen, 1971) and experimental animals (Smith, 1938). Alternatively, they may act as neuronal stem cells at a very early stage of epithelial differentiation. In light of these results, alternatives to the previous, morphologically based terminlogy are being used that take into account the biochemical and functional categorization of basal cells. Hence the term "horizontal basal cell" has been used (and is used here in Table 4.1) to describe the keratin (+) basal cells, which have a very low proliferation rate (if any), while the term "globose basal cell" is used to describe the proliferating population of cells that express neither NCAM nor keratin (Schwartz et al., 1990 meeting of the Assoc. Chemorec. Sci.). What remains to be determined is the correlation between this terminology and the morphological types described by Graziadei and Monti Graziadei. 1979).

Markers Selective for Immature Olfactory Neurons

Subsequent to final mitosis, newly born olfactory neurons must complete a sequence of morphological and functional differentiation. The morphological steps in this differentiation will include the initial formation of an

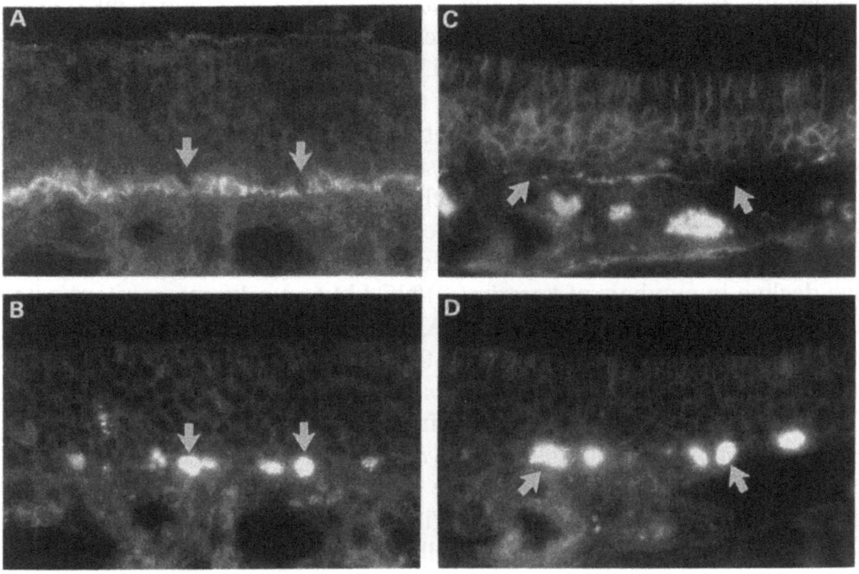

FIGURE 4.1. Double-labeling experiments show that proliferating basal cells are keratin (−) and NCAM (−). (A) Olfactory epithelium was stained with antikeratin antiserum to demonstrate the horizontal basal cell population (Texas Red fluorescence). (B) On the same section, the distribution of proliferating basal cells was demonstrated by using a monoclonal antibody against the thymidine analog bromodeoxyuridine (BrdU) (fluorescein fluorescence); this animal had been injected with BrdU 2 hr prior to sacrifice. Comparison of the two photomicrographs demonstrates that actively proliferating basal cells do not contain keratin (examples of cells identifiable in both photos of the same section are indicated by the arrowheads). (C) Staining with anti-NCAM antiserum (Texas Red fluorescence). (D) Anti-BrdU-labeled basal cells (fluorescein fluoresence). Likewise, the proliferating basal cells do not express NCAM (arrowheads signify the same cell in the two photos).

apical dendrite and a basal axon, and the extension of these processes to the surface of the epithelium and to the olfactory bulb, respectively. In the terminal phase of the differentiation, chemoreceptive cilia are elaborated by the olfactory knob, and the olfactory axons form synapses in the glomeruli of the olfactory bulb. The pulse-chase thymidine labeling data described above (Graziadei and Monti Graziadei, 1979) and experimental manipulations inducing the synchronization of olfactory neuron development by treatment with antimitotic agents (Lidow et al., 1986) both suggest that the maturation of olfactory neurons is paralleled by a progressive migration of the differentiating neuron from deep to superficial in the neuronal zone of the olfactory epithelium in adult animals. Accordingly, those neurons located more deeply in the epithelium can be identified with

TABLE 4.1. Biochemical markers of the differentiation of olfactory neurons

Horizonatal basal cells	*Globose basal cells*	*Immature neurons*	*Immature and mature neurons*	*Mature neurons*
Keratin (1, 2, 3)	BCL-1 (?)	GAP-43/B50 (3, 6)	NCAM (2, 3)	OMP (10)
BCL-1 (4) (?)	BCL-2 (?)	E-NCAM (3)	NEU-4 (7, 8)	114G12 (9)
BCL-2 (4) (?)	6B7 (?)		NEU-5 (7)	
6B7 (5) (?)			NEU-9 (7, 8)	
			112D5 (R2D5) (9)	

(?) signifies that the marker in question cannot be definitively assigned to this cell type
References: 1, Vollrath et al., 1985; 2, Calof and Chikaraishi, 1989; 3, Schwob, previously unpublished results; 4, Hempstead and Morgan, 1985a; 5, Akeson and Haines, 1989; 6, Verhaagen et al., 1989a, b; 7, Hempstead and Morgan, 1985b; 8, Carr et al., 1989; 9, Onoda, 1988b; 10, Margolis, 1988.

some reliability as more immature (it should be noted that this does not hold true at the early stages of embryonic development; Smart, 1971; Cuschieri and Bannister, 1975). This morphological feature can be exploited when trying to define biochemical markers or to characterize immature versus mature neurons. In addition, there are two alternative means of studying a more homogeneous population of immature neurons, which can then be compared to the normal adult olfactory epithelium and its relatively larger population of mature neurons. The first alternative is to study the embryonic olfactory epithelium at a stage when the olfactory neurons have not elaborated extensive cilia (Menco and Farbman, 1985a and b) and when they respond to odorants in a nonspecific manner (Gesteland et al., 1982). The second alternative is to study the adult olfactory epithelium that has been deprived of its synaptic target by removal of the olfactory bulb (Schwob et al., 1988; Schwob and Szumowski, 1989).

Olfactory bulbectomy has been known (and utilized) since the time of Le Gros Clark and Warwick (1946) to produce a thinning of the olfactory epithelium as a consequence of the decreased number of neurons. More recently, two phases in the response of the epithelium have been distinguished: acute and chronic (Costanzo and Graziadei, 1983). Following the acute devastation due to retrograde neuronal degeneration that occurs within a week after bulbectomy, there is a partial recovery in the thickness of the epithelium and in the number of olfactory neurons during the next few weeks until the epithelium becomes stable, but reduced in thickness, by about one month survival (Costanzo and Graziadei, 1983); this partial recovery occurs as a consequence of increased neurogenesis (Camara and Harding, 1984; Schwartz et al., 1989). What is responsible for the failure to achieve full recovery with long survivals after bulbectomy, i.e., long after the phase of acute degeneration as a consequence of direct damage is past? In the chronically bulbectomized epithelium, morphological evidence would suggest that the majority of the olfactory neurons present are imma-

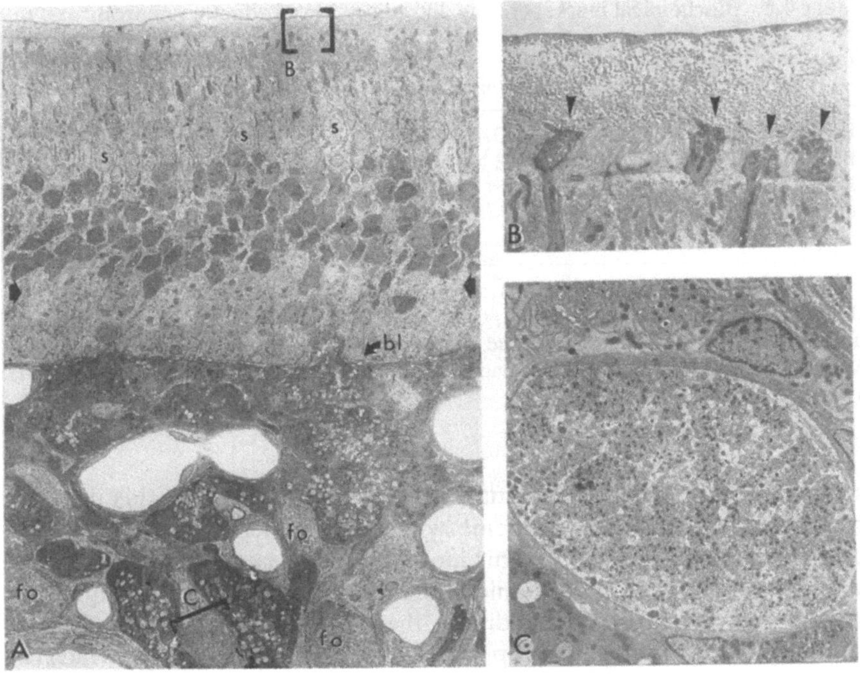

FIGURE 4.2. The septal olfactory epithelium of the unoperated side of a bulbectomized rat. (A) At low power, the epithelium resembles published descriptions (Graziadei and Monti Graziadei, 1979). The arrow marks the upper limit of the immature neuron/basal cell zone. Bracketed areas indicate the approximate region of the areas illustrated at higher power to the right (which are taken from a different thin section from this block). S—supporting cells. (B) Apical surface of the olfactory epithelium, illustrating the ciliated olfactory knobs (arrowheads) and the dense ciliary mat suspended in the mucus layer. (C) The fila olfactoria (fo) are composed of uniform, small-diameter, unmyelinated axons.

ture. At the electron microscopic level, it is apparent that the olfactory neurons in the chronically bulbectomized epithelium have not elaborated the extensive ciliary mat that is apparent on the unoperated side (Schwob et al., 1988; cf. Figures 4.2 and 4.3); indeed, most knobs lack cilia entirely, and the basal bodies are concentrated in the center of the knob and not around its rim as on the nonablated side. In addition, the somata of olfactory neurons on the ablated side are more electron lucent than the somata of superficially located, mature neurons on the unoperated side, but they do resemble the more immature neurons (or globose basal cells) that are situated deep in the unoperated epithelium just above the basal cell layer (also noted by Monti Graziadei, 1983). It has been suggested that the expansion of the population of immature neurons on the ablated side, and the profound reduction in mature neurons, is a consequence of accelerated

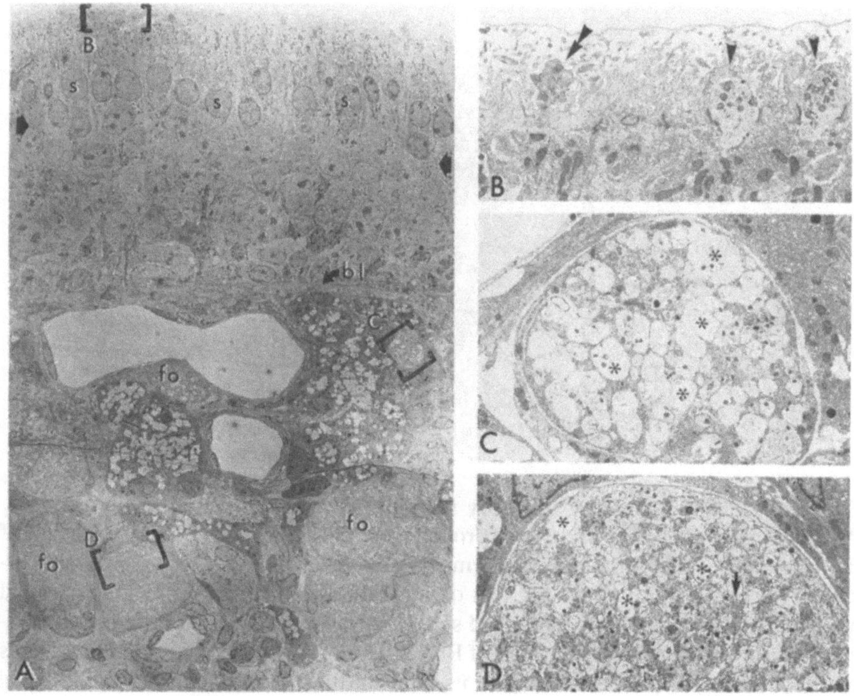

FIGURE 4.3. The septal olfactory epithelium of the lesioned side one month postbulbectomy. (A) At low power, the epithelium is thinner than on the control side (Figure 4.2), but the zone occupied by immature neurons/basal cells (from the basal lamina up to the arrows) is thicker. Bracketed areas indicate the approximate region of the areas illustrated at higher power to the right (which are taken from a different thin section from this block). (B) At the apical surface, the ciliary mat is absent, and most of the knobs have their basal bodies concentrated in the center without elaborating cilia (arrowheads). The double arrowhead indicates a knob that has elaborated stubby cilia. (C), (D) The fila olfactoria (fo) contain swollen and clearly degenerating axons (asterisks). Some axons are more normal in diameter (arrows).

neuronal turnover (Monti Graziadei, 1983). More recently, our laboratory and others have obtained direct evidence to indicate that, indeed, turnover is accelerated. For example, close morphological evaluation indicates that a substantial number of axons in the fila olfactoria are swollen and degenerating at long survivals after ablation (cf. Figures 4.2 and 4.3; Schwob et al., 1988). Furthermore, the life span of neurons was directly examined using a thymidine analog to label a cohort of newly postmitotic neurons and was shown to be shorter on the chronically ablated side as compared to the unoperated side. During the chronic phase (i.e., at long intervals) after bulb ablation, the proliferation rate (labeling index) of neuronal precursors

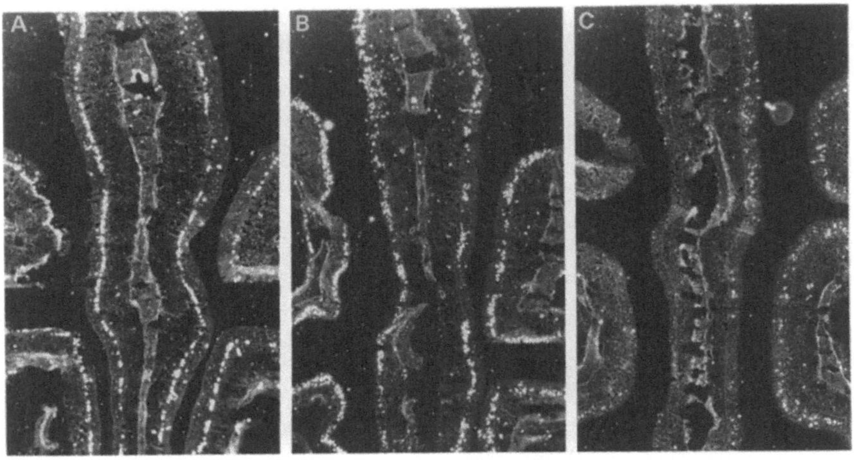

FIGURE 4.4. The life span of olfactory neurons is foreshortened after bulbectomy. Proliferating basal cells and their progeny are identified after incorporation of BrdU by anti-BrdU immunohistochemistry. Each dark-field photomicrograph includes the nasal septum and the tips of several endoturbinates at a long survival after bulbectomy. The bulbectomized side is on the left; the right remains unoperated. (A) Two hours after injection of BrdU. The rate of proliferation of basal cells on the ablated side is increased (counts indicate that the number of labeled basal cells is 50 percent higher). (B) Five days after BrdU. The increase in labeled cells on the ablated side persists as they migrate out from the basal cell zone and differentiate into neurons. On the ablated side (left), the outer limit of the band of labeling extends to the upper edge of the neuronal zone of the epithelium, whereas on the unoperated side (right), labeled neurons remain more deeply placed. (C) Two weeks after BrdU injection. The ratio of labeled cells on the ablated vs. the unoperated side has flipped and relatively *fewer* neurons have survived on the *ablated* side.

on the ablated side is increased about 1.5-fold relative to the unoperated epithelium (Figure 4.4; Schwob and Szumowski,1989; Schwartz et al., 1989). Up to a week after the labeling of the cohort, as the postmitotic cells are "chased" into the neuronal population, the ablated epithelium maintains a relatively *greater* number of labeled neurons compared to the control (Figure 4.4). However, on the ablated side, labeled neurons have progressed through the thickness of the epithelium to reach the outermost (apical) edge of the neuronal zone, while on the unoperated side, labeled neurons are still concentrated at the deep edge of the neuronal zone. By two weeks, virtually all of the labeled cohort of neurons have disappeared from the ablated epithelium, while most of them are still found in the control epithelium (Figure 4.4). Therefore, during the chronic phase after bulb ablation, more neurons are still being born on the ablated side, resulting in the expansion of the population of immature neurons relative to control, but in the absence of their synaptic target they die more quickly than in the control epithelium; i.e., neuronal turnover is accelerated. We interpret these results as an indication that olfactory neurons are trophically dependent on the olfactory bulb for their prolonged survival.

From the preceding discussion, it is apparent that immature olfactory neurons can be identified as: (1) those olfactory neurons adjacent to the basal layer in normal epithelium, (2) those neurons present at early embryonic stages, and (3) the vast majority of the neurons in the acutely or chronically ablated olfactory epithelium and their axons in the olfactory nerve. By using these criteria to classify neurons as mature or immature, a number of known proteins and monoclonal antibody markers can be shown to label immature olfactory neurons. These include NCAM (Calof and Chikaraishi, 1989; Carr et al., 1989; Miragall et al., 1988), monclonal antibody NEU-5 that recognizes an NCAM-like protein or an NCAM isoform (Hempstead and Morgan, 1985a and b; Carr et al., 1989), and several others listed in Table 4.1. Many of these are found on both immature and mature olfactory neurons (mature neurons can be recognized by the converse of the criteria for immaturity, see "Markers Selective for Mature Olfactory Neurons," below). However, there is one marker, namely GAP-43 (Growth Associated Protein of 43 kDa apparent molecular weight), that is *selective* for immature neurons, i.e., does not also label mature olfactory neurons, and another marker that also appears to be selective for immature olfactory neurons, namely E-NCAM, although on this point the evidence is still preliminary.

The protein GAP-43 (also known as B50) was discovered on the basis of its increased axoplasmic transport in the growing axons of retinal ganglion cells during optic nerve development and regeneration (Skene, 1989). During phases of axonal outgrowth, GAP-43 is preferentially concentrated in the growth cone (Meiri et al., 1986, 1988); however, at least at the early stages of axon outgrowth, GAP-43 is also found in the plasma membrane of the cell soma (Meiri et al., 1991). After the synaptic target is innervated (or reinnervated), the synthesis of this protein by retinal ganglion cells significantly decreases. The function of GAP-43 in the growth cone is presently unknown; however, it is subject to regulation by phosphorylation in response to several physiological factors that influence the direction or speed of axonal growth (Meiri et al., 1989). The pattern of GAP-43 labeling of the olfactory epithelium indicates that maturing olfactory neurons also pass through a stage when they transiently express this protein in cell body and axon. Staining of the normal adult rat olfactory epithelium with polyclonal antiserum (Verhaagen et al., 1989a) or monoclonal antibodies (Meiri et al., 1991) to GAP-43/B50 labels the somata and dendrites of a thin row of neurons immediately superficial to the basal layer (Schwob, submitted for publication; Figure 4.5). In sections doubly stained with antiserum to olfactory marker protein (OMP) and monoclonal anti-GAP-43 antibody, there is little or no overlap in the expression of these proteins;

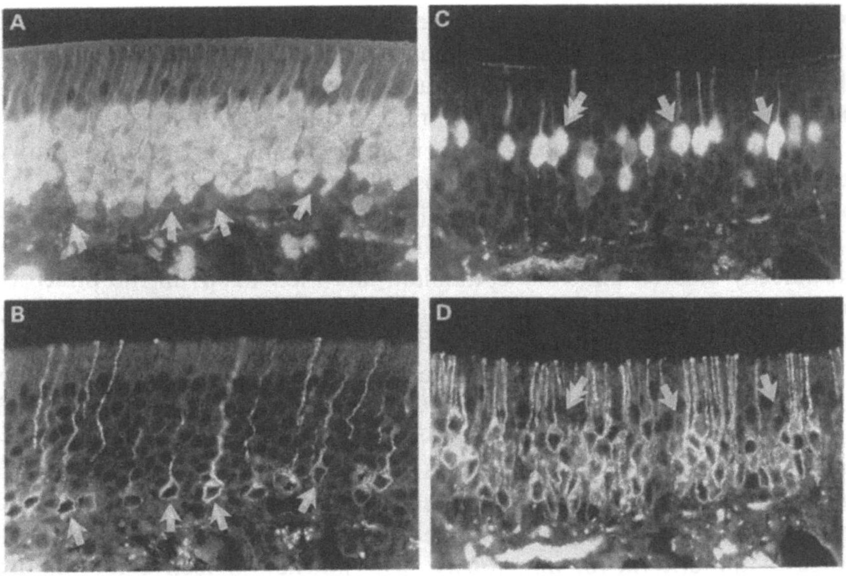

FIGURE 4.5. The number of OMP (+) olfactory neurons falls while the number of GAP-43 (+) olfactory neurons increases after bulbectomy. (A) and (B) illustrate a section of unoperated olfactory epithelium, doubly labeled with anti-OMP and anti-GAP-43, respectively. (C) and (D) illustrate a section of chronically bulbectomized olfactory epithelium also doubly labeled with anti-OMP and anti-GAP-43, respectively. Arrows indicate the same neurons in the two pairs of photos. The expression of GAP-43 and OMP is largely, but not exclusively reciprocal: Most GAP-43 (+) neurons do not contain OMP on either the ablated or the unoperated side (and vice versa), although other GAP-43 (+) neurons are weakly labeled with anti-OMP. For example, in the cluster of OMP (+) cells in the ablated epithelium indicated by the doubled arrow in (C), the middle one is the most weakly stained with OMP; it is also stained with anti-GAP-43.

neurons are either GAP-43 (+) or OMP (+), but seldom both (Figure 4.5). Similar double-labeling with antiserum to NCAM suggests that onset of GAP-43 and NCAM expression is coordinate; that is, NCAM does not appear to precede expression of GAP-43 during differentiation of the olfactory neurons, since few if any NCAM (+)/GAP-43 (−) cells are observed in the basal cell layer (Schwob, manuscript in preparation). These findings indicate that GAP-43 is expressed at an early stage in the differentiation of neurons from basal cells and is limited to immmature cells. Further evidence for this hypothesis can be cited. The proportion of GAP-43 (+) to GAP-43 (−) neurons is very much increased in the olfactory epithelium of progressively younger animals (Verhaagen et al., 1989a) and in adult rats subject to olfactory bulbectomy (Schwob and Szumowski, 1989; Figure 4.5). The selective expression of GAP-43 in immature neurons is reflected in the pattern of labeling of the axonal projection from the epithelium onto the bulb. The proportion of GAP-43 labeled to unlabeled olfactory axons in the olfactory nerve and bulb parallels the labeling of neuronal cell bodies; the olfactory nerve is most densely labeled in neonatal rats (Verhaagen et al., 1989a) and in animals that are bulbectomized (Figure 4.5; Schwob and Szumowski, 1989; Verhaagen et al., 1989b), while in normal adult rats, there is GAP-43 labeling of some but not all axons in the fine fiber bundles in the submucosa and in the large fascicles of the olfactory nerve in route to the olfactory bulb (Figure 4.5). Likewise, in the intact bulb the density of glomerular staining with anti-GAP-43 is a function of age (Verhaagen et al., 1989a; Schwob, unpublished observations). As the proportion of immature, GAP-43 (+) axons drops as the animal ages, a nonuniform pattern of glomerular labeling begins to emerge; i.e., the degree of labeling of glomeruli with the GAP-43 antibody varies from one glomerulus to another across the surface of the bulb in adult animals. For example, some glomeruli are completely filled with GAP-43-containing axons, even in animals as old as seven months that have been raised in a protective, barrier-isolated environment, while in others only a thin rim of staining or no staining is observed (Schwob, unpublished observations). Despite the correspondence between the numbers of GAP-43 labeled cell bodies and the density of axonal labeling in the olfactory nerve and bulb, we cannot say whether GAP-43-containing axons derive exclusively from GAP-43-containing neuron cell bodies. For example, it could be the case that axonal labeling persists for a time after the synthesis and accumulation of the protein ceases in the cell body. Additional experimentation will be required to elucidate the relationship between cell body and axon terminal labeling with anti-GAP-43.

In addition to GAP-43, the embryonic form of NCAM (E-NCAM) also appears to be a marker for the cell somata of newly born, immature neurons. This form of NCAM is largely absent from the adult nervous system, although it is abundant during development, and it contains a large amount of polysialic acid, which is the basis for its selective identification (Rutishauser and Jessel, 1988). It is unlikely that proliferating cells express E-NCAM, as had been previously suggested (Miragall et al., 1988), since both our laboratory (Figure 4.1) and Calof and Chikaraishi (1989) were unable to demonstrate NCAM (+) proliferating cells. Moreover, we do have the following preliminary data substantiating E-NCAM as a marker of immature olfactory neurons. First, most of the E-NCAM stained cells have the bipolar shape that is also seen with GAP-43. Second, the population of E-NCAM (+) cells is markedly expanded after bulb ablation in parallel with the increase in GAP-43 (+) cells. Accordingly, the E-NCAM (+) cell somata observed near the basal zone in the adult olfactory epithelium by Miragall et al. (1988) are more likely newly postmitotic neurons rather than proliferating globose basal cells as originally thought. As would be expected from the results in the epithelium, antibodies specific for E-

NCAM stain only a subset of olfactory axons in the olfactory nerve or in the bulb (Miragall et al., 1988; Schwob, unpublished observations); this may also be taken as evidence that E-NCAM is expressed in olfactory neurons for some portion of their life cycle. Functionally, the presence of polysialyated NCAM on growing olfactory axons would provide an increased "slipperiness" to the process as it maneuvers among the other axons already present in the olfactory nerve and would probably facilitate the formation of new axonal connections (Rutishauser and Jessel, 1988). No other markers selective for immature olfactory neurons have been identified to date. Although a panel of monoclonal antibodies raised against the adult olfactory epithelium has been used to stain the bulbectomized epithelium (Hempstead and Morgan, 1985b) and the embryonic epithelium (Carr et al., 1989), none has been identified for which staining is intensified in the bulbectomized epithelium or in the fetus, relative to the normal adult.

Markers Selective for Mature Olfactory Neurons

The converse of the criteria used to identify immature olfactory neurons may be used to classify olfactory neurons as mature. Hence, a marker that labels olfactory neurons located in the upper part of the epithelium, that cannot be found in the epithelium until late in the course of embryonic development, and that is absent or substantially decreased in the bulbectomized epithelium may be identified as a marker for mature olfactory neurons. Based on the preceding discussion (Menco and Farbman, 1985; Schwob et al., 1988), a *morphological* marker for maturity would be the elaboration of chemoreceptive cilia. The classic example of a *biochemical* marker that is present in mature olfactory neurons—indeed, that is largely limited to olfactory neurons and absent from other neurons and tissues in the body—is, of course, the Olfactory Marker Protein (OMP). The discovery and investigation of this protein have recently been reviewed by Margolis (1988). Since its isolation, it has been subject to extensive biochemical characterization, including cDNA cloning and the isolation of the gene (reviewed by Margolis, 1985; Danciger et al., 1989). It is known to be a cytoplasmic protein of 19 kDa molecular weight that lacks significant homology with any other known protein. Unfortunately, and despite substantial effort, OMP remains functionally anonymous. However, it is a boon to those interested in studying neuronal differentiation in the olfactory epithelium, since its expression is clearly limited to mature neurons. In the adult olfactory epithelium, Menco (1989) has recently completed a careful EM immunohistochemical study demonstrating that OMP expression in olfactory dendrites and dendritic knobs is limited to those that bear cilia; others with basal bodies but no cilia were not stained. It is likely that this difference in staining extends throughout the remainder of the cell body, since all or virtually all OMP (+) neurons have dendrites that extend to the apical epithelial surface (Monti Graziadei et al., 1977). Likewise, the finding that only a few olfactory neurons are OMP (+) after bulbectomy (Figure 4.5; Monti Graziadei, 1983) is indicative that OMP expression is limited to mature neurons. In particular, the number of OMP (+) neurons on the ablated side parallels the number of knobs that have elaborated cilia, rather than the total number of knobs (Schwob et al., 1988). This result, along with Menco's, would suggest that OMP expression is linked with ciliogenesis. In the case of the bulbectomized epithelium, as described above, the acceleration in neuronal turnover is likely to result in the death of the newly born olfactory neurons before or shortly after they have differentiated sufficiently to produce OMP and rudimentary cilia. As a second indicator of its tie to neuronal maturation, staining with antiserum to OMP is first observed in the olfactory epithelium just prior to the time that olfactory axons begin to make synapses in the embryonic olfactory bulb (Farbman and Margolis, 1980; Allen and Akeson, 1985; Farbman, 1986), and the numbers of OMP (+) neurons (and of olfactory knobs with cilia) in explants of the embryonic olfactory epithelium are clearly facilitated by inclusion of the olfactory bulb in the explant (Chuah and Farbman, 1983); whether the effect of the bulb on the explant is to promote neuronal survival, as it does in the adult olfactory epithelium, or whether it is promoting differentiation remains to be determined.

The same criteria can be used to identify other markers that are limited to mature olfactory neurons, and these other markers have been monoclonal antibodies (MAb's). The antigens in the olfactory neuron that these antibodies recognize have not been extensively characterized. However, these MAb's may prove useful in the future in improving our understanding of the differentiation of immature into mature olfactory neurons. The first of these, monoclonal antibody 114G12, is the best characterized to date (Onoda, 1988a and b; Onoda and Fujita, 1988). 114G12 recognizes a protein of 26 kDa molecular weight in both the olfactory epithelium and bulb of the rabbit, and staining with this antibody is found in other parts of the nervous system besides the primary olfactory projection; other stained structures include mitral cells in the bulb, layers I, II, and III of the piriform cortex (which encompasses both bulbar and intracortical synaptic zones), spinal cord, and cerebellum. However, within the nose, 114G12 recognizes only neurons of the main olfactory epithelium and their axons; there is no staining of vomeronasal epithelium or the vomeronasal nerve. These features distinguish it from OMP (Onoda and Fujita, 1988). Nonetheless, the pattern of labeling during embryonic development, the distribution of labeled neurons in the olfactory epithelium, and the changes in that distribution in the bulbectomized epithelium are highly reminescent of the expression of OMP under these conditions. That is, expression of the 114G12 antigen in the embryonic epithelium lags that of a marker that recognizes both immature and mature olfactory neurons (monoclonal antibody 112D5 or R2D5; Onoda, 1988a; Mori et al., 1985), it is limited to those neurons in the more superficial part of the neuronal zone of the epithelium in the normal adult rabbit, and the number of 114G12 (+) neurons as well as the density of staining of the olfactory nerve are markedly reduced in the acute and chronic stages after bulbectomy. Although no direct comparison of OMP and 114G12 expression in individual olfactory neurons has been made, the evidence to date would perhaps suggest that these two proteins are coordinately expressed by olfactory neurons as they make the transition from immature to mature stages in their life cycle.

The number of labeled olfactory neurons and the intensity of olfactory nerve staining with two other antibodies, NEU-4 and NEU-9, has been reported as decreased following bulbectomy in adult rats (Hempstead and Morgan, 1985b). The corresponding antigens for both of these are found on the surface of soma, dendrite, and axon of olfactory neurons (Carr et al., 1989). However, in contrast to the late embryonic expression of 114G12, both NEU-4 and NEU-9 label olfactory neurons a day or more prior to the appearance of OMP and prior to the elaboration of several morphological features that may be considered mature, such as multiciliated knobs and the formation of synapses in the olfactory bulb (Carr et al., 1989). In the case of NEU-4 and NEU-9, the data suggest that these antigens are not exclusively associated with mature olfactory neurons (at least not during embryological development). There may be a shift in the stage of olfactory neuronal differentiation at which these antigens are expressed in the adult animal, and potentially, the function of these molecules may also be altered in adults as compared to embryos. Alternatively, they may be marking a stage in differentiation that is intermediate between an immature (for example, GAP-43-containing) and a fully mature (OMP-containing) olfactory neuron, and that is particularly prominent in the fetus. Further experimentation will be required to clarify our understanding of the place of these antigens in the differentiation sequence.

Summary

In this section of the chapter, I have focused the discussion on those molecular species whose expression can clearly serve to mark the differentiation stage of an olfactory neuron (Table 4.1). This search is motivated by the perception that morphological criteria are inadequate for distinguishing functional states during olfactory neuron differentiation. In particular, I have concentrated on markers that are especially valuable for the study of the adult olfactory epithelium. This limitation in scope reflects certain prejudices: that the one truly remarkable advantage of this system is the opportunity it affords to study neurogenesis and neuronal differentiation (including the onset of sensory function) in a setting, namely the adult, that is more amenable to experimental manipulation and analysis than is the intact embryonic nervous system. Nonetheless, it is important to recognize that the embryonic olfactory epithelium may offer other advantages that offset its practical disadvantages. For example, it may be possible to distinguish a greater number of intermediate stages in the differentiation of olfactory neurons, since there is a temporal rather than a spatial disparity between these steps. Indeed, axon outgrowth begins around E13 in the rat, while the elaboration of multiciliated knobs is delayed until E16, which in turn precedes synaptogenesis in the bulb at E18 (Farbman, 1986). The developmental events have been carefully correlated with the sequential expression of a variety of markers by Farbman and his group (Carr et al., 1989). For a number of markers, expression is initially seen at some early stage in olfactory neuron differentiation and then persists in the mature neuron; these would include NCAM (and its probable homologue NEU-5), carnosine synthetase, and several of the markers selective for subgroups of olfactory neurons, such as 2B8, RB-8, and the blood group antigens H and B (all of which will be described in greater detail below). This group of markers may be useful for studying the transition from progenitor cell to immature neuron, but there are fewer clues to the functional role for the molecular species that are expressed both early and late. While those that are limited to immature forms only (such as GAP-43 and E-NCAM) are more likely to have a developmental role and, conversely, those that are associated with maturation of cilia and/or synaptogenesis are more likely to be related to onset and continuation of sensory function, it is more difficult to suggest or assign *a priori* the functional role of antigens that are present both early and late. Early and late antigens may serve a developmental function that is ongoing, but it is also possible that they serve mainly in a "housekeeping" role.

Molecular Correlates of Growth in Olfactory Neurons

The evidence discussed in the first section of this chapter would suggest that olfactory neurons, like other types of central and peripheral neurons, have a "growth phase" during which they undergo the early stages of neuronal differentiation; this would include a period of axonal growth and then synaptogenesis. Under this scheme, the difference between olfactory and other neurons is the presence of immature olfactory neurons in the adult animal. That is, in the adult olfactory epithelium, there are both neurons that are morphologically immature and also express markers, such as GAP-43 and E-NCAM, that are typical of immature or "juvenile" neurons elsewhere in the nervous system. However, the resemblence of olfactory neurons to other immature neurons persists beyond the immature phase of the life cycle of the olfactory neuron. *Mature* olfacory neurons in adult animals also phenotypically resemble "juvenile" central and peripheral neurons in that *mature* olfactory neurons *continue* to express several markers that are observed in the early embryonic and postnatal development of the nervous system ("juvenile" traits), and they *fail* to express other molecular species that are typical of the mature nervous system. In this case, olfactory neurons stand out as truly unique, and their uniqueness raises the question of whether the mature olfactory neuron has additional capacity or readiness for axonal growth despite the fact that neuronal differentiation has proceeded to the point where synapses have formed, OMP has been expressed, and sensory transduction has begun. In addition to those "juvenile" markers that are clearly being retained by mature olfactory neurons, there are other markers that are characteristic of juvenile neurons elsewhere in the nervous system that are also present in the adult olfactory epithelium and olfactory nerve at high levels, but the issue of whether their expression persists in mature olfactory neurons or whether it is limited to the immature neurons that are present in the adult remains open. If they are limited to the immature stage of olfactory neuron differentiation, then these juvenile markers may serve as additional examples of traits that are limited to the immature phase of olfactory neuron development (as are GAP-43 and E-NCAM). They would not, however, buttress the case that a unique set of phenotypic traits, which resemble some of the growth characteristics of other neurons, are retained in mature olfactory neurons.

Of the juvenile traits that persist in mature olfactory neurons, several can be classified under the broad heading of the expression of cytoskeletal proteins associated with immaturity. They may be of particular interest, because the organization and elaboration of the cytoskeleton during axonal outgrowth is obviously a key element in the process (Mitchison and Kirschner, 1988). The first category to be considered are those proteins that are subunits of intermediate filaments. Intermediate filaments are a common cytoskeletal element that are intermediate in size (measuring 10 to 12 nm in diameter) between microtubules and microfilaments. They are found in virtually all cell types in the body (except for cells of hematological lineage); however, the protein subunits that make up the filament vary among different cell types. For example, glial fibrillary acidic protein (GFAP) is characteristic of astrocytes in adult CNS, and the neurofilament triplet proteins are characteristic of neurons in the adult nervous system (Lazarides, 1980). However, the type of intermediate filament subunit that is expressed changes during the differentiation of these cells. For example, the neuronal progenitors of the neural tube and the neural crest contain the intermediate filament protein vimentin, which is typical of mesenchymal cells and many other types of immature epithelia (Lazarides, 1980). In these cases, vimentin is also observed in the newly postmitotic neuron for a brief period of time, concurrent with the beginning of process outgrowth, before neurofilament expression becomes manifest, and the vimentin disappears (Cochard and Paulin, 1984). At this stage, only the light and medium subunits of the neurofilament triplet, weighing 68 and 150 kDa, respectively, are present. The heavy neurofilament subunit (200 kDa) becomes expressed only later in development when synapses form and the axon becomes stabilized (Willard et al., 1985).

In contrast, mature olfactory neurons show two deviations from the sequence of events that is characteristic of intermediate filament expression in other neurons; in both cases the more juvenile phenotype is retained. First, mature olfactory neurons in many species of adult animals contain vimentin or a very highly vimentinlike protein, and second, they fail to express neurofilament proteins, with the exception of a minute subpopulation of mature neurons (Schwob et al., 1986). The first of these two findings, the retention of vimentin by mature olfactory neurons, may be considered somewhat controversial, since contradictory reports are present in the literature. However, the sum of evidence supports the conclusion that vimentin is present in mature olfactory neurons in some species of adult animals, particulary in rodents, may be summarized as follows. Several different types of antivimentin antibodies were used to demonstrate staining of the olfactory nerve and, to a lesser extent, of neuronal cell bodies in the olfactory epithelium of the rat. For each reagent, there is dense staining of the fascicles of the olfactory nerve extending along their length into the olfactory bulb, where vimentin-containing olfactory axons can be identified with both the light microscope and the electron microscope (in the case of monocolonal antivimentin antibody SBV-21) entering the glomerular neuropil (Schwob et al., 1986; and unpublished observations). The quality of labeling of neuronal cell bodies in the olfactory epithelium varies among the reagents and requires rigorous attention to fixation conditions; in our hands, the best labeling of cells in the olfactory epithleium and of axons in the olfactory nerve is obtained with cryostat sections of tissue fixed in organic solvents and stained with either of two monoclonal antivimentin antibodies SBV-21 or V9 (Osborn et al., 1984). The presence of labeled processes extending from the olfactory nerve layer into the underlying glomeruli is another powerful diagnostic indicator that olfactory axons contain vimentin (Figure 2 of Schwob et al., 1986). In addition, axons growing out from explants of the fetal olfactory epithelium also contain vimentin and can be individually identified in tissue culture (Figure 4.6). Moreover, the vimentinlike immunoreactive material in the olfactory nerve comigrates with authentic vimentin in Western blots stained with either polyclonal antiserum (unpublished results) or monoclonal antibody (Schwob et al., 1986). Some of our results in rats have recently been confirmed by Reinhard et al. (1988). These levels of analysis are absent from the paper that suggests that vimentin is not present in rat olfactory neurons (Vollrath et al., 1985); their observations are confined to the olfactory epithelium proper, and the authors make no attempt to identify fascicles of the olfactory nerve in the lamina propria or to say whether or not some of the vimentin-labeled structures in the lamina propria are indeed fascicles, as they appear to be (see Fig. 4j of Vollrath et al., 1985). In other species, the situation is less clear. Staining of olfactory nerve fascicles with some antivimentin antibodies is reported in frog and cow (Ophir and Lancet, 1988), but is absent in humans (Ophir and Lancet, 1988; Talamo et al.,

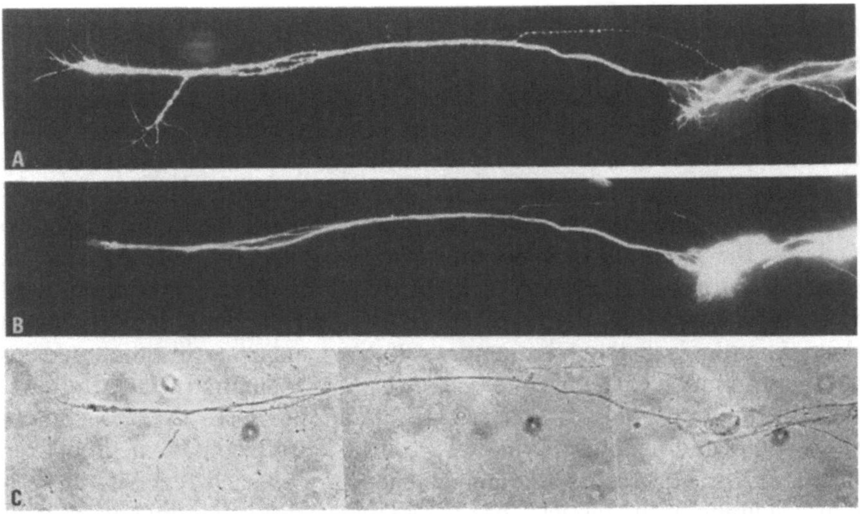

FIGURE 4.6. Vimentin is expressed in olfactory axons *in vitro*. These axons, at least four in number, have grown out as a bundle from an explant of E16 olfactory epithelium. The culture was doubly stained with anti-NCAM antiserum (A) and SBV-21, a monoclonal antibody against vimentin (B), and was visualized with phase-contrast optics (C). While NCAM extends through the growth cones to the tips of the filopodia, vimentin is limited to the shafts of the axons. The two vimentin-containing cells are fibroblasts.

1989). However, vimentin may be modified by phosphorylation and possibly other forms of posttranslational modification, and several antibodies to vimentin will recognize only a selectively expressed form found in one or at most a few types of cells, but will not react with all cells that express the vimentin gene (Dulbecco et al., 1983). As a consequence, immunohistochemical analyses that fail to demonstrate vimentin in olfactory neurons may reflect the choice of immunological reagent. Therefore, *in situ* hybridization with a cDNA probe to identify cells expressing the vimentin gene may be required to decide, in a definitive manner, the question of whether vimentin is expressed in olfactory neurons of all species.

A second issue related to vimentin in the olfactory epithelium regards the stage of olfactory neuron differentiation at which it is expressed. In rodents, two pieces of evidence suggest that vimentin is found both in immature and mature olfactory neurons. First, staining of the normal olfactory epithelium is found both in the zone that is occupied by immature neurons (see the section on GAP-43 above), located just superficial to the basal cell layer, and in the more superficial zone that is occupied by mature, OMP-containing neurons (Schwob et al., 1986). Second, in bulbectomized animals where immature olfactory neurons predominate, there is no real change (either increase or decrease) in the degree of staining of either the olfactory nerve or the epithelium as a consequence of the ablation (Schwob, unpublished observations). As a consequence, the expression of vimentin qualifies as an example of a juvenile marker that truly is retained in mature olfactory neurons.

A further example of the retention of a juvenile characteristic by mature olfactory neurons is the lack of neurofilament protein expression in the olfactory nerve or axon. Putting aside the human olfactory epithelium for the moment, there is general agreement that neurofilament proteins are absent from olfactory neurons, both cell soma and axon, in rats (Schwob et al., 1986; Vollrath et al., 1985), frogs (Ophir and Lancet, 1988), and cows (Ophir and Lancet, 1988). Rarely, bipolar neurons in the olfactory epithelium contain the medium and light subunits of the neurofilament triplet of proteins (but not the heavy subunit), and scattered neurofilament-positive axons are observed in the olfactory nerve layer of the bulb (Schwob et al., 1986). These are limited to the lateral and posterior part of the epithelium and number no more than 0.01 percent of the total olfactory neuronal population. Thus, the reciprocal relationship beteen vimentin and neurofilament expression that is seen in other neurons is maintained in olfactory neurons: neurofilament proteins and vimentin are not coexpressed. However, olfactory neurons fail to make the usual transition from vimentin to neurofilament protein expression, since they retain vimentin and do not make the neurofilament proteins. Even in those very few olfactory neurons that do make the transition and express neurofilament proteins, their pattern of expression is "immature," since they do not express the heavy subunit (Willard et al., 1985). The significance of this neurofilament-expressing subpopulation of olfactory neurons is uncertain; they may in some fashion be related to the long-lived olfactory neurons observed by Hinds' laboratory (Hinds et al., 1984). In this case, perhaps the transition to neurofilament expression can occur in olfactory neurons if they survive for a very long period of time after they are generated from the basal cells. In humans, the evidence regarding neurofilament expression is somewhat confusing, since Ophir and Lancet (1988) fail to demonstrate NFs in fetal olfactory epithelium, whereas Talamo et al. (1989) do report (but do not illustrate) some staining in autopsy material from nondemented aged controls with a particular MAb that recognizes the medium subunit. However, they do not see staining with other antibodies that recognize individual subunits (either light, medium, or heavy), nor with ones that recognize all three subunits at the same time. The findings in humans may be of particular import, since autopsy material from patients with Alzheimer's disease demonstrates neurofilament proteins, particularly phosphorylated epitopes, in fascicles of the olfactory nerve and in abnormal accumulations of neurites found at the base of the epithelium and in the lamina propria. These neuritic masses resemble the neuritic plaques and tangles found in the CNS of patients with Alzheimer's disease, since the olfactory accumulations also contained Tau protein (a microtubule-associated protein) and the epitope recognized by the MAb ALZ50 (Talamo et al., 1989). Therefore, olfactory neurons in human olfactory epithelium, like those in lower mammals, normally fail to make neurofilament proteins, but this pattern of gene expression may be altered in very interesting ways in disease states like Alzheimer's-type dementia. Because it is possible to harvest parts of the human olfactory epithelium by intranasal biopsy (Lovell et al., 1982), olfactory neurons may become a very useful system for analyzing the changes in neuronal cell biology that take place during the course of Alzheimer's disease.

In addition to the intermediate filament proteins, the microtubule associated proteins (MAPs) provide another example where the components of the olfactory system, including the mature olfactory neurons, retain proteins that are charcteristic of juvenile neurons elsewhere in the nervous system. MAPs are a class of proteins that copurify with tubulin by the repeated polymerization-depolymerization of microtubules from the brain (reviewed by Matus, 1988), and that appear to play a role in the stabilization of microtubules during the elongation and maintenance of axonal and dendritic processes. The distribution of individual MAPs is both spatially and developmentally regulated, that is, some MAPs are found only in dendrites, and others only in axons. While the expression of some MAPs is developmentally up-regulated, others are more or less exclusively found in the developing nervous system. In the olfactory system, olfactory neurons in adult animals continue to express high levels of both the phosphorylated and nonphosphorylated forms of MAP 5 (also known as MAP 1B and as MAP 1X), which in most other parts of the nervous system is found at high levels and is phosphorylated only during development (Schoenfeld et al., 1989; Viereck et al., 1989). Monoclonal antibodies to MAP5 label olfactory cell somata and dendrites as well as olfactory axons (including their terminal portions in the olfactory nerve and glomerular layers of the bulb), and the distribution of labeled cells throughout the width of the olfactory epithelium indicates that both immature and mature olfactory neurons contain this protein (Viereck et al., 1989). In similar fashion, the juvenile form of the Tau protein, which is another MAP, is found in a subset of olfactory axons as well as the whole of the vomeronasal nerve. The relative expression of Tau in immature vs. mature olfactory neurons remains uncertain. Interestingly, it would appear that the synaptic targets of the olfactory neurons, namely the mitral and tufted cells of the olfactory bulb, also retain the juvenile forms of several MAPs, including MAP5, MAP2, and Tau (Viereck et al., 1989). While the functional role of the juvenile MAPs in terms of the dynamic regulation of microtubules is unclear, and a detailed understanding of how shifts in microtubule chemistry relate to the processes of axonal stabilization or synapse formation is lacking, it is nonetheless intriguing that juvenile forms of MAPs are retained by both olfactory axon and target, given the capacity for synaptogenesis between these neuronal partners in adult animals.

Finally, the surfaces of olfactory neurons and their axons also retain a juvenile phenotype in several respects. For example, the cell surface glycoprotein Thy-1, which is a member of the immunoglobulin-cell adhesion molecule superfamily of proteins, increases nearly 100-fold during the development of the nervous system and is apparently expressed by most neurons in the adult PNS and CNS (reviewed by Morris, 1985). In contrast, most olfactory neurons, both immature and mature, in adult rats lack Thy-1 immunoreactivity (Morris and Barber, 1983; Schwob, unpublished observations). A very small percentage of olfactory neurons do express Thy-1; these are particularly found at the most superficial aspect of the neuronal zone of the epithelium (Schwob, unpublished observations). Therefore, in the case of Thy-1, as in the case of the expression of neurofilament proteins, some olfactory neurons are capable of making the transition from a juvenile phenotype to one that is typical of the adult nervous sytem elsewhere. The reason for these exceptions is unknown, but the Thy-1-positive olfactory neurons could potentially correspond to the subset of very long-lived olfactory neurons described by Hinds' group (Hinds et al., 1984). In this case, Thy-1-negative olfactory neurons, which are in the majority, may not survive long enough to permit the gene to be expressed. Alternatively, Thy-1 may be lacking from olfactory neurons for functional reasons. In the immune system, Thy-1 may mediate some forms of intercellular communication that result in elevated intracellular levels of calcium (Kroczek et al., 1986). Although the function of Thy-1 in the nervous system has not been established, the major increase in the expression of the protein during synaptogenesis has led to the suggestion that the molecule is involved in the stabilization of synapses, and its structural similarity to other cell-adhesion molecules would certainly be consistent with this hypothesis.

Several other juvenile markers are reported to be increased in the adult primary olfactory projection relative to other parts of the adult nervous system. These include the axonal cell-surface protein NILE/L1/Ng-CAM (Stallcup et al., 1985), an uncharacterized protein concentrated in neuronal growth cones that is recognized by a monoclonal antibody (Wallis et al., 1985), and the protease activator known as nexin (Reinhard et al., 1988). All three of these are likely to be involved in the process of axon outgrowth elsewhere in the nervous system, based on the pattern of expression or actual function-blocking assays in vitro. Unfortunately, we do not know whether the expression of these proteins is limited to the population of immature neurons (as GAP-43 and E-NCAM are) or whether these truly represent juvenile markers whose expression persists in mature olfactory neurons (as does vimentin). Our lack of information on this point is a consequence of the exclusively axonal localization of NILE and the growth-cone-associated antigen. It will be difficult to determine directly the neurons that express these axonal proteins and whether they are mature or immature olfactory neurons. However, this issue can be indirectly addressed by ablating the olfactory bulb. Should these other "juvenile" markers be expressed at a higher level after bulb ablation, then they, like GAP-43, are limited to immature olfactory neurons. Should the level of expression remain the same, then they, like vimentin, are probably also found in mature olfactory neurons.

Summary

The findings described in this section document the expression of "juvenile" markers in the adult olfactory epithelium. Furthermore, in the case of vimentin, neurofilament, and Thy-1, the juvenile pattern of expression has been shown to be characteristic of mature olfactory neurons. The cellular localization of others of these markers is unclear. At issue is the biological reason for the retention of a juvenile phenotype by mature olfactory neurons. Two alternatives are immediately apparent. The first would hypothesize that there is a functional correlate to the expression of juvenile markers. That is, mature olfactory neurons may have a continued capacity for growth based on the retention of a juvenile cytoskeletal (namely vimentin and the juvenile MAPs) and cell surface configuration (namely the absence of Thy-1). While speculative, this hypothesis could be tested, for example, by partially lesioning the olfactory nerve and assaying the numbers of synapses formed by indivivdual olfactory axons to ascertain whether intraglomerular sprouting might occur. The second alternative would hypothesize that the unique phenotypic features of olfactory neurons are a consequence of their embryological origin from olfactory placode as a lineage distinct from the neural tube. Certainly, that NCAM is absent from the surface of the neuronal progenitors (see above) is consistent with the notion that distinct programs of gene expression are established for the two types of tissue. However, this hypothesis neglects several similarities both in protein expression (namely, those proteins involved in the propagation of a nerve impulse and others involved in the formation of synapses) and common embryological origin from the neural *plate* (as distinct from the neural *tube*; Couly and Ledouarin, 1985). At present there is no definitive information available that would allow us to choose among the two alternatives mentioned, or any additional hypotheses.

Biochemically Defined Subclasses of Olfactory Neurons

The first two sections of this chapter described biochemical features that are common to all olfactory neurons or to more immature stages in their differentiation from progenitor cells. In this last section, the focus shifts to defining differences in phenotype that can be used to group olfactory neurons into subclasses. Several types of biochemical markers have been used to detect neuronal subclasses. These include lectins (which bind to defined carbohydrate structures on glycoproteins or glycolipids), monoclonal antibodies (which bind to a single epitope that has been biochemically defined in only a few instances), and histochemical assays for enzyme. The most useful classification of subclass-specific markers is a spatial one. Thus far, the markers that have been described can be divided into one of two classes: First, those that are expressed on a subpopulation of neurons scattered throughout the olfactory epithelium, without any restriction to one or another part, which I will term spatially *dispersed* markers; and second, those that are expressed by neurons that are limited to a particular part of the epithelium, which I will term spatially *restricted* markers. For the latter, most if not all of the neurons in that part of the epithelium express the marker. A few spatially restricted markers are also restricted to a portion of the neurons in this area, and this would be an example of a combined restriction. As a consequence of the axonal projection pattern from the olfactory epithelium to the main olfactory bulb, which is roughly quadrant-to-quadrant (Astic and Saucier, 1986; Astic et al., 1987; Clancy et al., 1985; Clark, 1957; Costanzo and Mozell, 1976; Costanzo and O'Connell, 1978; Stewart and Pedersen, 1987), spatially restricted epithelial markers will also have a limited distribution in the bulb. Therefore, we can use these markers to define zones in the epithelium and in the bulb that differ on a biochemical basis. The intersection of this anatomical-biochemical data with the issue of the functional organization of the peripheral olfactory system is of particular interest. Evidence from measures of summed activity such as EOG or multiunit recording indicates that odorant encoding by the epithelium has a spatial component that reflects both *inherent* differences in odorant sensitivity between different parts of the epithelium and differences in responsivity between different areas of the epithelium that are *imposed* by differential sorption of odorants as the air stream crosses the mucosa (reviewed by Moulton, 1976). As counterpoint to the level of spatial organization indicated by the inherent differences in odorant sensitivity, within any area of the epithelium that is sampled by the EOG, it is also likely that sensory neurons will differ in their sensitivity to odorants and be intermingled. For example, single units in frogs display a wide range of response patterns to a panel of odorants, and units varying enormously in their response to these odorants can be found at similar locations in the epithelium (Sicard, 1985; Sicard and Holley, 1984). These organizational features are not necessarily contradictory, since the inherent patterns of sensitivity, which are broadly rather than sharply defined, can be explained by differences in the proportion of units sensitive to odorant set A versus odorant set B at any particular point in the epithelium. I will return to the issue of the functional significance of these subclasses after considering their anatomical distribution and biochemical nature.

A listing of all known markers of olfactory neuronal subclasses, divided into the two types, spatially dispersed versus spatially restricted, and an

Table 4.2. Biochemically defined olfactory neuronal subclasses: spatially restricted markers

Market name	*Species*	*Distribution*	*Biochemical nature*
RB-8 (Schwob and Gottlieb, 1986, 1988)	Rat	Neurons of ventrolateral OE and axonal projection to ventrolateral OB. Stains all of the VMN and all AOB glomeruli.	MAb: binds to 125-kDa glycoprotein (cell-surface). PAb: replicates pattern.
R4B12 (Mori et al., 1985)	Rabbit	Similar to RB-8 in main OE and OB. Limited to rostrolateral AOB.	MAb: epitope uncharacterized.
CC2 (Schwarting and Crandall, 1989)	Rat	Complementary to RB-8 in main OE and OB, i.e., limited to dorsomedial OE and OB. Stains all AOB glomeruli.	MAb: binds to glycolipid with terminal α-fucose.
R5A10 (Mori et al., 1985)	Rabbit	Complementary to R4B12 in AOB, i.e., limited to caudomedial glomeruli.	Mab: epitope uncharacterized.
CC1 (Schwarting and Crandall, 1989)	Rat	Limited to rostral AOB.	Mab: binds to globoside.
Soybean agglutinin (Key and Giorgi, 1986a and b)	Rat, frog	Glomeruli in ventral OB (not all glomeruli in rat).	Lectin: binds to α-GalNAc, μ-GalNAc.
NADPH-diaphorase (Croul-Ottman and Brunjes, 1988)	Rat	Glomeruli in dorsomedial and dorsolateral OB stain heavily. In remainder of OB some variable glomerular staining.	Enzyme: catalyzes transfer of reducing equivalents. Neuron-selective distribution in CNS. Physiological role unknown.
Rat 204 (Pedersen et al., 1987)	Rat	Olfactory axons from dorsomedial OE and from VMO in fetal and neonatal pups.	MAb: uncharacterized epitope. Cell surface antigen.

Abbreviations: MAb, monoclonal antibody; PAb, polyclonal antiserum; OB, main olfactory bulb; OE, main olfactory epithelium; AOB, accessory olfactory bulb; VMN, vomeronasal nerve; VMO, vomeronasal organ (Jacobson's organ).

anatomical characterization of the neuronal subset that each defines is provided in Tables 4.2 and 4.3. Rather than repeating this information in the body of the text, I will elaborate on some of the relevant points and try to place the individual findings into a more general framework. For both spatially dispersed and spatially restricted markers, the results are of interest from several perspectives: (1) the anatomical relationship between the neuronal subsets in the olfactory epithelium and the distribution of their

TABLE 4.3. Biochemically defined olfactory neuronal subclasses: spatially distributed markers

Market name	*Species*	*Distribution*	*Biochemical nature*
Ulex europeus agglutinin I (Barber, 1989; Riggott and Scott, 1989)	Rat	Variable staining of adjacent glomeruli in OB. No regional differences noted. Olfactory specific.	Lectin: binds to α-L-fucose
2B8 (Allen and Akeson, 1985a and b)	Rat	Binds to about 30% of OSNs and axons. Variable staining of adjacent glomeruli in OB. No regional differences noted.	MAb: binds to cell surface glycoproteins of different m.w. as a function of developmental age.
2C5 (Mori, 1987)	Rabbit	Binds to 10–20% of OSNs and axons in OE. Variable staining of adjacent glomeruli in OB. No regional differences noted. Labels most neurons of VMO, uniform glomerular staining in AOB.	MAb: binds to Gal[α1 → 3]Gal[β1 → 4]GlcNAc → R.
4C9 (Mori, 1987)	Rabbit	Weakly labels subset of VMO neurons, part of VMN, and glomeruli in rostrolateral AOB. Corresponds to R4B12 (+) zone (see Table 4.2). No staining of OE and OB.	MAb: binds to Fuc → (Gal[β1 → 4]GlcNAc)$_n$ (trivial name: poly N-acetyl lactosamine)
Blood group H (Astic et al., 1989; Mollicone et al., 1985)	Rat	Function of developmental age: increasing number of neurons from E17 (50%) to P6 ("most"), then decreasing until P30. No description of nonuniform distribution.	PAb: binds to terminal α-L-fucose attched to precursor chain. On multiple proteins.
Blood group B (Astic et al., 1989; Mollicone et al., 1985)	Rat	Function of developmental age: increasing number of neurons from E17 ("a few") to P1 (still a low number) to P6 ("many") and continuing to increase in older animals. During fetal and neonatal period, there is variation in labeling among glomeruli, but this variation disappears in older rats.	PAb: binds to terminal α-D-galactose attached to H antigen. On multiple proteins.
Carbonic anhydrase (Brown et al., 1984)	Rat	Scattered, intensely (+) OSNs and axons in ON. Remainder weakly positive. Bulb not assessed.	Enzyme: catalyzes interconversion of CO_2 and HCO_3-. Typical of mitochondria-rich cells in transporting epithelium.

Abbreviations: Same as in Table 4.2. OSNs, olfactory sensory neurons of main olfactory epithelium.

axonal terminals in the olfactory bulb; (2) the biochemical nature of the marker that is differentially expressed; (3) the developmental mechanism responsible for the generation of the neuronal subsets and, in turn, the role of subset markers during the development of the olfactory epithelium and its axonal projection. For each of the two types of markers, these perspectives will be considered in turn. In the very last part, the implications of these two types of neuronal subclasses for sensory function will be considered together.

Spatially Restricted Subsets

For each of the spatially restricted epithelial markers, the distribution of labeled glomeruli is the same as would be expected from the quadrant-to-quadrant nature of the axonal projection from the epithelium onto the bulb (which had been determined using conventional tract-tracing techniques; Astic and Saucier, 1986; Astic et al., 1987; Clancy et al., 1985; Clark, 1957; Costanzo and Mozell, 1976; Costanzo and O'Connell, 1978; Stewart and Pedersen, 1987). For example, both RB-8 and R4B12, which appear to have identical distributions limited to the ventral and lateral olfactory epithelium of rats and rabbits, respectively, label glomeruli only in the ventral and lateral olfactory bulb to which the ventrolateral zone of the epithelium send its axons; in this zone, both antibodies seem to label all glomeruli heavily (Mori et al., 1985; Schwob and Gottlieb, 1986). Likewise, the antigen recognized by MAb CC2 is expressed by the dorsomedial olfactory epithelium of rats, its axons, and the glomeruli in the dorsomedial bulb to which neurons in this part of the epithelium project (Schwarting and Crandall, 1989). It is striking that the distribution of CC2 immunoreactivity is complementary to that of RB-8 in both the bulb and the epithelium. Indeed, the RB-8-defined boundary and the CC2-defined boundary appear to coincide both on the tips of the turbinates and along the nasal septum; however, the precise degree of overlap (if any) between the two antibodies' distributions, i.e., the degree of coexpression of both antigens, is presently unknown and will need to be determined by double labeling experiments on the same animal. Although the correspondence between the distribution of glomerular labeling and the staining in the epithelium is an expected result that would be predicted by tract-tracing results, it has implications beyond this simple relationship. For example, the density of RB-8 staining (and likewise R4B12 staining in the rabbit) in ventrolateral glomeruli and the identity of its distribution with OMP in these glomeruli signifies that a vast majority, if not all, of the neurons in this ventrolateral epithelium express this protein (Schwob and Gottlieb, 1986). Indeed, this pattern should be contrasted with the distribution of labeling in glomeruli of the ventral olfactory bulb by the lectin soybean agglutinin (SBA; Key and Giorgi, 1986b). In this case, while labeling is also limited to the ventral olfactory bulb, there is substantial variation in staining from one glomerulus to the next (Key and Giorgi, 1986b), indicating that not all neurons of the ventrolateral epithelium express the carbohydrate group recognized by the lectin. (Accordingly, SBA would be classified as a mixed marker showing both spatial restriction and selectivity for some neurons within that zone.) These numerical relationships could not have been easily predicted from the distribution of labeling in the periphery, since neuronal cell bodies do not stain intensely with either RB-8 or SBA.

It is apparent from these results that a pattern of contiguously staining glomeruli in the main olfactory bulb is sufficient to indicate that a marker is spatially restricted in the epithelium. However, the same relationship does not hold in the case of the accessory olfactory bulb. For example, monoclonal antibody 4C9 labels glomeruli limited to the rostrolateral accessory olfactory bulb (Mori, 1987). However, in the periphery, 4C9 labels neurons that are interspersed throughout the vomeronasal epithelium. This result would be predicted from direct tract-tracing evidence indicating that the axonal projection to the accessory olfactory bulb is diffuse and lacking in spatial organization (Mori et al., 1987). Therefore, the observation that a marker is restricted to glomeruli in a particular part of the accessory bulb does not indicate a similar restriction in the vomeronasal epithelium, in contrast to the situation with the main olfuctory bulb. Based on these findings, we can infer that two other antibodies, R4B12 and R5A10, that label glomeruli limited to the rostrolateral versus caudomedial accessory olfactory bulb, respectively, are expressed by neurons dispersed throughout the vomeronasal epithelium. It is worth noting that the lack of spatial restriction in the vomeronasal epithelium for R4B12 contrasts with its restricted distribution in the main olfactory epithelium. The reason for this altered pattern of expression is not clear.

The biochemical nature of one of these spatially restricted antigens, namely the one recognized by the monoclonal antibody RB-8, has been extensively described (Schwob and Gottlieb, 1986, 1988). The antibody recognizes a noncarbohydrate epitope of a 125-kDa glycoprotein exposed on the surface of olfactory axons. This protein has been purified to homogeneity, which has allowed us to determine a partial N-terminal amino acid sequence and to generate two independent polyclonal antisera. The patterns of labeling with the two antisera are identical to the labeling observed with the original monoclonal antibody. The identity of the patterns with antisera and monoclonal antibody strongly implies that the 125-kDa protein as a whole is being differentially expressed by the two zones of the epithelium, presumably as a consequence of differential gene regulation, and excludes the alternative explanation that the differential expression of a posttranslational modification of a common protein backbone can account for the difference between ventrolateral and dorsomedial epithelium. Along this same line, it is interesting to note that the monclonal antibody CC2, which has a complementary labeling distribution to that of

RB-8, recognizes a complex glycolipid (the nature of the particular epitope on this molecule is currently undefined) (Schwarting and Crandall, 1989). Therefore, the expression of more than one gene is being differentially regulated during the establishment of this dorsomedial-ventrolateral division of the main olfactory epithelium.

The identification of spatially restricted subsets of olfactory neurons raises two distinct developmental questions. First, what is the mechanism responsible for dividing the epithelium into these different zones? Second, what is the role of these spatially regulated markers in the subsequent development of the epithelium or its axonal projection onto the bulb? Unfortunately, nothing is known in answer to the first query. Translated into embryological terminology, at issue is the establishment of nonequivalent fates for the cells of the nasal placode. One hypothesis would predict that the fate of olfactory neurons, with respect to the establishment of these biochemically homogenous zones, is determined by the imposition of a chemical gradient across the placode that provides each cell with the necessary information about its position in the placode. There is good evidence to show that this type of mechanism for establishing positional information is responsible for determining cell fate in the vertebrate limb bud (rev. by Brockes, 1989). A competing hypothesis would predict that cell fate in the nasal placode is determined as a consequence of lineage from a very early stage in the development of the embryo, such as has been described for many invertebrate animals, like *C. elegans* (Purves and Lichtman, 1985). Experiments designed to alter cell fate in the nasal placode, such as rotation of the placode or changes in its size, could be attempted using the organ culture techniques pioneered by Farbman and colleagues or by experimental manipulation of amphibian or avian embryos *in situ* (Farbman, 1977). Since the divisions of the olfactory epithelium described above are defined by the expression of marker molecules, alterations in their size, placement, or boundaries of such biochemically defined zones can be easily assayed in response to these types of manipulations. On the second issue, namely the role of spatially restricted markers in the establishment of the projection pattern of the epithelium on to the bulb, little direct evidence is available. However, with respect to one of these markers, namely the RB-8 antigen, the time course of expression has been determined both during embryological development and during the differentiation of olfactory neurons from stem cells in the epithelium. We have preliminary evidence to indicate that the epithelium demonstrates a restricted pattern of RB-8 expression as early as E15 (using E1 to equal the first day after conception) (Figure 4.7). This is after the first axons have grown from the epithelium into the vicinity of the anlage of the bulb, but is prior to the formation of the first synapses in the bulb (Farbman, 1986). Likewise, the pattern of RB-8 expression has been determined after olfactory bulb ablation. As described in the first section, the vast majority of neurons in the bulbectomized epithelium are immature by both morphological and biochemical

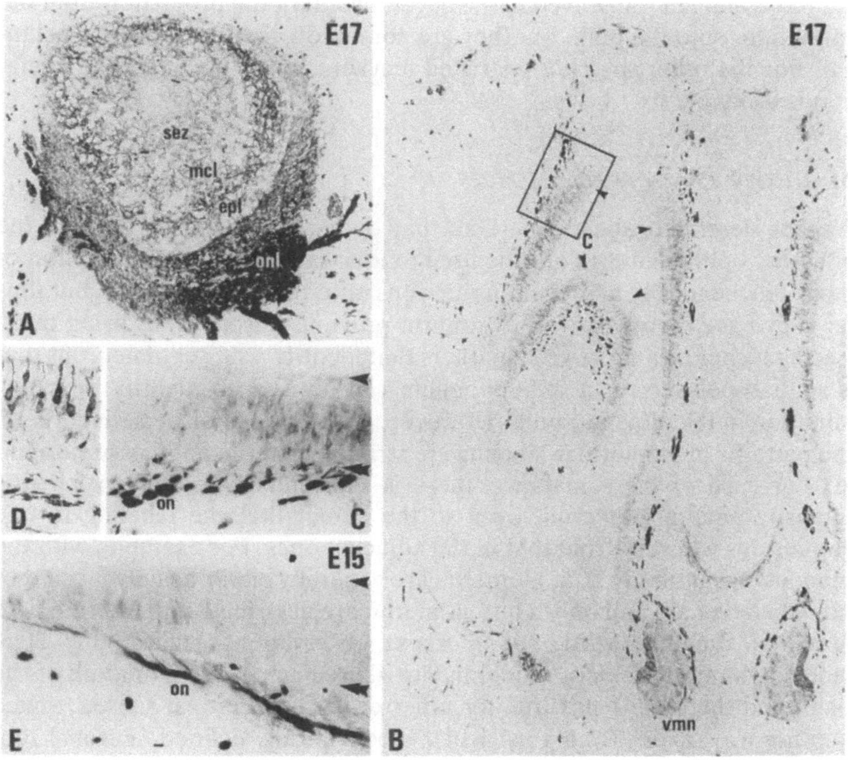

FIGURE 4.7. The RB-8 antigen is expressed during axon outgrowth prior to the time of synaptogenesis in the olfactory bulb. (A) Olfactory axons are segregated into RB-8 (+) and RB-8 (−) zones in the olfactory nerve layer at E17. Ventral is down, the midline is at the left edge of the photomicrograph. (B) Expression of the RB-8 antigen in the olfactory epithelium is spatially restricted at this time. Dorsal is up, the midline is to the right. There is staining of the ventral epithelium and associated axon fascicles. The boundaries between RB-8 (+) and (−) territories are indicated by the arrowheads. One of these (the boxed area) is shown at higher power in (C). (D) An adjacent section of dorsal epithelium stained with anti-OMP indicates that RB-8 (−) epithelium has achieved an equivalent stage of differentiation as ventral epithelium, i.e., there is no disparity in maturation between dorsal and ventral, with respect to this marker. (E) The first evidence of RB-8 staining was observed at E15, when there is weak staining of ventral olfactory epithelium and more prominent staining of fascicles of olfactory axons. VMN—vomeronasal organ.

criteria, and lack a normal target for synaptogenesis. Despite this, the RB-8 antigen is expressed but restricted to the ventrolateral olfactory epithelium as in the normal epithelium, which indicates that the protein is expressed in immature neurons during the phase of axonal growth (Schwob et al., 1988). Both of these results are consistent with the hypothesis that the RB-8 antigen is involved in the development of the projection from the epithelium onto the bulb, but they are admittedly indirect and circumstantial. For the other spatially restricted markers, even this limited information is lacking.

Spatially Dispersed Subsets

As was described above, markers that define a contiguous zone in the olfactory epithelium can also be used to illustrate the axonal projection of those biochemically homogenous neurons onto the olfactory bulb, but thus far they have merely served to confirm previous observations using tract-tracing techniques. Likewise, markers that identify a subset of neurons that is spatially dispersed in the epithelium can be used to identify their termination in the olfactory bulb. However, in the case of dispersed markers, the patterns of glomerular labeling are striking, and the findings are original. For each of these markers, there is substantial variation in labeling intensity among glomeruli, even to the extent that the labeling in one glomerulus will vary from that in the adjacent ones. For example, with the monoclonal antibody 2C5, glomeruli are generally either heavily labeled or not labeled at all, and only a few glomeruli are described with intermediate levels of staining (Mori, 1987). These observations suggest that 2C5-labeled glomeruli are not randomly distributed across the glomeruli of the bulb, and the same holds true for other dispersed neuronal subsets; stated another way, axons from a particular biochemically defined neuronal subset tend to project together to the same subset of glomeruli. At present, this conclusion should be considered as tentative, since no quantitative or statistical analysis has been reported. The developmental and functional implications of these results will be dealt with below.

Most of the dispersed neuronal subsets have been identified on the basis of reagents, either lectins or antibodies, that recognize carbohydrate groups that are covalent modifications of membrane-bound proteins or lipids. For many of these reagents, the carbohydrate chain to which they bind has not been precisely determined. This point can be illustrated with reference to lectins, which are usually classified according to the monosaccharide that will displace them from their binding site. However, other components of the carbohydrate chain clearly affect the binding as well; for example, the group of lectins that are dislodged by α-fucose differ among themselves in their ability to agglutinate red blood cells (Beeley, 1985) and in their staining pattern in the brain: *Ulex europeus* agglutinin-l labels only olfactory neurons in the rat nervous system (Barber, 1989), while staining with *Lotus tetragonolobus* agglutinin is more broadly distributed in the CNS (Schwob, unpublished results). Furthermore, the monoclonal antibody 4C9, which binds to a fucose-capped polylactosamine (also likely to be recognized by one or both of these lectins), fails to stain neurons in the main olfactory epithelium of rabbits and recognizes only a subset of vomeronasal neurons. Hence, we cannot say how these neuronal subsets differ from one another in exact biochemical terms. Nor has it been shown that these carbohydrate group differences are significant in a functional sense. Nonetheless, it has been suggested that carbohydrate structures play a major role in the process by which different subsets of olfactory axons reach their appropriate target, since some of them are expressed on the cell surface from an early stage in embryological development (for example, the epitope recognized by the monclonal antibody 2B8; Allen and Akeson, 1985b), and/or that they parallel differences between olfactory neurons in sensistivity to odorants. To some extent, this hypothesis is based on analogy to findings in the dorsal root ganglia (DRG), where the neuronal population has also been subdivided on the basis of differences in carbohydrate structures using a large panel of monoclonal antibodies that bind to different members of the lactoseries and the globoseries of carbohydrates (Dodd et al., 1984; Dodd and Jessell, 1985). In contrast to the limited information available for the olfactory system, the DRG neurons of a carbohydrate-defined subclass have been shown to share other common features that correlate with function, including embryonic lineage (small dorsomedial versus large ventrolateral population), the peptide neurotransmitters they express, and, most importantly, the somatosensory modality they subserve (putatively identified on the basis of the different lamina of the dorsal horn in which their axons terminate). Moreover, the existence of a class of soluble, endogenous lectins that bind selectively to these glycoconjugates is indirect evidence that the carbohydrate group that is exposed on the cell surface may help determine or guide the formation of axonal connections during the development of the dorsal root projection into the dorsal horn (Hynes et al., 1989). While the generation of these neuronal subsets in the DRG is not understood at a mechanistic level, the structures of the carbohydrate groups that distinguish them suggest that the subsets are generated in successive stages: neurons differ according to their carbohydrate backbone (lactose vs. globose) and the progressive modification of these backbones by the sequential addition of sugars. There are preliminary indications that subsets of olfactory neurons may also be identifiable on the basis of this same panel of anticarbohydrate antibodies (Dodd and Jessell, 1985). Unfortunately, a complete evaluation of the olfactory system using these reagents has not been published.

Functional Implications

For the DRG, the results described above suggest that biochemically defined subsets of DRG neurons apparently subserve different functions by conveying different somatosensory modalities, i.e., DRG neurons expressing lactoseries carbohydrates apparently encode nocioceptive stimuli rather than light touch. In the case of the olfactory system, the question, although it is phrased in terms of restricted sensitivity to odorants, remains essentially the same: do biochemically defined subsets of olfactory neurons, having either a spatially restricted or spatially dispersed distribution in the olfactory epithelium, differ with regard to olfactory sensitivity such that we can predict a neuron's sensitivity to odorants based solely on its biochemical phenotype? No clear answers have been found as yet. Indeed, there have been no published reports comparing sensory function with these biochemical divisions of the epithelium. However, several pieces of evidence suggest that these biochemically defined subclasses do have functional relevance. Much of the data derives from experiments that map the accumulation of radioactive 2-deoxyglucose (2-DG) in response to short periods of odorant exposure. In addition, recordings of the EOG in the epithelium or cribriform plate, or single-unit recordings from mitral cells of the bulb have been utilized, but these studies are fewer in number.

The evidence for a functional correlate is least direct for markers, like carbohydrate epitopes, that identify neurons scattered throughout the epithelium. However, these markers are potentially very interesting, because each neuronal subset projects onto the bulb in such a strikingly nonrandom, uneven way, filling some glomeruli, but avoiding others that are nearby or even adjacent. One might predict that a group of neurons with similar olfactory sensitivity would also be distributed throughout the epithelium, since sensitivity to an odorant is seldom sharply restricted, yet project in a punctate manner to individual glomeruli rather than regions of the glomerular layer, since the evidence from 2-DG studies would suggest that glomeruli tend to respond to odorants as a unit and often in isolation from surrounding glomeruli (Jourdan et al., 1980; Lancet et al., 1982; Benson et al., 1985). These results are suggestive, though indirect, evidence in support of the hypothesis that different carbohydrate-defined subsets will also show different sensitvity to odorants (Allen and Akeson, 1985a). However, the obverse relationship, namely that all the glomeruli labeled by a single carbohydrate marker are activatable by exactly the same set of odorants, is not likely to be true, since the number of glomeruli activated by a single odorant (as judged by 2-DG accumulation) is smaller than any of the carbohydrate-identified groupings of glomeruli that have been described to date (compare Jourdan et al., 1980 with Mori, 1987; Allen and Akeson, 1985a). For example, each carbohydrate marker will typically label 10 percent or more of the glomeruli on its own, but the number of glomeruli labeled with 2-DG after exposure to a single odorant is generally much lower than that. Hence, a carbohydrate-defined subset of neurons is not likely to be functionally homogenous, i.e., will not respond uniformly to the universe of odors, purely on the basis of the larger size of the subset as compared to the 2-DG results. This is not an unexpected result, since to group neurons into a few large functional subsets that each include 10 percent or more of olfactory neurons would not fit with the results of single-unit recordings in the frog olfactory epithelium, which indicate that the number of functional types of olfactory neurons far exceeds 20 (Gesteland et al., 1963; Sicard, 1985; Sicard and Holley, 1984). Does this mean that carbohydrate epitopes have no relation to *functional* subdivisions of the population of olfactory neurons, given the large number of glomeruli that are marked by any one of these carbohydrate reagents? This does not necessarily follow, since one potential mechanism to generate a greater number of groups would be to express more than one type of marker; under this mechanism, a subset would be defined by the combined expression of more than one type of marker (i.e., neurons that express both carbohydrate A and carbohydrate B would be classified together, while others that expressed A or B but not both would belong to other groups). As a consequence, the groups would be smaller, and all the neurons in the group theoretically could be sensitive to the same odorants. Evidence for or against this hypothesis is now lacking, but it would be useful to map the distribution of pairs of markers and compare these patterns of labelin with 2-DG accumulations in response to odorants. A related issue is the cell biological function of these markers that are limited to neurons whose axons tend to cluster in the same glomeruli. It may be, as was suggested for the neurons of the DRG (Dodd and Jessell, 1985), that these epitopes serve to direct axonal projections during development so that axons synapsing in the same glomerulus share these epitopes. Direct experiments utilizing reagents to remove or block these carbohydrate groups *in vitro* or *in vivo* would be required to prove that they actually do serve a role during the development of the projection.

For markers that define a limited spatial zone in the epithelium and its projection onto the bulb, there is some evidence that these zones differ in their sensitivity to odorants. Indeed, it is noteworthy, and may be taken as an indication of its potential importance, that this division of the epithelium into ventrolateral and dorsomedial parts is preserved in separate taxonomic orders. Three types of physiological evaluation have demonstrated that sensitivity to odorants is regionally restricted. First, both EOG (Mackay-Sim et al, 1982; Edwards et al., 1988) and voltage-sensitive dye recordings of odorant-evoked activity (Kent, 1990) demonstrate some degree of inherent spatial patterning of odorant responsivity in the olfactory epithelium of lower vertebrates and in rats. Second, 2-DG accumulation across the glomerular surface of the bulb in response to odorants is generally restricted to a few foci. Third, the changes in mitral cell morphology that occur in response to prolonged odorant exposure are limited to certain parts of the bulb. In most cases, the hypothesis that the biochemically defined divisions of the primary olfactory projection differ in odorant sensitivity has not been subjected to direct test, since the physiological results demonstrating a regional selectivity to odorants have seldom been compared with the anatomical findings that delimit the epithelium and bulb into biochemically homogeneous zones. However, for some of the spatially restricted markers, in particular RB-8 (Schwob and Gottlieb, 1986; 1988) and R4B12 (Mori et al., 1985), the division of the epithelium into a ventrolateral area that expresses the marker and a dorsomedial one that does not is sufficiently consistent from animal to animal to permit some indirect comparison with the three types of functional information mentioned above. An association between odorant sensitivity and expression of these markers would best be shown by finding that responsivity to some odorants is strictly limited to one zone or the other. Nonetheless, the hypothesis that marker-defined subsets of olfactory neurons are functionally relevant would be supported even if relative, rather than absolute, regional selectivity in response to odorants is found to correlate with the distribution of the marker. In fact, there is physiological evidence from both the olfactory epithelium and the olfactory bulb to indicate that RB-8 (−) and RB-8 (+) zones are differentially activated by odorants. For example, in the rat olfactory epithelium, EOG recordings were taken from 12 points across the tips of the endoturbinates in response to the odorants cineole, nicotine, i-pentanoic acid, and i-pentyl acetate (Edwards et al., 1988). All areas showed some response to four odorants, but there were regional differences in the amplitude of response for these odorants. Discriminant analysis was used to describe mathematically the response of each of the recording positions to the test odorants. With this type of analysis, the characteristics of the response of epithelium on the ventrally placed turbinates, which is RB-8 (+) territory, cluster together in the graphical representation of the analysis, away from the response characteristics of the recording positions on the tips of the dorsal turbinates, which is RB-8 (−) territory (Figure 4.8). Likewise, the patterns of 2-DG accumulation in the olfactory bulb suggest that RB-8 (+) territory is selectively responsive to some odorants that do not activate RB-8 (−) epithelium. For example, glomeruli activated by the odors amyl acetate, camphor, peppermint, and ethyl acetoacetate tend to cluster on the ventrolateral and ventromedial portions of the bulb, which is RB-8 (+) territory (in rats, Stewart et al., 1979; Jourdan, 1982; Jourdan et al., 1980; Coopersmith et al., 1986; in mice, Royet et al., 1987; Figure 4.9). In contrast, isovaleric acid activates glomeruli in the dorsomedial bulb in mice, which would be equivalent to RB-8 (−) territory in rats (Royet et al., 1987). In addition, morphological changes as a consequence of chronic odorant exposure also differentiate the two territories. For example, in rats exposed to cyclohexanone from postnatal day 14 for a period of 11 weeks, the dorsomedial bulb is spared the mitral cell shrinkage that is observed in the ventrolateral portions of the bulb (Laing et al., 1985; Panhuber and Laing, 1987). Mitral cell degeneration is more widespread in animals that are chronically exposed to cyclohexanone as adults, and degenerative changes now extend into the dorsomedial bulb, although there is still more sparing in this region than in the ventrolateral part (Panhuber et al., 1987). The age-related decrease in mitral cell sparing may reflect the progressive restriction of glomerular activation in response to other odorants as a function of age (Astic and Saucier, 1983). Eventually this process would culminate in the punctate pattern of 2-DG accumulation that is observed in response to odorants in the adult bulb (Jourdan et al., 1980). Nonetheless, these data are a further indica-

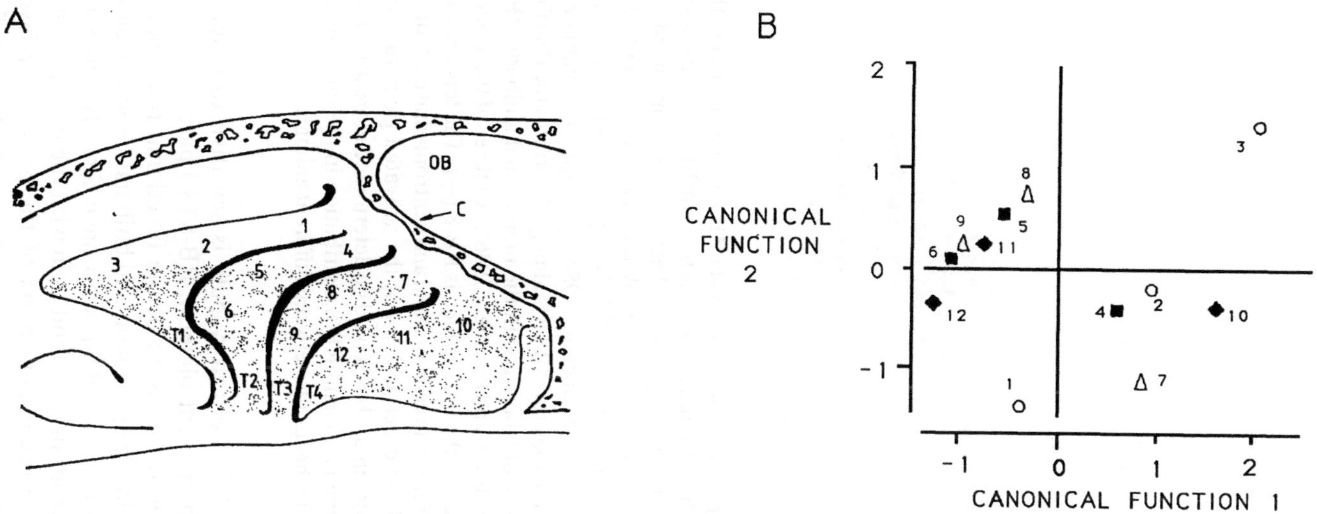

FIGURE 4.8. The response of RB-8 (−) parts of the olfactory epithelium to a panel of odorants is different from the response of RB-8 (+) parts. (A) EOGs were recorded in response to nicotine, i-pentyl acetate, i-pentanoic acid, and cineole from each of the 12 points indicated on the tips of the turbinates. The RB-8 (+) part of the epithelium is indicated by the shading. (B) A discriminant analysis calculated from the normalized EOG mathematically describes the response of each point on the epithelium to the test odorants. Functions were calculated to generate the maximum discrimination between the normalized EOGs, and the two most discriminating functions are shown. Therefore, a point on the graph represents the averaged response to all odorants from each of the 12 recording sites. A clustering of points together indicates that their response to the four odorants is similar, while a greater separation of points indicates a dissimilar response. Those points in RB-8 (−) territory (1–4) or adjacent to it (7, 10) are clearly separated from the responses of points in RB-8 (+) epithelium (5, 6, 8, 9, 11, 12), which form a fairly tight cluster. (Modified, by permission, from Edwards et al., 1988.)

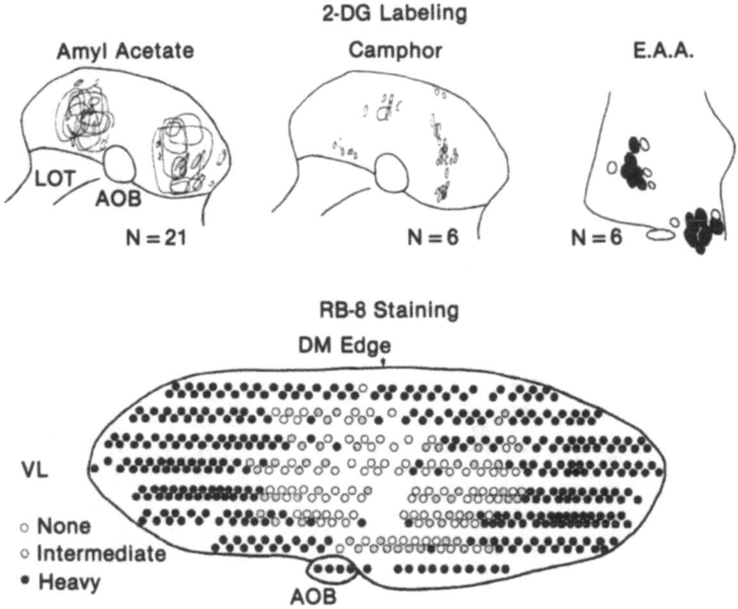

FIGURE 4.9. Some odors appear preferentially to activate RB-8 (+) glomeruli in the bulb. "Amyl acetate" and "camphor" summarize the patterns of 2-DG accumulation in the glomenular layer of the bulb in response to these odorants (modified, by permission, from Stewart et al., 1979. Copyright by Wiley-Liss, a division of John Wiley and Sons, 1979). "E.A.A." summarizes the 2-DG response to ethyl acetoacetate (modified, by permission, from Jourdan et al., 1980). (n's equal the number of bulbs analyzed to construct each composite). "R8-8 Staining" illustrates the distribution of glomeruli and their density of labeling with the monoclonal antibody RB-8 (modified, by permission, from Schwob and Gottlieb, 1986). All three of these odors fail to activate the dorsalmost part of the bulb, which is RB-8 (−). The tendency for these odorants to activate glomeruli in the ventrolateral and ventromedial parts of the bulb, which are RB-8 (+), and the exact position of the activated glomeruli with respect to the RB-8-defined border in the bulb could be determined more accurately by directly assessing 2-DG accumulation and RB-8 distribution in the same animal rather than attempting to translate the results between different animals.

tion that dorsomedial, RB-8 (−) olfactory bulb is sensitive to different odors than is the complementary ventrolateral, RB-8 (+) part.

The most direct evaluation of the issue of the functional relevance of biochemically defined zones in the epithelium and bulb has been presented in preliminary form by Mori et al. (1990), who have directly compared single-unit recordings of neurons in the rabbit bulb with the distribution of the R4B12 antigen in the corresponding parts of the glomerular layer. Although the number of units that has been studied is small, the data illustrate several differences between the dorsal and ventral olfactory bulb. By exposing the animal to a panel of 75 odorants, they have demonstrated that units in the dorsal, R4B12 (−) portion of the bulb respond to a smaller subset of the odorants, on average, than do units in the ventral, R4B12 (+) portion of the bulb. Moreover, there were differences between the dorsal and ventral bulb in terms of the odorants to which each portion of the bulb responded. For example, many of the units in the dorsal bulb were activated by caproic acid, whereas none of those located in the ventral bulb responded to these odorants. Conversely, the terpen-alcohols exclusively activated neurons in the ventral, R4B12 (+) bulb, whereas aldehydes and ketones were generally found to stimulate neurons in both divisions. From these data one can draw some tentative conclusions concerning both regional selectivity to odorants and its relationship to the division of the epithelium on the basis of RB-8/R4B12 expression. First, although the sample is small and statistical analysis is lacking, at least one of the odorants is specific for RB-8/R4B12 (−) territory, whereas others are specific for the complementary RB-8/R4B12 (+) zone. Taken in combination with the quadrant-to-quadrant projection of the epithelium onto the bulb, the conclusion would follow that some odorants are specifically transduced by one or the other of the R4B12/RB-8 defined zones of the epithelium, or alternatively that only the sensory neurons from that zone are able to excite the bulbar neurons. Second, the response to other odorants does not respect this biochemically defined boundary, and bulbar and epithelial neurons in both zones will be activated by these odorants (including units that responded to the region-specific odorants). However, it is not required that the sensitivity to all odorants be predicted by the RB-8/R4B12 border for this or other biochemically defined boundaries to be considered functionally relevant. Indeed, there may be other biochemically defined divisions of the epithelium and its axonal projection (as yet undiscovered) whose boundaries would be respected by these other odorants.

Taken together, the results correlating the physiological response to odorants with the biochemically defined subsets, both spatially restricted and spatially dispersed, of olfactory sensory and bulbar neurons provide suggestive evidence that these neuronal subsets differ in their sensitivity to odorants, i.e., that these divisions of the primary olfactory projection have functional relevance. The data, however, are incomplete and indirect, and, as a consequence, this conclusion is only preliminary. However, the data are sufficient to encourage further efforts whose aim is to study more directly the relationship between olfactory neuronal subsets and olfactory neuronal function.

General Conclusions and Synthesis

A biochemical approach to the olfactory system holds great promise for improving our understanding of the process of neuronal differentiation and the organization of the olfactory system for sensory function. The data considered in this chapter give clear indication of this promise and, just as clearly, illuminate the present deficiencies in our understanding. As a way of summarizing the discussion contained herein, I would like to suggest some areas where work needs to be done.

While much progress has been made in terms of identifying the stages that accompany the differentiation of olfactory neurons from their basal cell precursors, a much better definition of these stages will be required to enable us to understand the mechanism of differentiation at a molecular level. For example, the composition and regulation of the proliferating neuronal population has yet to be determined, largely because the means of identifying these cells in the epithelium is lacking. Once available, these markers could be used to study the dynamics of proliferation during recovery from injury, normal development, and the plasticity that occurs as a consequence of olfactory experience (Wysocki et al., 1989; Coopersmith et al., 1986; Panhuber and Laing, 1987). In addition, a better definition of the changes in protein composition that occur as a neuron makes a transition through immature to mature stages will be required prepatory to the application of molecular genetic techniques to deciphering the mechanism of neuronal differentiation.

The second focus of the chapter was the expression of juvenile phenotypic characteristics by otherwise mature olfactory neurons that have apparently stopped growing an axon. These features are surprising, and their significance is completely unknown. However, it is striking that many of these juvenile markers are structures or proteins that are intimately involved with axonal growth: intermediate filaments (e.g. vimentin), microtubule-associated proteins, and Thy-1 (whose expression parallels the cessation of active growth and the formation of synapses in other parts of the nervous system). Accordingly, these characteristics may be an indication of an ongoing capacity for growth by olfactory axons that persists even after the axons have contacted their target. This could take the form of localized, intraglomerular sprouting in response to a partial denervation of a glomerulus (which might result from ongoing neuronal turnover). At present there is no evidence for such a phenomenon, but, in this case, the biochemical results would suggest that this type of phenomenon is "worth a look."

Finally, the identification and characterization of biochemically defined subsets of olfactory neurons offer an excellent opportunity to elucidate functional relationships in the olfactory system. For the reasons outlined above, it is likely that neuronal subsets have specific restricted sensitivity to odorants, but this has not been definitely established. Direct comparisons of physiological measures of sensory activity with these biochemical classifications is an obvious strategy to elucidate the parallels between the neuronal subsets and olfactory function, and this type of approach is being pursued, but not widely. Moreover, in every case, these subset markers remain anonymous, i.e., none of them can be assigned a cell biological function. This is likely to be the most difficult aspect of the problem to solve, even though much of the data would suggest that these subset markers play a role during the development of the axonal projection of the epithelium onto the bulb. A priori, it is difficult to predict the type of experiment that could be successfully used to assay function, since the two so obviously go hand in hand. As a consequence, these experiments will be laborious and risky, but important nonetheless.

Despite these caveats, one can safely predict that molecular genetic and protein biochemical techniques, as they become widely available, will accelerate the pace at which functional insights into neuronal differentiation and the mechanism of olfactory function are achieved.

Acknowledgements. I would like to thank Dr. Kensaku Mori for sharing his data in advance of publication. Dr. Frank Margolis was exceptionally generous with his gifts of anti-OMP antibody, as were Dr. Jon Covault with the anti-NCAM antiserum, Dr. Stephen Blose with the antivimentin monoclonal antibody, and Dr. Urs Rutishauser with the anti-E-NCAM monoclonal antibody. I owe a special debt of gratitude to Audrey Stasky and Karen Szumowski for their outstanding technical assistance in accomplishing much of the work presented here that originated from our laboratory. I would also like to thank Dr. Karina Meiri for her comments on the manuscript and the gift of the monoclonal antibodies to GAP-43. The preparation of this chapter was supported by NIH grants R29 DC 00467 and P01 DC 00220, a grant from the March of Dimes Birth Defects Foundation, and a research fellowship from the Alfred P. Sloan Foundation.

References

Akeson, R.A., and S.L. Haines (1989). Rat olfactory cells and a central nervous system neuronal subpopulation share a cell surface antigen. *Brain Res.*, *488*: 202–212.

Allen, W.K., and R. Akeson (1985a). Identification of a cell surface glycoprotein family of olfactory receptor neurons with a monoclonal antibody. *J. Neurosci.*, *5*: 284–296.

Allen, W.K., and R. Akeson (1985b). Identification of an olfactory receptor neuron subclass: Cellular and molecular analysis during development. *Dev. Biol.*, *109*: 393–401.

Astic, L., and D. Saucier (1983). Ontogenesis of the functional activity of guinea-pig olfactory bulb: Autoradiography study with the 2-deoxyglucose method. *Dev. Brain Res.*, *10*: 257–263.

Astic, L., and D. Saucier (1986). Anatomical mapping of the neuroepithelial projection to the olfactory bulb in the rat. *Brain Res. Bull.*, *16*: 445–454.

Astic, L., D. Saucier, and A. Holley (1987). Topographical relationships between olfactory receptor cells and glomerular foci in the rat olfactory bulb. *Brain Res.*, *424*: 144–152.

Astic, L., J. LePendu, R. Mollicone, D. Saucier, and R. Oriol (1989). Cellular expression of H and B Antigens in the rat olfactory system during development. *J. Comp. Neurol.*, *289*: 386–394.

Barber, P.C. (1989). *Ulex Europeus* Agglutinin I binds exclusively to primary olfactory neurons in the rat nervous system. *Neuroscience*, *30*: 1–9.

Beeley, J.G. (1985). *Glycoprotein and Proteoglycan Techniques*. in R.H. Burdon and P.H. van Knippenberg (Eds), *Laboratory Techniques in Biochemistry and Molecular Biology*. New York: Plenum.

Benson, T.E., G.D. Burd, C.A. Greer, D.M.D. Landis, and G.M. Shepherd (1985). High-resolution 2-deoxyglucose autoradiography in quick frozen slabs of neonatal rat olfactory bulbs. *Brain Res.*, *339*: 67–78.

Brockes, J.P. (1989). Retinoids, homeobox genes, and limb morphogenesis. *Neuron*, *2*: 1285–1294.

Brown, D., L.-M. Garia-Sequra, and L. Orci (1984). Carbonic anhydrase is present in olfactory receptor cells. *Histochemistry*, *80*: 307–309.

Calof, A.L., and D.M. Chikaraishi (1989). Analysis of neurogenesis in a mammalian neuroepithelium: Proliferation and differentiation of an olfactory neuron precursor in vitro. *Neuron*, *3*: 115–127.

Camara, C.G., and J.W. Harding (1984). Thymidine incorporation in the olfactory epithelium of mice: Early exponential response induced by olfactory neurectomy. *Brain Res.*, *308*: 63–68.

Carr, V.M., A.I. Farbman, M.S. Lidow, L.M. Colletti, J.L. Hempstead, and J.I. Morgan (1989). Developmental expression of reactivity to monoclonal antibodies generated against olfactory epithelia. *J. Neurosci.*, *9*: 1179–1198.

Chuah, M.I., and A.I. Farbman (1983). Olfactory bulb increases marker protein in olfactory receptor cells. *J. Neurosci.*, *3*: 2197–2205.

Clancy, A.N., T.A. Schoenfeld, and F. Margolis (1985). Topographic organization of peripheral input to the hamster main olfactory bulb. *Chem. Senses*, *10*: 399–400.

Cochard, P., and D. Paulin (1984). Initial expression of neurofilaments and vimentin in the central and peripheral nervous system of the mouse embryo in vivo. *J. Neurosci.*, *4*: 2080–2094.

Coopersmith, R., S.R. Henderson, and M. Leon (1986). Odor specificity of the enhanced neural response following early odor experience in rats. *Brain Res.*, *392*: 191–197.

Costanzo, R.M., and M.M. Mozell (1976). Electrophysiological evidence for a topographical projection of the nasal mucosa onto the olfactory bulb of the frog. *J. Gen. Physiol.*, *68*: 297–312.

Costanzo, R.M., and R.J. O'Connell (1978). Spatially organized projections of hamster olfactory nerves. *Brain Res.*, *139*: 327–332.

Costanzo, R.M., and P.P.C. Graziadei (1983). A quantitative analysis of changes in the olfactory epithelium following bulbectomy in hamster. *J. Comp. Neurol.*, *215*: 370–381.

Couly, G.F., and N.M. LeDouarin (1985). Mapping of the early neural primordium in quail-chick chimeras. I. Developmenlal relationships between placodes, facial ectoderm, and prosencephalon. *Dev. Biol.*, *110*: 422–439.

Croul-Ottman, C.E., and P.C. Brunjes (1988). NADPH diaphorase staining within the developing olfactory bulbs of normal and unilaterally odor-deprived rats. *Brain Res.*, *460*: 323–328.

Cuschieri, A., and L.H. Bannister (1975). The development of the olfactory mucosa in the mouse: Light microscopy. *J. Anat.*, *119*: 277–286.

Dodd, J., D. Solter, and T.M. Jessell (1984). Monoclonal antibodies against carbohydrates differentiation antigens identify subsets of primary sensory neurons. *Nature*, *311*: 469–472.

Dodd, J., and T.M. Jessell (1985). Lactoseries carbohydrates specify subsets of dorsal root ganglion neurons projecting to the superficial dorsal horn of rat spinal cord. *J. Neurosci.*, *5*: 3278–3294.

Dulbecco, R., R. Allen, S. Okada, and M. Bowman (1983). Functional changes of intermediate filaments in fibroblastic cells revealed by a monoclonal antibody. *Proc. Natl. Acad. Sci. USA*, *88*: 1915–1918.

Edelman, G.M., S. Hoffman, C.-M. Chuong, J.-P. Thiery, R. Brackenbury, W.J. Gallin, M. Grumet, M.E. Greenberg, J.J. Hemperly, C. Cohen, and B.A. Cunningham (1983). Structure and modulation of neural cell adhesion molecules in early and late embryogenesis. Cold Spring Harbor. *Symp. Quant. Biol.*, *48*: 515–526.

Edwards, D.A., R.A. Mather, and G.H. Dodd (1988). Spatial variation in response to odorants of the rat olfactory epithelium. *Experientia*, *44*: 208– 211.

Farbman, A.I. (1986). Prenatal development of mammalian olfactory receptor cells. *Chem. Senses*, *11*: 3–18.

Farbman, A.I., and F.L. Margolis (1980). Olfactory marker protein during ontogeny: immunohistochemical localization. *Dev. Biol.*, *74*: 111–121.

Gesteland, R.C., R.A. Yancey, and A.I. Farbman (1982). Development of olfactory receptor neuron selectivity in the rat fetus. *Neuroscience*, *7*: 3127–3136.

Graziadei, P.P.C., and G.A. Monti Graziadei (1979). Neurogenesis and neuron regeneration in the olfactory system of mammals. 1. Morphological aspects of differentiation and structural organization of the olfactory sensory neurons. *J. Neurocytol.* *8*: 1–18.

Hempstead, J.L., and J.I. Morgan (1985a). A panel of monoclonal antibodies to the rat olfactory epithelium. *J. Neurosci.*, *5*: 438–449.

Hempstead, J.L., and J.I. Morgan (1985b). Monoclonal antibodies reveal novel aspects of the biochemistry and organization of olfactory neurons following unilateral olfactory bulbectomy. *J. Neurosci.* *5*: 2382–2387.

Hinds, J.W., P.L. Hinds, and N.A. McNelly (1984). An autoradiographic study of the mouse olfactory epithelium: Evidence for long-lived receptors. *Anat. Rec.*, *210*: 375–383.

Hynes, M.A., L.B. Buck, M. Gitt, S. Barondes, J. Dodd, and T.M. Jessell (1989). Carbohydrate recognition in neuronal development: structure and expression of surface oligosaccharides and beta-galactoside-binding lectins. *CIBA. Found. Symp.*, *148*: 189–210.

Jourdan, F. (1982). Spatial dimension in olfactory coding: a representation of the 2-deoxyglucose patterns of glomerular labeling in the olfactory bulb. *Brain Res.*, *240*: 341–344.

Jourdan, R., A. Duveau, L. Astic, and A. Holley (1980). Spatial distribution of [14C]2-deoxyglucose uptake in the olfactory bulbs of rats stimulated with two different odours. *Brain Res.*, *188*: 139–154.

Kent, P.F. (1990). The recording of odorant induced activity patterns from the olfactory mucosa using a voltage sensitive dye. Unpublished Ph.D. Thesis, SUNY Health Science Center, Syracuse.

Key, B., and P.P. Giorgi (1986a). Selective binding of soybean agglutinin to the olfactory system of xenopus. *Neuroscience*, *18*: 507–515.

Key, B., and P.P. Giorgi (1986b). Soybean agglutinin binding to the olfactory systems of the rat and mouse. *Neurosci. Lett.*, *69*: 131–136.

Kroczek, R.A., K.C. Gunter, R.N. Germain, and E.M. Shevach (1986). Thy-1 functions as a signal transduction molecule in T-lymphocytes and transfected B-lymphocytes. *Nature*, *322*: 181–184.

Laing, D.G., H. Panhuber, E.A. Pittman, M.E. Willcox, and G.K. Eagleson (1985). Prolonged exposure to an odor or deodorized air alters size of mitral cells in the olfactory bulb. *Brain Res.*, *336*: 81–87.

Lancet, D., C.A. Greer, J.S. Kauer, and G.M. Shepard (1982). Mapping of odor-related neuronal activity in the olfactory bulb by high-resolution 2-deoxyglucose autoradiography. *Proc. Natl. Acad. Sci. USA*, *79*: 670–674.

Lazarides, E. (1980). Intermediate filaments as mechanical integrators of cellular space. *Nature*, *283*: 249–256.

LeGros Clark, W. (1957). Inquiries into the anatomical basis of olfactory discrimination. *Proc. R. Soc. Med.*, *146*: 299–319.

Lidow, M.S., S.J. Kleene, and R.C. Gesteland (1986). Suppression of mitotic acitvity and synchronization of cell development in olfactory epithelium. *Brain Res.*, *393*: 145–162.

Lovell, M.A., B.W. Jafek, D.T. Moran, and J.C. Rowley, III (1982). Biopsy of human olfactory mucosa: An instrument and a technique. *Arch. Otolarygol.*, *108*: 247–249.

Mackay-Sim, A., P. Shaman, and D.G. Moulton (1982). Topographic coding of olfactory quality: Odorant-specific patterns of epithelial responsivity in the salamander. *J. Neurophysiol.*, *48*: 584–596.

Margolis, F.L. (1985). Olfactory marker protein: from PAGE band to cDNA clone. *Trends Neurosci.*, *8*: 542–546.

Margolis, F.L. (1988). Molecular cloning of olfactory-specific gene products. in F.L. Margolis and T.V. Getchall (Eds.) *Molecular Neurobiology of the olfactory system*. New York: Plenum, pp. 237–265.

Matus, A. (1988). Microtubule-associated proteins: Their potential role in determining neuronal morphology. *Ann. Rev. Neurosci.*, *11*: 29–44.

Meiri, K.F., K.H. Pfenninger, and M. Willard (1986). Growth associated protein, GAP-43, is a component of growth cones and corresponds to pp46, a major polypeptide of a subcellular fraction that is enriched in growth cones. *Proc. Natl. Acad. Sci. USA*, *83*: 3537–3541.

Meiri, K.F., M. Willard, and M.I. Johnson (1988). Distribution and phosphorylation of the growth associated protein GAP-43 in regenerating sympathetic neurons in culture. *J. Neurosci.*, *8*: 3537–3542.

Meiri, K.F., and P.R. Gordon-Weeks (1990). GAP-43 in growth cones is associated with areas of membrane that are tightly bound to substrate and is a component of a membrane skeleton subcellular fraction. *J. Neurosci.*, *10*: 256–266.

Meiri, K.F., and L.E. Bickerstaff, and J.E. Schwob (1991). Monoclonal antibodies show that Kinase C phosphorylation of GAP-43 during axonogenesis is both spatially and temporally restricted in vivo. *J. Cell Biol.*, *112*: 991–1005.

Menco, B.M. (1989). Electron-microscopic demonstration of olfactory-marker protein with protein G-gold in freeze-substituted, Lowicryl K11M-embedded rat olfactory-receptor cells. *Cell Tiss. Res.*, *256*: 275–281.

Menco, B.M., and A.I. Farbman (1985). Genesis of cilia and microvilli of rat nasal epithelia during pre-natal development. *J. Cell Sci.*, *78*: 283–310.

Miragall, F., G. Kadmon, M. Husmann, and M. Schachner (1988). Expression of cell adhesion molecules in the olfactory system of the adult mouse: Presence of the embryonic form of N-CAM. *Dev. Biol.*, *129*: 516–531.

Mitchison, T., and M. Kirschner (1988). Cytoskeletal dynamics and nerve growth. *Neuron*, *1*: 761–772.

Mollicone, R., J. Trojan, and R. Oriol (1985). Appearance of H and B antigens in primary sensory cells of the rat olfactory apparatus and inner ear. *Dev. Brain Res.*, *17*: 275–279.

Monti Graziadei, G.A. (1983). Experimental studies on the olfactory marker protein. III. The olfactory marker protein in the olfactory neuroepithelium lacking connections with the forebrain. *Brain Res.*, *262*: 303–308.

Monti Graziadei, G.A., F.L. Margolis, J.W. Harding, and P.P.C. Graziadei (1977). Immunocytochemistry of the olfactory marker protein. *J. Histochem. Cytochem.*, *25*: 1311–1316.

Mori, K. (1987). Monoclonal antibodies (2C5 and 4C9) against lactorseries carbohydrates identify subsets of olfactory and vomeronasal receptor cells and their axons in the rabbit. *Brain Res.*, *408*: 215–221.

Mori, K., S.C. Fujita, K. Imamura, and K. Obata (1985). Immunohistochemical study of subclasses of olfactory nerve fibers and their projections to the olfactory bulb in the rabbit. *J. Comp. Neurol.*, *242*: 214–229.

Mori, K., K. Imamura, S.C. Fujita, and K. Obata (1987). Projections of two subclasses of vomeronasal nerve fibers to the accessory olfactory bulb in the rabbit. *Neuroscience*, *20*: 259–278.

Mori, K., K. Imamura, and N. Onoda (1990). Signal processing in the rabbit olfactory bulb. *Proceedings, ISOT X*. In press.

Morris, R.J. (1985). Thy-1 in developing nervous tissue. *Dev. Neurosci.*, *7*: 133–160.

Morris, R.J., and P.C. Barber (1983). Fixation of Thy-1 in nervous tissue for immunohistochemistry: A quantitative assessment of the effect of different fixation conditions upon retention of antigenicity and the cross-linking of Thy-1. *J. Histochem. Cytochem.*, *31*: 263–274.

Moulton, D.G. (1976). Spatial patterning of response to odors in the peripheral olfactory system. *Physiol. Rev.*, *56*: 578–593.

Naessen, R. (1971). An enquiry of the morphological characteristics and possible changes with age in the olfactory region of man. *Acta Otolaryngol.*, *71*: 49–62.

Nakashima, T., C.P. Kimmelman, and J.B. Snow, Jr. (1984). Structure of human fetal and adult olfactory neuroepithelium. *Arch. Otolaryngol.*, *110*: 641–646.

Ophir, D., and D. Lancet (1988). Expression of intermediate filaments and desmoplakin in vertebrate olfactory mucosa. *Anat. Rec.*, *221*: 754–760.

Onoda, N. (1988a). Monoclonal antibody immunohistochemistry of rabbit olfactory receptor neurons during development. *Neuroscience*, *26*: 1003–1012.

Onoda, N. (1988b). Monoclonal antibody immunohistochemistry of degenerative and renewal pattern in rabbit olfactory receptor neurons following unilateral olfactory bulbectomy. *Neuroscience*, *26*: 1013–1022.

Onoda, N., and S.C. Fujita (1988). Monoclonal antibody immunohistochemistry of adult rabbit olfactory structures. *Neuroscience*, *26*: 993–1002.

Osborn, M., E. Debus, and K. Weber (1984). Monoclonal antibodies specific for vimentin. *Eur. J. Cell. Biol.*, *34*: 137–143.

Panhuber, H., and D.G. Laing (1987). The size of mitral cells is altered whem rats are exposed to an odor from their day of birth. *Dev. Brain Res.*, *34*: 133–140.

Panhuber, H., A. Mackay-Sim, and D.G. Laing (1987). Prolonged odor exposure causes cell shrinkage in the adult rat olfactory bulb. *Dev. Brain Res.*, *31*: 307–311.

Pedersen, P.E., B. Friedman, R. Smith, C.A. Greer, G.M. Shepherd, and S. Hockfield (1987). Correlations between 2DG activity and monoclonal antibody staining in the olfactory system of the fetal rat. *Soc. Neurosci. Abstr.*, *13*: 363.

Reinhard, E., R. Meier, W. Halfter, G. Rovelli, and D. Monard (1988). Detection of glia-derived nexin in the olfactory system of the rat. *Neuron*, *1*: 387–394.

Riggot, M.J., and J.W. Scott (1989). Lectin labeling of rat olfactory bulb glomeruli. *Soc. Neurosci. Abstr.*, *15*: 926.

Royet, J.P., G. Sicard, C. Souchier, and F. Jourdan (1987). Specificity of spatial patterns of glomerular activation in the mouse olfactory bulb: computer-assisted image analysis of 2-deoxyglucose autoradiogams. *Brain Res.*, *417*: 1–11.

Rutishauser, U., and T.M. Jessell (1988). Cell adhesion molecules in vertebrate neural development. *Physiol. Rev.*, *68*: 819–857.

Schoenfeld, T.A., L. McKerracher, R. Obar, and R.B. Vallee (1989). MAP 1A and MAP 1B are structurally related microtubule associated proteins with distinct developmental proteins in the CNS. *J. Neurosci.*, *9*: 1712–1730.

Schwarting, G.A., and J.E. Crandall (1989). Monoclonal antibodies to carbohydrate antigens identify subsets of olfactory and vomeronasal axons in the rat. *Soc. Neurosci. Abstr.*, *15*: 590.

Schwartz, M.A., D.M. Chikaraishi, and J.S. Kauer (1989). Characterization of the precursor population in mouse olfactory epithelium after bulbectomy. *Soc. Neurosci. Abstr.*, *15*: 749.

Schwob, J.E., and D.I. Gottlieb (1986). The primary olfactory projection has two chemically distinct zones. *J. Neurosci.*, *6*: 3393–3404.

Schwob, J.E., N.B. Farber, and D.I. Gottlieb (1986). Neurons of the olfactory epithelium in adult rats contain vimentin. *J. Neurosci.*, *6*: 208–217.

Schwob, J.E., and D.I. Gottlieb (1988). Purification and characterization of an antigen that is spatially segregated in the primary olfactory projection. *J. Neurosci.*, *8*: 3470–3480.

Schwob, J.E., K.E. Szumowski, and K.B. Brodie (1988). Differentiation of olfactory sensory neurons in the absence of their postsynaptic target. *Soc. Neurosci. Abstr.*, *14*: 1169.

Schwob, J.E., and K.E. Szumowski (1989). Olfactory sensory neurons are trophically dependent on the olfactory bulb for their prolonged survival. *Soc. Neurosci. Abstr.*, *15*: 749.

Sicard, G. (1985). Olfactory discrimination of structurally related molecules: Receptor cell responses to camphoraceous odorants. *Brain Res.*, *326*: 203–212.

Sicard, G., and A. Holley (1984). Receptor cell responses to odorants: similarities and differences among odorants. *Brain Res.*, *292*: 283–296.

Skene, J.H.P. (1989). Axonal growth-associated proteins. *Ann. Rev. Neurosci.*, *12*: 127–156.

Smart, I.H.M. (1971). Location and orientation of mitotic figures in the developing mouse olfactory epithelium. *J. Anat.*, *109*: 243–251.

Smith, C.G. (1936). Pathologic change in olfactory nasal mucosa of albino rats with "stunted" olfactory bulbs. *Arch. Otolaryngol.*, *25*: 131–143.

Stallcup, W.B., L.L. Beasley, and J.M. Levine (1985). Antibody against nerve growth factor—inducible large external (NILE) glycoprotein labels nerve fiber tracts in the developing nervous system. *J. Neurosci.*, *5*: 1090–1101.

Stewart, W.B., J.S. Kauer, and G.M. Shepherd (1979). Functional organization of rat olfactory bulb analysed by the 2-deoxyglucose method. *J. Comp. Neurol.*, *185*: 715–734.

Stewart, W.B., and P.E. Pedersen (1987). The spatial organization of olfactory nerve projections. *Brain Res.*, *411*: 248–258.

Talamo, B.R., R. Rudel, K.S. Kosik, V.M.-Y. Lee, S. Neff, L. Adelman, and J.S. Kauer (1989). Pathological changes in olfactory neurons in patients with Alzheimer's disease. *Nature*, *337*: 736–739.

Verhaagen, J., A.B. Oestreicher, W.H. Gispen, and F.L. Margolis (1989a). The expression of the growth associated protein B50/GAP 43 in the olfactory system of neonatal and adult rats. *J. Neurosci.*, *9*: 683–691.

Verhaagen, J., A.B. Oestricher, M. Grillo, Y.-S. Khew-Goodall, W.H. Gispen, and F.L. Margolis (1989b). Differential post-lesion expression of B-50/GAP-43 and olfactory marker protein in the olfactory system. *Soc. Neurosci. Abstr.*, *15*: 318.

Viereck, C., R.P. Tucker, and A. Matus (1989). The adult rat olfactory system expresses microtubule-associated proteins found in the developing brain. *J. Neurosci.*, *9*: 3547–3557.

Vollrath, M., M. Altmannsberger, K. Weber, et al. (1985). An ultrastructural and immunohistological study of the rat olfactory epithelium: Unique properties of olfactory sensory cells. *Differen.* *29*: 243–253.

Wallis, I., L. Ellis, K. Suh, and K.H. Pfenninger (1985). Immunolocalization of a neuronal growth-dependent membrane glycoprotein. *J. Cell Biol.*, *101*: 1990–1998.

Willard, M.B., K. Meiri, and M. Glicksman (1985). Changes of state during neuronal development: Regulation of axon elongation. In G.M., Edelman, W.E. Gall, and W.M. Cowan (Eds.), *Molecular Bases of Neural Development*, New York: Wiley, pp. 341–362.

Wysocki, C.J., K.M. Korries, and G.K. Beauchamp (1989). Ability to perceive androstenone can be acquired by ostensibly anosmic people. *Proc. Natl. Acad. Sci. USA*, *86*: 7176–7980.

5 Neuroanatomical Substrates of Olfaction

JOHN H. McLEAN and MICHAEL T. SHIPLEY

Introduction

In mammals, the olfactory system is important for reproduction, emotional responses, neuroendocrine regulation, and the recognition of conspecifics, predators, and prey. The linkages between olfaction and other neural functions present interesting questions. For example, the perception of flavor is a combination of olfactory and gustatory signals. What is the circuitry that integrates olfaction and taste to produce the integral conscious awareness of flavor? Odors are known to evoke strong memories; what are the neural substrates of this phenomenon? In some species, the odor of an estrous female is absolutely necessary to trigger sexual behavior in males. In other species, odors are important but not essential. What are the substrates linking olfactory stimuli to behavioral and neuroendocrine control circuits, and what differences in neural circuitry underlie differential dependence on odors for sexual behavior? An essential framework for understanding these issues is the neuroanatomy and neurochemical organization of the olfactory system and its linkages to other parts of the brain.

Before embarking on a more detailed account of olfactory circuitry, an overview of its major neural channels is presented. The first relay of olfactory information from the nose to the brain is in the olfactory bulb. The bulb contains output cells that convey olfactory information to higher-order olfactory structures and to other brain systems. The relay from the nose via the output cells of the bulb to other parts of the brain is influenced by local circuitry within the bulb and by centrifugal inputs from other regions of the central nervous system. Higher-order olfactory structures after the bulb include, from rostral to caudal, the olfactory peduncle (the anterior olfactory nucleus), the piriform cortex, the olfactory tubercle, the entorhinal cortex, and some amygdaloid nuclei. A more detailed account of the projections will follow. From these regions, further connections are made to regions that integrate the olfactory information so that (1) responses to olfaction can be elicited at the conscious memory level, (2) the information is processed so that we may distinguish odors or appreciate the flavor of something tasted, or (3) behavioral and/or visceral responses to odors are initiated. At first blush, this schematic of olfactory circuits appears to be relatively simple; thus it might seem that our understanding of the physiology of olfactory circuits may be comparable to that for the other sensory systems. However, this is not the case. There are a number of critical gaps in our knowledge of olfaction that have prevented the kind of integrative analysis of structure and function that has so greatly expanded our understanding of the visual, somatic sensory, and auditory systems. Foremost among these is our almost total ignorance of the nature of the "olfactory code." What are the relevant dimensions of olfactory stimuli that are extracted and processed by the olfactory system? Is there anything like finite "classes" of odors comparable to primary colors? At the present time this ignorance extends to the process of transduction itself. Are there specialized, odor-selective receptor sites on primary olfactory neurons, and, if so, how selective for different odors are such receptors? Are some olfactory receptor neurons concerned with general properties of odors (analogous to luminance and contrast for rods in the retina) while other receptor neurons are more specialized for classes of odors? Are the olfactory receptor neurons arranged in some orderly and specific way in the olfactory epithelium, and what is the relation between the location of primary neurons and their terminal projection in the bulb, i.e., is there any topographic organization of functions at any level of the olfactory system, or has the olfactory system evolved novel circuitry designed to analyze stimuli that do not impinge upon us in any intuitively obvious spatially packaged format?

This partial list of fundamental but unanswered questions gives some indication of the magnitude of the task of making sense of the anatomical wiring and chemical makeup of the olfactory system. At the present time, the best that can be hoped is that an understanding of the organizational principles of central olfactory circuits may provide some clues about, or impose some constraints upon, the ways in which multiple candidate "olfactory codes" might be accommodated by the actual hardware of the system. On the other hand, there is much to suggest that olfactory circuit organization and development has much in common with that of other neural systems, especially the cerebral cortex. The relatively simple (or primitive) cortical organization of the phylogenetically ancient olfactory system may contain important clues about the most basic principles of cortical organization. How such organization develops and the properties that have been conserved in more recently evolved cortical structures may provide insights about the hardware underlying fundamental computational features that have driven a selective expansion of cortical structures in the evolution of the mammalian brain.

This chapter reviews the connectional and chemical anatomy of the circuitry involved in olfaction. We concentrate more on some structures, such as the olfactory bulb, because that is where the important first olfactory relays take place and because it is the most thoroughly studied of all central olfactory structures. The discussion of connections is focused largely on those fairly directly related to olfactory circuitry. A detailed discussion of all secondary and tertiary olfactory connections would make the chapter unwieldy.

The input from the olfactory epithelium to the bulb is mediated by the olfactory receptor neuron (ORN). This cell is both the transducer and first order in olfaction. These neural receptor cells are discussed in greater detail elsewhere in this book. A final, but important, qualification is that most of the information presented in this chapter is based on data obtained from rats and hamsters, which have been the most thoroughly studied by experimental methods.

Cyto- and Chemoarchitecture of the Olfactory Bulb

In macrosmatic mammals (animals that rely on the sense of smell for survival), two subdivisions of the olfactory system are recognized, the main and accessory olfactory systems. These two subdivisions are parallel, yet, in many ways, separate in their integration of specific odor types. For example, ORNs in the olfactory epithelium transduce mostly volatile odors and transmit this information to the main olfactory bulb. By contrast, other ORNs located in the vomeronasal organ are exposed to odors during the activation of a norepinephrine-activated pump. Axons of vomeronasal neurons project exclusively to the accessory olfactory bulb, located at the dorsocaudal limit of the MOB. The central connections of these MOB and AOB to higher-order olfactory structures are parallel but essentially nonoverlapping, as will be discussed later. In contrast to macrosmatic mammals, microsmatic mammals such as humans either have no identifiable VNO-AOB or the VNO-AOB is only transiently present during fetal development (Humphrey, 1940; Macchi, 1951). Some mammals are anosmic (e.g., porpoises) and lack an olfactory bulb (Breathnach, 1960; Jacobs, McFarland, and Morgane, 1979). Indeed, the relative size of olfactory-related structures reflects the importance of olfaction to the animal. Thus, the olfactory bulb in humans is relatively small compared to the rest of the brain, whereas the rat, which depends heavily on olfaction for survival, has a relatively large olfactory bulb.

Main Olfactory Bulb

Olfactory Nerve Layer

In mammals, the main olfactory bulb is a highly laminated allocortex with several distinct layers that are easily visible in Nissl-stained sections. The most superficial layer is the olfactory nerve layer (onl), which contains axons from the primary olfactory neurons (Figure 5.1) and glial cells. The

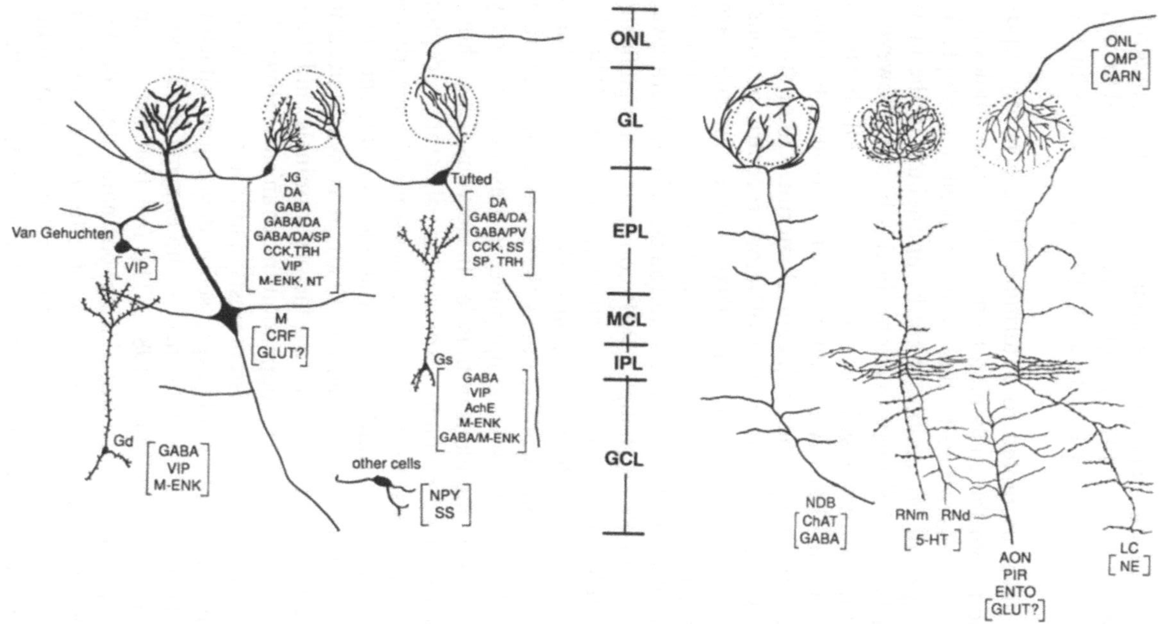

FIGURE 5.1. Left side. Diagram depicting some of the major neuronal types making up the main olfactory bulb and the putative neurotransmitters/neuropeptides found in these neurons (abbreviations found in Table 5.2). Right side. Diagram showing the major afferent inputs to the bulb and the constituent neurotransmitter/neuropeptides of these axons. Abbreviations: AON, anterior olfactory nucleus; CARN, carnosine; ChAT, choline acetyltransferase; DRN and MRN, dorsal and median raphe nuclei; ENTO, entorhinal cortex; epl, external plexiform layer; gcl, granule cell layer; GLUT, glutamate; gl, glomerular layer; ipl, internal plexiform layer; LC, locus coeruleus; mcl, mitral cell layer; NDB, nucleus of the diagonal band; NE, norepinephrine; OMP, olfactory marker protein; onl, olfactory nerve layer; PIR, piriform cortex; 5-HT, serotonin.

glial cells are thought to be derived from both the peripheral and central nervous system. At the olfactory nerve–glomerular interface, a specialized astrocyte, termed the "ensheathing cell" (Doucette, 1988), provides an incomplete wrapping that separates the glomeruli from the periglomerular region (Pinching and Powell, 1971b). The axons of primary olfactory neurons are thin, unmyelinated, and may terminate in several glomeruli (Cajal, 1911; Pinching and Powell, 1971b). Olfactory axons are said to only give off, but not receive (Pinching and Powell, 1971b), synapses in the glomeruli.

Numerous studies have attempted to determine if there is any topography in the projection from the epithelium to the bulb. These studies are reviewed in greater detail elsewhere in this book (Nickell and Shipley). There is some topography of projections from the epithelium to the bulb (see references in Nickell and Shipley's chapter in this book). However, there is also considerable diffuseness in the projection as well. Neurons from a specific region of the olfactory epithelium project strongly to a few glomeruli and diffusely to many other adjacent glomeruli. Other studies have explored the possible transmitter(s) of olfactory receptor neurons, but there is no definitive transmitter to date. Two soluble molecules are relatively specific to ORNs. These molecules include olfactory marker protein (OMP) (Margolis, 1980) and carnosine (Sakai, Kani, Karasawa, Yoshida, and Nagatsu, 1988). Although OMP constitutes a significant proportion of the protein content of the olfactory epithelia cell (Hirsch, Grillo, and Margolis, 1978; Hirsch and Margolis, 1979), there is no established physiological role for this protein, nor is there any indication that OMP is released by ORNs. Carnosine, a dipeptide, is present in a variety of tissues and is particularly abundant in muscle. In nervous tissue, however, carnosine appears to be uniquely expressed in olfactory receptor neurons (Ferriero and Macrides, 1975). Carnosine has some characteristics of a neurotransmitter candidate (Burd, Davis, Macrides, Grillo, and Margolis, 1982; Gonzalez-Estrada and Freeman, 1980): it is synthesized in ORNs, it is transported by rapid transport to olfactory terminals, and there is evidence that it is released by Ca^{++}-dependent mechanisms. However, to date, there is no consensus for a consistent postsynaptic action of carnosine. Adenosine deaminase is transiently expressed in developing olfactory epithelial cells, but the functional significance of the enzyme in ORNs is unknown (Senba, Daddona, and Magy, 1987). The fact that the enzyme decreases in olfactory-related structures during postnatal development suggests that it could be related to maturation of olfactory circuitry. In summary, although there are several selective markers for olfactory epithelial input to the bulb, definitive identification of a neurotransmitter is lacking.

Markers for some established transmitters have been observed in axons in the olfactory nerve layer. Occasional fibers containing CGRP (Rosenfeld et al., 1983) or substance P (Baker, 1986) have been reported in the onl, but the sources of these axons are not known. Axons containing substance P are found in the olfactory nerve (Baker, 1986) but these axons do not derive from the raphe nuclei where they might be expected to originate (Magoul, Oblin, Calas, and Araneda, 1988). In contrast, some axons from the raphe nuclei have been observed to course through the olfactory nerve to enter the olfactory bulb in tract-tracing studies, although the fibers were not labeled with antibodies to serotonin, a major transmitter of raphe neurons (McLean and Shipley, 1987a). Since the trigeminal ganglion contains neurons that colocalize CGRP and substance P (Uddman, Edvinsson, Ekman, Kingman, and McCulloch, 1985), it is possible that the substance P and CGRP fibers observed in the olfactory nerve layer are trigeminal axons, perhaps associated with blood vessels. Taken together, these studies indicate that although the olfactory nerve is composed predominantly of input from the olfactory epithelium to the olfactory bulb, the nerve may contain a few other fibers originating from other parts of the CNS or from PNS neurons.

Glomerular Layer

The next deepest layer of MOB, the glomerular layer, is one of the most distinctive in the bulb. The glomeruli are composed of cell-poor regions that are surrounded by a distinctive shell of small neurons and glia cells. The glomeruli are generally ovoid and range from 80 to 160 μm in diameter.

Glomerular Shape

It has been suggested that the shape of the glomeruli is influenced by the epithelial input (Monti Graziadei and Graziadei, 1984), since glomerular-shaped structures can be formed in brain sites where they normally are not found if chunks of olfactory epithelium are transplanted into host brain tissue. The Graziadeis suggest that the formation and shape of a glomerulus is a consequence of interactions among olfactory axons and that there is little influence from host tissue (Graziadei and Monti Graziadei, 1986). However, recent studies support the idea that glia may be very important in the formation and shape of glomeruli: in the absence of glia, glomeruli in the silk moth, *Manduca sexta*, are not formed (Oland, Tolbert, and Mosmann, 1988). Further, in mammals there is evidence that the condensation of olfactory axons into glomeruli is preceded by the formation of a protoglomerular structure by immature astrocytes (Bailey, Poston, and Shipley, 1989). Earlier in development, before the astrocytes form protoglomeruli, olfactory axons grow past the glomerular layer and ramify with no apparent order in the olfactory bulb. Thus, some factors of the target appear to be important in glomerular formation.

TABLE 5.1. Distribution of cells in MOB

Layer	Neuron types	Sizes	Cell density	Reference
GL	Periglomerular	5–8 μm	High	Pinching and Powell (1971b)
	Short axon	8–12 μm	Low	Pinching and Powell (1971b)
EPL	External tufted	10–15 μm	Moderate	Pinching and Powell (1971b)
	Middle tufted	15–18 μm	Low	Pinching and Powell (1971b)
	Internal tufted	18–25 μm	Low	Pinching and Powell (1971b)
	Van Gehuchten	12–17 μm	Low	Segal and Landis (1974)
MCL	Mitral	x = 22 μm	High	Segal and Landis (1974)
	Granule			
	Van Gehuchten	12–17 μm		Segal and Landis (1974)
IPL	Blanes	Small	Low	Rall, Shepherd, Reese, and Brightman (1966)
	Horizontal cell of Cajal			
GCL	Granule	10–16 μm	High	Segal and Landis (1974)
	Blanes	16–23 μm	Low	Segal and Landis (1974)
	Golgi	12–22 μm	Low	
EZ	Proliferating cells		High	

Abbreviations: GL, glomerular layer; EPL, external plexiform layer; MCL, mitral cell layer; IPL, internal plexiform layer; GCL, granule cell layer; EZ, ependymal zone.

Glomerular Number

Most estimates of glomerular number are similar, around 2000 glomeruli/bulb for rabbits (Allison and Warwick, 1949) and mice (Allison, 1953; Brunjes, 1983; Royet, Souchier, Jourdan, and Ploye, 1988; White, 1972). The number of glomeruli in rats has been estimated at 3000 (Meisami and Safari, 1981). This means that several million olfactory receptor cells send axons to a total of 2000 to 3000 glomeruli. Thus the glomerulus, the site where the first integration of olfactory information takes place in the CNS, is a site of convergence of primary olfactory axons (see also Nickell and Shipley book chapter).

Glomerular Cell Types and Circuitry

The glomeruli are surrounded by a thin wall of neurons and glial cells. The core of the glomeruli contains a rich plexis of axons and dendrites. Three principal neurons are present in the glomerular walls—periglomerular cells, slightly larger external tufted cells, and short axon cells (Pinching and Powell, 1971a). The short axon cells are often grouped with the periglomerular cells (Table 5.1). Together, the neurons that surround glomeruli are sometimes referred to as juxtaglomerular cells. The observations of Blanes (1898), Golgi (1875), and Cajal (1911) showed the distribution of juxtaglomerular cell dendrites and axons. These studies were in agreement that axons of external tufted cells did not penetrate glomeruli but, rather, ended between glomeruli. The small periglomerular cells have short axons that ramify within the same glomerulus or project to periglomeruli positions a few glomeruli away (Cajal, 1911). The larger juxtaglomerular cells (external tufted) project to other regions of the ipsilateral olfactory bulb (Schoenfeld, Marchand, and Macrides, 1985). The dendrites of juxtaglomerular neurons either ramify in glomeruli or surround glomeruli in the periglomerular region (Cajal, 1911; Macrides and Davis, 1983; McLean, Shipley, and Bernstein, 1989). Pinching and Powell (1971a) performed a thorough analysis of glomerular cell types and circuitry using Golgi-stained material and serial section ultrastructural analysis. They determined that the dendrites of small (5–8 μm) periglomerular cells penetrate more than one glomerulus but usually are preferential to one. The dendrites rarely fill the glomeruli and usually have many spinelike appendages. The short axon cells are somewhat larger than the periglomerular cells (8–12 μm) and are distinguished in Golgi material because their dendrites are entirely periglomerular. The axons of periglomerular or short axon cells usually go into or along the edge of up to 2 to 4 glomeruli. The external tufted cells are 10 to 15 μm in long axis, lie in the periglomerular region, and have usually one primary dendrite that arborizes in a segment of a single glomerulus. The dendrites are varicose but do not have typical spines. The axons of these cells generally give off several collaterals to the periglomerular region and then may pass to the external plexiform layer. Further connections of the external tufted cells are discussed later. There are several other neuronal components to glomeruli besides the dendrites of juxtaglomerular cells and axons of periglomerular cells and primary olfactory neurons. These include dendrites of deeper tufted and mitral cells and axons from central centrifugal sources (detailed later). The studies of Pinching and Powell (1971b) determined that there is direct contact between axons of primary olfactory neurons (PON) and juxtaglomerular cells, although further studies are required to determine whether olfactory axons terminate differentially on different classes of periglomerular cells that contain different transmitters. Olfactory axons also synapse densely upon the dendrites of mitral and tufted cells. The dendrites of mitral/tufted cells and periglomerular cells have very special synaptic relationships in the glomeruli. Reconstructions of serial sections revealed that the mitral/tufted dendrites, which are indistinguishable in normal EM material, make reciprocal synaptic contacts with the dendrites and gemmules (spinelike processes) of periglomerular cells. These specialized reciprocal synapses are often closely associated with each other (Pinching and Powell, 1971b). Our understanding of synaptic organization in the glomeruli is far from complete because in the past decade several neurotransmitters/ neuromodulators have been identified in juxtaglomerular cells. This means that the synaptic organization is probably more complicated than portrayed by original EM studies of synapses in the glomeruli because neurons of different transmitter phenotypes may differentially synapse with mitral/ tufted cells, olfactory axons, or other periglomerular neurons. Next, therefore, we review the different types of cells in MOB based on their neurochemical makeup as inferred from immunocytochemical, histochemical, and in situ hybridization studies.

A list of immunocytochemically identified cells in the main olfactory bulb is provided in Table 5.2 and are represented in diagram form in Figure 5.1. Starting with the glomerular layer, many juxtaglomerular cells are dopaminergic (Halász, Johansson, Hökfelt, Ljundahl, and Goldstein, 1981; Davis and Macrides, 1983) or GABAergic (Ribak, Vaughn, Saito, Barber, and Roberts, 1977). In the hamster, about 70 percent of the DA neurons are reported to also contain GABA, while about 45 percent of GABAergic cells contain dopamine (Kosaka, Hataguchi, Hama, Nagatsu, and Wu, 1985). In the rat, there also appears to be immunocytochemical colocalization of GABA and tyrosine hydroxylase in the juxtaglomerular cells (Gall, Hendry, Seroogy, and Jones, 1985), although at least one group has not found evidence for colocalization of the two markers in juxtaglomerular cells (Mugnaini, Oertel, and Wouterlood, 1984). Almost all substance P immunoreactive neurons in MOB of hamsters have been reported to contain both GABA and dopamine based on the presence of immunocytochemical markers for these transmitters in the same cells (Kosaka, Hama, Nagatsu, Wu, and Kosaka, 1988). Thus, some juxtaglomerular cells may contain both a catecholamine (dopamine), an amino acid inhibitory transmitter (GABA), and probably also an excitatory neuropeptide (substance P). In addition, at least one group has reported that met-enkephalin is present in periglomerular cells in the rat (Davis, Burd, and Macrides, 1982). A few juxtaglomerular cells contain vasoactive intestinal polypeptide (Gall, Seroogy, and Brecha, 1986; Sanides-Kohlrausch and Wahle, 1990). Some periglomerular cells with short axons that project to the deeper granule cell layer contain NADPH-diaphorase, neuropeptide-Y, and somatostatin, which Scott and collaborators (Scott, McDonald, and Pemberton, 1987) suggested may provide a direct route for periglomerular cells to influence granule cells. Also reportedly present in juxtaglomerular cells are cholecystokinin (Seroogy, Brecha, and Gall, 1985; Matsutani, Senba, and Tohyama, 1988), aspartic acid (Halasz, 1987), thyrotropin-releasing hormone (Merchenthaler, Csernus, Csontos, Petrusz, and Mess, 1988; Tsuruo, Hokfelt, and Visser, 1988), and protein kinase C (Saito, Kikkawa, Nishizuka, and Tanaka, 1988). In addition, acetylcholinesterase histochemistry has revealed a subpopulation of potential cholinoceptive neurons in the juxtaglomerular region (Nickell and Shipley, 1988).

At the present time, therefore, there are several subclasses of juxtaglomerular neurons. These neurochemical markers serve to identify potentially different cell types, but little is known of the functional significance of these different transmitters in olfactory processing. Because most juxtaglomerular cells are very small (5–12 μm) and because most of them have only local circuit connections, their physiological characteristics have been difficult to study.

Transient Expression of Peptides/Enzymes During Development

Some markers for some transmitters in juxtaglomerular cells are transiently expressed during olfactory bulb development. For example, Matsutani, Senba, and Tohyama (1988) showed that neurotensin-IR (NT-IR) periglomerular cells in MOB are most densely stained prenatally and for the most part are not visibly stained after postnatal day 10 (PND 10). The NT-IR cells are concentrated around a specialized region termed the modified glomerular complex (Teicher, Stewart, Kauer, and Shepherd, 1980). The transient expression of NT-IR in these neurons is interesting because the modified glomerular complex is thought to be involved in suckling, and the NT-IR label disappears around the time that suckling ceases (Matsutani, Senba, and Tohyama, 1988).

Substance P-IR staining of cells in the output cell layer of AOB also appears to have a transient expression during development that disappears after PND 10. It is interesting that mRNA for substance P is present in the mitral cells of MOB in adult rats (Warden and Young, 1988), but substance P product in mitral cells is not detectable by immunocytochemistry in the rat at any age. Perhaps the transcripts for substance P are produced without the actual expression of substance P in the cell or there is extensive posttranslation modification of the peptide, or perhaps the peptide is rapidly released so that there is no reserve pool to be detected by immunocytochemical staining. Substance P *is* observed by immunocytochemistry in periglomerular neurons in the hamster olfactory bulb (Baker, 1986), but not in the rat.

External Plexiform Layer

Immediately subjacent to the glomeruli is a layer of relatively low cell density but with a very rich neuropil, the external plexiform layer (epl). Golgi-stained sections reveal that most of the dendrites located in this layer belong to mitral/tufted and granule cells. The cells that are located in epl are superficial, middle, and deep tufted cells, named according to their relative depth in epl (Table 5.1). Tufted cells are larger than most periglomerular cells but generally smaller than mitral cells. There appears to be a gradual increase in tufted cell size from the superficial to the deep parts of the epl (Pinching and Powell, 1971a; Switzer, de Olmos, and Heimer, 1985). Cajal originally described the tufted cells as displaced mitral cells, since they generally had the appearance of mitral cells (shaped like a bishop's mitre). The middle and deep tufted cells have projections similar to those of the mitral cells (Schoenfeld and Macrides, 1984); thus, they may functionally be considered part of one group, the output cells of the bulb. The axons of the superficial (external) tufted cells project mainly to other sites in the same olfactory bulb. Middle and deep tufted (in epl near mitral cells) and mitral cells also have local collaterals in the ipsilateral bulb, but most of them appear to project out of the olfactory bulb to the

TABLE 5.2. Putative neurotransmitter/peptide content of cells in MOB

Peptide/ neurotransmitter	Cell type (or location)	Cell size	Cell number	Species	Reference
Aspartate/glutamate	Small mitral	90 μm^2	Few	Rat	1, 2
Cholecystokinin (CCK)	Juxtaglomerular (JG)	11	Few	Rat	3
	External tufted	17	Many	Rat	3
	Middle tufted	19	Many	Rat	3
	Internal tufted		Few	Rat	3
	Deep short axon		Rare	Rat	3
	Vertical cells of Cajal		Rare	Rat	3
Dopamine (DA)	Juxtaglomerular	8–13	Many	Rat, hamster, turtle, mice	4–11
met-enkephalin (M-ENK)	GL—PG	Small	Many	Hamster	12, 13
	GCL—granule cells	Small	<GL		13
GABA	GL—PG	Small	Many	Rat, hamster	14, 15
	GCL—granule cell	8–10 μm	Many	Rat	16
GABA/DA	PG	Small	Several	Rat	17, 18, not 19
GABA/parvalbumin	EPL—external tufted	Small	Few	Rat	20
GABA/enkephalin	GCL—granule cell	Small	Rare	Rat	20
GABA/SP/DA	PG	Small	<GABA/DA	Hamster	17
NAAG	Mitral	No report	Many	Rat	21–23

NADPH-diaphorase	Short axon			Rat	24
NPY	Deep GCL—short axon, possibly commissural	14–32 μm	Few	Rat, human, cat, marmoset	15, 25, 26
	PG—short axon	14–32 μm	Few	Rat, human	15, 25
SP	External tufted (juxtaglomerular and EPL)	12–17	Many	Hamster, but none in rat, mouse, guinea pig, rabbit, cat	12, 27
TRH	PG, tufted	10 μm	10–15% of PG cells	Rat	28, 29
VIP	PG	<10 μm	Few	Cat	26
	EPL—middle tufted (vert. cells of Cajal) (Van Gehuchten) GCL, IPL	12 μm	Few	Rat, cat	25, 26

*Immunoreactivity decreases with postnatal development.

Abbreviations: PG, periglomerular; EPL, external plexiform layer; GCL, granule cell layer; GABA, β-aminobutyric acid; NAAG, *N*-acetylaspartylglutamate; NPY, neuropeptide Y; SP, substance P, TRH, thyrotropin-releasing hormone; VIP, vasoactive intestinal polypeptide.

References: 1, Fuller and Price (1988); 2, Halász (1987); 3, Shepherd (1972); 4, Baker (1986); 5, Baker et al. (1983); 6, Baker et al. (1984); 7, Baker et al. (1988); 8, Davis and Macrides (1983); 9, Gall et al. (1985); 10, Halasz et al. (1981); 11, McLean and Shipley (1988); 12, Davis et al. (1982); 13, Matsutani et al. (1989a); 14, Jacobs et al. (1979); 15, Ohm et al. (1988); 16, Ribak et al. (1977); 17, Kosaka et al. (1988); 18, Kosaka et al. (1985); 19, Mugnalni et al. (1984); 20, Kosaka et al. (1987); 21, Blakely et al. (1987); 22, ffrench-Mullen et al. (1985); 23, Allison and Warwick (1949); 24, Scott et al. (1987); 25, Gall et al. (1986); 26, Sanides-Kohlrausch and Wahle (1990); 27, Kream et al. (1984); 28, Merchenthaler et al. (1988); 29, Tsuruo et al. (1988).

anterior olfactory nucleus and other rostral olfactory cortical structures (Schoenfeld and Macrides, 1984; Schoenfeld and Scott, 1986; Marchand and Macrides, 1985). The dendritic morphologies of tufted cells vary, but generally they have at least one dendrite that enters a glomerulus and other secondary dendrites that run tangentially in the epl (Scott, 1986). The secondary dendrites of tufted cells are thought to form reciprocal synapses with granule cells (Shepherd, 1972).

A second class of neurons in the epl are the Van Gehuchten cells. These cells are characterized by two or more thick primary dendrites that remain in the epl. In addition, axons from these cells terminate around mitral and tufted cells. Many of these Van Gehuchten cells stain positively for vasoactive intestinal polypeptide (Sanides-Kohlrausch and Wahle, 1990) and calcium binding protein (Shipley, in preparation).

With the advent of immunocytochemical markers over the last decade, several neurochemically distinct classes of tufted cells have been identified. Table 5.2 shows the different transmitter/peptides observed in tufted cells. A brief description of different neurochemical markers found in tufted cells is included below with the description of mitral cells.

Mitral Cell Layer

Deep to the external plexiform layer is the mitral cell layer. This is a thin layer that contains the somata of mitral cells (25–35 μm diameter) arranged in almost a monolayer. These cells are the principal output cells of the bulb and, with some minor species differences (cf. Scott, 1988), have one apical dendrite that enters a single glomerulus, branches extensively there, and is contacted by olfactory axons or other central inputs (Shepherd, 1972). As it courses through the EPL, the apical dendrite of mitral and tufted cells appears to receive very few synapses, and the main shaft of the apical dendrite appears to be insulated by glial sheaths (Shipley and Zahm, 1990). The secondary dendrites of mitral cells may extend up to 2 mm in the EPL and are oriented parallel to the surface of the bulb (see also Nickell and Shipley, this book). These secondary dendrites engage in dendro-dendritic synapses with dendrites of granule cells and possible other inputs (Jackowski, Parnavelas, and Lieberman, 1978; Rall, Shepherd, Reese, and Brightman, 1966). The primary dendrites of mitral cells appear to enter only one glomerulus; the dendrites of approximately 25 mitral cells enter each glomerulus (Cajal, 1911). The physiological significance of this anatomical arrangement is discussed in the chapter on physiology of the olfactory bulb (Nickell and Shipley). A proposed transmitter for mitral cells is N-acetyl-aspartyl-glutamate (NAG) (Blakley, Ory-Lavolle, Grzanna, Koller, and Coyle, 1987; ffrench-Mullen, Koller, Zaczek, Coyle, Hori, and Carpenter, 1985), which has been observed exclusively in mitral cells using immunocytochemistry (Blakley, Ory-Lavolle, Grzanna, Koller, and Coyle, 1987). However, a recent neurophysiology study has cast doubt on a transmitter role for NAG (Whittemore and Koerner, 1989). Recently, corticotropin releasing factor (CRF) has been demonstrated in mitral and some tufted cells using both immunocytochemistry and in situ hybridization in the rat (Imaki, Nahon, Sawchenko, and Vale, 1989). CRF fibers were also observed in the molecular layer of the piriform cortex. This finding is consistent with CRF being a releasable neural peptide in mitral cells, since mitral cells synaptically terminate in the molecular layer of the piriform cortex. A similar localization of CRF has been reported in the squirrel monkey, suggesting that this peptide may be used as a transmitter/modulator in many mammals (Bassett, Foote, and Shipley, 1989). A few unusually small mitral cells appear to contain aspartate and project to the piriform cortex (Fuller and Price, 1988).

Many of the middle tufted cells are reported to contain vasoactive intestinal polypeptide in the rat (Gall, Seroogy, and Brecha, 1986), but in the cat, this peptide appears to be in van Gehuchten cells and not tufted cells (Sanides-Kohlrausch and Wahle, 1990). A large population of tufted cells (superficial to deep) contain the neuropeptide cholecystokinin (CCK) (Seroogy, Brecha, and Gall, 1985). One study using in situ hybridization detected substance P transcripts in some external tufted and up to half of the mitral cells in MOB of rat (Warden and Young, 1988), but to date no studies using immunocytochemistry have detected substance P in mitral cells in any species (Baker, 1986; Inagaki et al., 1982; Schultz, Quirion, Chronwall, Chase, and O'Donahue, 1984). Even in the hamster, which has many substance P juxtaglomerular cells in MOB, substance P is not present in the mitral/tufted cells, at least as detected by immunocytochemistry (Baker, 1986).

Internal Plexiform Layer

Immediately subjacent to the mitral cell layer is the internal plexiform layer (ipl). The ipl is another layer with a low density of cells but with many axons and dendrites (Table 5.1). Golgi studies indicate that the ipl contains the axons of mitral cells, dendrites of granule cells, and axons from other unidentified sources. Recent immunocytochemical/tracing studies show that at least some of the axons in ipl originate from raphe nuclei (McLean and Shipley, 1987), locus coeruleus (McLean et al., 1989), and the diagonal band (Shipley et al., in preparation) (see "Neurochemistry of Organization and Central ('Centrifugal') Afferents to the Olfactory Bulb" for further discussion).

Granule Cell Layer

The granule cell layer is the deepest neuronal layer in the bulb and contains many small (8–10 μm in long axis) granule neurons arranged in rowlike clusters of tightly packed somata. Granule cells have also been found mixed with mitral cells in the mitral cell layer. Golgi studies indicate that granule cells have no axons. They have basal dendrites that ramify in the gcl and a larger apical dendritic formation that enters and ramifies extensively in the EPL. Generally, the more superficial the granule cell is located in the gcl, the more superficially located are its apical dendritic ramifications in the external plexiform layer, while granule cells deeper in gcl have dendrites that ramify in the deeper parts of epl (Mori, Kishi, and Ojima, 1983; Orona, Scott, and Rainer, 1983; Scott, 1986). However, not all granule cells follow this pattern. The exceptions include (1) granule cells having distal dendrites that project deeper, toward the center of the bulb (Schneider and Macrides, 1978), (2) granule cells with dendrites that do not reach the external plexiform layer, and (3) large, nongranule cells in the granule cell layer; these cells are called Blanes cells (Table 5.1).

At the reciprocal synapse between the dendrite of the granule cell and the secondary dendrites of mitral/tufted cells, the mitral (and presumably tufted) cell excites the granule cell dendrite (Shepherd, 1972). Most of the granule cells contain GABA (Ribak, Vaughn, Saito, Barber, and Sato, 1977), which inhibits the mitral and tufted cells (Jahr and Nicoll, 1982; Shepherd, 1972). Some peptides have also been localized in other cells in the granule cell layer. For example, a few cells containing neuropeptide-Y (NPY) are found in this layer in the rat (Gall, Seroogy, and Brecha, 1986) and human (Ohm, Braak, Probst, and Weindl, 1988). In the rat, the axons of NPY cells appear to ramify in the more superficial layers such as the glomerular layer (Gall, Seroogy, and Brecha, 1986). These neurons are probably short-axon cells as opposed to granule cells. Some inframitral cells may be cholinoceptive, i.e., they contain acetylcholinesterase, but they are not cholinergic (Nickell and Shipley, 1988) because they do not possess choline acetyltransferase, the synthetic enzyme for acetylcholine. Thus, these cells may receive cholinergic input and hydrolyse ACh but may not synthesize acetylcholine themselves. These candidate cholinoceptive cells are located just deep to the mitral cell layer in the internal plexiform layer and in the superficial parts of the pranule cell layer.

There is evidence that inframitral cholinoceptive neurons are excited by cholinergic synaptic inputs from the basal forebrain (described later in this chapter) and that this cholinoceptive interneuron excites granule cells, which leads to mitral cell inhibition (Nickell and Shipley, 1988). Many of these cholinoceptive cells are large and similar in size and location to Blane cells (Macrides and Davis, 1983; Cajal, 1911).

Ependymal Zone

The deepest layer in the main olfactory bulb is the subependymal/ ependymal zone, a cell-poor region in the adult. The cells in this layer line the ventricle (if present), and during development the progenitors of many olfactory bulb cells are found in this zone, although other olfactory bulb neurons (granule cells) may be generated in more caudal forebrain regions and then migrate into the bulb in what has been termed "the rostral migratory stream" (Altman, 1969; Kishi, 1987; Shimada, 1966).

Influence of the Olfactory Nerve on Bulb Neurons

Input from the olfactory epithelium is important for the normal function of many types of glomerular neurons. For example, the olfactory nerve influences the expression of the DA phenotype in dopaminergic juxtaglomerular neurons. If the olfactory epithelium is destroyed by detergent or $ZnSO_4$, or if functional activity in the pathway is perturbed by closure of the nares, then tyrosine hydroxylase (TOH) immunoreactivity (DA marker) is lost in juxtaglomerular neurons. This phenomenon of transneuronal regulation of transmitter phenotype has been demonstrated in adult rats (Kawano and Margolis, 1982; Baker, Kawano, Margolis, and John, 1983; Baker, Kawano, Albert, Joh, Reis, and Margolis, 1984), mice (Nadi, Head, Grillo, Hempstead, Grannot-Reisfeld, and Margolis, 1981; Baker, Kawano, Margolis, and Joh, 1983), dogs (Nadi, Head, Grillo, Hempstead, Grannot-Reisfeld, and Margolis, 1981), hamsters (Kream, Davis, Kawano, Margolis, and Macrides, 1984), and developing rats (McLean and Shipley, 1988). The loss of TOH immunoreactivity is not necessarily due to cell death because the neurons can be detected with antibodies to another DA enzyme (Baker et al., 1984). Thus, these cells appear to require the epithelial input in order to maintain TOH and, hence, express their DA phenotype. Epithelial input is also required in order to trigger the initial developmental expression of TOH (McLean and Shipley, 1988). Thus, both the developmental induction and maintenance of the DA phenotype depends on the presence and normal function of the olfactory nerve.

It is not known if the influence of olfactory epithelial input on the DA neurons is by direct synaptic contact or through less direct means such as transneuronal regulation of mitral/tufted cell dendrites, which could then influence dopamine expression in the PG cells. As mentioned above, Pinching and Powell (1971c) showed that there is direct contact of PON axons with PG cells; however, it is not known whether olfactory axons synapse on dopaminergic juxtaglomerular cells. Recently it has been demonstrated that the predominant DA receptor in the bulb is of the D_2 subtype. Surprisingly, this receptor is localized not only in the glomerular layer as expected, but equally densely in the olfactory nerve layer (Nickell, Norman, Wyatt, and Shipley, 1990). When the olfactory epithelium is destroyed by $ZnSO_4$, all D_2 receptor binding is eliminated in both the olfactory nerve layer and the glomeruli (Shipley, unpublished observations). These results raise the interesting possibility that the DA neurons are presynaptic to olfactory nerve terminals. While it has been presumed for two decades that olfactory nerve terminals do not receive presynaptic contacts, it is possible that contacts by dendrites releasing DA onto olfactory terminals might not exhibit a synaptic morphology and thus might have escaped detection in the classical EM studies of Pinching and Powell (1971b). If the D_2 receptors are predominantly localized on olfactory nerve endings, the loss of tyrosine hydroxylase in juxtaglomerular DA neurons following deafferentation might reflect a down regulation of DA synthesis related to the DA neurons having lost their postsynaptic targets.

Although the phenotypic expression of juxtaglomerular DA cells and external tufted substance P-IR cells in the hamster (Kream, Davis, Kawano, Margolis, and Macrides, 1984) is clearly influenced by olfactory nerve lesions, the expression of GABA in periglomerular neurons is not affected (Baker, Towle, and Margolis, 1988). Since it is claimed that about 70 percent of the DA periglomerular cells also contain GABA (Kosaka et al., 1987), it is possible that the DA synthetic pathway of the cell is preferentially dependent on the olfactory nerve. New results indicate that the influence of the olfactory nerve also extends to the expression of CCK in tufted cells and to CRF in mitral cells (Shipley et al., in preparation).

The influence of the olfactory nerve on other cell types in the bulb has been studied in another way. Laing and colleagues (1985) found that exposing animals to pure air resulted in decreased size of mitral cells, while exposure to specific odors over prolonged periods caused regional increases in mitral cell size. These findings were interpreted to suggest that odor quality may be coded at the level of the mitral cells (Panhuber and Laing, 1987).

Light and dark granule cells have been described (Struble and Walters, 1982), and these two subpopulations differ in their reaction to unilateral nares occlusion (Skeen, Due, and Douglas, 1985; Frazier and Brunjes, 1988); the lighter granule cells found in the deeper regions of gcl are first to be affected by nares closure. Frazier and Brunjes (1988) speculate that the lighter cells are affected first because they arrive first in gcl. Nares occlusion reduces the number of both light and dark granule cells, and it has been speculated that the reduction is due to cell death rather than decreased cell proliferation (Kaplan, McNelly, and Hinds, 1985; Frazier and Brunjes, 1988), since the granule cells continue to be born in the proliferative subependymal zone (Frazier and Brunjes, 1988).

Accessory Olfactory Bulb

The accessory olfactory bulb (AOB) has some similar cytoarchitectural features to the MOB, but is much smaller. It is located at the caudal-dorsal end of the MOB. The vomeronasal nerve transmits information from the VNO to the glomeruli of the AOB. The glomerular layer (agl) is less distinct in the AOB than in the MOB because the AOB glomeruli are fewer and smaller. In addition, the periglomerular cells are far fewer than in the MOB, with the result that the glomeruli are not so neatly delineated by a shell of cell bodies. The term "periglomerular" is less accurate in the AOB than in the MOB because in the AOB, the few periglomerular cells tend to

TABLE 5.3. Putative neurotransmitter/peptide content of cells in AOB

Peptide/ neurotransmitter	Cell type	Cell size	Cell number	Species	Reference
Aspartate	Output	Medium	Many	Rat	1
Dopamine (DA)	JG cell	Small	Rare	Rat, hamster, rabbit, cat, guinea pig, mouse	2, 3
GABA	PG-granule	Small	Many	Mouse	4
met-enkephalin	GL-GCL-small	Small	Fewer than in MOB	Rat, hamster	3, 5
Neurotensin	Output	Medium	Many	Guinea pig	6
Substance P	Output	10–15 μm	Several	Rat*	5
	Output		Many	Hamster	2
	GCL-granule	Medium 9 μm	Numerous	Rat, hamster	2, 3, but not 7
VIP	Mitral/AEPL (Van Gehuchten)	12 μm	Few	Cat	8

*Cell number decreases during postnatal development.
Abbreviations: JG, juxtaglomerular; PG, periglomerular; GCL, granule cell layer; VIP, vasoactive intestinal polypeptide.
References: 1, Fuller and Price (1988); 2, Baker et al. (1983); 3, Matsutani et al. (1989a); 4, Benjamin et al. (1982); 5, Carson (1984b); 6, Matsutani et al. (1989b); 7, Macrides and Davis (1983); 8, Sanides-Kohlrausch and Wahle (1990).

stay on either side of the glomeruli rather than surrounding them. The external plexiform layer of the AOB (aepl) and the mitral cell layer are also less distinct than the corresponding layers of the MOB, mainly because the "mitral" cells of the AOB are multilayered and mixed in with aepl. The AOB internal plexiform layer (aipl) is unremarkable and is situated between the mitral cell layer and the lateral olfactory tract. The granule cell layer (agcl) of the AOB, located deep to the lateral olfactory tract, contains the same small type of cell as in the MOB granule cell layer. Despite being called the mitral cell layer by many authors, the output cells in the AOB are much more polymorphic than their counterparts in the MOB. The transmitter phenotype of the output cells of the AOB has been studied recently (Table 5.3). One suspected transmitter is aspartate. ^{3}H-aspartate was injected into the AON and other caudal regions of olfactory cortex. This amino acid is thought to be preferentially taken up by synaptic endings that release aspartate as a transmitter, and the labeled aspartate is transported retrogradely from the injection site back to the cell body. Many more of the mitral cells in the AOB appear to be aspartatergic than the mitral cells in the MOB (Fuller and Price, 1988). Many mitral cells in the AOB of the guinea pig contain neurotensin (Matsutani, Senba, and Tohyama, 1989), while in the rat, mitral cells transiently express substance P, but the expression in these output cells gradually diminishes after postnatal day 10. Interestingly, substance P-IR granule cells increase in number at the time when mitral cell expression is decreasing (Matsutani, Senba, and Tohyama, 1989). The functional significance of this decrease in substance P expression has not been determined, although one might speculate that the decrease may be tied to changes in the pup responses to certain pheromones during postnatal development.

There are several important anatomical differences between the AOB and the MOB that may reflect differences in processing of olfactory inputs. The outermost layer of the AOB contains the preterminal axons of the vomeronasal nerve, which terminates in the glomeruli of the AOB. The few "periglomerular cells" in the AOB are neurochemically different from those in the MOB. The most obvious difference is the lack of dopaminergic periglomerular cells in the AOB. Also lacking are the substance P-containing external tufted cells that are abundant in the MOB of some species (Baker, 1986). GABAergic periglomerular and granule cells are present in the AOB (Baker, Towle, and Margolis, 1988). Substance P-containing cells are most prominent in the agcl of rats; in contrast, there are fewer substance P-IR cells in agcl in the rabbit, guinea pig, cat, and hamster, and in mice these cells appear to be absent (Baker, 1986).

It is clear, then, that AOB glomeruli lack some of the neurochemical neuronal components that are present in MOB glomeruli. As we shall see later, the differences between the MOB and the AOB are not confined to intrinsic cells; there are also distinct differences in the central connection of these two structures. The functional significance of the differences in intrinsic AOB-MOB neurons is still unclear; periglomerular cells in both the AOB and the MOB are difficult to study physiologically because they are very small (5–10 μm). Nonetheless, the obvious differences in the neurochemistry of periglomerular cells in the MOB and the AOB suggest that there are significant differences in the processing of olfactory information at the first level of synaptic transfer between the epithelium (olfactory or vomeronasal) and other central structures.

The growth of the AOB is influenced by steroids produced by the testes (Roos, Roos, Schaeffer, and Aron, 1988). The AOB in the male rat is statistically larger than that in the female, but if the male is castrated, the AOB development is decreased to the size of the female AOB. These findings in the AOB correlate well with sexual dimorphisms in other regions (e.g., hypothalamic nuclei) that are known to govern sexual behavior, have been found to be sexually dimorphic (cf. Arnold and Gorski, 1984 for review), and receive direct or indirect projections from the AOB (Scalia and Winans, 1975; Simerly, Young, Capozza, and Swanson, 1989).

Outputs of the Olfactory Bulb

Outputs of the MOB

As discussed earlier, the primary output of the MOB is through mitral cells, although there is also a large contribution from middle and deep tufted cells (Schoenfeld and Macrides, 1984). The output of the MOB has some degree of organization. For example, neurons in the dorsolateral quadrant of the MOB project to the dorsal part of the external subdivision of the AON, whereas output cells of the ventral half of the MOB project to the lateral subdivision (Schoenfeld and Macrides, 1984). Output neurons of the MOB that project to more caudal regions (e.g., piriform cortex) are evenly distributed in the MOB.

The axons of the mitral cells give off collaterals within the bulb in the internal plexiform and granule cell layers (Kishi, Mori, and Ojima, 1984). The main branches course predominantly, although not exclusively, in the lateral olfactory tract (which forms at the level of the AOB). These caudally directed axons give off collaterals in AON and other regions of olfactory cortex. Intracellular injections of HRP into mitral cells have been used to show the arborizations of individual axons. The axons often have many collateral terminal arbors in AON and fewer in the more caudally located piriform cortex (Ojima, Mori, and Kishi, 1984). The terminal arbors have a patchy anterior-posterior distribution in layer la of AON and piriform cortex, which according to Ojima and colleagues (1984) is reminiscent of the patchy distribution of thalamic input to the visual cortex. Some mitral cells branch to project to both the olfactory cortex and the olfactory tubercle.

Injection of anterograde tracers into the MOB results in labeling of many structures of the ipsilateral hemisphere. Only a slight contralateral projection appears at all, and that is to the AON (Shipley and Adamek, 1984). The MOB projects to the superficial plexiform layer of the ipsilateral anterior olfactory nucleus. The projection is extremely dense and confined to the superficial half of layer Ia (Broadwell, 1975; Shipley and Adamek, 1984). The label is continuous with other more caudal structures, which include the taenia tecta, anterior hippocampal continuation, indusium grisium, olfactory tubercle, and the lateral and medial entorhinal area. Collectively, the regions directly innervated by the output of the bulb have been referred to as primary olfactory cortex (de Olmos, Hardy, and Heimer, 1978). Most of these projections have been reported in several species (for review, cf. Shipley and Adamek, 1984). A recent study has also shown a direct MOB projection to the supraoptic nucleus (Smithson, Weiss, and Hatton, 1989). The functional significance of this projection may be in its regulation of water retention. The supraoptic nucleus contains neurons that secrete vasopressin into the circulatory system via the

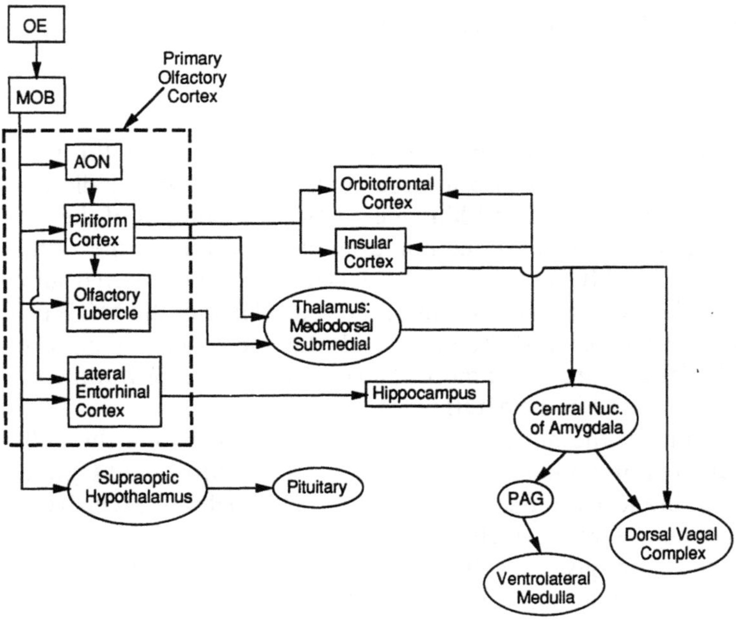

FIGURE 5.2. Diagram summarizing the major efferent connections of the main olfactory system. The regions within the thick dashed lines are within the primary olfactory cortex. In this and the following figures, cortical regions are indicated within rectangles and subcortical regions within ovals. Abbreviations: AON, anterior olfactory nucleus; MOB, main olfactory bulb; OE, olfactory epithelium, PAG, periaqueductal gray.

pituitary. It is thus speculated that excitatory input from the MOB to vasopressin neurons in the supraoptic nucleus increases vasopressin release from the pituitary and increases body tissue retention of water (Smithson, Weiss, and Hatton, 1989). The primary efferent pathways of the main olfactory system are summarized in Figure 5.2.

Outputs of the AOB

Although it has been reported that MOB neurons project to the anterior cortical and posterolateral cortical amygdaloid nuclei (Broadwell, 1975; Scalia and Winans, 1975), Shipley and Adamek (1984) found that only injections including the accessory olfactory bulb (AOB) project to the amygdala in the mouse. Thus, it is suspected that the AOB but not the MOB has direct projections to the amygdala. This pathway may be involved in the processing of pheromonal information. Evidence from different approaches has produced several theories about how this pathway and further secondary connections help to produce emotional and behavioral

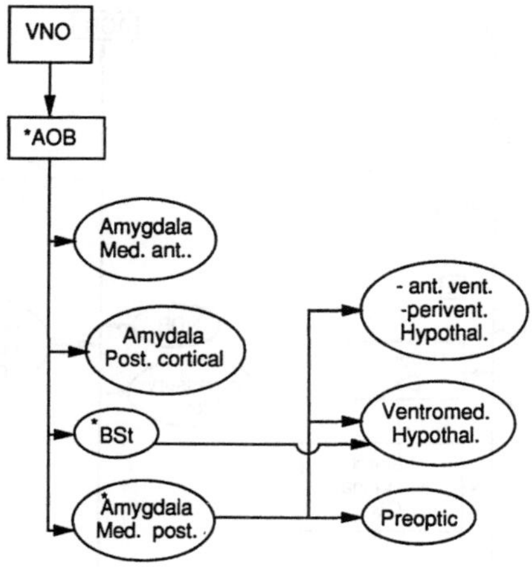

FIGURE 5.3. The efferent pathways of the accessory olfactory system. Note that many of the connections pass via the amygdala to the hypothalamic nuclei. Abbreviations: AOB, accessory olfactory bulb; BSt, bed nucleus of stria terminalis; VNO, vomeronasal organ.

responses to pheromones. These secondary connections, especially to the amygdala, may be acted upon directly by hormones, and their proposed functions are described in a later section. The primary connections of the accessory olfactory system are summarized in Figure 5.3.

Organization and Neurochemistry of Central ("Centrifugal") Afferents to the Olfactory Bulb

Afferents to the MOB

The centrifugal afferent inputs to the olfactory bulb are very dense, come from many sources (Figure 5.4), and must play important roles in controlling or modulating neural processing in the olfactory bulb. These inputs are numerous and diverse but display some degree of organization and neurochemical heterogeneity. The centrifugal afferents may be divided into two groups: afferents from olfactory related structures and nonolfactory subcortical modulatory afferents to the olfactory bulb. These are distinguished because the olfactory-related projections are almost certainly strongly linked to olfactory sensory and association functions, while the modulatory afferents are systems that have widespread CNS projections and appear to influence CNS neurons across system lines. Olfactory-related centrifugal afferents arise from many sources, including the AON, piriform cortex, periamygdaloid cortex, entorhinal cortex, nucleus of the

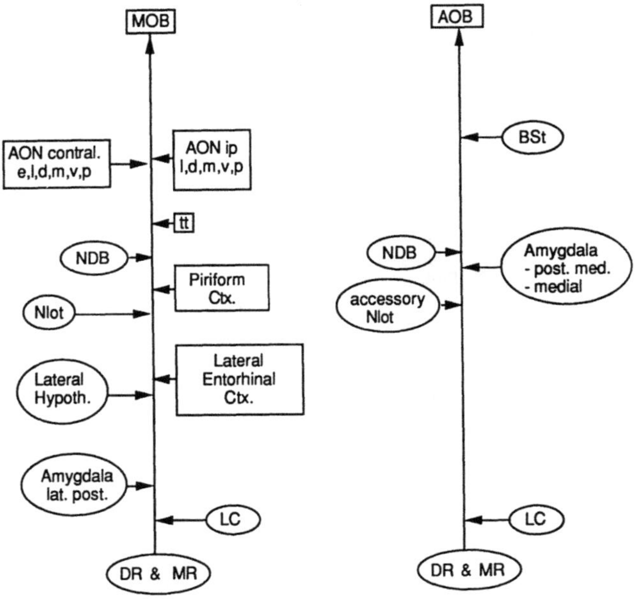

FIGURE 5.4. Centrifugal inputs to the main and accessory olfactory bulbs. Note the restricted sources of afferents to the AOB compared to the MOB. Abbreviations: AOB, accessory olfactory bulb; AON contral e, l, d, m, v, p, contralateral anterior olfactory nucleus—pars externa, lateralis, dorsalis, medialis, ventralis, posterior; AON ip, ipsilateral AON; BSt, bed nucleus of stria terminalis; DR, dorsal raphe nucleus; LC, locus coeruleus; MR, median raphe nucleus; NDB, nucleus of the diagonal band including horizontal and vertical limbs; Nlot, nucleus of the lateral olfactory tract; tt, taenia tecta.

lateral olfactory tract, and amygdala. Subcortical modulatory afferents originate in the brain stem and basal forebrain.

Anterior Olfactory Nucleus (AON)

Caudal to the olfactory bulb is a distinctive cortical structure, the anterior olfactory nucleus (AON). Herrick (1924), who first coined the term "anterior olfactory nucleus," regarded it as a nucleus. However, the AON is laminated, consisting of a plexiform layer and a rather homogeneous layer of tightly packed cells except in the center.

Golgi and Nissl studies have shown that the AON is laminated and contains pyramidal cells (Haberly and Price, 1978b). Thus, recent workers have been more inclined to consider the AON a cortical rather than a nuclear structure. Golgi studies by Haberly and Price (1978b) reveal that many of the output cells in the different subdivisions of the AON are pyramidal, save for neurons in the external subdivision. The presence of pyramidal cells and the laminar organization of afferent inputs to the subdivisions of the AON lead Haberly and Price (1978b) to suggest that the AON is, in fact, cortically organized. The AON has been divided into several subdivisions based on cytoarchitecture and connections (Haberly and Price, 1978b; de Olmos, Hardy, and Heimer, 1978). Since the extrinsic connections of the AON are better understood than its intrinsic circuitry or transmitters, further description of the AON is postponed until a later section. All subdivisions of the ipsilateral AON project to both the ipsilateral and contralateral main olfactory bulbs except the external division (pars externa pE), which projects only to the contralateral MOB (de Olmos, Hardy, and Heimer, 1978). The AON contains the largest number of neurons projecting to the bulb from any one source (Carson, 1984a). Thus, there is bilateral representation of olfactory information at the level of the AON. Kucharski and Hall (1987) have shown that the AON can access and recall existing olfactory memories stored in the contralateral AON or bulb.

The AON does not appear to project to the AOB (de Olmos, Hardy, and Heimer, 1978). There is some degree of laminar topography of the terminal projections to the MOB from different AON subdivisions, piriform cortex, and NLOT as observed in the rat (Luskin and Price, 1983) and hamster (Davis and Macries, 1981). In both species, the ventral and posterior subdivisions of the AON project bilaterally to the granule cell layer (mainly superficial) and the deep third of the glomerular layer. In the hamster the lateral and dorsal divisions of the AON project mainly to the superficial gcl and gl (ipsilateral), with no projection to the glomeruli, while in the rat the dorsal division terminates evenly in the gcl and lightly in the external plexiform layer but not in the glomerular layer. The external subdivision of the AON in both species terminates heavily in a thin band just deep to the internal plexiform layer.

In this chapter we do not exhaustively review all the cytoarchitectural subdivisions of the AON, but for interested readers these subdivisions have been described recently (de Olmos, Hardy, and Heimer, 1978; Macrides and Davis, 1983; Shipley and Adamek, 1984; Haberly and Price, 1978a).

Aspartate has been proposed as a transmitter of neurons mainly in the dorsal and external divisions of the AON based on selecting retrograde transport of ^{3}H aspartate (Watanabe and Kawana, 1984; Fuller and Price, 1988). Fewer neurons in other subdivisions of the AON contain aspartate, and no neurons in other known afferents to the bulb contain aspartate. Met-enkephalin and somatatinergic neurons in the AON appear to project to the olfactory bulb (Davis, Burd, and Macrides, 1982).

Primary Olfactory Cortex: (Piriform/Entorhinal/Transitional Cortices)

The caudal part of the AON is continuous through transitional zones with the primary olfactory cortex laterally and with the anterior hippocampal medially. The primary olfactory cortex includes the piriform cortex, periamygdaloid, and transition cortex and the lateral entorhinal cortex. Haberly and Price (1978a) divided the piriform cortex into three layers that are further subdivided on the basis of cytoarchitecture and afferent connections. Layer I, the superficial plexiform layer, is divided into Ia and Ib, which receive different types of afferents; Layer Ia receives afferents from the ipsilateral olfactory bulb, while Layer Ib receives association fibers from the AON and from other parts of the primary olfactory cortex. Layer II, the superficial compact cell layer, is also divided into two sublayers; the more superficial zone has a lower cell density and the deeper division has a higher cell density. Layer III is the deepest cell layer. Deep to the piriform cortex is the endopiriform nucleus, which was recognized by Loo (1931) in the opossum and was further described by Haberly and Price (1978a). In the piriform cortex, the projection cells in layers IIb and III have extensive basal dendrites, while those in layer IIa lack basal dendrites (Haberly and Price, 1978) and are reminiscent of dentate granule cells in the hippocampus. Transitional zones between the AON and the rostral piriform cortex have been defined by de Olmos and collaborators (1978) where the cells gradually become one specialized structure. The lateral transitional area (towards the piriform cortex) contains many cells that project to the MOB; the dorsal transition (between the AON and the frontal cortex) and the medial transition (towards the anterior hippocampus-archicortex) contain relatively fewer cells that project to the bulb.

Caudal to the piriform and transitional cortex is the entorhinal cortex. This cortex is divided into medial, lateral, and intermediate divisions and has six layers. The lateral entorhinal cortex is further subdivided into three divisions based upon position and relative density of cells in each layer. As noted by de Olmos, Hardy, and Heimer (1978), there are several transitional regions between the piriform cortex and the entorhinal cortex, AON, and piriform cortex and between olfactory cortices and neocortex. These transitional regions have been described in detail previously (de Olmos, Hardy, and Heimer, 1978), and the reader is referred to that paper for further information.

The ipsilateral piriform cortex, lateral entorhinal cortex, and transitional areas project to the bulb. The projections are heavier from the rostral than from the caudal parts of the primary olfactory cortex in the rat and mouse (Shipley and Adamek, 1984). Other cortical areas projecting to the MOB include taenia tecta and a small projection from the anterior hippocampal continuation (Shipley and Adamek, 1984; de Olmos, Hardy, and Heimer, 1978). A few cells in the posterolateral and medial cortical amygdaloid areas may project to the MOB (Shipley and Adamek, 1984). The transmitter(s) of these olfactory cortical projections to the bulb are not known, although glutamate would be suspected because that putative excitatory amino acid is found in many cells in layers II and III of the piriform and entorhinal cortex (Kaneko and Mizuno, 1988). It is thought that glutamate excites inhibitory GABAergic granule cells in the MOB, which in turn inhibit firing of mitral cells (Nicoll, 1971). Sources of the glutamate could be from mitral cells via reciprocal dendrodendritic synapses between mitral and granule cell dendrites (Halász and Shepherd, 1983), or possibly from afferents to the bulb from the olfactory cortex, although there is no direct evidence that these afferents are glutamatergic.

In the hamster, neurons in the rostral to caudal levels of the piriform cortex terminate from superficial to deep in the gcl of MOB, respectively. However, this gradual shift in termination is not as apparent in the rat. The periamygdaloid cortex and NLOT terminate in the deep gcl (Luskin and Price, 1983). In summary, it would appear that most of the afferent projections to the MOB from the AON project most heavily to the ipsilateral gcl.

"Nonolfactory" Modulatory Inputs to the Olfactory Bulb

In addition to centrifugal inputs from olfactory cortical and subcortical structures, the olfactory bulb is significantly targeted by afferents from nonolfactory subcortical modulatory inputs. These inputs arise from three sources: the nucleus of the diagonal band, dorsal and median raphe nuclei, and the locus coeruleus. The nucleus of the diagonal band (NDB) is a component of the basal forebrain magnocellular system, which includes the NDB, nucleus basalis, and the medial septum. These basal forebrain neurons innervate most regions of the neocortex, the hippocampus, and many other forebrain regions, including the amygdala and the thalamus. The nucleus locus coeruleus and the dorsal and median raphe nuclei innervate cortical and subcortical structures throughout the CNS.

Nucleus of the Diagonal Band (NDB)

In the mouse about 3.5 percent of all neurons that project to the bulb originate in the horizontal limb of the diagonal band (Carson, 1984a); far fewer originate in the vertical limb of the diagonal band (Carson, 1984a; Shipley and Adamek, 1984). Many NDB cells are cholinergic (Carson, 1984b; Macrides, Davis, Youngs, Nadi, and Margolis, 1981), but a recent double label study (Zaborszki, Carlsen, Brashear, and Heimer, 1986) showed that there are at least two distinct populations of NDB neurons projecting to the olfactory bulb. About 20 percent of the NDB neurons that project to the bulb are cholinergic; most of these cells are concentrated in the rostromedial portion of the horizontal limb of the NDB. At least as many NDB bulbopetal neurons are GABAergic and and are preferentially localized mainly in the lateral-caudal regions of the horizontal limb of the NDB (Zaborszky, Carlsen, Brashear, and Heimer, 1986). Thus, there are at least two transmitter candidates in NDB projections to the bulb.

The termination organization of NDB projections to the bulb is, thus, potentially complicated by the mixture of cholinergic/GABAergic neurons for the NDB. Accordingly, although the two types of neurons are somewhat segregated in the NDB, injections of anterograde tracers to label NDB to projections to the bulb (Luiten, Gaykema, Traber, and Spenser, 1988) may not distinguish the cholinergic from GABAergic input. Cholinesterase is one marker for the cholinergic axons. AChE label in the bulb is preferentially concentrated in the internal plexiform layer, gcl, the inner third of external plexiform layer, and the glomerular layer (gl). Some glomeruli are more densely stained for AChE and correspond to regions of LHRH innervation (Zheng, Caldani, and Jordan, 1988). The source of LHRH in these specialized glomeruli is unknown, although Zheng, Caldani, and Jordon (1988) suggested VLDB as a possible source. The glomeruli containing dense AChE and LHRH labeling are perhaps the modified glomerular complex as defined by Greer, Teicher, and collaborators (1980, 1982). It has been suggested that these glomeruli may be areas of specialized olfactory processing during development (Teicher, Stewart, Kauer, and Shepherd, 1980).

AChE staining in the MOB is partially confounded by the presence of cholinoceptive neurons in the bulb (see above). A more suitable marker for cholinergic axons is choline acetyltransferase (ChAT), the requisite enzyme for acetylcholine production. Antibodies to ChAT generally do not robustly stain cholinergic axons. However, we have been able to obtain excellent labeling using a polyclonal antibody to ChAT (generously donated by F. Eckenstein). ChAT-stained axons are located in regions similar to those described for AChE and are very fine in diameter. The glomeruli of the AOB lack both ChAT and AChE staining. Thus, this represents another example in which the glomerular innervation differs between the AOB and the MOB.

The GABAergic projection from the NDB (Zaborszki, Carlsen, Brashear, and Heimer, 1986) is more difficult to characterize than the cholinergic because the intrinsic GABAergic periglomerular and granule cells in the bulb provide such a massive intrinsic GABAergic innervation of the bulb.

Raphe Nuclei

The dorsal and median raphe provide strong inputs to the olfactory bulb. In the rat, about 1,000 dorsal and 300 median raphe neurons innervate the bulb. These neurons are serotonergic and do not contain tyrosine hydroxylase (McLean and Shipley, 1987a) or substance P (Magoul, Oblin, Calas, and Araneda, 1988). The raphe input enters the MOB via the olfactory peduncle and perhaps via the olfactory nerve layer (McLean and Shipley, 1987a). Thick serotonergic fibers preferentially innervate the glomeruli of the MOB, while thinner serotonergic axons preferentially innervate inframitral layers (McLean and Shipley, 1987a). In neocortex, thin axons originate in the dorsal raphe and thick axons from median raphe neurons (Mamounas and Molliver, 1988), so one may assume that the same type of arrangement occurs in the MOB.

The density of serotonergic axons in the glomeruli continues to increase throughout postnatal development (McLean and Shipley, 1987b). The reason for this continued increase of serotonin innervation is not known but may reflect a nontransmitter role for serotonin. There is continuous turnover and replacement of primary olfactory neurons. New olfactory axons grow into the glomeruli (Costanzo, 1974; Graziadei, 1973; Graziadei and Monti Graziadei, 1979). Serotonin may influence the growth or formation of synapses by these new olfactory axons, since serotonin has been shown to inhibit synaptogenesis elsewhere (Haydon, McCabb, and Kater, 1984). Serotonergic axons do not innervate the glomeruli of the AOB, just as cholinergic axons avoid this layer. Since the AOB has far fewer PG cells than the MOB, the paucity of 5HT and ACh innervation of the AOB glomerular layer suggests that serotonergic and cholinergic inputs target PG cells in the MOB; the relative absence of PG neurons in the AOB, thus, might account for the lack of serotonergic/cholinergic input to that layer in the AOB. Interestingly, the serotonergic innervation of MOB glomeruli in primates is considerably reduced (Azmitia and Gannon, 1986; Takeuchi, Kimura, and Sano, 1982) compared with that in other mammals that have been studied, such as rats (McLean and Shipley, 1987a), cats (Takeuchi, Kimura, and Sano, 1982), rabbits (personal observation), and hamsters (Kream, Davis, Kawano, Margolis, and Macrides, 1984). The specific functions of the 5HT and ACh inputs are unknown; however, their widespread distribution in the CNS has suggested that they may modulate global aspects of neural processing.

Locus coeruleus

A significant modulatory input to the bulb is from the pontine nucleus, the locus coeruleus (LC). Shipley, Halloran, and de la Torre (1985) estimated that up to 40 percent of LC neurons (400–600 of a total of 1600 LC neurons) project to the bulb in the rat. The axons of neurons in the LC project mainly to the infraglomerular layers of the bulb, particularly the internal plexiform and granule cell layers. The external plexiform and mitral cell layers are moderately innervated, while the glomerular layer is nearly devoid of NE input. This highly specific laminar innervation pattern, unusual for LC terminal fields, is observed in both the MOB and the AOB. In the AOB, the internal plexiform layer is, in fact, sharply demarcated by the dense DBH-IR fibers running through it just deep to the multicellular output cell layer (mitral cell layer) (McLean, Shipley, Nickell, Aston-Jones, and Reyher, 1989). Based on these light microscopic studies, it is likely that the major target of the NE input is granule cells (McLean et al., 1989). Several studies have examined the physiological actions of noradrenaline in the bulb. Salmoiraghi, Bloom, and Costa (1964) and McLennan (1971) made iontophoretic applications of NE and NE agonists; both studies reported decreased mitral cell firing. This result was interpreted to mean that NE causes granule cells to increase GABA release, thus inhibiting mitral cells. McLennan showed that NE failed to decrease mitral cell firing when bucuculin, a specific antagonist in to the $GABA_A$ receptor, was coapplied with NE. However, using a slice preparation, Jahr and Nicholl (1982) found that NE application caused increased mitral cell firing. Additional studies are needed to understand the role of endogenously released NE in the adult olfactory bulb. LC activation and release of NE have been implicated in a heightened state of vigilance (Aston-Jones and Bloom, 1981). There are several lines of evidence that lend support to this theory in the bulb. We have suggested (McLean, Shipley, Nickell, Aston-Jones, and Reyher, 1989) that NE is released onto granule cells, resulting in increased granule cell release of GABA onto the lateral dendrites of mitral and tufted cells. We speculate that this decreases spontaneous, but not sensory-driven, activity of mitral cells because the apical dendrite of the mitral cell receives sensory input in the glomeruli and the glomeruli are not innervated by NE fibers. NE has been reported to decrease spontaneous versus sensory-evoked activity in other cortical neurons (Waterhouse, Moises, and Woodward, 1986). The result of selective inhibition of mitral/tufted lateral dendrites might be to increase the signal-to-signal noise of olfactory stimuli to other parts of the brain during heightened vigilance so that salient stimuli (food or predator) may be more readily detected and acted upon.

Other recent findings suggest that noradrenaline may be necessary for newborn rats to form learned odor preferences (Sullivan and Leon, 1987; Wilson and Leon, 1988). Noradrenergic fibers arrive in the bulb before birth and increase in density (McLean and Shipley, 1991) at a time when olfactory bulb circuits are still being established in the bulb. The timing of noradrenergic axon arrival in the MOB (McLean and Shipley, 1991) correlates with pharmacological evidence of noradrenergic influence on mitral cell excitability in the immature bulb (Wilson and Leon, 1988).

Afferents to the AOB

There are substantial and potentially important differences between centrifugal inputs to the MOB and the AOB (Figure 5.4). The differences appear to be in two respects: (1) The inputs to the AOB are from far fewer brain regions than inputs to the MOB—the major inputs to the AOB are from the amygdala, while the primary olfactory cortex is the major input to the MOB—and (2) brain regions that project to both the AOB and the MOB appear to have dissimilar termination patterns in the two targets. First, let us consider the disparate inputs to the AOB and the MOB. Studies utilizing anterograde and retrograde tracers show that the major afferent inputs to the AOB are from the bed nucleus of the stria terminalis, the nucleus of the accessory olfactory tract, the medial amygdala nucleus, and the posteromedial cortical amygdala nucleus (de Olmos, Hardy, and Heimer, 1978; Shipley and Adamek, 1984). Obviously, from this description the

AOB does not share the massive olfactory peduncle and olfactory cortex projections that innervate the MOB. The bed nucleus of the stria terminalis and the medial amygdala are limbic structures that could mediate pheromonal/neuroendocrine function. The precise terminal organization of these inputs to the AOB is not known. Some centrifugal afferents are common to both the MOB and the AOB. These afferents include the subcortical modulatory systems: diagonal band, dorsal and median raphe nuclei, and locus coeruleus. The distribution of these common inputs differs between the AOB and the MOB, especially with respect to the cholinergic and serotonergic inputs. In the AOB, both the cholinergic and serotonergic inputs absolutely avoid the glomeruli, whereas they heavily innervate the glomeruli of the MOB. The cholinergic-serotonergic inputs to the AOB are mainly to the granule cell layer and internal plexiform layer (McLean and Shipley, 1987a). The noradrenergic projection from the locus coeruleus is mainly confined to inframitral layers in both the MOB and the AOB. Noradrenergic input to the AOB appears to have a very interesting function in mice. Bruce (1960) found that recently mated female mice will abort if presented with the odors of a strange male mouse that is not the mate. This effect is blocked if the NE input to the female's AOB is removed immediately after mating, presumably before olfactory memories of the mate are formed (Keverne and de la Riva, 1982; Rosser and Keverne, 1985). Thus, NE appears to be important in strengthening the memory of the odor of the "husband". NE has also been shown to be necessary for other olfactory memories, such as maternal recognition in sheep (Pissonier et al., 1985) and odor preference in young rats (Sullivan, Wilson, and Leon, 1989). The mechanism of memory formation has been examined by Kaba, Rosser, and Keverne (1989) in the context of pregnancy block. They suggest that the dendro-dendritic synapse between granule cells and mitral cells in the AOB may be critical for the memory formation and that NE, by enhancing the inhibition of a subset of mitral cells for several hours following mating, may facilitate a selective odor memory. They further suggest that, as a consequence of such neural activity, presenting the pregnant female with the stud male would produce activity of mitral cells matching that produced around the time of mating, while strange males would produce different patterns of mitral cell activity leading to neuroendocrine responses that abort the pregnancy.

Secondary Olfactory Connections and Integration

The olfactory system communicates with other regions of the brain to integrate olfactory information with other sensory systems and higher brain functions. There are several pathways that are involved in the integration of olfactory information. The pathways are summarized by diagram in Figures 5.2 and 5.3.

Olfaction and Motor Activity

Several pathways link olfactory-related structures to what Heimer and Wilson (1975) have referred to as the ventral striatum. These connections are proposed to provide a means by which limbic (and possibly olfactory) information is integrated with the motor control regions of the striatum so that visceral and somatic effectors may be controlled by these pathways (Newman and Winans, 1980). Olfactory linkages to the ventral striatum are mediated by parallel projections from the MOB, the AON, and the piriform cortex to the olfactory tubercle. Neurons in the olfactory tubercle and some in piriform cortex project to the nucleus accumbens (part of the ventral striatum), which in turn projects to the ventral pallidum and substantia nigra, pars reticulata (Newman and Winans, 1983). A possible candidate transmitter in the projection from the MOB to the olfactory tubercle is CCK. Recent immunocytochemical and tracing studies suggest that CCK immunoreactive terminals, arising from the MOB, possibly from CCK-containing tufted cells, densely innervate the lateral portion of the olfactory tubercle (Matsutani, Senba, and Tohyama, 1989). The chemical neuroanatomy of other projections from the bulb, the AON, and the primary olfactory cortex to the ventral striatum are is not known, but one may guess that the pyramidal cells of the piriform and entorhinal cortex contribute excitatory amino acid-mediated projections to the nucleus accumbens because many cells in the AON and primary cortex contain glutamate (Kaneko and Mizuno, 1988) and project to the olfactory tubercle and nucleus accumbens (Newman and Winans, 1988). The neurochemistry of the accumbens/olfactory tubercle projections are only open to speculation at this time. It would be interesting if the neurochemical circuitry parallels the cortex (glutamate) ⟶ neostriatum (GABA, enkephalin, substance P) ⟶ pallidal (GABA) ⟶ thalamus (glutamate) ⟶ cortex loop of the basal ganglia. The transmitters/peptides and precise circuitry of those pathways have recently been reviewed (Albin, Young, and Penney, 1989), but the equivalent relationships in the ventral striatal pathways remain to be determined. Recent immunocytochemical studies show that the following neuroactive markers are present in neurons of the olfactory tubercle: gonadotropin-releasing hormone associated peptide (Merchanthaler, Culler, Petrusz, Flerko, and Negro-Vilar, 1989) and cholecystokinin-8 (Pfister, Shade, and Ott, 1989). It is not yet known if these neurons receive direct input from the olfactory bulb and/or project to other olfactory processing regions.

Olfaction and Memory

The entorhinal cortex receives a substantial input from the MOB (Broadwell, 1975; de Olmos, Hardy, and Heimer, 1978; Shipley and Adamek, 1984). In turn, the medial and lateral enthorhinal cortex project to the dentate gyrus and CA fields of the hippocampus. Recent studies show that MOB projections to the entorhinal cortex make direct contact with stellate cells located in layer II that in turn project via the perforant path to the hippocampus (Schwerdtfeger, Buhl, and Germroth, 1990). In addition, the primary olfactory cortex has direct connections to the entorhinal cortex. Because the hippocampus is important in memory function, these olfactory-entorhinal-hippocampal circuits may be important for establishment or recall of olfactory memories formed or associated with other events. The chemical neuroanatomy of these connections is not known. It should be noted that important olfactory memories and olfactory learning may take place at the level of the olfactory bulb (as discussed above in "Afferents to the AOB").

Sex and Affective Responses to Odors

Olfaction plays an important role in sexual behavior in many animals. Macrosmatic animals have a highly developed ability to use olfaction for identifying sexual partners, enemies, and food, that is, these animals use olfaction for survival and continuation of the species. The sexual linkage may not be as strong in humans, but we may still possess the neural hardware tying odors to sexual arousal, and certainly the profit and loss statements of the fragrance industry attest to some role for olfaction in human sex drives. In macrosmatic animals, the accessory olfactory bulb is believed to be involved in processing pheromones that are initially transduced by way of the vomeronasal organ, and via the AOB, to the posterior cortical and anteromedial cortical nuclei of the amygdala (de Olmos, Hardy, and Heimer, 1978; Scalia and Winans, 1975; Shipley and Adamek, 1984). These amygdaloid nuclei have projections to other amygdaloid and hypothalamic nuclei (cf. Shioska et al., 1983, for review). Some of these secondary olfactory connections are known to strongly influence sexual drive, and the neurons involved in the connections may contain steroid receptors and release peptides that mediate the sexual responses. For example, neurons in the posterodorsal part of the medial nucleus of the amygdala (MeAp) contain neurons that project to four cell groups that are known to be sexually dimorphic and differ in their roles in reproduction. The medial preoptic nucleus (MPN) is one of the sexually dimorphic targets of MeAp, and lesions of MPN decrease male copulatory behavior (cf. Simerly, Young, Capozza, and Swanson, 1989, for review). Estrogen regulates the expression of cholecystokinin at the mRNA level in cells of MeAp; many cells containing cholecystokinin in MeAp project to the medial preoptic nucleus. In the female rat, cholecystokinin injection in the medial preoptic region enhances luteinizing hormone secretion (Kimura et al., 1987), although its action on male sexual responses is not known. Nevertheless, this pathway *could* provide the neuroendocrine and anatomical substrates to regulate copulatory behavior in male rats. More research is needed to determine the precise actions of peptides on sexual behavior and how the expression of the peptides is regulated by olfactory stimulation.

The accessory olfactory pathways that pass through the amygdala to the hypothalamic nuclei could also mediate the animal's visceral responses to stressful olfactory-induced behavior. Whether labeled lines of transmitter specific, spatially ordered circuits are present for pheromonal input is not known yet.

Cognicance of Odors

In contrast to the affective sexual and endocrine responses to odor mediated by relays with the amygdala, other pathways are thought to be involved in olfactory cognicance (meaning conscious discrimination of odors). Obviously, there are several sites where this discrimination could take place, since virtually all of the primary olfactory regions are cortical; the olfactory bulb, anterior olfactory cortex, piriform cortex, and entorhinal cortex are all allocortical, having less than six layers. However, evidence is lacking for the conscious discrimination of odors in these allocortical structures. On the other hand, olfactory information eventually makes its way to the neocortex via the thalamus. Physiological evidence in monkeys has shown that some degree of odor discrimination may take place in the lateral and posterior areas of the orbitofrontal cortex (cf. Takagi, 1984, for review). Further evidence suggests that the olfactory information is relayed either through the mediodorsal thalamus or through some other nonthalamic route (cf. Takagi, 1984, for review). There are several studies showing the potential anatomical path of olfactory information to the thalamus. For example, the olfactory tubercle, insular cortex, and piriform cortex receive input from mitral cell axons (Broadwell, 1975; de Olmos, Hardy, and Heimer, 1978; Ojima and Kishi, 1984; Shipley and Adamek, 1984). In turn, the olfactory tubercle and the piriform cortex project to the dorsomedial thalamus (Benjamin, Jackson, Golden, and West, 1982; Powell, Cowan, and Raisman, 1965; Price, 1985) and possibly the submedial nucleus (Price and Slotnick, 1983), although the projection from the piriform cortex (but not the insular cortex) has recently been questioned (Motokizawa, Yasuda, and Shimomura, 1988). The dorsomedial thalamus projects to the posterior orbitofrontal cortex, so one might argue that this pathway mediates the olfactory discrimination. Physiological evidence, however, suggests that the dorsomedial thalamus projection is stronger to the centroposterior portion of orbitofrontal cortex, which, according to Takagi (1984), is more involved in integrating odor sensations than in discriminating odors because individual cells in that region respond to several odors. In contrast, Takagi (1984) argues that the lateral posterior orbitofrontal cortex is the more definitive site for odor discrimination because individual cells located there commonly respond to only one odor.

The transmitters involved in the pathways discussed above are not known, although one might suspect that the excitatory amino acid glutamate is involved, since it is found in all thalamic nuclei and projection cells of the piriform cortex (Kaneka and Mizuno, 1988).

Olfaction and Taste/Visceral Integration

Olfactory stimuli can activate visceral response and autonomic adjustments, such as gastric secretions, salivation, and changes in heart rate. The circuits that mediate these functions are becoming known. One possibility is that the MOB and piriform cortex connections with the insular cortex might be involved in these functions (Saper, 1982; Shipley, 1982; Shipley and Sanders, 1982; Shipley and Geinisman, 1984; Ruggiero, Mraovitch, Granata, Anwar, and Reis, 1987). Recent studies in the mouse show that a portion of the granular insular cortex is a site of significant overlap between olfactory and visceral information (Shipley and Geinisman, 1984), and this has been confirmed in the rat (Krushel and van der Kooy, 1988). In addition, the medial frontal cortex may be an area of motor control of visceral activity (Neafsey, Hurley-Gius, and Arvanitis, 1986). There are direct projections from the MOB to the ventral part of the medial frontal cortex, and there are reciprocal connections between the insular and the medial frontal cortex. Olfactory pathways may act through the insular cortex and medial frontal cortex to influence autonomic and visceral function via direct projections to cardiovascular regions of the ventral medulla and the solitary nucleus (Ruggiero et al., 1987). Alternatively, the insular cortex may influence cardiovascular regions by less direct routes. For example, the central nucleus of the amygdala, which receives a dense projection from the insular cortex (Shipley and Sanders, 1982), projects to brainstem autonomic centers such as the periaqueductal gray and dorsal vagal complex (Hopkins and Holstege, 1978; Hopkins, McLean, and Takeuchi, 1981; Rizvi, Ennis, Behbehani, and Shipley, 1991). Portions of the periaqueductal gray project to the ventral lateral medulla (Van Bockstaele, Pieribone, and Aston-Jones, 1989) and may be involved in pressor and depressor responses of the cardiovascular system (Carriur, Bandler, and Damporey, 1987). Another region that may be involved in the integration of various senses is the posterolateral orbitofrontal cortex. This cortex receives input from the insular cortex (Wiggins, Rolls, and Baylis, 1987), and many neurons in the orbitofrontal cortex of the primate react to both taste and smell (Wiggins, Rolls, and Chisholm, 1988) and even visual inputs (Rolls, 1989). Thus, the orbitofrontal region may be an area where higher-level integration of multiple sensory modes (taste, smell, and vision) takes place.

As can be appreciated from the above description, relatively little is known about olfactory function and integration with other modalities. Research into these questions will undoubtedly provide interesting and valuable information.

Acknowledgments. Dr. McLean would like to thank Memorial University audiovisual services for assistance with the illustrations, and its secretarial staff for helping to type the manuscript. We would like to acknowledge grant support for portions of this chapter that refer to studies by the authors. Grant support to J.H.M is from M.R.C. of Canada #MT-10931, and grant support to M.T.S is from NIDCDP01DC00347 and DOD DAMD17-86-C-6005.

References

Alberts, J.R. (1976). Olfactory contributions to behavioral development in rodents. In R.L. Doty (Ed.), *Mammalian Olfaction, Reproductive Processes and Behavior*. New York: Academic, pp. 67–95.

Alberts, J.R. (1981). Ontogeny of olfaction: Reciprocal roles of sensation and behavior in the development of perception. In R.N. Aslin, J.R. Alberts, and M.R. Petersen (Eds.), *Development of Perception; Psychobiological Perspectives*, Vol. 1. New York: Academic, pp. 322–357.

Albin, R.L., A.B., Young, and J.B. Penney (1989). The functional anatomy of basal ganglia disorders. *Trends Neurosci.*, *12*, 366–375.

Allison, A.C. (1953). The morphology of the olfactory system in the vertebrates. *Biol. Rev. the Cambridge Philosophical Soc.*, *28*, 195–244.

Allison, A.C., and R.T.T. Warwick (1949). Quantitative observations on the olfactory system of the rabbit. *Brain*, *72*, 186–197.

Altman, J. (1969). Autoradiographic and histological studies of postnatal neurogenesis. IV. Cell proliferation and migration in the anterior forebrain, with special reference to persisting neurogenesis in the olfactory bulb. *J. Comp. Neurol.*, *126*, 433–458.

Arnold, A.P., and R.A. Gorski (1984). Gonadal steroid induction of structural sex differences in the central nervous system. *Ann. Rev. Neurosci.*, *7*, 413–442.

Astic, L. and D. Saucier (1986). Anatomical mapping of the neuroepithelial projection to the olfactory bulb in the rat. *Brain Res. Bull.*, *16*, 445–454.

Aston-Jones, G., and F.E. Bloom (1981). Activity of norepinephrine-containing locus coeruleus neurons in behaving rats anticipates fluctuations in the sleep-waking cycle. *J. Neurosci*, *1*, 876–886.

Azmitia, E.C., and P.J. Gannon (1986). The primate serotonergic system: A review of human and animal studies and a report on *Macaca jascicularis*. In S. Fahn et al. (Eds.), *Advances in Neurology*, Vol 43, *Myoclonus*, New York: Raven Press, pp. 407–468.

Bailey, M.S., M.R. Poston, and M.T. Shipley (1989). Unique morphology of olfactory bulb radial glial: role in glomerular development? *Soc. Neurosci. Abstr.*, *15*, 690.

Baker, H. (1986). Species differences in the distribution of substance P and tyrosine hydroxylase immunoreactivity in the olfactory bulb. *J. Comp. Neurol.*, *252*, 206–226.

Baker, H., T. Kawano, F.L., Margolis, and T.H. Joh (1983). Transneuronal regulation of tyrosine hydroxylase expression in olfactory bulb of mouse and rat. *J Neurosci.*, *3*, 69–78.

Baker, H., T. Kawano, V. Albert, T.H. Joh, D.J. Reis, D.J. and F.L. Margolis

(1984). Olfactory bulb dopamine neurons survive deafferentation-induced loss of tyrosine hydroxylase. *Neuroscience*, *11*, 605–615.

Baker, H., A.C. Towle, and F.L. Margolis (1988). Differential afferent regulation of dopaminergic and GABAergic neurons in the mouse main olfactory bulb. *Brain Res.*, *450*, 69–80.

Bassett, J.L., S.L. Foote, and M.T. Shipley (1989). Localization of corticotropin-releasing factor (CRF)-like immunoreactivity in monkey (*Saimiri aciureus*) olfactory bulb and its efferents. *Soc. Neurosci. Abstr.*, *15*, 561.

Benjamin, R.M., J.C. Jackson, G.T. Golden, and C.H.K. West (1982). Sources of olfactory inputs to opposum mediodorsal nucleus identified by horseradish peroxidase and autoradiographic methods. *J. Comp. Neurol.*, *207*, 358–368.

Blakely, R.D., L. Ory- Lavollee, R. Grzanna, K.J. Koller, and J.T. Coyle (1987). Selective immunocytochemical staining of mitral cells in rat olfactory bulb with affinity purified antibodies against N-acetyl-aspartyl-glutamate. *Brain Res.*, *402*, 373– 378.

Breathnach, A.S. (1960). The cetacean central nervous system. *Biol. Rev.*, *35*, 187–230.

Broadwell, R.D. (1975). Olfactory relationships of the telencephalon and diencephalon in the rabbit. I. An autoradiographic study of the efferent connections of the main and accessory olfactory bulbs. *J. Comp. Neurol.*, *163*, 329–346.

Broadwell, R.D. (1975). Olfactory relationships of the telencephalon and diencephalon in the rabbit. II. An autoradiographic and horseradish peroxidase study of the efferent connections of the anterior olfactory nucleus. *J. Comp. Neurol.*, *164*, 389–410.

Bruce, H.M. (1960). A block to pregnancy in the mouse caused by proximity of strange males. *J. Reproduction Fertility*, *1*, 96–103.

Brunjes, P.C. (1983). Olfactory bulb maturation in *Acomys cahirinus*: Is neural growth similar in precocial and altricial murids? *Dev. Brain Res.*, *8*, 335–341.

Brunjes, P.C., L.K. Smith-Crafts, and R. McCarty (1985). Unilateral odor deprivation: Effects on the development of olfactory bulb catecholamines and behavior. *Brain Res.*, *354*, 1–6.

Burd, G.D., B.J. Davis, F. Macrides, M. Grillo, and F.L. Margolis (1982). Carnosine in primary afferents of the olfactory system: an autoradiographic and biochemical study. *J. Neurosci.*, *2*, 244–255.

Cajal, S.R. (1911). *Histologie du Système Nerveux de l'Hommes dt des Vertébrés*. Tome 2, Madrid: Institut Ramon y Cajal, 1972.

Carriur, P., R. Bandler, and R.A.L. Damporey (1987). Viscerotopic control of regional vascular buds by discrete groups of neurons in the midbrain periaqueductal gray. *Brain Res.*, *493*, 385–390.

Carson, K.A. (1984a). Quantitative localization of neurons projecting to the mouse main olfactory bulb. *Brain Res. Bull.*, *12*, 629–634.

Carson, K.A. (1984b). Localization of acetylcholinesterase-positive neurons projecting to the mouse main olfactory bulb. *Brain Res. Bull.*, *12*, 635–639.

Cooper, H.M., G. Mick, and M. Magnin (1989). Retinal projection to mammalian telencephalon. *Brain Res.*, *477*, 350–357.

Costanzo, R.M. (1984). Comparison of neurogenesis and cell replacement in the hamster olfactory systems with and without a target (olfactory bulb). *Brain Res.*, *307*, 295–301.

Davis, B.J., and F. Macrides (1981). The organization of centrifugal projections from the anterior olfactory nucleus, ventral hippocampal rudiment, and piriform cortex to the main olfactory bulb in the hamster: an autoradiographic study. *J. Comp. Neurol.*, *203*, 475–493.

Davis, B.J., G.D. Burd, and F. Macrides (1982). Localization of methionine-enkephalin, substance P, and somatostatin immunoreactivities in the main olfactory bulb of the hamster. *J. Comp. Neurol.*, *204*, 377–383.

Davis, B.J., and F. Macrides (1983). Tyrosine hydroxylase immunoreactive neurons and fibers in the olfactory system of the hamster. *J. of Comp. Neurol.*, *214*, 427–440.

de Olmos, J., H. Hardy, and L. Heimer (1978). The afferent connections of the main and the accessory olfactory bulb formations in the rat: an experimental HRP-study. *J. Comp. Neurol.*, *181*, 213–244.

Doucette, R. (1988). Development of the nerve fiber layer in the olfactory bulb of mouse embryos. *J. Comp. Neurol.*, *285*, 514–527.

Ferriero, D., and F.L. Margolis (1975). Denervation in the primary olfactory pathway of mice. I. Effects on carnosine and other amine compounds. *Brain Res.*, *94*, 75–86.

ffrench-Mullen, J.M., K. Koller, R. Zaczek, J.T. Coyle, N. Hori, and D.O. Carpenter (1985). N-acetylaspartylglutamate: possible role as the neurotransmitter of the lateral olfactory tract. *Proc. Nat. Acad. Sci. USA.*, *82*, 3897–3900.

Frazier, L.L., and P.C. Brunjes (1988). Unilateral odor deprivation: early postnatal changes in olfactory bulb cell density and number. *J. Comp. Neurol.*, *269*, 355- 370.

Fuller, T.A., and J.L. Price (1988). Putative glutamatergic and/or aspartatergic cells in the main and accessory olfactory bulbs of the rat. *J. Comp. Neurol.*, *276*, 209–218.

Gall, C., S.H.C. Hendry, K.B. Seroogy, E.G. Jones, and J.W. Haycock (1985). Evidence for coexistence of GABA and dopamine in neurons of the rat olfactory bulb. *J. Comp. Neurol.*, *266*, 307–318.

Gall, C., K.B. Seroogy, and N. Brecha (1986). Distribution of VIP- and NPY-like immunoreactivities in rat main olfactory bulb. *Brain Res.*, *374*, 389–394.

Gonzalez-Estrada, M.T., and W.J. Freeman (1980). Effects of carnosine on olfactory bulb EEG, evoked potentials and DC potentials. *Brain Res.*, *202*, 373–386.

Graziadei, P.P.C. (1973). Cell dynamics in the olfactory mucosa. *Tiss. Cell*, *5*, 113–131.

Graziadei, P.P.C., and G.A. Monti Graziadei (1979). Neurogenesis and neuron regeneration in the olfactory system of mammals. I. Morphological aspects of differentiation and structural organization of the olfactory sensory neurons. *J. Neurocytol.*, *8*, 1–18.

Graziadei, P.P.C., and G.A. Monti Graziadei (1986). Principles of organization of the vertebrate olfactory glomerulus: an hypothesis. *Neuroscience*, *9*, 1025–1035.

Greer, C.A., W.B. Stewart, M.H. Teicher, and G.M. Shepherd (1982). Functional development of the olfactory bulb and a unique glomerular complex in the neonatal rat. *J. Neurosci.*, *2*, 1744–1759.

Guevara-Aguilar, R., L.P. Solano-Flores, O.A. Donatti-Albarran, and U. Aguilar-Baturoni (1982). Differential projections from locus coeruleus to olfactory bulb and olfactory tubercle: an HRP study. *Brain Res Bull.*, *8*, 711–719.

Haberly, L.B., and J.L. Price (1978a). Association and commissural fiber systems of the olfactory cortex of the rat. I. Systems originating in the piriform cortex and adjacent areas. *J. Comp. Neurol.*, *178*, 711–740.

Haberly, L.B., and J.L. Price (1978b). Association and commissural fiber systems of the olfactory cortex of the rat. II. Systems originating in the olfactory peduncle. *J. Comp. Neurol.*, *181*, 781–808.

Halász, N. (1987). Autoradiographic localization of high affinity uptalie sites for ^{3}H-D-aspartatein the rat olfactory bulb. *Acta Biol. Hungaria*, *38*, 383–392.

Halász, N., O. Johansson, T. Hökfelt, A. Ljundahl, and M. Goldstein (1981). Immunoreactive identification of two types of dopamine neuron in the rat olfactory bulb as seen by serial sectioning. *J. Neurocytol.*, *10*, 251–259.

Halász, N., and G.M. Shepherd (1983). Neurochemistry of the vertebrate olfactory bulb. *Neuroscience*, *10*, 579–619.

Harlan, R.E., B.D. Shivers, S.R. Fox, K.A. Kaplove, B.S. Schachter, and D.W. Pfaff (1989). Distribution and partial characterization of immunoreactive prolactin in the rat brain. *Neuroendocrinology*, *49*, 7–22.

Haydon, P.G., D.P. McCabb, and S.B. Kater (1984). Serotonin selectively inhibits growth cone mobility and synaptogenesis of specific identified neurons. *Science*, *226*, 561–564.

Heimer, L., and R.D. Wilson (1975). The subcortical projections of the allocortex: Similarities in the neural connections of the hippocampus, the piriform cortex, and the neocortex. In M. Santini (Ed.), Golgi Centennial Symposium, *Perspectives in Neurobiology*. New York: Raven Press, pp. 177–193.

Herrick, C.J. (1924). The nucleus olfactorius anterior of the opposum. *J. Comp. Neurol.*, *37*, 217–259.

Hirsch, J.D., M. Grillo, and F.L. Margolis (1978). Ligand binding studies in the mouse olfactory bulb: Identification and characterization of a L-[^{3}H]carnosine binding site. *Brain Res.*, *158*, 407–422.

Hirsch, J.D., and F.L. Margolis (1979). L-[^{3}H]carnosine binding in the olfactory bulb. II. Biochemical and biological studies. *Brain Res.*, *174*, 81–94.

Hopkins, D.A., and G. Holstege (1978). Amygdaloid projections to the mesencephalon, pons and medulla oblongata in the cat. *Exp. Brain Res.*, *32*, 529–547.

Hopkins, D.A., J.H. McLean, and Y. Takeuchi (1981). Amygdalotegmental projections: light and electron microscopic studies utilizing anterograde degeneration and anterograde and retrograde transport of horseradish peroxidase (HRP). In Y. Ben-Ari (Ed.), *The Amygdaloid Complex*. Amsterdam: Elsevier. pp. 133–147.

Humphrey, T. (1940). The development of the olfactory and the accessory olfactory formation in human embryos and fetuses. *J. Comp. Neurol.*, *73*, 431–468.

Imaki, T., J.L. Nahon, P.E, Sawchenko, and W. Vale (1989). Widespread expression of corticotropin-releasing factor messenger RNA and immunoreactivity in the rat olfactory bulb. *Brain Res.*, *496*, 35–44.

Inokuchi, A., J.P. Restrepo, and J.B. Snow, Jr. (1987). Effect of stimulation of the horizontal limb of the diagonal band on rat olfactory bulb activity. *Am. J. Otolaryngol.*, *8*, 205–210.

Jackowski, A., J.G. Parnavelas, and A.R. Lieberman (1978). The reciprocal synapse in the external plexiform layer of the mammalian olfactory bulb. *Brain Res.*, *195*, 17–28.

Jacobs, M.S., W.L. McFarland, and P.J. Morgane (1979). The anatomy of the brain of the bottlenose dolphin (*Tursiops truncatus*) rhinic lobe (rhinecephalon): the archicortex. *Brain Res. Bull.*, *4*, Suppl. 1, 1–108.

Jahr, C.E., and R.A. Nicoll (1982). Noradrenergic modulation of dendrodendritic inhibition of the olfactory bulb. *Nalure (London)*, *297*, 227–228.

Kaba, H., A. Rosser, and B. Keverne (1989). Neural basis of olfactory memory in the context of pregnance block. *Neuroscience*, *32*, 657–662.

Kanecko, T., and N. Mizuno (1988). Immunohistochemical study of glutaminase-containing neurons in the cerebral cortex and thalamus of the rat. *J. Comp. Neurol.*, *267*, 590–602.

Kaplan, M.S., N.A. McNelly, and J.W. Hinds (1985). Population dynamics of adult-formed granule neurons of the rat olfactory bulb. *J. Comp. Neurol.*, *239*, 117–125.

Kawano, T., and F.L. Margolis (1982). Transynaptic regulation of olfactory bulb catecholamines in mice and rats. *J. Neurochem.*, *39*, 342–348.

Keverne, E.B., and C. de la Riva (1982). Pheromones in mice: reciprocal interaction between the nose and brain. *Nature*, *296*, 148–150.

Kimura, F., N. Mitsugi, J. Arita, T. Akema, and K. Yoshida (1987). Effects of preoptic injections of gastrin, cholecystokin, secretin, vasoactive intestinal peptide and PHI on the secretion of luteinizing hormone and prolactin in ovariectomized estrogen-primed rats. *Brain Res.*, *410*, 315–322.

Kishi, K. (1987). Golgi studies on the development of granule cells of the rat olfactory bulb with reference to migration in the subependymal layer. *J. Comp. Neurol.*, *258*, 112–124.

Kishi, K., K. Mori, and H. Ojima (1984). Distribution of local axon collaterals of mitral, displaced mitral, and tufted cells in the rabbit olfactory bulb. *J. Comp. Neurol.*, *225*, 511–526.

Kosaka, T., Y. Hataguchi, K. Hama, I. Nagastsu, and J.-Y. Wu (1985). Coexistence of immunoreactivities for glutamate decarboxylase and tyrosine hydroxylase in some neurons in the periglomerular region of the rat main olfactory bulb: possible coexistence of gamma-aminobutyric acid (GABA) and dopamine. *Brain Res.*, *343*, 166–171.

Kosaka, T., K. Kosaka, C.W. Heizmann, I. Nagatsu, J.-Y. Wu, N. Yanaihara, and K. Hama (1987). An aspect of the organization of the GABAergic system in the rat main olfactory bulb: laminar distribution of immunohistochemically defined subpopulations of GABAergic neurons. *Brain Res.*, *411*, 373–378.

Kosaka, T., K. Hama, I. Nagatsu, J.-I. Wu, and T. Kosaka (1988). Possible coexistence of amino acid (gamma-aminobutyric acid), amine (dopamine) and peptide (substance P); neurons containing immunoreactivities for glutamic acid decarboxylase, tyrosine hydroxylase and substance P in the hamster main olfactory bulb. *Exp. Brain Res.*, *71*, 633–642.

Kream, R.M., B.J. Davis, T. Kawano, F.L. Margolis, and F. Macrides (1984). Substance P and catecholaminergic expression in neurons of the hamster main olfactory bulb. *J. Comp. Neurol.*, *222*, 140–154.

Krushel, L.A., and D. van der Kooy (1988). Visceral cortex: integration of the mucosal senses with limbic information in the rat agranular insular cortex. *J. Comp. Neurol.*, *270*, 39–54.

Kucharski, D., and W.G. Hall (1987). New routes to olfactory memory. *Science*, *238*, 786–788.

Laing, D.G., H. Panhuber, E.A. Pittman, M.E. Willcox, and G.K. Eagleson (1985). Prolonged exposure to an odor or deodorized air alters the size of mitral cells in the olfactory bulb. *Brain Res.*, *336*, 81–87.

Li, C.-S., H. Kava, H. Saito, and K. Seto (1989). Excitatory influence of the accessory olfactory bulb on tuberoinfundibular arcuate neurons of female mice and its modulation by oestrogen. *Neuroscience*, *29*, 201–208.

Loo, Y.T. (1931). The forebrain of the opposum, *Didelphys virginiana*. II. Histology. *J. Comp. Neurol.*, *52*, 1–148.

Lorente de Nó, R. (1933). Studies on the structure of the cerebral cortex. I. The area entorhinalis. *J. Psychol. Neurol.*, *45*, 381–438.

Luiten, P.G.M., R.P.A. Gaykema, J. Traber, and D.G. Spenser, Jr. (1987). Cortical projection patterns of magnocellular basal nucleus subdivisions as revealed by anterogradely transported *Phaseolus vulgaris* leucoagglutin. *Brain Res.*, *413*, 229–250.

Luskin, M.B., and J.L. Price (1983). The topographic organization of associational fibers of the olactory system in the rat, including centrifugal fibers to the olfactory bulb. *J. Comp. Neurol.*, *216*, 264–291.

Macchi, G. (1951). The ontogenetic development of the olfactory telencephalon in man. *J. Comp. Neurol.*, *95*, 245–303.

Macrides, F., B.J. Davis, W.M. Youngs, N.S. Nadi, and F.L. Margolis (1981). Cholinergic and catecholaminergic afferents to the olfactory bulb in the hamster: A neuroanatomical, biochemical, and histological investigation. *J. Comp. Neurol.*, *203*, 495–514.

Macrides, F., and B.J. Davis (1983). The olfactory bulb. In P.C. Emson, (Ed.), *Chemical Neuroanatomy*. New York: Raven Press, pp. 391–426.

Magoul, R., A. Oblin, A. Calas, and S. Araneda (1988). Serotonergic projections to the spinal cord but not those to the olfactory bulb also contain substance P. A combined immunocytochemical and autoradiographic study following retrograde axonal transport of [^{3}H]serotonin labeled products. *Neuroscience*, *26*, 959–969.

Mamounas, L.A., and M.E. Molliver (1988). Evidence for dual serotonergic projections to Deocortex: axons from the dorsal and median raphe are differentially vulnerable to the neurotoxin p-chloroamphetamine (PCA). *Exp. Neurol.*, *102*, 23–36.

Margolis, F.L. (1980). A marker protein for the olfactory chemoreceptor neuron. In R.A. Bradshaw and D.M. Schneider (Eds.), *Proteins of the Nervous System*. New York: Raven Press, pp. 59–84.

Matsutani, S., E. Senba, and M. Tohyama (1988). Neuropeptide- and neurotransmitter-related immunoreactivities in the developing rat olfactory bulb. *J. Comp. Neurol.*, *272*, 331–342.

Matsutani, S., E. Senba, and M. Tohyama (1989a). Distribution of neuropeptide-like immunoreactivities in the guinea pig olfactory bulb. *J. Comp. Neurol.*, *280*, 577–586.

Matsutani, S., E. Senba, and M. Tohyama (1989b). Terminal field of cholecystokinin-8-1ike immunoreactive projection neurons of the rat main olfactory bulb. *J. Comp. Neurol.*, *285*, 73–82.

McLean, J.H., and M.T. Shipley (1987a). Serotonergic afferents to the rat olfactory bulb: I. Origins and laminar specificity of serotonergic inputs in the adult rat. *J. Neurosci.*, *7*, 3016–3028.

McLean, J.H., and M.T. Shipley (1987b). Serotonergic afferents to the rat olfac-

tory bulb. II. Changes in fiber distribution during development. *J. Neurosci.*, *7*, 3029–3039.
McLean, J.H., and M.T. Shipley (1988). Postmitotic, postmigrational expression of tyrosine hydroxylase in olfactory bulb dopaminergic neurons. *J. Neurosci.*, *8*, 3658–3669.
McLean, J.H., M.T. Shipley, and D.I. Bernstein (1989). Golgi-like, transneuronal retrograde labelling with CNS injections of Herpes simplex virus type 1. *Brain Res. Bull.*, *22*, 867–881.
McLean, J.H., M.T. Shipley, W.T. Nickell, G. Aston-Jones, and C.K.H. Rehyer (1989). Chemoanatomical organization of the noradrenergic input from the locus coeruleus to the olfactory bulb of the adult rat. *J. Comp. Neurol.*, *258*, 339–349.
McLean, J.H., and M.T. Shipley (1991). Postnatal development of the noradrenergic projection from locus coeruleus to the olfactory bulb in the rat. *J. Comp. Neurol.*, *304*, 467–477.
McLennan, H. (1971). The pharmacology of inhibition of mitral cells in the olfactory bulb. *Brain Res.*, *29*, 177–184.
Meisami, E., and L. Safari (1981). A quantitative study of the effects of early unilateral olfactory deprivation on the number and distribution of mitral and tufted cells and of the glomeruli in the Rat olfactory system. *Brain Res.*, *221*, 81–107.
Merchenthaler, I., V. Csernus, C. Csontos, P. Petrusz, and B. Mess (1988). New data on the immunocytochemical localization of thyrotopin-releasing hormone in the rat central nervous system. *Am. J. Anat.*, *181*, 359–376.
Monti Graziadei, G.A., and P.P.C. Graziadei (1984). The olfactory organ. Neural transplantation. In J.R. Sladek, Jr., and D.M. Gash (Eds.), *Neural Transplants*. New York: Plenum, pp. 167–186.
Mori, K., K. Kishi, and H. Ojima (1983). Distribution of dendrites of mitral, displaced mitral, and granule cells in the rabbit olfactory bulb. *J. Comp. Neurol.*, *226*, 339–355.
Morrison, E.E., and P.P.C. Graziadei (1983). Transplants of olfactory mucosa in the rat brain. I. A light microscopic study of transplant organization. *Brain Res.*, *279*, 241–245.
Motokizawa, F., N. Yasuda, and F. Shimomura (1988). The piriform cortex is not a direct olfactory relay to the mediodorsal thalamic nucleus in cats and rabbits. *Neurosci. Lett. 84*, 41–45.
Mugnaini, E., W.H. Oertel, and F.F. Wouterlood (1984). Immunocytochemical localization of GABA neurons and dopamine neurons in the rat main and accessory olfactory bulb. *Neurosci. Lett.*, *47*, 221–226.
Nadi, N.S., R. Head, M. Grillo, J. Hempstead, N. Grannot-Reisfeld, and F.L. Margolis (1981). Chemical deafferentation of the olfactory bulb: plasticity of the levels of tyrosine hydroxylase, dopamine and norepinephrine. *Brain Res.*, *213*, 365–377.
Neafsey, E.J., K.M. Hurley-Guis, and D. Arvanitis (1986). The topographical organization of neurons in the rat medial frontal, insular and olfactory cortex projecting to the solitary nucleus, olfactory bulb, periaqueductal gray and superior colliculus. *Brain Res.*, *377*, 261–270.
Newman, R., and S.S. Winans (1980). An experimental study of the ventral striatum of the golden hamster. II. Neuronal connections of the olfactory tubercle. *J. Comp. Neurol.*, *191*, 193–212.
Nicoll, R.A. (1971). Pharmacological evidence for GABA as the transmitter in

granule cell inhibition in the olfactory bulb. *Brain Res.*, *35*, 137–149.

Nickell, W.T., and M.T. Shipley (1988). Two anatomically specific classes of candidate cholinoceptive neurons in the rat olfactory bulb. *J. Neurosci.*, *8*, 4482–4491.

Nickell, W.T., A.B. Norman, L.M. Wyatt, and M.T. Shipley (1990). Localization of dopamine receptor subtypes in the olfactory bulb: [^{3}H]spirerone (D2) binding in the glomerular and nerve layers. *Soc. Neurosci Abstr.*, *16*, 101.

Ohm, T.G., E. Braak, A. Probst, and A. Weindl (1988). Neuropeptide Y-like immunoreactive neurons in the human olfactory bulb. *Brain Res.*, *451*, 295–300.

Ojima, H., K. Mori, and K. Kishi (1984). The trajectory of mitral cell axons in the rabbit olfactory cortex revealed by intracellular HRP injection. *J. Comp. Neurol.*, *230*, 77–87.

Oland, L.A., L.P. Tolbert, and K.L. Mossman (1988). Radiation-induced reduction of the glial population during development disrupts the formation of olfactory glomeruli in an insect. *J. Neurosci.*, *8*, 353–367.

Orona, E., J.W. Scott, and E.C. Rainer (1983). Different granule cell populations innervate superficial and deep regions of the external plexiform layer in rat olfactory bulb. *J. Comp. Neurol.*, *217*, 227–237.

Orona, E., E.C. Rainer, and J.W. Scott (1984). Dendritic and axonal populations of mitral and tufted cells innervate the rat olfactory bulb. *J. Comp. Neurol.*, *226*, 346–356.

Panhuber, H., and D.G. Laing (1987). The size of mitral cells is altered when rats are exposed to an odor from their day of birth. *Brain Res.*, *431*, 133–140.

Pfister, C., R. Shade, and T. Ott (1989). Cholecystokinin-8-1ike immunoreactive structures in the olfactory bulb of the rat. *J. für Hirnforshung (Berlin)*, *30*, 257–264.

Pinching, A.J., and T.P.S. Powell (1971a). The neuron types of the glomerular layer of the olfactory bulb. *J. Cell. Sci.*, *9*, 305–345.

Pinching, A.J., and T.P.S. Powell (1971b). The neuropil of the glomeruli of the olfactory bulb. *J. Cell. Sci.*, *9*, 347–377.

Pinching, A.J., and T.P.S. Powell (1971c). The neuropil of the periglomerular region of the olfactory bulb. *J. Cell. Sci.*, *9*, 379–409.

Pissonier, D., J.C. Thiery, C. Fabre-Nse, P. Poindron, and E.B. Keverne (1985). The importance of olfactory bulb noradrenalin for maternal recognition in sheep. *Physiol. Behav.*, *35*, 361–363.

Powell, T.P.S., W.M., Cowan, and G. Raisman (1965). The central olfactory connexions. *J. Anat.*, *99*, 791–813.

Price, J.L. (1985). Beyond the primary olfactory cortex: olfactory-related areas in the neocortex, thalamus and hypothalamus. *Chem. Senses*, *10*. 239–258.

Price, J.L., and B.M. Slotnick (1983). Dual olfactory representation in the rat thalamus: an anatomical and electrophysiological study. *J. Comp. Neurol.*, *215*, 63–77.

Rall, W., G.M. Shepherd, T.S. Reese, and M.W. Brightman (1966). Dendrodendritic synaptic pathway for inhibition in the olfactory bulb. *Exp. Neurol.*, *14*, 44–56.

Ribak, C.E., J.E. Vaughn, K. Saito, R. Barber, and E. Roberts (1977). Glutamate decarboxylase localization in neurons of the olfactory bulb. *Brain Res.*, *126*, 1–18.

Rizvi, T.A., M. Ennis, M. Behbehani, and M.T. Shipley (1991). Connections between the central nucleus of the amygdala and the midbrain periaqueductal gray:

topography and reciprocity. *J. Comp. Neurol.*, *303*, 121–132.
Rolls, E.T. (1989). Information processing in the taste system of primates. *J. Exp. Biol.*, *146*, 141–164.
Roos, J., M. Roos, C. Schaeffer, and C. Aron (1988). Sexual differences in the development of accessory olfactory bulbs in the rat. *J. Comp. Neurol*, *270*, 121–131.
Rosenfeld, M.G., J.-J. Mermond, S.G. Amara, L.W. Swanson, P.E. Sawchenko, J. River, W.W. Vale, and R.M. Evans (1983). Production of a novel neuropeptide encoded by the calcitonin gene via tissue-specific RNA processing. *Nature*, *304*, 129–135.
Rosser, A.E., and E.B. Keverne (1985). The importance of central noradrenergic neurones in the formation of an olfactory memory in the prevention of pregnancy block. *Neuroscience*, *15*, 1141–1147.
Royet, J.P., C. Souchier, F. Jourdan, and H. Ploye (1988). Morphometric study of the glomerular population in the mouse olfactory bulb: numerical density and size distribution along the rostrocaudal axis. *J. Comp. Neurol.*, *270*, 559–568.
Ruggiero, D.A., S. Mraovitch, A.R. Granata, M. Anwar, and D.J. Reis (1987). A role of insular cortex in cardiovascular function. *J. Comp. Neurol.*, *257*, 189–207.
Saito, N., U. Kikkawa, Y. Nishizuka, and C. Tanaka (1988). Distribution of protein kinase C-like immunoreactive neurons in rat brain. *J. Neurosci.*, *8*, 369–382.
Sakai, M., K. Kani, N. Karasawa, M. Yoshida, and I. Nagatsu (1988) Carnosine-like immunoreactivity in the olfactory bulb of the rat: an electron microscopic study. *Brain Res.*, *438*, 335–338.
Salmoiraghi, G.C., F.E. Bloom, and E. Costa (1964). Adrenergic mechanisms in rabbit olfactory bulb. *Am. J. Physiol.*, *207*, 1417–1424.
Sanides-Kohlrausch, C., and P. Wahle (1990). VIP- and PHI-immunoreactivity in olfactory centers of the adult cat. *J. Comp. Neurol.*, *294*, 325–339.
Sanides-Kohlrausch, C., and P. Wahle (1990). Morphology of neuropeptide Y-immunoreactive neurons in the cat olfactory bulb and olfactory peduncle: post-natal development and species comparison. *J. Comp. Neurol.*, *291*, 468–489.
Saper, C.B. (1982). Convergence of autonomic and limbic connections in the insular cortex of the rat. *J. Comp. Neurol.*, *210*, 163–173.
Scalia, F., and S.S. Winans (1975). The differential projections of the olfactory bulb and accessory olfactory bulb in mammals. *J. Comp. Neurol.*, *161*, 31–56.
Schneider, S.P., and F. Macrides (1978). Laminar distribution of interneurons in the main olfactory bulb of adult hamster. *Brain Res. Bull.*, *3*, 73–82.
Schoenfeld, T.A., and F. Macrides (1984). Topographic organization of connections between the main olfactory bulb and pars externa of the anterior olfactory nucleus in the hamster. *J. Comp. Neurol.*, *227*, 121–135.
Schoenfeld, T.A., J.E. Marchand and F. Macrides (1985). Topographic organization of tufted cell axonal projections in the hamster main olfactory bulb: an intrabulbar associational system. *J. Comp. Neurol.*, *235*, 503–518.
Schwerdtfeger, W.K., E.H. Buhl, and P. Germroth (1990). Disynaptic olfactory input to the hippocampus mediated by stellate cells in the entorhinal cortex. *J. Comp. Neurol.*, *292*, 163–177.
Schwob, J.E., and J.L. Price (1984). The development of axonal connections in the central olfactory system of rats. *J. Comp. Neurol.*, *223*, 177–202.
Schwob, J.E., N.B. Farber, and D.I. Gottlieb (1986). Neurons in the olfactory epithelium in adult rats contain vimentin, *J. Neurosci.*, *6*, 208–217.

Scott, J.W., (1986). The olfactory bulb and central pathways. *Experenlia*, *42*, 223–232.

Scott, J.W., J.K. McDonald, and J.L. Pemberton (1987). Short axon cells of the rat olfactory bulb display NADPH-diaphorase activity, neuropeptide-Yimmunoreactivity, and somatostatin-like immunoreactivity. *J. Comp. Neurol.*, *260*, 378–391.

Segal, M., and S. Landis (1974). Afferents to the hippocampus of the rat studied with the method of retrograde transport of horseradish peroxidase. *Brain Res.*, *78*, 1–15.

Senba, E., P.E. Daddona, and J.I. Nagy (1987). Adenosine deaminase-containing neurons in the olfactory system of the rat during development. *Brain Res. Bull.* *18*, 635–648.

Seroogy, K.B., N. Brecha, and C. Gall (1985). Distribution of cholecystokinin-like immunoreactivity in the rat main olfactory bulb. *J. Comp. Neurol.*, *239*, 373–383.

Shepherd, G.M. (1972). Synaptic organization of the mammalian olfactory bulb. *Physiol. Rev.*, *52*, 864–917.

Shimada, M. (1966). Cytokinetics and histogenesis of early postnatal mouse brain as studied by H^3-thymidine autoradiography. *Arch. Histologicum Japonicum*, *4*, 413–437.

Shiosaka, S., M. Sakanaka, S. Inagaki, E. Senba, Y. Hara, K. Takatsuki, H. Takagi, Y. Kawai, and M. Toyhama (1983). Putative neurolransmitters in the amygdaloid complex with special reference to peptidergic pathways. In P.C. Emson (Ed.), *Chemical Neuroanatomy*. New York: Raven Press, pp. 359–389.

Shipley, M.T. (1982). Insular cortex projections to the nucleus of the solitary tract and brainstem visceromotor nuclei in the mouse. *Brain Res. Bull.* *8*, 139–148.

Shipley, M.T., and M.S. Sanders (1982). Special senses are special: Evidence for a reciprocal bilateral pathway between insular cortex and nucleus parabrachialis. *Brain Res. Bull.*, *8*, 493–501.

Shipley, M.T., and G.D. Adamek (1984). The connections of the mouse olfactory bulb: a study using orthograde and retrograde transport of wheat germ agglutinin conjugated to horseradish peroxidase. *Brain Res. Bull.*, *12*, 669–688.

Shipley, M.T., and Y. Geinisman (1984). Anatomical evidence for convergence of olfactory, gustatory, and visceral afferent pathways in mouse cerebral cortex. *Brain Res.*, *12*, 221–226.

Shipley, M.T., F.J. Halloran, and J. de la Torre (1985). Surprisingly rich projection from locus coeruleus to the olfactory bulb in the rat. *Brain Res.*, *329*, 294–299.

Skeen, L.C., B.R. Due, B.R. and F.E. Douglas (1985). Effects of early anosmia on two classes of granule cells in developing mouse olfactory bulb. *Neurosci. Lett.*, *54*, 301–306.

Simerly, R.B., B.J. Young, M.A. Capozza, and L.W. Swanson (1989). Estrogen differentially regulates neuropeptide gene expression in a sexually dimorphic olfactory pathway. *Proc. Nat. Acad. Sci. USA*, *86*, 4766–4770.

Smithson, K.G., M.L. Weiss, and G.I. Hatton (1989). Supraoptic nucleus afferents from the main olfactory bulb. 1. Anatomical evidence from anterograde and retrograde tracers in rat. *Neuroscience*, *31*, 277–287.

Struble, R.G., and C.P. Walters (1982). Light microscopic differentiation of two populations of rat olfactory bulb granule cells. *Brain Res.*, *236*, 237–251.

Sullivan, R.M., D.A. Wilson, and M. Leon (1989). Norepinephrine and learning-

induced plasticity in infant rat olfactory system. *J. Neurosci.*, *9*, 3998–4006.
Switzer, R.C., J. de Olmos, and L. Heimer (1985). Olfactory system. In G. Paxinos (Ed.), *The Rat Nervous System*, Vol. 1, *Forebrain and Midbrain*. Australia: Academic, pp. 1–37.
Takeuchi, Y., H. Kimura, and Y. Sano (1982). Immunohistochemical demonstration of serotonin nerve fibers in the olfactory bulb of the rat, cat and monkey. *Histochemistry 75*, 461–471.
Teicher, M.H., W.B. Stewart, J.S. Kauer, and G.M. Shepherd (1980). Suckling pheromone stimulation of a modified glomerular region in the developing rat olfactory bulb revealed by the 2-deoxyglucose method. *Brain Res.*, *194*, 530–535.
Tsuruo, Y., T. Hökfelt, and T.J. Visser (1988). Thyrotropin-releasing hormone (TRH)-immunoreactive neuron populations in the rat olfactory bulb. *Brain Res.*, *447*, 183–187.
Uddman, R., L. Edvinsson, R. Ekman, T. Kingman, and J. McCulloch (1985). Innervation of the feline cerebral vasculature by nerve fibers containing calcitonin gene-related peptide: Trigeminal origin and coexistence with substance P. *Neurosci. Lett. 62*, 131–136.
Van Bockstaele, E.J., V.A. Pieribone, and G. Aston-Jones (1989). Diverse afferents converge on the nucleus paragigantocellularis in the rat ventrolateral medulla: Retrograde and anterograde studies. *J. Comp. Neurol.*, *290*, 561–584.
van Ooteghem, S.A., and M.T. Shipley (1984). Factors affecting the sensitivity and consistency of the Koelle-Friedenwald histochemical method for localization of acetylcholinesterase. *Brain Res. Bull.*, *12*, 543–553.
Warden, M.K., and W.S. Young, III (1988). Distribution of cells containing mRNAs encoding substance P and Neurokinin B in rat central nervous system. *J. Comp. Neurol.*, *272*, 90–113.
Watanabe, K., and E. Kawana (1984). Selective retrograde transport of tritiated D-aspartate from the olfactory bulb to the anterior olfactory nucleus, pyriform cortex and nucleus of the lateral olfactory tract. *Brain Res.*, *296*, 148–151.
Waterhouse, B.D., H.C. Moises, and D.J. Woodward (1986). Interaction of serotonin with somatosensory cortical neuronal responses to afferent inputs and putative neurotransmitters. *Brain Res. Bull. 17*, 507–518.
White, E.L. (1972). Synaptic organization in the olfactory glomerulus of the mouse. *Brain Res.*, *37*, 69–80.
Whittemore, E.R., and J.F. Koerner (1989). An explanation for the purported excitation of piriform cortical neurons by N-acetyl-L-aspartyl-L-glutamic acid (NAAG). *Proc. Nal. Acade. Sci. USA*, *86*, 9602–9605.
Wiggins, L.L., E.T. Rolls and G.C. Baylis (1987). Afferent connections of the caudolateral orbitofrontal cortex area of the primate. *Soc. Neurosci. Abstr.*, *13*, 1406.
Wiggins, L.L., E.T. Rolls, and T. Chisholm (1988). Multimodal convergence within and beyond the secondary taste cortex in primates. *Soc. Neurosci. Abstr.*, *14*, 1186.
Wilson, D.A,, and M. Leon (1988). Noradrenergic modulation of olfactory bulb excitability in the postnatal rat. *Brain Res.*, *470*, 69–75.
Witter, M.P., A.W. Griffioen, B. Jorritsma-Byham, and J.L.M. Krijnen (1988). Entorhinal projections to the hippocampal CA1 region in the rat: an underestimated pathway. *Neurosci. Lett.*, *85*, 193–198.
Zaborszky, L., J. Carlsen, H.R. Brashear, and L. Heimer (1986). Cholinergic and

GABAergic afferents to the olfactory bulb in the rat with special emphasis on the projection neurons in the nucleus of the horizontal limb of the diagonal band. *J. Comp. Neurol.*, *243*, 488–509.

Zheng, L.M., M. Caldani and F. Jourdan (1988). Immunocytochemical identification of luteinizing hormone-releasing hormone-positive fibres and terminals in the olfactory system of the rat. *Neuroscience*, *24*, 567–578.

6
Neurophysiology of the Olfactory Bulb

W.T. NICKELL and MICHAEL T. SHIPLEY

Introduction

The olfactory bulb receives neural signals from a sheet of sensory epithelium located in the nasal cavity and transforms this information into output signals that are transmitted to central cortical structures. The olfactory system is thus similar to other sensory systems in that a two-dimensional array of receptors projects upon a two-dimensional cortex. It differs, however, in having no obvious inherent relation between odor characteristics and the receptive surface. One hypothesis for the neural coding of odor quality is that there are a finite number of olfactory receptor subtypes, each responsive to some particular property of odorous molecules. Different odors would generate distinctive patterns of responses across the several receptor types. These hypothetical receptor types *might* be segregated into separate regions of the epithelium, and this segregation *might* be conveyed topographically to the olfactory bulb, but there is no necessity of this.

Although the hypothesis of discrete olfactory receptor types is widely favored and is the basis of much current research, the existence of primary olfactory neurons highly selective for particular chemical classes of odors has not been clearly demonstrated. Thus, the alternative hypothesis, that olfactory receptor neurons are responsive to many odors and that odor discrimination arises as the result of physical segregation mechanisms, remains viable and is also the basis of current research.

This ignorance of the neural coding of olfactory quality sets the background for the study of olfactory bulb physiology. Different approaches to discrimination and coding of odor quality presumably require different kinds of processing by the bulb. Thus, the function of olfactory bulb circuits is inextricably linked with the question of how olfactory information is coded.

We will first review the classically described anatomy and physiology of the olfactory bulb. Then, we consider how different modes of olfactory coding might be influenced by bulb circuitry. Finally, we discuss the role of known transmitter-effector systems in the olfactory bulb circuits. Other

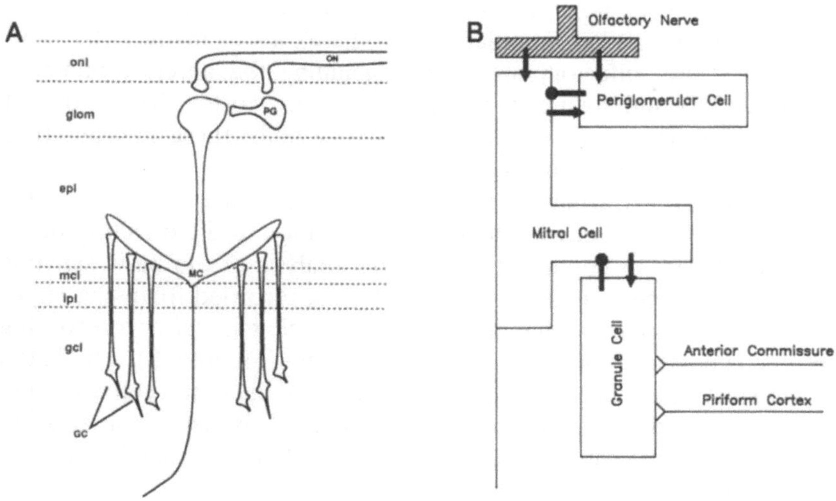

FIGURE 6.1. Diagrammatic representations of olfactory bulb cellular elements and connections. A. Schematic representation of major cellular elements. B. Diagrammatic representation of major synaptic interconnections. Arrows represent excitatory synapses; balls represent inhibitory synapses. Abbreviations: onl, olfactory nerve layer; glom, glomerular layer; epl, external plexiform layer; mcl, mitral cell layer; ipl, internal plexiform layer; gcl, granule cell layer; ON, olfactory nerve; PG, periglomerular cell; MC, mitral cell; GC, granule cell.

recent reviews of olfactory bulb function and anatomy may be of interest (Shepherd, 1972a and b, 1979; Halasz and Shepherd, 1983; Macrides and Davis, 1983; Scott and Harrison, 1987; Takagi, 1989; Shepherd and Greer, 1990).

Basic Circuitry of the Olfactory Bulb

The basic cell types and synaptic connections of the olfactory bulb are illustrated in Figure 6.1. The cell bodies of the main output neurons of the bulb, the mitral cells, lie in a narrow band in the interior of the bulb (the mitral cell layer, mcl). Each mitral cell sends a single primary (or apical) dendrite into the superficial layer of the bulb, the glomerular layer (gl), and several secondary dendrites into the external plexiform layer (epl). In the glomerular layer the mitral cell primary dendrite arborizes within a single glomerulus to receive synaptic contacts from terminals of the olfactory nerve. Each mitral cell sends a single axon out of the bulb toward more central olfactory structures. These axons collect in a discrete bundle, the lateral olfactory tract (LOT), which travels along the surface of the piriform cortex. A second class of output neurons, the tufted cells (not pictured), are morphologically similar to mitral cells but have their soma located in the epl and glomerular layer. The axons of many tufted cells project to rostral olfactory structures, including the anterior olfactory nucleus (AON) and rostral parts of the piriform cortex. The axons of other tufted cells remain within the olfactory bulb; these neurons thus function as local circuit neurons.

The study of extracellular field potentials generated by antidromic activation of the LOT or by orthodromic activation of the olfactory nerve (ON) has played an important role in the analysis of the basic synaptic events in the olfactory bulb. These waveforms, recorded following either LOT or ON stimulation, change in a characteristic manner as a recording electrode is passed through the layers of the bulb (Figure 6.2). The initial event is a fast potential that reflects the invasion of the mitral cell soma and possibly of the primary and secondary dendrites. This fast potential is followed by a slower wave, positive when the recording electrode is in the gcl and negative when it is above the mcl.

Following a single shock to the LOT or the olfactory nerve, there is a period of inhibition of the spontaneous activity of mitral cells. This inhibition is accompanied by a reduction in amplitude of the response to a test shock to the ON or the LOT (Phillips et al. 1963; Shepherd, 1963a and b).

Electrophysiological analysis, combined with ultrastructural studies of the synaptic connections of the epl, provided an explanation for these observations (Phillips et al., 1963; Rall et al., 1966; Rall and Shepherd, 1968; Review: Shepherd, 1972a and b). The positive wave in the gcl must result from synaptic currents flowing into cells that send processes across the mcl. The granule cells are the only large population of neurons fitting this description. It was considered that the positive wave might result from inhibitory currents arising in synapses below the mcl; however, it was argued that inhibitory currents are usually much smaller than excitatory currents and would not be large enough to generate the observed field potential. Thus, the positive potential is the result of excitatory currents flowing into granule cells from synapses on their dendrites in the epl, above the mitral cell layer (Rall et al., 1966; Phillips et al., 1963; Rall and Shepherd, 1968; Review: Shepherd, 1972a and b). These conclusions were confirmed by intracellular recordings from mitral cells (Mori and Takagi, 1978a).

Ultrastructural studies (Rall et al., 1966) demonstrated that mitral cell secondary dendrites form asymmetric synapses with the dendrites of granule cells in the epl, accounting for the synaptic current flowing into granule cell dendrites following stimulation of the LOT. The presumably excitatory synapse from mitral secondary dendrites to granule cell dendrites was frequently associated with a reciprocal synapse from the granule cell back to the mitral cell secondary dendrite.

Later work (Ribak et al., 1977) confirmed that granule cells are GABAergic. The inhibition associated with activation of the granule cells

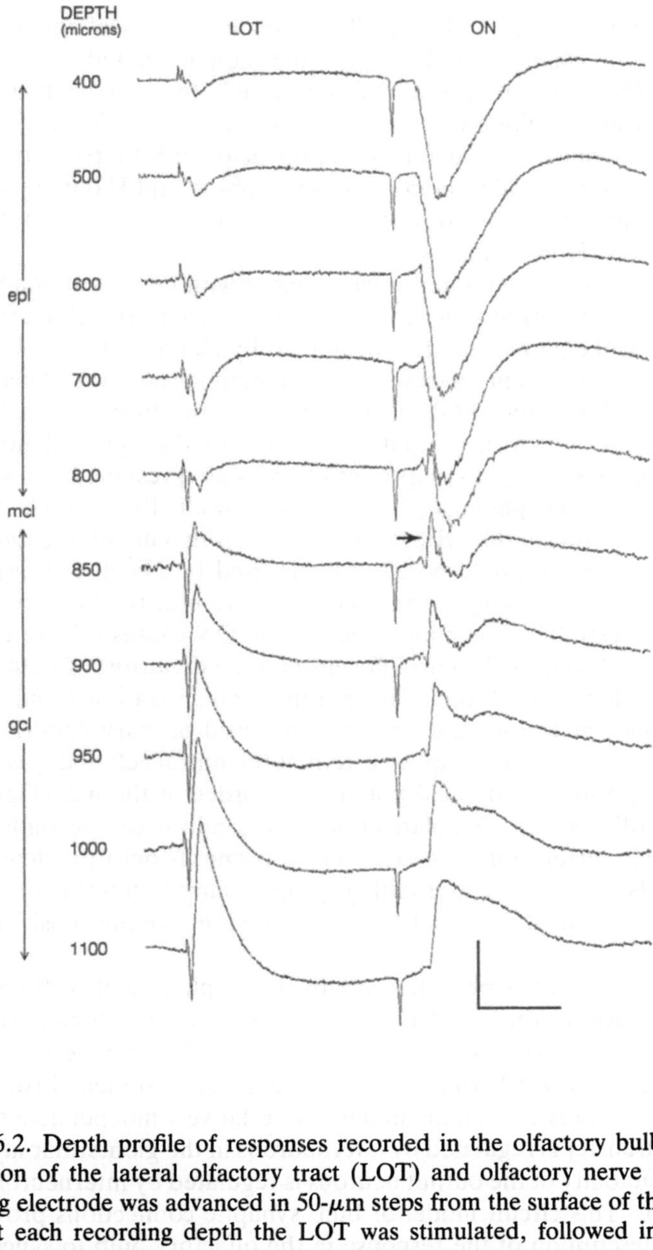

FIGURE 6.2. Depth profile of responses recorded in the olfactory bulb following stimulation of the lateral olfactory tract (LOT) and olfactory nerve (ON). The recording electrode was advanced in 50-μm steps from the surface of the olfactory bulb. At each recording depth the LOT was stimulated, followed in 50 ms by stimulation of the ON. Data were recorded by the authors, but are similar to those reported by Powell et al. (1963) and Nicoll (1972a). Calibration: 3 mV, 20 ms, positive upward. Abbreviations: epl, external plexiform layer; mcl, mitral cell layer; gcl, granule cell layer.

is blocked by picrotoxin and bicuculline (Nicoll, 1971; McLennan, 1971), as expected of classical GABAergic inhibition mediated by $GABA_A$-receptors. The transmitter used by mitral cells has not been definitely identified (ffrench-Mullen et al., 1985; Whittemore and Koerner, 1989), although antagonists of excitatory amino acid (EAA) receptors block transmission from mitral to granule cells (Jacobson and Hamberger, 1986; Jacobson et al., 1986) and from the terminals of LOT fibers to piriform cortex neurons (Hori et al., 1982).

Incoming olfactory nerve terminals synapse with the apical dendrites of mitral and tufted cells in spherical regions of neuropil, the glomeruli. The perimeters of the glomeruli are surrounded by glia and the cell bodies of interneurons that send processes into the glomerular neuropil. These interneurons [periglomerular (PG) cells] receive synapses from terminals of the olfactory nerve and form output synapses with the apical dendrites of mitral/tufted cells. Output synapses of PG cells are presumed inhibitory on the basis of their morphology (Pinching and Powell, 1971a and b) and use of GABA as a transmitter (Ribak et al., 1977; Mugnaini et al., 1984).

Analysis of evoked field potentials was used by Nicoll (1972a) to describe the events following orthodromic activation of the bulb by a single shock to the ON. The incoming volley in the ON causes release of transmitter from ON terminals onto PG cells and the dendrites of mitral/tufted cells in the glomerular layer. The resulting depolarization of mitral dendrites is conducted down the unbranched mitral primary dendrite to the mitral soma. Exit of the synaptic current at the mitral cell body layer (mcl) produces a positive-going field potential recorded at the mcl (Figure 6.2, arrow; Nicoll, 1972a). Depolarization of the mitral cell secondary dendrites by this current releases excitatory transmitter onto the dendrites of granule cells in the epl. The resulting synaptic current entering the granule cells produces a field potential that is negative above and positive below the mcl.

These classic studies provide a relatively simple model of the synaptic connections within the olfactory bulb. Transfer of olfactory information through the bulb involves a single synapse from olfactory nerve terminals to one of the two kinds (mitral and tufted cells) of principal (or output) neurons. This transfer is modulated by two relatively independent systems: (1) Initial transfer is regulated by interneurons in the glomerular layer, and (2) the excitability of the output neurons is regulated by interneurons in the gcl and epl. This circuit model of bulb synaptic connections provides an excellent description of the response in the olfactory bulb to synchronous activation of input and output pathways. The role of these connections in processing olfactory information is, however, poorly understood. The impulses conveyed to the olfactory bulb from primary olfactory neurons in response to odor stimulation are patterned both spatially and temporally. The response of the bulb to this input will depend upon both the spatial and temporal structure of this signal and characteristics of bulb circuitry that are not addressed in this standard model.

The Olfactory Code and the Anatomy of the Olfactory Nerve

Analysis of the functions of the olfactory bulb would be greatly simplified if the coding of information in primary olfactory axons were known. Unfortunately, however, widely differing concepts of the nature of this coding can be supported by available data and plausible argument. Since different assumptions about the organization of the input to the bulb lead to significantly different interpretations of experimental results, it is useful to discuss different models of olfactory coding that have been developed in the literature.

Broad Topographic Mapping

Although there is no obvious inherent relation between odor qualities and the surface of olfactory receptors, a mapping might be created either by segregation of receptor subtypes sensitive to classes of odors into different regions of the epithelium (Kauer and Moulton, 1974; Mackay-Sim et al., 1982) or by physical mechanisms that segregate odorants according to their physical properties onto regions of the epithelium (Mozell, 1964, 1966, 1970; Moulton, 1976). Either of these mechanisms would generate spatial and temporal patterns that might be conveyed to the olfactory bulb by topographic ordering of olfactory nerve axons. Topographic order in the projection from the olfactory epithelium to the bulb is crude relative to that of other sensory systems (e.g., Land and Shepherd, 1974; Land, 1973; Astic and Saucier, 1986; Saucier and Astic, 1986; Costanzo and O'Connell, 1980; MacKay-Sim and Nathan, 1984; Dubois-Dauphin, 1981; Duncan et al., 1990). This modest topographic ordering might, however, be sufficient to map a small number of epithelial regions onto the bulb surface; topographic ordering of odor properties might be sharpened by reorganization of olfactory axons within the olfactory nerve before termination in the bulb (Kauer, 1987).

Thus, coding of olfactory quality either by discrete receptor types or by physical properties might result in a mapping of odor properties onto the surface of the olfactory bulb.

There is no consensus as to the number of discrete odor qualities or the degree of resolution that might be achieved by physical segregation mechanisms, although neither the physical mechanisms available for segregating odors on the epithelium nor the degree of topography observed anatomically supports the idea that very large numbers of odor "qualities" are mapped onto the bulb surface. Relatively few "qualities" are required, however: 7 qualities discriminated into 8 levels of activation would, as an arbitrary example, define 8^7 or 2,097,152 distinct odors. Thus, one plausible model of olfactory coding is that the surface of the bulb is divided into a relatively small number of regions, each of which represents a single odor "quality."

Figure 6.3A is a diagrammatic representation of this concept. It is assumed (for illustrative purposes only) that there are seven classes of olfactory receptor neurons and that the axons of these neurons segregate into seven discrete regions of the bulb surface. Through some combination of intrinsic properties and physical segregation mechanisms, each of these classes of neurons is presumed to respond to a single odor "quality." A particular odor would activate each of the seven bulb regions proportionally to its activation of the class of primary olfactory neuron. Presumably, some odors will strongly activate one or a few bulb regions; other odors should produce more even activation of the bulb surface.

Experimental support for this model is provided by the observation that exposure of rats to high concentrations of certain odors results in regions of high metabolic activity in the bulb, as measured by the uptake of 2-deoxyglucose; for a given odor, the locations of these regions are similar in animals of the same species (Jourdan et al., 1980; Lancet et al., 1982; Royet et al., 1987; Sharp et al., 1975; Stewart et al., 1979).

Glomeruli as Functional Units

The broad topographic model of olfactory coding paces little emphasis on the role of glomeruli as functional units. An alternative concept, inspired by the organization of the olfactory nerve near the bulb, assumes an important role for the glomeruli.

Axons that travel from the epithelium in parallel fascicles break into a dense plexus after passage through the cribriform plate. Individual glomeruli receive fascicles entering from different directions (Le Gros Clark, 1951, 1956). A similar plexus of fibers is seen in amphibians (Scalia, 1976; Kauer, 1981; Daston et al., 1990; Duncan et al., 1990). The early observations of Le Gros Clark were confirmed by later studies using degeneration or tracers to label the fibers arising from restricted regions of the epithelium (Land and Shepherd, 1974; Land, 1973; Adamek et al., 1986; Stewart and Pedersen, 1987). Within the broad regions of the bulb labeled after focal injections or lesions in the epithelium, densely labeled glomeruli are often adjacent to lightly labeled glomeruli. Fascicles of labeled fibers can often be seen traveling significant distances through the nerve layer to enter a particular glomerulus (Land and Shepherd, 1974; Land, 1973).

Within the nerve layer on the surface of the bulb, the spread of fibers is limited. A cut across the nerve layer results in a relatively narrow denervated band caudal to the cut (Wellis and Scott, 1989); mitral cells beneath the denervated region were not responsive to stimulation. Freeman (1972c, 1974d) reached a similar conclusion by mapping the extent of field potentials and of excited mitral cells following focal electrical stimulation of the nerve layer.

The analysis of this anatomical arrangement by Le Gros Clark (1951, 1956) cannot be much improved: "One possibility is that there are different

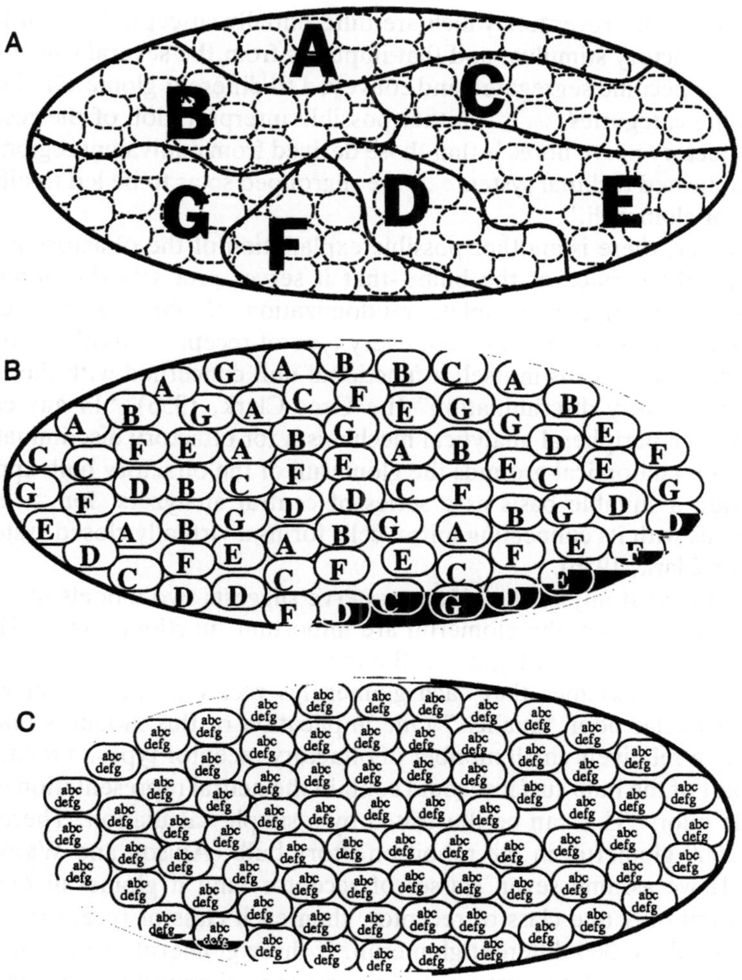

FIGURE 6.3. Possible mapping of odor characteristics onto the surface of the olfactory bulb. It is assumed that the olfactory epithelium responds to seven characteristics of odors. The three diagrams illustrate three logically possible ways in which these characteristics may be mapped onto the surface of the olfactory bulb.

A. Broad topography.
The fibers representing the different odor characteristics are presumed to segregate into broad regions of the bulb surface.

B. Glomerular selectivity.
Fibers representing the different odor characteristics are presumed to segregate into single glomeruli. Small regions of the bulb will then contain the representation for all possible odors.

C. Universal glomeruli.
Fibers representing the different odor characteristics are presumed to enter glomeruli randomly. Each glomerulus then contains the representation of all possible odors.

types of olfactory receptors which are differentially susceptible to various forms of olfactory stimulus, and that impulses from the several categories of receptor become segregated and conveyed to different glomeruli of corresponding categories. . . . Another possible interpretation of the resorting of olfactory nerve fibres is that those derived from equivalent regions of the complicated turbinal system . . . are regrouped so as to be led to different sets of glomeruli."

"However, there is another possible explanation of the olfactory nerve plexus on the surface of the bulb—that it serves precisely the opposite purpose of ensuring a complete randomization of fibres so that each glomerulus receives impulses from every type of receptor; in other words, that each glomerulus is in itself a functional unit concerned with the total range of olfactory discrimination" (Le Gros Clark, 1956). "In any case, whatever the peripheral analytical mechanism for olfactory discrimination may be, on anatomical grounds the glomeruli of the olfactory bulb appear to provide a suitable basis as a series of central analyzers, for . . . each glomerulus with its connecting mitral cells forms a virtually closed system" (Le Gros Clark, 1956).

Thus, the anatomy of the olfactory nerve suggests two models of olfactory coding in which the glomeruli are important functional units. These models are illustrated in Figure 6.3B and C.

Figure 6.3B assumes that each glomerulus receives fibers specific for some particular odor "quality." One implication of this model is that, if there are a relatively small number of olfactory receptor types, a relatively small part of the bulb surface would contain the neural representation of all possible odors. This can be illustrated by a simple calculation. There are about 2000 glomeruli in the rabbit olfactory bulb (Allison and Warwick, 1949). If we assume seven classes of receptor, and if glomeruli receive axons from only one class of receptor, then each class of receptor would project to about 300 different glomeruli. If these glomeruli were scattered randomly over the bulb surface, and the average glomerulus were 200 μm in diameter, then on average a region of the bulb surface of about 0.2 mm^2 (0.5 mm diameter) would contain one of each type of glomerulus. Larger numbers of receptor types would require a proportionally larger part of the bulb surface to represent all possible odors. In the limit of a very large number of odor qualities, the entire bulb would be required to discriminate between odors.

The second model suggested by Le Gros Clark, in which each glomerulus receives a sample of all possible fibers, is illustrated in Figure 6.3C. Random distribution of individual fibers would result in a nearly equal distribution of fiber types among the glomeruli (Le Gros Clark, 1956). Each glomerulus would be capable of discriminating all possible odors. In this case, as with the model illustrated in Figure 6.3B, small portions of the bulb surface would be capable of representing all possible odors; however, glomeruli would be identical.

Evidence that a relatively small part of the bulb is capable of representing odor quality is provided by the observation of Slotnick et al. (1987) that a learned olfactory discrimination survives ablation of most of the bulb.

Several kinds of experimental evidence indicate that glomeruli are heterogeneous functional units, as suggested in the illustration of Figure 6.3B. Metabolic activity in individual glomeruli, determined by 2-deoxyglucose (Lancet et al., 1982) or by cytochrome oxidase activity (Shipley and Costanzo, 1984; Costanzo et al., 1984; Weinberg and Meisami, 1989; Mouradian and Scott, 1988), is constant within a glomerulus but varies with no apparent pattern from glomerulus to glomerulus. Also consistent with the importance of glomeruli as functional units is the observation that a slow electrical potential recorded across a single glomerulus showed a different pattern of response to a battery of nine odorants for each glomerulus (Leveteau and MacLeod, 1966).

The models presented in Figure 6.3 represent three very different ways in which olfactory quality might be represented on the bulb surface. In the first model (Figure 6.3A), the olfactory system functions somewhat like an eye with a cloudy lens. Each odor produces a blurred pattern on the bulb surface. In analogy with the retina, lateral inhibitory connections within the bulb might "sharpen" these patterns by enhancing differences between regions receiving different levels of excitation. The models that emphasize the role of the glomerulus, on the other hand, show less resemblance to the vertebrate retina. In the model of Figure 6.3B, in which individual glomeruli receive fibers representing a single odor "quality," active and inactive glomeruli would be randomly distributed over the bulb surface; integration of activity across these separate glomeruli would involve extensive connections between the mitral/tufted cells associated with the active glomeruli. Models such as those suggested in Figure 6.3C, in which each glomerulus receives input from many classes of fibers, might require extensive synaptic processing within the glomerulus.

An important constraint on the mapping of odor quality onto the bulb surface is set by the fact that primary olfactory neurons undergo turnover and replacement throughout life (Moulton, 1974; Graziadei and Metcalf, 1971). Unless the replacement axons are able to reestablish similar connections, perceived odor quality would be altered and the animal would lose the ability to remember odors. This suggests that the olfactory pathway has the ability to establish appropriate connections long after embryonic development. This might be accomplished by the maintained expression of biochemical markers that guide new axons of a particular type to their appropriate termination. An alternative hypothesis is suggested by the observation that regenerating olfactory neurons exhibit increased odor selectivity after establishing contact with the olfactory bulb (Adamek et al., 1984; Mair et al., 1982). Primary olfactory neurons might have unspecified odor selectivity until their terminals reach the glomerular layer of the olfactory bulb; selectivity could be specified by molecular signals retrogradely transferred from cells in the bulb to neurons in the epithelium.

The different models illustrated in Figure 6.3 emphasize the fact that there are several plausible ways in which odor qualities could be "mapped" onto the surface of the bulb. An interesting question is whether these "mappings" should be termed topographic. In the case of the retina, cochlea, and skin, contiguous regions of receptors map into the brain as contiguous regions of termination. This is the usual sense of the term "topographical organization." However, if like elements in a receptor array are randomly distributed but become associated when they map onto the brain, the resulting map is no less specific. If like receptors are randomly placed in the epithelium in order to increase their exposure to odorants, then the convergence of like receptors at the level of the bulb would serve the same function as mapping contiguous regions in sensory systems where receptor contiguity is an essential component of the coding.

The connections and operations carried out by circuits of the olfactory bulb are presumably optimized to process the mapping actually used by the olfactory system. Thus, analysis of the specific temporal and spatial characteristics of bulb circuitry may provide useful clues about the nature of the olfactory code.

Neural Processing in the Glomerular Layer

Summation of Inputs

The firing rate of a primary olfactory neuron is presumably proportional to the concentration in its vicinity of the odorants to which it is sensitive. Even for an ideal transducer, however, there is a necessary minimum of noise arising from the random variation of odorant concentration in a small region (Schild, 1988). An imperfect transduction mechanism might add additional noise. There are, however, almost certainly a very large number of olfactory neurons associated with each class of odorant, and the sum of the activity of all or part of these neurons would be a far more precise indicator of odorant concentration. Summation of the activity of many primary neurons would appear necessary in order to make use of the large epithelial surface provided in macrosmatic mammals. Thus, one plausible function of the initial synaptic connection in the bulb is the derivation of an average activity of input fibers.

If there are 50,000,000 olfactory receptors and 2000 glomeruli (Allison and Warwick, 1949), then each glomerulus receives 25,000 olfactory axons. If there are 25 mitral cells associated with each glomerulus (Allison and Warwick, 1949), then the average number of axons terminating on each mitral cell is about 1000. Electrotonic conduction within the apical dendrite of a single mitral cell provides an adequate mechanism for summation of the inputs to that neuron. The reduction in noise in the summated activity is $\sqrt{1000}$ or about 30-fold (Holley and Doving, 1977; Schild, 1988). If a glomerulus receives axons of a single class, noise in the summated activity of these axons would be reduced by the $\sqrt{25,000}$ or 160-fold from the noise present in a single axon. It is not clear, however, how much of this enormous potential increase in sensitivity is realized. There is no evidence that the mitral cells associated with a single glomerulus project to common targets (Scott et al., 1980; Scott, 1981; Ojima et al., 1984). Hence, there appears to be no opportunity for additional convergence of the mitral cells from a single glomerulus onto a relatively small number of target neurons, as would appear necessary for effective summation of their activity.

Inhibitory Systems of the Glomerular Layer: Periglomerular (PG) Neurons

In addition to summation of inputs from olfactory nerve terminals, a second plausible function of the circuitry of the glomerular layer is regulation of transmission from the olfactory nerve to mitral and tufted cells. Such regulation might, for example, act to maintain the excitation of mitral/tufted cells relatively constant over a wide range of odor intensities in order to increase the dynamic range of the olfactory system. Ultrastructural studies of the synaptic connections of the glomerular layer provide an anatomical substrate for this kind of regulation.

The standard wiring diagram of the glomerular layer is primarily based upon Golgi and ultrastructural studies of synaptic connections of this region (Pinching and Powell, 1971a and b; Schneider and Macrides, 1978; Macrides et al., 1985). Terminals of the olfactory nerve synapse both with the apical dendrites of mitral and tufted cells and with periglomerular cells. Periglomerular cells, which send processes primarily into a single glomerulus, form reciprocal synapses with the apical dendrites of mitral and tufted cells. The synapses leaving PG cells are symmetrical and, therefore, are considered inhibitory. There is thus a "feedforward" pathway for inhibition of second-order neurons. The mitral/tufted side of the reciprocal synapse is asymmetrical and is, therefore, considered excitatory. Thus there is also a pathway for "feedback" inhibition from mitral and tufted cells back to PG cells. Several smaller classes of neurons send axons to neighboring glomeruli (Schneider and Macrides, 1978).

A schematic diagram of these connections is presented in Figure 6.1. The predictions arising from this schematic diagram have been partially confirmed by physiological experiments.

Responses of PG Cells to Electrical and Odor Stimulation

Extracellularly recorded units in the glomerular layer fire one or a few spikes following electrical stimulation of the olfactory nerve (Getchell and

Shepherd, 1975b; Freeman, 1974b). Intracellular recordings from a small population of PG neurons showed that ON stimulation produces an EPSP of about 30 ms duration (Wellis and Scott, 1990). In some cells, larger EPSPs resulted in fewer spikes, presumably because of depolarization block. In extracellular recordings this depolarization block would be interpreted as inhibition.

PG neurons are facilitated following weak shocks to the olfactory nerve; at longer stimulus intervals or with stronger stimuli, these neurons are inhibited (Getchell and Shepherd, 1975b). Facilitation and inhibition were inferred from changes in response latency, response threshold, or the number of spikes generated by a constant stimulus. Intracellular recordings from PG cells revealed an IPSP of 200 ms duration produced by ON stimulation (Wellis and Scott, 1990).

Reciprocal dendrodendritic synapses between mitral/tufted apical dendrites and PG cell dendrites were also found in EM studies of the glomerular layer (Pinching, 1970; Pinching and Powell, 1971a and b). Antidromic activation of mitral cells by LOT stimulation should result in depolarization of the apical dendrites and activation of these synapses; however, LOT stimulation appears to have little effect on the activity of glomerular layer neurons (Getchell and Shepherd, 1975b; Wellis and Scott, 1990).

Excitatory Systems of the Glomerular Layer

There is also evidence for excitatory interactions within the glomerular layer. Electrical stimulation of the olfactory nerve activates some units that should be outside the spread of the activated fibers, and some units are activated after long latencies suggestive of polysynaptic pathways (Getchell and Shepherd, 1975b; Wellis and Scott, 1990). In turtle bulb, orthodromic but not antidromic stimulation activates a slow excitatory current (Nowycky et al., 1981).

Freeman (Freeman, 1974a and b; Martinez and Freeman, 1984) has argued that PG cells are primarily excitatory. The most direct evidence is the demonstration of a potential difference indicating current flow between the glomerular and external plexiform layers. This potential is activated by orthodromic (olfactory nerve) but not by antidromic (LOT) stimulation and is of relatively long duration (30–50 ms). It is argued that this potential is the result of excitatory input to the apical dendrites of mitral cells.

There are several morphological cell types within the glomerular layer (Schneider and Macrides, 1978; Macrides et al., 1985); any of these might provide excitatory inputs to mitral and tufted cells, although there are no available data on the synaptic connections of these subtypes. In summary, the available evidence suggests that both excitatory and inhibitory actions are present within the glomerular layer; further work is needed to define the roles of these events in signal processing in the glomeruli.

Responses of PG Cells to Odors

The intracellular responses of periglomerular cells to odors provide the only direct evidence of the actions of odors on PG cells (Wellis and Scott, 1990). All cells (3) responded to odor stimulation with a burst of spikes associated with a depolarization. Two of the three cells showed a reduction in response with successive "sniffs," although there was no apparent IPSP.

Glomerular Inhibition

Because olfactory nerve stimulation activates both glomerular and mitral/granule synapses, it is difficult to separate the relative contribution of these two inhibitory systems to mitral cell inhibition. In some preparations, however, stimulation of the olfactory nerve (orthodromic stimulation) results in a longer-lasting inhibition of mitral cells than stimulation of the LOT (antidromic stimulation). Freeman (1974a) reported that at stimulation rates above 2 pulses/s, the response to ON stimulation [average evoked potential (AEP) recorded with surface electrodes] progressively diminishes, whereas the response to antidromic activation of LOT maintains a constant amplitude with stimulus rates of 15 pulses/s. The inhibition of the ON response varied significantly between experiments (from 2 to 15 pulses/s being required for inhibition of the orthodromic response). Strong inhibition of the ON response was associated with a surface negative shift in the baseline of the oscillatory response evoked by the weak ON stimulus. A conditioning shock to the ON resulted in inhibition of the response to a test shock which recovered with an exponential time course of hundreds of milliseconds. The period of inhibition is substantially longer than the duration of spike activity generated in PG neurons by ON stimulation (Freeman, 1974b). This inhibition was not activated by antidromic (LOT) stimulation (Freeman, 1974a), indicating that it is not associated with the mitral/granule inhibitory system, which is as effectively activated by antidromic as by orthodromic stimulation. Thus, activation of interneurons in the glomerular layer may cause a relatively slow inhibitory process that reduces transmission from olfactory nerve terminals to mitral/tufted cells.

A slow inhibitory process in the glomerular layer is also present in the isolated turtle olfactory bulb. A conditioning ON shock inhibits the response to a test ON shock as long as 1 s. This inhibition was not accompanied by a detectable change in membrane resistance, nor was it generated by antidromic activation of mitral cells through the LOT (Mori et al., 1984).

Possible Mechanisms of Glomerular Inhibition

Additional clues about the functions of glomerular circuitry are provided by the localization of transmitters and receptor subtypes within the bulb:

PG cells contain the transmitters GABA (Ribak et al., 1977; Mugnaini et al., 1984) and dopamine (Halasz et al., 1978, 1981; Davis and Macrides, 1983; McLean and Shipley, 1988). GABA and DA may be colocalized in some PG cells (Gall et al., 1987; Kosaka et al., 1985), but these observations are not confirmed by other workers (Mugnaini et al., 1984).

Receptor subtypes for these transmitters that probably operate through slow "second messenger" mechanisms are specifically localized to the glomerular layer. Binding of the $GABA_B$ agonist baclofen in the olfactory bulb is segregated in the glomerular layer (Bowery et al., 1987). DA receptors in the bulb appear to be the D2 subtype (Dawson et al., 1988; Closse et al., 1988; Boyson et al., 1986; Gehlert and Wamsley, 1985). The binding of D2 ligand spiperone in the rat olfactory bulb is located in the glomerular layer and also in the olfactory nerve layer (Mansour et al., 1990; Nickell et al., 1990, 1991). In other parts of the CNS, $GABA_B$ and D2 receptors have been shown to inhibit neurons by similar second messenger-mediated mechanisms (Pinnock, 1984; Gahwiler and Brown, 1985; Innis and Aghajanian, 1987; Lacey et al., 1988).

Topical application of baclofen inhibits the field potential response to ON stimulation in the frog (Potapov, 1985; Potapov and Trepakov, 1986) and in the rat (Nickell and Shipley, 1988c), suggesting a role for $GABA_B$ receptors in control of transmission from the olfactory nerve to mitral and tufted cells. In isolated turtle bulb, DA agonists had different actions than baclofen, regulating the strength of inhibition of the response to a test olfactory shock by a conditioning shock (Nowycky et al., 1983).

Thus, pharmacological data are consistent with the physiological and anatomical observations described above in suggesting that PG cells containing GABA or dopamine modulate transmission from the nerve to mitral tufted cells via slowly acting inhibitory receptors. It is probable that the more rapidly acting $GABA_A$ receptors are also present in the glomerular layer; however, physiological demonstration of their actions is difficult because their time course and pharmacology are presumably identical to those of the $GABA_A$ receptors mediating inhibition at the mitral/granule synapse in the epl.

Inhibition of transmission from olfactory nerve terminals to mitral and tufted cells might be mediated postsynaptically by shunting of currents in the apical dendrites or presynaptically through some mechanism that reduces the release of transmitter from olfactory nerve terminals, or both. Because classical ultrastructural studies (Pinching and Powell, 1971a and b) did not find classical synapses on olfactory nerve terminals, it has usually been considered that transmitter mediated control of transmitter release from these terminals is improbable (Freeman, 1974a; Mori et al., 1984).

There is, however, some evidence for a presynaptic locus of glomerular inhibition. The slow time course of the inhibition suggested nerve terminal inhibition by buildup of some substance such as K^+ in the glomeruli (Freeman, 1974a; Mori et al., 1984; Jahr and Nicoll, 1981). Jahr and Nicoll

(1981) used a sucrose gap technique to monitor currents flowing into the terminals of the frog olfactory nerve. Electrical stimulation of the nerve produces a prolonged depolarization of the terminals. Jahr and Nicoll considered this effect to be analogous to the depolarization associated with presynaptic inhibition in dorsal roots of the spinal cord (Review, Nicoll and Alger, 1979). The presence of spiperone binding in both the glomerular and nerve layers (Mansour et al., 1990; Nickell et al., 1990, 1991) suggests a presynaptic localization of D_2 receptors on olfactory nerve terminals, and thus a direct presynaptic action of dopamine on these terminals.

Significance of the Glomerular Layer for Signal Processing

A slow inhibitory system in the glomerular layer activated by the olfactory nerve and regulating transmission from the olfactory nerve to secondary neurons would be ideally situated to function as an "automatic volume control of input" (Freeman, 1974b). This circuitry would maintain excitation of mitral/tufted cells relatively constant over a wide range of odor intensities; increased activity in olfactory nerve terminals would recruit additional inhibition. This would act to increase the dynamic range of second-order neurons.

The processes of some juxtaglomerular interneurons appear to extend over a distance corresponding to about 2 to 3 glomeruli (Macrides et al., 1985; Schneider and Macrides, 1978); thus, an "automatic volume control" might respond to the input level over a region comprising 4 to 9 glomeruli. Within a glomerulus, there is presumably strong electrotonic coupling of synaptic currents between parts of individual neurons; coupling between glomeruli is presumably via axons and, hence, requires action potentials (Getchell and Shepherd, 1975b).

The Functional Connection between the Glomerular Layer and the Deeper Layers of the Olfactory Bulb

The poorly understood integrative functions of juxtaglomerular interneurons result in a net current transferred to the apical dendritic tufts of the mitral and tufted cells. The only connection between this apical tuft and the mitral soma is the single unbranched primary dendrite. In the monkey, the primary dendritic shaft of some mitral cells may be myelinated (Pinching, 1971). In the rat, the primary dendrite shafts, identified by Golgi impregnation in EM sections, receive only a small number of synaptic contacts and are surrounded over most of their length by a series of glial sheaths (Shipley and Zahm, 1990). Thus, the mitral dendritic shaft may be electrically insulated and thus electrotonically short, providing a relatively tight coupling between the branched part of the apical dendrite and the cell body.

In intracellular recordings from turtle (Mori et al., 1982) and rabbit (Mori and Takagi, 1975) mitral cell somata, orthodromic stimulation generates a fast prepotential on the rising phase of the EPSP, and in some cells orthodromic stimulation produces an action potential in the soma without a preceding EPSP. These observations suggest that an action potential is generated somewhere in the apical dendrite or in the dendritic shaft.

The Mitral/Granule Inhibitory System

Much experimental work has focused on the reciprocal synaptic connections between the secondary dendrites of mitral/tufted cells and the granule cells because the density of these synaptic interactions suggests their fundamental importance to bulb function.

Morphology of Mitral and Tufted Cells

The morphologies of mitral and tufted cells have been studied in detail using several methods and species. Macrides and Schneider (1982, hamster) characterized the distributions of dendrites of mitral and tufted cells in Golgi-impregnated material. Intracellular filling of impaled neurons with HRP was used to describe the detailed structure of individual neurons (Mori et al., 1983; Kishi et al., 1984; rabbit; Mori et al., 1981b; turtle). In mammals, each mitral cell sends a single primary dendrite to a single glomerulus. (In the turtle bulb, mitral cells typically send a primary dendrite to two glomeruli; Mori et al., 1981b.) In addition to this primary dendrite, each mitral cell also sprouts several secondary dendrites. These secondary dendrites extend for very long distances through the epl, often covering 2 mm or one-fourth of the bulb's circumference. Reconstruction of mitral cells did not reveal any preferential orientation of the dendrites; the arbors tended to be symmetrical.

Scott's laboratory studied a sample of mitral and tufted cells (Orona et al., 1983; rat) using extracellular iontophoretic deposits of HRP among the dendrites of mitral/tufted cells. Dendrites passing through the injection site took up the tracer, which was transported throughout the neuron, providing Golgi-like labeling of a small population of neurons. These workers describe two classes of mitral cells. Secondary dendrites of the most numerous class (Type I) ramify almost entirely in the deep part of the external plexiform layer. A second class of mitral cells (Type II) sends dendrites almost exclusively to more superficial parts of the epl. The distributions of the dendrites of these two types of mitral cell corresponds with the laminar distribution of cytochrome oxidase activity in the epl (Mouradian and Scott, 1988).

The superficially located dendrites of Type II mitral cells extend over a smaller area than the deeper dendrites of Type I mitral cells. The population studied by Orona et al. may not be a random sample, since visualization of neurons was dependent upon the placement of the iontophoretic injection; however, in their population 79 of 274 mitral cells (29 percent) were classed as "Type II."

Some calculations based on the dimensions of secondary dendrites provide insight into their possible functions. Each Type I mitral cell sustained an average of 6.2 secondary dendrites, each of which branches several times (Orona et al., 1983). There are approximately 25 mitral cells projecting into a single glomerulus (Allison and Warwick, 1949). If the secondary dendrites of these mitral cells extended radially outward equally in all directions, there would be 150 secondary dendrites from each glomerular unit, providing 1 secondary dendrite per 2.4 degrees. At 2 mm from the center of the glomerular unit, secondary dendrites from one glomerulus would be separated by about 80 μm. Since each secondary dendrite branches several times, dendrites may be even more closely spaced. A similar conclusion can be reached from the estimate of Mori et al. (1983) of the mean total length of the secondary dendrites of a single mitral cell as 15,000 μm; 25 mitral cells would have a total dendritic length of 375,000 μm (375 mm). If these dendrites were cut into 2-mm lengths and distributed evenly over a 2-mm square the average spacing would be 11 μm. Since the diameter of the entire dendritic arbor of a granule cell is 100 to 200 μm (Orona et al., 1983), each granule cell could be contacted by dendrites of several mitral cells from each glomerulus. Thus, if the mitral cells of a single glomerulus are activated or inhibited as a unit and the secondary dendrites of these mitral cells sustain action potentials, their activation could create a uniform excitation of granule cells over a large part of the bulb.

Recurrent Collaterals of Mitral and Tufted Cells

Both mitral and tufted cells have axon collaterals that ramify exclusively in the gcl (Kishi et al., 1984). These collaterals extend over greater distances than the secondary dendrites, but the axons tend to project in only one direction.

If the reciprocal synapses between mitral secondary dendrites and granule cells are suppressed by a conditioning antidromic shock to the LOT, then the response to a test LOT shock is positive in the gcl (Nicoll, 1970), as would be expected from activation of mitral axon collaterals synapsing on granule cells in the gcl. These connections may thus provide a pathway for mitral cell activation of granule cells that is not affected by inhibition of mitral cell secondary dendrites.

Tufted Cells

The group of cells anatomically classified as tufted cells is in fact a heterogeneous collection of cells with different projection patterns and transmitters (Macrides et al., 1985). Some tufted cells project out of the bulb to the anterior olfactory nucleus and the anterior piriform cortex (Scott et al., 1980; Scott, 1981). Tufted cells also project via the internal plexiform layer to a region on the opposite side of the olfactory bulb (Schoenfeld et al., 1985). A third class of tufted cells resides near or within the glomerular layer and sends axons to other nearby glomeruli (Macrides et al., 1985). Tufted cells respond more reliably and more vigorously to odors or to electrical stimulation of the olfactory nerve than mitral cells (Schneider and Scott, 1983).

Granule Cells

Freeman (1972a) noted that Golgi-impregnated granule cells tend to be of the same length; thus, deeply situated granule cells will tend to interact with mitral dendrites deep in the epl, while superficially situated granule cells will interact principally with mitral and tufted cells that have dendrites in the superficial parts of the epl. This laminar segregation of the granule cell connections was studied more rigorously by Orona et al. (1983), again using the technique of localized extracellular injection of HRP. According to their reconstructions, each granule cell soma sends one or two dendrites toward deeper parts of the gcl, about 50 to 100 μm, and a single apical dendrite extends upward toward the epl. This dendrite generally does not branch until it reaches the epl; it may not branch extensively until it reaches the superficial epl. The dendritic arbor of the granule cell extends over an area of diameter 100 to 200 μm.

This laminar distribution of granule cells and their apical dendrites, combined with the segregation of mitral and tufted cell secondary dendrites into different depths of the epl (Orona et al., 1984), implies that mitral and tufted cells interact with different populations of granule cells. Tufted cells that project from the bulb terminate primarily in the anterior olfactory nucleus (AON) and anterior piriform cortex (Scott et al., 1980; Scott, 1981; Schneider and Scott, 1983). Reciprocal projections from the AON, the anterior piriform cortex, and the anterior commissure terminate primarily in the ipl and superficial gcl and thus primarily influence superficial granule cells, whose dendrites most probably contact tufted cells. Mitral cells project to more caudal levels of piriform cortex, and the reciprocal projections from posterior piriform cortex terminate more deeply in the gcl (Davis and Macrides, 1981; Luskin and Price, 1983). Thus, Orona et al. (1983) propose that there may be two parallel and partially independent olfactory processing systems in the bulb: the mitral cells and their associated granule cells, and the tufted cells, with their separate population of granule cells. It is not clear, however, to what extent the segregation of granule cell contacts is common to all mammals. In the rabbit, Mori et al. (1983) describe a class of granule cells with short dendrites; these cells have their cell bodies near the mcl and project into the deep parts of the epl.

Macrides et al. (1985), using Golgi impregnations in the hamster olfactory bulb, agree with Mori et al. that granule cells are heterogeneous in length. Thus, the generality of the conclusions of Freeman (1972a) and Orona et al. (1983) is not clear.

Granule cells appear to be electrotonically short (Rall and Shepherd, 1968; Shepherd and Brayton, 1979); thus, depolarization of any part of the cell spreads relatively effectively to all parts. Most granule cells appear not to be spontaneously active in the absence of olfactory or electrical stimulation, as passage of a recording electrode through the gcl ordinarily encounters no active units; however, spontaneously active units that may be granule cells can be isolated near the mcl (McLennan, 1971). Granule cells can produce spikes, however, during excitation by antidromic (LOT) or orthodromic (anterior commissure) stimulation (Mori and Kishi, 1982). In a population of 20 granule cells studied with intracellular recording, 8 responded to odor stimulation with spikes, 7 responded with excitatory responses, 4 showed no response, and 1 cell was inhibited (Wellis and Scott, 1990).

A recent study using dye coupling, EM analysis, and antibodies to gap junctions demonstrated that granule cell bodies are coupled by tight junctions (Reyher et al., 1991). There are no electrical measurements of the degree of electrical coupling; however, a possible function of these junctions is to link groups of granule cells into synchronously active units.

Lateral Inhibition versus Self-inhibition

The responses of the mitral/granule system to single or paired orthodromic and antidromic volleys are well understood. But the function of this reciprocal synaptic mechanism in processing odor-evoked activity depends on its response to spatially and temporally patterned activity in the olfactory nerve.

One possible function of the mitral/granule system is to mediate some form of lateral inhibition that would sharpen the differences between active and inactive regions of the bulb, a function analogous to similar operations in the retina and other sensory systems. In the classical descriptions of the mitral/granule system, however, mitral cells are described as primarily self-inhibitory (Rall et al., 1966; Rall and Shepherd, 1968; Rall, 1972; Shepherd, 1972a). Self-inhibition would predominate if there are many reciprocal synapses near the mitral soma or if the secondary dendrites are electrotonically short (so that synaptic currents occurring anywhere on the dendrite are conducted with relatively little decrement to the soma). If, on the other hand, the secondary dendrites are electrotonically long, there are relatively few reciprocal synapses near the soma, and the dendrites sustain action potentials, inhibitory currents generated in distal parts of the dendrite would have less effect on the soma, but action potentials in the soma might propagate out the dendrite, activating the reciprocal synapses with granule cells (Figure 6.1). In this case an active mitral cell (or group of cells) would preferentially inhibit other mitral cells in the more remote reaches of its dendritic arborization. To the degree that one or the other of these two modes of inhibition predominates during the processing of odor signals, very different kinds of functions could be performed by the mitral/granule system.

As noted above, the secondary dendrites of mitral cells extend for extraordinary distances, on the order of 2 mm. With reasonable values of membrane resistance, the secondary dendrites would be electrotonically long; activity in the reciprocal synapses would decrease length constants further. Thus, voltage changes in distal parts of the dendrites are probably only weakly coupled to the soma. Inhibitory activity in more proximal parts of the dendrite soma would be more strongly coupled to the soma. Both self-inhibition and lateral inhibition may be generated by this anatomical arrangement. The relative importance of the two kinds of inhibition can be assessed by intracellular recordings from mitral cells, and data from such recordings are available from several studies.

The inhibitory potentials (IPSPs) recorded intracellularly from mitral cells following LOT stimulation increase in a graded manner as LOT stimulation strength is increased from threshold. The appearance of a spike in the impaled cell does not cause a large increment in the IPSP (Yamamoto et al., 1962; Mori and Takagi, 1978), suggesting that most of the IPSP results from activity in other mitral cells. In the isolated turtle olfactory bulb, individual mitral cells were impaled and stimulated by injection of current through the recording pipette (Jahr and Nicoll, 1982a). In nearly all cells tested, stimulation of the impaled cell alone produced a smaller IPSP than either orthodromic (ON activation) or antidromic (LOT) stimulation. There was considerable variability, however: in some cells direct stimulation of the impaled neuron produced a large IPSP, but in many others only a small IPSP was produced.

These observations suggest that both self-inhibition and lateral inhibition are present in the mitral/granule system. The rate of spikes from the soma of a mitral cell probably reflects the summation of excitatory currents from the primary dendrite and inhibitory currents from proximal parts of the secondary dendrites. An action potential generated in the soma probably propagates out the secondary dendrites, possibly as a calcium spike (Jahr and Nicoll, 1982a). This propagated spike will release excitatory transmitter, but the quantity of transmitter released and the extent of spread of the spike will depend upon the local level of inhibition. The IPSP (in the dendrite) following the spike blocks further release of transmitter for a period of several milliseconds. The activity of other mitral cells and of centrifugal inputs to the bulb (see below) may also set independent levels of inhibition in the distal parts of the secondary dendrites. Thus, quite complex temporal and spatial patterns could be generated.

The effect of the mitral/granule system on inputs to the bulb can be indirectly assessed by determining the pattern of excitation and inhibition caused by localized stimulation of the olfactory nerve. Freeman (1974c) mapped the pattern of average evoked potential (AEP) and single unit responses following focal stimulation of the olfactory nerve. This study concluded that ON stimulation results in an initial excitatory focus which is replaced by a larger inhibitory focus, not in an excitatory focus surrounded by an inhibited region, as expected for pure lateral inhibition.

Oscillatory Responses of the Mitral/Granule System

The studies reviewed above demonstrate that the possible modes of operation of the mitral/granule system are too complex to be adequately described by simple labels such as self-inhibition or lateral inhibition. An adequate description of the behavior of this system may require the development of models with appropriate simplifying assumptions. One such model is based on the ability of the mitral/granule system to develop oscillations.

Oscillatory activity of about 50 Hz was a prominent feature of early descriptions of the response of the olfactory bulb to odors (Adrian, 1950). Depth recordings demonstrate that the oscillations recorded at the bulb surface are the result of synaptic currents flowing into granule cells following mitral cell activation (Freeman, 1972a). Stimulation of the LOT or ON at low stimulus amplitudes results in oscillatory averaged responses rather than the monophasic responses observed after large shocks (Freeman, 1972b). Antidromic activation of mitral cells by stimulation of the LOT at frequencies near this natural oscillation results in alternation of large and small responses (Mori et al., 1977; Westecker, 1970).

The generation of these oscillations can be rationalized from the classical descriptions of the mitral/granule system (Shepherd, 1972a and b, 1979; Rall, 1972). An alternative viewpoint is provided by noting that the mitral/granule system constitutes a negative feedback loop (Freeman, 1972b). All such feedback systems tend to oscillate at the frequency corresponding to the total of delays around the loop; thus, the mitral/granule system may be expected to oscillate at a frequency corresponding to the duration of the IPSP plus the delay in the EPSP that generated it. The duration of the IPSP is strongly influenced by anesthetics; in unanesthetized preparations, or with certain anesthetics, this period is approximately 20 msec, corresponding to the experimentally observed oscillation frequency of 50 Hz. Barbiturate anesthesia, which increases the delay around the loop by prolongation of the IPSP (Stewart and Scott, 1976; Scott and Stewart, 1979; Nicoll, 1972b) reduces the frequency of the spontaneous oscillations (Adrian, 1950; Scott and Aaron, 1977).

This oscillatory behavior provides one model for the operation of the mitral/granule system. If a spatially patterned input is applied to the bulb, groups of neighboring mitral cells receiving sufficient excitatory input might break into oscillation; interactions mediated by the long secondary dendrites would tend to synchronize these groups of neighboring cells over a large portion of the bulb surface. Mitral cells receiving insufficient excitation for oscillation would be inhibited by the activity of the oscillating cells.

This mode of operation could accommodate either the broad topographic or the glomerular models of olfactory input. If olfactory information is coded on broad regions of the bulb (Figure 6.3A), then the mitral cells within a region receiving sufficient excitation might oscillate; the long secondary dendrites might serve to inhibit other regions receiving less excitatory input. Alternatively, if olfactory input is coded into individual glomeruli (Figure 6.3B), the long secondary dendrites might function to synchronize oscillations between active but spatially separated glomeruli.

Responses of Bulb Neurons to Odors and Electrical Stimulation

The "product" of the connections of the glomerular and deep layers of the bulb is activity of output neurons, the mitral and tufted cells. The spatial and temporal patterns of mitral/tufted cell activity constitute the information passed on to more central olfactory structures. However, our ignorance of the kinds of patterns generated by olfactory nerve inputs renders interpretation of recordings from individual mitral and tufted cells difficult. Since we do not know the location of the neurons actually involved in the response to a particular odor, there is no assurance that the individual mitral cell contacted by the recording electrode is involved in any important way in the response to any particular odor. Thus, searching for patterns in the responses of bulb neurons to odors is somewhat analogous to recording from the visual system while allowing the monkey to hold the light.

Nevertheless, it should be possible to distinguish between broad classes of activity patterns. If odorants produce broad zones of activity (as illustrated in Figure 6.3A), then it would be expected that different mitral cells within a region would exhibit similar patterns of response to application of the same odorant. On the other hand, if the mitral cells associated with each glomerulus function as a unit (Figure 6.3B), then the responses of mitral cells positioned within 100 to 200 μm of one another might be similar, while the responses of slightly more separated mitral cells would be independent.

Responses of Mitral and Tufted Cells to Odors

Responses of mitral cells to odors are typically complex: during odor exposure, units may be initially excited then inhibited, may be inhibited then excited, or may exhibit more complex responses. The character of these responses may alter with odor concentration. Individual units may more reliably discriminate between odors if unit activity is recorded in relation to an artificial sniff cycle (Macrides and Chorover, 1972). Recent authors have emphasized the necessity of testing the response of each cell over a range of odor concentrations (Meredith, 1986; Harrison and Scott, 1986). Testing with several odors, each at several different concentrations, showed that a significant number of cells respond differently to at least two odors at all odor concentrations (Wellis et al., 1989). Similar results in the salamander led to the concept of "concentration tuning" of bulb units: individual cells appeared to respond best to a particular concentration of each odorant (Kauer, 1974).

If an odor causes strong excitation in a particular region of the bulb surface (Figure 6.3A), then there ought to be lateral inhibition of surrounding cells. Meredith (1986) found a few pairs of cells, separated by relatively large distances, that gave approximately opposite responses to odor stimulation. Pairs of this type appear not to be numerous, however. On the other hand, closely neighboring mitral cells, presumed to connect with the same glomerulus, showed strong correlations in their responses to odors (Buonoviso and Chaput, 1990).

Effect of Anesthetics

In interpreting the significance of responses of individual mitral cells to odorants, it is important to reemphasize the potent effects of anesthetics on the mitral/granule inhibitory system (Stewart and Scott, 1976; Scott and Stewart, 1979; Nicoll, 1972b). Most studies discussed here were performed in animals anesthetized with pentobarbital. This anesthetic profoundly increases the duration and strength of inhibition in the bulb. Thus, mitral cells in an awake animal may be appreciably more active and responsive to odors than is suggested by available data.

Connections between the Piriform Cortex, AON, and Olfactory Bulb

The olfactory bulb receives heavy centrifugal afferent inputs from most of its target structures, resulting in a complex feedback system. A simplified diagram of these connections is presented as Figure 6.4. The axons of most tufted cells that project out of the bulb terminate in the anterior olfactory nucleus (AON) and anterior piriform cortex. Except for the pars externa (PE) portion of this nucleus, neurons in all the parts of the AON send axons back to the ipsilateral bulb. In addition, most of these axons (63 percent; Alheid et al., 1984; Mori et al., 1979) bifurcate and send a projection by way of the anterior limb of the anterior commissure (AC) to the contralateral bulb. Axons projecting to the contralateral olfactory bulb

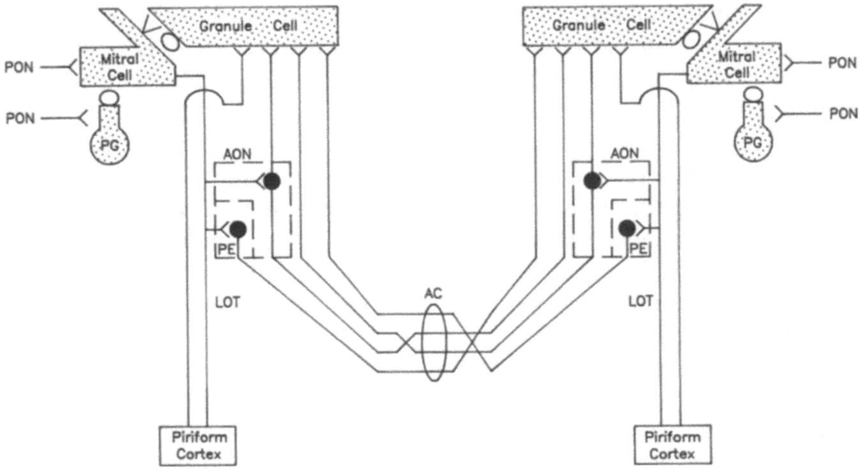

FIGURE 6.4. Diagrammatic representation of commissural and centrifugal connections between the two main olfactory bulbs. Abbreviations: AON, anterior olfactory nucleus; PE, anterior olfactory nucleus pars externa; PG, periglomerular cell; LOT, lateral olfactory tract; AC, anterior commissure; PON, primary olfactory nerve.

have a higher conduction velocity than those projecting to the ipsilateral bulb, suggesting that the axon projecting over the longer path is of larger diameter (Moyano and Molina, 1980). Most AON fibers terminate in the deeper parts of the gcl of both bulbs, although fibers from a portion of the AON (ventroposterior) reach the glomerular layer (Luskin and Price, 1983; Davis and Macrides, 1981). Projections from a given subdivision of the AON generally have the same laminar distribution in the two bulbs, but the projection to the contralateral bulb is less dense (Luskin and Price, 1983). There is no obvious topographic ordering in the ipsi- or contralateral projections of the AON neurons, except for the pars externa (PE).

PE neurons project only to the contralateral bulb (Alheid et al., 1984), terminating primarily in the ipl (Luskin and Price, 1983; Davis and Macrides, 1981). PE neurons receive branches of ipsilateral mitral axons and terminal branches of tufted cells (Scott et al 1985; Ojima et al., 1984). Thus, PE neurons receive major input from the mitral/tufted cells in the ipsilateral bulb and send axons only to the contralateral bulb. The connection between the two bulbs passing through the pars externa appears to be topographically organized in the circumferential dimension: injections into restricted portions of the PE result in labeling of complementary longitudinal strips in both bulbs (Luskin and Price, 1983; Schoenfeld and Macrides, 1984). The PE projection, thus, is anatomically suited to convey topographically ordered information between the two bulbs.

The axons of mitral cells terminate along the length of the piriform cortex; the piriform cortex sends significant projections back to the olfactory bulb and AON. These centrifugal projections from the piriform cortex terminate primarily in the deeper parts of the gcl (Haberly and Price, 1978; Luskin and Price, 1983).

Physiological Studies of Commissural and Ipsilateral Association Systems

High-frequency stimulation of the AC depresses odor-induced activity in the bulb. Section of the AC augments this activity (Kerr and Hagbarth, 1955; von Baumgarten et al., 1962); however, section of the AC also reduces olfactory acuity (Bennett, 1968). Application of an odor to one epithelium a few milliseconds before application of the same odor to the contralateral epithelium is reported to suppress a response recorded across the glomerular layer of the contralateral bulb (Leveteau and MacLeod, 1969).

Both AC and piriform cortex (PC) stimulation produce a negative field potential recorded in the gcl (Walsh, 1959; Nakashima et al., 1978), as expected if excitatory currents are flowing into granule cells in that layer. Intracellular recordings from mitral cells (Yamamoto et al., 1963; Mori and Takagi, 1978b) demonstrate that AC stimulation produces an IPSP in mitral cells that has an appreciably slower rise time and longer duration than the IPSP generated by antidromic stimulation of the LOT. These results indicate that AC and PC stimulation causes excitation of granule cells that release GABA onto mitral cells, resulting in an IPSP in the mitral cells. The response to AC stimulation summates with repeated shocks: single shocks produced no visible IPSP, but three or more shocks at 100 Hz caused a significant hyperpolarization accompanied by a block of antidromic invasion of the impaled cell. Preceding AC stimulation blocks the IPSP that normally follows antidromic stimulation of the LOT (Yamamoto et al., 1963), suggesting that the commissural system is able to modulate the strength of lateral inhibition in the mitral/granule system.

Modulatory Control of Synaptic Processes in the Olfactory Bulb

The olfactory bulb resembles other cortical structures in receiving neuromodulatory projections, norepinephrine, serotonin, and acetylcholine, from brainstem and basal forebrain structures. These modulatory inputs may significantly influence signal processing by bulb circuits; hence it is appropriate to consider the functional anatomy and physiology of these inputs.

Monoaminergic Transmitters

The existence of monoaminergic projections to the olfactory bulb was initially demonstrated using a variety of markers and pharmacological manipulations (Halasz et al., 1978, 1981). Unambiguous description of these projections required the use of selective immunohistochemical techniques (McLean and Shipley, 1987; McLean et al., 1989). These later studies demonstrate that the termination of these projections is closely related to the laminar structure of the olfactory bulb.

Norepinephrine (NE)

Norepinephrine-containing fibers from the locus coeruleus are primarily concentrated in the gcl and the inner part of the epl and do not reach the glomerular layer (McLean et al., 1989; Figure 6.5B). Electrophysiological studies using single unit recording and iontophoretic application of drugs suggest that NE acts upon the granule cell, although the direction of the reported effects is disputed. In the rat, iontophoresis of adrenergic agonists results in inhibition of firing of mitral cells (Bloom et al., 1964; Salmoiraghi et al., 1964). This inhibition is blocked, however, by simultaneous application of the $GABA_A$ antagonist bicuculline, suggesting that the mitral inhibition is mediated indirectly through excitation of granule cells by NE (McLennan, 1971). Contrary results have been reported by Jahr and Nicoll (1982b) using isolated turtle olfactory bulb. In this preparation, bath application of NE decreased the strength of mitral cell inhibition.

A different kind of evidence concerning the actions of NE in the olfactory bulb is provided by biochemical measurements of the effect of adrenergic agonists on cAMP turnover in olfactory bulb slices (Stone and Herrera, 1986). Application of either a β_1 agonist (which is directly stimulatory to adenylcyclase) or an α_1 agonist (which indirectly blocks the inhibitory subunit of adenylcyclase through activation of phosphatidyl inositol) results in a moderate increase in AC turnover of about 4 times control. Combination of the two agonists results in a 16-fold increase in AC (Stone and Hererra, 1986). This synergistic action of the two receptors is consistent with interaction of their effector mechanisms within single cells, presumably granule cells.

This evidence indicates that there are both α_1 and β_1 adrenergic receptors on granule cells and that the NE input activates adenylate cyclase in the granule cells. Although it does not appear to have been measured directly, the actions of the α_1 agonist imply that the phosphatidyl inositol system is also activated by NE. Both these systems are presumably "excitatory" and result in increased release of transmitter (GABA) for a given stimulatory input; however, biophysical measurements of changes of membrane properties caused by these agents are not available.

Thus, the bulk of the evidence suggests that noradrenergic fibers from the locus coeruleus (LC) terminate on granule cells and modulate their

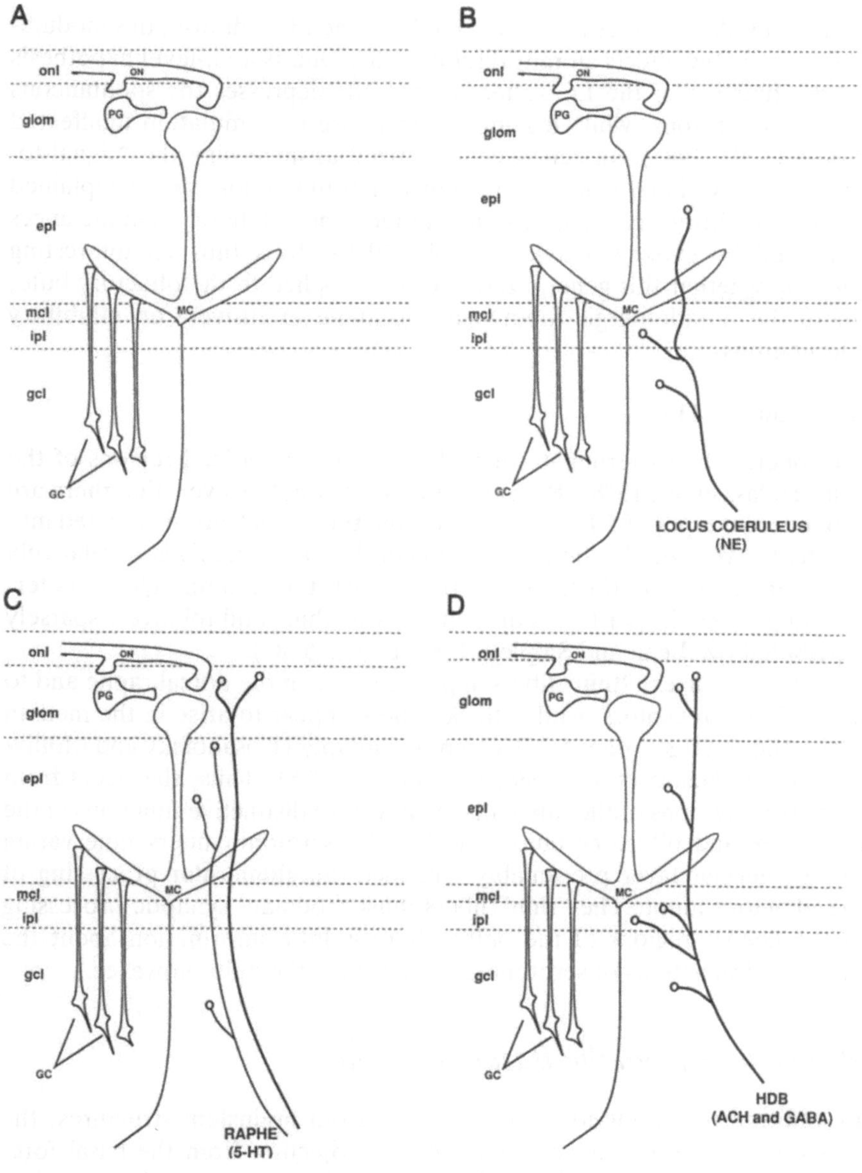

FIGURE 6.5. Schematic representation of distribution of modulatory inputs to the olfactory bulb. A. Schematic representation of bulb synaptic connections. B. Noradrenergic projection from locus coeruleus. C. Serotonin (5-HT) projection from raphe nuclei. "Thick" fibers terminate in the glomerular layer, while "thin" fibers terminate in the deeper layers of the olfactory bulb. D. Cholinergic projection from horizontal limb of the diagonal band (HDB). Abbreviations: onl, olfactory nerve layer; glom, glomerular layer; epl, external plexiform layer; mcl, mitral cell layer; ipl, internal plexiform layer; gcl, granule cell layer; ON, olfactory nerve; PG, periglomerular cell; MC, mitral cell; GC, granule cell.

release of GABA. Since the projection from the LC is diffuse, this modulation presumably affects all mitral/tufted cells. One system-level hypothesis for the function of the LC system is that NE depresses the spontaneous activity of neurons, while leaving the response to stimulation unaffected (Foote et al., 1983), an action characterized as increasing the "signal-to-noise ratio." Cellular studies have shown that this action can be explained by "disinhibitory" mechanisms that reduce the inhibitory conductances generated by evoked activity (Nicoll, 1988). It is thus an interesting question whether this generalization can be applied to the olfactory bulb, where the noradrenergic input may terminate exclusively on inhibitory interneurons.

Serotonin (5- HT)

Serotonergic fibers terminate in both deep and superficial regions of the bulb (Halasz et al., 1978). Recent work has shown, however, that there are two different types of 5-HT fibers, thick and thin, which are segregated into different layers of the bulb. The glomerular layer receives a relatively heavy projection of "thick" fibers. In contrast, the serotonergic fibers terminating in the deeper layers of the bulb are "thin" and relatively sparsely distributed (McLean and Shipley, 1987; Figure 6.5C).

In the neocortex, "thin" fibers appear to arise in the dorsal raphe and to target 5-HT_2 receptors, while "thick" fibers appear to arise in the median raphe and to target other 5-HT receptor subtypes (Kosakofsky and Molliver, 1987; O'Hearn et al., 1988; Blue et al., 1988). Thus, the fibers from these two divisions of the raphe nuclei may have distinctive functions in the neocortex and olfactory bulb. The "thick" serotonin fibers innervating the glomerular layer presumably influence the glomerular processing of the olfactory input. The "thin" fibers must modulate synaptic processing in the deeper regions of the bulb. There is little information about the physiological actions of serotonergic agonists in the bulb, however.

Projections from the Basal Forebrain

In addition to the monoaminergic inputs from brainstem structures, the olfactory bulb also receives an extensive projection from the basal forebrain (Macrides et al., 1981). This projection is analogous with the projections from medial septum to hippocampus and from nucleus basalis to neocortex. These projections have long been known to contain acetylcholine; more recent work has demonstrated the presence of other transmitters as well (Rye et al., 1984; Brashear et al., 1986; Zaborsky et al., 1986).

With a recording electrode in the gcl of the olfactory bulb, electrical stimulation of the horizontal limb of the diagonal band (HDB) produces a biphasic field potential, an initial positive deflection followed by a slower negative wave. Repetitive stimulation at 10 Hz results in a marked potentiation of the negative portion of the response. Analysis of the depth profile of the field potential suggested that the negative (in gcl) wave results from excitatory currents flowing into granule cells. This is confirmed by the observation that mitral cells are inhibited during potentiation of the field potential (Nickell and Shipley, 1988b).

Acetylcholine (ACh)

The cholinergic input to the olfactory bulb, which arises in the horizontal limb of the diagonal band (HDB; Macrides et al., 1981), terminates in both the deep and superficial regions of the bulb. The deeper of these bands is concentrated in the outer part of the gcl, the ipl, and the deep part of the epl. The superficial band of ACh input is located in the glomerular layer and the superficial part of the epl. The localization of AChE staining, of immunohistochemically stained ChAT fibers (Shipley et al., in preparation), of ChAT activity (Godfrey et al., 1980), and of tracers transported from HDB (Macrides et al., 1981) are consistent with this conclusion. The location of the deep band of ACh fibers is consistent with termination upon any of several cell types, including mitral cells, granule cells, or nongranule cell interneurons located in the gcl (Schneider and Macrides, 1978; Nickell and Shipley, 1988a and b; Figure 6.5D), or the terminals of afferent inputs.

There are two populations of interneurons in the bulb that express AChE and not ChAT; these interneurons may be cholinoceptive (Nickell and Shipley, 1988a). One population of interneurons is situated just below the mitral cell layer. A second population is a subset of the PG interneurons in the glomerular layer. Thus, the two populations of putative cholinoceptive neurons are located in the two zones of heaviest cholinergic input.

Muscarinic agonists activate phosphatidylinositol (PI) turnover in the olfactory bulb (Gonzales and Crews, 1985). Maximal effects of NE and carbachol on PI activity are additive in the bulb, suggesting that the receptors for the two transmitters may be located on different cells. Since NE receptors are most probably located on granule cells, this may suggest that muscarinic receptors coupled to phosphatidylinositol metabolism are not located on granule cells.

Other Transmitters in HDB

Of the HDB neurons projecting to the olfactory bulb, about 30 percent contain GAD, the synthetic enzyme for GABA (Rye et al., 1984; Brashear et al., 1986; Zaborszky et al., 1986). Since there are intrinsic GABAergic neurons in all parts of the olfactory bulb (Macrides and Davis, 1983), it is difficult to determine whether the GABA fibers from HDB are distributed in the same way as the cholinergic input from this nucleus. Similarly, determination of the physiological role of the extrinsic GABAergic projection is complicated by the presence of intrinsic GABAergic neurons.

The identified cholinergic and GABAergic neurons in HDB projecting to the bulb account for 40 to 50 percent of HDB neurons. Immunohistochemical procedures may not have labeled all ChAT- or GAD-containing cells; however, it is also possible that the remaining HDB neurons projecting to the bulb contain one or more other transmitters.

Conclusion

The olfactory bulb occupies a similar place in the olfactory pathway as the retina in the visual pathway. This has led to a search for analogies between the olfactory bulb and the better understood visual system. The olfactory bulb shares several similarities in circuitry with the retina; thus, the bulb may perform similar kinds of transformations (e.g., edge enhancement, lateral inhibition) on its input data (Shepherd, 1972a and b, 1979). There are, however, significant differences between the olfactory bulb and the retina that limit the usefulness of this analogy. First, there is no requirement that odorants form coherent spatial patterns of activity on the bulb surface. Second, the output of the bulb is not topographically organized. Finally, unlike the retina, the olfactory bulb receives extensive reciprocal projections from its target structures. The bulb is in this regard more similar to higher cortical structures, which receive extensive "association" projections from other cortical areas.

The cortical structure most closely associated with the olfactory bulb, the piriform cortex, has recently been the subject of numerical modeling efforts (Haberly and Bower, 1989). These studies demonstrate that networks that have a plausible resemblance to the synaptic connections of the piriform cortex are able to store and retrieve patterns that lack any spatial organization. In this regard, the olfactory cortex is suggested to have a functional resemblance to association neocortex, which also must process inputs with no apparent spatial or topographic organization (Haberly, 1985). The close reciprocal connections between the olfactory bulb and the piriform cortex suggest that it may be necessary to consider the bulb and piriform cortex as a single system; one recent modeling study makes use of postulated connections between the bulb and the cortex (Ambrose-Ingerson et al., 1990). Thus, realistic, testable models of olfactory processing may require understanding of the coding of olfactory information onto the surface of the olfactory bulb and of the transformations accomplished by bulb circuitry.

Acknowledgments. The authors were partially supported during preparation of this manuscript by: NIDCD8 P01 DC00347, NINDS NS29218, NINDS NS24698, and DAMD17-91-C-1071.

We wish to think David Wellis and John Scott for allowing us to cite observations from an unpublished manuscript, J.W. Scott for critical comments on the manuscript, and H.J. Duncan for comments on the manuscript and for assistance with the figures.

References

Adamek, G.D., R.C. Gesteland, R.G. Mair, and B. Oakley (1984). Transduction physiology of olfactory receptor cilia. *Brain Res.*, *310*, 87–97.

Adamek, G.D., W.T. Nickell, and M.T. Shipley (1986). Evidence for diffuse and focal projections from the olfactory epithelium to the bulb. *Chem. Senses*, *11*, 575.

Adrian, E.D. (1950). The electrical activity of the olfactory bulb. *EEG Clin. Neurophysiol.*, *2*, 377–388.

Alheid, G.F., J. Carlsen, J. DeOlmos, and L. Heimer (1984). Quantitative determination of collateral anterior olfactory nucleus projections using a fluorescent tracer with an algebraic solution to the problem of double retrograde labelling. *Brain Res.*, *292*, 17–22.

Allison, A.C., and R.T.T. Warwick (1949). Quantitative observations on the olfactory system of the rabbit. *Brain*, *72*, 186–197.

Ambros-Ingerson, J., R. Granger, and G. Lynch (1990). Simulation of paleocortex performs hierarchical clustering. *Science*, *247*, 1344–1347.

Andrade, R., R.C. Malenka, and R.A. Nicoll (1986). A G protein couples serotonin and $GABA_B$ receptors to the same channels in hippocampus. *Science*, *234*, 1261–1265.

Astic, L., and D. Saucier (1986). Anatomical mapping of the neuroepithelial projection to the olfactory bulb in the rat. *Brain Res. Bull.*, *16*, 445–454.

Bennett, M.H. (1968). The role of the anterior limb of the anterior commissure in olfaction. *Physiol. Behav.*, *3*, 507–515.

Bloom, F.E., E. Costa, and G.C. Salmoiraghi (1964). Analysis of individual rabbit olfactory bulb neuron responses to the microelectrophoresis of acetylcholine, norepinephrine and serotonin synergists and antagonists. *J. Pharmacol. Exp. Therp.*, *146*, 16–23.

Blue, M.E., K.A. Yagaloff, L.A. Mamounas, P.R. Hartig, and M.E. Molliver (1988). Correspondence between 5-HT_2 receptors and serotonergic axons in rat neocortex. *Brain Res.*, *453*, 315–328.

Bowery, N.G., A.L. Hudson, and G.W. Price (1987). $GABA_A$ and $GABA_B$ receptor site distribution in the rat central nervous system. *Neuroscience*, *20*, 365–383.

Boyson, S.J., P. McGonigle, and P.B. Molinoff (1986). Quantitative autoradiographic localization of the D1 and D2 subtypes of dopamine receptors in rat brain. *J. Neurosci.*, *6*, 3177–3188.

Brashear, H.R., L. Zaborszky, and L. Heimer (1986). Distribution of GABAergic and cholinergic neurons in the rat diagonal band. *Neuroscience*, *17*, 439–451.

Buonoviso, N., and M.A. Chaput (1990). Response similarity to odors in olfactory bulb output cells presumed to be connected to the same glomerulus: Electrophysiological study using simultaneous single-unit recordings. *J. Neurophysiol.*, *63*, 447–454.

Closse, A., M. Camps, A. Wanner, and J.M. Palacios (1988). *In vivo* labeling of brain dopamine D2 receptors using the high-affinity specific D2 agonist [3H] CV 205–502. *Brain Res.*, *440*, 123–132.

Costanzo, R.M., and R.J. O'Connell (1980). Receptive fields of second-order neurons in the olfactory bulb of the hamster. *J. Gen. Physiol.*, *76*, 53–68.

Costanzo, R.M., M.T. Shipley, and S. Van Ooteghem (1984). Ontogeny and lesion

induced changes in cytochrome oxidase (CO) levels in the olfactory system. *Neurosci. Abstr.*, *10*, 118.
Daston, M.M., G.D. Adamek, and R.C. Gesteland (1990). Ultrastructural organization of receptor cell axons in frog olfactory nerve. *Brain Res.*, *537*, 69–75.
Daval, G., and J. Leveteau (1974). Electrophysiological studies of centrifugal and centripetal connections of the anterior olfactory nucleus. *Brain Res.*, *78*, 395–410.
Davis, B.J., and F. Macrides (1981). The organization of centrifugal projections from the anterior olfactory nucleus, ventral hippocampal rudiment, and piriform cortex to the main olfactory bulb in the hamster: An autoradiographic study. *J. Comp. Neurol.*, *203*, 475–493.
Davis, B.J., and F. Macrides (1983). Tyrosine hydroxylase immunoreactive neurons and fibers in the olfactory system of the hamster. *J. Comp. Neurol.*, *214*, 427–440.
Dawson, T.M., P. Barone, A. Sidhu, J.K. Wamsley, and T.N. Chase (1988). The D1 dopamine receptor in the rat brain: quantitative autoradiographic localization using an iodinated ligand. *Neuroscience*, *26*, 83–100.
Dubois-Dauphin, M., E. Tribolet, and J.J. Dreifuss (1981). Relations somatotopiques entre la muquese olfactive et le bulbe olfactif chez le triton. *Brain Res.*, *219*, 269–287.
Duncan, H.J., W.T. Nickell, M.T. Shipley, and R.C. Gesteland (1990). Organization of projections from olfactory epithelium to olfactory bulb in the frog, *Rana pipiens*. *J. Comp. Neurol.*, *299*, 299–311.
Foote, S.L., F.E. Bloom, and G. Aston-Jones (1983). Nucleus locus ceruleus: New evidence of anatomical and physiological specificity. *Physiol. Rev.*, *63*, 844–914.
ffrench-Mullen, J.M.H., K. Koller, R. Zaczek, J.T. Coyle, N. Hori, and D.O. Carpenter (1985). N-Acetylaspartylglutamate: Possible role as the neurotransmitter of the lateral olfactory tract. *Proc. Natl. Acad. Sci.* USA, *82*, 3897–3900.
Freeman, W.J. (1972a). Depth recording of averaged evoked potential of olfactory bulb. *J. Neurophysiol.*, *35*, 780–796.
Freeman, W.J. (1972b). Measurement of oscillatory responses to electrical stimulation in olfactory bulb of cat. *J. Neurophysiol.*, *35*, 762–769.
Freeman, W.J. (1972c). Spatial divergence and temporal dispersion in the primary olfactory nerve of cat. *J. Neurophysiol.*, *35*, 733–744.
Freeman, W.J. (1974a). Attenuation of transmission through glomeruli of olfactory bulb on paired shock stimulation. *Brain Res.*, *65*, 77–90.
Freeman, W.J. (1974b). Relation of glomerular neuronal activity to glomerular transmission attenuation. *Brain Res.*, *65*, 91–107.
Freeman, W.J. (1974c). Average transmission distance from mitral-tufted cells to granule cells in the olfactory bulb. *EEG Clini. Neurophysiol.*, *36*, 609–618.
Freeman, WJ. (1974d). Topographic organization of primary olfactory nerve in cat and rabbit as shown by evoked potentials. *EEG Clin. Neurophysiol.*, *36*, 33–45.
Gahwiler, B.H., and D.A. Brown (1985). $GABA_B$- receptor-activated K^+ current in voltage-clamped CA3 pyramidal cells in hippocampal cultures. *Proc. Natl. Acad. Sci. USA*, *82*, 1558–1562.
Gall, C.M., S.H. Hendry, K.B. Seroogy, E.G. Jones, and J.W. Haycock (1987). Evidence for coexistence of GABA and dopamine in neurons of the rat olfactory bulb. *J. Comp. Neurol.*, *266*, 307–318.
Gehlert, D.R. and J.K. Wamsley (1985). Dopamine receptors in the rat brain: A

quantitative autoradiographic analysis. *J. Neurosci.*, *8*, 2352–2365.

Getchell, T.V., and G.M. Shepherd (1975a). Synaptic actions on mitral and tufted cells elicited by olfactory nerve volleys in the rabbit. *J. Physiol.*, *251*, 497–522.

Getchell, T.V., and G.M. Shepherd (1975b). Short-axon cells in the olfactory bulb: dendrodendritic synaptic interactions. *J. Physiol.*, *251*, 523–548.

Godfrey, D.A., C.D. Ross, A.D. Herrmann, and F.M. Matschinsky (1980). Distribution and derivation of cholinergic elements in the rat olfactory bulb. *Neuroscience*, *5*, 273–292.

Gonzales, R.A., and F.T. Crews (1985). Cholinergic- and adrenergic-stimulated inositide hydrolysis in brain: Interaction, regional distribution and coupling mechanisms. *J. Neurochemistry*. *45*, 1076–1084.

Graziadei, P.P.C., and J.F. Metcalf (1971). Autoradiographic and ultrastructural observations on the frog's olfactory mucosa. *Z. Zellforsch. Microsk. Anat.*, *116*, 305–318.

Haberly, L.B. (1985). Neuronal circuitry in olfactory cortex: Anatomy and functional implications. *Chem. Senses*, *10*, 219–238.

Haberly, L.B., and J.L. Price (1978). Association and commissural fiber systems of the olfactory cortex of the rat: Systems originating in the piriform cortex and adjacent areas. *J. Comp. Neurol.*, *178*, 711–740.

Haberly, L.B., and J.M. Bower (1989). Olfactory cortex: model circuit for study of associative memory? *TINS*, *12*, 258–264.

Halasz, N., A. Ljungdahl, and T. Hokfelt (1978). Transmitter histochemistry of the rat olfactory bulb. II. Fluorescence histochemical, autoradiographic and electron microscopic localization of monoamines. *Brain Res.*, *154*, 253–271.

Halasz, N., O. Johansson, T. Hokfelt, A. Ljungdahl, and M. Goldstein (1981). Immunohistochemical identification of two types of dopamine neuron in the rat olfactory bulb as seen by serial sectioning. *J. Neurocytol.*, *10*, 251–259.

Halasz, N., and G.M. Shepherd (1983). Neurochemistry of the vertebrate olfactory bulb. *Neuroscience*, *10*, 579–619.

Hamilton, K.A., and J.S. Kauer (1985). Intracellular potentials of salamander mitral/tufted neurons in response to odor stimulation. *Brain Res.*, *338*, 181–185.

Hamilton, K.A., and J.S. Kauer (1988). Responses of mitral/tufted cells to orthodromic and antidromic electrical stimulation in the olfactory bulb of the tiger salamander. *J. Neurophysiol.*, *59*, 1736–1755.

Harrison, T.A., and J.W. Scott (1986). Olfactory bulb responses to odor stimulation: Analysis of response pattern and intensity relationships. *J. Neurophysiol.*, *56*, 1571–1589.

Holley, A., and K.B. Doving (1977). Receptor sensitivity, acceptor distribution, convergence and neural coding in the olfactory system. In J. LeMagnen and P. MacLeod (Eds.), *Olfaction and Taste Vl. Paris*. London: IRL, pp. 113–133.

Hori, N., C.R. Auker, D.J. Braitman, and D.O. Carpenter (1982). Pharmacologic sensitivity of amino acid responses and synaptic activation in *in vitro* prepiriform neurons. *J. Neurophysiol.*, *48*, 1289–1301.

Innis, R.B., and G.K. Aghajanian (1987). Pertussis toxin blocks autoreceptor-mediated inhibition of dopaminergic neurons in rat substantia nigra. *Brain Res.*, *411*, 139–143.

Jacobson, I., and A. Hamberger (1986). Effects of kynurenic acid on evoked extracellular field potentials in the rat olfactory bulb *in vivo*. *Brain Res.*, *386*, 389–392.

Jacobson, I., S. Butcher, and A. Hamberger (1986). An analysis of the effects of excitatory amino acid receptor antagonists on evoked field potentials in the olfactory bulb. *Neuroscience*, *19*, 267–273.

Jahr, C.E., and R.A. Nicoll (1981). Primary afferent depolarization in the *in vitro* frog olfactory bulb. *J. Physiol.*, *318*, 375–384.

Jahr, C.E., and R.A. Nicoll (1982a). An intracellular analysis of dendrodendritic inhibition in the turtle *in vitro* olfactory bulb. *J. Physiol.*, *326*, 213–234.

Jahr, C.E., and R.A. Nicoll (1982b). Noradrenergic modulation of dendrodendritic inhibition of the olfactory bulb. *Nature (Lond.)*, *287*, 227–228.

Jourdan, F., A. Duveau, L. Astic, and A. Holley (1980). Spatial distribution of [^{14}C] 2-deoxyglucose uptake in the olfactory bulbs of rats stimulated with two different odours. *Brain Res.*, *188*, 139–154.

Kauer, J.S. (1974). Response patterns of amphibian olfactory bulb neurones to odour stimulation. *J. Physiol.*, *243*, 695–715.

Kauer, J.S. (1981). Olfactory receptor staining using horseradish peroxidase. *Anat. Rec.*, *200*, 331–336.

Kauer, J.S. (1987). Coding in the olfactory system: In T.E. Finger and W.L. Silver (Eds.), *Neurobiology of Taste and Smell*. New York: Wiley, pp. 205–231.

Kauer, J.S., and D.G. Moulton (1974). Responses of olfactory bulb neurones to odour stimulation of small nasal areas in the salamander. *J. Physiol.*, *243*, 717–737.

Kerr, D.I.B., and K.E. Hagbarth (1955). An investigation of the olfactory centrifugal system. *J. Neurophysiol.*, *18*, 362–374.

Kishi, K., K. Mori, and H. Ojima (1984). Distribution of local axon collaterals of mitrals, displaced mitrals, and tufted cells in the rabbit olfactory bulb. *J. Comp. Neurol.*, *225*, 511–526.

Kosaka, T., Y. Hataguchi, K. Hama, I. Nagatsu, and J.-Y. Wu (1985). Coexistence of immunoreactivities for glutamate decarboxylase and tyrosine hydroxylase in some neurons in the periglomerular region of the rat main olfactory bulb: possible coexistence of gamma-aminobutyric acid (GABA) and dopamine. *Brain Res.*, *343*, 166–171.

Kosakofsky, B.E., and M.E. Molliver (1987). The serotonergic innervation of cerebral cortex: different classes of axon terminals arise from dorsal and median raphe nuclei. *Synapse*, *1*, 153–168.

Lacey, M.G., N.B. Mercuri, and R.A. North (1988). On the potassium conductance increase activated by $GABA_B$ and dopamine D2 receptors in rat subsantia nigra neurones. *J. Physiol.*, *401*, 437–453.

Lancet, D., C.A. Greer, J.S. Kauer, and G.M. Shepherd (1982). Mapping of odor-related neuronal activity in the olfactory bulb by high-resolution 2-deoxyglucose autoradiography. *Proc. Natl. Acad. Sci. USA*, *79*, 670–674.

Land, L.J. (1973). Localized projection of olfactory nerves to rabbit olfactory bulb. *Brain Res.*, *63*, 153–166.

Land, L.J., and G.M. Shepherd (1974). Autoradiographic analysis of olfactory receptor projections in the rabbit. *Brain Res.*, *70*, 506–510.

Le Gros Clark, W.E. (1951). The projection of the olfactory epithelium on the olfactory bulb in the rabbit. *J. Neurol. Neurosurg. Psychiatry*, *14*, 1–10.

Le Gros Clark, W.E. (1956). Inquiries into the anatomical basis of olfactory discrimination. *Proc. R. Soc. of London Series B*, *146*, 299–319.

Leveteau, J., and P. MacLeod (1966). Olfactory discrimination in the rabbit olfac-

tory glomerulus. *Science*, *153*, 175–176.
Leveteau, J., and P. MacLeod (1969). Reciprocal inhibition at glomerular level during bilateral olfactory stimulation. In C. Pfaffman (Ed.), *Olfaction and Taste, III*. New York: Rockefeller University Press, pp. 212–215.
Luskin, M.B. and J.L. Price (1983). The topographic organization of associational fibers of the olfactory system of the rat, including centrifugal fibers to the olfactory bulb. *J. Comp. Neurol.*, *216*, 264–291.
Mackay-Sim, A., P. Shaman, and D.G. Moulton (1982). Topographic coding of odor quality; odorant-specific patterns of epithelial responsivity in the salamander. *J. Neurophysiol.*, *48*, 584–596.
Mackay-Sim, A., and M.H. Nathan (1984). The projection from the olfactory epithelium to the olfactory bulb in the salamander, *Ambystoma tigrinum*. *Anat. Embryol.*, *170*, 93–97.
Macrides, F., and S.L. Chorover (1972). Olfactory bulb units: Activity correlated with inhalation cycles and odor quality. *Science*, *175*, 84–87.
Macrides, F., B.J. Davis, W.M. Youngs, N.S. Nadi, and F.L. Margolis (1981). Cholinergic and catecholaminergic afferents to the olfactory bulb in the hamster: A neuroanatomical, biochemical and histochemical investigation. *J. Comp. Neurol.*, *203*: 495–514.
Macrides, F., and S.P. Schneider (1982). Laminar organization of mitral and tufted cells in the main olfactory bulb of the adult hamster. *J. Comp. Neurol.*, *208*, 419–430.
Macrides, F., and B.J. Davis (1983). The olfactory bulb. In P.C. Emson (Ed.), *Chemical Neuroanatomy*. New York: Raven Press, pp. 391–426.
Macrides, F., T.A. Schoenfeld, J.E. Marchand, and A.N. Clancy (1985). Evidence for morphologically, neurochemically and functionally heterogeneous classes of mitral and tufted cells in the olfactory bulb. *Chem. Senses*, *10*, 175–202.
Mair, R.G., R.C. Gesteland, and D.L. Blank (1982). Changes in morphology and physiology of olfactory receptor cilia during development. *Neuroscience*, *7*, 3091–3103.
Mansour, A., J.H. Meador-Woodruff, J.R. Bunzow, O. Civelli, H. Akil, and S.J. Watson (1990). Localization of dopamine D2 receptor mRNA and D1 and D2 receptor binding in the rat brain and pituitary: An *in situ* hybridization-receptor and autoradiographic analysis. *J. Neurosci.*, *10*, 2587–2600.
Martinez, D.P., and W.J. Freeman (1984). Periglomerular cell action on mitral cells in olfactory bulb shown by current source density analysis. *Brain Res.*, *308*, 223–233.
McLean, J.H., and M.T. Shipley (1987). Serotonergic afferents to the rat olfactory bulb. I. Origins and laminar specificity of serotonergic inputs in adult rat. *J. Neurosci.*, *7*, 3016–3028.
McLean, J.H., and M.T. Shipley (1988). Postmitotic, postmigrational expression of tyrosine hydroxylase in olfactory bulb dopaminergic neurons. *J. Neurosci.*, *8*, 3658–3669.
McLean, J.H., M.T. Shipley, W.T. Nickell, G. Aston-Jones, and C.K.H. Reyher (1989). Chemoanatomical organization of the noradrenergic input from locus coeruleus to the olfactory bulb of the adult rat. *J. Comp. Neurol.*, *285*, 339–349.
McLennan, H. (1971). The pharmacology of inhibition of mitral cells in the olfactory bulb. *Brain Res.*, *29*, 177–184.
Meredith, M. (1986). Patterned response to odor in mammalian olfactory bulb:

the influence of intensity. *J. Neurophysiol.*, *56*, 572–597.
Mori, K. and S.F. Takagi (1975). Spike generation in the mitral cell dendrite of the rabbit olfactory bulb. *Brain Res.*, *100*, 685–689.
Mori, K., S. Kogure, and S.F. Takagi (1977). Alternating responses of olfactory bulb neurons to repetitive lateral olfactory tract stimulation. *Brain Res.*, *133*, 150–155.
Mori, K., and S.F. Takagi (1978a). An intracellular study of dendrodendritic inhibitory synapses on mitral cells in the rabbit olfactory bulb. *J. Physiol. (Lond.)*, *279*, 569–588.
Mori, K., and S.F. Takagi (1978b). Activation and inhibition of olfactory bulb neurones by anterior commissure volleys in the rabbit. *J. Physiol. (Lond.)*, *279*, 589–604.
Mori, K., M. Satou, and S.F. Takagi (1979). Axonal projection of anterior olfactory nuclear neurons to the olfactory bulb bilaterally. *Exp. Neurol.*, *64*, 295–305.
Mori, K., M.C. Nowycky, and G.M. Shepherd (1981a). Analysis of a long-duration inhibitory potential in mitral cells in the isolated turtle olfactory bulb. *J. Physiol. (Lond.)*, *314*, 311–320.
Mori, K., M.C. Nowycky, and G.M. Shepherd (1981b). Electrophysiological analysis of mitral cells in the isolated turtle olfactory bulb. *J. Physiol. (Lond.)*, *314*, 281–294.
Mori, K., and K. Kishi (1982). The morphology and physiology of the granule cells in the rabbit olfactory bulb revealed by intracellular recording and HRP injection. *Brain Res.*, *247*, 129–133.
Mori, K., M.C. Nowycky, and G.M. Shepherd (1982). Impulse activity in presynaptic dendrites: analysis of mitral cells in the isolated turtle olfactory bulb. *J. Neurosci.*, *2*, 497–502.
Mori, K., K. Kishi, and H. Ojima (1983). Distribution of dendrites of mitral, displaced mitral, tufted, and granule cells in the rabbit olfactory bulb. *J. Comp. Neurol.*, *219*, 339–355.
Mori, K., M.C. Nowycky, and G.M. Shepherd (1984). Synaptic excitatory and inhibitory interactions at distal dendritic sites on mitral cells in the isolated turtle olfactory bulb. *J. Neurosci.*, *4*, 2291–2296.
Moulton, D.G. (1974). Dynamics of cell populations in the olfactory epithelium. *Ann. NY Acad. Sci.*, *237*, 52–61.
Moulton, D.G. (1976). Spatial patterning of response to odors in the peripheral olfactory system. *Physiol. Rev.*, *56*, 578–593.
Mouradian, L.E., and J.W. Scott (1988). Cytochrome oxidase marks dendritic zones of the rat olfactory bulb external plexiform layer. *J. Comp. Neurol.*, *271*, 507–518.
Moyano, H.F., and J.C. Molina (1980). Axonal projections and conduction properties of olfactory peduncle neurons in the rat. *Exp. Brain Res.*, *39*, 241–248.
Mozell, M.M. (1964). Evidence for sorption as a mechanism of the olfactory analysis of vapors. *Nature*, *203*, 1181–1182.
Mozell, M.M. (1966). The spatio-temporal analysis of odor at the level of the olfactory receptor sheet. *J. Gen. Physiol.*, *50*, 25–41.
Mozell, M.M. (1970). Evidence for a chromatographic model of olfaction. *J. Gen. Physiol.*, *56*, 46–63.
Mugnaini, E., W.H. Oertel, and F.F. Wouterlood (1984). Immunocytochemical localization of GABA neurons and dopamine neurons in the rat main and acces-

sory olfactory bulbs. *Neurosci. Lett.*, *47*, 221–226.

Nakashima, M., K. Mori, and S.F. Takagi (1978). Centrifugal influence on olfactory bulb activity in the rabbit. *Brain Res.*, *154*, 301–316.

Nickell, W.T., and M.T. Shipley (1988a). Two anatomically specific classes of candidate cholinoceptive neurons in the rat olfactory bulb. *J. Neurosci.*, *8*, 4482–4491.

Nickell, W.T., and M.T. Shipley (1988b). Neurophysiology of magnocellular forebrain inputs to the olfactory bulb in the rat: Frequency potentiation of field potentials and inhibition of output neurons. *J. Neurosci.*, *8*, 4492–4502.

Nickell, W.T., and M.T. Shipley (1988c). Pharmacology of inhibition in the glomerular layer of the olfactory bulb. *Chem. Senses*, *13*, 724.

Nickell, W.T., A.B. Norman, L.M. Wyatt, and M.T. Shipley (1990). Localization of dopamine receptor subtypes in the olfactory bulb: [3H]Spiperone (D2) binding in the glomerular and nerve layers. *Neurosci. Abstr.*, *16*, 101.

Nickell, W.T., A.B. Norman, L.M. Wyatt, and M.T. Shipley (1991). Olfactory bulb DA receptors may be located on terminals of the olfactory nerve. *Neuroreport*, *2*, 9–12.

Nicoll, R.A. (1970). Recurrent excitatory pathways of anterior commissure and mitral cell axons in the olfactory bulb. *Brain Res.*, *19*, 491–493.

Nicoll, R.A. (1971). Pharmacological evidence for GABA as the transmitter in granule cell inhibition in the olfactory bulb. *Brain Res.*, *35*, 137–149.

Nicoll, R.A. (1972a). Olfactory nerves and their excitatory action in the olfactory bulb. *Exp. Brain Res.*, *14*, 185–197.

Nicoll, R.A. (1972b). The effects of anaesthetics on synaptic excitation and inhibition in the olfactory bulb. *J. Physiol. (Lond.)*, *223*, 803–814.

Nicoll, R.A. (1988). The coupling of neurotransmitter receptors to ion channels in the brain. *Science*, *241*, 545–551.

Nicoll, R.A., and B.E. Alger (1979). Presynaptic inhibition: Transmitter and ionic mechanisms. *Int. Rev. Neurobiol.*, *21*, 217–258.

Nowycky, M.C., K. Mori, and G.M. Shepherd (1981). Blockade of synaptic inhibition reveals long-lasting synaptic excitation in turtle olfactory bulb. *J. Neurophysiol.*, *46*, 649–658.

Nowycky, M.C., N. Halasz, and G.M. Shepherd (1983). Evoked field potential analysis of dopaminergic mechanisms in the isolated turtle olfactory bulb. *Neuroscience*, *8*, 717–722.

O'Hearn, E., G. Battaglia, E.B. De Souza, M.J. Kuhar, and M.E. Molliver (1988). Methylenedioxyamphetamine (MDA) and methylenedioxymethamphetamine (MDMA) cause ablation of serotonergic axon terminals in forebrain: immunocytochemical evidence. *J. Neurosci.*, *8*, 2788–2803.

Ojima, H., K. Mori, and K. Kishi (1984). The trajectory of mitral cell axons in the rabbit olfactory cortex revealed by intracellular HRP injection. *J. Comp. Neurol.*, *230*, 77–87.

Orona, E., J.W. Scott, and E.C. Rainer (1983). Different granule cell populations innervate superficial and deep regions of the external plexiform layer in rat olfactory bulb. *J. Comp. Neurol.*, *217*, 227–237.

Orona, E., E.C. Rainer, and J.W. Scott (1984). Dendritic and axonal organization of mitral and tufted cells in the rat olfactory bulb. *J. Comp. Neurol.*, *226*, 346–356.

Phillips, C.G., T.P.S. Powell, and G.M. Shepherd (1963). Responses of mitral cells

to stimulation of the lateral olfactory tract in the rabbit. *J. Physiol. (Lond.)*, *168*, 65–88.
Pinching, A.J. (1970). Synaptic connexions in the glomerular layer of the olfactory bulb. *J. Physiol. (Lond.)*, *210*, 14P–15P.
Pinching, A.J. (1971). Myelinated dendritic segments in the monkey olfactory bulb. *Brain Res.*, *29*, 133–138.
Pinching, A.J., and T.P.S. Powell (1971a). The neuropil of the glomeruli of the olfactory bulb. *J. Cell. Sci.*, *9*, 347–377.
Pinching, A.J., and T.P.S. Powell (1971b). The neuropil of the periglomerular region of the olfactory bulb. *J. Cell. Sci.*, *9*, 379–409.
Pinnock, R.D. (1984). Hyperpolarizing action of baclofen on neurons in the rat substantia nigra slice. *Brain Res.*, *322*, 337–340.
Potapov, A.A. (1985). Baclofen inhibition of synaptic transmission in frog olfactory bulb glomeruli. *Nierofiziologiia*, *17*, 834–837.
Potapov, A.A., and V.V. Trepakov (1986). Two types of GABA-receptors in the intact olfactory bulb and primordial hippocamp of frogs: pharmacological data. *Biull-Eksp-Biol-Med.*, *101*, 317–320.
Price, J.L., and T.P.S. Powell (1970). The synaptology of granule cells of the olfactory bulb. *J. Cell. Sci.*, *7*, 125–155.
Rall, W. (1972). Dendritic neuron theory and dendrodendritic synapses in a simple cortical system. In F.O. Schmitt (Ed.), *The Neurosciences: Second Study Program*. New York: Rockefeller University Press, pp. 552–565.
Rall, W., G.M. Shepherd, T.S. Reese, and M.W. Brightman (1966). Dendrodendritic synaptic pathway for synaptic interactions in the olfactory bulb. *Exp. Neurol.*, *14*, 44–56.
Rall, W., and G.M. Shepherd (1968). Theoretical reconstruction of field potentials and dendrodendritic synaptic interactions in olfactory bulb. *J. Neurophysiol.*, *31*, 884–915.
Reyher, C.K.H., J. Lubke, W.J. Larsen, G.M. Hendrix, M.T. Shipley, and H.G. Baumgarten (1991). Olfactory bulb granule cell aggregates: Morphological evidence for interperikaryal electrotonic coupling via gap junctions. *J. Neurosci.*, *11*, 1485–1495.
Ribak, C.E., J.E. Vaughn, K. Saito, R. Barber, and E. Roberts (1977). Glutamate decarboxylase localization in neurons of the olfactory bulb. *Brain Res.*, *126*, 1–18.
Royet, J.P., G. Sicard, C. Souchier, and F. Jourdan (1987). Specificity of spatial patterns of glomerular activation in the mouse olfactory bulb: computer-assisted image analysis of 2-deoxyglucose autoradiograms. *Brain Res.*, *417*, 1–11.
Rye, D.B., B.H. Wainer, M.-M. Mesulam, E.J. Mufson, and C.B. Saper (1984). Cortical projections arising from the basal forebrain: A study of cholinergic and noncholinergic components employing combined retrograde tracing and immunohistochemical localization of choline acetyltransferase. *Neuroscience*, *13*, 627–643.
Salmoiraghi, G.C., F.E. Bloom, and E. Costa (1964). Adrenergic mechanisms in the rabbit olfactory bulb. *Am. J. Physiol.*, *207*, 1417–1424.
Saucier, D., and L. Astic (1986). Analysis of the topographic organization of olfactory epithelium projections in the rat. *Brain Res. Bull.*, *16*, 455–462.
Scalia, F. (1976). Structure of olfactory and accessory olfactory systems. In R. Llinas and W. Precht (Eds.), *Frog Neurobiology*. Berlin: Springer-Verlag, pp. 213–233.

Schild, D. (1988). Principles of odor coding and a neural network for odor discrimination. *Biophysiol. J.*, *54*, 1001–1011.
Schneider, S.P., and F. Macrides (1978). Laminar distribution of interneurons in the main olfactory bulb of the adult hamster. *Brain Res. Bull.*, *3*, 73–82.
Schneider, S.P., and J.W. Scott (1983). Orthodromic response properties of rat olfactory bulb mitral and tufted cells correlate with their projection patterns. *J. Neurophysiol.*, *50*, 358–378.
Schoenfeld, T.A., and F. Macrides (1984). Topographic organization of connections between the main olfactory bulb and pars externa of the anterior olfactory nucleus in the hamster. *J. Comp. Neurol.*, *227*, 121–135.
Schoenfeld, T.A., J.E. Marchand, and F. Macrides (1985). Topographic organization of tufted cell axonal projections in the hamster main olfactory bulb: An intrabulbar associational system. *J. Comp. Neurol.*, *235*, 503–518.
Scott, J.W. (1981). Electrophysiological identification of mitral and tufted cells and distributions of their axons in olfactory system of rat. *J. Neurophysiol.*, *46*, 918–931.
Scott, J.W. and D.D. Aaron (1977). Averaged induced waves in the olfactory bulb of the rat during odor stimulation. *Exp. Neurol.*, *55*, 654–665.
Scott, J.W., and W.B. Stewart (1979). Mechanisms of augmented field potential responses in the rat olfactory bulb. *Brain Res.*, *163*, 21–32.
Scott, J.W., R.L. McBride, and S. P. Schneider (1980). The organization of projections from the olfactory bulb to the piriform cortex and olfactory tubercle in the rat. *J. Comp. Neurol.*, *194*, 519–534.
Scott, J.W., E.C. Ranier, J.L. Pemberton, E. Orona, and L.E. Mouradian (1985). Pattern of rat olfactory bulb mitral and tufted cell connections to the anterior olfactory nucleus pars externa. *J. Comp. Neurol.*, *242*, 415–424.
Scott, J.W., and T.A. Harrison (1987). The olfactory bulb: Anatomy and physiology. In T.E. Finger and W.L. Silver (Eds.), *Neurobiology of Taste and Smell.* New York: Wiley, pp. 151–178.
Sharp, F.R., J.S. Kauer, and G.M. Shepherd (1975). Local sites of activity-related glucose metabolism in rat olfactory bulb during olfactory stimulation. *Brain Res.*, *98*, 596–600.
Shepherd, G.M. (1963a). Responses of mitral cells to olfactory nerve volleys in the rabbit. *J. Physiol. (Lond.)*, *168*, 89–100.
Shepherd, G.M. (1963b). Neuronal systems controlling mitral cell excitability. *J. Physiol. (Lond.)*, *168*, 101–117.
Shepherd, G.M. (1972a). Synaptic organization of the mammalian olfactory bulb. *Physiol. Rev.*, *52*, 864–917.
Shepherd, G.M. (1972b). The olfactory bulb as a simple cortical system: Experimental analysis and functional implications. In F.O. Schmitt (Ed.), *The Neurosciences: Second Study Program.* New York: Rockefeller University Press, pp. 539–552.
Shepherd, G.M. (1979). Olfactory bulb. In G.M. Shepherd, *The Synaptic Organization of the Brain*, 2nd ed. New York: Oxford University Press, pp. 152–183.
Shepherd, G.M., and R.K. Brayton (1979). Computer simulation of a dendrodendritic synaptic circuit for self- and lateral-inhibition in the olfactory bulb. *Brain Res.*, *175*, 377–382.
Shepherd, G.M., and C.A. Greer (1990). Olfactory bulb. In G.M. Shepherd (Ed.), *The Synaptic Organization of the Brain.* 3rd ed. New York: Oxford University Press, pp. 133–169.

Shipley, M.T., and R. Costanzo (1984). Olfactory bulb cytochrome oxidase (CO) staining patterns suggest that glomeruli are functional units. *Neurosci. Abstr.*, *10*, 118.

Shipley, M.T., and D.S. Zahm (1990). Differential synaptic processing on apical versus lateral mitral/tufted cell dendrites. *Chem. Senses*, *15*, 638.

Slotnick, B.M., G. Graham, D.G. Laing, and G.A. Bell (1987). Detection of proprionic acid vapor by rats with lesions of olfactory bulb areas associated with high 2-DG uptake. *Brain Res.*, *417*, 343–346.

Stewart, W.B. (1985). Labelling of olfactory bulb glomeruli following horseradish peroxidase lavage of the nasal cavity. *Brain Res.*, *347*, 200–203.

Stewart, W.B., and J.W. Scott (1976). Anesthetic-dependent field potential interactions in the olfactory bulb. *Brain Res.*, *103*, 487–499.

Stewart, W.B., J.S. Kauer, and G.M. Shepherd (1979). Functional organization of rat olfactory bulb analyzed by the 2-deoxyglucose method. *J. Comp. Neurol.*, *185*, 715–734.

Stewart, W.B., and P.E. Pedersen (1987). The spatial organization of olfactory nerve projections. *Brain Res.*, *411*, 248–258.

Stone, E.A., and A.S. Herrera (1986). Alpha-adrenergic modulation of cyclic AMP formation in rat CNS: Highest level in olfactory bulb. *Brain Res.*, *384*, 401–403.

Takagi, S. (1989). *Human Olfaction*. Tokyo: University of Tokyo Press.

von Baumgarten, R., J.D. Green, and M. Mancia (1962). Slow waves in the olfactory bulb and their relation to unitary discharges. *EEG. Clin. Neurophysiol.*, *14*, 621–634.

Walsh, R.R. (1959). Olfactory bulb potentials evoked by electrical stimulation of the contralateral bulb. *Am. J. Physiol.*, *196*, 327–329.

Weinberg, C., and E. Meisami (1989). Comparison of staining patterns of rat olfactory bulb glomeruli by cytochrome oxidase and acid phosphatase. *Chem. Senses*, *14*, 759.

Wellis, D. P., and J.W. Scott (1989). Olfactory receptor axons run in parallel on the lateral face of the rat olfactory bulb. *Chem. Senses*, *14*, 759.

Wellis, D. P., J.W. Scott, and T.A. Harrison (1989). Discrimination among odorants by single neurons of the rat olfactory bulb. *J. Neurophysiol.*, *61*, 1161–1177.

Wellis, D.P., and J.W. Scott (1990). Intracellular responses of identified rat olfactory bulb interneurons to electrical and odor stimulation. *J. Neurophysiol.*, *64*, 932–947.

Westecker, M.E. (1970). Alternating characteristics of the evoked potential in the olfactory bulb in response to repetitive stimulation of the lateral olfactory tract. *Brain Res.*, *17*, 142–144.

Whittemore, E.R., and J.F. Koerner (1989). An explanation for the purported excitation of piriform cortical neurons by N-acetyl-L-aspartyl-L-glutamic acid (NAAG). *Proc. Natl. Acad. Sci. USA*, *86*, 9602–9605.

Yamamoto, C., T. Yamamoto, and K. Iwama (1963). The inhibitory systems in the olfactory bulb studied by intracellular recording. *J. Neurophysiol.*, *26*, 403–415.

Zaborszky, L., J. Carlsen, H.R. Brashear, and L. Heimer (1986). Cholinergic and GABAergic afferents to the olfactory bulb in the rat with special emphasis on the projection neurons in the nucleus of the horizontal limb of the diagonal band. *J. Comp. Neurol.*, *243*, 488–509.

7
Olfactory Learning and Memory in the Rat: A "Model System" for Studies of the Neurobiology of Memory

TIM OTTO and HOWARD EICHENBAUM

Introduction

The study of olfactory learning and memory in the rat not only is an interesting research area in its own right, but additionally, and perhaps more importantly, provides a means by which a comprehensive, biologically based theory of memory can be developed. The purpose of this chapter is to review the behavioral, anatomical, and physiological data supporting this view, and in doing so, to determine the extent to which the rodent olfactory system serves as a model that can be used to extend our understanding of the neurobiological bases of human memory.

Among the more conspicuous features of human memory are rapid acquisition, large capacity, longevity, and flexibility. These capacities are demonstrated in everyday life by our ability to retain events heard, read, or seen, and in our ability to flexibly reflect on and integrate our representations of those events. Investigations of human amnesia have demonstrated that it is just these capacities that are lost after damage to certain limbic targets of central olfactory projections (see Squire, 1987).

The importance of limbic structures to the encoding of certain types of new information became apparent in neuropsychological studies of the patient H.M., in whom the medial temporal lobes were removed bilaterally in an effort to treat intractable epilepsy (Scoville and Milner, 1957). Subsequent to this surgical procedure, H.M. could recall in great detail and reflect upon events prior to his operation, but was unable to form any new, lasting memories. The range of impaired memory capacity in H.M. extends widely to retention of most everyday events, but it is not complete; he demonstrates spared learning on a limited variety of tasks, specifically tasks in which retention can be revealed primarily by a mere repetition of the act of learning (Moscovitch, 1984). Together with results of similar studies of other human amnesics with more limited telencephalic damage (Squire, Shimamura, and Amaral, 1989), the data suggest that only certain types of memory representation depend on limbic telencephalic pathways. Specifically, memories that are rapidly acquired and can subsequently be reflected upon and integrated for use in new ways are particulary susceptible to telencephalic damage. Further, specific limbic structures within those pathways are critical to the creation, but not the storage, of the memory representation. Several authors have proposed that memory storage itself occurs in the cortical association areas associated with sensory and motor processing relevant to the learning experience (Squire, Cohen, and Nadel, 1984; Teyler and DiScenna, 1986).

On the Need for a "Model System" for Studies on the Neurobiology of Learning and Memory

Recent attempts to characterize the anatomical and physiological substrates of learning and memory have met with considerable success. Anatomical investigations have consistently illustrated the decidedly prominent role of particular limbic and cortical areas in memory processing and storage. Damage to these areas in both humans and laboratory animals has convincingly demonstrated the existence of multiple memory systems, apparently subserved by discrete neural pathways. Furthermore, it has become increasingly clear that the brain exhibits a remarkable degree of plasticity observable at the physiological, anatomical, and biochemical levels of analysis; indeed, synaptic plasticity has become the backbone of current neurobiological theories of learning and memory.

Unfortunately, there exists no generally accepted, comprehensive theory of memory integrating the "systems" and "physiological-biochemical" levels. This is due in part to the diversity of behavioral paradigms used by investigators across levels. For example, while the systems approach has provided a wealth of important data demonstrating the significance of various limbic structures to visual memory in primates, the complexity of the visual system precludes readily interpretable investigations into the physiological correlates of information storage. Similarly, while physiological studies in the rat have identified behavioral correlates of the activity of single neurons, these data are not easily accommodated into systems-level theories of memory. For example, it has been demonstrated that individual hippocampal cells fire preferentially when the animal is in a defined spatial location (Ranck, 1973), but we do not as yet understand the significance of these correlates in the encoding of spatial memories (Eichenbaum and Cohen, 1988).

Progress in a comprehensive approach to the biological underpinnings of memory depends on the development of a "model system" that would allow anatomical, physiological, and behavioral data to be collected on the same species and integrated across levels of analysis (see Otto and Etchenbaum, 1992). A true model system would additionally permit generalizations beyond the species under investigation; implicit in its development is the assumption that this system can be used to investigate human memory processes. A model system would thus necessarily be based on an appropriate choice of a subject and behavioral paradigm in which "humanlike" memory processing is observed. In addition, to facilitate physiological investigations, the stimulus modality employed in these behavioral paradigms must involve known, accessible sensory pathways into the limbic structures implicated in human memory processing. In the following sections we will review the data suggesting that the rodent olfactory system is a "model system" for studying the neurobiological bases of learning and memory. We will then discuss ways in which this system can be exploited to provide new information critical to our general understanding of learning and memory processes. Specifically, we will attempt to establish what investigations of rodent olfactory learning can tell us about the anatomical substrates of multiple memory systems and about the physiological plasticity underlying memory formation.

Rodent Olfaction as a "Model System" for Studies on the Neurobiology of Learning and Memory

Nauta once suggested that the olfactory system provides a "flow diagram for telencephalic processing in the mammalian brain" (1972). Converging evidence from behavioral, anatomical, and physiological studies suggests that the rodent olfactory system is indeed an ideal sensory modality for investigations of mammalian learning and memory dependent on telencephalic pathways. Consistent with the attributes of a "model system" described above, behavioral studies have consistently demonstrated that laboratory animals, including rats, exhibit remarkably "humanlike" learning in tasks using olfactory cues. Further, anatomical investigations suggest that olfactory-limbic pathways are fully evolved and parallel the relevant neocortical-limbic pathways of dominant sense modalities in primates. In this section, we will review the behavioral and anatomical data supporting the hypothesis that the rodent olfactory system is a "model system" for studies of human learning and memory.

Olfactory Learning and Memory in the Rat: Behavioral Studies

Human and nonhuman primates were traditionally thought to be unique among mammals in their ability to rapidly acquire vast amounts of information, to retain that information indefinitely, and to integrate that information for use in novel circumstances. However, evidence accumulated in recent years indicates that rats are capable of demonstrating each of these performance characteristics of human learning and memory provided the tasks employ olfactory stimuli.

One of the more commonly employed measures of learning ability is the speed with which "learning sets" are acquired. Learning set formation is characterized by a progressive improvement in an organism's ability to solve new problems of a similar class, and the relative rate of learning set acquisition has been used as a tool with which to rank species in intelligence (Warren, 1973). Early attempts to scale organisms in intelligence in this way met with severe criticism, primarily for their failure to consider the relevance of the stimuli employed to the organism tested. Indeed, the importance of choosing stimuli appropriate to the dominant sensory modality of the species under investigation is perhaps most apparent when considering the rat: While rats are incapable of acquiring learning sets with visual or auditory stimuli, they do so with considerable facility when presented with olfactory stimuli (Nigrosh, Slotnick, and Nevin, 1975). This is perhaps not surprising considering the dominant role played by olfaction in rodent social and sexual behavior (reviewed by Doty, 1986).

The phenomenon of rat olfactory learning set formation has been demonstrated and replicated in several laboratories (Jennings and Keefer, 1969; Slotnick and Katz, 1974; Eichenbaum et al., 1986; Otto, Schlotter, Staubli, Eichenbaum, and Lynch, 1991). Invariably, when presented with a series of session-unique odor discriminations, rats rapidly improve in their discrimination performance until at asymptote they are able to discriminate new odor pairs after only a few trials. Rats can form olfactory learning sets in paradigms in which they are presented with two odors simultaneously and must choose the reinforced odor (Eichenbaum et al., 1988) and in paradigms in which two odors are presented individually and sequentially across trials, and the response demand is to execute or not execute a single response ("go/no-go" discrimination; Otto, Schlotter, Staubli, Eichenbaum, and Lynch, 1991). Asymptotic performance in these tasks demonstrates not only that, within the appropriate stimulus modality, rats can learn complex abstract rules, but that they are able to acquire new information after very few exposures.

In addition to learning sets, rats can acquire other abstract concepts using olfactory stimuli, including "oddity" (Langworthy and Jennings, 1972) transitive inference (Wible, Eichenbaum, and Otto, 1990) and delayed nonmatching to sample (Otto, Eichenbaum, and Wible 1990), a task demanding both a capacity for short-term memory and a conditional logic (matching) operation. The visual analog of olfactory delayed nonmatching to sample has become a benchmark test of primate memory (Mishkin and Delacour, 1975; Gaffan 1974).

Investigation of the longevity of olfactory memories has been problematic because well-trained rats can acquire novel discriminations as rapidly as they can demonstrate savings on familiar discriminations. Thus, to assess the retention of odor memories, some investigators have employed "reversal tests" in which pairs of odors that were presented previously are again presented, but with reversed reward valence; the strength of memory is reflected in the extent to which the animal responds in accordance with the originally trained valence assignments (Otto, Schlotter, Staubli, Eichenbaum, and Lynch, 1991; Eichenbaum et al., 1986; Staubli et al., 1987). Using this method, it has been shown that olfactory information is retained for up to 65 days (the longest retention interval tested, Otto, Schlotter, Staubli, Eichenbaum, and Lynch, 1991) and apparently remains intact despite several intervening discriminations of new odor pairs (Staubli et al., 1987) or limited training (Otto, Schottler, Staubli, Eichenbaum, and Lynch 1991). Collectively, the olfactory learning set and retention data indicate that rats are capable of rapidly forming lasting representations of olfactory information.

Rapid and long-lasting olfactory learning in rats can also be observed in more ethologically valid contexts. Rats quickly learn to avoid odorized foods that predict later illness (Supak, Macrides, and Chorover, 1971). These aversions are learned even more rapidly when conjoint olfactory-gustatory cues are used (Palmerino, Rusiniak, and Garcia, 1980). Even fetal rats exhibit rapid acquisition of odor aversion (Smotherman, 1982) and odor preference (Blass and Teicher, 1980). Indeed, odor-guided learning begins quite early in development in a wide range of species, including rats and humans (reviewed in Leon, 1983).

Considered together, the behavioral data on olfactory learning indicate that rats can rapidly acquire new olfactory information in a variety of contexts, can retain that information for significant periods, and can demonstrate a seemingly limitless olfactory memory capacity. The learning set data further suggest that with olfactory stimuli, rats are capable of integrating acquired information and using it in novel situations in a manner apparently indistinguishable from that of primates. These are the qualities of learning and memory that are often lost following insult to certain limbic structures in humans. Thus, combined with our ability to produce well-circumscribed lesions in the rat, olfactory learning paradigms permit detailed analyses of the specific role of various limbic structures in learning and memory in rats; these results should be interpretable in a context relevant to human amnesia. Hence, with regard to behavioral performance, rat olfactory learning is an ideal assay from which to generalize to human learning.

Olfactory Learning and Memory in the Rat: Anatomical Substrates

As discussed previously, whether the rodent olfactory system provides an adequate model for studying human learning and memory depends in part upon olfactory anatomy. In this section, we will review the olfactory pathways that might support the behavioral performance described above. Specifically, we will examine the extent to which olfactory-limbic projections parallel the sensory-limbic pathways demonstrated to be critical to primate memory. A more comprehensive description of olfactory anatomy appears elsewhere in this volume (McLean and Shipley).

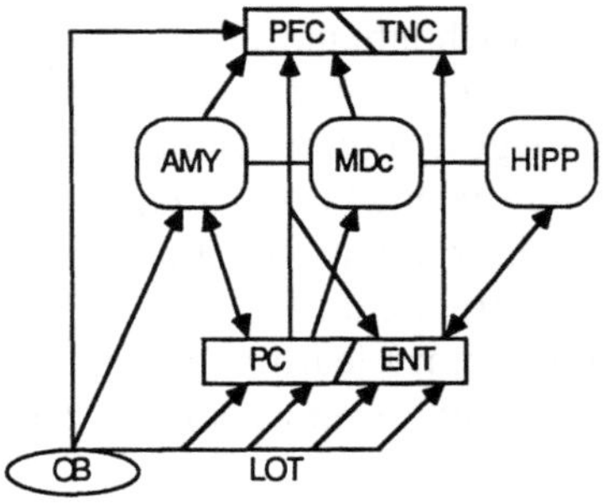

FIGURE 7.1. Anatomical pathways between the olfactory system and the limbic areas in the rat. Depicted here are connections to the limbic areas known to be critical for some types of memory formation in the primate. Double-headed arrows indicate reciprocal connections. Abbreviations: OB, olfactory bulb; LOT, lateral olfactory tract; PC, piriform cortex; ENT, entorhinal cortex; AMY, amygdala; MDc, central division of the dorsomedial nucleus of the thalamus; HIPP, hippocampus; PFC, prefrontal cortex; TNC, temporal neocortex.

The neocortical association areas that process complex visual, auditory, and tactile stimuli for learning and memory in primates project to three main limbic targets, specifically, the amygdala (Nauta, 1972), the dorsomedial nucleus of the thalamus (MD) and its associated prefrontal cortical target (Leonard, 1972), and the hippocampus (Witter, VanHoesen, and Amaral, 1989). Damage to these areas has been strongly implicated in memory disorders in human and nonhuman primates (see Squire, 1987).

The efferents of the rat olfactory bulb and olfactory cortex have precisely the same limbic targets. Via the lateral olfactory tract (LOT), the olfactory bulbs project monosynaptically to layer 1A of the piriform (primary olfactory) cortex and the lateral subdivision of adjacent entorhinal cortex; most of the olfactory-limbic projections arise from these cortical areas. As illustrated in Figure 7.1, the rat olfactory-limbic pathways can be divided into the same three secondary subsystems as primate neocortical-limbic pathways, each of which will be discussed individually.

Amygdala

Several amygdaloid nuclei (anterior amygdaloid area and anterior and posterolateral cortical amygdaloid nuclei) receive inputs primarily from layer 2 and the deep layer of the anterior piriform cortex. In addition, layer 2 of the posterior piriform cortex projects to the lateral amygdaloid nucleus and the basolateral, basomedial, and rostral central amygdaloid nuclei. Connections between piriform and most of these amygdaloid nuclei are reciprocal (Switzer, DeOlmos, and Heimer, 1985). Amygdalar inputs arise also from the entorhinal cortex (Aggleton, Burton, and Passingham, 1980) and the olfactory bulb (Price, 1985). The amygdala projects to prefrontal cortex (Krettek and Price, 1977a) and to entorhinal cortex (Insausti, Amaral, and Cowan, 1987). Odorants have been shown to elicit excitatory responses in single units of the amygdala in both rats (Cain and Bindra, 1972) and monkeys (Tanabe, Iino, and Takagi, 1975).

Mediodorsal Nucleus of the Thalamus

One segment of MD (the central nucles, hereafter referred to as MDc) receives a direct projection from deep cells in the piriform cortex (Switzer et al., 1985; Price and Slotnick, 1983; Krettek and Price, 1977b). MDc projects to the orbital subdivision of the prefrontal cortex, with fibers terminating in layers 1 and 2 (Krettek and Price, 1977b). Direct input to the prefrontal cortex from the olfactory bulb and piriform cortex has also been reported (Cinelli, Ferreyra-Moyano, and Barragan, 1987; Gerfen and Clavier, 1972). While single units in MDc of cats (Motokizawa, 1974) and rabbits (Imamura, Onoda, and Takagi, 1984) respond to both biologically relevant and biologically irrelevant odors, units in the prefrontal cortex respond almost exclusively to biologically significant odors such as urine, feces, and food (Onoda and Iino, 1980).

Hippocampus

The lateral subdivision of entorhinal cortex receives direct olfactory input from the olfactory bulb as well as from the piriform cortex, and it is from this region that the primary olfactory efferents to the hippocampus arise. Axons of entorhinal cortex layer 2 stellate neurons, forming the perforant path, terminate in the dentate gyrus and in area CA1 of the hippocampus proper (for review see Eichenbaum and Buckingham, 1989). The hippocampus projects back to the deep layers of entorhinal cortex (Sorensen and Shipley, 1979), a zone from which projections to temporal neocortex (Kosel, VanHoesen, and Rosene, 1982) and prefrontal cortex (Swanson, 1981) arise. Hippocampus also projects back to piriform cortex (Switzer et al., 1985). Odorants appear to elicit primarily inhibitory responses in hippocampal single units in the anesthetized, tracheotomized rat (Cain and Bindra, 1972).

Summary

The rat olfactory system is intimately linked to limbic structures known to play a prominent role in primate learning and memory, namely the amygdala, the dorsomedial nucleus of the thalamus, and the hippocampus. These olfactory-limbic pathways are fully evolved in rodents and parallel the sensory-limbic pathways that are relevant to memory processing in primates. Considered together with the "primatelike" behavioral aspects of olfactory learning and memory in rats, the evidence strongly suggests that the rat olfactory system has both the functional and anatomical properties that characterize a model system for studying the neurobiological bases of learning and memory.

How, then, can we take advantage of this model system to investigate mammalian learning and memory? Investigations of olfactory memory have begun to provide important insights into the anatomical and physiological bases of memory processing in two ways. First, lesion and single-unit recording studies of rats solving olfactory tasks have helped distinguish the roles of the olfactory-limbic pathways in memory processing, and have significantly advanced our understanding of the anatomical bases of multiple memory systems. Second, investigations into learning-induced synaptic plasticity have addressed the potential role of olfactory cortical structures in memory storage. These issues will be discussed separately in the following two sections.

Discrete Limbic Pathways are Critical for Dissociable Olfactory Memory Processes

Using the rat olfactory system as our "model," we can explore the mnemonic functions of each of the three limbic targets of olfactory information, and inquire further whether these anatomically separate systems subserve different memory functions. Delineating the role of specific brain structures in learning and memory depends on our ability to demonstrate that a given structure is required for a particular aspect of memory processing and further, that those memory processes are reflected in the physiological activity within that structure. These kinds of evidence can be derived, respectively, from studies of the effects of discrete brain lesions on olfactory learning and memory, and from investigations of physiological (single-unit) correlates of olfactory information processing. In this section, we will review the results of experiments employing both strategies in each of the olfactory-limbic targets and their cortical projection areas.

Historical Overview

Few studies have systematically examined the behavioral effects of damage to the primary or secondary olfactory projection targets. Early investigations explored the effects of large cortical and subcortical lesions on olfactory learning. Swann (1934) tested rats' postoperative retention of odors in an appetitively motivated simultaneous discrimination task following bilateral lesions in specific cytoarchitectural areas throughout the cortex. He found that lesions to these areas beyond the olfactory bulb or anterior LOT did not produce a consistent deficit in discrimination performance. Included his analyses were subjects with considerable, although incomplete damage to the amygdala, hippocampus, or frontal lobes. Shortly thereafter, Allen (1940, 1941) investigated the effects of damage to most of the same secondary olfactory targets in dogs. His findings were also negative; ablation of the frontal lobes, hippocampus, or piriform cortex and amygdala did not prevent development of aversively reinforced conditioned responses to odor stimuli. Brown and Ghiselli (1938) also reported unimpaired olfactory learning following lesions to the thalamus and pretectal areas and concluded that, consistent with Lashley's views on the equipotentiality of cortical involvement in memory storage, olfactory habits could not be localized to any particular structure, and that their representation might be widely distributed in the brain.

Investigations of humans and nonhuman primates have been more successful in demonstrating an important role for specific cortical areas in higher olfactory processing, including odor discrimination and identification. A series of studies on monkeys (Santibanez and Pinto Hamuy, 1957; Brown, 1963; Brown et al., 1963) implicated the temporal association cortex, and in particular the anterior temporal pole, in discrimination learning guided by odor cues in the Wisconsin General Test Apparatus. More recently, Tanabe et al. (1975) demonstrated that lesions of a posterior lateral portion of the orbitofrontal cortex in monkeys resulted in impaired retention of odor-cued discrimination of edible versus quinine-tainted foods.

Consistent with these findings, evidence from human patients indicates that orbitofrontal and temporal cortex are selectively involved in higher-order aspects of olfactory processing. Unilateral cortical excisions in either area result in little or no alteration of odor thresholds but cause significant deficits in odor discrimination (Potter and Butters, 1980; Eskenazi et al., 1983), identification (Eskenazi et al., 1983; Henkin et al., 1977), and recognition memory (Eskenazi et al., 1983; Rausch et al., 1977). A similar pattern of relatively intact olfactory detection and impaired olfactory discrimination, identification, and recognition memory is observed in patients with Korsakoff's psychosis, which is characterized by damage to medial thalamic areas in the secondary olfactory pathways (see Mair and Flint, this volume; Jones et al., 1978). Furthermore, the patient H.M., who underwent bilateral medial-temporal lobe resection, demonstrates a striking dissociation between detective and discriminative functions; he has a normal capacity for detecting and discriminating odor stimuli by intensity, as determined by a variety of measures, but is completely unable to differentiate, recognize, or identify odors by quality (Eichenbaum, Morton, Potter, and Corkin, 1983).

The lesion approach to identifying the role of specific limbic structures in rodent learning and memory has been employed with renewed enthusiasm in recent years, due largely to the discovery of dissociable memory systems in human amnesics and nonhuman primates with cortical or limbic system damage (see above). Many studies have sought not only to identify which brain areas are important to the storage or retrieval of information, but to determine whether these areas encode different "types" of memory, as they seem to in human and nonhuman primates.

With regard to olfactory memory, recent investigations on animals have focused on developments of behavioral assays of olfactory detection and discrimination in rodents. Studies over the last 10 years, employing a variety of behavioral paradigms, collectively suggest that separate secondary olfactory pathways play dissociable roles in olfactory learning and memory. Thus secondary olfactory pathways via the amygdala, MDc/prefrontal cortex, and hippocampus appear to utilize olfactory cues for different kinds of odor memory representations.

Amygdala

Slotnick (1985), using more refined behavioral methods than those in previous studies, found that damage to the anterior amygdala transecting the posterior LOT had no effect on learning and retention of successive odor discriminations and discrimination reversals, and further, did not affect olfactory thresholds. Eichenbaum et al. (1986) confirmed and extended these findings, reporting unimpaired learning of successive odor discriminations following amygdala lesions alone or in combination with damage to the hippocampal system.

Although apparently uninvolved in odor detection or discrimination, the amygdala does seem to be critical to some types of odor-guided behavior, including behavior dependent on odor memory. Destruction of the amygdala or its local anesthesia via intracranial novocaine application does not produce anosmia, but does disrupt odor- and taste-potentiated odor aversions (Bermudez-Rattoni et al., 1982, 1983) and odor preference learning (Beaulieu, Morris, and Petrides, 1988). Furthermore, amygdala lesions or transection of the LOT just anterior to the amygdala disrupt odor-guided sexual behavior (Devor, 1973; Macrides et al., 1976). These effects do not appear to be secondary to disruption of hippocampal inputs (Sapolsky and Eichenbaum, 1980). Thus, while the amygdala is not critical to odor perception or to arbitrary associations of odors with reward, its processing may be crucial to the development of species-typical responses dependent on odor memories.

Dorsomedial Nucleus of the Thalamus and Prefrontal Cortex

In contrast to the findings on the amygdala, several studies have demonstrated the importance of the thalamo-prefrontal pathway in odor discrimination learning. Eichenbaum et al. (1980) preoperatively trained rats on odor detection tasks involving differential responses to an odor versus a clean air stream and on a series of successive odor discriminations. After lesions of the MDc, rats performed as well as sham-operated subjects in reacquiring the odor detection and had normal thresholds as determined by the method of descending limits. They further exhibited retention of a preoperatively acquired problem involving a pair of odors that were easily discriminable by humans. In contrast, MDc lesions impaired the postoperative acquisition and retention of a novel discrimination involving a pair of odors that humans find difficult to differentiate, even though the rats themselves have no difficulty distinguishing these odors.

Other results of this study further indicated that the learning impairment associated with MDc lesions reflects a "cognitive" rather than a "sensory" deficit. Sham-operated rats required more trials to learn a difficult discrimination problem if it was preceded by an easy problem. However, rats with MDc lesions required about the same number of trials to acquire the difficult problem irrespective of the difficulty of the preceding problem. Thus the MDc appears to be involved in higher-order aspects of the development of odor associations, in this case an experience-dependent "contrast effect," and not in the sensory aspects of detection or differentiation of odors. Other investigations have shown that rats with MD lesions demonstrate impaired olfactory reversal learning set acquisition (Slotnick and Kaneko, 1981) and little savings in relearning olfactory discriminations (Staubli et al., 1987). The learning impairment associated with MD damage, like that of Korsakoff patients, is multimodal (Knoth, 1988).

Recent studies, using both rats and primates, suggest that orbital prefrontal cortex, a projection target of both piriform cortex and MDc, is critical to some types of odor-guided learning (Tanabe et al., 1975; Eichenbaum, Clegg, and Fealey, 1983). Lesions of this area in rats has recently been shown to impair acquisition of delayed non-match to sample (Otto and Eichenbaum, 1991). Furthermore, lesions of orbital prefrontal cortex result in greater impairment in odor discrimination acquisition and retention than those observed following lesions of MDc, and the deficit is proportional to the amount of cortical damage (Eichenbaum et al., 1980). Unlike lesions in the medial thalamus (Knoth and Mair, 1991), destruction of the orbital prefrontal cortex results in a modality-specific deficit in odor-guided learning. Eichenbaum, Clegg, and Fealey (1983) demonstrated a double dissociation of learning deficits in odor discrimination and spatial delayed alternation following selective ablations of the orbital and medial components, respectively, of the prefrontal cortex of rats. Furthermore, the observed impairments in discrimination performance were attributable to an increase in perseverative responses within each respective task, rather than a general behavioral disinhibition or sensory defect. These observations of modality-specific and response-dependent impairments after selective prefrontal lesions are analogous to those observed in nonhuman primates (for review, see Rosenkilde, 1979). It has been suggested that the evolution of prefrontal cortex coincides with the observation of excellent learning capacity within specific sensory modalities across species, and may thus serve as the neocortical substrate for expression of "dominant sensory modalities" in learning and memory (Eichenbaum, Morton, Potter, and Corkin, 1983).

Hippocampus

Hippocampal system dysfunction also results in distinct effects on olfactory learning and memory, but the evidence suggests that the hippocampal system may play a different role in memory processing from that of the thalamo-prefrontal pathway. Like MD or prefrontal damage, insults to the hippocampal system do not interfere with odor perception, and impair odor discrimination learning under certain conditions. Specifically, by modifying the manner in which the same discriminative stimuli are presented so as to encourage or hinder their direct comparison, a profound impairment or facilitation in performance is observed as a consequence of hippocampal system damage (see Figure 7.2). These data are reviewed below.

Other studies have shown that lesions of the entorhinal cortex disrupt rats' performance in odor-guided delayed non-matching to sample when long retention intervals are used, but have no effect when those intervals are brief (Otto and Eichenbaum, 1991). This pattern of results is strikingly similar to those obtained using visually-guided tasks with monkeys and humans (Squire, 1987). Following hippocampal system damage produced by either fornix transection or entorhinal lesion, rats are impaired in discrimination learning and in the acquisition of an olfactory learning set if the odor stimuli are presented simultaneously (Eichenbaum et al., 1988); this impairment is exacerbated by long intertrial intervals (Staubli et al., 1984). In addition, hippocampal denervation in pretrained rats blocks the learning of new simultaneous discriminations while leaving intact the memories of odor cues learned preoperatively (Staubli et al., 1984). This dissociation of impaired postoperative acquisition and spared retention of preoperatively acquired olfactory memories in rats is strikingly similar to the pattern of impaired and spared memory capacities reported in patient H.M. Notably, however, H.M. is surprisingly incapable of discriminating or identifying common odors, even though his olfactory detection thresholds and ability to discriminate and identify the same preoperatively experienced objects by other sensory modalities is normal (Eichenbaum, Morton, Potter, and Corkin, 1983). His inability to discriminate or identify even common odors may be due to the extent of extra-hippocampal damage, which included, among other areas, damage to a considerable portion of the olfactory projection cortex.

→

FIGURE 7.2. Hippocampal system dysfunction results in either an impairment or facilitation of olfactory discrimination learning depending upon the manner in which the discriminative stimuli are presented. A. Simultaneous-cue, go-left/go-right odor discrimination. Two odors are presented simultaneously, and the rat must choose the correct odor for water reinforcement. Each problem is composed of a unique odor pair. Subjects with sham lesions improve across problems, exhibiting acquisition of learning set. In contrast, subjects with lesions of the fornix, a

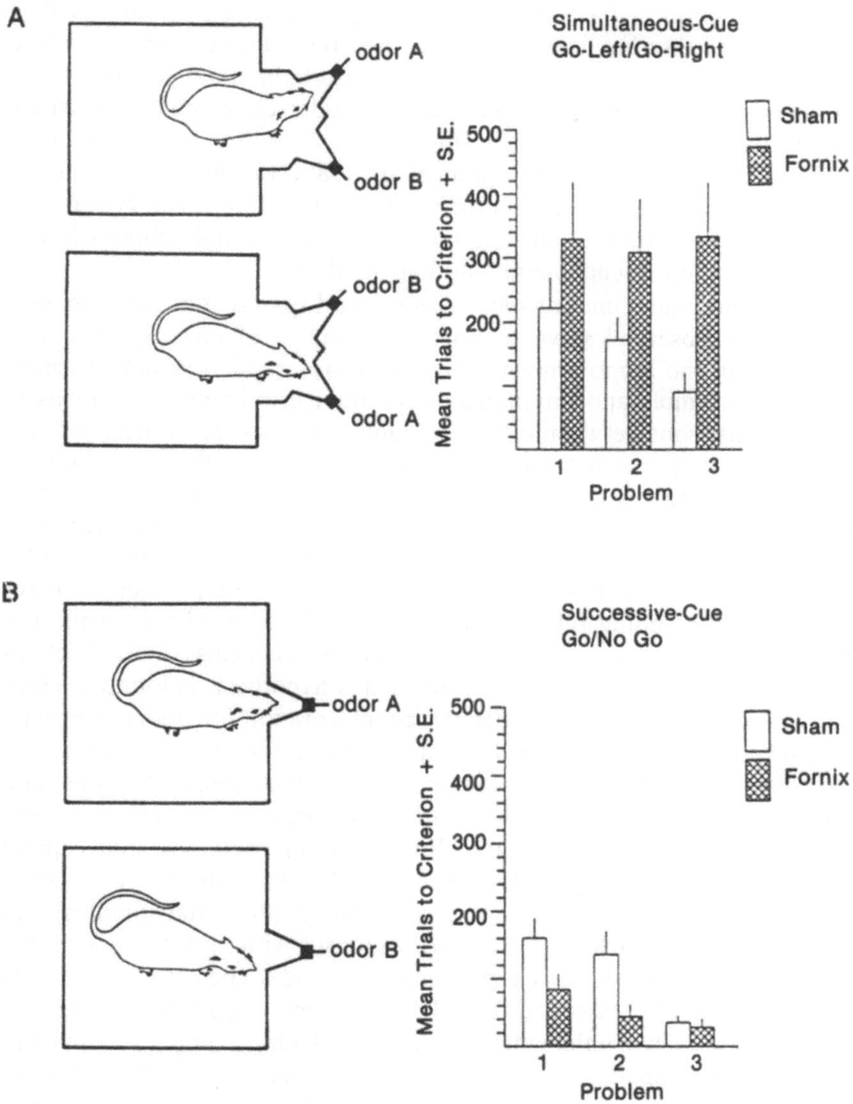

major output pathway of the hippocampus, are dramatically impaired on the initial discrimination and do not acquire the learning set. B. Successive-cue, go/no-go odor discrimination. This task also requires subjects to discriminate problem-unique pairs of odors, but within a problem the two odors are presented individually across trials in a random fashion. The subject is required to execute a nose-poke response in the presence of the positive odor for water reinforcement, and to withold this response in the presence of the negative odor. Note that while both groups improve across problems, illustrating learning set acquisition, subjects with fornix lesions acquire problems more rapidly than their sham counterparts. From Eichenbaum et al. (1988). (Copyright 1987 by the American Psychological Association, reprinted by permission.)

Under different task demands, hippocampal damage does not result in impaired odor discrimination. Rather, when the odor cues are presented individually in a discrete trials, "go, no-go" discrimination task, hippocampal system damage results in a *facilitation* of odor discrimination learning and of olfactory learning set acquisition (Otto, Schottler, Staubli, Eichenbaum, and Lynch, 1991; Eichenbaum et al., 1988). Furthermore, when the discriminative stimuli are presented in this format, the memory for odor valence remains intact at least as long as that in normal subjects (Otto, Schottler, Staubli, Eichenbaum, and Lynch, 1991).

The circumstances under which impaired and spared odor discimination learning are observed serve to clarify the nature of memory processes subserved by the hippocampus. The procedures explicit in delayed nonmatching to sample and simultaneous-cue odor discrimination emphasize direct comparisons between odor cues, and performance in these tasks is impaired by hippocampal system damage. Conversely, the procedures in successive-cue odor discrimination hinder direct cue comparison, and performance in this task is facilitated by hippocampal system damage. The contrast in learning abilities of rats with hippocampal system damage under varying task demands has led to the proposal that the hippocampus is responsible for encoding relations among multiple odor stimuli, while the encoding of individual stimuli and their valence occurs elsewhere (Eichenbaum et al., 1988). Evidence in favor of this hypothesis is found in comparisons of the nature of the memory representation for odors in normal rats and rats with hippocampal system damage. Using the simultaneous odor discrimination task, Eichenbaum et al. (1989) found that rats with hippocampal system damage, while severely impaired on most problems, could occasionally acquire some discrimination problems. Subsequent analyses indicated that rats with hippocampal system damage, even when performing as successfully as normal rats, had an abnormal stimulus sampling strategy by which they executed responses without investigating the separate cues. Further, their representation of odor cues was inflexible in that it could not be extended usefully to tests involving novel pairings of familiar odors. Thus, unlike intact rats, rats with hippocampal system damage apparently failed to distinguish, compare, and encode the relations among odor cues. Instead, they acquired an individual response to each unique odor compound experienced during acquisition of a discrimination, reproducing the representational strategy that is so successful for them in successive odor discrimination.

Complementary evidence from unit recording studies supports the view that the hippocampus is involved in the representation of relations among odors. These studies have revealed a category of hippocampal neurons that fire selectively during the period in which a rat is sampling odors prior to its behavioral response during odor discrimination learning. These cells are active both in a task that critically depends on hippocampal function and in a task where hippocampal system damage results in facilitation of performance, and their activity reflects the relevant relations among odor cues in each task. Thus, in simultaneous odor discrimination, some neurons fire selectively during the presentation of particular left-right configurations of the odor cues (Wiener et al., 1989). In successive odor discrimination, cells of the same category fire selectively depending on the sequence of odor presentation (see Figure 7.3). In each case the activity of these cells seems to reflect comparisons among odor cues presented simultaneously or sequentially, consistent with the hypothesis that the hippocampus participates in a memory representation of relations among odor cues (Eichenbaum and Cohen, 1988; Eichenbaum, Mathews, and Cohen, 1989). Consistent with this interpretation are the data showing that, during the odor-sampling period of delayed non-matching to sample performance, many hippocampal cells exhibit differential firing on "match" versus "non-match" trials (Otto, et al., 1990).

Chief among the conclusions regarding the role of the hippocampus in olfactory learning and memory is that, given appropriate task demands, hippocampal damage can produce striking effects on learning and retention with no concomitant loss in perceptual ability. Further, since the hippocampus does not appear to be necessary for the discrimination of previously learned cues, it is not a storage site for olfactory memories. These behavioral deficits closely parallel those following similar damage in human and nonhuman primates. Furthermore, neuropsychological and unit-recording studies on olfactory memory in rats have contributed to a more detailed description of the nature of hippocampal participation in memory processes. Together, these studies suggest that the hippocampus is critical to the construction of a representation of relevant relations among odor cues.

Summary

The history of studies on the localization of olfactory memory has in general paralleled that of other studies focusing on "localization of the engram." Early efforts indicated that no particular cortical or subcortical structure was required for storage of olfactory "habits." Recently, more sophisticated lesion and behavioral methods have demonstrated that specific areas in the frontal and temporal lobes constitute different olfactory memory "systems," anatomically distinct pathways that are responsible for functionally distinct odor memory representations. The amygdala is preferentially involved in odor-guided, species-specific behavioral adaptations. The dorsomedial nucleus of the thalamus appears to mediate odor discrimination performance based on cue similarity and experience. Finally, the hippocampal system appears to be involved in the encoding of relations among distinct odor cues. None of the pathways appears to be necessary to odor perception or to long-term storage of olfactory memories.

It is important to note that convergence of these representations is likely

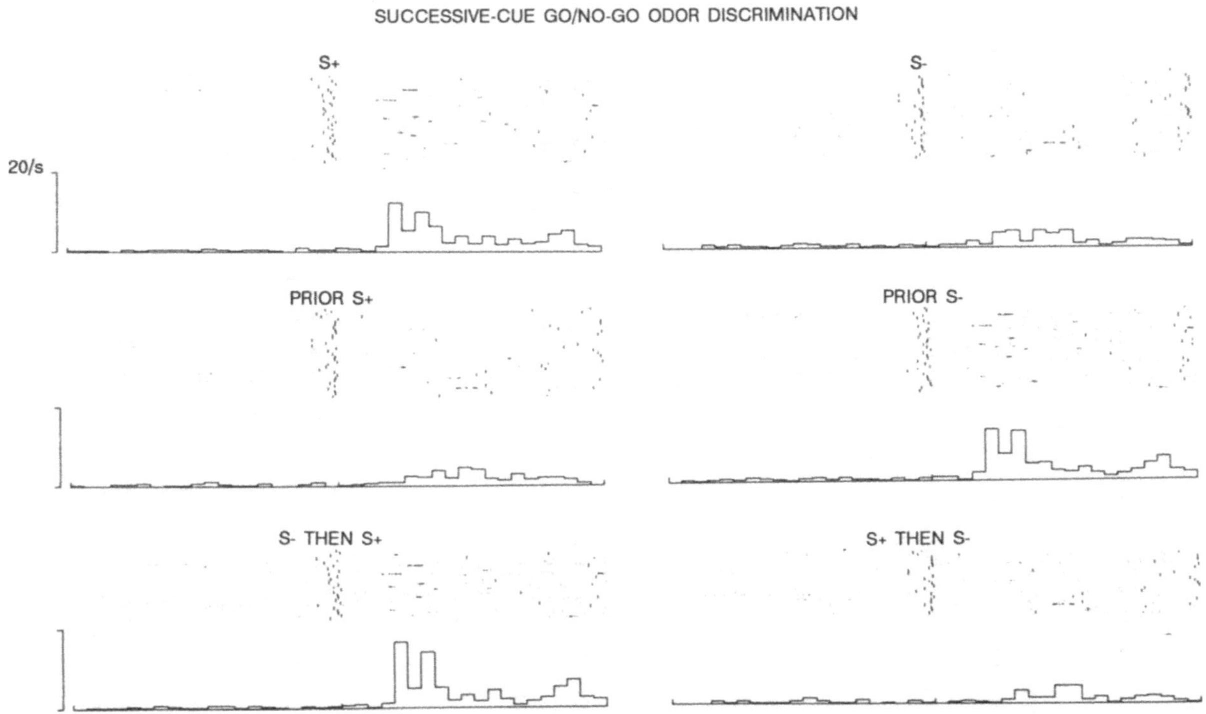

FIGURE 7.3. Raster displays and cumulative histograms of spikes from single hippocampal neurons activated during successive-cue, go/no-go discrimination learning. Unit activity is time-locked to the peak of the first inhalation after odor presentation. S+ and S−, respectively, refer to the positive and negative odors presented. The cell was significantly more active during sampling of S+ than S−, and was more active if the odor presented on the immediately preceding trial was S− than S+, regardless of the odor currently being sampled. This unit was *most* active when the rat sampled the positive odor having just previously sampled the negative odor. These data suggest that hippocampus may participate in a representation of the temporal relations among odor cues. (Reprinted, by permission, from Eichenbaum et al. 1987.)

at the nodal points where these anatomical pathways meet. Both the hippocampus, which in addition to its primary olfactory afferents receives input indirectly from the prefrontal cortex and directly from the amygdala, and the orbital prefrontal cortex, which receives input from the olfactory bulbs, amygdala, medial-thalamus, and hippocampus, appear to be involved in higher-order aspects of olfactory learning *and* in odor-guided species-typical social and ingestive behaviors (Sapolsky and Eichenbaum, 1980; Lasiter et al., 1985; Miller, Nonneman, Kelly, Neisewander, and Isaac, 1986). Thus the functional dissociations described above are currently only suggestive, and are based on comparisons of the effects of circumscribed lesions in different behavioral paradigms. More direct comparisons of the roles of these structures in olfactory learning and memory will undoubtedly disclose additional functional interactions.

Piriform Cortex as a Storage Site for Olfactory Memories

The three main limbic targets of olfactory efferents encode distinct types of olfactory information, but the existing evidence strongly suggests that these limbic areas are not sites of permanent olfactory memory storage. This conclusion is consistent with indications that in humans these same structures perform critical processing functions necessary to, but are not sites of, memory storage. Squire et al. (1984) and others (Teyler and Discenna, 1986; Rolls, 1989), have suggested that memory storage itself occurs in those cortical areas processing the sensory and motor components of the remembered event, and that this storage is subserved by changes in synaptic efficacy among a network of neurons within that area of cortex. The suggestion that changing synaptic weights constitutes the physiological substrate of memory storage is not new; Sechenov (1863) offered a theoretical description of neural "changeability" by which memories and sensations were "preserved in a latent state by the nervous apparatus." A similar, more explicit theoretical account of the selective strengthening of synaptic transmission underlying learning and memory was advanced by Hebb (1949). In this section, we will review the evidence suggesting that the rat olfactory system provides a uniquely suitable cortical domain in which to assess the relevance of cortical synaptic plasticity to memory formation.

Synaptic Plasticity

Learning-dependent neural plasticity has been reported in the olfactory bulb. For example, Leon and colleagues (see Leon, 1987, for review) have demonstrated increases in the responsiveness and alterations in the funtional topography of olfactory bulb principal neurons following odor preference learning in rat pups. Freeman and colleagues (Gray et al., 1986;

Bressler and Freeman, 1980) have demonstrated marked alterations in EEG patterns in olfactory bulbs following olfactory learning. These studies provide important evidence of learning-induced plasticity in the olfactory sensorium, but they do not address specifically the question of cortical olfactory memory storage, nor do they suggest explicitly a physiological mechanism that might subserve memory storage.

One form of synaptic plasticity suggested as a candidate physiological mnemonic device is the long-term potentiation (LTP) of synaptic efficacy following brief bouts of tetanic afferent stimulation. LTP reflects an increase in synaptic strength, and is typically characterized by an increase in amplitude and slope of monosynaptic field potentials evoked by orthodromic stimulation. Since its discovery, LTP has quickly become the leading candidate as a physiological mechanism for memory storage, primarily because of its similarity in induction and maintenance characteristics to behaviorally defined memory (see Teyler, 1986; Lynch and Baudry, 1984). Specifically, long-lasting memories are acquired rapidly, are induced by brief behavioral events, and are strengthened by repetition; LTP is by nature long-lasting, is induced easily by brief bouts of naturally occurring patterns of synaptic activity, and can be strengthened by repeated episodes of afferent stimulation. Furthermore, hippocampal LTP induction is blocked by some pharmacological agents which also interfere with forms of learning dependent on hippocampal function (Morris et al., 1986).

The existence of LTP has been well documented in the hippocampus and other structures (Racine, Milgram, and Hafner, 1983; Teyler, 1989). While in the hippocampus LTP can be elicited over a wide range of stimulation parameters, recent studies have revealed three stimulation parameters optimal for the induction of LTP: high-frequency bursts of orthodromic stimulation, activity 100 to 200 ms prior to a burst (Larson, Wong, and Lynch, 1986; Rose and Dunwiddie, 1986), and burst delivery near the positive peak of the ongoing dentate theta rhythm (Pavlides, Greenstein, Grudman, and Winson, 1988). Otto, Eichenbaum, Wiener, and Wible (1991) have recently shown that each of these three stimulation parameters is reflected in the endogenous firing patterns of CA1 pyramidal cells in rats during mnemonic processing. Specifically, CA1 pyramidal cells discharge in high-frequency bursts in rats performing a simultaneous odor discrimination task. Further, these bursts occur near the positive peak of the ongoing dentate theta rhythm and are preceeded by neural activity at latencies of 100 to 200 ms. Importantly, the occurrence of these bursts is time-locked to periods of stimulus sampling and analysis. Thus, at least in the hippocampus, the conditions appropriate for inducing LTP commonly occur during olfactory learning.

Until very recently, the demonstration of synaptic potentiation has not been pursued in the cortex as intensively as it has been in the hippocampus for two reasons. First, long-lasting increases in synaptic efficacy following intense afferent stimulation were first reported not in the cortex but in the hippocampus (Bliss and Garner-Medwin, 1971; Lomo, 1971). This exceedingly important finding was naturally replicated and extended in several laboratories in both the intact rat and the *in vitro* hippocampal slice preparation, the latter of which has become the standard paradigm for studies of the biochemical bases of hippocampal long-term potentiation (see Lynch, Muller, Seubert, and Larson, 1988). Second, and perhaps more important, entorhinal cortex monosynaptically innervates the dentate gyrus granule cells in the hippocampus via the densely packed perforant path, which provides a relatively accessible and isolatable route by which to selectively stimulate the hippocampus. Additionally, the simplified laminar architecture of the hippocampus (see Teyler, 1986) permits relatively easy recording of monosynaptic field potentials associated with orthodromic stimulation of both extrinsic and intrinsic fibers.

In order to address the question of cortical synaptic plasticity and its relationship to memory storage, one needs a cortical system in which learning-induced changes, if present, are likely to be detected. The ability to observe these changes in cortical synaptic efficacy depends largely upon the anatomy of the chosen area of cortex and its sensory afferents. Thus, like those of the hippocampus, the primary afferents of the chosen area of cortex must consist of an isolated and accessible pathway, and these afferents must innervate cortex in a monosynaptic fashion. Similarly, in order to facilitate recording of monosynaptic field potentials associated with stimulation of these afferents, the cytoarchitecture within the cortex itself must be relatively simple and laminar, and its fundamental intrinsic interactions well understood.

Piriform Cortex Anatomy

Unlike other sensory systems, primary olfactory cortex receives monosynaptic input from the sense organ, and is the source of most olfactory-limbic pathways. Because of its generous and reciprocal connections with many limbic areas known to play a role in memory processing, piriform cortex seems to be a likely candidate for the integration of olfactory and limbic processes and thus for olfactory memory storage.

Despite the wealth of its extrinsic connections, piriform cortex has a relatively simple anatomical organization. Piriform is composed of no more than three distinct layers: Layer 1, the superficial plexiform layer, is composed of the apical dendrites of the pyramidal cells composing layers 2 and 3. Axons of the mitral and tufted cells within the olfactory bulb, composing the lateral olfactory tract (LOT), project ipsilaterally to the entire rostral-caudal length of piriform, establishing synapses in layer 1A (Switzer et al., 1985). These bulbar projections monosynaptically innervate piriform in an apparently random, nontopographic fashion (Luskin and Price, 1982). Layer 1B appears to receive input from collateral fibers of neighboring pyramidal cells, composing an elaborate intracortical associational system (Haberly and Price, 1978). Layers 2 and 3 represent the superficial and deep soma layers, respectively (Haberly, 1983).

The anatomy of piriform cortex and its sensory afferents has led several authors to conclude that piriform cortex serves as a useful model for studying cortical systems in general (Switzer et al., 1985), and cortical information storage in particular (Lynch, 1986; Haberly and Bower, 1989). Specifically, piriform cortex provides the ideal anatomical domain in which to investigate the role of cortical LTP in memory storage. The LOT provides a densely packed, accessible source of cortical afferents, monosynaptically innervating piriform cortex in a known fashion. Further, the intrinsic architecture of piriform cortex permits collection and analysis of excitatory postsynaptic potentials (EPSPs) generated by electrical stimulation of these afferents. Thus, piriform cortex appears to be uniquely suited to studies of whether physiological plasticity similar to that observed in the hippocampus and other areas of cortex can be induced here, and if so, whether that plasticity is associated with learning and memory.

"Electrical Odors" and Cortical LTP

Most studies of LTP have focused on its biochemical bases and physiological induction parameters in the *in vitro* hippocampal slice preparation. As we have seen, however, the hippocampus does not appear to be a site of permanent memory storage. If LTP is truly a physiological substrate of memory, it must be inducible in cortical storage areas by afferent stimulation composing the remembered event. As described above, the anatomy of piriform cortex and its sensory afferents renders this cortical area especially suitable to studies of this type.

Stimulation of the LOT has been reported to produce LTP in piriform cortex *in vitro* (Kanter and Haberly, 1989). However, to determine if LTP represents the physiological basis of olfactory information storage, it is critical to demonstrate that electrical stimulation of the cortical afferents produces LTP in the piriform cortex of the intact, behaving rat. We must further be able to demonstrate that the same stimulation serves as a learning cue, and that this cue is remembered.

Electrical Odors

The use of electrical stimulation of olfactory structures in place of odorous stimuli was originally reported by Mouly, Vigouroux, and Holley (1985), who demonstrated that stimulation of discrete patches of rat olfactory bulb served effectively as distinct, discriminable stimuli in an olfactory learning task. Roman et al. (1987) extended these findings by demonstrating that LOT stimulation produced robust sniffing reactions in rats, and could serve as a learning cue in an operant discrimination task. These studies did not, however, address directly the issue of whether stimulation of the LOT is, to the rat, perceptually similar to naturally odorous stimuli. More recent evidence indicates that LOT stimulation is indeed perceived by rats as an odor, that this stimulation can be used effectively as a discriminative stimulus by rats during olfactory discrimination learning, and that this information is remembered as are odors (Otto et al., 1988). These data are reviewed below.

In order to determine whether repetitive bursts of stimulation of the LOT were perceived by rats as odorous stimuli, a test of the relative transfer of learning from chemical odorants to LOT stimulation or to two auditory cues (tones) was assessed. Electrical stimuli were composed of 2-s trains of electrical stimulation (400 μA, 4 pulse bursts at 100 Hz repeated at 5 Hz), and are hereafter referred to as "theta pattern" stimulation because of their similarity to the cellular bursting associated with the hippocampal theta rhythm. Transfer from chemical odorants to LOT stimulation or to tones was assumed to reflect the extent to which each class of stimuli was perceived by the rats as similar to chemical odorants. If indeed theta pattern stimulation of the LOT was perceived by the rats as an odor, one would expect nearly perfect transfer of learning from odors to LOT stimulation.

Prior to the transfer test, subjects were well trained on 10 to 15 odor pair discriminations in the successive-cue, go, no-go version of the two-odor discrimination task. Following learning set acquisition and establishment of asymptotic performance, adult rats were implanted bilaterally with bipolar stimulating electrodes in the LOT. Following recovery and retraining using chemical odorants, one group of subjects was required to discriminate between two tones with clean air being continuously ejected from the odor port. Another group was required to discriminate between electrical stimulation of one LOT plus clean air (positive cue) and clean air alone (negative cue); this method of stimulus presentation required subjects to use LOT stimulation as the relevant cue. Subjects were run to a criterion of 18/20 correct responses or 200 trials maximum. Data from the transfer test appear in Figure 7.4. As shown, rats transfer much more readily from chemical odorants to LOT stimulation than to tones. These data are viewed as supporting the notion that theta pattern stimulation of the LOT is perceived by rats as an odor cue; that is, electrical stimulation of the LOT when used as a cue in olfactory discrimination paradigms constitutes behaviorally meaningful "electrical odors."

In order to determine whether "electrical odor" information was retained by subjects as is that of chemical odorants, those subjects trained with "electrical odor" stimuli were presented with a reversal of cue valence 24 hours following training. A separate group of unoperated controls, trained on the traditional odor version of this task, were presented with an odor reversal. Results appear in Figure 7.5. As illustrated, the performance of both groups was affected dramatically by reversing the valence of previously learned cues, as one might expect if the significance of those cues were retained.

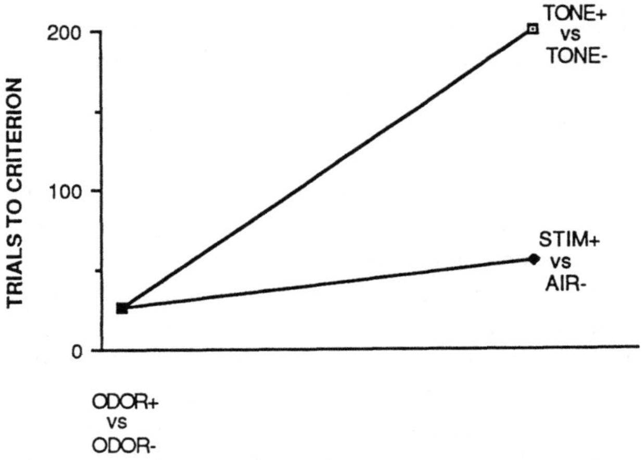

FIGURE 7.4. Transfer from discriminations between two odorants to discriminations between two tones or to discriminations between an "electrical odor" and clean air. Performance is plotted as trials to a criterion of 18 correct responses in 20 consecutive trials. Animals not reaching criterion in 200 trials were assigned a score of 200. The extent of transfer is assumed to be a measure of the similarity of each class of stimuli (tones and electrical stimulation of the lateral olfactory tract) to real odorants. These data suggest that electrical stimulation of the LOT is perceived by rats as an odor.

Cortical LTP Induced by Electrical Odors

So "electrical odors" appear to be effective olfactory cues and are remembered as well as real odorants. The question remains as to whether these "electrical odors" induce plasticity in piriform cortex, and if so, whether this plasticity is correlated with memory formation. Roman et al. (1988) originally reported LTP in piriform cortex following only behaviorally relevant LOT stimulation. Similar analyses were conducted by Otto et al. (1988) following electrical odor training.

Prior to electrical odor training, subjects were implanted bilaterally with bipolar stimulating electrodes in the LOT and monopolar recording electrodes in the superficial plexiform layer (la) of piriform cortex. Twenty dendritic population EPSPs evoked in piriform cortex by single-pulse stimulation of the ipsilateral LOT were collected for both hemispheres immediately before (baseline) and at several time points following electrical odor training. The current applied to each stimulating electrode during this procedure was adjusted so as to produce an EPSP approximately 50 percent of the maximum attainable. Initial slope and peak amplitude were recorded for each response. After collection of the pretraining baseline, one stimulating electrode was used to deliver theta pattern stimulation during "electrical odor" training, which, as described above, required a discrimination between LOT stimulation and clean air. Subjects were trained to

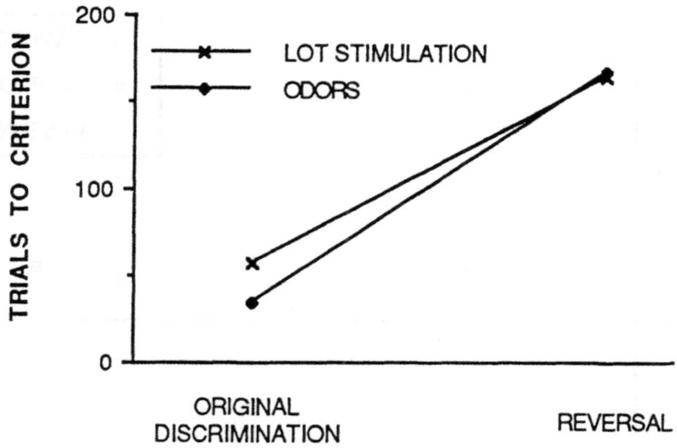

FIGURE 7.5. Performance of rats presented with a reversal of cue valence following initial discrimination training in rats trained with odorants or with electrical stimulation of the LOT. Poor performance in the reversal test indicates intact memory of the previously learned cues. These data indicate that the valence of both real and "electrical" odors is retained by rats.

a performance criterion of 18 correct responses in 20 consecutive trials. Immediately after reaching this criterion, an opaque barrier was placed between the rat and the odor and water ports, preventing the subject from seeing or interacting with the operanda, and the same number of 2-s theta trains as delivered during "electrical odor" training were then delivered to the contralateral electrode. Thus each hemisphere received the same number and type of stimuli, but in one case these stimuli were behaviorally relevant (trained) and in the other they were behaviorally irrelevant (control).

As illustrated in Figure 7.6, behaviorally irrelevant theta stimulation of the control electrode induced synaptic potentiation in ipsilateral piriform cortex which decayed within 24 hours. In marked contrast, the potentiation observed following "electrical odor" training was both more robust and persistent, strongly suggesting that synaptic modifications of sufficiently long duration to account for behavioral memory require that the stimulus be behaviorally meaningful. Although these data are based on small sample sizes and are therefore considered preliminary, they indicate that cortical LTP may indeed be the physiological substrate of olfactory memory formation.

Summary

Piriform cortex has intimate and reciprocal connections both with limbic areas implicated in learning and memory and with olfactory sensory fibers; it thus seems to be a likely storage site for at least some types of olfactory

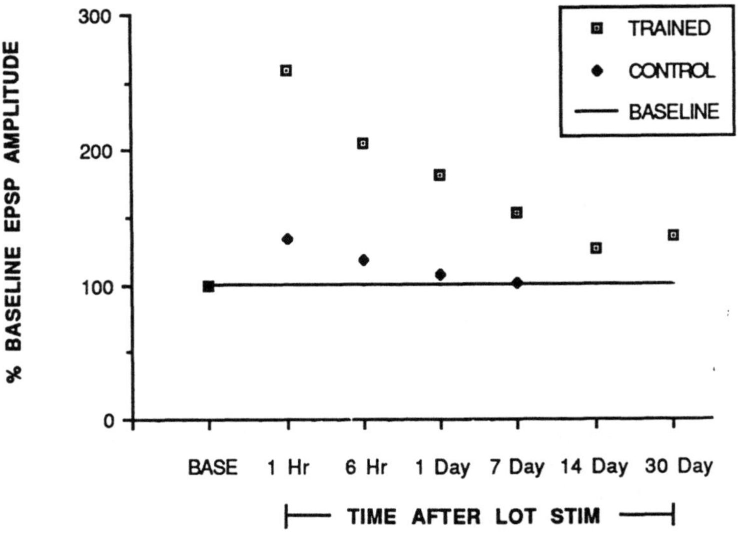

FIGURE 7.6. LTP in piriform cortex following "electrical odor" training or behaviorally irrelevant stimulation of the lateral olfactory tract. LTP decays within 24 hours following behaviorally irrelevant stimulation, while LTP associated with odor training is far more robust and persistent.

information. The results of studies using patterned electrical stimulation of the LOT as discriminative stimuli in discrimination tasks suggest that these "electrical odors" are perceived and remembered by rats as if they were natural odors. Furthermore, this stimulation, when used as a learning cue, can induce in piriform cortex long-term potentiation, a candidate physiological mnemonic device whose induction characteristics are similar to those of behavioral memory. These findings suggest that synaptic efficacy in cortex can indeed be altered by behaviorally relevant electrical stimulation, and further provide a paradigm with which to integrate the study of memory storage across the physiological and behavioral levels.

Conclusions

The goal of this chapter has been to review the behavioral, anatomical, and physiological data pertaining to rodent olfactory learning and memory, and, based on these data, to advance the hypothesis that the rodent olfactory system serves as a model for studying the neurobiological bases of human learning and memory. As discussed at the outset, a "model system" for studies of the neurobiology of learning and memory would allow integration of anatomical, physiological, and behavioral data collected on the same species, and would additionally permit generalizations of these data to humans.

Several aspects of this review support our contention that the rat olfactory system does indeed constitute a "model system." First, rats solve olfactory tasks and encode olfactory information in a manner comparable to the way in which human and nonhuman primates learn and remember visual information. Rats rapidly acquire specific olfactory information, retain that information for extended periods, and form olfactory "learning sets." Second, the rat olfactory-limbic pathways are fully developed and parallel relevant sensory-limbic pathways critical to memory in primates. Each of these limbic targets, the amygdala, dorsomedial nucleus of the thalamus, and hippocampus, is directly innervated by primary olfactory efferents. Third, as demonstrated by both physiological (unit recording) and anatomical (lesion) studies, the same limbic structures implicated in human learning and memory are critical to higher-order processing, storage, and retrieval of olfactory information in rodents. Moreover, these discrete structures play dissociable roles in olfactory memory; that is, separate olfactory-limbic pathways appear to encode different types of information. Supporting this conclusion are studies showing that lesions of the hippocampal system impair rats' ability to solve simultaneous olfactory discriminations, but leave intact rats' ability to learn and remember discriminations in which the olfactory cues are presented successively. Characteristics of this dissociation of "impaired vs. spared" memory are similar to the hypothesized multiple memory systems in humans, for which there is considerable data. Fourth and finally, the anatomical organization of olfactory cortex and its sensory afferents permits examination of the physiological plasticity presumed to underlie cortical information storage. Preliminary evidence suggests that synapses within piriform cortex do exhibit significant plasticity following behaviorally relevant stimulation. Assimilation of these findings leads inevitably to the conclusion that the rodent olfactory system is indeed a model system for studies of learning and memory in general.

The utility of any model can be evaluated by not only assessing the extent to which it provides a means of addressing unanswered questions, but additionally by determining whether it permits new types of analyses not possible in studies of other species and sensory systems. As we have illustrated, studies of rodent olfactory learning and memory have already begun to provide valuable insights into the anatomical and physiological bases of mammalian memory; however, many important questions remain unanswered. In our view, the primary advantage of the olfactory "model" system is that it can be used to address these questions, providing information previously unattainable. For instance, apart from those involving the hippocampus, there exist scant single-unit data from limbic and cortical structures in awake, behaving animals solving tasks requiring memory processing. Examinations of the behavioral correlates of hippocampal unit firing in rats solving olfactory tasks have been useful in developing a theory regarding the role of this structure in memory processes at large; there is every reason to expect that similarly revealing data can be obtained from other memory-related limbic structures. Also, while the results of existing lesion studies strongly suggest a dissociation of memory function among the three limbic areas discussed here, this conclusion is only allusive. Parametric studies assessing the effects of discrete brain lesions across all varieties of olfactory learning and memory tasks will advance our understanding of multiple memory systems and their anatomical substrates. Finally, and perhaps most importantly, the connectional organization and cytoarchitecture of the olfactory system affords the singular opportunity to examine learning-dependent physiological plasticity in cortex. The anatomical and behavioral features of rodent olfaction provide a domain in which these data can not only be accumulated, but synthesized into a unified account of mammalian memory storage.

Acknowledgments. Preparation of this manuscript was supported in part by NIH grant FS32-NS08136 to T.O. and NSF grants BNS-8721157 and BNS-8810095 to H.E.

References

Allen, W.F. (1940). Effect of ablating the frontal lobes, hippocampi, and occipito-parieto-temporal (excepting pyriform areas) lobes on positive and negative olfactory conditioned reflexes. *Am. J. Physiol.*, *128*, 754–771.

Allen, W.F. (1941). Effect of ablating the pyriform-amygdaloid areas and hippocampi on positive and negative olfactory conditioned reflexes and on conditioned olfactory differentiation. *Am. J. Physiol.*, *132*, 81–92.

Aggleton, J.P., M.J. Burton, and R.E. Passingham (1980). Cortical and subcortical afferents to the amygdala of the rhesus monkey (Macaca mulatta). *Brain Res.*, *190*, 347–368.

Beaulieu, N., R. Morris, and M. Petrides (1988). Lesions of the basolateral amygdala and of the pyriform cortex impair acquisition of a conditioned odor preference *Soc. Neurosci. Abstr.*, *14*, 1226.

Bermudez-Rattoni, F., S.W. Keifer, C.V. Grijalva, and J. Garcia (1982). Basal and central amygdala involvement in the acquisition of taste and odor aversions. *Soc. Neurosci. Abstr.*, *4*, 501.

Bermudez-Rattoni, F., K.W. Rusiniak, and J. Garcia (1983). Flavor-illness aversions: Potentiation of odor by taste is disrupted by application of novocaine to the amygdala. *Behav. Neural Biol.*, *37*, 61–75.

Blass, E.M., and M.H. Teicher (1980). Suckling. *Science*, *210*, 15–22.

Bliss, T.V.P., and A.L. Garner-Medwin (1971). Long-lasting increases of synaptic influence in the unanesthetized hippocampus. *J. Physiol.*, *216*, 32–33.

Bressler, S.L., and Freeman, W.J. (1980). Frequency analysis of olfactory system EEG in cat, rabbit, and rat. *EEG Clin. Neurophysiol.*, *50*, 19–24.

Brown, C.W., and Ghiselli, E.E. (1938). Subcortical mechanisms in learning. IV. Olfactory discrimination. *J. Comp. Physiol. Psychol.*, *26*, 109–120.

Brown, T.S. (1963). Olfactory and visual discrimination in the monkey after selective lesions of the temporal lobe. *J. Comp. Physiol. Psychol.*, *56*, 764–768.

Brown. T.S, H.E. Rosvold, and M. Mishkin (1963). Olfactory discrimination after temporal lobe lesions in monkeys. *J. Comp. Physiol. Psychol.*, *56*, 190–195.

Cain, D.P., and D. Bindra (1972). Responses of amygdala single units to odors in the rat. *Exp. Neurol.*, *35*, 98–110.

Cinelli, A.R., H. Ferreyra-Moyano, and E. Barragan (1987). *Brain Res. Bull.*, *19*, 651–661.

Deacon, T.W., H. Eichenbaum, P. Rosenberg, and K.W. Eckmann (1983). Afferent connections of the perirhinal cortex in the rat. *J. Comp. Neurol.*, *220*, 168–190.

Devor, M. (1973). Components of mating behavior dissociated by lateral olfactory tract transection in male hamsters. *Brain Res.*, *64*, 437–441.

Doty, R.L. (1986). Odor-guided behavior in mammals. *Experientia*, *42*, 257–271.

Eichenbaum, H., K.W. Eckmann, and K.J. Shedlack (1980). Thalamocortical mechanisms in odor guided behavior. I. The effects of lesions on the mediodorsal nucleus and frontal cortex on olfactory discrimination in the rat. *Brain Behav. Evolution*, *17*, 255–275.

Eichenbaum, H., R.A. Clegg, and A. Feeley (1983). A re-examination of functional subdivisions of the rodent prefrontal cortex. *Exp. Neurol.*, *79*, 434–451.

Eichenbaum, H., T. Morton, H. Potter, and S. Corkin (1983). Selective olfactory deficits in case H.M. *Brain*, *106*, 459–472.

Eichenbaum, H., T. Parikh, and N.J. Cohen (1985). Delayed non-match to sample with trial-unique odor stimuli in intact and fornix-damaged rats: A new test for recognition memory and model of temporal-lobe amnesia. *Soc. Neurosci. Abstr.*, *11*, 333.

Eichenbaum, H., A. Fagan, and N.J. Cohen (1986). Normal olfactory discrimination learning set and facilitation of reversal learning after medial-temporal damage in rats: Implications for an account of preserved learning abilities in amnesia. *J. Neurosci.*, *6*, 1876–1884.

Eichenbaum, H., M. Kuperstein, A. Fagan, and J. Nagode (1987). Cue-sampling and goal-approach correlates of hippocampal unit activity in rats performing an odor-discrimination task. *J. Neurosci.*, *7*, 716–732.

Eichenbaum, H., and N.J. Cohen (1988). Representation in the hippocampus: what do hippocampal neurons code? *Trends Neurosci.*, *11*, 244–248.

Eichenbaum, H., A. Fagan, P. Mathews, N.J. Cohen (1988). Hippocampal system dysfunction and odor discrimination learning in rats: Impairment or facilitation depending on representational demands. *Behav. Neurosci.*, *102*, 331–339.

Eichenbaum, H., P. Mathews, and N.J. Cohen (1989). Further studies of hippocampal representation during odor discrimination learning. In press.

Eichenbaum, H., S.I. Wiener, M.L. Shapiro, and N.J. Cohen (1989). The organization of spatial coding in the hippocampus: A study of neural ensemble activity. *J. Neurosci.*, *9*, 2764–2775.

Eichenbaum, H., and J. Buckingham (1991). Studies on hippocampal processing: Experiment, theory, and model: In M. Gabriel and J. Moore (Eds.) *Neurocomputation and Learning: Foundations of Adaptive Networks*, Cambridge: M.I.T. Press. pp. 171–231.

Eskenazi, B., W.S. Cain, R.A. Novelly, and K.B. Friend (1983). Olfactory functioning in temporal lobectomy patients. *Neuropsychologia*, *24*, 553–562.

Gaffan, D. (1974). Recognition impaired and association intact in the memory of monkeys after transection of the fornix. *J. Comp. Physiol. Psychol.*, *86*, 1100–1109.

Gerfen, C.R., and R.M. Clavier (1979). Neural inputs to the prefrontal agranular insular cortex in the rat: horseradish peroxidase study. *Brain Res. Bull.*, *4*, 347–353.

Gray, C.M., W.J. Freeman, and J.E. Skinner (1986). Chemical dependencies of learning in the rabbit olfactory bulb: Acquisition of the transient spatial pattern change depends on norepinephrine. *Behav. Neurosci.*, *100*, 585–596.

Haberly, L.B. (1983). Structure of the piriform cortex of the opossum. I. Description of neuron types with Golgi methods. *J. Comp. Neurol.*, *213*, 163–187.

Haberly, L.B., and J.L. Price (1978). Association and commissural fiber systems of the olfactory cortex of the rat. I. Systems originating in the piriform cortex and adjacent areas. *J. Comp. Neurol.*, *178*, 711–740.

Haberly, L.B., and J.M. Bower (1989). Olfactory cortex: model circuit for study of associative memory? *Trends Neurosci.*, *12*, 258–264.

Hebb, D.O. (1949). *The Organization of Behavior.* New York: Wiley.

Henkin, R.I., H. Comiter, P. Fedio, and D. O'Doherty (1977). Defects in taste and smell following temporal lobectomy. *Trans. Am. Neurol. Association*, *102*, 146–150.

Imamura, K., N. Onoda, and S.F. Takagi (1984). Odor response characteristics of thalamic mediodorsal nucleus neurons in the rabbit. *Jap. J. Physiol.*, *34*, 55–73.

Insausti, R., D.G. Amaral, and W.M. Cowan (1987). The entorhinal cortex of the monkey. III. Subcortical afferents. *J. Comp. Neurol.*, *264*, 396–408.

Jennings, J.W., and L.H. Keefer (1969). Olfactory learning set in two varieties of domestic rat. *Psychol. Rep.*, *24*, 3–15.

Jones, B.P., N. Butters, H. Moskowitz, and K. Montgomery (1978). Olfactory and gustatory capacities of alcoholic Korsakoff patients. *Neuropsychologia*, *16*, 332–337.

Kanter, E.D., and L.B. Haberly (1989). APV dependent induction of long-term potentiation in piriform (olfactory) cortex slices. *Soc. Neurosci. Abstr.*, *15*, 929.

Knoth, R.L., and R.G. Mair (1991). Response latency and accuracy on a pretrained nonmatching-to-sample task in rats recovered from pyrithiamine-induced thiamine deficiency. *Behav. Neurosci.*, *105*, 375–385.

Kosel, K.C., G.W. VanHoesen, and D.L. Rosene (1982). Non-hippocampal cortical projections from the entorhinal cortex in the rat and and rhesus monkey. *Brain Res.,* 244, 210–213.

Krettek, J.E., and J.L. Price (1977a). Projections from the amygdaloid complex to the cerebral cortex and thalamus in the rat and cat. *J. Comp. Neurol.*, *172*, 687–722.

Krettek, J.E., and J.L. Price (1977b). The cortical projections of the mediodorsal nucleus and adjacent thalamic nuclei in the rat. *J. Comp. Neurol.*, *171*, 157–192.

Langworthy, R.A., and J.W. Jennings (1972). Odd ball, abstract, olfactory learning in laboratory rats. *Psychol. Rec.*, *22*, 487–490.

Larson, J., D. Wong, and G. Lynch (1986). Patterned stimulation at the theta frequency is optimal for the induction of hippocampal long-term potentiation. *Brain Res.*, *441*, 111–118.

Lasiter, P.S., D.A. Deems, and J. Garcia (1985). Involvement of the anterior insular gustatory neocortex in taste-potentiated odor aversion learning. *Physiol. Behav.*, *34*, 71–77.

Leon, M. (1983). Chemical communication in mother-young interactions. In J. Vandenbergh (Ed.), *Pheromones and Reproduction in Mammals.* New York: Academic, pp. 39–77.

Leon, M. (1987). Plasticity of olfactory output circuits related to early olfactory learning. *Trends Neurosci.*, *10*, 434–438.

Leonard, C.M. (1972). Connections of the mediodorsal nuclei. *Brain Behav. Evolution*, *6*, 524–541.

Lomo, T. (1971). Potentiation of monosynaptic EPSPs in the perforant path-dentate granule cell synapse. *Exp. Brain Res.*, *12*, 46–63.

Luskin, M.B., and J.L. Price (1982). The distribution of axon collaterals from the olfactory bulb and the nucleus of the horizontal limb of the diagonal band to the olfactory cortex demonstrated by double retrograde labeling techniques. *J. Comp. Neurol.*, *209*, 249–263.

Lynch, G. (1986). *Synapses, Circuits, and the Beginnings of Memory.* Cambridge, MA: MIT Press.

Lynch, G., and M. Baudry (1984). The biochemistry of memory: a new and specific hypothesis. *Science, 224*, 1057–1063.

Lynch, G., D. Muller, P. Seubert, and J. Larson (1988). Long-term potentiation: Persisting problems and recent results. *Brain Res. Bull.*, *21*, 363–372.

Macrides, F., A.C. Firl, S.P. Schneider, A. Bartke, and D.G. Stein (1976). Effects of one stage or serial transections of the lateral olfactory tracts on behavior and plasma testosterone levels in male hamsters. *Brain Res.*, *109*, 97–109.

Macrides, F., H.B. Eichenbaum, and W.B. Forbes (1982). Temporal relationship between sniffing and the limbic theta rhythm during odor discrimination reversal learning. *J. Neurosci.*, *2*, 1705–1717.

Mair, R.G., C. Capra, W.J. McEntee, and T. Engen (1980). Odor discrimination and memory in Korsakoff's psychosis. *J. Exp. Psychol., Human Percept. Perform*, *6*, 445–458.

Mair, R.G., R.L. Doty, K.M. Kelly, C.S. Wilson, P.J. Langlais, W.J. McEntee, and T.A. Vollmecke (1986). Multimodal sensory discrimination deficits in Korsakoff's psychosis. *Neuropsychologia*, *24*, 831–839.

Miller, J.S., A.J. Nonneman, K.S. Kelly, J.L. Neisewander, and W.L. Isaac (1986). Disruption of neophobia, conditioned odor aversion, and conditioned taste aversion in rats with hippocampal lesions. *Behav. Neural Biol.*, *45*, 240–253.

Mishkin, M., and J. Delacour (1975). An analysis of short-term visual memory in the monkey. *J. Exp. Psychol.: Animal Behav. Processes*, *1*, 326–334.

Morris, R.G.M., E. Anderson, G.S. Lynch, and M. Baudry (1986). Selective impairment of learning and blockade of long-term potentiation by an N-methyl-D-aspartate receptor antagonist, AP5. *Nature*, *319*, 774–776.

Moscovitch, M. (1984). The sufficient conditions for demonstrating preserved memory in amnesia: a task analysis. In L.R. Squire and N. Butters (Eds.), *Neuropsychology of Memory.* New York: Guilford, pp. 104–114.

Motokizawa, F. (1974). Olfactory input to the thalamus: Electrophysiological evidence. *Brain Research*, *67*, 334–337.

Mouly, A.M., M. Vigouroux, and A. Holley (1985). On the ability of rats to discriminate between microstimulations of the olfactory bulb in different locations. *Behav. Brain Res.*, *17*, 45–58.

Nauta, W.J.H. (1972). Neural associations of the frontal cortex. *Acta Neurobiol.*

Exp., *32*, 125–140.

Nigrosh, B.J., B.M. Slotnick, and J.A. Nevin (1975). Olfactory discrimination, reversal learning, and stimulus control in rats. *J. Comp. Physiol. Psychol.*, *89*, 285–294.

Onoda, N., and M. Iino (1980). Selective responses to odors of animal products in the neocortex neurons of rabbits. *Proc. Jap. Acad.*, *56*, 300–305.

Otto, T., U. Staubli, and G. Lynch (1988). Theta stimulation of the lateral olfactory tract serves as a cue in a successive-cue olfactory discrimination task. *Soc. Neurosci. Abstr.*, *14*, 568.

Otto, T., H. Eichenbaum, and C.G. Wible (1990). Behavioral correlates of hippocampal unit activity in an odor-guided delayed non-matching to sample task. *Soc. Neurosci. Abstr.*, *16*, 263.

Otto, T., H. Eichenbaum (1991). Dissociable roles of hippocampus and orbitofrontal cortex in an odor-guided delayed nonmatch to sample task. *Soc. Neurosci. Abstr.* 17.

Otto, T., H. Eichenbaum, S.I. Wiener, and C.G. Wible (1991). Learning-related patterns of CA1 spike trains parallel stimulation parameters optimal for inducing hippocampal long-term potentiation. *Hippocampus* 1:181–192.

Otto, T., F. Schottler, U. Staubli, H. Eichenbaum, and G. Lynch (1991). The hippocampus and olfactory discrimination learning: Effects of entorhinal cortex lesions on learning-set acquisition and on odor memory in a successive-cue go/no-go task. *Behav. Neurosci.* 105:111–119.

Otto, T., H. Eichenbaum (1992). Toward a comprehensive account of hippocampal function: Studies of olfactory learning permit an integration of data across multiple levels of neurobiological analysis. In L.R. Squire and N. Butters (Eds.), *Neuropsychology of Memory, 2nd Edition.* New York: Guilford.

Palmerino, C.C., K.W. Rusiniak, and J. Garcia (1980). Flavor-illness aversions: The peculiar roles of odor and taste in memory for poison. *Science*, *208*, 753–755.

Potter, H., and N. Butters (1980). An assessment of olfactory deficits in patients with damage to the prefrontal cortex. *Neuropsychologia*, *18*, 621–628.

Price, J.L. (1985). Beyond the primary olfactory cortex: olfactory-related areas in the neocortex, thalamus and hypothalamus. *Chem. Senses*, *10*, 239–258.

Price, J.L., and B.M. Slotnick (1983). Dual olfactory representation in the rat thalamus: An anatomical and electrophysiological study. *J. Comp. Neurol.*, *215*, 63–77.

Racine, R.J., N.W. Milgram, and S. Hafner (1983). Long-term potentiation phenomena in the rat limbic forebrain. *Brain Res.*, *260*, 217–231.

Ranck, J.B. (1973). Studies on single neurons in dorsal hippocampal formation and septum in unrestrained rats. Part I. Behavioral correlates and firing repertoires. *Exp. Neurol.*, *41*, 462–531.

Rausch, R., E.A. Serafetinides, and P.H. Crandall (1977). Olfactory memory in patients with anterior temporal lobectomy. *Cortex*, *13*, 445–452.

Rolls, E.T. (1989). Functions of neuronal networks in the hippocampus and neocortex in memory. In J.H. Byrne and W.O. Berry (Eds.), *Neural Models of Plasticity: Experimental and Theoretical Approaches*. New York: Academic, pp. 240–265.

Roman, F., U. Staubli, and G. Lynch (1987). Evidence for synaptic potentiation in a cortical network during learning. *Brain Res.*, *418*, 221–226.

Rose, G.M., and T.V. Dunwiddie (1986). Induction of hippocampal long-term potentiation using physiologically-patterned stimulation. *Neurosci. Lett.*, *69*, 244–248.

Rosenkilde, C.E. (1979). Functional heterogeneity of the prefrontal cortex in the monkey: a review. *Behav. Neural Biol.*, *25*, 301–345.

Santibanez, G., and T. Pinto Hamuy (1957). Olfactory discrimination deficits in monkeys with temporal lobe ablations. *J. Comp. Physiol. Psychol.*, *50*, 472–473.

Sapolsky, R.M., and H. Eichenbaum (1980). Thalamocortical mechanisms in odor-guided behavior. II. Effects of lesions of the mediodorsal thalamic nucleus and frontal cortex on odor preferences and sexual behavior in the hamster. *Brain, Behav. Evolution*, *17*, 276–290.

Scoville, W.B., and B. Milner (1957). Loss of recent memory after bilateral hippocampal lesions. *J. Neurol., Neurosurg., Psychiatry*, *20*, 11–21.

Sechenov, I.M. (1863/1965). *Reflexes of the Brain.* Cambridge, MA: MIT Press.

Slotnick, B.M. (1985). Olfactory discrimination in rats with anterior amygdala lesions. *Behav. Neurosci.*, *99*, 956–963.

Slotnick, B.M., and H.M. Katz (1974). Olfactory learning-set formation in rats. *Science*, *185*, 796–798.

Slotnick, B.M., and N. Kaneko (1981). Role of mediodorsal thalamic nucleus in olfactory discrimination learning in rats. *Science*, *214*, 91–92.

Smotherman, W.P. (1982). Odor aversion learning by the rat fetus. *Physiol. Behav.*, *29*, 769–777.

Sorensen, K.E., and K.T. Shipley (1979). Projections from the subiculum to the deep layers of the ipsilateral presubicular and entorhinal cortices in the guinea pig. *J. Comp. Neurol.*, *188*, 313–334.

Squire, L.R. (1987). *Memory and Brain.* New York: Oxford University Press.

Squire, L.R., N.J. Cohen, and L. Nadel (1984). The medial temporal region and memory consolidation: A new hypothesis. In H. Weingartner and E.S. Parker (Eds.), *Memory Consolidation.* Hillsdale, NJ: Lawrence Erlbaum, pp. 185–210.

Squire, L.R., A.P. Shimamura, and D.G. Amaral (1989). Memory and the hippocampus. In J.H. Byrne and W.O. Berry (Eds.), *Neural Models of Plasticity.* San Diego, CA: Academic, pp. 208–239.

Staubli, U., G. Ivy, and G. Lynch (1984). Hippocampal denervation causes rapid forgetting of olfactory information in rats. *Proc. Nat. Acad. of Sci. USA*, *81*, 5885–5887.

Staubli, U., D. Fraser, R. Faraday, and G. Lynch (1987). Olfaction and the "data" memory system in rats. *Behav. Neurosci.*, *101*, 757–765.

Staubli, U., F. Schlotter, and D. Nejat-Bina (1987). Role of the dorsomedial thalamic nucleus and piriform cortex in processing olfactory information. *Behav. Brain Res.*, *25*, 117–129.

Supak, T.D., F. Macrides, and S.L. Chorover (1971). The bait-shyness effect extended to olfactory discrimination. *Communications Behav. Biol.*, *5*, 321–324.

Swann, H.G. (1934). The function of the brain in olfaction. II. The results of destruction of olfactory and other nervous structures upon the discrimination of odors. *J. Comp. Neurol.*, *59*, 175–201.

Swanson, L.W. (1981). A direct projection from Ammon's horn to prefrontal cortex in the rat. *Brain Res.*, *217*, 150–154.

Switzer, R.C., J. de Olmos, and L. Heimer (1985). Olfactory system. In G. Paxinos (Ed.), *The Rat Nervous System,* Vol. 1, *Forebrain and Midbrain.* Orlando, FL: Academic.

Tanabe, T., M. Iino, and S.F. Takagi (1975). Discrimination of odors in olfactory bulb, piriform-amygdaloid areas, and orbitofrontal cortex of the monkey. *J. Neurophysiol.*, *38*, 1284–1296.

Tanabe, T., H. Yarita, M. Iino, Y. Ooshima, and S.F. Takagi (1975). An olfactory projection area in the orbitofrontal cortex of the monkey. *J. Neurophysiol.*, *38*, 1269–1283.

Teyler, T.J. (1986). Memory: Electrophysiological analogs. In J.L. Martinez and R.P. Kesner (Eds.), *Learning and Memory: A Biological View*. San Diego, CA: Academic, pp. 237–266.

Teyler, T.J. (1989). Comparative aspects of hippocampal and neocortical long-term potentiation. *J. Neurosci. Methods*, *28*, 101–108.

Teyler, T.J., and P. DiScenna (1986). The hippocampal memory indexing theory. *Behav. Neurosci.*, *100*, 147–154.

Warren, J.M. (1973). Learning in vertebrates. In D.A. Dewsbury and D.A. Rethlingshafer (Eds.), *Comparative Psychology: A Modern Survey*. New York: McGraw-Hill.

Wible, C.G., H., Eichenbaum, and T. Otto (1990). A task designed to demonstrate a declarative memory representation of odor cues in rats. *Soc. Neurosci. Abstr.*, *16*, 605.

Wiener, S.I., C.A. Paul, and H. Eichenbaum (1989). Spatial and behavioral correlates of hippocampal neuronal activity. *J. Neurosci.*, *9*, 2737–2763.

Witter, M.P., G.W. Van Hoesen, and D.G. Amaral (1989). Topographical organization of the entorhinal projection to the dentate gyrus of the monkey. *J. Neurosci.*, *9*, 216–228.

8
Sex Differences in Olfaction in Mammals

KATHLEEN M. DORRIES

Introduction

Olfaction plays a major role in mammalian reproduction. Chemical cues can affect reproductive physiology or directly evoke behaviors essential to reproduction. Olfactory stimuli emitted by many species of mammals modulate conspecifics' sexual, aggressive, and maternal behavior. Such responses to intraspecific chemical signals are often dimorphic; that is, they are seen only, or to a greater degree, in one sex. Hormonal determinants of reproductive behaviors mediated by olfactory stimuli have been explored extensively. Investigation of the effects of sex hormones on the sensory systems involved in the reception and processing of those cues, however, lags far behind. The aim of this chapter is to demonstrate that research in this area may enhance our understanding of both sexually dimorphic behavioral responses and olfactory processing itself. The study of sex differences in processing of taste stimuli to which animals respond dimorphically has been advocated for these same reasons (Di Lorenzo and Monroe, 1989), and recent results appear promising (Di Lorenzo et al., 1989; Di Lorenzo and Monroe, 1990).

The chapter will focus on mammals, although I will cite examples from other classes where useful. It will also be concerned in particular with the main and accessory olfactory systems, because virtually all research on sex differences in mammalian olfaction has focused on these systems. After a brief summary of the role of olfaction in reproductive behavior in mammals [the literature is extensive and more comprehensive reviews may be found elsewhere—see, for example, Vandenbergh (1983)], I will review recent research on sexual dimorphism in olfactory structures in rodents, and then results of studies on sex differences in olfactory sensitivity. Investigation of sex differences in olfactory sensitivity has followed two approaches: (1) the study of sex differences in general olfactory sensitivity, and (2) the study of sex differences in sensitivity to specific "biologically significant" stimuli. I will discuss implications and problems in the application of each of these approaches, and describe two potential models for the study of sex differences in olfaction in mammals.

Olfaction and Reproduction

Odors that mediate mammalian behavior are often referred to as pheromones, although the use of this term has been criticized (Bronson, 1976). Because of its original usage in insect literature to describe the capacity of specific chemicals emitted by an insect to elicit stereotyped responses in conspecifics (Karlson and Butenandt, 1959), the term "pheromone" discounts the importance of experience and multiplicity of sensory stimuli in the behavior of mammals. Nonetheless, the concept of pheromones is used widely in referring to olfactory stimuli involved in chemical communication in mammals (see Vandenbergh, 1983). The pheromonal functions of olfactory stimuli may be divided generally into two types: priming and signaling (see Vandenbergh, 1983). Priming pheromones are olfactory stimuli that modulate the physiology of the receiver. Pheromones have been shown to regulate the onset of puberty (for example, Nishimura, Utsumi, Yuhara, Fujitani, and Iritani, 1989; Pearce and Hughes, 1987); affect expression, duration, and synchrony of ovarian cycles (see McClintock, 1983); and terminate or block pregnancy (Elefthériou, Bronson, and Zarrow, 1962; Parkes and Bruce, 1962).

Signaling pheromones mediate the expression of a wide variety of behaviors. Olfactory cues, often in combination with stimuli from other modalities, initiate attraction and investigatory behaviors in reproduction (Darby, Devor, and Chorover, 1975; Johnston, 1974), as well as eliciting receptive postures in females (Signoret, 1970) and mounting or intromission in males (Singer, Macrides, Clancy, and Agosta, 1986). Odors are important in maternal responses (Stern, 1990); for example, they facilitate recognition of young in sheep (Baldwin and Shillito, 1974), goats (Klopfer and Gamble, 1966), and swine (Signoret, Baldwin, Fraser, and Hafez, 1975), and determine whether pups will elicit aggressive or maternal responses in rats (Fleming, Vaccarino, Tambosso, and Chee, 1979) and hamsters (Marques, 1979). Olfactory involvement in sex-related territorial and aggressive behavior has also been reported. For example, male mice emit odors that promote aggression in other males, and females emit odors that deter aggression (see Lee, 1976, for review).

Sexual Dimorphism in Olfactory Structures

Because of the importance of olfaction in many facets of mammalian reproduction, it seems likely that olfactory systems may be tuned to specific stimuli, a phenomenon that occurs in audition. The auditory system of some species of both frogs and crickets is most responsive to sound frequencies included in their own species-specific mating calls (Capranica, Frishkopf, and Nevo, 1973; Hoy, Hahn, and Paul, 1977; Narins and Capranica, 1976). Many olfactory stimuli are of much greater significance to one sex than to the other, or affect animals of different sexes or reproductive conditions differentially, initiating specific behaviors or changes in physiology in one but not another. One might expect to find sexual dimorphism in the olfactory system that processes these stimuli and mediates such changes. Dimorphic structures have been found in the olfactory systems of insects that show sex-specific responses to pheromones in courtship (Masson and Mustaparta, 1990). A striking example of this was demonstrated by Schneiderman, Matsumoto, and Hildebrand (1982) in the moth *Manduca sexta*. Females with precursors of male antennae (imaginal discs) grafted in place of their own developed male antennae, as well as a male-specific brain region, the macroglomerular complex, in the ipsilateral antennal lobe of the brain. A relationship between this change in structure and a specific behavioral response to a pheromone was established by an experiment showing that females with at least one male antenna sometimes displayed orienting behavior toward the pheromone source, whereas females with no male antennae oriented only toward odors from egg-laying sites (tobacco leaves; Schneiderman, Hildebrand, Brennan, and Tumlinson, 1986).

In mammals, olfaction may include up to five different chemosensory systems, depending on the species (Meredith, 1983). Thus far, however, only the main and accessory (vomeronasal) olfactory systems have been examined for sexual dimorphism in peripheral and central structures; the trigeminal system, the nervous terminalis, and the septal organ have not yet been examined. While both the main olfactory and vomeronasal systems mediate responses to pheromones in mammals, it has been hypothesized, based on differences in both effective stimuli and central connections of the two systems, that the functions of the main and accessory olfactory systems differ (Scalia and Winans, 1975). For example, attraction to and investigation of odors may be mediated by the main olfactory system (Johnston, 1985), and reception and processing of primer pheromones may be the domain of the vomeronasal system (Meredith, 1983).

Receptor cells from the main olfactory epithelium synapse in the main olfactory bulb (MOB), which has primary projections to numerous areas. These include the anterior olfactory nucleus, the nucleus of the lateral olfactory tract, the olfactory tubercle, the pyriform cortex, entorhinal cortex, and the anterior cortical and posterior cortical nuclei of the amygdala (Johnston, 1985; Scalia et al., 1975). Receptor cells from the sensory epithelium of the vomeronasal organ synapse in the accessory olfactory bulb (AOB). The primary projections of the AOB are much more restricted than those of the MOB: these include the nucleus of the accessory olfactory tract, the bed nucleus of the stria terminalis, and the medial amygdaloid nucleus and the posteromedial sector of the cortical amygdaloid nucleus (Scalia et al., 1975; Scalia and Winans, 1976).

Dimorphism in mammalian olfactory organs and their central nervous system pathways has been studied mainly in rats (Table 8.1). In the main olfactory system, Balboni and Vanelli (1982) found differences in the olfactory epithelium between prepubertal and postpubertal rats, and, in postpubertal rats, found that several measures of the olfactory epithelium in

TABLE 8.1. Sexual dimorphism in the main and accessory olfactory systems

Measure	Direction of differences	Reference
	MAIN OLFACTORY SYSTEM	
Main olfactory epithelium		
Basal cell number[1]	M, dF > pM, pF > eF	Balboni and Vanell, 1982
Stratefied nuclei number	pM, pF > M, dF > eF	Balboni and Vanell, 1982
Epithelium height	pM, pF > M, dF > eF	Balboni and Vanell, 1982
	ACCESSORY OLFACTORY SYSTEM	
Vomeronasal organ		
VNO volumes	M, tpF > F, cM	Segovia and Guillamon, 1982
Vomeronasal epithelium		
Volume	M, tpF > F, cM	Segovia and Guillamon, 1982
Height[2]	F, M > cM, oF	Segovia, Paniagua, Nistal, and Guillamon, 1984
Receptor number	M, tpF > F, cM	Segovia and Guillamon, 1982
Cell nuclear size	F, cM > M, tpF	Segovia and Guillamon, 1982
Cell nuclear size[2]	F > M > cM, oF	Segovia, Paniagua, Nistal, and Guillamon, 1984
Accessory olfactory bulb		
AOB volume, total	M, tpF > F, cM	Segovia, Orensanz, Valencia, and Guillamon, 1984
Mitral layer, volume	M, tpF > F, cM	Segovia, Orensanz, Valencia, and Guillamon, 1984
Cell number	M, tpF > F, cM	Valencia, Segovia, and Guillamon, 1986
Granular layer, volume	M, tpF > F, cM	Segovia, Orensanz, Valencia, and Guillamon, 1984
Light cell number	M, tpF > F, cM	Valencia, Segovia, and Guillamon, 1986
Dark cell number	M, tpF > F, cM	Valencia, Segovia, and Guillamon, 1986
Glomerular layer, volume	M > cM	Segovia, Orensanz, Valencia, and Guillamon, 1984
Plexiform layer, volume	M, tpF > F, cM	Segovia, Orensanz, Valencia, and Guillamon, 1984
Bed Nucleus of the Stria Terminalis		
Medial anterior volume	F, cM > M	Del Abril, Segovia, and Guillamon, 1987
Medial posterior volume	M, tpF > F, cM	Del Abril, Segovia, and Guillamon, 1987
Cell number	M, tpF > F, cM	Guillamon, Segovia, and del Abril, 1988
Encapsulated portion volume[2]	M, cM > F, oF	Hines, Davis, Coquelin, Goy, and Gorski, 1985

KEY: c, castrated; o, ovariectomized; tp, treated with testosterone propionate; p, prepubertal; e, estrous; d, diestrous.

[1] All treatments at neonatal day 1 unless otherwise noted.

[2] Treatment as an adult.

estrous females differed from those in males and diestrous females. Prepubertal rats of both sexes had a greater epithelium height and number of stratified nuclei than adult males and females, whereas they were intermediate between diestrous and estrous females for basal cell number. On all three measures, adult males and diestrous females did not differ from each other, but exceeded estrous females. These data suggest that the structure of main olfactory epithelium, perhaps including epithelial surface area and access of odorants to receptors, changes developmentally and also varies as a function of estrous cycle stage in rats.

Structures of the accessory olfactory system have been more extensively studied. In six-month-old (sexually mature) rats, total vomeronasal organ volume in males exceeds that in females by approximately 40 percent. Epithelium volume and receptor number are 33 percent and 20 percent greater, respectively, in males (Segovia and Guillamón, 1982). Castration of males and treatment of females with testosterone propionate (TP), both at neonatal day one, reverse the direction of the dimorphism: castrated males do not differ from control females, and TP-treated females do not differ from control males. Segovia et al. (1982) found that receptor cell nuclei were significantly larger (by 3 to 4 percent) in intact females and castrated males than in TP-treated females and intact males. Gonadectomy in adulthood also affected receptor cell nucleus size: castration reduced nucleus size in males by 5 percent, while ovariectomy decreased size by almost 20 percent (Segovia, Paniagua, Nistal, and Guillamón, 1984). Vomeronasal epithelium height was reduced by 20 percent in gonadectomized rats, with no difference between sexes (Segovia, Paniagua, Nistal, and Guillamón, 1984). As in the main olfactory system, variation in cell size and number in the vomeronasal epithelium might affect odorant access to receptor cell surfaces. Such peripheral changes might cause differences in olfactory sensitivity in general, or to specific compounds if regions of the epithelium or classes of cells were affected differentially.

Many accessory olfactory bulb measures are also affected by testicular hormones, all favoring intact males over castrated males. Castration at birth resulted in a 43 percent decrement in total AOB volume in three-month-old (postpubertal) rats (Segovia, Orensanz, Valencia, and Guillamón, 1984). Volumes of the plexiform, mitral, granular, and glomerular cell layers of the AOB were reduced by 38 to 50 percent (Segovia, Orensanz, Valencia, and Guillamón, 1984). Castration at birth was also associated with a reduction in the number of cells in the mitral layer (Valencia, Segovia, and Guillamón, 1986) and in two different populations of granular layer cells (light and dark; Segovia, Valencia, Calés, and Guillamón, 1986). Cell numbers in the plexiform and glomerular layers have not been measured. No statistically significant sex differences were found for glomerular layer volume, but for all other measures, intact females resembled castrated males, and TP-treated females resembled normal males.

Finally, the volumes of several regions of the bed nucleus of the stria terminalis (BNST) that form part of the accessory olfactory pathway may also be dimorphic. Rats were again treated at birth and sacrificed at three months of age (del Abril, Segovia, and Guillamón, 1987; Guillamón, de Blas, and Segovia, 1988). Although no sex differences were found in the total volume of the BNST or of the medial, lateral, or intermediate portions (del Abril et al., 1987), sex differences were found in regions within the medial portion. In the medial posterior region, both volume (del Abril et al., 1987) and cell number (Guillamón et al., 1988) showed differences in the same direction as the majority of measurements in the accessory olfactory bulb, with males exceeding females by 15 percent and 24 percent, respectively. In contrast, the medial anterior region was larger in intact females and castrated males than in intact males (del Abril et al., 1987). Intact males' medial anterior region volume was approximately 75 percent of the females' and 66 percent of the castrated males'. Hines, Davis, Coquelin, Goy, and Gorski (1985) reported that the encapsulated portion of the BNST in guinea pigs was greater in volume in males than in females (by approximately 25 percent), and that gonadectomy in adulthood had no effect on this sex difference.

Thus, several regions of the accessory olfactory pathway are reported to be under sex hormone control. Most of the differences described favor males or TP-treated females, suggesting that endogenous androgen levels during development or adulthood affect olfactory structures, and, more specifically, their volume and cell number. The larger olfactory structures in males may permit a greater sensitivity to or more complex processing of olfactory stimuli. Male rats can distinguish receptive from nonreceptive females by smell (Le Magnen, 1952), and castration is known to disrupt their preference for estrous over nonestrous odors (Brown, 1977; Le Magnen, 1952).

The studies reported here represent preliminary work, however, and more thorough investigation is necessary to draw definitive conclusions. Except for the Hines et al. (1985) report, all of these studies have been done in the rat, and virtually all of the work has been done in one laboratory. Furthermore, results reported by Segovia and Guillamón and their colleagues were apparently not corrected for body or brain size, nor were comparisons made with control areas outside of the olfactory system. It would therefore be of great interest to confirm these results. Then we could ask: How might sex hormone effects on olfactory structures be related to sexual dimorphism in behavioral responses to olfactory stimuli?

Sex Differences in Olfactory Sensitivity

The study of sex differences in olfactory sensitivity has been approached in two ways, each based on a separate set of assumptions. In both approaches, comparisons are made between individuals with different sex hormone levels as adults, during development, or both. The hypotheses tested differ, however, and this affects both the design of the study in terms of stimuli selected, and interpretations of results.

General Olfactory Sensitivity

The first approach tests the hypothesis that olfaction as a unitary entity is affected by hormonal condition in adulthood or during development. Subjects are tested with one or more stimuli, and the subject's sensitivity to those stimuli is assumed to reflect that individual's olfactory sensitivity as a whole. Since only a small set of stimuli are used, ideally the set should be selected to be representative, in some way, of the total range of possible olfactory stimuli. This logic is very often followed in taste testing (for example, see Brosvic and McLaughlin, 1989; Di Lorenzo et al., 1989, 1990). Investigators use four stimuli representing four primary tastes (sweet, sour, salty, and bitter), implicitly or explicitly assuming that there are four corresponding taste subsystems, again unitary in nature. Although the view that there are in some sense four such primaries seems widely accepted, the unitary nature of those subsystems has been questioned (Jackinovich and Sugarman, 1988; Schiffman, 1988).

The concept of primaries in olfaction was advanced in the 1970s, based on evidence from studies of specific anosmias (Amoore, 1977). One study of effects of sex hormones on olfaction did use seven odorants representing primaries suggested by Amoore (Herberhold, Genkin, Brändle, Leitner, and Wöllmer, 1982). Evidence now suggests, however, that there is not a small set of olfactory primaries arising from seven or so specific receptor types, but in fact hundreds of olfactory receptor types may exist (Buck and Axel, 1991). The stimuli used in olfactory testing are therefore often chosen for practical reasons: they are readily available, they do not stick to surfaces of the odor delivery apparatus as much as other chemicals might, and/or they have been used by others (for example, Doty and Ferguson-Segall, 1989). In some cases, no explicit justification for using specific odorants is stated at all (Phillips and Vallowe, 1975). The use of a few stimuli appears to be based on the assumption that measuring sensitivity to any one odorant or set of odorants adequately measures the sensitivity of the overall system. While it is the case that testing with a truly comprehensive set of stimuli is impractical or impossible, the use of one or a few stimuli under the assumption that they represent all odorants may present methodological or theoretical problems, which I will discuss later in this section.

In spite of such problems, an approach based on the assumption that olfactory sensitivity varies in a unitary manner has been used quite frequently for investigation of sex differences in olfactory sensitivity in both humans and rodents (Table 8.2a). Human subjects have generally been tested with psychophysical threshold measures, using forced choice or signal detection paradigms. Either psychophysical methods using oper-

TABLE 8.2. Sex hormone effects on olfactory sensitivity[1]

	a. Neutral Odors		b. Biologically significant odors	
	Rodent	Human	Rodent	Human
Sex difference	M > F rat[2] M > F mouse[3]	F > M F = M		F > M musks 16–androstenes testosterone NE androstenone
Estrous/ menstrual cycle effect	+ peak sensitivity at estrus—rat + peak sensitivity at proestrus—mouse	+ peak midcycle + peak midcycle and menses NE		+ peak midcycle—pentadecalactone NE pentadecalactone
Pregnancy/ parturition effect	+ late pregnancy/parturition—mouse − pseudopreg., cyclicity elimin.—rat			+ late pregnancy—pentadecalactone
Gonadectomy effect	+ ovariectomy, cyclicity elim.—rat NE castration—rat		NE castration: estrous vs nonestrous odors—rat	− ovariectomy—pentadecalactone
Sex hormone treatment effect	+ TP: ovariectomized female + E: ovariectomized female − P: ovariectomized female rat			+ E: normal man + E: hypogonadal women pentadecalactone
Developmental change		Sex difference in adults but not children		Sex difference in adults but not children 16-androstenes musks

KEY: +, sensitivity increases; −, sensitivity decreases; NE, no effect or difference between groups; F, female; M, male; E, estradiol; TP, testosterone propionate; P, progesterone.

[1] See text for references.

[2] Peak sensitivity occurs during estrus.

[3] Peak sensitivity occurs during proestrus.

ant tasks or neural recording in the olfactory bulb have been used to determine olfactory thresholds in rodents.

Experiments with rodents indicate that endogenous sex hormone levels do affect general olfactory sensitivity. Olfactory sensitivity was reported to vary with estrous cycle stage in both rats (Phillips et al., 1975; Pietras and Moulton, 1974) and mice (Rittner and Schmidt, 1982). Rats and mice differed, however, in the cycle phase at which peak sensitivity occurred: in rats the peak occurred during proestrus, and in mice during estrus. Threshold measurements in mice based on recordings of neural response in the olfactory bulb (Schmidt, 1978; Schmidt and Schmidt, 1980; Schmidt, Schmidt, and Breipohl, 1982) also indicated cyclic variation, with olfactory sensitivity peaking during estrus. In both rats (Pietras et al., 1974) and mice (Schmidt et al., 1980), males had lower thresholds than females, except females at peak sensitivity. Olfactory sensitivity increased late in gestation in the mouse (Schmidt et al., 1982), while it decreased in the rat during pseudopregnancy induced with estradiol (Pietras et al., 1974). In ovariectomized rats, cyclic variation in olfactory sensitivity was eliminated, and sensitivity was greater than in diestrous sham-operated females (Phillips et al., 1975; Pietras et al., 1974). Olfactory sensitivity in males was not affected by gonadectomy. Both estrogen and androgen injections increased sensitivity of ovariectomized female rats, while progesterone treatment decreased sensitivity (Phillips et al., 1975). Thus, olfactory sensitivity in rats and mice varied with both natural and experimentally induced fluctuations in sex hormone levels.

The results of most studies on variation in olfactory sensitivity with menstrual cycle stage in humans are in accord with rodent estrous cycle results. All but one study (Herberhold et al., 1982) reported cyclic variation, with peaks in sensitivity at midcycle (Henkin, 1974; Schneider and Wolf, 1955) or midcycle and during the menses (Doty, Hall, Flickinger, and Sondheimer, 1982; Doty, Snyder, Huggins, and Lowry, 1981). Mair, Bouffard, Engen, and Morton (1978) reported cyclic variation in thresholds to relatively involatile odorants but not to volatile chemicals. They suggest that this is due to hormonal effects on the composition of nasal mucus: for example, increased mucous viscosity may reduce access of less volatile chemicals to receptors, but have little effect on the passage of volatile compounds that have shorter retention times.

In contrast with rodent results, studies comparing olfactory sensitivity of men and women have found that women have lower thresholds than men (Deems and Doty, 1987; Koelega and Köster, 1974; Schneider et al., 1955), or that women and men are equally sensitive (Le Magnen, 1952b; Venstrom and Amoore, 1968). In these studies, women's menstrual cycle stage was not determined. Testosterone treatment of hypogonadal women was associated with a decrease in olfactory sensitivity, whereas those treated with estrogen increased in sensitivity (Schneider, Costiloe, Howard, and Wolf, 1958).

In general, estrogens appear to increase olfactory sensitivity per se, while progesterone decreases sensitivity (Phillips et al., 1975; Pietras et al., 1974; Schmidt et al., 1982). These conclusions are supported by rodent cycle data, the elimination of peak sensitivity in ovariectomized females, and pseudopregnancy results [though pseudopregnancy is induced by estrogen injection, it is characterized by high progesterone levels (Pietras et al., 1974)], as well as by the hormone treatment results. The effect of androgens is less straightforward. Testosterone treatment increased sensitivity in ovariectomized female rats, but decreased sensitivity in hypogonadal women. Castration of male rats appears not to have any effect on olfactory sensitivity. It is also unclear why male rodents in general have lower olfactory thresholds than females, while in humans females are equally sensitive or more sensitive than males. This species difference is not surprising, however, in light of species differences in olfactory-mediated behaviors.

In most of the human cycle studies, determination of menstrual cycle stage was imprecise, based on days from menses and occasionally basal body temperature records. Strong conclusions about relationships between endogenous sex hormone levels and olfactory sensitivity therefore cannot be drawn from these data. Exceptions include experiments by Doty and coworkers (1981, 1982) in which cycle stage was determined by hormone titre in regularly taken blood samples. These experiments, however, found relationships between olfactory sensitivity and cardiovascular parameters, while sensitivity fluctuation was not dependent on sex hormone levels. Thresholds varied cyclically in women taking oral contraceptives whose estrogen, progesterone, and luteinizing hormone levels were stable. These results suggest that the menstrual variation in olfactory sensitivity may be related to some aspect of central arousal, rather than directly to sex hormone levels.

As I suggested earlier, interpretation of results of studies of sex hormone effects on olfaction may be confounded when a small set of stimuli is used to measure overall sensitivity on the assumption that the set represents all odors. Two problems will be discussed here. First, it follows from this assumption that there is no such thing as a control odor. Results reviewed here are interpreted to mean that sex hormone levels in adulthood or during development affect olfactory sensitivity per se. But do the differences reported truly reflect changes in olfactory sensitivity, or are they caused by variation in some other aspect of performance on the tasks, such as motivation? This is a possibility, given that sex differences in performance on operant tasks involving no olfactory stimuli of any kind have been reported (Calés, Segovia, Enríquez, and Guillamón, 1988). Thus control groups must be added to the design in order to rule out this possibility. Controls would have to learn the same task but with nonolfactory stimuli because there can be no control odor if sensitivity of the olfactory system is assumed to vary as a whole. In fact, none of the studies summarized above included such controls. Measurements of neural thresholds in the olfactory bulb do suggest, however, that olfactory sensitivity itself is affected. A second solution is the use of a signal detection paradigm, which is designed to eliminate influences of criterion shifting on threshold measures. This paradigm has been used by Doty and his colleagues in both human (Doty et al., 1981, 1982) and rodent (Doty et al., 1989) experiments and by Mair et al. (1978) in their menstrual cycle study. Using signal detection analysis, Doty et al. (1989) found effects of castration in rats on task performance, but not on ethyl acetate threshold. However, Mair et al. (1978) and Doty et al. (1981, 1982) did find menstrual variation in olfactory sensitivity using this method.

Second, the validity of the assumption that overall olfactory sensitivity can be determined by measuring sensitivity to a set of odors may be questionable. The relative sensitivity of subjects to one or some odorants does not necessarily predict the relative sensitivity to other odorants. In the most extreme case, the subject is completely unable to detect a particular chemical in spite of having a normal sense of smell (specific anosmia; Amoore, 1977). Moreover, odorant-specific changes in threshold within subjects have been reported: long-term periodic exposure to 5α-androst-16-en-3-one (androstenone) and amyl acetate increased sensitivity to the former but not to the latter (Wysocki, Dorries, and Beauchamp, 1989). Thus, while studies with both humans and rodents demonstrate effects of sex hormones on olfactory sensitivity, one must be careful in assuming that sensitivity to individual odors represents sensitivity of the olfactory system as a whole. It is possible that sex differences exist in sensitivity to specific olfactory stimuli.

Sensitivity to Biologically Significant Odors

The second approach to studying sex differences in olfaction tests the hypothesis that olfactory sensitivity to sex-specific "biologically significant" odors is determined by sex or sex hormone condition. A biologically significant stimulus would be one that an animal normally encounters in its environment and that conveys information valuable to the animal's survival or reproduction. Koelega et al. (1974) refer to these as having "biological (sexual) significance," and to other odors as "neutral." Le Magnen (1952) advanced the hypothesis that olfactory sensitivity to biological odors is determined by sex hormone levels, while sensitivity to neutral odors is not. Since then, there have been a number of studies of sex hormone effects on sensitivity to biological odors (Table 8.2b). Compounds of greatest interest would be odors responded to primarily by individuals of one sex, including odors with priming or signaling pheromonal activity. Such chemicals are likely to be species-specific. For example, a male hamster might be more sensitive than a female hamster to vaginal odors from conspecific females, and female sheep might be more sensitive than males to lamb odors or odors from the urine of conspecific males.

Most of the investigation of sex hormone effects on sensitivity to specific odors has focused on human sensitivity to 16-androstene steroids or pentadecalactone musk. Le Magnen (1952) reported that sensitivity to exaltolide, a musk, varied with menstrual cycle stage, peaking at midcycle, while sensitivity to neutral odors did not. Several investigators confirmed his finding of menstrual variation of exaltolide sensitivity (Good, Geary, and Engen, 1976; Vierling and Rock, 1967), although one study found no cyclic variation (Amoore, Popplewell, and Whissell-Buechy, 1975). Le Magnen (1952) also reported that ovariectomy in human females resulted in decreased sensitivity to exaltolide, and estrogen treatment of ovariectomized females increased sensitivity. The latter finding was confirmed by Good et al. (1976). Sensitivity to the musk also increased late in pregnancy in humans (Good et al., 1976).

Specific anosmia to androstenone, a 16-androstene steroid, occurs in a large percentage of human subjects tested (ranging from approximately 8 to 50 percent), with a sex difference in its occurrence. Women are more likely than men to smell the odor (Dorries, Schmidt, Beauchamp, and Wysocki, 1989; Griffiths and Patterson, 1970; Koelega et al., 1974; except see Gower, Bird, Sharma, and House, 1985), and of those who can detect the odor, women have lower thresholds than men (Dorries et al., 1989). This sex difference appears during adolescence (Dorries et al., 1989; Koelega et al., 1974); almost all prepubertal children appear to be able to detect the odor (Schmidt and Beauchamp, 1988).

Only two studies have examined effects of sex hormones on sensitivity to biologically significant odors in rodents. Both investigated the effect of castration of male rats on response to estrous or nonestrous odors of female rats. Carr and Caul (1962) determined that both intact and castrated males were able to discriminate between air drawn over estrous and diestrous female rats. Carr, Solberg, and Pfaffmann (1962) measured absolute olfactory detection threshold for estrous female urine in intact and castrated male rats, and found no significant difference between the two groups.

While in principle the study of sex differences in sensitivity to specific biologically significant stimuli may be more useful in increasing our understanding of the relationship between olfaction and behavior, thus far the application of this approach has been flawed in several ways. First, many investigators have again used no procedures to control for performance confounds (Carr and Caul, 1962; Carr, Solberg, and Pfaffmann, 1962; Good et al., 1976; Griffiths et al., 1970; Vierling et al., 1967), even though here use of control odorants is feasible. Consequently, it is still not clear in those studies whether olfactory sensitivity per se is affected by sex hormones, or, if it is, whether the effects of sex hormones are specific to the biologically significant odor. Le Magnen (1952), Koelega et al. (1974), and Dorries et al. (1989) did use both "biological" and "neutral" odors. The variations in olfactory sensitivity reported by Le Magnen and Dorries et al. were specific to exaltolide and androstenone, respectively. However, Koelega et al. (1974) measured olfactory sensitivity to two musks and seven 16-androstenes, as well as five "neutral" odors, and found that women were more sensitive to both biological and neutral odors, except for one neutral odor (amyl acetate) for which there was no sex difference.

Second, in using this approach, investigators have interpreted "biologically significant" very broadly. Stimuli selected are often synthetic musks. Though musklike compounds may be used for chemical communication by some species (Johnston, 1983), we cannot assume that musks are biologically meaningful to all mammals. In fact, Pietras et al. (1974) used pentadecalactone as a stimulus in their studies and stated that none of their stimuli were "known to have particular biological significance for the rat." Furthermore, if a species did communicate using a musk, it is likely that only a specific musk (or set of musks) would be significant to that species, rather than all musks.

Are 16-androstenes biologically significant stimuli for humans? 16-Androstenes have been isolated in human male sweat (Brooksbank, Brown, and Gustafsson, 1974) and urine (Brooksbank and Haslewood, 1961). Female humans also produce these steroids, but at much lower levels (Brooksbank et al., 1961). Possible pheromonal effects of androstenol and androstenone have been investigated in humans (Benton, 1982; Black and Biron, 1982; Cowley, Johnson, and Brooksbank, 1977; Filsinger, Monte, Braun, and Linder, 1984; Kirk-Smith and Booth, 1980; Kirk-Smith, Booth, Carrol, and Davies, 1978). Although several authors reported that the steroids affected mood, attraction, or judgments of other people, none gave clear, convincing results. The experimenters who employed the most ecologically valid design (androstenol applied to the bodies of confederates who interacted with subjects) found no effect of androstenol on ratings of attractiveness of confederates wearing the compound (Black et al., 1982). While 16-androstenes may be involved in chemical communication in humans, no consistent behavioral responses to them have yet been identified.

Although rodent studies did not test responses to specific chemicals, but rather used the odors of entire females (estrous and nonestrous) or their urine, they did include odors that are clearly biologically relevant for investigating possible sex hormone effects. Sex differences in sensitivity would be most expected for odors that are meaningful or perhaps essential to the reproductive success of one sex but not the other. Very few specific odors of this kind have yet been identified in mammals (see Johnston, 1983). The exceptions include dimethyl disulfide in hamsters (Singer et al., 1976), androstenone in pigs (Reed, Melrose, and Patterson, 1974), and 2-*sec*-butyl-4,5-dihydrothiazole and 3,4-dehydro-*exo*-brevicomin in mice (Novotny, Harvey, Jemiolo, and Alberts, 1985). Until recently, none of the chemicals had been investigated for sex differences in sensitivity to the odor.

At this time, not enough is known about the effects of sex hormones on sensitivity to biologically significant odors to draw any general conclusions about mechanisms. However, in the long run, this direction of research may prove more useful than the study of sex differences in general olfactory sensitivity for expanding our understanding of processing of olfactory information. The fact that variation in sensitivity associated with sex hormone levels has also been reported in several other sensory modalities [including taste, hearing, and touch (Doty et al., 1982; Henkin, 1974)] suggests that there might be a common underlying basis for these phenomena. Thus, elucidating the mechanisms involved in sex differences in general olfactory sensitivity may tell us little about olfaction per se. Studying stimulus-specific sex differences in sensitivity, on the other hand, might reveal how information from a single stimulus or set of stimuli is processed in the central nervous system, by determining how hormone-dependent changes in the olfactory system affect responses to one stimulus but not another. The study of hormone effects in, for example, the songbird song system has dramatically increased our understanding of the development of the central nervous system and neuronal plasticity in adulthood (Alvarcz-Buylla, Kirn, and Nottebohm, 1990; Arnold, 1985; DeVoogd and Nottebohm, 1981). Using this approach, we may also gain tremendous insight into necessary and sufficient conditions for perceiving and responding to a specific olfactory stimulus.

Model Systems for Studying Sex Differences in Olfaction

An ideal system for the study of effects of sex hormones on olfaction would satisfy certain conditions: the presence of an identified olfactory stimulus or stimuli, and a corresponding sexually dimorphic and well-characterized behavioral response. In addition, an understanding of the development of the dimorphic behavior would be extremely useful in forming and testing hypotheses about sex-hormone-mediated developmental changes in sensitivity to the odor or odors. Similarly, knowledge of the olfactory system involved in the reception and processing of the odor (i.e., main olfactory, vomeronasal, etc.) would be a great advantage in looking for sex hormone effects on corresponding anatomy and/or physiology in the central nervous system. A model of this kind would allow us to learn how sensory systems and neural substrates of dimorphic behaviors interact: how specific stimuli elicit or modulate behavioral responses in one individual but not in a conspecific of a different age, sex, or hormonal condition.

Two species meet many of the criteria for an ideal model (Table 8.3). These are the domestic pig (*Sus scrofa*) and the golden hamster (*Mesocricetus auratus*). For both, stimuli have been identified that elicit or facilitate the display of specific behaviors. The behaviors are sexually dimorphic, seen almost exclusively in one sex and not the other under natural conditions, and are affected by manipulations of sex hormone levels.

TABLE 8.3. Model systems[1]

	Pig		Hamster	
Behavior	Attraction of estrous females to boars	Immobile mating posture by estrous sows in the presence of boar	Male attraction to and investigation of estrous females	Male mounting of estrous females and intromission
Relation to estrous cycle	Estrus or proestrus only	Estrus only		
Hormonal influences during development:			sex difference appears at puberty	
Gonadectomy	+ pp. castration NE a. castration NE ovariectomy	+ pp. castration NE a. castration NE ovariectomy	− a. castration	− pp. castration − a. castration
Hormone treatment during development	−EB: pp. castratess & o. females	−EB: pp. castratess & o. females	+ TP: pp. males pp. females	+ TP: pp. castrates
Hormone treatment during adulthood	+ EB: females & pp. castrates	+ EB: females & pp. castrates + TP: females & pp. castrates	+ TP: a. castrates a. ovariectomized females	NE TP: pp. castrates + TP: a. castrates
Chemical stimulus		Androstenone	Dimethyl disulphide	Aphrodisin
Olfactory system			Main olfactory system—contact not necessary NE VNX − deafferentation	Vomeronasal system—contact necessary − VNX
Dimorphic sensitivity		a. females more sensitive than a. males		

KEY: +, positive effect on display of behavior; −, negative effect; NE, no effect; a, adult; pp, prepubertal; VNX, vomeronasal organ removal; TP, testosterone propionate; EB, estradiol benzoate.
[1] See text for references.

Domestic Pig

A sow in estrus or late proestrus is strongly attracted to intact males (boars): in a choice test the sow spends significantly more time near a male than a female (Signoret, 1967). During estrus, sows display an immobile mating stance in the presence of a boar (Signoret, 1970; Signoret et al., 1975). Estrogens play both an activational and an organizational role in the display of attraction and standing behaviors. A pig that experiences high estrogen levels during puberty is defeminized, reducing the expression of attraction and standing in adulthood (organization; Adkins-Regan, Orgeur, and Signoret, 1989). In adulthood, expression of both behaviors is dependent on the high levels of estrogens that occur during late proestrus and estrous (activation; Figure 8.1). A single estradiol benzoate (EB) injection will induce both attraction and standing behavior in nonestrous or ovariectomized sows (Signoret, 1967; Signoret, 1970). An EB injection in males castrated prepubertally, but not postpubertally, will also induce both attraction to a boar (Ford, 1983) and the immobile stance (Berry and Signoret, 1984; Ford, 1982, 1983). Testosterone propionate (TP) has the same activational effect on standing behavior as EB. This similarity between androgen and estrogen is probably due to aromatization of the androgen (conversion to estrogen), as nonaromatizable dihydrotestosterone does not induce standing (Ford, 1982). EB administered throughout the pubertal period reduces attraction and standing in both females and castrated males (Adkins-Regan et al., 1989). Pubertal treatment with TP also reduces the percentage of females and castrated males that show the standing response (Ford and Christenson, 1987). The effect of TP on attraction has not been investigated.

The cues that elicit attraction and standing in estrous sows include acoustic and tactile stimuli in addition to olfactory stimuli (Signoret, 1970). Olfactory cues alone, however, can elicit these behaviors. Estrous sows are attracted to intact adult males but not castrated males even when both males are anesthetized and not visible (Signoret, 1967), and removal of the olfactory bulbs disrupts attraction (Signoret, 1974). The standing response can be elicited by putting an estrous sow into a pen previously occupied by a boar (Signoret, 1970). Several 16-androstene steroids isolated from boar submaxillary saliva (Patterson, 1968) have been demonstrated to facilitate the display of the immobile stance in estrous sows (Melrose, Reed, and Patterson, 1971). Androstenone is the most effective compound, eliciting standing in more sows than any other single compound or combination tested (Melrose et al., 1971; Reed et al., 1974).

To date, there are no published experiments demonstrating sex hormone effects on attraction and standing responses using androstenone alone as the stimulus. Thus, although the behavior that androstenone facilitates has been shown to be dimorphic, it cannot yet be said that responses to androstenone per se are dimorphic. However, because the behavioral responses

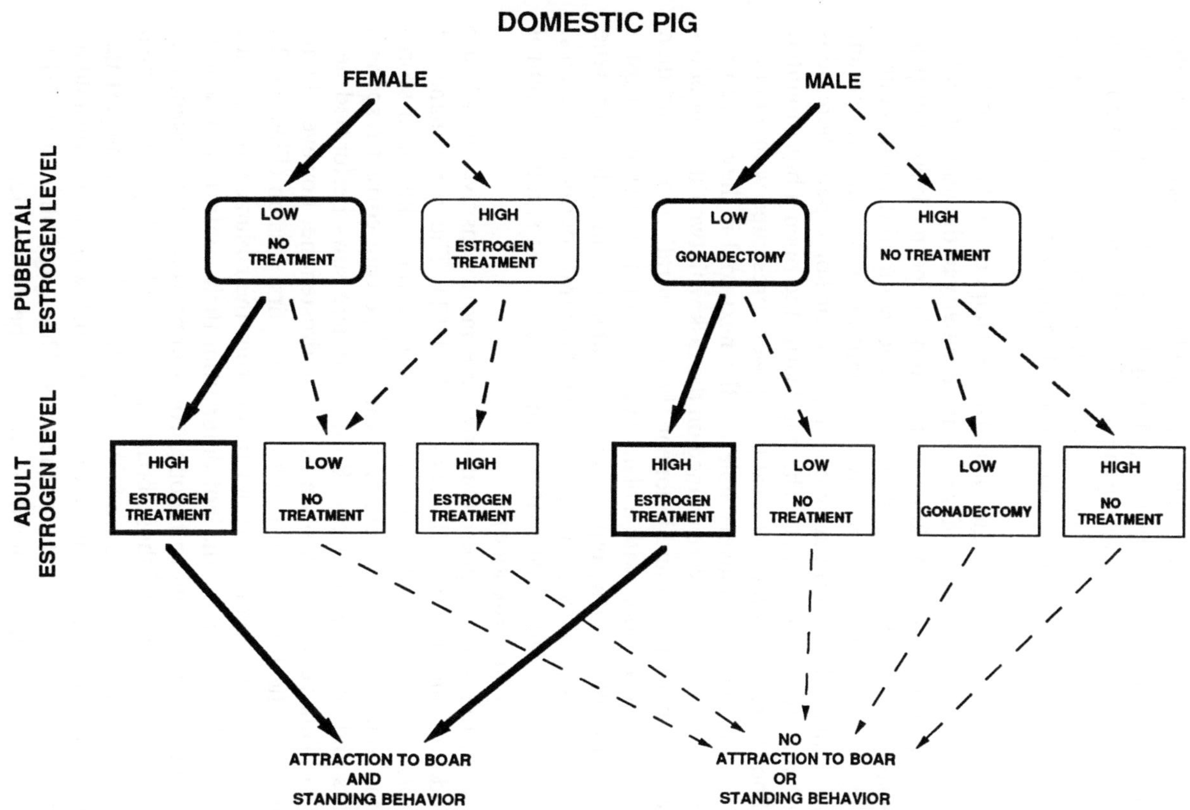

FIGURE 8.1. Organizational (prepubertal) and activational (adult) effects of estrogen on attraction to boars and receptive standing behavior in pigs. Solid lines indicate paths that result in the display of both behaviors in adulthood. Broken lines represent paths that eliminate both behaviors in adult animals. See text for references.

androstenone elicits are under hormonal control, it is reasonable to hypothesize that olfactory sensitivity to the steroid might also show a sex difference. If olfactory sensitivity to androstenone is under organizational control by sex hormones in the way that the behavioral response to the boar is, prepubertal male and female pigs should be equally sensitive to the odor, and males should lose sensitivity during the pubertal period, resulting in dimorphic sensitivity favoring the female in adulthood. Credibility is lent to this hypothesis by the fact that the same developmental pattern of sensitivity to androstenone has been reported in humans (Dorries et al., 1989).

With Elizabeth Adkins-Regan and Bruce Halpern, I have recently begun an investigation of sex differences in sensitivity to the odor of androstenone in pigs (Dorries, Adkins-Regan, and Halpern, 1990; Dorries, Adkins-Regan, and Halpern, 1991). We have compared androstenone sensitivity of five adult males and five pregnant females using an operant task. We trained pigs to sniff at an opening in their home pen where odorized or blank air was presented. The pigs learned to open a box mounted on the floor to receive a food reward if they were presented with an odor, and to wait at the odor delivery hole for the next trial if they were presented with a blank. They were trained on the task with two different odorants successively (geraniol and limonene, both "neutral" odors), and then tested with a control odor, with a blank (versus blank), and with androstenone. The control odor was amyl acetate for half of the animals and pinene for the others. Both of these are presumably "neutral" odors for the pigs. Testing consisted of four 100-trial sessions for each test stimulus (control odor, blank, and androstenone).

Results of this study suggest that sows are more sensitive than boars to the odor of androstenone (Dorries et al., 1991). The performance of boars and sows did not differ for the control odorant: the mean scores across the four test sessions were 85.6 and 87.5 percent correct responses for boars and sows, respectively. Both boars and sows performed near chance in blank versus blank sessions. On androstenone, however, boars performed significantly worse than sows. In the first session (Figure 8.2), boars' androstenone scores did not differ from their blank scores, while sows' androstenone scores did not differ from their control odor scores. Four of the five boars' androstenone scores improved across sessions, as did the scores of three of those boars when tested with pinene at near threshold concentrations (Figure 8.3). We believe these data suggest that the concentration of androstenone we used was below or near threshold for the boars, but above threshold for the sows. Thus, pigs' olfactory sensitivity to the odor of androstenone does appear to be dimorphic.

We are currently carrying out a study to replicate our results with nonpregnant sows, and to expand our investigation to include effects of developmental stage and castration on androstenone sensitivity. It will also be essential to determine which olfactory system mediates the behavioral

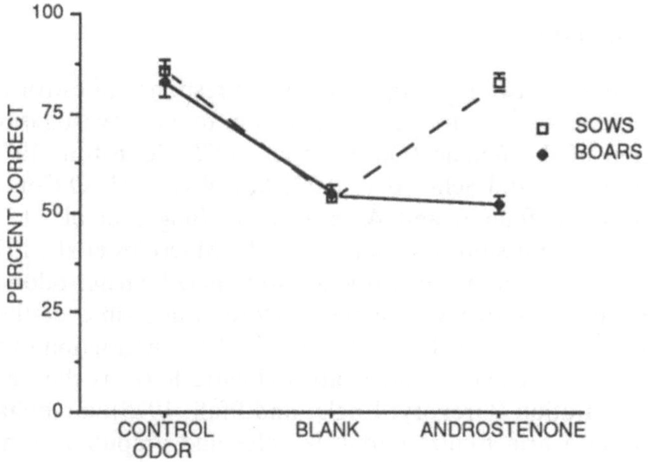

FIGURE 8.2. Mean (± se) percent correct responses for sows ($n = 5$) and boars ($n = 5$) in the first of four 100-trial test sessions with androstenone versus blank, control odor (either pinene or amyl acetate) versus blank, and blank versus blank trials. Chance is 50 percent correct. (Data from Dorries et al. 1991.)

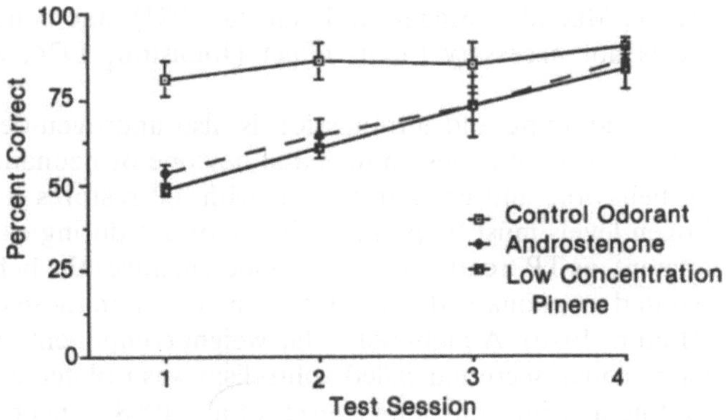

FIGURE 8.3. Mean (± se) percent correct responses for boars ($n = 3$) in four 100-trial test sessions with control odor versus blank, low-concentration pinene versus blank, and androstenone versus blank. Chance is 50 percent correct. (Data from Dorries et al. 1991.)

response to androstenone in order to use this system as an effective model. The next steps will then be to determine whether there is dimorphism in the anatomy or physiology of olfactory structures corresponding to behavioral and apparent olfactory sex differences, and to explore connections between olfactory sensitivity and dimorphic response to odors.

Golden Hamster

Male hamsters show two types of behaviors in response to estrous females that are mediated by olfactory cues: attraction to and investigation of the genital region of the female (Darby et al., 1975; Johnston, 1974, 1975; Macrides, Johnson, and Schneider, 1977; Murphy, 1973; O'Connell, Singer, Macrides, Pfaffmann, and Agosta, 1978; Singer et al., 1976), and mounting and intromission (Darby et al., 1975; Macrides et al., 1977; Murphy, 1973). A sex difference in attraction to female hamster odors appears around the time of puberty in males, before which time neither sex is attracted to the odor (Johnston and Coplin, 1979). Attraction of males to female odor is under androgenic control (Figure 8.4), as the behavior is reduced by castration (Gregory, Engle, and Pfaff, 1975) and enhanced by TP injection in ovariectomized adult females and prepubertal males and females (Johnston et al., 1979). Dimethyl disulfide (DMDS), a component of estrous female hamster vaginal secretions, has been shown to be responsible in large part for the attractive quality of the secretion (O'Connell et al., 1978; Singer et al., 1976). The effect of DMDS appears to be mediated by the main olfactory system, as deafferentation of both main and accessory olfactory systems eliminates the attractiveness of vaginal secretion (Powers and Winans, 1975), but removal of the vomeronasal organ does not (Clancy, Macrides, Singer, and Agosta, 1984), and contact with the substance is not necessary for its effect (Johnston, 1974; Murphy, 1973).

Expression of mounting and intromission is also androgen-dependent (Figure 8.4). Castration of males on neonatal day one or neonatal day 15 abolishes the behavior, and early treatment with TP restores it (Eaton, 1970). Androgen levels must be particularly important during early postnatal development, as TP treatment in adulthood enhances the behavior of hamsters castrated on neonatal day 15, but has no effect on those castrated on day one (Eaton, 1970). A high-molecular-weight component of estrous female hamster vaginal secretion called aphrodisin was isolated and found to elicit mounting and intromission (Henzel et al., 1988; Singer, Clancy, Macrides, and Agosta, 1984; Singer et al., 1986). Aphrodisin is currently believed to be an odorant binding protein for an as yet unknown stimulus molecule (Singer and Macrides, 1990). Because contact with vaginal secretions is necessary for display of mounting and intromission (Singer et al., 1984) and removal of the vomeronasal organ abolishes the behavior in inexperienced males (Clancy et al., 1984), the accessory olfactory system is implicated in mediation of response to aphrodisin.

As with pigs, the effects of sex and of hormonal manipulation on response to the isolated compounds have never been tested. Nor have sex differences in sensitivity to either DMDS or aphrodisin been investigated. Yet hamsters may be extremely promising for further investigation of the effects of sex hormones on olfaction in mammals. Two compounds have been isolated, one associated with the main olfactory system and the other with the vomeronasal system, each of which mediates the expression of a separate, sexually dimorphic aspect of reproductive behavior. Along with the pig, the hamster thus invites study not only of sex differences in sensitivity to specific odors, but also of the neural correlates of sensitivity differences and of connections between olfactory structures and neural substrates for the dimorphic behaviors these stimuli mediate.

Applications of the Model

There are several areas of research in olfactory processing that may profit from the use of the type of model described here. First, results of Segovia et al. (1982) suggest that sex hormones could alter sensitivity to specific chemicals by modulating the number or types of receptors expressed in olfactory epithelia. Variation across the estrous cycle has been reported for serotonin receptor density in neurons of the rat (Biegon, Bercovitz, and Samuel, 1980), and testosterone has been shown to increase acetylcholine receptor number in rat muscle fibers (Bleisch, Harrelson, and Luine, 1982). Olfactory receptor molecules might be modulated in a similar way. It has been hypothesized that specific anosmia results from a lack of receptors for an odor (Amoore, 1977), and that increases in sensitivity after exposure to that odor may be due to a proliferation of those receptors (Wysocki et al., 1989). Specific receptors for androstenone have been sought in the olfactory epithelium of the sow (Gennings, Gower, and Bannister, 1977; Hancock, Gennings, and Gower, 1985). Though these attempts have been unsuccessful thus far, further progress might be made employing more sophisticated steroid binding techniques (McEwen et al., 1982; Schumacher, 1990). Rosenblum, Sorensen, Stacey, and Peter (1991) have recently succeeded in demonstrating specific binding of a steroid pheromone to olfactory membranes in gold fish.

Two lines of investigation using biologically significant stimuli may contribute to our understanding of the functional organization of the olfactory bulb. Woo, Coopersmith, and Leon (1987) reported that specific olfactory bulb glomeruli are active during stimulation with particular odorants, as measured by C^{14} 2-deoxyglucose (2DG) uptake, and that exposure to an odor under conditions simulating maternal contact increases the area of the corresponding glomeruli. Sex hormones may likewise affect olfactory processing by modifying olfactory bulb glomeruli size or organization. This hypothesis could be tested by examining 2DG uptake in, for example, hamsters of different sexes or sex hormone treatments after exposure to DMDS.

Recording from neurons in the main or accessory olfactory pathways while stimulating with biologically significant odors can provide further information about areas involved in the processing of those stimuli. Pfaff and Gregory (1971) found no effect of castration on olfactory bulb and preoptic

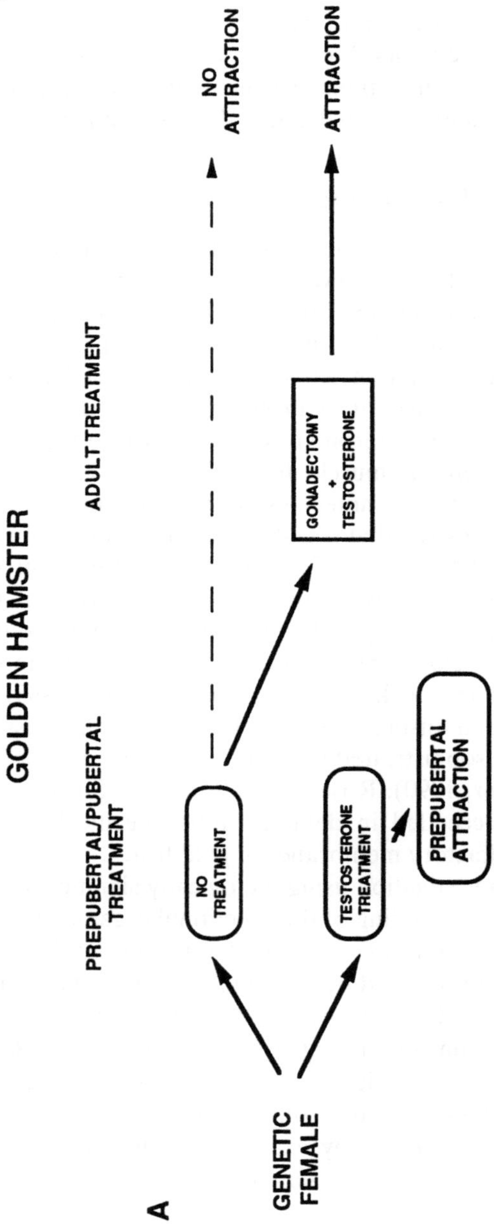
A
GOLDEN HAMSTER
GENETIC FEMALE
PREPUBERTAL/PUBERTAL TREATMENT
ADULT TREATMENT
NO TREATMENT
TESTOSTERONE TREATMENT
PREPUBERTAL ATTRACTION
GONADECTOMY + TESTOSTERONE
NO ATTRACTION
ATTRACTION

B

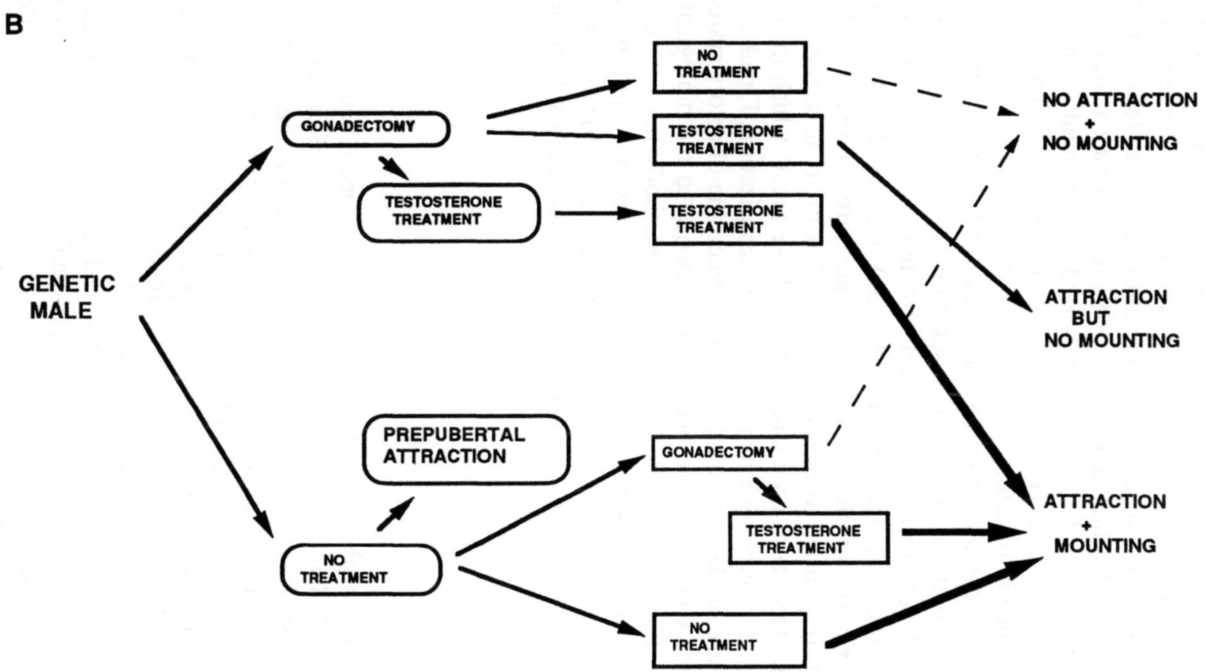

FIGURE 8.4. Effects of sex hormone manipulation on (A) attraction to female odors in female golden hamsters, and (B) attraction to female odors and mounting in male golden hamsters. Thick solid lines indicate paths that lead to expression of both attraction and mounting in adults. Medium solid lines indicate paths that lead to attraction to female odors only. Thin broken lines represent paths that eliminate both behaviors in adults. See text for references.

area neural responses to urine odors in rats, but suggested that this may be because they used whole urine, rather than "the relevant volatile chemicals" in urine, as the stimulus. MacLeod, Rheinhardt, and Ellendorff (1979) recorded from the mitral layer of the pig olfactory bulb during exposure to androstenone and testosterone. Some cells tested were responsive to both steroids, some to neither, and some to either androstenone or testosterone but not both. Although MacLeod et al. (1979) tested intact males, castrated males, and diestrous females, they reported that the number of cells tested was too small to analyze results for sex differences. Given the reported sex difference in olfactory sensitivity to androstenone (Dorries et al., 1991), further research in this area may be particularly rewarding.

In the song system of songbirds, where sex differences have been extensively investigated, sex hormones affect cell size and number, and the extent of the dendritic arbor (DeVoogd, 1986; DeVoogd et al., 1981). We may examine sex hormone effects on the same measures in regions of the olfactory system responsive to biological odors. Sensitivity to sex-specific stimuli might be dependent on connections in those areas, which in turn may depend on cell number or extent of dendritic arborization. Finally, we may also use these models to examine links between the olfactory system and brain regions controlling the behavioral response to the biologically relevant odors. For these studies, tract-tracing methods may be used to determine how specific areas in the olfactory pathway responsive to androstenone, for example, are linked to neural substrates for the standing response, and how these connections are affected by sex hormone treatment.

Conclusions

The existence of dimorphism in olfactory systems has been demonstrated at both the anatomical and the behavioral level. Beyond the fact that these sex differences exist, however, very little is known. What is the functional significance of the dimorphism of the accessory olfactory system of the rat? What are the anatomical or physiological correlates of sex hormone–associated variations in olfactory sensitivity reported in both rodents and humans? To what extent, if at all, do sex differences in olfaction underlie sex differences in olfactory-mediated behaviors?

I have outlined here characteristics that an ideal model for investigating such questions would require, and suggested two animals in which many of these criteria are met. In the first, the domestic pig, we have demonstrated a sex difference in sensitivity to a chemical that elicits a dimorphic behavioral response. The hamster appears equally promising. Using the pig and the hamster as models, we may begin to be able to answer the questions posed above, as well as questions about how specific chemical stimuli are processed by the main and accessory olfactory systems.

Acknowledgments. The author's research was funded in part by a grant-in-aid of research from Sigma Xi, The Scientific Research Society. I thank the biopsychology research group of the Psychology Department at Cornell University for comments on the manuscript. I am especially indebted to Elizabeth Adkins-Regan and Bruce Halpern for discussion and encouragement.

References

Alvarez-Buylla, A., J.R. Kirn, and F. Nottebohm (1990). Birth of projection neurons in adult avian brain may be related to perceptual or motor learning. *Science*, *249*, 1444–1446.

Adkins-Regan, E., P. Orgeur, and J.P. Signoret (1989). Sexual differentiation of reproductive behavior in pigs: Defeminizing effects of pubertal estradiol. *Hormones Behav.*, *23*, 290–303.

Amoore, J.E. (1977). Specific anosmia and the concept of primary odors. *Chem. Senses*, *2*, 267–281.

Amoore, J.E., J.R. Popplewell, and D. Whissel-Buechy (1975). Sensitivity of women to musk odor: no menstrual variation. *J. Chem. Ecol.*, *1*, 291–297.

Arnold, A.P. (1985). Gonadal steroid-induced organization and reorganization of neural circuits involved in bird song. In C.W. Cotman (Ed.), *Synaptic Plasticity*. New York: Guilford, pp. 263–285.

Balboni, G.C., and G.B. Vannelli (1982). Morphological features of the olfactory epithelium in prepubertal and postpubertal rats. In W. Breipohl (Ed.), *Olfaction and Endocrine Regulation*. London: IRL, pp. 285–297.

Baldwin, B.A., and E.E. Shillito (1974). The effects of ablation of the olfactory bulbs on parturition and maternal behavior in Saoy sheep. *Animal Behav.*, *22*, 220–223.

Benton, D. (1982). The influence of androstenol—a putative human pheromone—on mood throughout the menstrual cycle. *Biol. Psychol.*, *15*, 249–256.

Berry, M., and J.P. Signoret (1984). Sex play and behavioral sexualization in the pig. *Reproduction Nutr. Dev.*, *24*, 507–513.

Black, S.L., and C. Biron (1982). Androstenol as a human pheromone: No effect on perceived physical attractiveness. *Behav. Neural Biol.*, *34*, 326–330.

Biegon, A., H. Bercovitz, and D. Samuel (1980). Serotonin receptor concentration during the estrous cycle of the rat. *Brain Res.*, *187*, 221–225.

Bleisch, W.V., A.L. Harrelson, and V.N. Luine (1982). Testosterone increases acetylcholine receptor number in the "levator ani" muscle of the rat. *J. Neurobiol.*, *13*, 153–161.

Bronson, F.H. (1976). Urine marking in mice: causes and effects. In R.L. Doty (Ed.), *Mammalian Olfaction, Reproductive Processes, and Behavior*. New York: Academic, pp. 8–28.

Brooksbank, B.W.L., and G.A.D. Haselwood (1961). The estimation of androst-16-en-3α-ol in human urine. *Biochem. J.*, *80*, 488–496.

Brooksbank, B.W.L., R. Brown, and J.A. Gustafsson (1974). The detection of 5α-androst-16-en-3-ol in human male axillary sweat. *Experientia*, *39*, 790–792.

Brosvic, G.M., and W.W. McLaughlin (1989). Quality specific differences in human taste detection thresholds as a function of stimulus volume. *Physiol. Behav.*, *45*, 15–20.

Brown, R.E. (1977). Odor preference and urine-marking scales in male and female rats: effects of gonadectomy and sexual experience on responses to conspecific odors. *J. Comp. Physiol. Psychol.*, *91*, 1190–1206.
Buck, L. and R. Axel (1991). A novel multigene family may encode odorant receptors: a molecular basis for odor recognition. *Cell*, *65*, 175–187.
Calés, J.M., S. Segovia, P. Enríquez, and A. Guillamón (1988). Effects of early postnatal gonadal steroids on aquisition and extinction of a single alternation response in discrete-trials lever pressing. *Physiol. Behav.*, *43*, 747–753.
Capranica, R.R., L.S. Frishkopf, and E. Nevo (1973). Encoding of geographic dialects in the auditory system of the cricket frog. *Science*, *182*, 1272–1275.
Carr, W.J., and W.F. Caul (1962). The effect of castration in rat upon the discrimination of sex odours. *Animal Behav.*, *10*, 20–27.
Carr, W.J., B. Solberg, and C. Pfaffmann (1962). The olfactory threshold for estrous female urine in normal and castrated male rats. *J. Comp. Physiol. Psychol.*, *55*, 415–417.
Clancy, A.N., F. Macrides, A.G. Singer, and W.C. Agosta (1984). Male hamster copulatory responses to a high molecular weight fraction of vaginal discharge: effects of vomeronasal organ removal. *Physiol. Behav.*, *33*, 653–660.
Cowley, J.J., A.L. Johnson, and B.W.L. Brooksbank (1977). The effect of two odorous compounds on perfomance in an assessment-of-people test. *Psychoneuroendocrinology*, *2*, 159–172.
Darby, E.M., M. Devor, and S.L. Chorover (1975). A presumptive sex pheromone in the hamster. *J. Comp. Physiol. Psychol.*, *88*, 496–502.
Deems, D.A., and R.L. Doty (1987). Age-related changes in the phenyl ethyl alcohol odor detection threshold. *Trans. Penn. Acad. Opthalmol. Otolaryngol.*, *39*, 646–650.
Del Abril, A., S. Segovia, and A. Guillamón (1987). The bed nucleus of the stria terminalis in the rat: regional sex differences controlled by gonadal steroids early after birth. *Dev. Brain Res.*, *32*, 295–300.
DeVoogd, T.J. (1986). Steroid interactions with structure and function of avian song control regions. *J. Neurobiol.*, *17*, 177–201.
DeVoogd, T.J., and F. Nottebohm (1981). Gonadal hormones induce dendritic growth in the adult avian brain. *Science*, *214*, 202–204.
Di Lorenzo, P.M., and S. Monroe (1989). Taste responses in the parabrachial pons of male, female and pregnant rats. *Brain Res. Bull.*, *23*, 219–227.
Di Lorenzo, P.M., and S. Monroe (1990). Taste responses in the parabrachial pons of ovariectomized rats. *Brain Res. Bull.*, *25*, 741–748.
Dorries, K.M., H.J. Schmidt, G.K. Beauchamp, and C.J. Wysocki (1989). Changes in sensitivity to the odor of androstenone during adolescence. *Dev. Psychobiol.*, *22*, 423–435.
Dorries, K.M., E. Adkins-Regan, and B.P. Halpern (1990). Domestic pig: possible model for study of specific anosmia to androstenone. *Chem. Senses*, *15*, 567.
Dorries, K. M., E. Adkins-Regan, and B.P. Halpern (1991). Sex difference in olfactory sensitivity to the boar chemosignal, androstenone, in the domestic pig. *Animal Behav.*, *42*, 403–411.
Doty, R.L., P.J. Snyder, G.R. Huggins, and L.D. Lowry (1981). Endocrine, cardiovascular, and physiological correlates of olfactory sensitivity changes during the human menstrual cycle. *J. Comp. Physiol. Psychol.*, *95*, 45–60.
Doty, R.L., J.W. Hall, G.L. Flickinger, and S.J. Sondheimer (1982). Cyclical

changes in olfactory and auditory sensitivity during the menstrual cycle: no attenuation by oral contraceptive medication. In W. Breipohl (Ed.), *Olfaction and Endocrine Regulation.* London: IRL, pp. 285–297.

Doty, R.L., and M. Ferguson-Segall (1989). Influence of adult castration on the olfactory sensitivity of the male rat: a signal detection analysis. *Behav. Neurosci.*, 103, 691–694.

Eaton, G. (1970). Effect of a single prepubertal injection of testosterone propionate on adult bisexual behavior of male hamsters castrated at birth. *Endocrinology*, *87*, 934–940.

Elefthériou, B.E., F.H. Bronson, and M.X. Zarrow (1962). Interaction of olfactory and other environmental stimuli on implantation in the deer mouse. *Science*, *137*, 764.

Filsinger, E.E., W.C. Monte, J.J. Braun, and D.E. Linder (1984). Human (*Homo sapiens*) response to the pig (*Sus scrofa*) sex pheromone 5 alpha-androst-16-en-3-one. *J. Comp. Psychol.*, *98*, 219–222.

Fleming, A., F. Vaccarino, L. Tambosso, and P. Chee (1979). Vomeronasal and olfactory system modulation of maternal behavior in the rat. *Science*, *203*, 372–374.

Ford, J.J. (1982). Testicular control of defeminization in male pigs. *Biol. Reproduction*, *27*, 425–430.

Ford, J.J. (1983). Postnatal differentiation of sexual preference in male pigs. *Hormones Behav.*, *17*, 152–162.

Ford, J.J., and R.K. Christenson (1987). Influences of pre- and postnatal testosterone treatment on defeminization of sexual receptivity in pigs. *Biol. Reproduction*, *36*, 581–587.

Gennings, J.N., D.B. Gower, and L.H. Bannister (1977). Studies on the receptors to 5α-androst-16-en-3-one and 5α-androst-16-en-3α-ol in sow nasal mucosa. *Biochim. Biophys. Acta*, *496*, 547–556.

Good, P.R., N. Geary, and T. Engen (1976). The effect of estrogen on odor detection. *Chem. Senses*, *2*, 45–50.

Gower, D.B., S. Bird, P. Sharma, and F.R. House (1985). Axillary 5α-androst-16-en-3-one in men and women: relationships with olfactory acuity to odorous 16-androstenes. *Experientia*, *41*, 1134–1136.

Gregory, E., K. Engle, and D. Pfaff (1975). Male hamster preference for odors of female hamster vaginal discharge: studies of experiential and hormonal determinants. *J. Comp. Physiol. Psychol.*, *89*, 442–446.

Griffiths, N.M., and R.L.S. Patterson (1970). Human olfactory responses to 5α-androst-16-en-3-one—principal component of boar taint. *J. Sci. Food Agr.*, *21*, 4–6.

Guillamón, A., S. Segovia, and A. del Abril (1988). Early effects of gonadal steroids on the neuron number in the medial posterior region and the lateral division of the bed nucleus of the stria terminalis in the rat. *Dev. Brain Res.*, *40*, 306–310.

Hancock, M.R., J.N. Gennings, and D.B. Gower (1985). On the existence of receptors to the pheromonal steroid 5α-androst-16-en-3-one, in porcine nasal epithelium. *FEBS Lett.*, *181*, 328–334.

Henkin, R.I. (1974). Sensory changes during the menstrual cycle. In M. Ferin, F. Halberg, R.M. Richart, and R.L. Vande Wiele (Eds.), *Biorhythms and Human Reproduction.* New York: Wiley, pp. 277–285.

Henzel, W.J., H. Rodriguez, A.G. Singer, J.T. Stults, F. Macrides, W.C. Agosta, and H. Niall (1988). The primary structure of aphrodisin. *J. Biol. Chem.*, *263*, 16682–16687.

Herberhold, C., H. Genkin, L.W. Brändle, H. Leitner, and W. Wöllmer (1982). Olfactory threshold and hormone levels during the human menstrual cycle. In W. Breipohl (Ed.), *Olfaction and Endocrine Regulation*. London: IRL, pp. 343–351.

Hines, M., F.C. Davis, A. Coquelin, R.W. Goy, and R.A. Gorski (1985). Sexually dimorphic regions in the medial preoptic area and the bed nucleus of the stria terminalis of the guinea pig brain: a description and investigation of their relationship to gonadal steroids in adulthood. *J. Neurosci.*, *5*, 40–47.

Hoy, R.R., J. Hahn, and R.C. Paul (1977). Hybrid cricket auditory behavior: evidence for genetic coupling in animal communication. *Science*, *195*, 82–83.

Jackinovich, Jr., W., and D. Sugarman (1988). Sugar taste reception in mammals. *Chem. Senses*, *13*, 13–31.

Johnston, R.E. (1974). Sexual attraction function of golden hamster vaginal secretion. *Behav. Biol.*, *12*, 111–117.

Johnston, R.E. (1975). Sexual excitation function of hamster vaginal secretion. *Animal Learning Behav.*, *3*, 161–166.

Johnston, R.E. (1983). Chemical signals and reproductive behavior. In J.G. Vandenbergh (Ed.), *Pheromones and Reproduction in Mammals*. New York: Academic, pp. 3–38.

Johnston, R.E. (1985). Olfactory and vomeronasal mechanisms of communication. In D.W. Pfaff (Ed.), *Taste, Olfaction, and the Central Nervous System*. New York: Rockefeller University Press, pp. 323–346.

Johnston, R.E., and B. Coplin (1979). Development of responses to vaginal secretion and other substances in golden hamsters. *Behav. Neural Biol.*, *25*, 473–489.

Karlson, P, and A. Butenandt (1959). Pheromones (ectohormones) in insects. *Ann. Rev. Entomol.*, *4*, 39–58.

Kirk-Smith, M., D.A. Booth, D. Carroll, and P. Davies (1978). Human social attitudes affected by androstenol. *Res. Communications Psychol., Psychiatry, and Behav.*, *3*, 379–384.

Kirk-Smith, M., and D.A. Booth (1980). Effect of androstenone on choice of location in others' presence. In H. Van der Starre (Ed.), *Olfaction and Taste VII*. London: IRL, pp. 397–400.

Klopfer, P.H., and L. Gamble (1966). Maternal "imprinting" in goats: the role of chemical senses. *Zeitschrift für Tierpsychologie*, *23*, 588–592.

Koelega, H.S., and E.P. Köster (1974). Some experiments on sex differences in odor perception. *Ann. NY Acad. Sci.*, *237*, 234–246.

Lee, C.-T. (1976). Agonistic behavior, sexual attraction, and olfaction in mice. In R.L. Doty (Ed.), *Mammalian Olfaction, Reproductive Processes, and Behavior*. New York: Academic, pp. 161–180.

Le Magnen, J. (1952). Les phenomenes olfacto-sexuels chez le rat blanc. *Archives des Sciences Physiologiques*, *6*, 295–332.

MacLeod, N., W. Rheinhardt, and F. Ellendorff (1979). Olfactory bulb neurons of the pig respond to an identified steroidal pheromone and testosterone. *Brain Res.*, *164*, 323–327.

Macrides, F., P.A. Johnson, and S.P. Schneider (1977). Responses of the male golden hamster to vaginal secretion and dimethyl disulfide: attraction versus sexual behavior. *Behav. Biol.*, *20*, 377–386.

Mair, R.G., J.A. Bouffard, T. Engen, and T.H. Morton (1978). Olfactory sensitivity during the menstrual cycle. *Sensory Processes*, *2*, 90–98.

Marques, D.M. (1979). Roles of the main and vomeronasal systems in the response of the female hamster to young. *Behav. Neural Biol.*, *26*, 311–329.

Masson, C., and H. Mustaparta (1990). Chemical information processing in the olfactory system of insects. *Physiol. Rev.*, *70*, 199–245.

McClintock, M. (1983). Pheromonal regulation of the ovarian cycle: enhancement, suppression, and synchrony. In J.G. Vandenbergh (Ed.), *Pheromones and Reproduction in Mammals.* New York: Academic, pp. 95–112.

McEwen, B.S., A. Biegon, P.G. Davis, L.C. Krey, V.N. Luine, M.Y. McGinnis, C.M. Paden, B. Parsons, and T.C. Rainbow (1982). Steroid hormones: humoral signals which alter brain cell properties and functions. *Recent Prog. Brain Res.*, *38*, 41–83.

Melrose, D.R., H.C.B. Reed, and R.L.S. Patterson (1971). Androgen steroids associated with boar odour as an aid to the detection of oestrus in pig artificial insemination. *Br. Veterinary J.*, *127*, 497–501.

Meredith, M. (1983). Sensory physiology of pheromone communication. In J.G. Vandenbergh (Ed.), *Pheromones and Reproduction in Mammals.* New York: Academic, pp. 95–112.

Murphy, M.R. (1973). Effects of female hamster vaginal discharge on the behavior of male hamsters. *Behav. Biol.*, *9*, 367–375.

Narins, P.M., and R.R. Capranica (1976). Sexual differences in the auditory system of the tree frog Eleutherodactylus coqui. *Science*, *192*, 378–380.

Nishimura, K., K. Utsumi, M. Yuhara, Y. Fujitani, and A. Iritani (1989). Identification of puberty-accelerating pheromones in male mouse urine. *J. Exp. Zool.*, *251*, 300–305.

Novotny, M., S. Harvey, B. Jemiolo, and J. Alberts (1985). Synthetic pheromones that produce inter-male aggression in mice. *Proc. Natl. Acad. Sci. USA*, *82*, 2059–2061.

O'Connell, R.J., A.G. Singer, F. Macrides, C. Pfaffmann, and W.C. Agosta (1978). Responses of the male golden hamster to mixtures of odorants identified from vaginal discharge. *Behav. Biol.*, *24*, 244–255.

Parkes, A.S., and H.M. Bruce (1962). Pregnancy-block in female mice placed in boxes soiled by males. *J. Reproduction Fertility*, *4*, 303–308.

Patterson, R.L.S. (1968). 5α-androst-16-en-3-one: Compound responsible for taint in boar fat. *J. Sci. Food Agr.*, *19*, 31–37.

Pearce, G.P., and P.E. Hughes (1987). The influence of boar component stimuli on puberty attainment in the gilt. *Animal Production*, *44*, 293–302.

Pfaff, D.W., and E. Gregory (1971). Olfactory coding in olfactory bulb and medial forebrain bundle of normal and castrated male rats. *J. Neurophysiol*, *34*, 208–215.

Phillips, P.D., and H.H. Vallowe (1975). Cyclic fluctuations in odor detection by female rats and the temporal influences of exogenous steroids on ovariectomized rats. *Proc. Penn. Acad. Sci.*, *49*, 160–164.

Pietras, R.J., and D.G. Moulton (1974). Hormonal influences on odor detection in rats: changes associated with the estrous cycle, pseudopregnancy, ovariectomy, and administration of testosterone proprionate. *Physiol. Behav.*, *12*, 475–491.

Powers, J.B., and S.S. Winans (1975). Vomeronasal organ: critical role in mediating sexual behavior of the male hamster. *Science*, *187*, 961–963.

Reed, H.C., D.R. Melrose, and R.L.S. Patterson (1974). Androgen steroids as an aid to the detection of oestrus in pig artificial insemination. *Br. Veterinary J.*, *130*, 61–66.

Rittner, M., and U. Schmidt (1982). The influence of the sexual cycle on the olfactory sensitivity of wild female house mice (*Mus musculus domesticus*). *Zeitschrift für Säugetierkunde*, *47*, 47–50.

Rosenblum, P.M., P.W. Sorensen, N.E. Stacey, and R.E. Peter (1991). Binding of the steroidal pheromone 17α, 20β-dihydroxy-4-pregnen-3-one to gold fish (*carassius auratus*) olfactory epithelium membrane preparations. *Chem. Senses*, *16*, 143–154.

Scalia, F., and S.S. Winans (1975). The differential projections of the olfactory bulb and accessory olfactory bulb in mammals. *J. Comp. Neurol.*, *161*, 31–56.

Scalia, F., and S.S. Winans (1976). New perspectives on the morphology of the olfactory system: olfactory and vomeronasal pathways in mammals. In R.L. Doty (Ed.), *Mammalian Olfaction, Reproductive Processes, and Behavior.* New York: Academic, pp. 8–28.

Schiffman, S.S. (1988). Taste transduction and modulation. *News in Physiol. Sci.*, *3*, 109–112.

Schmidt, C. (1978). Olfactory threshold and its dependence on the sexual status in the female laboratory mouse. *Naturwissenschaften*, *65*, 601.

Schmidt, C., and U. Schmidt (1980). Changes of olfactory sensitivity during the estrus cycle in female laboratory mice. *Chem. Senses*, *5*, 359–365.

Schmidt H.J., and G.K. Beauchamp (1988). Adult-like odor preferences and aversions in 3-year-old children. *Child Dev.*, *59*, 1136–1143.

Schmidt, U., C. Schmidt, and W. Breipohl (1982). Olfactory sensitivity changes during estrus cycle, gestation and lactation in mice. In W. Breipohl (Ed.), *Olfaction and Endocrine Regulation.* London: IRL, pp. 323–332.

Schneider, R.A., and S. Wolf (1955). Olfactory perception thresholds for citral utilizing a new type olfactorium. *J. Appl. Physiol.*, *8*, 337–342.

Schneider, R.A., J.P. Costiloe, R.P. Howard, and S. Wolf (1958). Olfactory perception thresholds in hypogonadal women: changes accompanying administration of androgen and estrogen. *J. Clin. Endocrinol. Metab.*, *18*, 379–390.

Schneiderman, A.M., S.G. Matsumoto, and J.G. Hildebrand (1982). Trans-sexually grafted antennae influence development of sexually dimorphic neurones in moth brain. *Nature*, *298*, 844–846.

Schneiderman, A.M., J.G. Hildebrand, M.M. Brennan, and J.H. Tumlinson (1986). Trans-sexually grafted antennae alter pheromone-directed behavior in a moth. *Nature*, *323*, 801–803.

Schumacher, M. (1990). Rapid membrane effects of steroid hormones: an emerging concept in neuroendocrinology. *Trends Neurosci.*, *13*, 359–361.

Segovia, S., and A. Guillamón (1982). Effects of sex steroids on the development of the vomeronasal organ in the rat. *Dev. Brain Res.*, *5*, 209–212.

Segovia, S., L.M. Orensanz, A. Valencia, and A. Guillamón (1984). Effects of sex steroids on the development of the accessory olfactory bulb in the rat: a volumetric study. *Dev. Brain Res.*, *16*, 312–314.

Segovia, S., R. Paniagua, M. Nistal, and A. Guillamón (1984). Effects of postpubertal gonadectomy on the neurosensorial epithelium of the vomeronasal organ in the rat. *Dev. Brain Res.*, *14*, 289–291.

Segovia, S., A. Valencia, J.M. Calés, and A. Guillamón (1986). Effects of sex steroids on the development of two granule cell subpopulations in the rat accessory olfactory bulb. *Dev. Brain Res.*, *30*, 283–286.

Signoret, J.P. (1967). Attraction de la femmelle en estrus par le mâle chez les porcins. *Rev. Comptes Animales*, *4*, 10–22.

Signoret, J.P. (1970). Swine behavior in reproduction. In *Effects of Disease and Stress on Reproductive Efficiency in Swine.* Symposium Proceedings 70-0, Extension Service, University of Nebraska College of Agriculture, pp. 28–45.

Signoret, J.P. (1974). Rôle des différentes informations sensorielles dans l'attraction de la femelle en oestrus par le mâle chez les porcins. *Ann. Biol. Animale Biochim. Biophys.*, *14*, 747–755.

Signoret, J.P., B.A. Baldwin, D. Fraser, and E.S.E. Hafez (1975). The behaviour of swine. In E.S.E. Hafez (Ed.), *The Behaviour of Domestic Animals.* Baltimore: Williams & Wilkins, pp. 295–329.

Singer, A.G., W.C. Agosta, R.J. O'Connell, C. Pfaffmann, D.V. Bowen, and F.H. Field (1976). Dimethyl disulphide: an attractant pheromone in hamster vaginal secretion. *Science*, *191*, 948–950.

Singer, A.G., A.N. Clancy, F. Macrides, and W.C. Agosta (1984). Chemical studies of hamster vaginal discharge: male behavioral responses to a high molecular weight fraction require physical contact. *Physiol. Behav.*, *33*, 645–651.

Singer, A.G., F. Macrides, A.N. Clancy, and W.C. Agosta (1986). Purification and analysis of a proteinaceous aphrodisiac pheromone from hamster vaginal discharge. *J. Biol. Chem.*, *261*, 13323–13326.

Singer, A.G., and F. Macrides (1990). Aphrodisin: pheromone or transducer? *Chem. Senses*, *15*, 199–203.

Stern, J.M. (1990). Multisensory regulation of maternal behavior and masculine sexual behavior: a revised review. *Neurosci. Biobehav. Rev.*, *14*, 183–200.

Valencia, A., S. Segovia, and A. Guillamón (1986). Effects of sex steroids on the development of the accessory olfactory bulb mitral cells in the rat. *Dev. Brain Res.*, *24*, 287–290.

Vandenbergh, J.G. (Ed.) (1983). *Pheromones and Reproduction in Mammals.* New York: Academic.

Venstrom, D., and J.E. Amoore (1968). Olfactory threshold in relation to age, sex or smoking. *J. Food Sci.*, *33*, 264–265.

Vierling, J.S., and J. Rock (1967). Variations in olfactory sensitivity to Exaltolide during the menstrual cycle. *J. Appl. Physiol.*, *22*, 311–315.

Woo, C.C., R. Coopersmith, and M. Leon (1987). Functional changes in olfactory bulb morphology following early learning. *J. Comp. Neurol.*, *263*, 113–125.

Wysocki, C.J., K.M. Dorries, and G.K. Beauchamp (1989). Ability to perceive androstenone can be acquired by ostensibly anosmic people. *Proc. Natl. Acad. Sci. USA*, *86*, 7976–7978.

Part III
Clinical Issues in the Study of Olfaction

9
Techniques in the Quantitative Study of Human Olfaction

William S. Cain, J. Enrique Cometto-Muñiz, and
René A. de Wijk

Introduction

This chapter will illustrate some techniques and procedures relevant to the study of the functional properties of human olfaction. Rather than deal with methodology in detail, we will deal largely with the outcome of psychophysical work on the premise that outcome illustrates methodology best. Although olfaction may be the least well understood of the five classical senses, there do exist solid data on various of its functions. We will cover only a few here and will occasionally highlight techniques that may prove of interest to a person who wishes to evaluate the modality clinically. First, however, we will look briefly at matters of stimulus control.

Stimulus Control and Measurement

The difficulty of controlling and delivering the olfactory stimulus is in some measure responsible for the slow pace of understanding of olfaction. The complications of stimulus control and measurement undoubtedly inhibit nonexperts from entering the field and bringing new perspectives. It can appear to the nonexpert that stimulus choice and control must conform to arcane rules known only to a few. To add to the burden, almost no devices for stimulus control are commercially available. One must not only choose a stimulus, one must find a way to control it that often entails construction of a one-of-a-kind device. Some of the more elaborate olfactometers have required years to build, only to serve for one investigation and then to languish and fill lab space for years thereafter.

Static Olfactometry

Methods to generate odorants and deliver them to the nose divide themselves relatively well into the *static*, where odoriferous vapor exists in an enclosed or semienclosed volume, or the *dynamic*, where vapor flows continuously, carried by a dilution gas—generally odorless air—toward the

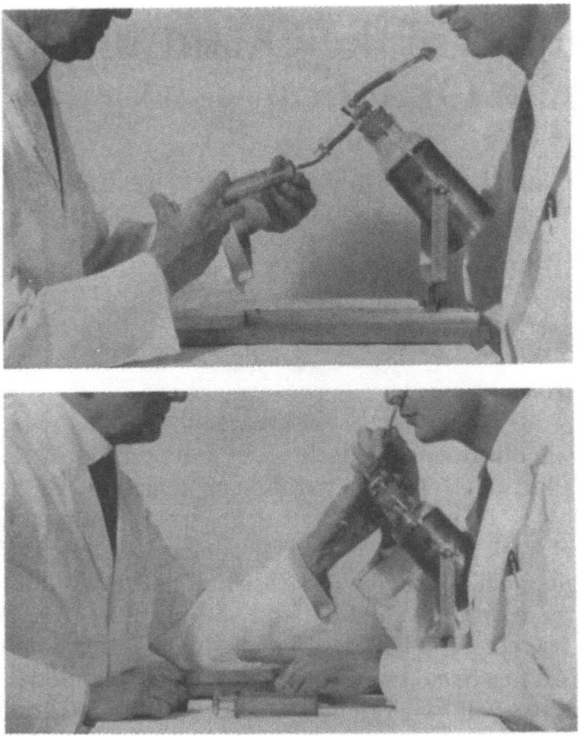

FIGURE 9.1. The arrangement for the blast injection technique. A: "Loading" the bottle for testing. B: Upon signal from the subject—who raises his finger while holding his breath—the experimenter presses the valve to release the blast of odorized air. (Reprinted, by permission, from Elsberg and Levy, 1935.)

nose of a subject. Glass bottles containing various dilutions of odorant in liquid provide the prototype for the static method, whereas olfactometers of many varieties provide examples of the dynamic method.

Elsberg and Levy (1935) developed a sort of hybrid technique, called the blast injection method, for clinical use by neurologists (Figure 9.1). It took the static headspace above undiluted odorant and expelled it through a nosepiece. With the nosepiece closed, the operator would push a syringe connected to one side of the odor vessel to a desired setting (thereby taking advantage of the compressibility of air) and lock it in place. The pressure generated in the bottle would determine the flow rate and volume of air that exited when the operator opened the stopcock to the nosepiece placed just inside a patient's nostrils. Participants did not have to inhale to receive the odorant, which made the stimulation independent of inhalation pattern. The operator varied the intensity of stimulation through the amount of pressure applied to the bottle, i.e., by the degree of compression, read as number of cubic centimeters on the syringe.

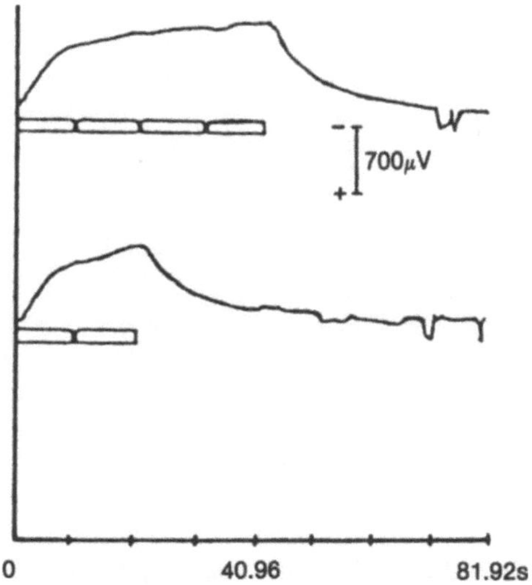

FIGURE 9.2. Human electro-olfactogram, a potential that reflects the activity of many receptors, elicited by hydrogen sulfide at two stimulus durations. (Adapted, by permission, from Kobal, 1981.)

Elsberg's technique had the flaws that it caused progressive drying of the nasal mucosa—eliciting pungent or irritant sensations with repetitive stimulation—and that it might also mistakenly measure air pressure thresholds rather than olfactory thresholds (Wenzel, 1948). Largely because of those concerns, it dropped out of the repertoire of clinical techniques. Nevertheless, it served as a forerunner of a technique used by Kobal (1981), who employs humidified air and adds odorant to a flowing airstream in such a manner as to preclude pressure artifacts. This recent development has proved useful for the measurement of the electro-olfactogram, a mass potential from the olfactory mucosa, and olfactory evoked potentials in humans (Figure 9.2). In due course, such potentials may provide truly objective means to assess the olfactory functioning.

Static olfactometry, generally the easiest form of stimulus control, customarily entails the use of a number of glass or odorless plastic vessels, each with a dilution of odorant in an odorless solvent. Choice of the solvent poses some challenge. Distilled and deionized water is one choice, but some odorants are not stable in it—e.g., esters slowly hydrolize to acids and alcohols—and many odorants have little or no water solubility. In those cases, mineral oil, propylene glycol U.S.P., or some low-vapor-pressure esters might be used.

The actual stimulus in static olfactometry is the concentration of the odorant in the *headspace above a solution*. In principle, headspace concentration varies proportionally with the concentration of the odorant *in the solution* by a factor known as the "activity coefficient." This factor differs

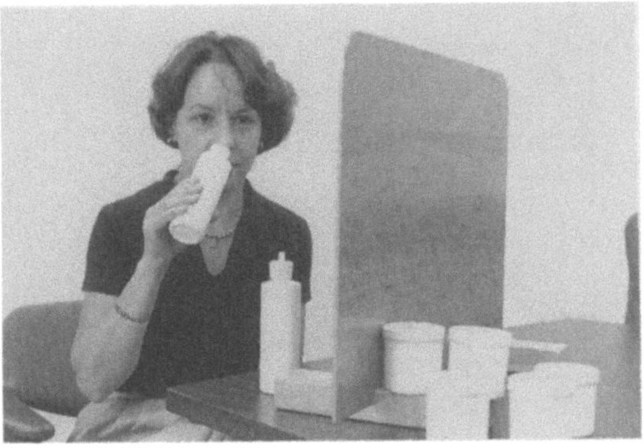

FIGURE 9.3. Squeezable bottles with pop-out spouts used for clinical testing. (Reprinted, by permission, from Cain, 1989.)

among odorants diluted with the same solvent, among solvents, and sometimes among concentrations of the same odorant-solvent pair. A means of checking vapor-phase concentration becomes the only safeguard against incorrect reliance on questionable assumptions about the relation between liquid-phase and vapor-phase concentrations. The majority of investigations of olfaction have unfortunately included no measurement of the concentration delivered to a subject's nose. Gas chromatography makes such calibration relatively simple.

When subjects open bottles containing odorant, bring the open end to their nose, and sniff from the headspace above the liquid solution, they also inhale a certain amount of surrounding room air, thereby diluting the stimulus. Squeezable sniff bottles with pop-out spouts that fit into one or the other nostril for monorhinic testing can circumvent the problem of dilution (see Figure 9.3) (Amoore and Ollman, 1983; Cain, 1989; Cain, Gent, Catalanotto, and Goodspeed, 1983; Cain, Gent, Goodspeed, and Leonard, 1988). Subjects place the spout into the specified nostril, then squeeze and sniff simultaneously. In this way, the odorant vapor enters the nasal cavity efficiently, and each nostril can be tested separately. The simplicity of the method, the ease of use and handling of the bottles, and the possibility for monorhinic testing make this procedure convenient in both clinical work (Cain, 1989) and basic research (Cometto-Muñiz and Cain, 1990; J.C. Stevens and Cain, 1987).

The headspace above an odorant solution requires time to regain its original—presqueeze—vapor concentration. If a bottle is left still, the duration would be on the order of minutes (around 30 min). Much faster reequilibration—on the order of fraction of a minute—occurs if the liquid is shaken in a circular pattern (Dravnieks, 1975). If this precaution is supplemented with the availability of two bottles for each concentration step and alternative use of each, the problem of reequilibration is essentially solved.

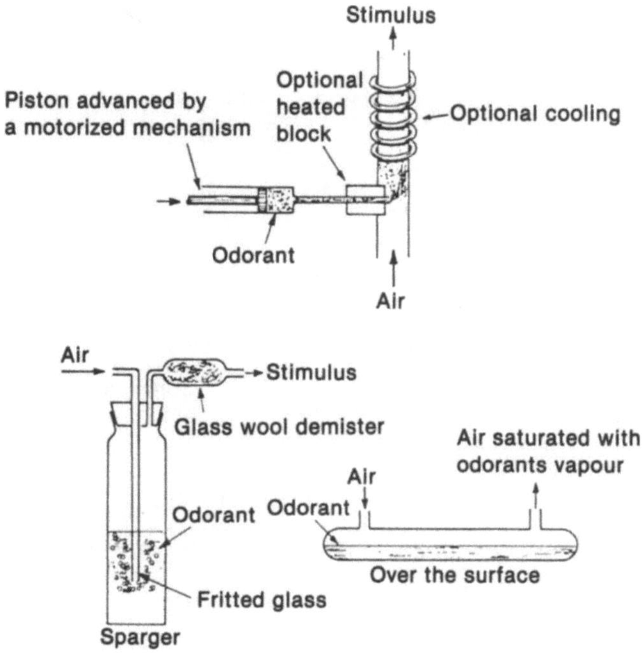

FIGURE 9.4. Various techniques for introducing odorant into a stream of non-odorous air. (Reprinted, by permission, from Dravnieks, 1975.)

Dynamic Olfactometry

As noted above, dynamic olfactometry involves a stream of odorized air—generally at vapor saturation for that temperature—that can be mixed in various proportions with odorless air or nitrogen acting as solvent or carrier. An array of tubing, valves, flowmeters—either rotameters or capillaries with manometers—saturating and mixing vessels, deodorizing and air conditioning—temperature, humidity—devices provide the necessary equipment for the generation and control of odorants.

Loading the carrier gas with vapors from liquid odorants can be achieved by direct vaporization into the carrier stream, by bubbling the odorless gas through the liquid, or by passing the gas over an extended surface of the odorant (Figure 9.4). The bubbling method has the disadvantage of forming aerosols, small droplets of the liquid suspended in the gas phase. These can be reduced by aerosol filters (typically glass wool) but may not always be completely eliminated.

Dravnieks (1975) described the details of various dynamic olfactometers used for animal and human research, including Köster's (1967, 1971), Sanders's (1970), and Moulton's (1972). (Dravnieks also included an annotated bibliography with the complete reference and a brief summary of olfactometers described as early as 1930.) Particularly interesting was his

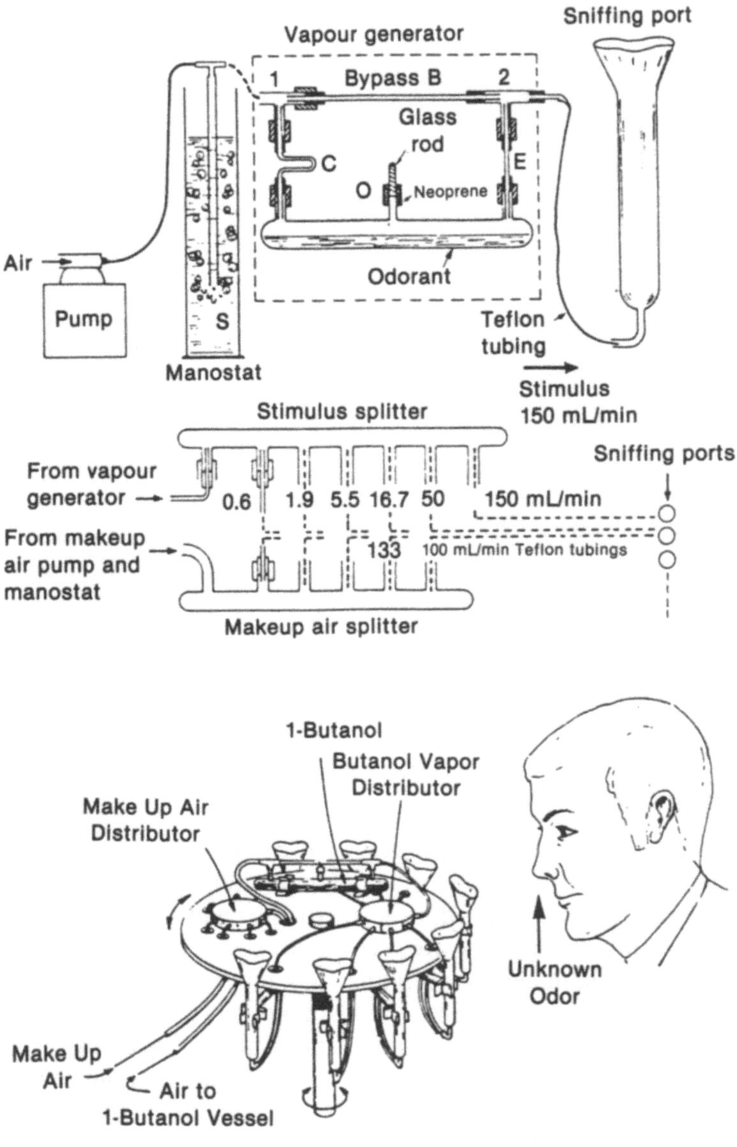

FIGURE 9.5. Top: Schematic representation of the Dravnieks Binary Dilution Olfactometer. (Reprinted, by permission, from Dravnieks, 1975.) Bottom: A perspective drawing of the olfactometer. (Reprinted, by permission, from Dravnieks, 1977.)

own flexible, simple, valveless device, the Dravnieks Binary Dilution Olfactometer (Dravnieks, 1975), which combined stability of concentration and portability with ease of use and maintenance (see Figure 9.5).

In the Binary Dilution Olfactometer, the odorous stimulus is generated by pumping air—at a pressure controlled by a water column manostat—

across the surface of the odorant. The odorant-saturated air is mixed in desired proportions with odorless air. This is achieved via a stimulus splitter to which stainless steel capillaries of varying lengths and internal diameters are attached, providing an orderly series of various flows of odorous air (each related to the next by a fixed factor). The stimulus splitter is complemented with a make-up (odorless) air splitter that brings up the total flow from each sniffing port to a common value. An operator needs merely to place odorant into a small saturator and check a couple of flow rates. All the rest occurs automatically. The olfactometer is made of glass, stainless steel, very short pieces of neoprene tubing (to interconnect the capillaries), Tygon tubing for the initial pumping of the air, and Teflon tubing to feed the sniffing ports.

The Dravnieks Binary Dilution Olfactometer, when allowed to run in a steady-state mode, essentially avoids the need for calibration through external analytical instruments, such as a gas chromatograph. Since the strongest stimulus is odorant-saturated air, the concentration of which may be calculated or looked up from tables, and the other stimuli are fixed dilutions (on a volume basis) of that vapor-saturated stream, actual concentrations are simple to compute.

Dravnieks also developed what is called a Dynamic Forced-Choice Triangle Olfactometer for measurement of thresholds (Dravnieks and Prokop, 1975; Dravnieks, Prokop, and Boehme, 1978). This instrument consists of six sniffing stations, each with three identical sniffing nozzles. One nozzle in each station delivers a certain dilution of the stimulus, whereas the other two deliver odorless air. The stimulus varies in concentration by a factor of 3 from station to station. During threshold testing, a subject starts with the first station (weakest stimulus) under instruction to choose the nozzle that smells different from (stronger than) the other two, and then goes to the next station (threefold more concentrated), where he seeks to do the same, and so on. The total flow rate emerging from each nozzle (blank or stimulus) is fixed at either 3 l/min or 0.5 l/min. The highest dilution achieved can also vary from 81-fold to 1700-fold. This device, like the Binary Dilution Olfactometer, is available commercially from IIT Research Institute, Chicago, IL

Environmental Chambers

Air-dilution olfactometers and static dilution devices may have limitations on the amount of air they deliver to the nose per unit time. Whether the limitations have any practical consequences remains largely unexplored. Work by Laing (1982, 1983) offers design end points of interest: An average human sniff lasts 0.4 s, has a volume of 200 cm^3, and reaches an instantaneous flow rate of 30 l/min. (Big sniffs do not necessarily increase odor magnitude over small sniffs.) Although suitably designed devices can accommodate these characteristics, an odor chamber offers the greatest freedom of human odor sampling. Such chambers provide a way to perform odor and pungency research with environmental realism. The results obtained permit an almost-in-the-field understanding of problems related to indoor air quality (Cain and Leaderer, 1982; Cain, Leaderer, Isseroff, Berglund, Huey, Lipsitt, and Perlman, 1983; Cain, See, and Tosun, 1986; Cain, Tosun, See, and Leaderer, 1987; Clausen, Fanger, Cain, and Leaderer, 1986), the perception of environmental fragrances (Schiet and Cain, 1990), masking agents, and gas warning agents.

The air in the typical environmental chamber can be precisely controlled and monitored for temperature, humidity, and flow rate of both recirculation and outside (fresh) air. The air flow through the chamber should provide good mixing of any stimulus delivered and a fairly quick renewal of room air with fresh air, without creating disturbing air currents. A useful added feature is an anteroom to minimize perturbances created by entering and exiting the chamber.

Heating and cooling systems, as well as desiccant and humidifying devices, condition the air entering the chamber according to the requirements of each experiment. Measurement of decays of tracer gases (e.g., carbon dioxide) that are easy to deliver and monitor establishes actual ventilation rates.

Thresholds

Odor thresholds have been measured since the middle of the nineteenth century (Cain, 1978). The corpus of published threshold data currently encompasses hundreds of materials (of hundreds of thousands of odorants), some of interest because of their flavors or their fragrancing effects, some because of their possible presence in polluted atmospheres, and some because olfactory researchers have arbitrarily chosen to study them. The corpus of data has unfortunately little thematic uniformity. A search through one or another published compilation may uncover thresholds for some members of a chemical series, but not for others, or for some isomers, but not others (Fazzalari, 1978; van Gemert and Nettenbreijer, 1977). The literature on smell contains disappointingly few prospective efforts to attack the molecular basis for odor potency through the collection of thresholds for many related chemicals.

Chemical Series and the Physicochemical Basis for Potency

Occasionally investigators have explored some members of one or another chemical series in a search for insight into the physicochemical basis for odor potency. Figure 9.6 shows thresholds obtained in such efforts for

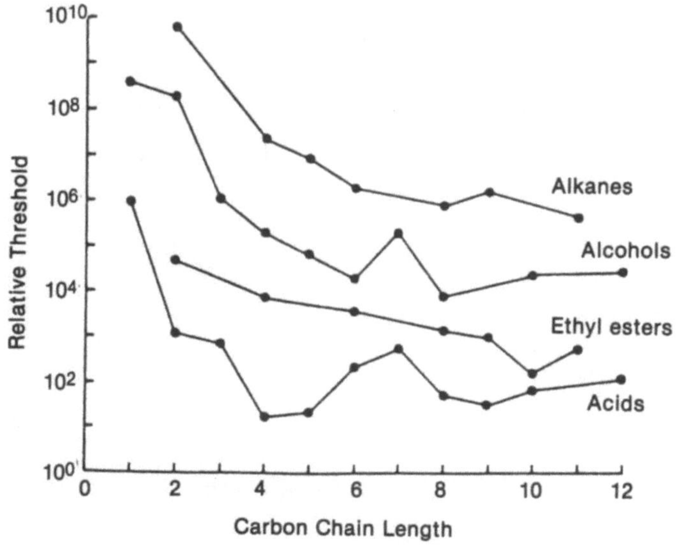

FIGURE 9.6. Odor thresholds for human subjects of four aliphatic series. (Reprinted, by permission, from Cain, 1988; copyright by John Wiley Sons, Inc., 1988; data from Laffort, 1969.)

various aliphatic series. The outcomes implied that within a series, odor potency grows with molecular size and with lipid solubility. Once such factors are taken into account, the enormous odorant-to-odorant range of thresholds of the sort seen in Table 9.1 foreshortens somewhat.

Research of Cometto-Muñiz and Cain (1990) illustrates the point, both for olfaction and for pungency (e.g., irritation). Figure 9.7 displays thresholds for aliphatic alcohols in normal (normosmic) subjects (odor thresholds) and in persons (anosmics) without a functional sense of smell (pungency or trigeminal thresholds). As expected from general character-

TABLE 9.1. Odor detection thresholds of selected compounds

Compound	Odor threshold concn. in air	
	(mg/m³)	(ppm, v/v)
Ethane	1.5×10^5	120,000
Methanol	6.6×10^2	500
Chloroform	3.2×10^2	65
Benzene	1.7×10^1	5.2
Camphor	1.1×10^0	0.17
Furfural	2.3×10^{-1}	0.059
Isoamyl acetate	3.8×10^{-2}	0.0071
5α-Androst-16-en-3-one	2.1×10^{-3}	0.00019
2-Methoxy-3-isobutylpyrazine	3.6×10^{-5}	0.00000054

Adapted from Amoore (1982).

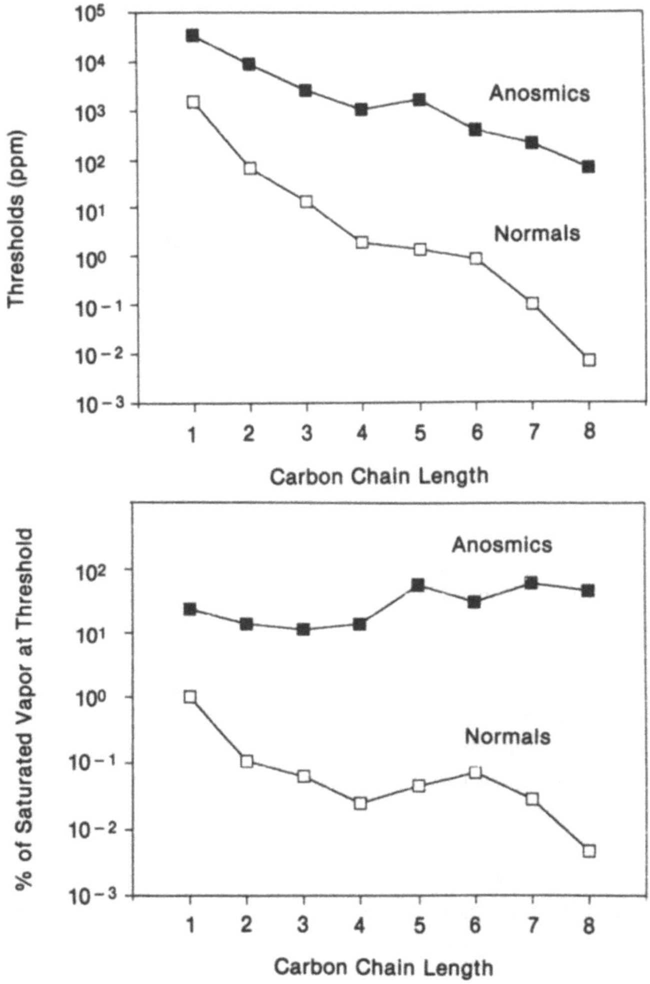

FIGURE 9.7. Top: Odor thresholds measured in normosmics (empty symbols) and pungency thresholds measured in anosmics (filled symbols) for the eight aliphatic alcohols from methanol to 1-octanol. Bottom: Thermodynamic activity at the odor threshold from normosmics (empty symbols) and at the pungency threshold from anosmics (filled symbols). The activity was calculated as the ratio between vapor concentration at threshold odor or pungency to saturated vapor concentration, multiplied by 100. (Reprinted, by permission, from Cometto-Muñiz and Cain, 1990.)

istics of olfaction and the common chemical sense, pungency thresholds lie well above odor thresholds, but both types of thresholds decline with chain length. When plotted in terms of what we might call incident vapor-phase concentration, measured via gas chromatography, the odor threshold changes by a factor of 2.1×10^5 from methyl to octyl alcohol, and the pungency threshold changes by a factor of 50. When plotted in terms of percent saturated vapor, an index of thermodynamic activity, the odor threshold varies by a factor of 210 (and remains rather constant between ethyl and heptyl alcohol) and pungency threshold by a factor of 5 (Figure 9.7).

The reduction in range with the threshold data plotted in terms of percent saturated vapor implies that much of the change in threshold over the series results from variation in the amount of material that entered the biophase (watery mucus and lipid cell membrane). This was particularly true for pungency. With nonreactive materials such as the aliphatic alcohols, their ability to elicit pungency apparently comes from a nonspecific effect derived from reaching a critical concentration in mucosal tissue. At equal concentrations in tissue, different alcohols cause the same degree of pungency.

As always, conclusions in olfaction rest upon only fragmentary data. Laffort and colleagues (Laffort, Patte, and Etcheto, 1974), however, used existing threshold data to build a reasonable model of odor potency from various molecular parameters.

Threshold Methodology

The odor threshold measured most often is the detection threshold, whereby subjects seek to distinguish the presence of odor from odorless air, and do not seek to specify or recognize odor quality. Compilations often include the so-called recognition threshold also, where subjects are required to specify quality. Such a threshold typically occurs at a concentration about threefold above the point of detection. It will come as no surprise that olfactory thresholds depend largely on the methodology used to gather them and that compilations suffer very much from methodological inconsistencies. Multiple entries for the same odorant may differ by as much as four orders of magnitude. Such differences can derive from such physical or chemical matters as the solvent used to dilute a material. Except for olfactometers arranged like the Dravnieks device, essentially every estimate of threshold is questionable without an assessment of vapor-phase concentration. Nevertheless, psychophysical methodology per se also plays its role.

To take the simplest case, odor thresholds will vary according to whether the experimenter presents concentrations in ascending, descending, or random order. An ascending series generally leads to a low threshold, a descending series to a high threshold, and a random series to an intermediate threshold (see Figure 9.8). Higher thresholds found for descending stimuli probably occur because of adaptation (Köster, 1975). Adaptation may also play a role with stimuli presented in random order, particularly with short interstimulus intervals. Intervals of 60 s generally lead to satisfactory results.

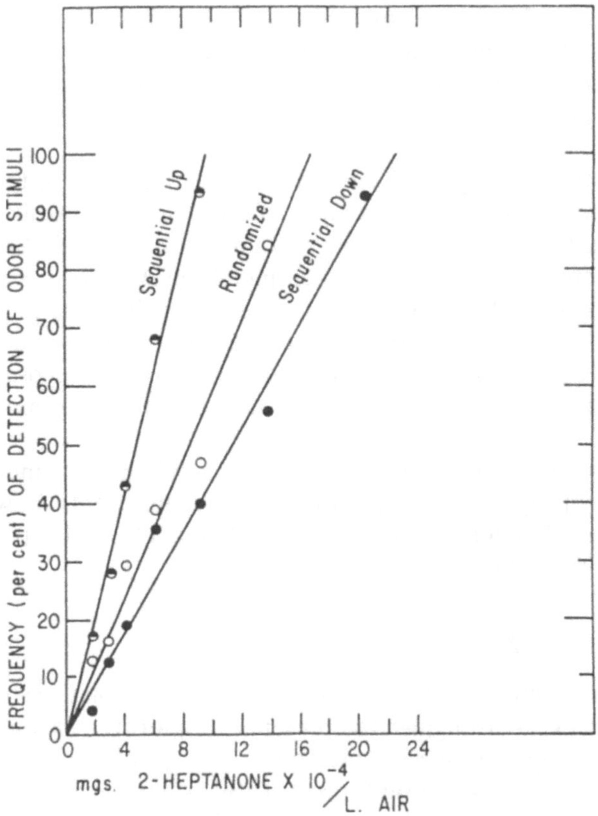

FIGURE 9.8. Psychometric functions for 2-heptanone obtained by three methods of stimulus presentation. (Reprinted, by permission, from Pangborn, Berg, Roessler and Webb, 1964.)

Time figures pervasively in determinations of odor thresholds. Slow recovery from the effects of stimulation force the experiments to proceed at an often agonizing pace. In the experiments described in Figure 9.8, subjects received no blanks. Unfortunately, then, the threshold could depend as much upon response criterion as upon olfactory sensitivity. Forced-choice methods, where subjects must choose between a stimulus and at least one blank, circumvent the problem, but only at the expense of time. Various experimenters have used an ascending method of limits in a forced-choice mode (two alternatives).

In a forced-choice variation used in clinical evaluation, a correct answer at any given concentration leads to re-presentation of that same concentration until a criterion, such as five correct in a row, is reached or until the subject makes an error, whereupon concentration increases by one step (Cain, 1989). In this variation, threshold is defined as the point of 100 percent detection rather than at the more conventional point halfway between chance and perfect performance. The procedure has limitations imposed by the inevitable occurrence of strings of correct answers by chance. Such limitations can be circumvented by increases in the number of correct answers required to quit testing, by repeat testing, and so on. Alternatively, another threshold procedure, such as the staircase method, could be employed (Doty, Gregor, and Settle, 1986). All such maneuvers take time, which may be plentiful in the laboratory, but not perforce in the clinic.

A somewhat less orthodox method for threshold measurement involves the extrapolation of threshold concentrations from psychophysical functions erected from ratings of perceived intensity. The procedure implies some knowledge of the form of the psychophysical function in the vicinity of the threshold, a matter that remains unsettled (Marks and J.C. Stevens, 1968). The approach has yielded satisfactory results in some cases (Berglund, Hogman, and Johansson, 1988; Overbosch, de Wijk, de Jonge, and Köster, 1989), but seems a risky substitute for measuring thresholds directly.

Random versus Systematic Variability

Olfactory thresholds exhibit notorious person-to-person variability, sometimes as much as five log units, or 100,000 to 1, but more commonly three log units (Brown, Maclean, and Robinette, 1968). Although various methods of threshold measurement and various definitions of threshold (50 percent detection versus 100 percent detection; detection versus recognition) can account for some variability from study to study, they fail to account for variation within a study. This presumably arises from poor stimulus control, unreliability of methods, differences in the experience of subjects, and undoubtedly some true individual differences. Cain and Gent (1990) found that reliability increased from one threshold test to another over the course of four days. During this time, thresholds declined progressively. Hence, subjects become more uniform and seemingly more sensitive with practice.

In another experiment of how thresholds changed with repeated testing, Rabin and Cain (1986) found that practice generalized from one odorant to another and from one nostril to another, and presumably depended upon something other than peripheral sensory factors. Thresholds were still declining after three sessions of three hours (Figure 9.9). De Wijk (1989) found stable individual threshold concentrations after subjects received approximately 30 hours of practice.

Aging has an influence on olfactory thresholds more or less irrespective of the stimulus employed. In single threshold measurements, groups of elderly adults typically yield thresholds two- to tenfold above those of young adults (Cain and J.C. Stevens, 1989; J.C. Stevens, Cain, and Weinstein, 1987) (see Figure 9.10). Cain and Gent (1991) found that the apparent effect of age increased with the reliability of threshold measurements.

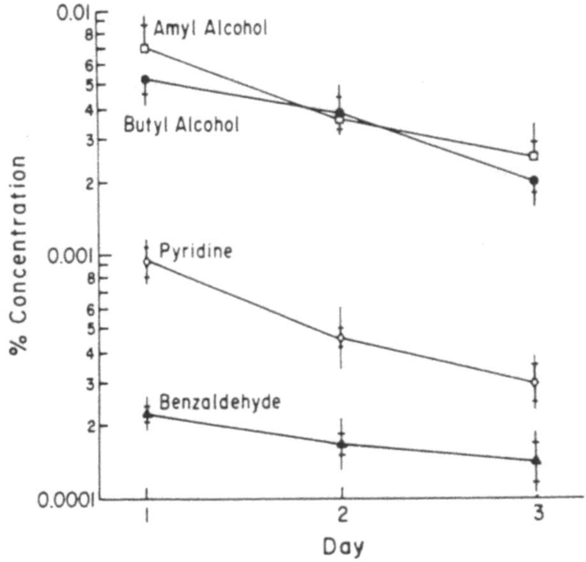

FIGURE 9.9. Odor thresholds measured over three days. Bars depict standard errors computed with and without normalization for a general factor of sensitivity across subjects. (Reprinted, by permission, from Rabin and Cain, 1986.)

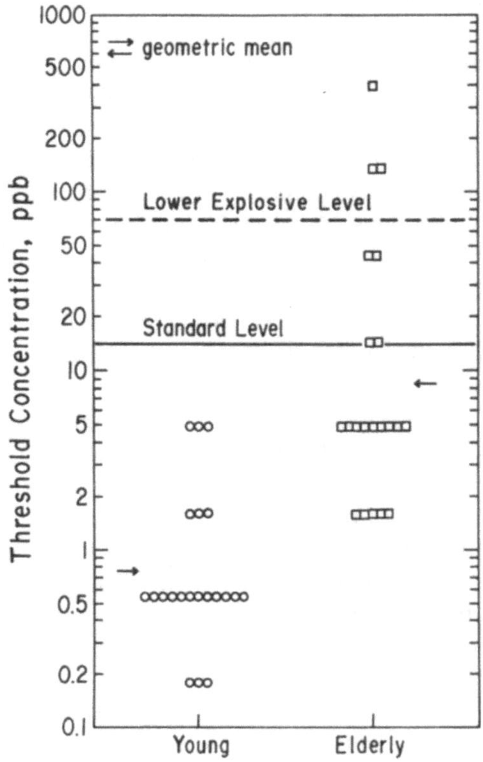

Whereas age accounted for 15 percent of the variance of measurements made on a single day, it accounted for 50 percent of the variance of measurements averaged over four days (Figure 9.11). Over a four-decade age range from the 20s to the 50s, threshold averaged over four days of testing increased more than 60-fold with age. Residual individual differences, after accounting for age-related influences, equaled about 30 to 1, which lies well below the traditional 1000 to 1 seen when age or other organismic variables are not taken into account.

Systematic influences of organismic and environmental factors on odor threshold often seem ephemeral. For example, one threshold study will uncover an effect of smoking and another will not (Hubert, Fabsitz, Feinleib, and Brown, 1980; Martin and Pangborn, 1970; Matzker, 1965; Venstrom and Amoore, 1968). One will find an advantage for females and another will not (Venstrom and Amoore, 1968; Koelega, 1970; Koelega and Köster, 1974; LeMagnen, 1952; Punter, 1983; Doty, Gregor, and Settle, 1986). Presumably, positive effects come closer to the truth than negative effects. The need seems clear: higher reliability will uncover the true effects more clearly and will give a more realistic picture of the true magnitude of an effect.

Temporal Integration and Adaptation

The threshold (or a comparable index of sensitivity) has often served as the workhorse of the psychophysical laboratory. If one wishes to study the effects of, say, time on the olfactory response, then a threshold study can usually reveal it. For instance, threshold studies have revealed that at the beginning of stimulation, detectability increases with duration, i.e., temporal integration occurs, but that after a brief time, detectability decreases, i.e., adaptation occurs. Stuiver (1958), who investigated the effects of temporal integration on olfactory thresholds, found that for very short stimulus durations (critical durations up to 160 to 200 ms), stimulus concentration and stimulus duration were completely interchangable (Figure 9.12). Detectability increased with stimulus durations up to 1.6 s. Longer stimulus durations, however, led to a decrease in detectability via adaptation (Figure 9.13). Eventually, the threshold concentration may rise to equal the concentration of the adapting stimulus. As a result, the adapting stimulus may no longer be perceived. The adaptation time required for the cessation of smell, or ATCS (de Wijk, 1989; Elsberg and Levy, 1935; Mullins, 1955;

FIGURE 9.10. Detection thresholds for the gas warning agent ethyl mercaptan in young and old subjects. The standard level refers to the concentration of ethyl mercaptan (14 ppb) that would be achieved when a leak of propane from a freshly filled cylinder achieves a concentration of 0.47 percent (one-fifth of the lower level of flammability). (Reprinted, by permission, from J.C. Stevens, Cain, and Weinstein, 1987. Copyright 1987, National Fire Protection Association, Quincy, MA 02269.)

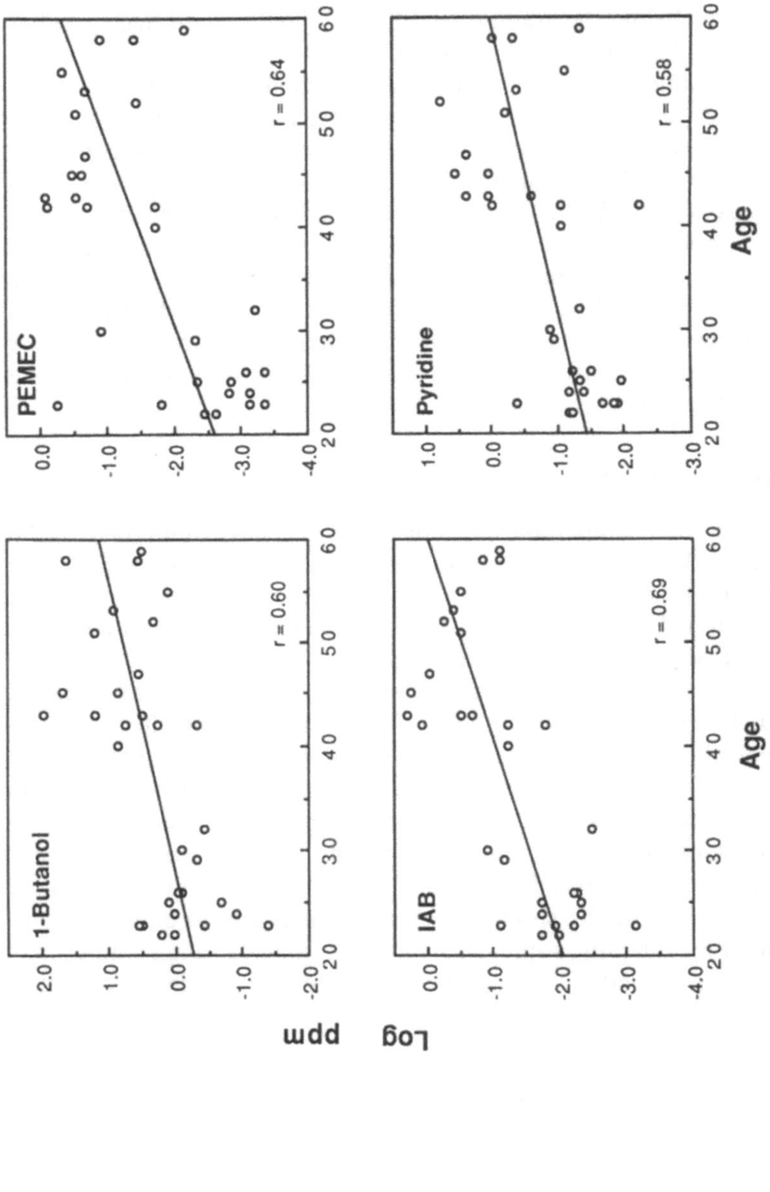

FIGURE 9.11. Relationship between threshold and age for 1-butanol, pyridine, isoamyl butyrate (IAB), and phenylethylmethylethyl carbinol (PEMEC). Each point represents a participant. (Reprinted, by permission, from Cain and Gent, 1990. Copyright 1990 by the American Psychological Association.)

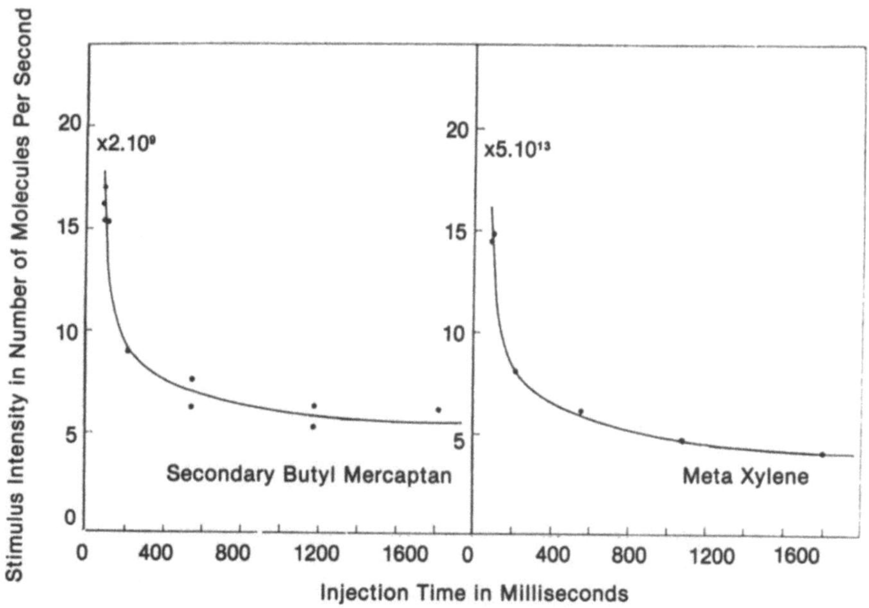

FIGURE 9.12. Detection thresholds, expressed in terms of number of molecules per second, as a function of stimulus injection time for *sec*-butyl mercaptan and *m*-xylene. Flow rate was 100 ml/s. (Reprinted, by permission, from Stuiver, 1958.)

Stuiver, 1958; Woodrow and Karpman, 1917), increases with stimulus concentration, and varies with the test odorant (Figure 9.14).

Through the decades, researchers have sought one or another psychophysical maneuver to uncover what we might call the relatedness of odors. The phenomenon of cross-adaptation has often seemed the most likely route to such understanding, though various obstacles have stood in the way. In the cross-adaptation paradigm, an experimenter exposes a subject to one odorant and tests sensitivity with another. Although logic suggests that molecules that share receptor cells or receptor sites should cross-adapt more strongly than those that do not, the outcome of the experiments has yielded little in the way of discernible patterns (e.g., Köster, 1971). This may have occurred for several reasons. Receptor cells in the olfactory mucosa may have specific and nonspecific receptor sites, and therefore every odorant may show some cross-adaptation with every other odorant. The amount of interaction, particularly that which arises from nonspecific interaction, may depend on how much material filters through to the relevant biophase. Alternatively, a given cell may possess unrelated specific receptor sites. Another possibility is that properties that have little to do with exactly which receptor sites a molecule stimulates, such as water solubility, may determine where the material deposits itself on the olfactory mucosa, and accordingly related molecules may have little opportunity to show

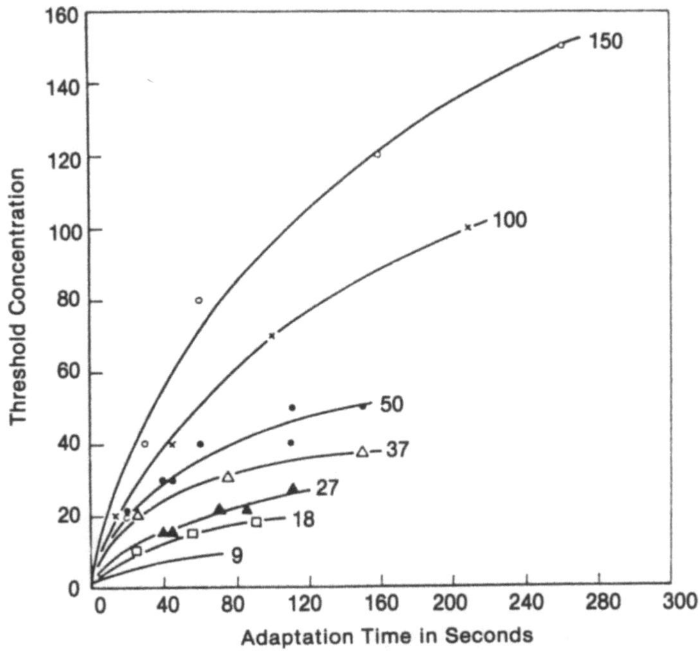

FIGURE 9.13. Increase of the threshold for 2-octanol during adaptation to various stimulus concentrations. Both adapting concentrations (parameter) and threshold concentrations are expressed in multiples of the unadapted threshold concentration. (Reprinted, by permission, from Stuiver, 1958.)

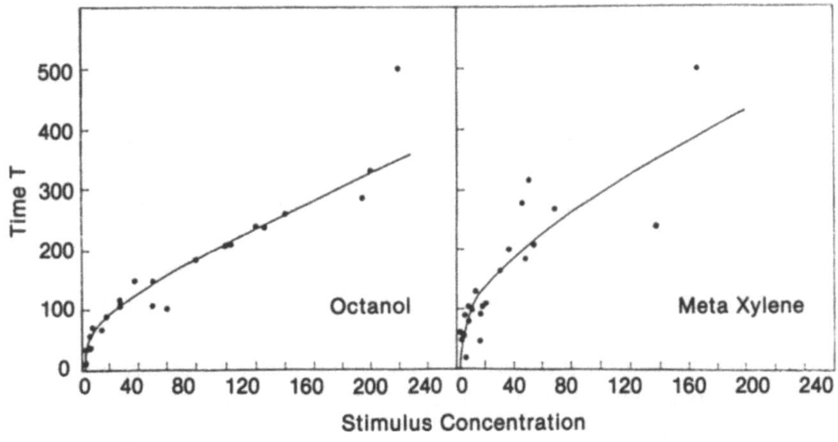

FIGURE 9.14. Relation between the concentration of the adapting stimulus (expressed in multiples of the unadapted threshold concentration) and the adaptation time required for the cessation of the smell sensation (ATCS) for 2-octanol and *m*-xylene. (Reprinted, by permission, from Stuiver, 1958.)

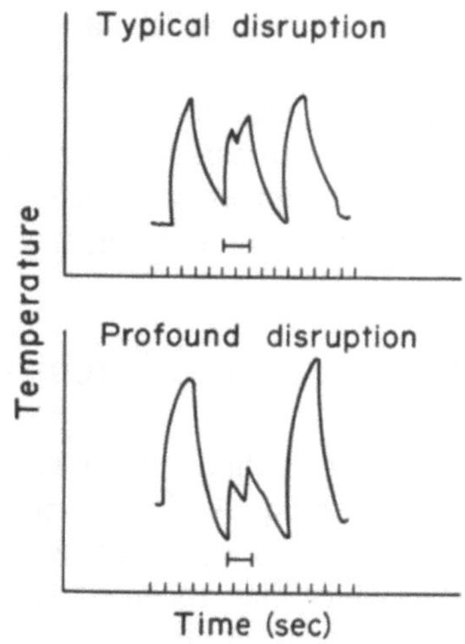

FIGURE 9.15. Breathing patterns detected by changes in temperature of a nasal thermocouple before, during, and after presentation of carbon dioxide at a concentration sufficient to disrupt breathing, eliciting a reflex, transitory apnea. Such measurements confirmed that smokers showed the threshold at a significantly higher carbon dioxide concentration (52.3 ± 2.2 percent) than nonsmokers (41.8 ± 2.6 percent). (Reprinted, by permission, from Cometto-Muñiz and Cain, 1982.)

interaction. Without some fundamental insights into the biophysics of the olfactory mucosa, cross-adaptation may continue to frustrate as a means of solving the odor quality issue psychophysically (Cain and Polak, 1990; Köster, 1971).

Common Chemical Sense

A final comment seems appropriate regarding thresholds for *nasal pungency* evoked by stimulation of the common chemical sense. Thresholds for such sensations as irritation, burning, prickling, tingling, freshness, stinging, piquancy, and the like can prove difficult to measure because of simultaneous odor sensations elicited by the chemicals employed to provoke them, and because of mutual interactions between odor and pungency (Cain and Murphy, 1980). The use of anosmic subjects (i.e., persons lacking a functional sense of smell) provides one way to address common chemical functioning (see Cometto-Muñiz and Cain, 1990). A technique that measures the *threshold* for a momentary *reflex* interruption of inhalation evoked by nasal pungency provides another way and holds promise as an *objective* index of the functional status of the nasal common chemical sense (Cometto-Muñiz and Cain, 1982; Dunn, Cometto-Muñiz, and Cain, 1982; García-Medina and Cain, 1982) (Figure 9.15). The reflex occurs well above the sensory threshold. Results from experiments on bilateral integration in the nose (García-Medina and Cain, 1982), nasal pungency responses in smokers and nonsmokers (Cometto-Muñiz and Cain, 1982), and temporal integration of nasal pungency (Cometto-Muñiz and Cain, 1984) showed excellent agreement between the threshold for the reflex and the psychophysical estimate of perceived pungency, suggesting that the threshold for the reflex occurs at a criterion level of pungency.

Measurement of Perceived Odor Intensity

All the phenomena of interest at threshold reflect themselves at suprathreshold levels, where the techniques of measurement differ from those used to study absolute sensitivity. In what follows, we will ignore the topic of differential sensitivity, which has received relatively little attention (Cain, 1977), but will instead focus on magnitude of sensation.

Fechner's Law

Researchers in psychophysics have long grappled with how to measure the basic input-output function of a sensory modality. Gustav Fechner, the father of psychophysics, sought the answer by construction of a scale of cumulative just noticeable differences (jnd's) in sensation (Fechner, 1860). The jnd thereby came to be treated as a unit of perceived magnitude. To construct such a scale empirically, that is, by the actual measurement and cumulation of jnd's along the dynamic range of a modality, would prove a formidable obstacle to progress. Fechner, however, found a way around the obstacle. He accepted that the magnitude of stimulation that would prove just resolvable, that is, would give a jnd, equaled a constant fraction of the reference stimulus. Hence, if it took a one-unit change in a stimulus of a magnitude of 10 units to resolve a difference, then it would take a 10-unit change in a stimulus of 100 units, a 100-unit change in a stimulus of 1000 units, and so on.

The rule of proportionality of resolving power is known as Weber's law. Fechner showed that joint acceptance of Weber's law and of the assumption that all jnd's had equal subjective size, that is, formed a true unit of subjective magnitude, led to a simple *logarithmic rule* relating sensation to stimulus magnitude. For almost a century, Fechner's logarithmic law prevailed for characterization of input-output.

The logarithmic law stayed alive in part because a simple method of judgment, known as category rating, often tended to yield logarithmic functions (Marks, 1968). Such a simple method of judgment, such as use of a 5-point or 7-point scale, for example, saw relatively little use in work designed to understand sensory processes, but often served in practical work. Figure 9.16 offers one example of many sets of odor/irritation functions derived from screening gas warning agents by the U.S. Bureau of

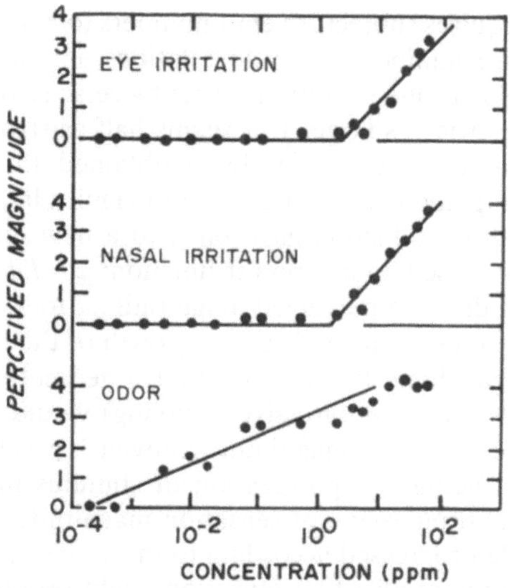

FIGURE 9.16. Psychophysical functions, derived from category scaling, for eye irritation, nasal irritation, and odor of benzyl mercaptan. (Reprinted, by permission, from Katz and Talbert, 1930.)

Mines (Katz and Talbert, 1930). The data illustrate that the same vapor that stimulates olfaction at low concentrations may stimulate both olfaction and the irritation sense, often called the common chemical sense (Cain, 1990), at higher concentrations. These results also reveal that both odor intensity and irritation intensity functions obtained through category rating can be described by logarithmic functions, but that irritation functions are much steeper than odor functions.

Ratio Scaling and the Power Law

About 40 years ago, category scaling came under attack, as did Fechner's law again, with the introduction of what are commonly called ratio scaling techniques. In the terminology of measurement theory, scales formed from category ratings fall into a class called interval scales. Such scales, exemplified in physical science by the Celsius and Fahrenheit scales of temperature, possess no true zero and do not allow statements regarding the ratios of scale values. On a seven-point category scale, a judgment of six does not equal twice as much as a judgment of three. Its interval properties do, however, allow such statements as that a scale value of six falls the same distance from a scale value of four as does a scale value of two.

The introduction of ratio scaling techniques by S.S. Stevens (e.g., 1957) and his colleagues at the Harvard Laboratory of Psychoacoustics injected the discipline of psychophysics with considerable life and with unparalleled activity. The most commonly used ratio judgment technique, called magnitude estimation, requires subjects to emit numbers to match sensations and to preserve in those numbers the ratio relations among the impressions (S.S. Stevens, 1956). If one sensation seems twice as strong as another, it deserves a number twice as large. If it seems half as strong, it deserves a number half as large. And so on. Data obtained through magnitude estimation and companion techniques called magnitude production, ratio estimation, and ratio production gave birth to a new formulation of the psychophysical law, based upon a power function: $\varphi = k\phi^{\beta}$, where ϕ refers to physical magnitude, φ to perceived magnitude, k to a constant of proportionality, and β to the shape and rate of growth of the function.

Whereas the logarithmic law implied that a geometric progression of stimulus magnitudes (e.g., progressive doubling) would lead to an arithmetic progression of sensation magnitudes, Stevens's psychophysical power law implied that a geometric progression of stimulus magnitudes would lead to a geometric progression of sensation magnitudes. For example, if stimulus magnitude progressed according to the series 1, 2, 4, 8, and 16, a logarithmic law would predict that sensation might progress in the following manner: 0.5, 1.0, 1.5, 2.0, and 2.5, that is, each multiple of stimulus magnitude might yield a constant half-unit increment in sensation. The power law would predict, however, that sensation would progress according to equal percentage increments in magnitude. In the case of a square-root relation, for example, sensation would grow by 41 percent for each doubling of stimulus magnitude. Hence, for the series above, sensation would progress: 0.5, 0.7, 1.0, 1.41, and 2.0.

A square-root relation may hold in some cases, but in actuality the exponent can have any value and varies substantially from one sensory continuum to another. For loudness, or sound pressure level, for instance, it equals two-thirds, and for brightness, it equals one-third. To some degree, an exponent will vary with the frequency of a sound and the wavelength of a light. In olfaction, the exponent of the power function commonly varies from one stimulus to another (Figure 9.17) (Cain, 1969). It may vary from under 0.10 to over 0.70. It virtually always falls below 1.0, which implies a compression of sensation magnitude over stimulus magnitude (Cain and Moskowitz, 1974). Often the compression is very marked, that is, the exponents are very low. As examples, let us consider exponents of 0.7 and 0.1. In the former case, a tenfold increase in concentration would lead to a fivefold increase in perceived odor intensity. In the latter case, the tenfold increase in concentration would lead to just a 25 percent increase in perceived odor intensity. Patte and colleagues (1975) have related variation of the exponent to various physicochemical properties.

Temporal Integration and Adaptation

Phenomena measured quantitatively at the threshold level often occur over a different time scale above the threshold. Whereas temporal integration

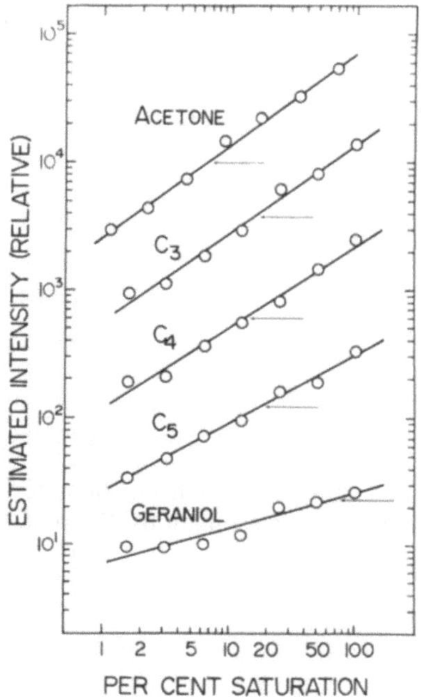

FIGURE 9.17. Magnitude estimation as a function of concentration for five odorants. The functions have been displaced in the vertical direction for clarity. The arrows indicate points of equal intensity across odorants and provide the means to relate the functions to one another. (Reprinted, by permission, from Cain, 1969.)

occurred over intervals up to about 2 s at the threshold, it occurs over intervals as long as 6 s above the threshold (Figure 9.18). Critical durations of such length occur infrequently in the sensory domain (de Wijk, 1989; von Békésy, 1964). For the common chemical sense, suprathreshold temporal integration can occur even over durations as long as an hour (Cain, See, and Tosun, 1986; Cain, Tosun, See, and Leaderer, 1987; Cometto-Muñiz and Cain, 1984).

Phenomena such as adaptation also hold as much or even more interest at suprathreshold levels as at the threshold. Ratio scaling techniques and the coincident formulation of the psychophysical power law also led to a coherent way to describe phenomena such as adaptation. Various investigators have asked how adaptation to a particular concentration of an odorant transforms the the psychophysical function for odor intensity.

Figure 9.19 gives an example of families of adaptation functions for two odorants with very similar earthy-chocolate odor qualities (Cain and Polak, 1991). The uppermost functions in each set describe perceived magnitude when subjects were adapted to room air before they judged the various test stimuli. The functions fitted to the data conform to a generalized version of the psychophysical power law, $\varphi = k(\phi - \phi_0)^{\beta}$, where ϕ_0 is

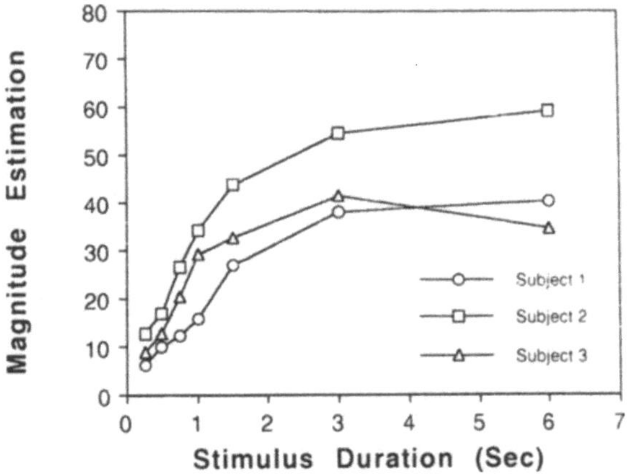

FIGURE 9.18. Individual temporal integration functions for *n*-hexane. Each point represents the mean of 50 magnitude estimations. From de Wijk, 1989.

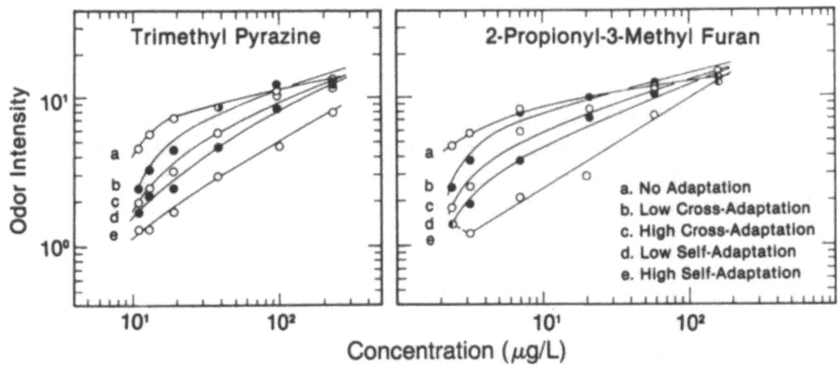

FIGURE 9.19. Families of psychophysical functions for the odor intensity of trimethyl pyrazine (TMP) and 2-propionyl-3-methyl furan (PMF) under two conditions of self-adaptation (low and high adapting concentrations) and two conditions of cross-adaptation (low and high) of one substance by the other. (Reprinted, by permission, from Cain and Polak, 1991.)

an estimated constant. With subjects adapted to a concentration of the test odorant (self-adaptation) before smelling the test stimuli, perceived magnitude fell, more in the case of adaptation to a stronger than to a weaker concentration. With subjects adapted to a concentration of the other odorant (cross-adaptation), perceived magnitude also fell, though not quite so severely as with self-adaptation. Severity of adaptation, generally more evident at low than at high test concentrations, revealed itself largely by increases in the exponents of the psychophysical functions. As Figure 9.20 shows, comparable conditions of self-adaptation and cross-adaptation led

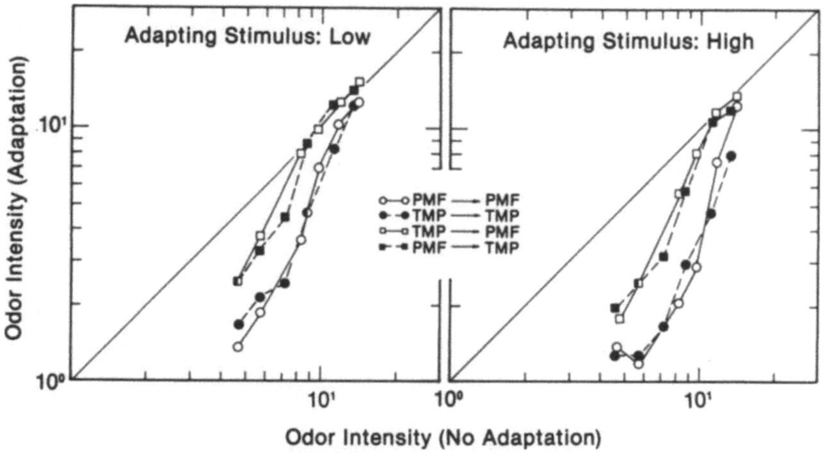

FIGURE 9.20. Same data as in Figure 9.19, but plotted as perceived odor intensity of TMP and PMF after odor adaptation vs. perceived odor intensity after adaptation with air. (Reprinted, by permission, from Cain and Polak, 1991. Copyright by Oxford University Press.)

to almost identical changes in perceived magnitude from the one odorant to the other. In this case, the data suggested that adapting stimuli matched in perceived intensity will produce equal degrees of self-adaptation and, at least in very similar smelling substances, equal degrees of mutual cross-adaptation.

Gender, Aging, and Smoking

A technique called magnitude matching, devised to compare absolute values of perceived magnitude across groups of subjects, has rather recently entered the psychophysical repertoire (J.C. Stevens and Marks, 1980). It has been used in the study of olfaction and nasal pungency in relation to aging (J.C. Stevens and Cain, 1986, 1987), gender (Cometto-Muñiz and Noriega, 1985), and smoking (Cometto-Muñiz and Cain, 1982).

The magnitude matching procedure requires subjects to assign numbers to perceived intensity in more than one sensory modality in a given session. Subjects are instructed to use a common scale across modalities. One modality serves as a reference. In the study of the influence of aging on the sense of smell, the taste modality has often served as reference because aging has little or no influence on perceived taste intensity. Subjects have judged both saltiness and the odor intensity of interest. Figure 9.21 shows functions for the odor intensity of isoamyl butyrate (Cain and J.C. Stevens, 1989). The vertical positions of the functions reflect odor intensity after normalization to taste intensity. The technique has consistently found that the threshold difference between young and elderly subjects translates itself rather uniformly up the concentration scale, so that elderly subjects

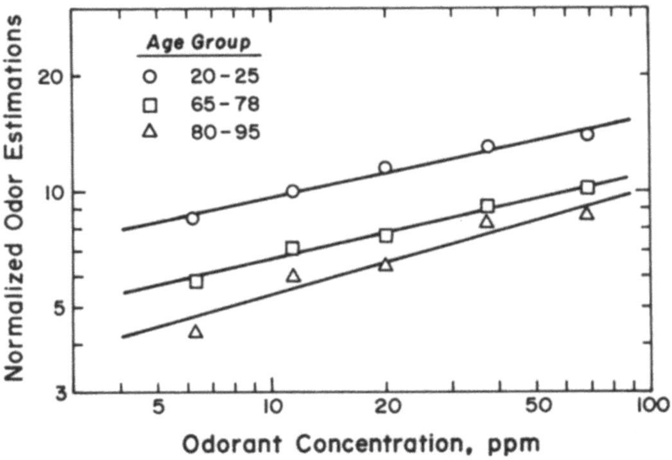

FIGURE 9.21. Magnitude matching functions for odor intensity of isoamyl butyrate in three age groups. The functions were normalized according to intensity of the salty taste of sodium chloride in the three groups (see text). (Reprinted, by permission, from Cain and J.C. Stevens, 1989.)

find both strong and weak odors comparably weakened. A similar parallel shift in the functions also characterizes nasal pungency perceived by young and elderly (J.C. Stevens, Plantinga, and Cain, 1982; J.C. Stevens and Cain, 1986), by nonsmokers and smokers, with smokers finding nasal pungency weaker than nonsmokers (Cometto-Muñiz and Cain, 1982), and by females and males, with males finding nasal pungency weaker than females (Cometto-Muñiz and Noriega, 1985).

Acknowledgment. Preparation supported by NIH Grant DC00284. Dr. Cometto-Muñiz is a member of the Carrera del Investigador Científico, Conseso Nacional de Investigaciones Científicas y Técnicas (CONICET), República Argentina.

References

Amoore, J.E. (1982). Odor theory and odor classification. In E.T. Theimer (Ed.), *Fragrance Chemistry: The Science of the Sense of Smell.* New York: Academic, pp. 27–76.

Amoore, J.E., and D. Venstrom (1966). Sensory analysis of odor qualities in terms of the stereochemical theory. *J. Food Sci.*, *31*, 118–128.

Amoore, J.E., and B. Ollman (1983). Practical test kits for quantitatively evaluating the sense of smell. *Rhinology*, *21*, 49–54.

Berglund, B., L. Hogman, and I. Johansson (1988). Reliability of odor measurements near threshold. *Reports from the Department of Psychology.* University of Stockholm, paper #682, *November.*

Brown, K.S., C.M. Maclean, and R.R. Robinette (1968). Sensitivity to chemical odors. *Human Biol.*, *40*, 456–472.

Cain, W.S. (1969). Odor intensity: Differences in the exponent of the psychophysical function. *Percept. Psychophys.*, *6*, 349–354.

Cain, W.S. (1977). Bilateral interaction in olfaction. *Nature*, *268*, 50–52.

Cain, W.S. (1978). History of research on smell. In E.C. Carterette and M.P. Friedman (Eds.), *Handbook of Perception*, Vol. 6A, *Tasting and Smelling*. New York: Academic, pp. 197–229.

Cain, W.S. (1988). Olfaction. In R. Atkinson, R. Herrnstein, G. Lindzey, and R. Luce (Eds.), *Stevens' Handbook of Experimental Psychology*. New York: Wiley, pp. 409–460.

Cain, W.S. (1989). Testing olfaction in a clinical setting. *Ear, Nose Throat J.*, *68*, 316–328.

Cain, W.S. (1990). Perceptual characteristics of nasal irritation. In B.G. Green, J.R. Mason, and M.R. Kare (Eds.), *Chemical Senses: Irritation*. New York: Marcel Dekker, pp. 43–58.

Cain, W.S., and H.R. Moskowitz (1974). Psychophysical scaling of odors. In A. Turk, J.W. Johnston, Jr., and D.G. Moulton (Eds.), *Human Responses to Environmental Odors*. New York: Academic, pp. 2–32.

Cain, W.S., and C.L. Murphy (1980). Interaction between the chemoreceptive modalities of odour and irritation. *Nature*, *284*, 255–257.

Cain, W.S., and B.P. Leaderer (1982). Ventilation requirements in occupied spaces during smoking and non-smoking occupancy. *Environ. Int.*, *8*, 505–514.

Cain, W.S., J. Gent, F.A. Catalanotto, and R.B. Goodspeed (1983). Clinical evaluation of olfaction. *Am. J. Otolaryngol.*, *4*, 252–256.

Cain, W.S., B.P. Leaderer, R. Isseroff, L.G. Berglund, R.J. Huey, E.D. Lipsitt, and D. Perlman (1983). Ventilation requirements in buildings: I. Control of occupancy odors and tobacco smoke odor. *Atmospheric Environ.*, *17*, 83–97.

Cain, W.S., L.C. See, and T. Tosun (1986). Irritation and odor from formaldehyde: Chamber Studies. *IAQ '86: Managing Indoor Air for Health and Energy Conservation*. Atlanta: American Society of Heating, Refrigerating and Air-Conditioning Engineers, pp. 126–137.

Cain, W.S., T. Tosun, L.C. See, and B.P. Leaderer (1987). Environmental tobacco smoke: Sensory reactions of occupants. *Atmospheric Environ.*, *21*, 347–353.

Cain, W.S., J.F. Gent, R.B. Goodspeed, and G. Leonard (1988). Evaluation of olfactory dysfunction in the Connecticut Chemosensory Clinical Research Center. *Larynoscope*, *98*, 83–88.

Cain, W.S., and J.C. Stevens (1989). Uniformity of olfactory loss in aging. *Ann. NY Acad. Sci.*, *561*, 29–38.

Cain, W.S., and J.F. Gent (1991). Olfactory sensitivity: Reliability, generality, and association with aging. *J. Exp. Psychol.: Hum. Percept. Perform.*, *17*, 382–391.

Cain, W.S., and E.H. Polak (1991). Olfactory adaptation as an aspect of odor similarity. *Chem. Senses*.

Clausen, G.H., P.O. Fanger, W.S. Cain, and B.P. Leaderer (1986). Stability of body odor in enclosed spaces. *Environ. Int.*, *12*, 201–205.

Cometto-Muñiz, J.E., and W.S. Cain (1982). Perception of nasal pungency in smokers and nonsmokers. *Physiol. Behav.*, *29*, 727–731.

Cometto-Muñiz, J.E., and W.S. Cain (1984). Temporal integration of pungency. *Chem. Senses*, *8*, 315–327.

Cometto-Muñiz, J.E., and G. Noriega (1985). Gender differences in the perception of pungency. *Physiol. Behav.*, *34*, 385–389.

Cometto-Muñiz, J.E., and W.S. Cain (1990). Thresholds for odor and nasal pungency. *Physiol. Behav.*, *48*, 719–725.

de Wijk, R.A. (1989). Temporal Factors in Human Olfactory Perception. Unpublished Doctoral thesis, University of Utrecht, Netherlands.

Doty, R.L., T.P. Gregor, and R.G. Settle (1986). Influence of intertrial interval and sniff-bottle volume on phenyl ethyl alcohol odor detection thresholds. *Chem. Senses*, *11*, 259–264.

Dravnieks, A. (1975). Instrumental aspects of olfactometry. In D.G. Moulton, A. Turk, and J.W. Johnston Jr. (Eds.), *Methods in Olfactory Research.* New York: Academic, pp. 1–58.

Dravnieks, A. (1977). Correlation of odor intensities and vapor pressures with structural properties of odorants. In R.A. Scanlan (Ed.), *Flavor Quality: Objective Measurement.* Washington, DC: American Chemical Society, pp. 11–28.

Dravnieks, A., and W.H. Prokop (1975). Source emission odor measurement by a dynamic forced-choice triangle olfactometer. *J. Air Pollution Control Assoc.*, *25*, 28–35.

Dravnieks, A., W.H. Prokop, and W.R. Boehme (1978). Measurement of ambient odors using dynamic forced-choice triangle olfactometer. *J. Air Pollution Control Assoc.*, *28*, 1124–1130.

Dunn, J.D., J.E. Cometto-Muñiz, and W.S. Cain (1982). Nasal reflexes: Reduced sensitivity to CO_2 irritation in cigarette smokers. *J. Appl. Toxicol.*, *2*, 176–178.

Elsberg, C.A., and I. Levy (1935). The sense of smell. A new and simple method of quantitative olfactometry. *Bull. Neurol. Inst. NY*, *4*, 5–19.

Fazzalari, F. (1978). *Compilation of Odor and Taste Threshold Values Data.* Philadelphia: American Society for Testing and Materials.

Fechner, G.T. (1860). *Elemente der Psychophysik*, Vols. I and II. Leipzig: Breitkopf & Haertel.

García-Medina, M.R., and W.S. Cain (1982). Bilateral integration in the common chemical sense. *Physiol. Behav.*, *29*, 349–353.

Hubert, H.B., R.R. Fabsitz, M. Feinleib, and K.S. Brown (1980). Olfactory sensitivity in humans: Genetic versus environmental control. *Science*, *208*, 607–609.

Katz, S.H., and E.J. Talbert (1930). Intensities of odors and irritating effects of warning agents for inflammable and poisonous gases. *US Department of Commerce, paper #480*, 1–37.

Kobal, G. (1981). *Elektrophysiologische Untersuchungen des Menslichen Geruchsinns (Electrophysiological Investigation of the Human Sense of Smell).* New York: Thieme-Copythek.

Koelega, H.S. (1970). Extraversion, sex, arousal and olfactory sensitivity. *Acta Psychol.*, *34*, 51–66.

Koelega, H.S., and E.P. Köster (1974). Some experiments on sex differences in odor perception. *Ann. NY Acad. Sci.*, *237*, 234–246.

Köster, E.P. (1967). Olfactometre modifié et automatisé. *Parf. Cosm. Sav.*, *10*, 509–513.

Köster, E.P. (1971). Adaptation and Cross-Adaptation in Olfaction. Unpublished Doctoral thesis, University of Utrecht, The Netherlands.

Köster, E.P. (1975). Human psychophysics in olfaction. In D.G. Moulton, A. Turk, and J.W. Johnston Jr. (Eds.), *Methods in Olfactory Research.* New York: Academic, pp. 345–374.

Laffort, P. (1969). A linear relationship between olfactory effectiveness and identified molecular characteristics, extended to fifty pure substances. In C. Pfaffmann (Ed.), *Olfaction and Taste III*. New York: Rockefeller University Press, pp. 150–157.
Laffort, P., F. Patte, and M. Etcheto (1974). Olfactory coding on the basis of physicochemical properties. *Ann. NY Acad. Sci.*, *237*, 193–208.
Laing, D.G. (1982). Characterization of human behaviour during odour perception. *Perception*, *11*, 221–230.
Laing, D.G. (1983). Natural sniffing gives optimum odour perception for humans. *Perception*, *12*, 99–117.
LeMagnen, J. (1952). Les phénomenes olfacto-sexuels chez l'homme. *Arch. Sci. Phys.*, *6*, 125–160.
Marks, L.E. (1968). Stimulus-range, number of categories, and the form of the category scale. *Am. J. Psychol.*, *81*, 467–479.
Marks, L.E., and J.C. Stevens (1968). The form of the psychophysical function near threshold. *Percept. Psychophys.*, *4*, 315–318.
Martin, S., and R.M. Pangborn (1970). A note on responses to ethyl alcohol before and after smoking. *Percept. Psychophys.*, *8*, 169–170.
Matzker, J. (1965). Riechen und Lebensalter—Riechen und Raucher. *Urnschan*, *65*, 673.
Moulton, D.G., G. Celebi, and R.P. Fink (1972). Olfaction in mammals—two aspects: proliferation of cells in the olfactory epithelium and sensitivity to odours. In G.E.W. Wolstenholme and J. Knight (Eds.), *Taste and Smell in Vertebrates*. London: Churchill, pp. 227–250.
Mullins, L.J. (1955). Olfaction. *Ann. NY Acad. Sci.*, *62*, 247–276.
Overbosch, P., R.A. de Wijk, T.J.R. de Jonge, and E.P. Köster (1989). Temporal integration and reaction times in human smell. *Physiol. Behav.*, *145*, 615–626.
Pangborn, R.M., H.W. Berg, E.B. Roessler, and A.D. Webb (1964). Influence of methodology on olfactory response. *Perceptual and Motor Skills*, *18*, 91–103.
Patte, F., M. Etcheto, and P. Laffort (1975). Selected and standardized values of suprathreshold odor intensities for 110 substances. *Chem. Senses Flav.*, *1*, 283–305.
Punter, P.H. (1983). Measurement of human olfactory thresholds for several groups of structurally related compounds. *Chem. Senses*, *7*, 215–235.
Rabin, M.D., and W.S. Cain (1986). Determinants of measured olfactory sensitivity. *Percept. Psychophys.*, *39*, 281–286.
Sanders, G.R., R.A. Umbraco, S.A. Twiss, and P.K. Mueller (1970). *The Measurement of Malodor in a Community by Dynamic Olfactometry*, presented at Conference on Methods for Measuring and Evaluating Odorous Air Pollutants at the Source and in the Ambient Air, Stockholm.
Schiet, F.T., and W.S. Cain (1990). Odor intensity of mixed and unmixed stimuli under environmentally realistic conditions. *Perception*, *19*, 123–132.
Stevens, J.C., and L.E. Marks (1980). Cross-modality matching functions generated by magnitude estimation. *Percept. Psychophys.*, *27*, 379–389.
Stevens, J.C., A. Plantinga, and W.S. Cain (1982). Reduction of odor and nasal pungency associated with aging. *Neurobiol. Aging*, *3*, 125–132.
Stevens, J.C., and W.S. Cain (1986). Aging and the perception of nasal irritation. *Physiol. Behav.*, *37*, 323–328.
Stevens, J.C., and W.S. Cain (1987). Old-age deficits in the sense of smell gauged by thresholds, magnitude matching, and odor identification. *Psychol. Aging*, *2*, 36–42.

Stevens, J.C., W.S. Cain, and D.E. Weinstein (1987). Aging impairs the ability to detect gas odor. *Fire Technology*, *23*, 198–204.

Stevens, S.S. (1956). The direct estimation of sensory magnitudes: loudness. *Am. J. Psychol.*, *29*, 1–25.

Stevens, S.S. (1957). On the psychophysical law. *Psychol. Rev.*, *64*, 153–181.

Stuiver, M. (1958). *Biophysics of the Sense of Smell.* Unpublished Doctoral thesis, University of Groningen, The Netherlands.

van Gemert, L.J., and A.H. Nettenbreijer (1977). *Compilation of Odour Threshold Values in Air and Water.* Voorburg, Netherlands: National Institute for Water Supply.

Venstrom, D., and J.E. Amoore (1968). Olfactory threshold in relation to age, sex or smoking. *J. Food Sci.*, *33*, 264–265.

von Békésy, G. (1964). Olfactory analogue to directional hearing. *J. Appl. Physiol.*, *19*, 369–373.

Wenzel, B.M. (1948). Techniques in olfactometry: A critical review of the last one hundred years. *Psychol. Bull.*, *45*, 231–247.

Woodrow, H., and B. Karpman (1917). A new olfactometric technique and some results. *J. Exp. Psychol.*, *2*, 431–447.

10 Stimulus Selection in the Design and Interpretation of Olfactory Studies

Avery N. Gilbert and Mark S. Greenberg

Introduction

Olfactory perception is one of the last great frontiers of contemporary neuroscience. While our understanding of receptor activation and stimulus transduction has expanded greatly in recent years, the central processing of olfactory information—its encoding, transformation, and intermodal integration—is known schematically at best.

Nevertheless, there is little doubt that the neurophysiological basis of odor perception will become increasingly clarified. Less certain is whether a neurophysiological account by itself will provide a satisfying explanation of the most characteristic phenomenological aspects of odor perception (Gilbert and Kare, 1991). Individual differences in odor hedonics, the ontogeny of odor preferences, the accessibility of odor memory, and the waning of odor sensitivity with normal and abnormal aging are central to our conception of olfaction.

Chief among the experiential features of olfaction is odor quality, and the closely allied topic of odor hedonics. Unlike gustation, which can be decomposed into four primary taste qualities whose neurophysiological and developmental bases can be traced independently, olfaction has resisted attempts at odor quality categorization.

Does the lack of consensus on odor primaries reveal a fundamental fact about smell function? Or is it a reflection of empirical techniques that have been at best driven by unsupported theory, and at worst simply arbitrary? How do experimentalists conceptualize the relations between odors? How do they determine which are appropriate to the experiment at hand? Has the choice of odor stimuli shaped the nature of the results? A survey of the recent experimental literature was undertaken to determine whether broad answers to these questions exist. As more investigators from other disciplines enter the olfactory field, they are faced with the problem of selecting appropriate stimuli. Another aim of this review was to provide them with a compendium of commonly used odors and the rationale behind them. The results of this overview are presented below, followed by a discussion of the trends found, and finally by the ways in which these trends impinge on the design and interpretation of odor experiments.

Survey of the Literature

Scope of the Survey

Experimental studies on olfaction were selected from six areas of the literature, spanning clinical and basic research studies and both normal and patient populations. The topic areas were children, memory, aging and dementia, other diseases and disorders, brain damage, and normal subjects. Studies that assessed only detection thresholds were not included. The selected studies span the period from the early 1970s through 1989. The rationale for stimulus selection, if provided by the original authors, is reported here. The odor characterizations in parentheses that follow the chemical names of stimuli are those provided by the original authors.

The studies reviewed here are neither a random nor an exhaustive catalogue of human olfactory research. Rather, they were chosen to be representative of work taking place in fields that use different subject populations and methodologies. The use of six fields reduces the impact that a bias in study selection in any one field might have on the survey as a whole.

Studies on Children

Self, Horowitz, and Paden (1972) examined response to four odors in newborn infants. They used anise, lavender, asafetida, and valerian, with water as a control. Asafetida and anise were chosen in order to replicate earlier findings by other authors; lavender and valerian were used to replace what the authors considered to be the trigeminally active odorants used previously.

Balogh and Porter (1986) tested olfactory preferences in neonates resulting from mere exposure to odors. They used two odors (cherry and white ginger) based on the authors' previous finding that infants showed no difference in responsiveness to these two smells.

Schmidt and Beauchamp (1988) examined odor hedonics in three-year-old children with nine odorants "selected to represent a wide range of odor qualities and hedonic values." They used C-16 aldehyde (strawberry), phenyl ethyl methyl ethyl carbinol (PEMEC) (floral), L-carvone (spearmint), methyl salicylate (wintergreen), eugenol (cloves), amyl acetate (banana), butyric acid (strong cheese/vomit), pyridine (spoiled milk), and androstenone (variable: sweaty, sandalwood, or no odor).

Strickland, Jessee, and Filsinger (1988) examined the ability of two- to five-year-old children to discriminate between pleasant and unpleasant odors. A training procedure used vanillin and butyric acid, and the actual test odors were benzaldehyde (almond) and dimethyl disulfide (sweaty). The rationale for selection was that in a previous study one of the authors found these to be the most and least preferred odors chosen by children.

Fabes and Filsinger (1986) looked at the relative contributions of odor and visual cues in the preferences of children aged three to five years. The four stimuli used were α-androstenone, dimethyl disulfide, benzaldehyde, and no odor. The odors were selected because they had been "successfully and safely used in previous research with adults" and because "they represent a broad range of odorants that differ in psychophysical qualities."

Laing and Clark (1983) studied the olfactory preferences of pre-, mid-, and postpubertal males. They tested the odors of 10 foods commonly consumed in Australia (the test country): roast beef, fish, chocolate, onion, Vegemite™, peanut butter, spearmint, chicken, coffee, and Parmesan cheese. The stimuli were intended to include staple food items as well as some known to elicit divided opinion (e.g., coffee).

Koelega and Köster (1974) examined sex differences in odor sensitivity among prepubescents, pubescents, and adults, using a forced choice procedure (three blank and one odorous flask). In various replications of the test, the stimuli were amyl acetate, pentadecanolide, oxahexadecanolide, pyridine, safrol, *m*-xylene, and seven androsten- compounds. The two -decanolides were described as "biological odors" (musk), and the amyl acetate and safrol as "neutral odors."

Summary

In the studies of children reviewed in this section, the selection rationale for odor stimuli frequently involved previous findings, such as the odors having produced equal responsiveness in subjects (Balogh and Porter, 1986) or extremes in response (Strickland et al., 1988). Earlier work was also cited in attempts at replication (Self et al., 1972) and for the previous safe and successful use of odors (Fabes and Filsinger, 1986).

Three studies made use of relatively large (≥9) stimulus sets. Of these, one was intentionally composed entirely of food-related odors (Laing and Clark, 1983), while another contained 78 percent food-related stimuli, even though the authors characterized the set as covering a "wide range" of odor quality and hedonic values (Schmidt and Beauchamp, 1988). The third study (Koelega and Köster, 1974) contained only 15 percent food-related odors, but the two stimuli the authors characterized as "neutral" were both food-related.

Studies on Memory

Engen, Kuisman, and Eimas (1973) studied short-term recognition memory using a set of 100 "diverse odorants" that represented "the odors of familiar household products such as garlic, vanilla, and floorwax; chemical compounds such as acetates, alcohols, and aldehydes; perfumed products such as neroli oil and Arpège perfume." A complete list of specific items was not provided.

Desor and Beauchamp (1974) tested naive subjects' ability to identify "whole odors of common, odorous objects." The 32 stimuli were coffee, beer, liver, paint, white bread, tomatoes, banana, human urine, crayons, chocolate, cheddar cheese, cat feces, cinnamon, sauerkraut, cigar, rubber, onion, honey, ham, rubbing alcohol, rose, pine sawdust, coconut, tunafish, mothballs, green pepper, vinegar, motor oil, leather, pine needles, popcorn, and cigarette butts. In a second experiment, the authors used an enlarged odor set of 64 stimuli, which included 38 new stimuli, of which 24 were food items.

Davis (1975) performed paired-associate learning studies comparing the acquisition of odors versus abstract figures to number responses. In a first experiment, the odorants were ethyl alcohol, amyl acetate, methyl salicylate, and cyclohexanone. They were chosen because they "would be commonly recognized as odors similar to liquor, bananas, wintergreen candy, and almond extract." In a second experiment, odorants were "selected based principally on their appearance in the work of a large number of other investigators. The combination of physical attributes such as stability, vapor pressure, molecular rigidity, as well as availability in highly refined grades, received secondary emphasis." Only 12 of the 20 test stimuli in this experiment were identified when they were used as stimuli in a third experiment: ethyl acetate, xylene, eugenol, methyl salicylate, safrole, citral, cyclopentanone, cyclohexanone, benzaldehyde, octyl acetate, heptyl acetate, and cyclohexanol.

Lawless and Cain (1975) used recognition tests to examine decay curves of olfactory memory. The odorants were a set of 22 compounds "chosen for distinctiveness and dissimilarity" and matched for perceptual intensity. They were musk oil, turpentine, butanone, eugenol (cloves), benzaldehyde, bourbon, salicaldehyde, isobutyric acid, isopropyl alcohol, Eau de Lanvin, ethyl-*N*-butylamine, phenyl ethyl acetate, lavandin oil, *p*-menthan-3-ol (menthol), vinegar, methyl salicylate (wintergreen), piperonal, 1-butanol, lemon oil, trimethylamine, pyridine, and amyl acetate.

Lawless (1978) compared memory decay functions for odors, simple shapes, and complex pictures. The 24 stimuli chosen were "common household odors," as one purpose of the study was to determine whether such odors displayed an advantage in recognition memory tasks. The stimuli were coffee, paint, honey, chocolate, aftershave, Pine-Sol, maple syrup, beer, cat litter, mothballs, peanut butter, hand soap, cheddar cheese, fish, lemon, Clorox bleach, cigars, vinegar, rye whiskey, cinnamon, perfume, onion, oregano, and shoe polish.

Eich (1978) investigated fragrances as cues for remembering words, and how verbal associations mediate retrieval. A total of 12 odors was used:

leather, pine, cinnamon, cherry, lime, cola, coconut, orange, pickle, perfume, chocolate, and spearmint. The 12 stimuli were selected from a larger set that was experimentally prescreened to generate correlation coefficients relating a noun with a fragrance cue. Half of the selected odors had high fragrance-to-word associations, and half had low associations.

Cain (1979) examined how the labeling and naming of odors modulates odor identification. The stimuli were 80 "commonly encountered, 'ecologically valid' substances." The substances were organized into categories: meats, fish, fruits, spices, snacks, condiments, beverages, confections, household products, medical products, raw materials, personal products, other. Specific items were not listed.

Cain and Krause (1979) conducted three experiments that demonstrated that typically poor performance on olfactory confrontation naming tasks will improve given the use of highly familiar stimuli and the use of certain test procedures. In experiment 1, 12 odors were chosen from a larger set of 80 "commonly encountered substances (foods, household products, etc.)," based on the following criteria: stable, highly identifiable, and highly familiar. The test stimuli selected were Johnson's baby powder, Juicy Fruit chewing gum, Baker's unsweetened chocolate, cinnamon, cloves, ground coffee, leather, mothballs, peanut butter, potato chips, Ivory bar soap, and wintergreen. In a second experiment, 11 of these odors and an additional 8 commercially available microencapsulated odors were included: banana, carnation, black cherry, chocolate, dill pickle, grape, lemon, and mint. The grounds for selection were that the "experimenters exercised their own judgment."

Rabin and Cain (1984) used the ability to name an odor as a means of investigating the association between recognition performance and adequacy of memory encoding. The odor stimuli were 80 "common substances selected for diversity of odor quality" and were divided into a target set and three distractor sets of 20 items each. The stimuli were almond extract, ammonia, apple, baby powder, banana, Band-Aid, barbecue sauce, beer, black pepper, bleach, bologna, bubble gum, burnt paper, burnt toast, bourbon, Brut aftershave, caramel, cat food (dry), cheese, chocolate, cigarette, cigar butt, clay, cleaning fluid, cinnamon, cloves, coffee, cough syrup, crayons, coconut, garlic, ginger, grape drink, herbal shampoo, honey, horseradish, ketchup, fruit gum, leather, lemon, lighter fluid, liverwurst, Lysol cleaner, maple syrup, molasses, mothballs, motor oil, mustard, nail polish remover, nutmeg, onion, orange extract, oregano, peanut butter, pencil shavings, peppermint extract, perfume, pine shavings, pipe tobacco, popcorn, potato chips, prune (dried), raspberry, rubber, rubbing alcohol, sardines, sherry, shoe polish, soap (bar), soy sauce, spearmint gum, strawberry, tea leaves, tuna, turpentine, vanilla extract, wintergreen extract, varnish, Vicks, and vinegar.

Lyman and McDaniel (1986) studied the effects of different encoding strategies on the long-term memory for odors. Their stimuli were those used previously by Rabin and Cain (1984), although these were arranged into a 30-item target (acquisition) set and a single 30-item distractor set.

Walk and Johns (1984) studied interference and facilitation effects on short-term memory for odors. They used 36 common food substances organized into six categories: condiments, fruit, herbs and spices, spices, sweets, and vegetables. As a rationale for stimulus selection, the authors stated that "food substances were used rather than chemical entities so that subjects would be familiar with at least their names."

Barker and Weaver (1983) had subjects estimate the intensity of a target smell, then, following a delaying interval, pick the target from a five-step dilution series of the same odor (pyridine).

Ehrlichman (1987) explored different methods of unilateral odor presentation to investigate the hemispheric lateralization of positive and negative affect. He used four odor stimuli selected to be pleasant (almond, orange) and unpleasant (thiophine, pyridine).

Ehrlichman and Halpern (1988) used pleasant and unpleasant odors to alter the current mood of subjects while asking for recall of a personal memory relating to affectively neutral words. The two odors were almond extract and pyridine (an unscented air stream was used as a control). The particular odors and concentrations "had been selected after pretesting revealed them to be reliably rated as pleasant and unpleasant, respectively, under the conditions of this study." Ehrlichman and Halpern note that odor makes a good paradigm for their work because of the hedonic aspect of its experience, which is far less studied than the physiological and behavioral aspects. Note that one of the two test odors was food-related.

Cann and Ross (1989) examined the effects of odor stimuli on memory processes and social judgments. Subjects were exposed to ambient odor during learning sessions and tested for recognition in the presence of the original or an alternative odor. Two ambient odors were used to "create a pleasant or an unpleasant olfactory context." Island Gardenia™ (Jovan) was selected based on its similarity to the fragrance used by Baron (1981) to investigate odor influences on social perceptions. Ammonium sulfide was selected based on its use as an unpleasant olfactory stimulus in a study of air pollution and aggression (Rotton et al., 1978).

Summary

From the studies reviewed in this section, it is clear that experiments on odor memory generally involve substantial numbers (≥12) of stimuli. A frequent criterion for selection is that the odors be those of "common" substances (Desor and Beauchamp, 1974; Lawless, 1978; Cain, 1979; Rabin and Cain, 1984) or that they be "familiar" (Cain and Krause, 1979; Walk and Johns, 1984). Yet only a single study reported quantitative prescreening to determine the degree of association between potential stimuli and their corresponding substances (Eich, 1978). In the remaining studies, the perception of commonness or familiarity was presumably, and in one instance expressly (Cain and Krause, 1979), left to the judgment of the experimenter. Examination of the actual odors used in these studies gives the impression that common and familiar is equated with food-related. From 50 to 70 percent of the odors used fell into this category [excluding the study of Walk and Johns (1984), which made use exclusively of food substances]. This bias may not have materially affected the conclusions to narrowly drawn experimental hypotheses. Yet broader speculations such as Cain's (1979) on the relation of odor nameability to memory may have been compromised (63 percent of his stimulus categories were food-related). Of the 80 items used in a subsequent study (Rabin and Cain, 1984), 51 (64 percent) were food-related.

Three memory studies used relatively few (≤4) stimuli. They were concerned with the role of affect on odor memory, and the stimuli were selected to represent a pleasant versus unpleasant dichotomy. In two instances (Ehrlichman, 1987; Ehrlichman and Halpern, 1988) the pleasant category was composed entirely of food-related odors; in the third it was a floral perfume (Cann and Ross, 1989).

Studies on Aging and Dementia

Waldton (1974) studied impaired cranial nerve function in senile dementia. He reported qualitative, longitudinal observation of the subjects' ability to detect and identify six odors (asafetida, camphor, citrus fruit, Eau de Cologne, menthol, and gasoline). No rationale was provided for stimulus selection.

Knupfer and Spiegel (1986) compared normal aged subjects with others having Alzheimer's or vascular dementia. They tested olfactory thresholds with eucalyptol, citral, and prunolide. To test recognition, identification, and naming, they used 20 odors, arranged in 10 pairs (five similar and five dissimilar). Odor pairs were pretested to confirm similarity/dissimilarity. The rationale for stimulus selection was that "the smells should be familiar from everyday life and that they should not change their qualitative characteristics with increasing concentration." The pairs selected were coconut and cloves, mothballs and camphor, eucalyptus and smoked bacon, marzipan and bitter almond, aniseed and orange, grass and hay, rum and licorice, lemon and flower, asparagus and pine needles, and peppermint and spearmint.

Koss (1986) examined olfactory function in Alzheimer's disease using a discrimination paradigm. A subject was presented with a target odor, then a second odor, and asked if the second was the same or different. The author used 10 stimuli selected to represent pleasant, unpleasant, and neutral qualities. The odors were ground coffee, lemon, coconut, vanilla, mint, phenethyl alcohol (rose), Stinson cheese, decaying meat, brass polish, and water.

Wood and Harkins (1987) examined odor identification ability as a function of age, stimulus selection, and type of retrieval cue. They screened potential subjects for sensory loss with 20 odorants selected from a set used previously to test memory (Rabin and Cain, 1984) that had received positive familiarity ratings by young women (Cain, 1979). The proportion of food-related odors (55 percent) was similar to that in the Rabin and Cain (1984) set. The actual test stimuli were 20 odorants "selected to be representative of the items identified by approximately 80 older persons as remembered from their childhood," although items of more recent origin were included. The test stimuli were asafetida, cured country ham, cured bright-leaf tobacco, sour milk, lye soap, peppermint, pork chitterlings, sage, mothballs, perfume, vinegar, vanilla extract, ammonia, coffee, tuna, glue, lighter fluid, hair conditioner, bubble gum, and turpentine.

Rezek (1987) compared Alzheimer's disease patients to age-matched controls on three measures of olfactory function. Detection threshold was determined for amyl alcohol and for cinnamon oil. Odor identification and recognition were tested with five stimuli: cinnamon oil, lemon, ground coffee, peppermint, and turpentine.

Moberg, Pearlson, Speedie, Lipsey, Strauss, and Folstein (1987) tested olfactory recognition in Huntington's and Alzheimer's disease patients. Their paradigm consisted of a presentation of 10 target odors, a five-minute delay, and then a presentation of 20 odors (the 10 targets and 10 distractors, including 5 similar to targets and 5 not similar to targets). The 10 target odors were onion, vanilla, almond, peppermint, lemon, cinnamon, pine, mothball, birch tar rectified (smoke), and isoamyl acetate (banana). The distractor odors were wintergreen, clove, black pepper, orange, ethyl brassylate (musk), apple, phenyl ethyl alcohol (floral), camphor, garlic, and butyric acid (rancid butter).

Kesslak, Cotman, Chui, Van Den Noort, Fang, Pfeffer, and Lynch (1988) studied olfactory response in three patient groups (Alzheimer's disease, Parkinson's disease, and multiple sclerosis) and age-matched controls. Smell identification ability was tested with the commercially available Smell Identification Test (SIT) (Doty, Shaman, and Dann, 1984). Nonanosmic patients were given 15 match-to-sample odor items. Each item was a different target odor, followed by three choices. The rationale for stimulus selection was that at least some of the stimuli be "uncommon odors (not readily identifiable)." The actual test odors were not specifically identified.

Wysocki and Gilbert (1989) examined the relation between age and olfactory perception in an analysis of the results of the *National Geographic* Smell Survey. In that survey, approximately 1.42 million readers of the *National Geographic* sampled six microencapsulated odor samples and returned an accompanying questionnaire intended to provide demographic data. The odors used were 5α-androst-16-en-3-one, isoamyl acetate, Galaxolide^{TM}, eugenol, mercaptans, and rose. They were selected "to include both pleasant and offensive odors, familiar as well as less familiar ones, food- and non-food-related odors, complex mixtures and unitary compounds, and odorants associated with specific anosmias."

Koss, Weiffenbach, Haxby, and Friedland (1987) characterized olfactory function in Alzheimer's disease patients, using pyridine (Amoore's commercially available dilution series) for detection thresholds and the SIT for odor recognition. Koss, Weiffenbach, Haxby, and Friedland (1988) compared olfactory detection and identification in early Alzheimer's disease and healthy controls using a protocol similar to that of Koss et al. (1987).

Doty, Reyes, and Gregor (1987) examined olfactory function in Alzheimer's disease. Detection thresholds were determined with phenyl ethyl alcohol. Odor identification was assessed with the SIT. Doty, Deems, and Stellar (1988) used the same protocol in examining olfactory function in Parkinson's disease.

Hurwitz, Kopala, Clark, and Jones (1988) examined odor recognition in schizophrenics and psychiatric and normal control populations using the SIT. Kopala, Clark, and Hurwitz (1989) studied sex differences in olfactory function in schizophrenic patients and normal control subjects. Detection thresholds were determined with *n*-butanol. Olfactory recognition was measured with the SIT.

Summary

In these studies of aging and dementia based on clinical populations, a twofold approach to olfactory assessment is apparent, in which detection threshold is measured in addition to odor identification ability (the latter most often with the SIT) (e.g., Koss et al., 1987, 1988; Doty et al., 1987, 1988; Hurwitz et al., 1988; Kopala et al., 1989). Because of its prevalence as a diagnostic instrument, the SIT deserves attention. Inspection of the 40 test items (licorice, menthol, soap, turpentine, lemon, bubble gum, cinnamon, mint, cherry, orange, chocolate, lilac, coconut, smoke, fruit punch, watermelon, rose, pine, wintergreen, paint thinner, cheddar cheese, banana, pizza, onion, strawberry, clove, gasoline, grape, motor oil, peanut, grass, root beer, lime, natural gas, dill pickle, pineapple, cedar, gingerbread, leather, and peach) shows that 27 (68 percent) are food-related (Doty et al., 1984). The predominance of food-related smells in studies of aging and dementia is not confined to those that use the SIT. When everyday items were used as odors, 33 to 70 percent of the stimuli were food-related (Waldton, 1974; Knupfer and Spiegel, 1986; Koss, 1986; Wood and Harkins, 1987; Rezek, 1987; Moberg et al., 1987; Wysocki and Gilbert, 1989). Whether such odors, being experienced over a lifetime as familiar, safe, comforting, and pleasant, are selectively retained in the face of age and dementia-related changes in memory is an open question.

Studies on Other Diseases and Disorders

Barton (1974) studied olfactory perception in patients with leprosy in rural India. The four stimuli were menthol in peppermint, methyl salicylate, oil of turpentine, and amyl acetate. They were selected on the basis of "differing as far as possible in character and, to a certain degree, in strength," and because they were colorless.

Weiffenbach and McCarthy (1984) compared olfactory function in cystic fibrosis patients with that in normal controls. PEMEC was used to test detection thresholds. Olfactory identification was assessed with the SIT. Warner, Peabody, and Berger (1988) used the SIT to test odor recognition in Down's syndrome patients. Neil-Dwyer and Marus (1989) examined smell and taste function in patients with mild to moderate hypertension who received ACE inhibitors. The SIT was used to test smell identification, and PEMEC (Amoore's dilution series) was used to determine detection thresholds.

Conrad, Corwin, Katz, Serby, LeFavour, and Rotrosen (1987) examined memory and olfactory performance in patients receiving chronic hemodialysis treatment. The stimuli were 20 odors selected from the SIT. Subjects were presented with the odor and a name, and asked if the odor name was correct. The selected stimuli were orange, lemon, grape, cherry, peanut, leather, fruit punch, chocolate, mint, banana, rose, bubble gum, pineapple, peach, onion, motor oil, turpentine, cinnamon, gasoline, and soap.

Corwin (1989) compared olfactory discrimination and response bias in hemodialysis patients, using as stimuli a subset of the 40 odors in the SIT. She selected 20 of the items in 10 pairs to "cover a range of odor qualities, and to be moderately to closely matched on profiles within pairs."

Murphy and Cain (1986) examined olfactory sensitivity and odor identification in blind versus sighted adults. Odor thresholds were determined with *n*-butyl alcohol. Odor identification was tested by confrontation naming of 80 "everyday odors," which on inspection (Fig. 3, p. 179) appear to be those used by Rabin and Cain (1984).

Smith (1988) offered guidance on the assessment of patients with taste and smell disorders, and outlined the University of Cincinnati Taste and Smell Center test protocol. This includes three tests of olfactory function: a butanol dilution series for threshold, the SIT for odor recognition, and an "Odor ID" test consisting of three trigeminal stimulants (unnamed) and seven odorants (baby powder, chocolate, cinnamon, coffee, mothballs, pèanut butter, and soap).

Wright (1987) attempted to go beyond threshold determinations to characterize in a standardized way the qualitative aspects of olfactory dysfunction. He proposed a confusion matrix method wherein 10 odors are presented, and the patient is requested to identify the odor from a list of the odor names. Wright used ammonia, Clorox, licorice, mothballs, peppermint, roses, turpentine, vanilla, Vicks, and vinegar. Stimuli were selected based on their being widely identifiable (by 80 percent of 50 pretested subjects), and on some being "regarded as almost exclusive stimulants for the olfactory nerve" while others had "strong trigeminal components."

A tabulation from seven neurology manuals (Cain and Krause, 1979) showed that of 17 odorants recommended for testing olfactory function, 8 were food-related. Cinnamon, cloves, and coffee were the most frequently recommended (in four manuals each).

Summary

In these studies of olfactory perception in various diseases and disorders, identification ability was frequently tested with the commercially available SIT (Weiffenbach and McCarthy, 1984; Warner et al., 1988; Neil-Dwyer and Marus, 1989; Smith, 1988) with its bias toward food-related odors. In two studies that used subsets of the SIT stimuli (Conrad et al., 1987; Corwin, 1989), food-related odors made up nearly the same proportion (70 percent) as in the full SIT (68 percent). Similarly high proportions of food-related stimuli were used in studies that did not use the SIT (64 percent, Murphy and Cain, 1986; 75 percent, Barton, 1974). In Wright's (1987) diagnostic system for olfactory dysfunction, 40 percent of the odors were food-related, while Smith's (1988) diagnostic protocol employs, in addition to a threshold measurement and the SIT, an identification test in which 57 percent of the stimuli are food-related.

Studies of Brain-Damaged Subjects

Gordon and Sperry (1969) studied the lateralization of olfactory perception in a series of commissurotomy patients, using a test set of 16 odors plus water. One rationale for stimulus selection was that the odor be "paired with an object that could easily be associated with the odor." The stimuli were lemon, pineapple, banana, tomato juice, apple juice, peanut butter, coffee, tobacco, Arpège perfume, aftershave lotion, mint, fish oil, hydrogen sulfide, butyric acid, garlic, and mentholatum.

Mair and Engen (1976) studied the relationship of aphasic lesions and the ability to associate odors with visual targets. The 14 odorants were "all common household products representing a wide variety of odor qualities." The stimuli were arranged in pairs based on similarity, with the author's similarity judgments validated by pretesting on 74 normal subjects. The odor pairs were Ivory soap and paraffin, Kiwi shoe polish (brown) and London oil paint (brown), Seagram's whiskey and Taylor's cream sherry, Fleishman's gin and Gordon's vodka, Ronson lighter fluid and kerosene, Durkee black pepper and Durkee nutmeg, Durkee cinnamon and Durkee rosemary.

Potter and Butters (1980) examined olfactory deficits in patients with prefrontal lesions, alcoholic Korsakoff's patients, and brain-damaged and normal controls. Detection thresholds were determined with *n*-butanol. Odor quality discrimination was tested with similarity ratings obtained from match and mismatch presentations of four odor pairs. The pairs were "selected to cover a broad range of difficulty, as judged from pilot work conducted with seven naive normal subjects." The pairs were allyl caproate and amyl proprionate, eugenol and guaiacol, methyl salicylate and guaiacol, and *n*-butyric acid and eugenol.

Eichenbaum, Morton, Potter, and Corkin (1983) tested the famous patient H.M. (bilateral medial temporal lobe resection) and found dissociation between "the perceptual phenomena of odour detection and discrimination." The investigators determined the detection threshold once with dilution series of almond, ethanol, lemon, and acetone, and again with *n*-butanol. Odor intensity discrimination was tested with the four odorants used in the first threshold test. Adaptation was tested with *n*-butanol and β-phenethyl alcohol. Odor quality discrimination was tested first in a signal detection method with three odor pairs (methyl salicylate and guaiacol, eugenol and guaiacol, and allyl caproate and amyl proprionate). Next, two concentrations each of *n*-butanol and β-phenethyl alcohol were used in a triangle match-to-sample test. Finally, an odor matching task used nine "common odors," coconut, mint, peppermint, almond, vanilla, orange, cloves, raspberry aldehyde, and β-phenethyl alcohol (rose).

Abraham and Mathai (1983) studied the effect of right temporal lobe lesions on a test involving the matching of smells. The four stimuli (menthone, fenchone, α-terpinol, and α-terpinylacetate) were selected on the basis of being "colourless, aromatic, non-irritant, non-volatile, stable, chemical liquids."

Levin, High, and Eisenberg (1985) described a study of olfactory naming and recognition in patients with closed head injury compared to age-matched normal controls. They used an "Olfactory Identification Test" consisting of commercially microencapsulated fragrance labels. Patients were presented the odor in a confrontation naming task, followed by a three-option multiple-choice naming task. The 12 odorants were selected to be "familiar scents which would be likely to evoke an association." Trigeminal irritants were excluded. Pilot testing with normal subjects confirmed the ease of detection and recognition of the test odorants. The scents were carnation, pine, cherry, grape, orange, strawberry, pickle, banana, chocolate, lemon, perfume no. 2 [sic], and rose.

Eskenazi, Cain, Novelly, and Mattson (1986) examined odor perception in patients with unilateral anterior temporal lobe resection, and in patients with unoperated temporal lobe epilepsy. Detection thresholds were determined for each nostril with 1-butanol. Odor recognition memory was tested by presenting odors of 14 "common objects" (seven per nostril), followed after a 10-minute delay by the same 14 odors interspersed with 14 distractors. Later, the 14 targets were presented again, interspersed with a different set of 14 distractors. The 42 stimuli (with the exceptions of paprika, Playdough [sic], and silver polish) were from those used by Rabin and Cain (1984).

Bellas, Novelly, Eskenazi, and Wasserstein (1988a) examined unilateral neglect in patients with right hemisphere damage. Detection thresholds were determined with butyl alcohol. The test procedure was an olfactory double simultaneous stimulation paradigm, with four stimulus odors: coffee, peanut butter, Ivory soap, and Juicy Fruit chewing gum. Vinegar was used in a subsequent assessment of trigeminal effects. Three criteria were used in stimulus selection: odors most easily identified by normal subjects, odors least confused with one another, and odors most easily identified by brain-damaged (temporal lobectomy) patients. In addition, Bellas et al. (1988b) examined unilateral displacement in the olfactory sense in patients with right hemisphere lesions, using the same four odorants as Bellas et al. (1988a).

Jones-Gotman and Zatorre (1988) investigated odor identification ability of patients with focal surgical brain lesions. Odor identification ability was tested with the SIT. Odor detection thresholds were determined with *n*-butyl alcohol.

Eskenazi, Cain, Lipsitt, and Novelly (1988) reported two case studies of olfactory function following callosotomy. Thresholds were determined with 1-butanol. Odor discrimination was tested presurgery with benzyl acetate, geraniol, methyl butyrate, benzaldehyde, and pyridine, and postsurgery with l-carvone, ethyl-*n*-butylamine, pentanedione, benzaldehyde, and pyridine. Recognition memory was assessed with 14 "everyday odorants" (chocolate, shoe polish, crayons, garlic, etc.). An odor identification task employed 10 odorants: baby powder, chewing gum, chocolate, cinnamon, coffee, mothballs, peanut butter, potato chips, soap, and wintergreen. An odor-visual matching task used "15 common everyday odorants," namely Band-Aid, garlic, bubble gum, popcorn, coconut, mustard, crayon, cigarette, black pepper, pencil, orange, apple, pipe tobacco, cherry, and olive. An odor-tactile matching task made use of 10 scented objects: bar of soap, block of wood, onion, newspaper, peanuts, tea bag, banana, rubber hose, egg, and lemon. With a few exceptions (cherry, olive, newspaper, and egg) the stimuli in the last two experiments were drawn from the set used by Rabin and Cain (1984).

Summary

Many of the studies of brain-damaged patients described in this section used relatively large (>9 items) stimulus sets, in which 57 to 89 percent of the stimuli were food-related odors. In those studies using relatively small

(3 to 6 items) stimulus sets, at least one food-related odor was always included, while in two studies (Bellas et al., 1988a and b) three of the four stimuli were food odors.

Studies on Normal Subjects

A number of investigators have made use of multidimensional scaling techniques to explore the psychological responses to odor of normal subjects. Berglund, Berglund, Engen, and Ekman (1973) used multidimensional scaling to examine the psychophysical dimensions of a set of odors matched in intensity but varying in quality. The 21 stimuli were selected from 28 used previously by the same authors (seven were discarded as being out of range on intensity). The rationale for selection of stimuli was that "they provide a variety of qualitatively different, pleasant and unpleasant odors." The stimuli were acetone (mildly pungent, somewhat aromatic), *n*-amyl acetate (apples, banana), anethole, benzaldehyde (bitter almonds), benzyl acetate (jasmine, pears), 1-butanol (weak fusel oil), *n*-butyric acid (pungent, rancid), ethyl acetate (fruity), ethyl methyl ketone (acetone-like), eugenol (clove), furfurole (almond), geranial (lemon, sweet rose), guaiacol (agreeable aromatic), *n*-heptane (gasoline), 1-heptanol, menthol (peppermint), methyl salicylate (strong wintergreen), nonane (candied ginger), *n*-octane, 2-phenyl ethyl alcohol (faint rose), isovaleric acid (disagreeable, rancid cheese).

Gregson and Mitchell (1974) investigated the generality of verbally based, multidimensional scaling of odors, using seven stimuli selected "for their diversity of odor quality" and because some were previously implicated as examples of primary odor qualities. The stimuli were isoamyl acetate, methyl proprionate, *n*-propanol, *d*-menthol, *l*-carvone, eugenol, and Exaltolide. Jones, Roberts, and Holman (1978) compared the ways in which recognition memory for odors is related to similarity judgments of the same odors quantified by multidimensional scaling. The stimuli were 11 dried herbs and spices: basil, bay, celery, marjoram, mint, oregano, parsley, rosemary, sage, tarragon, and thyme. Gilbert, Greenberg, and Beauchamp (1989) used multidimensional scaling of five stimulus odors (cedarwood, patchouli, rose, lemon, and orange) to investigate how sex, handedness, and side of nose modulate odor perception.

Lawless (1989) used multidimensional scaling to examine how subjects sorted three sets of six olfactory stimuli each. The sets were woody fragrances (pine oil, and isobornyl acetate, cedarwood oil, cedrene, sandalwood oil, and sandela), citrus fragrances (citral, limonene, dimyrcetol, lemon oil, orange oil, and lime oil), and partially woody/partially citrus fragrances (dihydromyrcenol, elemi oil, dihydroterpineol, dihydromyrcenyl acetate, tetrahydromugyl acetate, and mandarin oil). The fragrance types were selected to provide two distinctive stimulus sets, and a third whose odors were ambiguous with respect to membership in the other sets.

Other studies of normal subjects have dealt with various behavioral and physiological parameters of the response to odor. Gilbert, Fridlund, and Sabini (1987) examined the decoding of posed and spontaneous facial responses to odor. They used butyric acid, isovaleric acid, eugenol, phenylethyl alcohol, and a mineral oil blank as stimuli. Mester, Doty, Shapiro, and Frye (1988) investigated how body tilt (i.e., position in the sagittal plane) affects odor identification (assessed with the SIT) and various physiological parameters. Lorig and Schwartz (1988) recorded EEG activity during odor inhalation. In one experiment, the stimuli were spiced apple, eucalyptus, lavender, and a distilled water control. In a second experiment, the stimuli were "refined fine fragrances of similar odor," namely "five floral note perfumes prepared in an unscented detergent base," used in order to "assess the effects of different odors while minimizing their memorial and cognitive differences."

Trotter, Dallas, and Verdone (1988) studied the incorporation of olfactory stimuli into REM dreams. "Half of the olfactory stimuli were pleasant, and half were unpleasant," based on a preliminary odor survey of 50 students. In two nights of preliminary study, bananas as well as "hand lotion, lip gloss, and corn nuts" were the trial stimuli. Pleasant test stimuli included coffee, peanut butter, rose, cinnamon, chocolate, and lemon. Unpleasant test items included wood alcohol, dirty ashtray, match smoke, mold, dog feces, and onion.

Several investigations have focused on the social aspects of fragrance. Baron (1980) exposed subjects to a pleasant odor applied to experimental forms while manipulating levels of physical aggression. Two odors were tested: Jungle Gardenia perfume and a pine-scented aerosol. Both had been prescreened for level of pleasantness by a panel of five judges. Baron (1981) investigated how a perfume worn by a female confederate influenced the social perception of her by male subjects. Jungle Gardenia was selected on the basis of its being rated "both pleasant and potent" in pretesting by a panel of five undergraduate males. Baron (1983) presented an investigation into the social psychology of scent wearing. (See also the chapter by Ehrlichman in this volume.) Student subjects role played job interviewers, while experimental confederates pretended to role-play interviewees. Female confederates wore Jontue, and the males Brut. These stimuli were chosen based on pretesting of students with 11 popular perfumes and colognes. Kirk-Smith, Van Toller, and Dodd (1983) paired a stressful test situation with odor or no odor, and compared self-ratings of subjects when tested later in the presence of the same odor. Trimethylundecylenic aldehyde was selected as a stimulus because, in a pilot study, subjects "found it difficult to describe or make positive associations" to it. Filsinger, Fabes, and Hughston (1987) explored the relation of a personality measure (Eysenck's Introversion-Extraversion Scale) to ratings of strength, familiarity, and pleasantness of α-androstenone, α-androstenol, Exaltolide, and pyridine.

A number of experiments have examined the effects of putative human pheromones on sexual behavior and arousal, social attitudes and judgments, mood, and physical spacing. In most instances, the compounds tested were ones naturally produced by the human body that had been implicated as pheromones in studies on nonhuman mammals. The boar pheromone androstenol (5α-androst-16-en-3α-ol) was used in a study by Kirk-Smith, Booth, Carroll, and Davies (1978). Androstenol was tested along with Exaltolide (synthetic musk) and no odor controls by Black and Biron (1982). Benton (1982) and Benton and Wastell (1986) tested androstenol in ethanol against an ethanol control. Filsinger, Braun, and Monte (1985) examined androstenol along with androstenone, Exaltolide, and a no-odor control. Gustavson, Dawson, and Bonett (1987) compared androstenol to an androsterone (3α-hydroxy-5α-androstan-17-one) control. Morris and Udry (1978) examined the effects of a "synthetic hypothetical human female pheromone" composed of six short-chain aliphatic acids (acetic, propionic, isobutyric, *n*-butyric, isovaleric, and isocaproic).

Human body odors have been the stimuli in another set of investigations. Wallace (1977) examined the discriminability of odors from the hands of different individuals. Body odors absorbed on clean T-shirts were the means by which Porter, Cernoch, and Balogh (1985) examined olfactory similarity between mothers and offspring. Using a similar technique, Porter, Balogh, Cernoch, and Franchi (1986) examined the identifiability of infant body odors by adult kin.

The cross-cultural aspects of odor perception have also been the focus of study. Pangborn, Guinard, and Davis (1988) studied cross-cultural differences in the response to 21 odors (anise, banana, cedar, chocolate, cinnamon, garlic, grape, lemon, methyl salicylate, natural gas, onion, pickle, pineapple, pizza, peppermint, root beer, skunk, smoke, strawberry, and vanilla), and analyzed the results in the context of regional cuisine. Takagi (1989) reviewed the development of a standardized method of olfactometry in Japan. It was decided that candidate odors should be distinct from one another, easily perceptible to many people, stable over time, and of chemically pure, simple structures. On this basis, 10 odors were chosen: τ-undecalactone (canned peach), isovaleric acid (putrid, sweaty, etc.), cyclotene (burnt, carmel), skatole (vegetable garbage), β-phenyl ethyl alcohol (rose), Exaltolide (musk), phenol (disinfectants, hospitals), acetic acid (vinegar), diallyl sulfide (garlic), and *dl*-camphor. After preliminary testing, the first five odorants were selected for further quantitative olfactometry.

Several investigators have examined the interaction of nonolfactory stimuli features and the perception of odor. Davis (1981) studied the effects on odor identification of nonolfactory context cues (colors or color words) that were either relevant or irrelevant to the odor identity. The odorants were drawn from a commercially available library of microencapsulated fragrances (3M Company). Selection was based on the odors being "representative of real objects." They were lemon, coconut, chocolate, cherry, smoke, grass, pickle, cedar, cinnamon, banana, garlic, grape, skunk, carnation, clove, whisky, butter, pineapple, heptyl acetate, and onion.

Batic and Gabassi (1987) investigated the relative dominance of olfactory and visual cues in memory for odors by using a set of seven familiar fruit smells, with congruent or incongruent printed labels. The odors were extracts of apricot, pineapple, orange, coconut, strawberry, lemon, and pear. Gabassi and Batic (1987) studied whether the presentation of an odor with a congruent or incongruent odor name altered free recall for the odor names. The stimuli were nine fruit smells [apple and banana in addition to the seven used by Batic and Gabassi (1987)].

Multivariate experiments with normal subjects have examined responses to multiple odor stimuli and multiple measures of response. Moskowitz, Dravnieks, and Klarman (1976) had observers rate various concentrations of 32 odorants in relation to the odor of butanol, in order to determine the relation between odor intensity and pleasantness. The stimuli were chosen because they "represented a diversity of odor qualities and were of varying pleasantness and unpleasantness." They were benzene (solvent), propyl-*n*-butyrate (fruity), 1-hexanol (alcoholic, metallic), hexyalamine (rancid, fishy), benzaldehyde (almond, cherry), eugenol (spicy cloves), hexadienal (grassy), methyl salicylate (oil of wintergreen), pinene (turpentine), vanillin (vanilla), thiophene (rotten, sulfidic), cedrol (cedar oil), toluene (solvent), 2-hexanone (sweet, fruity), ethyl butyrate (estery), 1-propyl proprionate (ethereal, estery), hexanoic acid (fatty, goaty), cyclohexanol (musty), cyclohexylamine (fishy), 3-hexanol (alcoholic), methyl valerate (harsh, estery), cyclohexanone (sweet, alcoholic), cyclohexene (harsh, solvent), butanol (fusel oil), pyridine (rank, harsh), phenylethanol (rose), propionic acid (acid, putrid), dibutylmethyl carbinol acetate (flowery), guaiacol (burnt), hexaldehyde (grassy, lily), butyl acetate (banana), and methyl isovalerate (estery, harsh).

Moskowitz and Barbe (1977) studied how the sensory evaluation of five odors (methyl salicylate, caproic acid, isobutyl isobutyrate, methyl disulfide, and camphor) varied as a function of their various combinatoric mixtures. Pendse (1987) examined hemispheric asymmetry on a category judgment task using olfactory stimuli, namely five concentrations of *n*-butanol.

Eskenazi, Cain, and Friend (1986) used a variety of tests to characterize the olfactory aptitude of normal subjects. Quality discrimination was tested with *l*-carvone, ethyl-*n*-butylamine, 2,3-pentanedione, and pyridine. A recognition memory task used those odors along with 20 distractors (1-octanol, methyl butyrate, 2-butanone, isopentyl acetate, lavandin, 1-propanol, *d*-*p*-mentha,6-8-dien,2-one, linalool, benzyl acetate, geraniol, *p*-mentha,3-one, *d*-limonene, linalyl acetate, eugenol, dodecyl alcohol, hexane, phenethyl alcohol, acetophenone, turpentine, and anethole). For an odor-visual matching task, 15 "familiar odorants and their corresponding stimulus-objects" were used (mustard, tobacco, black pepper, maraschino cherry, green olive, Band-Aid, shredded coconut, garlic powder, bubble gum, popcorn, pencil shavings, orange juice concentrate, and fresh apple). For an odor-tactile matching task, the odorants were tea leaves, Ivory bar soap, wood shavings, onion powder, peanut butter, newspaper, banana, egg, lemon seasoning, and rubber. In one test of verbal identification, subjects had to select the names of the 10 target odors (baby powder, chewing gum, chocolate, cinnamon, coffee, mothballs, peanut butter, potato chips, soap, and wintergreen) from a list that included 10 distractors. In another test of verbal identification, subjects had to identify 20 stimuli "chosen because they seemed from pilot experiments relatively difficult to identify in a free recall situation." The odors were silver polish, sauerkraut, graham crackers, shoe polish, paprika, baby shampoo, witch hazel, cloves, Brut aftershave, clay, Lysol, beer, oregano, ginger, soy sauce, nutmeg, steak sauce, Playdough [sic], prunes, and cork.

Summary

A broad array of studies was reviewed in this section. Three groups of them are notable for using odor stimuli that are not food-related. Studies of the social aspects of odor made use of perfumes and other nonfood odors (Baron, 1980, 1981, 1983; Kirk-Smith et al., 1983; Filsinger et al., 1987). Studies on human pheromones (Kirk-Smith et al., 1978; Black and Biron, 1982; Benton, 1982; Benton and Wastell, 1986; Filsinger et al., 1985; Gustavson et al., 1987) and those on human body odors (Wallace, 1977; Porter et al., 1985, 1986) used no food-related odors.

In contrast, there were several investigations in which all the stimuli were food-related (Davis, 1975; Jones et al., 1978; Batic and Gabassi, 1987; Gabassi and Batic, 1987). In the remaining studies, 20 to 70 percent of the odor stimuli were food-related.

Discussion

A striking feature of olfactory investigations in a variety of disciplines is the predominant use of food-related odors as experimental stimuli. In the majority of studies reviewed here, reliance on these odors was not the result of an explicitly stated desire to narrow the range of stimuli. On the contrary, studies with a preponderance of food-related odors often claim that stimulus selection was intended to cover a wide range of odor types and hedonic values. Several implicit rationales may underlie this choice of odor category. Food odors are readily and inexpensively obtained, and are easily measured, stored, and disposed of. They are innocuous from a safety standpoint, and this may facilitate approval by human subjects committees and participation by volunteers. They are also among the most familiar, recognizable, and identifiable (i.e., nameable) of odors, which may be advantageous in certain experimental paradigms.

Nevertheless, there are a number of ways in which reliance on this category of odors may be problematic. Most critically, it may limit the generalizability of conclusions drawn from olfactory experiments. One basic consideration is the universality of response to food-related odors. Simply put, what smells like food to persons of one culture may not smell edible to those of another. Data from the *National Geographic* Smell Survey show that only about 6 percent of Americans report a willingness to eat something smelling like roses, compared to nearly 32 percent of respondents from India, a result consistent with historical patterns of regional cuisine (Gilbert and Wysocki, unpublished results). Subtle cross-cultural differences can exist even among odors defined as edible. For instance, Pangborn et al. (1988) found that liking for methyl salicylate was notably low in Switzerland and England, but high in Canada and California. These findings suggest that culture-bound differences in dietary preferences may influence response to food odors, and thereby limit the generalizability of results based on these odors.

This conclusion may not be shared by all investigators. Cain (1979), for one, claimed that the use of odors of "everyday" objects and substances lent "ecological validity" to his study of odor nameability, which used an 80-item set composed of 64 percent food-related odors (see also Cain and Krause, 1979; Rabin and Cain, 1984; Murphy and Cain, 1986; Lyman and McDaniel, 1986; Eskenazi, Cain, and Friend, 1986; Eskenazi et al., 1988). Such a claim raises three issues. First, by what measure can "everyday" odors be operationally defined as distinct from less typical ones? No investigator has proposed such a metric, while many have claimed such quotidian status for their stimuli. Second, even with such a metric, the cultural context of the subject population needs to be specified. After all, Manhattan's everyday odors are unlikely to be those of rural Wisconsin. Third, what is the "ecological validity" of presenting food odors without an accompanying taste? Sniffing a potential food item initiates a sequence of anticipated ingestion that typically includes conscious, unconscious, and physiological processes culminating in a consummatory behavior. It is somewhat bizarre to present subjects with a series of food odors and yet not provide the actual experience of ingestion.

Perhaps the predominance of food-related stimuli in olfactory experiments merely reflects the degree to which these smells are represented in a population's aggregate olfactory experience. Can this experience be characterized? Schleidt, Neumann, and Morishita (1988) surveyed a large sample of Germans and Japanese regarding associations to freely recalled unpleasant and pleasant odors. Schleidt et al. grouped the recalled odors into categories based on source: civilization, food and drink, nature, man, and a fifth catch-all category. The food-and-drink category included 26 percent of memories recalled by the German sample and 23 percent of those from the Japanese sample. From this it might appear that typical stimulus sets, often containing 50 percent or more food odors, are disproportionate with respect to aggregate experience.

Food-related odors may be an appropriate choice for testing certain experimental hypotheses, but their use can introduce artifact. Lorig and Schwartz (1987) recorded EEGs while subjects imagined food odors, and concluded that images of food reduced theta activity. If the presentation of food-related odors and task demands typically activates food-related imagery, and if such imagery reliably alters CNS activity, then such studies may be confounded. A subject's response may be specific to a food-related pattern of neurophysiological activity, and not representative of olfactory perception at large.

The majority of food-and-drink-related memories are pleasant (Schleidt et al., 1988). Negative attributions are largely confined to rotten or deteriorated foods (exceptions are garlic, onions, and alcoholic drinks). If similar trends are present in other cultures, it can be presumed that subjects enter olfactory experiments with a hedonic bias in favor of food odors as a class.

Food odors are typically ones we have experienced since infancy. As a consequence, our identification of them and our hedonic response to them may be overlearned or conditioned. As experimental stimuli, they may circumvent deeper processing channels and fail to yield insight into the cognitive processing of olfactory information. They may contribute to error variance by activating conditioned taste aversions in some subjects. Finally, they may represent a confounding variable, if not an outright provocation, for hungry subjects. It would behoove an experimenter to compare the response to food and nonfood odors as a procedural check.

Odors occupy a particular place and time—they are cultural artifacts. This becomes a consideration in assessing the olfactory abilities of the elderly. Wood and Harkins (1987) note that "older persons appear most likely to recognize common odors if the substances have been in use over a long period of time. The use of 'modern' odorants as test stimuli may well distort results in favor of younger cohorts." In their experiment, Wood and Harkins found that glue, lighter fluid, hair conditioner, and bubble gum (relatively recent product developments) were more readily identifiable by younger (18- to 35-year-olds) than by older subjects (63- to 90-year-olds).

Odors have been used as markers of particular commercial brands (Byrne-Quinn, 1988), and the more successful among them have become, with time, synonymous with and denotative of a product category. These cultural assumptions often surface in the scientific literature. Soap, for instance, is a frequently used odor stimulus, and when the specific brand is identified, it is invariably Ivory bar soap.

This may reflect a broader tendency for investigators to select the most prototypical exemplar of a particular odor category. For instance, when cheese is used as a stimulus it is most often cheddar rather than Camembert, Roquefort, or other highly fragrant varieties. A powerful preconception appears to be at work in stimulus selection.

Preference for the use of prototypical odors results in a narrowed field of potential stimuli. One way to avoid this is to use the perfumer's ability to create notes that are clearly category-typical (e.g., fruity), yet not prototypical or representational of any single category member (e.g., banana). An example might be a guava-watermelon-like note. This approach would have the benefit of expanding the range of experimental stimuli, increasing their novelty, and reducing reliance on standard odors.

Experimentalists could also find more diverse and novel alternative odors in the "nature" category (Schleidt et al., 1988) which has a high proportion of positive associations. Materials such as leaves and twigs (fir-balsam, petitgrain), roots (iris, vetiver, costus), wood (sandalwood, cedarwood, bois de rose, amyris), gums and resins (olibanum, benzoin, galbanum, tolu), moss (oakmoss, treemoss), grass (citronella), and hay (foin) are frequently used in perfumery, along with lesser-known floral scents (ylang ylang, mimosa, broom) (Priest, 1981).

Body odors, which are certainly important from social, medical, and psychological points of view, have been used infrequently as experimental stimuli. To date, they have figured largely in studies of kin recognition, and have seldom been included in stimulus sets used to test other hypotheses. They would seem to hold a good deal of promise for investigations of the affective elements of odor response.

The trend revealed in this literature survey is unsettling and merits the attention of experimentalists. While no one odor category will satisfy the demands of all experimental purposes, an awareness of the constraints on interpretation imposed by stimulus selection, and an openness to different olfactory dimensions, would be salutary. Failing that, we are creating a science of olfaction based on cinnamon and coffee.

References

Abraham, A., and K.V. Mathai (1983). The effect of right temporal lobe lesions on matching of smells. *Neuropsychologia*, *21*, 277–281.

Balogh, R.D., and R.H. Porter (1986). Olfactory preferences resulting from mere exposure in human neonates. *Infant Behav. Dev.*, *9*, 395–401.

Barker, L.M., and C.A. Weaver (1983). Rapid, permanent, loss of memory for absolute intensity of taste and smell. *Bull. Psychonomic Soc.*, *21*, 281–284.

Baron, R.A. (1980). Olfaction and human social behavior: Effects of pleasant scents on physical aggression. *Basic Appl. Soc. Psychol.*, *1*, 163–172.

Baron, R.A. (1981). Olfaction and human social behavior: Effects of a pleasant scent on attraction and social perception. *Personality Social Psychol. Bull.*, *7*, 611–616.

Baron, R.A. (1983). "Sweet smell of success"? The impact of pleasant artificial scents on evaluations of job applicants. *J. Appl. Psychol.*, *68*, 709–713.

Barton, R.P.E. (1974). Olfaction in leprosy. *J. Laryngol. Otol.*, *88*, 355–361.
Batic, N., and P.G. Gabassi (1987). Visual dominance in olfactory memory. *Percept. Motor Skills*, *65*, 88–90.
Bellas, D.N., R.A. Novelly, B. Eskenazi, and J. Wasserstein (1988a). The nature of unilateral neglect in the olfactory sensory system. *Neuropsychologia*, *26*, 45–52.
Bellas, D.N., R.A. Novelly, B. Eskenazi, and J. Wasserstein (1988b). Unilateral displacement in the olfactory sense: A manifestation of the unilateral neglect syndrome. *Cortex*, *24*, 267–275.
Benton, D. (1982). The influence of androstenol—a putative human pheromone—on mood throughout the menstrual cycle. *Biol. Psychol.*, *15*, 249–256.
Benton, D., and V. Wastell (1986). Effects of androstenol on human sexual arousal. *Biol. Psychol.*, *22*, 141–147.
Berglund, B., U. Berglund, T. Engen, and G. Ekman (1973). Multidimensional analysis of twenty-one odors. *Scand. J. Psychol.*, *14*, 131–137.
Black, S.L., and C. Biron (1982). Androstenol as a human pheromone: No effect on perceived physical attractiveness. *Behav. Neural Biol.*, *34*, 326–330.
Byrne-Quinn, J. (1988). Perfume, people, perceptions and products. In S. Van Toller and G.H. Dodd (Eds.), *Perfumery: The Psychology and Biology of Fragrance*. London: Chapman and Hall, pp. 205–216.
Cain, W.S. (1979). To know with the nose: Keys to odor identification. *Science*, *203*, 467–470.
Cain, W.S., and R.J. Krause (1979). Olfactory testing: Rules for odor identification. *Neurol. Res.*, *1*, 1–9.
Cann, A., and D.A. Ross (1989). Olfactory stimuli as context cues in human memory. *Am. J. Psychol.*, *102*, 91–102.
Conrad, P., J. Corwin, L. Katz, M. Serby, G. LeFavour, and J. Rotrosen (1987). Olfaction and hemodialysis: Baseline and acute treatment decrements. *Nephron*, *47*, 115–118.
Corwin, J. (1989). Olfactory identification in hemodialysis: Acute and chronic effects on discrimination and response bias. *Neuropsychologia*, *27*, 513–522.
Davis, R.G. (1975). Acquisition of verbal associations to olfactory stimuli of varying familiarity and to abstract visual stimuli. *J. Exp. Psychol., Human Learning Memory*, *104*, 134–142.
Davis, R.G. (1981). The role of nonolfactory context cues in odor identification. *Percept. Psychophys.*, *30*, 83–89.
Desor, J.A., and G.K. Beauchamp (1974). The human capacity to transmit olfactory information. *Percept. Psychophys.*, *16*, 551–556.
Doty, R.L., P. Shaman, and M. Dann (1984). Development of the University of Pennsylvania smell identification test: A standardized microencapsulated test of olfactory function. *Physiol. Behav.*, *32*, 489–502.
Doty, R.L., P.F. Reyes, and T. Gregor (1987). Presence of both odor identification and detection deficits in Alzheimer's disease. *Brain Res. Bull.*, *18*, 597–600.
Doty, R.L., D.A. Deems, and S. Stellar (1988). Olfactory dysfunction in parkinsonism: A general deficit unrelated to neurologic signs, disease stage, or disease duration. *Neurology*, *38*, 1237–1244.
Ehrlichman, H. (1987). Hemispheric asymmetry and positive-negative affect. Duality and unity of the brain: Unified functioning and specialisation of the hemispheres. *Wenner-Gren Symp.*, *47*, 194–206. New York: Plenum.

Ehrlichman, H., and J.N. Halpern (1988). Affect and memory: Effects of pleasant and unpleasant odors on retrieval of happy and unhappy memories. *J. Pers. Soc. Psych.*, *55*, 769–779.

Eich, J.E. (1978). Fragrances as cues for remembering words. *J. Verbal Learning Verbal Behav.*, *17*, 103–111.

Eichenbaum, H., T.H. Morton, H. Potter, and S. Corkin (1983). Selective olfactory deficits in case H.M. *Brain*, *106*, 459–472.

Engen, T., J.E. Kuisma, and P.D. Eimas (1973). Short-term memory of odors. *J. Exp. Psychol.*, *99*, 222–225.

Eskenazi, B., W.S. Cain, and K. Friend (1986). Exploration of olfactory aptitude. *Bull. Psychonomic Soc.*, *24*, 203–206.

Eskenazi, B., W.S. Cain, R.A. Novelly, and R. Mattson (1986). Odor perception in temporal lobe epilepsy patients with and without temporal lobectomy. *Neuropsychologia*, *24*, 553–562.

Eskenazi, B., W.S. Cain, E.D. Lipsitt, and R.A. Novelly (1988). Olfactory functioning and callosotomy: A report of two cases. *Yale J. Biol. Med.*, *61*, 447–456.

Fabes, R.A., and E.E. Filsinger (1986). Olfaction and young children's preferences: A comparison of odor and visual cues. *Perception Psychophys.*, *40*, 171–176.

Filsinger, E.E., J.J. Braun, and W.C. Monte (1985). An examination of the effects of putative pheromones on human judgments. *Ethol. Sociobiol.*, *6*, 227–236.

Filsinger, E.E., R.A. Fabes, and G. Hughston (1987). Introversion-extraversion and dimensions of olfactory perception. *Percept. Motor Skills*, *64*, 695– 699.

Gabassi, P.G., and N. Batic (1987). Interference processes in visual and olfactory stimulations. *Percept. Motor Skills*, *65*, 79–82.

Gilbert, A.N., A.J. Fridlund, and J. Sabini (1987). Social and hedonic determinants of facial displays to odor. *Chem. Senses*, *12*, 355–363.

Gilbert, A.N., M.S. Greenberg, and G.K. Beauchamp (1989). Sex, handedness, and side of nose modulate odor judgments. *Neuropsychologia*, *27*, 505–511.

Gilbert, A.N. and M.R. Kare (1991). Odour perception and its mechanisms. In P.M. Müller and D. Lamparsky (Eds.), *Perfumes: Art, Science and Technology*. London: Elsevier Applied Science.

Gordon, H.W., and R.W. Sperry (1969). Lateralization of olfactory perception in the surgically separated hemispheres of man. *Neuropsychologia*, *7*, 111–120.

Gregson, R.A.M., and M.J. Mitchell (1974). Odor quality similarity scaling and odor-word profile matching. *Chem. Senses*, *1*, 95–101.

Gustavson, A.R., M.E. Dawson, and D.G. Bonett (1987). Androstenol, a putative human pheromone, affects human (*Homo sapiens*) male choice performance. *J. Comp. Psychol.*, *101*, 210–212.

Hurwitz, T., L. Kopala, C. Clark, and B. Jones (1988). Olfactory deficits in schizophrenia. *Biol. Psychiat.*, *23*, 123–128.

Jones, F.N., K. Roberts, and E.W. Holman (1978). Similarity judgments and recognition memory for some common spices. *Percept. Psychophys.*, *24*, 2–6.

Jones-Gotman, M., and R.J. Zatorre (1988). Olfactory identification deficits in patients with focal cerebral excision. *Neuropsychologia*, *26*, 387–400.

Kesslak, J.P., C.W. Cotman, H.C. Chui, S. Van Den Noort, H. Fang, R. Pfeffer, and G. Lynch (1988). Olfactory tests as possible probes for detecting and monitoring Alzheimer's disease. *Neurobiol. Aging*, *9*, 399–403.

Kirk-Smith, M.D., D.A. Booth, D. Carroll, and P. Davies (1978). Human social attitudes affected by androstenol. *Res. Comm. Psychol. Psychiatry Behav.*, *3*, 379–384.

Kirk-Smith, M.D., C. Van Toller, and G.H. Dodd (1983). Unconscious odour conditioning in human subjects. *Biol. Psychol.*, *17*, 221–231.

Knupfer, L., and R. Spiegel (1986). Differences in olfactory test performance and vascular type dementia individuals. *Int. J. Geriatr. Psychiatry*, *1*, 3–14.

Koelega, H.S., and E.P. Köster (1974). Some experiments on sex differences in odor perception. *Ann. NY Acad. Sci.*, *237*, 234–246.

Kopala, L., C. Clark, and T.A. Hurwitz (1989). Sex differences in olfactory function in schizophrenia. *Am. J. Psychiatry*, *146*, 1320–1322.

Koss, E. (1986). Olfactory dysfunction in Alzheimer's disease. *Dev. Neuropsychol.*, *2*, 89–99.

Koss, E., J.M. Weiffenbach, J.V. Haxby, and R.P. Friedland (1987). Olfactory detection and recognition in Alzheimer's disease. *Lancet*, March 14, 622.

Koss, E., J.M. Weiffenbach, J.V. Haxby, and R.P. Friedland (1988). Olfactory detection and identification performance are dissociated in early Alzheimer's disease. *Neurology*, *38*, 1228–1232.

Laing, D.G., and P.J. Clark (1983). Puberty and olfactory preferences of males. *Physiol. Behav.*, *30*, 591–597.

Lawless, H.T. (1978). Recognition of common odors, pictures, and simple shapes. *Percept. Psychophys*, *24*, 493–495.

Lawless, H.T. (1989). Exploration of fragrance categories and ambiguous odors using multidimensional scaling and cluster analysis. *Chem. Senses*, *14*, 349–360.

Lawless, H.T., and W.S. Cain (1975). Recognition memory for odors. *Chem. Senses and Flavor*, *1*, 331–337.

Levin, H.S., W.M. High, and H.M. Eisenberg (1985). Impairment of olfactory recognition after closed head injury. *Brain*, *108*, 579–591.

Lorig, T.S., and G.E. Schwartz (1987). EEG activity during food and relaxation imagery. *Psychophysiology*, *24*, 599.

Lorig, T.S., and G.E. Schwartz (1988). Brain and odor: I. Alteration of human EEG by odor administration. *Psychobiology*, *16*, 281–284.

Lyman, B.J., and M.A. McDaniel (1986). Effects of encoding strategy on long-term memory for odours. *Q. J. Exp. Psychol.*, *38A*, 753–765.

Mair, R.G., and T. Engen (1976). Some effects of aphasic lesions on odor perception. *Sensory Processes*, *1*, 33–39.

Mester, A.F., R.L. Doty, A. Shapiro, and R.E. Frye (1988). Influence of body tilt within the sagittal plane on odor identification performance. *Aviation Space Environ. Med.*, *59*, 734–737.

Moberg, P.J.,G.D. Pearlson, L.J. Speedie, J.R. Lipsey, M.E. Strauss, and S.E. Folstein (1987). Olfactory recognition: Differential impairments in early and late Huntington's and Alzheimer's diseases. *J. Clin. Exp. Neuropsychol.*, *9*, 650–664.

Morris, N.A., and J.R. Udry (1978). Pheromonal influences on human sexual behaviour: An experimental search. *J. Biosocial Sci.*, *10*, 147–157.

Moskowitz, H.R., A. Dravnieks, and L.A. Klarman (1976). Odor intensity and pleasantness for a diverse set of odorants. *Percept. Psychophys.*, *19*, 122–128.

Moskowitz, H.R., and C.D. Barbe (1977). Profiling of odor components and their mixtures. *Sensory Processes*, *1*, 212–226.

Murphy, C., and W.S. Cain (1986). Odor identification: The blind are better. *Physiol. Behav.*, *37*, 177–180.
Neil-Dwyer, G., and A. Marus (1989). ACE inhibitors in hypertension: Assessment of taste and smell function in clinical trials. *J. Hypertens.*, *3*, 169–176.
Pangborn, R.M., J.-X. Guinard, and R.G. Davis (1988). Regional aroma preferences. *Food Quality and Preference*, *1*, 11–19.
Pendse, S.G. (1987). Hemispheric asymmetry in olfaction on a category judgement task. *Perceptual and Motor Skills*, *64*, 495–498.
Porter, R.H., J.M. Cernoch, and R.D. Balogh (1985). Odor signatures and kin recognition. *Physiol. Behav.*, *34*, 445–448.
Porter, R.H., R.D. Balogh, J.M. Cernoch, and C. Franchi (1986). Recognition of kin through characteristic body odors. *Chem. Senses*, *11*, 389–395.
Potter, H., and N. Butters (1980). An assessment of olfactory deficits in patients with damage to prefrontal cortex. *Neuropsychologia*, *18*, 621–628.
Priest, N.O. (1981). Natural materials and the practicing perfumer. *Perfumer and Flavorist*, *6*, 33–52.
Rabin, M.D., and W.S. Cain (1984). Odor recognition: Familiarity, identifiability, and encoding consistency. *J. Exp. Psychol. Learning Memory Cognition*, *10*, 316–325.
Rezek, D.L. (1987). Olfactory deficits as a neurological sign in dementia of the Alzheimer type. *Arch. Neurol.*, *44*, 1030–1032.
Rotton, J., T. Barry, J. Frey, and E. Soler (1978). Air pollution and interpersonal attraction. *J. Appl. Soc. Psychol.*, *8*, 57–71.
Schleidt, M., P. Neumann, and H. Morishita (1988). Pleasure and disgust: Memories and associations of pleasant and unpleasant odours in Germany and Japan. *Chem. Senses*, *13*, 279–293.
Schmidt, H.J., and G.K. Beauchamp (1988). Adult-like odor preferences and aversions in three-year-old children. *Child Dev.*, *59*, 1136–1143.
Self, P.A., F.D. Horowitz, and L.Y. Paden (1972). Olfaction in newborn infants. *Dev. Psychol.*, *7*, 349–363.
Smith, D.V. (1988). Assessment of patients with taste and smell disorders. *Acta Otolaryngol.*, Supplement 458, 129–133.
Strickland, M., P.O. Jessee, and E.E. Filsinger (1988). A procedure for obtaining young children's reports of olfactory stimuli. *Percept. Psychophys.*, *44*, 379–382.
Takagi, S.F. (1989). Standardized olfactometries in Japan—a review over ten years. *Chem. Senses*, *14*, 25–46.
Trotter, K., K. Dallas and P. Verdone (1988). Olfactory stimuli and their effects on REM dreams. *Psychiatr. J. Univ. Ottawa*, *13*, 94–96.
Waldton, S. (1974). Clinical observations of impaired cranial nerve function in senile dementia. *Acta Psychiatr. Scand.*, *50*, 539–547.
Walk, H.A., and E.E. Johns (1984). Interference and facilitation in short-term memory for odors. *Percept. Psychophys.*, *36*, 508–514.
Wallace, P. (1977). Individual discrimination of humans by odor. *Physiol. Behav.*, *19*, 577–579.
Warner, M.D., C.A. Peabody, and P.A. Berger (1988). Olfactory deficits and Down's syndrome. *Biol. Psychiat.*, *23*, 833–836.
Weiffenbach, J.M., and V.P. McCarthy (1984). Olfactory deficits in cystic fibrosis: Distribution and severity. *Chem. Senses*, *9*, 193–199.

Wood, J.B., and S.W. Harkins (1987). Effects of age, stimulus selection, and retrieval environment on odor identification. *J. Gerontol.*, *42*, 584–588.

Wright, H.N. (1987). Characterization of olfactory dysfunction. *Arch. Otolaryngol. Head Neck Surg.*, *113*, 163–168.

Wysocki, C.J., and A.N. Gilbert (1989). The National Geographic Smell Survey: Effects of age are heterogenous. In C. Murphy, W.S. Cain, and D.M. Hegsted (Eds.), *Nutrition and the Chemical Senses in Aging: Recent Advances and Current Research Needs. Ann. NY Acad. Sci.*, *561*, 12–28.

11
Assessing Olfaction: Cognitive and Measurement Issues

JUNE CORWIN

"The act of smelling something, anything, is remarkably like the act of thinking itself."

Lewis Thomas (1980).

Introduction

In this chapter, I will apply an information processing approach[1] to the examination of standard and novel olfactory assessment paradigms, with particular attention to the assessment of olfactory identification in cognitively impaired subjects (specifically patients with dementia or mental retardation) and in normal aged subjects.

Information can be presented in any sensory modality for processing and subsequent testing of the efficiency and accuracy of that processing. In the cognitive psychology literature, it has been shown that some kinds of information are easier to process than others, are better encoded by subjects, are more efficiently stored, and serve as better retrieval aids, and that information presented in one sensory modality often evokes representations of the information in another. Furthermore, the relative efficiency of a given information-containing cue probably depends quite strictly on what the experimenter asks the subject to do with it (see Snodgrass, Levy-Berger, and Haydon, 1985).

[1] For clarity of exposition, a three-stage model of memory (see Atkinson and Schiffrin, 1968) is assumed, with attention as the front end, leading to a temporally and quantitatively limited short-term memory buffer, in which information is retained by rehearsal for either output, further processing, and/or storage, and an infinite-capacity long-term memory system that holds representations of prior events (episodic memory) and information (semantic memory) in codes related to the various sensory modalities. We assume that external stimuli are processed by subjects to give rise to internal representations of the information available in the stimulus, and that the quality of the processing ("deep" or "shallow"; see Craik and Lockhart, 1972) affects the completeness of the generated internal representations.

This information processing approach is commonly assumed when auditory and visual stimuli are used for assessment of perception, learning, and memory, but is not often considered in the study of olfactory functions (see Corwin, Serby, and Rotrosen, 1986; Gilbert, 1986). Specifically, tasks that use olfactory stimuli have encoding, processing, storage, and retrieval demands similar to those of tasks in other sensory modalities, with the important additional consideration that olfactory stimuli appear not to automatically engender verbal and/or visual representations, but are more likely to be evocative of affective states and significant episodes (Cain, 1979; Engen, 1987). That is, the standard hierarchy of task demands applies to tasks using olfactory stimuli, with the addition that the overall difficulty of olfactory tasks is most likely greater than that of tasks using verbal or visual material because of the particular encoding difficulties (Engen, 1987). In this regard, it is interesting to note that there are no commonly used English words for inability to smell or for varieties of suboptimal function, as there are for inability to see (blind, nearsighted) or hear (deaf, hard of hearing).

As an example of these principles, consider the problem of assessing subjects' ability to identify odors (given that they can detect them). Note that exactly the same considerations apply to measuring olfactory sensitivity and memory for odors.

The most difficult identification task is odor name recall, in which the subject is presented an odor and asked to supply the name; normal people are quite poor at tasks of this type (Cain, 1979). From an information processing point of view, this is not surprising, because in odor name recall tasks subjects must generate verbal labels for the presented odor, generate or find odor representations of the verbal labels stored in semantic memory, search for the best match, and finally decide on and emit the verbal response. The process can fail at many levels: absence of or poor quality verbal labels in the lexicon, search processes, and generation of representations, storage of the representations in short-term memory, and/or decision rules. That is, failures can be due to olfactory, language-related, or memory-related problems, or to some summation or interaction among relatively minimal impairments in more than one of these domains. Thus, recall tasks are not appropriate for the clinic.

Forced-choice tasks are substantially easier. In these paradigms, both verbal labels and olfactory stimuli are presented, and the subject is required to match a supplied name with one of several odors or an odor with one of several supplied names. The University of Pennsylvania Smell Identification Test (UPSIT; Doty, Shaman, Dann, and Kimmelman, 1984) is the most widely used forced-choice olfactory identification test. The measurement of performance for the UPSIT is the number of correct choices on the 40 trials, each with one odor sample and four verbal labels. This score can be transformed to percent correct for comparison with other tasks with different numbers of trials or choices.

In the case of forced-choice paradigms, the problem of generating either representations or labels is reduced, but not eliminated. Here, subjects are required to generate an olfactory representation for each supplied verbal label and match it to the supplied olfactory stimulus, or vice versa. It appears more likely that subjects approach tasks of this sort in the former fashion, given that people are notoriously poor at naming odors (see the above discussion of recall tasks). As well, the number of alternatives among which the subject must choose affects difficulty. In a one-to-four paradigm, the subject must keep track of a maximum of 16 possible combinations of verbal label and olfactory stimulus or representation per test question (odor-name, name-odor, odor-odor, name-name, for each of the four name-odor pairs in a trial) and then make some decision about which pair yields the best fit. However, 7 ± 2 "chunks" of information is the average capacity of short-term memory (Miller, 1956), and information can be retained in this store for not more than 6 to 9 without rehearsal or further processing (Brown, 1958; Peterson and Peterson, 1959). Given the constraints on the quantitative and temporal capacity of short-term memory, subjects must deploy some strategy that they must keep in mind while doing the task. For example, a subject may choose to proceed as we were taught to do to optimize performance on standard multiple-choice tests like the MCAT or the GRE: process one of the four word-odor pairs at a time, discarding it from further consideration if the fit between the representations is sufficiently poor, or marking it in some way as a possible correct match; the subject is then left with a smaller choice set from which to make a final decision. Elderly subjects, small children, and patients with Alzheimer's disease, frontal lobe damage, or depression all have been shown to have difficulty effectively generating and deploying such metacognitive strategies (Botwinick and Storandt, 1974; Lezak, 1983; Poon, 1988).

Perhaps the simplest behavioral paradigm that one can use to address the problem of assessing olfactory identification is a yes/no task in which subjects are presented with a series of single names paired with a single odor and asked, "Is this odor X?" Word-odor pairs are chosen so that sometimes the correct answer is yes, and sometimes no. Here, the memory and processing requirements are reduced to their absolute minimum: The decision is binary and simpler than in even a one-to-two-alternative forced-choice task (yes/no? versus A:X? or A:Y?); only a maximum of four representations need be generated and processed; the short-term memory load is minimized. Thus, it seems likely that failures on this task are more accurate reflections of failures in olfactory processes rather than in memory or language-related processes. The primary measure of performance on yes/no tasks reflects the ability of subjects to tell the two criterial classes of items apart, that is, discrimination. Additionally, yes/no tasks offer the advantage of allowing the tester to assess the nature of the decision rule that subjects use when they are uncertain about whether or not the word-odor pair matches, that is, response bias. Bias of this type cannot be measured easily in forced-choice paradigms in which subjects do not have the opportunity to say they don't know (as in the UPSIT).

Because I am arguing that yes/no paradigms may be preferred when assessing olfactory function in the clinic for both theoretical and pragmatic reasons, and to indicate that the same kind of task analysis could (and often should) be performed for other types of olfactory paradigms, I now present the assumptions underlying measurement of performance on yes/no tasks and the derivations of the measures associated with two approaches. In addition, I will demonstrate that the theoretical approach taken to calculating the measures of performance even within a single task can affect the results. A more complete discussion of the theoretical bases of measurement is found in Snodgrass and Corwin (1988).

Following the section on measurement of performance on yes/no identification tasks, I will turn to some data that show the real-life consequences of varying task requirements on olfactory identification performance.

Measuring Performance on Yes/No Tasks

Several factors must be taken into account when discrimination and response bias measures are calculated, especially when clinical populations are tested: The measures must be able to register performance at or below chance; the discrimination measure must not be spuriously affected by changes in response bias; and the response bias measure must not be affected by differences in discrimination. There are two theoretical approaches that meet these requirements: Two-High Threshold theory (2HT) and a variant of Signal Detection Theory with the bias measure C rather than the more commonly used β (SDT-C). Both theories use the same raw data: the number of HITS (yes responses to items for which the correct answer is yes, that is, yes responses to TARGETS); the number of MISSES (no responses to targets); the number of FALSE ALARMS (yes responses to items for which the correct response is no, that is, yes responses to DISTRACTORS); and the number of CORRECT REJECTIONS (no responses to distractors). Performance on yes/no tasks can be summarized in a stimulus-response matrix as shown in Figure 11.1.

Raw numbers of the different response types are usually transformed into conditionalized rates. Because the measures of performance under Signal Detection Theory discussed later are undefined for hit rates of 1 or false alarm rates of 0, we suggest routinely applying a linear transformation in calculating hit and false alarm rates (this has been done for all yes/no data presented here). The transformation, achieved by adding 1 to each cell of the stimulus/response matrix [as suggested for log-linear models by Upton (1978)], produces the following equations for hit and false alarm rates:

	TARGETS	DISTRACTORS	
	HITS	FALSE ALARMS	YES
RESPONSES			
	MISSES	CORRECT REJECTIONS	NO

FIGURE 11.1. Stimulus/response matrix for YES/NO tasks.

$$\text{Hit rate (HR)} = \text{Number of hits} + 0.5/\text{Number or match trials} + 1$$

$$\text{False alarm rate (FAR)} = \text{Number of false alarms} + 0.5/\text{Number of no-match trials} + 1$$

Note that hits and false alarms completely summarize performance; the miss rate is the complement of the hit rate, and the correct rejection rate is the complement of the false alarm rate. The next problem is to combine the response type rates into measures of the two aspects of performance. To do this, one must have a theory of the psychological processes generating responses; all measures of performance are based implicitly or explicitly on a theory.

Signal Detection Theory with C

In the application of SDT-C to olfactory identification, it is assumed that matching word-odor pairs (e.g., target pairs like "onion" label–onion odor) lie along a continuum of similarity. Some pairs will be judged to give rise to highly similar representations, some moderately similar representations, and some quite dissimilar representations. Mismatched pairs (e.g., pairs with distractors, such as "onion"–rose) will also be distributed along the similarity axis, but with a lower mean level of similarity (see Figure 11.2). These similarity distributions correspond to the "old" (target) and "new" (distractor) familiarity distributions for recognition memory SDT and the "signal plus noise" (target) and "noise alone" (distractor) sensation strength distributions for sensory SDT.

It is assumed that subjects cannot *directly* know if a word-odor pair matches or does not match, they cannot know which distribution is being sampled in a given trial. The only information of which the subject is aware

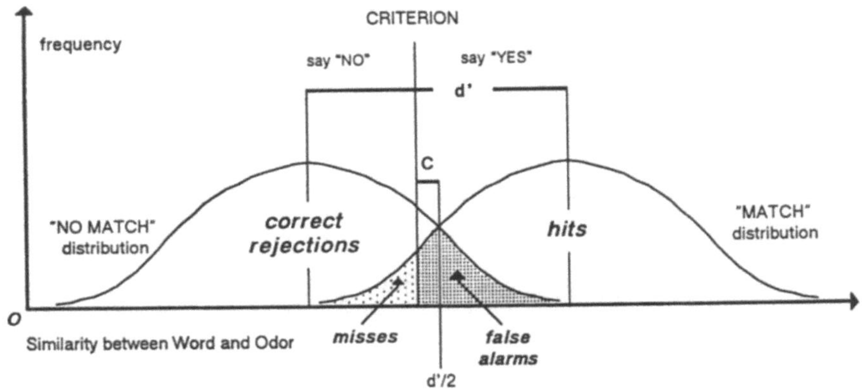

FIGURE 11.2. Assumptions underlying Signal Detection Theory. The discrimination measure d' is the distance between the means of the match and no-match distributions; the bias measure C is the distance of the criterion from the intersection of the match and no-match distributions, which is set equal to zero by dividing d' by 2.

is the perceived level of similarity between the representations evoked by the verbal label and the supplied odor sample. Therefore, the subject must establish a criterion value of similarity above which he or she says that the odor matches the word and below which he or she says that the pair does not match. This criterion is under the subject's control, and can be explicitly manipulated by the experimenter by changing the costs and rewards of different types of responses (the payoff matrix) or manipulating the ratio of target to distractor trials (the presentation probabilities).

In any version of SDT, the discrimination measure is the distance between the means of the match and no-match distributions; response bias is indexed by a measure locating the criterial value, a determination that may be done in several ways. The discrimination measure we will use is d'_L, the bias measure is C_L. These measures, when expressed as d' and C, may be calculated assuming unit normal curves for both distributions, but this is a computationally unwieldy procedure. We therefore suggest assuming logistic rather than unit normal distributions, as suggested originally by Noreen (1977). It has been shown that bias and discrimination measures calculated from logistic distributions behave identically to those calculated from normal distributions, and are substantially easier to compute. Details of the derivations are given in Snodgrass and Corwin (1988) and Noreen (1977).

The discrimination measure d'_L is the distance between the means of the match and no-match distributions in units of x, which is a standard score comparable to the z-score derived from the normal distribution. The x-score for a point on the logistic distribution may easily be calculated because the natural log of the ratio of the area above the x-score to the area below the x-score produces the x-score itself. The areas are estimated by the appropriate hit and miss and false alarm and correct rejection rates as for SDT with normal distributions. X-scores are used to derive the following d'-like measure, where ln is the natural logarithm:

$$d'_L = \ln \{ [HR(1 - FAR)]/[(1 - HR)FAR] \}$$

The problem of locating the criterion above which value the subject says "Yes the name matches the odor" and below which he or she says "no" may be solved in several ways. The most common method (SDT-β) locates the criterion by taking the ratio of the density (height) of the signal + noise distribution to the density of the noise distribution at this point, a procedure that yields β when applied to normal distributions. A similar ratio measure may be computed using logistic distributions, but both are not independent of their respective discrimination measures. Additionally, if the level of similarity varies from word-odor pair to word-odor pair (e.g., "onion"–garlic versus "onion"–rose), in order to compute B the subject must know a priori specifically which pairs of curves the word-odor pair come from in order to use them to decide if a given word-odor pair has a similarity value lying above or below the criterion. This operation appears to require perfect prior knowledge of the source of a pair's similarity value, that is, the distributions from which it comes—the very decision we are asking subjects to make as the test of performance.

It appears more reasonable to assume that the subject knows only how similar the word-odor pair is, and that match and no-match distributions for highly and less similar word-odor pairs are ordered in a mirrorlike fashion on the similarity axis (see Glanzer and Adams, 1985). In this way it is possible for subjects to locate the criterion with reference to the similarity axis alone rather than to the overlying pairs of distributions. The calculation of the bias measure based on these assumptions may be performed by setting the zero point on the similarity axis at the intersection of the match and no-match distributions, and calculating the distance of the criterion from this zero point. The procedure yields the criterion location C_L, in x-score units from the logistic distribution as follows:

$$C_L = 0.5 \{\ln [(1 - FAR)(1 - HR)]/[(HR)(FAR)]\}$$

Using this formula, a neutral criterion has C_L of zero; conservative nay-saying biases produce positive values of C_L and liberal biases negative values.

Two-High Threshold Theory

Threshold theories, in contrast to signal detection models, do not assume a continuously distributed set of word-odor similarities, but rather posit discrete states that are separated by thresholds. These thresholds are the esti-

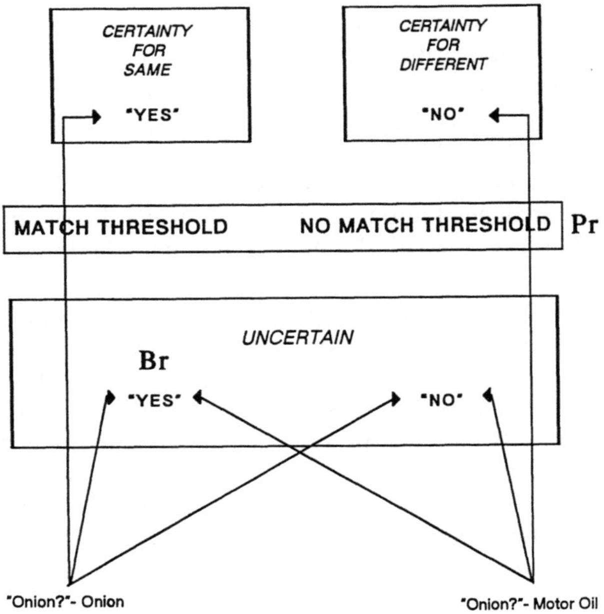

FIGURE 11.3. Assumptions underlying Two-High Threshold Theory. The discrimination measure Pr is the threshold above which subjects are in a certain state (that word and odor stimuli either match or don't match); the bias measure Br is the probability of guessing yes when in the uncertain state.

mates of discrimination. In the Two-High Threshold model, there are two thresholds that define three states, as shown in Figure 11.3. If a word-odor pair crosses the Match threshold, the subject is certain that the stimuli match and perforce must say "yes, the odor matches the word"; if the pair crosses the No Match threshold, the subject must say "no, the stimuli are different"; if the word-item pair fails to cross either threshold, the subject is in the uncertain state and will guess either yes or no with some probability. From a single pair of hit and false alarm rates, it is not possible to generate separate estimates of the Match and No Match thresholds; we therefore assume that they are equal, and call this common threshold Pr. The assumption of equal thresholds is supported by the mirror effect in recognition memory; this effect is the finding that as hits increase across manipulations, false alarms decrease in an inverse fashion (Glanzer and Adams, 1985).

Thus, the hit rate has two components: true identifications and a proportion of the lucky guesses from the uncertain state. False alarms in this model can be generated only from the uncertain state, and therefore the FAR is a direct estimate of the probability of saying yes when uncertain. Because of these relationships, the hit rate equals the number of items crossing Pr (which estimates Pr itself) plus the FAR. Rearranging these terms gives an estimate of Pr as:

$$\mathrm{Pr} = \mathrm{HR} - \mathrm{FAR}$$

the familiar correction for guessing, first suggested by Woodworth in 1938. Note as well that Pr is a linear transform of percent correct,[2] in this way discrimination performance on a forced-choice and a yes/no paradigm can be compared directly.

The bias measure in 2HT is the likelihood of guessing yes from the uncertain state. For a false alarm response, two events must occur. the subject fails to identify the pair as matching (which occurs with probability 1 − Pr), and the subject guesses yes when uncertain [which occurs with probability Br (the bias estimate)]. In this fashion, the FAR is related to Br, the bias estimate by:

$$\mathrm{FAR} = (1 - \mathrm{Pr})(\mathrm{Br})$$

Rearranging terms and substituting for Pr its estimate HR − FAR yields the following equation for Br expressed in terms of the hit and false alarm rates:

$$\mathrm{Br} = \mathrm{FAR}/[1 - (\mathrm{HR} - \mathrm{FAR})]$$

Neutral response bias yields Br = 0.5; values greater than 0.5 indicate liberal bias, less than 0.5, conservative bias.

---

[2]The transformation of percent correct to Pr is given by the following, if one assumes equal numbers of targets and distractors such that the total number of trials is given by 2*N*.

$$\text{Percent Correct} = \text{Hits} + \text{Correct Rejections}/2N$$
$$\mathrm{Pr} = (\text{Hits}/N) - (\text{False Alarms}/N)$$
$$\text{Correct Rejections} = \#\ \text{distractors} - \text{False Alarms}$$

Rearranging terms gives:

$$\text{False Alarms} = \#\ \text{Distractors} - \text{Correct Rejections}$$

Substituting for False Alarms in the equation for Pr and rearranging terms gives:

$$\begin{aligned}\mathrm{Pr} &= (\text{Hits}/N) - (N/N - \text{Correct Rejections}/N)\\ &= \text{Hits}/N - 1 + \text{Correct Rejections}/N\\ &= [(\text{Hits} + \text{Correct Rejections})/N] - 1\end{aligned}$$

Substituting 1/2 Percent Correct for [(Hits + Correct Rejections)/*N*] and rearranging terms gives:

$$\text{Percent Correct} = (\mathrm{Pr} + 1)/2$$

Thus, any time that Percent Correct is calculated, Two-High Threshold Theory is assumed to underlie the generation of Hits and False Alarms.

Comparison of the Two Approaches

It is not possible to decide which of the two theories is preferable on theoretical grounds or from examining the fits between the theories and actual data, and both approaches meet our desiderata for good measures of discrimination and bias. Furthermore, because each theory defines different sets of hit and false alarm rate pairs that represent equal discrimination and equal bias (i.e., different isodiscrimination and isobias curves), it is possible to get statistical differences between groups under one theory but not under the other. Consider the following example from a task in which there are 7 targets and 7 distractors:

					Discrimination		Bias	
Subject	*Hits*	*False alarms*	*HR*	*FAR*	*Pr*	d'_L	*Br*	C_L
1	2	0	0.313	0.063	0.25	1.92	0.08	1.75
2	3	1	0.438	0.188	0.25	1.22	0.25	0.86
3	4	2	0.563	0.313	0.25	1.04	0.42	0.27
4	5	3	0.688	0.438	0.25	1.04	0.58	−0.27
5	6	4	0.813	0.563	0.25	1.92	0.75	−0.86

Note that under 2HT each of the subjects showed identical discrimination in Pr, whereas under SDT, these five subjects yielded three different values of d'_L. For the bias measures, the range of values is different for the two theories, and no two subjects demonstrate exactly the same strategy. The most conservative subject (# 1) produces Br = 0.08 and $C_L = 1.75$, the most liberal subject (# 5) guesses yes when uncertain 75 percent of the time (Br = 0.75), and his criterion is located 0.86 x-score units to the left of the intersection of the match and no-match distributions. The difference between the most conservative and most liberal subject is 0.67 under 2HT and −2.61 under SDT-C. Thus, the same data summarized under the two theories can produce different results, although clearly there will be high correlations between the measures from SDT-C and those from 2HT. If the differences between groups are large, they will appear under both approaches. Our general rule of thumb is therefore to use both 2HT and SDT measures when summarizing performance.

However, on practical grounds one approach may be preferred. For example, if yes/no and four-alternative forced-choice tasks are to be compared, 2HT should be used. Percent correct is a linear transform of Pr, and conversion of four-alternative forced-choice scores to d' is unwieldy and possibly incorrect (Williams, 1980).

Let us now turn to some real data drawn from two studies, the first assessing olfactory identification abilities in normal elderly and Alzheimer's disease patients (Corwin et al., 1989), and the second looking at the same functions in children and young adults with idiopathic mental retardation and Down's syndrome compared to their controls (Hemdal and Corwin, 1989).

Olfactory Identification in Aging and Alzheimer's Disease

We administered the UPSIT and a YES/NO identification paradigm using UPSIT stimuli to 93 cognitively normal adults from age 24 to 84 and 16 mildly and 13 moderately impaired Alzheimer's disease patients. The YES/NO task used 20 of the 40 UPSIT stimuli, chosen to cover the range of olfactory qualities used in the UPSIT, arranged in 10 pairs so that members of the pairs were moderately similar in quality according to the Doty, Shaman, and Dann ratings (1984). There were 1 target and 1 distractor trial per item, using the other item in the pair as the distractor; e.g., for the pair rose–bubblegum, the subject was presented the rose stimulus twice; on the target trial he or she was asked if it was rose, and on the distractor trial, he or she was asked if it was bubblegum. This procedure yields 20 target and 20 distractor trials, for a total of 40 trials (the same number of trials as the UPSIT). The normal subjects had normal memory for age and WAIS vocabulary as determined by the Guild Memory Test or MiniMental Status Exam plus the Global Deterioration Scale (Crook, Ferris, and Gilbert, 1980; Folstein, Folstein, and McHugh, 1975; Reisberg, Ferris, deLeon, and Crook, 1982) and no condition known to affect olfactory function (Schiffman, 1983). Alzheimer's disease (AD) patients met standard criteria for probable AD (McKhann et al., 1983) and primary degenerative dementia (American Psychiatric Association, 1983). They were divided into mildly and moderately impaired subgroups based on either Guild Memory Test scores or MiniMental State exam scores (Folstein, Folstein, and McHugh, 1975), (GDS 3 and 4 or MMS $>$ 19 for mild). The cognitively normal subjects were divided into subgroups by decade of age. The UPSIT was scored as percent correct; the YES/NO task was scored as percent correct [(Pr + 1)/2], Pr and d'_L, and Br and C_L.

Figure 11.4A below shows UPSIT and YES/NO percent correct scores for all subjects. As can be seen in the figure, although the UPSIT and YES/NO scores show the same general pattern, the curves are by no means parallel. Figure 11.4B plots this relationship on a within-subject basis; it is apparent that the relationship between UPSIT and YES/NO scores worsens as performance worsens. In fact, Pearson's r is 0.57 ($p < 0.0001$) between the pairs of scores for the normal subjects and 0.21 (ns) for the AD patients.

Two mixed-design ANOVAs using subject group as the between-subject factor and test score (UPSIT and YES/NO percent correct) as levels of a

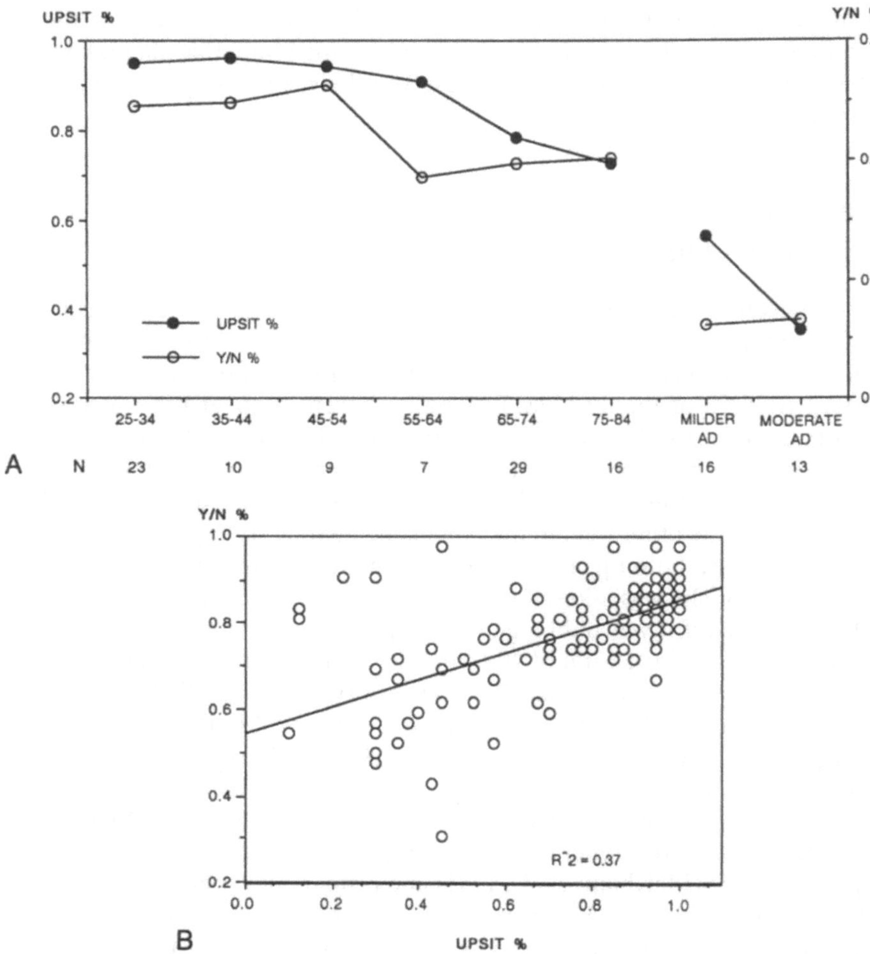

FIGURE 11.4. A. Performance by normal and Alzheimer's disease subjects on the UPSIT and YES/NO olfactory identification tasks. B. Correlation between UPSIT and YES/NO scores.

repeated factor were calculated for the normal subjects alone (6 levels), and for the two groups of AD patients with their age-matched controls (65 to 74-year-olds). In the analyses of untransformed data, there was an interaction between subject group and measure [normal subjects: $F(5, 88) = 8.60, p < 0.0001$; AD patients versus controls: $F(2, 55) = 10.07, p < 0.0001$]. Because the ranges of scores and levels of chance performance are different on the two tasks, similar calculations were performed on z-transformed scores with identical results.

To more closely examine the source of the interactions between group and measures, we calculated separate one-way ANOVAs for UPSIT and YES/NO percent correct. In the normal subjects, UPSIT scores yielded

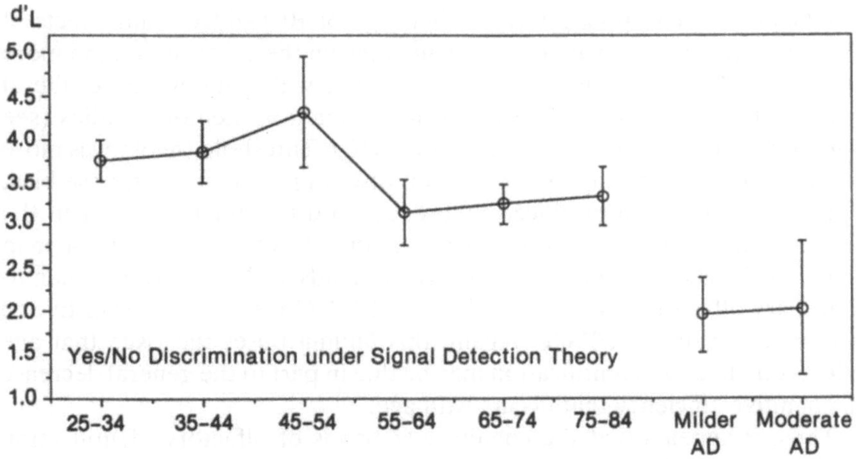

FIGURE 11.5. Discrimination scores of normal and Alzheimer's disease subjects under Signal Detection Theory on the YES/NO olfactory identification task.

a main effect of subject group at $F(5, 88) = 9.86$, $p = 0.0001$, with 6 significant pairwise differences (of a possible 15) in the expected directions. This finding replicates the pattern of results of the normative UPSIT study (Doty, Shaman, Applebaum, Giberson, Siksorski, and Rosenberg, 1984).

The main effect of (normal) subject group in the analysis of percent correct from the YES/NO task was significantly weaker than the UPSIT [$F(5, 88) = 2.35$, $p = 0.05$], with fewer (3) significant pairwise differences. In the analyses for the AD patients versus their controls, UPSIT percent correct yielded $F(2, 55) = 23.59$, $p < 0.0001$, with all pairwise differences significant. On the other hand, YES/NO percent correct yielded $F(2, 55) = 7.54$, $p = 0.001$, with significant pairwise differences between each patient group and the controls but not between the milder and moderately impaired patient groups. Note as well in the figure that the AD patients do relatively better on the YES/NO task than on the UPSIT, as predicted by the task demand hypothesis.

We next looked at the remaining (Br, d'_L and C_L) performance measure on the YES/NO task for the normal sample and for patients versus their controls; d'_L is plotted in Figure 11.5. For d'_L in the normal subjects, in contrast to the analysis of YES/NO percent correct, there was no main effect of group [$F(5, 88) = 1.36$, $p = 0.25$] and a single significant pairwise difference (between the 45–55 and 65–74 year olds, an effect that appeared as well in the analysis of percent correct). Here we see the effect of choice of theory on results.

For the patients versus their controls, d'_L did differentiate among groups [$F(2, 55) = 3.43$, $p = 0.04$], but as in the analysis of YES/NO percent correct, the pairwise differences between the more impaired AD patients versus their controls and between the two groups of patients did not reach significance. For response bias, neither C_L nor Br yielded main effects of diagnosis. But the pairwise comparison between the controls and the more impaired AD patients was significant for Br, with patients more liberal than controls, a result also obtained in recognition memory studies (see Snodgrass and Corwin, 1988). Thus, Two-High Threshold theory was more sensitive to age and diagnosis effects on discrimination and response bias, but neither approach produced differences in discrimination between the mild and moderate AD groups, suggesting that the effect of severity seen in the UPSIT may be due to the cognitive demands of the task rather than to a primary olfactory deficit. As well, the YES/NO task was less sensitive to age effects than the UPSIT. Again, this finding raises the issue that age effects on olfactory identification may be due in part to the general decrease in cognitive efficiency that occurs with age.

Thus, it appears that the cognitive demands of olfactory identification tasks influence the results of studies with cognitively suboptimal subjects. One could argue that the subjects who performed most poorly were not even able to understand the task requirements, or that the choice of the particular subset of UPSIT items used in the YES/NO task influenced performance. We next present results from a study that explicitly controlled for such possibilities and test the generalities of the findings of the first study to two other cognitively impaired samples. As well, in this study we more closely examined the relationships between response bias in the YES/NO task and discrimination in both tasks.

Olfactory Identification in Down's Syndrome and Idiopathic Mental Retardation

These data are selected from Hemdal and Corwin, 1989. In this study, 20 Down's syndrome (DS) patients (mean age 23.1, mean WISC or WISC-R IQ 45.85), 15 subjects with idiopathic mental retardation (IMR) (mean age 20.6, mean IQ 49.87), and 21 age-matched controls of normal intelligence (mean age 23.8) were tested with the 20 most familiar UPSIT items (modified UPSIT, M-UPSIT), a yes/no olfactory task with the identical items, and two control tasks using the one-to-four forced-choice and yes/no procedures with common objects presented to the touch as the stimulus material. No subject was tested with the olfactory stimuli unless he or she achieved scores of at least 80 percent correct on both tactile tasks, thus ensuring that all subjects could understand and comply with the task requirements.

The mean percent correct on the M-UPSIT and YES/NO tasks are plotted in Figure 11.6. Again we see that the differences between groups are smaller for the YES/NO task than for the M-UPSIT.

A mixed-design ANOVA with subject group (DS, IMR, and controls) as the between-subject variable and M-UPSIT and YES/NO percent correct as levels of a repeated factor was calculated as before. As in the previous

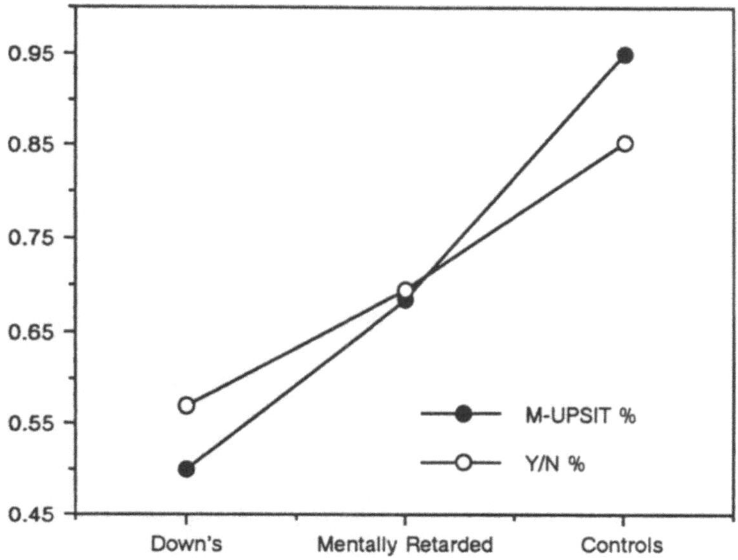

FIGURE 11.6. Performance by normal controls, Down's syndrome (DS), and Idiopathic Mentally Retarded (IMR) subjects on the UPSIT and YES/NO Olfactory identification tasks.

experiment, there were main effects for subject group [$F(2, 52) = 93.15$, $p < 0.0001$], no effect for measure [$F(2, 52) < 1$], and a significant group by measure interaction [$F(2, 52) = 6.76$, $p < 0.003$]. The interaction is due to significantly better performance by the controls on the M-UPSIT than on the YES/NO task, worse performance on the M-UPSIT than on the YES/NO task by the Down's subjects, and no difference between tests for the Idiopathic Mental Retardation group (see Figure 11.6). The interaction appears as well in analyses of z-transforrned data. Here again we see the effects of cognitive processes on olfactory task performance: the normals benefit from the presence of alternatives, whereas the DS subjects perform as though the alternatives interfere with selection of correct responses.

The one-way analyses of variance on each test separately with planned comparisons show that the DS group performed the most poorly, the IMR group produced intermediate performance, and the controls scored highest on all three measures of discrimination [UPSIT percent correct, Pr and d'_L; minimum $F(2, 53) = 45.52$, $p = 0.0001$ for d'_L].

Similar analyses were calculated for the bias measures C_L and Br. These results are plotted in Figure 11.7; bias under both theories was more liberal in the cognitively impaired samples than in controls [minimum $F(2, 53) = 10.27$, $p < 0.0001$ for C_L]. Furthermore, the correlations between response bias under both theories and M-UPSIT performance were significant for the DS subjects in the direction that the better the UPSIT performance, the more liberal the response bias [for Br, $r(18) = 0.45$, $p = 0.05$], but this relationship did not hold in the other two groups despite

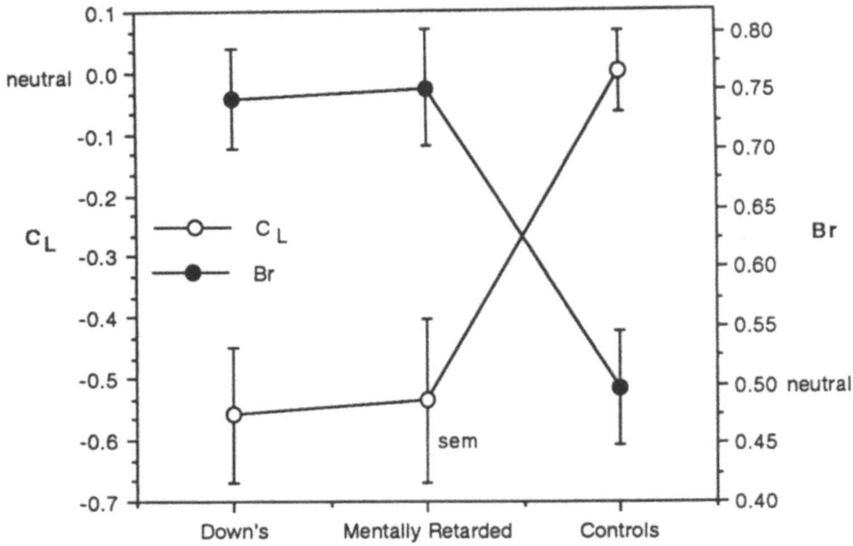

FIGURE 11.7. Response bias under Signal Detection and Two-High Threshold Theories for controls, Down's syndrome (DS), and Idiopathically Mentally Retarded (IMR) subjects on the YES/NO task.

abnormally liberal bias in the IMR subjects [maximum $r(13) = 0.45$, ns for C_L in the IMR group). Given the significant correlation between bias and M-UPSIT scores in the DS sample, we used Br as a covariate in an analysis of M-UPSIT scores to see whether accounting for bias effects in M-UPSIT performance affected the magnitude of the differences among groups; this procedure did reduce the F ratio from 67.2 to 55.31.

How does abnormal bias on the YES/NO procedure relate to bias on the M-UPSIT? We can examine bias on the UPSIT as a function of the position of alternatives on the list of four possible choices, treating incorrect choices of items in a given position as false alarms and correct choices at the same position as hits to calculate C_L and/or Br for each position. These calculations also permit us to see whether performance on the M-UPSIT meets one of the major criteria for the use of forced-choice paradigms, namely that subjects choose each of the alternative positions equally when uncertain (Williams, 1980). The resulting measures are graphed in Figure 11.8.

Note that the normal subjects guess that the correct choice is in position 2 over the other list positions, and that both the DS and IMR subjects show a preference for guessing the last position when uncertain. It turns out that there are more correct choices in position 2 for the 20 most familiar original UPSIT items selected for the M-UPSIT (8 of 20); the normal subjects were able to use this information to deploy a probability-matching strategy based on the actual presentation probabilities, whereas the DS and IMR subjects could not. In addition, position 4 was chosen more often by the

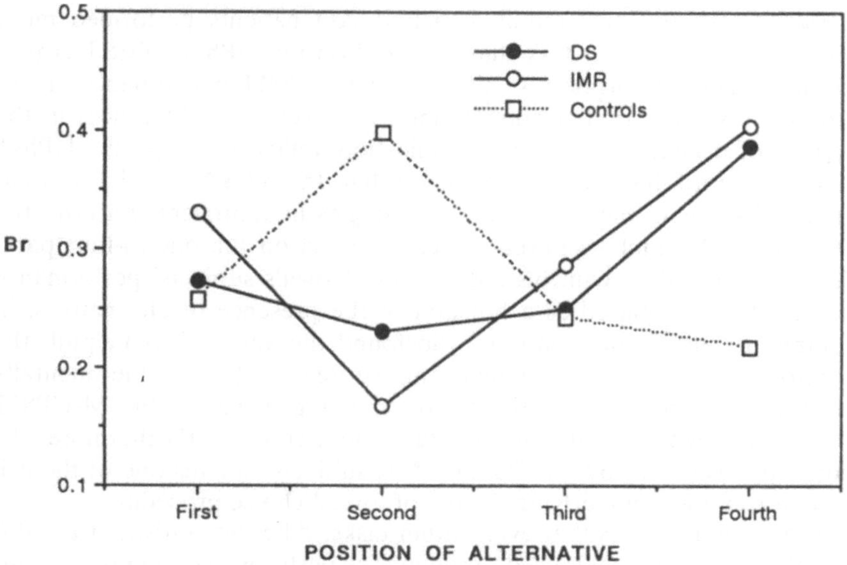

FIGURE 11.8. Response bias under Two-High Threshold Theory on the UPSIT by position of response alternative.

DS and IMR subjects when they were uncertain; that is, the impaired subjects showed a significant recency effect. The combination of these two effects produced a diagnosis by position interaction in the analysis of Br at $F(6, 51) = 6.53, p < 0.001$. Again, this suggests that the cognitive demands of the one-to-four alternative task affect performance: it appears as though the impaired subjects use a suboptimal approach, processing one word-odor pair at a time and, if it is rejected, moving on to the next pair until the last is reached and then settling for that, making little recursion through the choices. Importantly, these findings suggest that the performance of cognitively abnormal subjects may violate one of the assumptions underlying forced-choice procedures; that is, the DS and IMR subjects demonstrate a clear bias towards the last choice in the list rather than distributing their guesses randomly.

Summary

In this chapter, I have raised and attempted to address several issues in the study of olfactory processes in cognitively impaired subjects. The first, and most important, is that the cognitive demands of the olfactory tasks used clearly affect performance in a variety of cognitively impaired groups. The most dramatic examples of the cognitive demand effect in the present data are the differences between UPSIT and YES/NO task performance in AD patients: the one-to-four alternative task produced a severity effect, the

YES/NO task did not; mildly impaired AD patients performed more poorly on the YES/NO task than they did on the UPSIT; YES/NO performance did not correlate significantly with UPSIT performance in this group. Less dramatic but equally important were the differences in the pattern of decline in olfactory identification ability with age: the UPSIT yielded a more linear decline with age than the YES/NO task. As well, impaired subjects may use different strategies in approaching a task: the DS and IMR samples showed a recency effect on the one-to-four paradigm, whereas their controls did not; the Down's subjects' performance was relatively impaired as a function of the presence of alternatives, in contrast to that of the controls, who found the alternatives helpful; the controls used a probability-matching strategy, whereas the mentally retarded subjects did not; the discrimininating power of the M-UPSIT decreased when YES/NO bias was taken into account. Furthermore, the position bias shown by the DS and IMR subjects violates one of the primary assumptions underlying the use of forced-choice procedures.

The second point is that even within tasks, different statistics from different theoretical models used to summarize performance can produce different results: in these and other studies (see Snodgrass and Corwin, 1988), Two-High Threshold Theory was more sensitive than SDT-C to diagnosis and age effects.

What is the researcher or clinician to do? Given that olfactory stimuli have qualitatively different properties than visual and auditory stimuli, the issues raised in this chapter are particularly important to address. On pragmatic and theoretical grounds, I suggest using the simplest task that will give an interpretable result, or, if possible, more than one task to reveal different aspects of performance. It is important as well to ensure that cognitively impaired subjects understand and can comply with task requirements; that is, use control tasks in other sensory modalities when possible.

Can this approach be applied to studies of odor detection? The standard signal detection sensitivity paradigm employs a set of stimuli of varying intensity plus odorless blanks. To get a stable estimate of individual sensitivity to each intensity, many trials are needed, but the olfactory system habituates quickly (see Meiselman and Rivlin, 1986). Forced-choice paradigms have the advantage of requiring relatively few trials, but the drawbacks are task demand confounds and the problem that single-point thresholds do not well characterize sensitivity [the issue that motivated the development of SDT methods in psychophysics (see Galanter, 1962; Green and Swets, 1974)].

To get around the problem of habituation in yes/no detection paradigms, normative studies can be performed to establish those intensities around the limits of normal sensitivity, so that reasonable numbers of yes/no trials can be administered to patient samples (we have been able to use a 60-trial sensitivity task with AD patients, albeit with some difficulty). As well, an average d'_L or Pr across the presented stimulus intensities can be calculated for each subject to get a more stable estimate of performance that still reflects some psychophysical "truth"; additionally, the variances around an average value of d' can be compared (Gourevitch and Galanter, 1967). At this stage in the study of abnormal olfaction, investigators are looking primarily for group rather than individual or condition-wise differences, so the use of larger samples than is common in "normal" psychophysics is warranted, especially given the inherent variance in symptom type and severity among patients with a particular diagnosis.

Acknowledgments. I thank Paula Hemdal, Gay Snodgrass, and Adam Reeves for their helpful suggestions. This work was supported by the Veterans Administration.

References

American Psychiatric Association (1983). *Diagnostic and Statistic Manual of Mental Disorders*, 3rd ed. Washington, DC: American Psychiatric Association.

Atkinson, R.C., and R.M. Shiffrin (1968). Human memory: A proposed system and its control processes. In K.W. Spence and J.T. Spence (Eds.), *The Psychology of Learning and Motivation*, Vol. 2. New York: Academic P, pp. 89–195.

Botwinick, J., and M. Storandt (1974). *Memory Related Functions and Age.* Springfield, IL: Charles C Thomas.

Brown, J.A. (1958). Some tests of the decay theory of immediate memory. *Q. J. Exp. Psychol.*, *10*, 12–21.

Cain, W.S. (1979). To know with the nose: Keys to odor identification. *Science*, *203*, 467–470.

Corwin, J., M. Serby, and J. Rotrosen (1986). Olfactory deficits in AD: What we know about the nose. *Neurobiol. Aging*, *7*(6), 580–581.

Corwin, J., P. Larson, A.S. Ward, and M. Serby (1990). Olfactory identification deficits in aging and depression: A function of task demands? *J. Clin. Exp. Neuropsychol. 12*, 90.

Craik, F.I.M. and R.S. Lockhart (1972). Levels of processing: A framework for memory research. *J. Verbal Learning Verbal Behav.*, *11*, 268–294.

Crook, T., J.G. Gilbert, and S. Ferris (1980). Operationalizing memory impairment for elderly persons: The Guild Memory Test. *Psychol. Rep.*, *46*, 1316–1318.

Engen, T. (1987). Remembering odors and their names. *Am. Sci.*, *75*, 497–504.

Folstein, M.F., S.E. Folstein and P.R. McHugh (1975). MiniMental State: A practical method of grading the cognitive state of patients for the clinician. *J. Psychiatr. Res.*, *12*, 189–198.

Galanter, E. (1962). Contemporary psychophysics. In T. Newcomb (Ed), *New Directions in Psychology*. New York: Holt, Reinhart and Winston, pp. 87–156.

Glanzer, M., and J. Adams (1985). The mirror effect in recognition memory. *Memory Cognition*, *13*, 8–20.

Gourevitch, V., and E. Galanter (1967). A significance test for one parameter isosensitivity functions. *Psychometrika*, *32*, 25–33.

Doty, R.L., P. Shaman, S.L. Applebaum, R. Giberson, L. Sikorski, and L. Rosenberg (1984). Smell identification ability: Changes with age. *Science*, *226*, 1441–1443.

Doty, R.L., P. Shaman, M. Dann and C. Kimmelman (1984). University of Pennsylvania Smell Identification Test: A rapid olfactory function test for the clinic. *Laryngoscope*, *94*(2), 176–178.

Engen, T. (1987). Remembering odors and their names. *Am. Sci.*, *75*, 496–503.

Gilbert, A.N. (1986). The neuropsychology of olfaction in Alzheimer's disease. *Neurobiol. Aging*, *7*(6), 578–579.

Green, D.M., and J.A. Swets (1974). *Signal Detection Theory and Psychophysics*. Huntington, NY: Krieger.

Hemdal, P., and J. Corwin (1989). Olfactory dysfunctions in Down's syndrome. *J. Clin. Exp. Neuropsychol.*, *11*(1), 65.

Lezak, M.L. (1983). *Neuropsychological Assessment*, 2nd ed. New York: Oxford University Press.

McKahnn, G. D. Drachman, M. Folstein, R. Katzman, D. Price, and E.M. Stadlan (1983). Clinical diagnosis of Alzheimer's disease; Report of the NINCDS-ADRDA work group under the auspices of Department of Health and Human Services Task Force on Alzheimer's Disease. *Neurology* (Cleveland), *34*, 939–944.

Meiselman, H.L. and R.S. Rivlin (Eds.) (1986). *Clinical Measurement of Taste and Smell*. New York: Macmillan.

Miller, G.A. (1956). The magical number seven plus or minus two: Some limits on our capacity for processing information. *Psychol. Rev.*, *63*, 81–97.

Noreen, D.L. (1986). Relations among some models of choice. Paper presented at the Tenth Annual Mathematical Psychology Meetings, University of California, Los Angeles, CA, August.

Peterson, L.R., and M.J. Peterson (1959). Short term retention of individual verbal items. *J. Exp. Psychol.*, *58*, 193–198.

Poon, L. (Ed.) (1986). *Handbook for Clinical Memory Assessment of Older Adults*. Washington, DC: American Psychological Association.

Reisberg, B., S.H. Ferris, M. DeLeon, and T. Crook (1982). The Global Deterioration Scale for the assessment of primary degenerative dementia. *Am. J. Psychiatry*, *139*, 1136–1139.

Schiffman, S. (1983). Taste and smell in disease. *N. Engl. J. Med.*, *308*, 1275–1279.

Snodgrass, J.G., G. Levy-Berger, and M. Haydon (1985). *Human Experimental Psychology*. New York: Oxford University Press.

Snodgrass, J.G. and J. Corwin (1988). Pragmatics of measuring recognition memory: Applications to dementia and amnesia. *J. Exp. Psychol.*, *117*(1), 34–50.

Thomas, L. (1980). Notes of a biology watcher: On smell. *N. Eng. J. Med.*, *302*, 731–733.

Upton, G.J.G. (1978). *The Analysis of Cross Tabulated Data*. Chichester, England: Wiley.

Williams, M.V. (1980). Receiver Operating Characteristics: The effect of distribution on between-group comparisons. In L.W. Poon (Ed.), *Aging in the 1980's: Psychological issues*. Washington, DC: American Psychological Association, pp. 558–567.

Woodworth, R.S. (1938). *Experimental Psychology*. New York: Holt.

12
A Neurolinguistic Perspective of the Study of Olfaction

Karen L. Chobor

Introduction

It is widely accepted that naming odors is a difficult task. In normal subjects, it is common to smell an odor and to recognize that it is familiar and belongs to a general class or category, but to be unable to come up with a specific label for it. This is known as the tip-of-the-nose phenomenon (Lawless and Engen, 1977). Several factors play a role here: (1) The encoding of an odor is highly experience-specific, so retrieval out of context is an arduous task; (2) it is difficult to "imagine" an odor, that is, there are no odor images as there are visual images; (3) there is no universally accepted classification system for odors (i.e., pungent, spicy, etc.), leading to greater reliance on specific item associations; (4) the associations made with odors tend not to be lexical, that is, they are made with a context and/or an object within that context.

In cognitively impaired subjects for whom naming in the standard (auditory and visual) modalities is difficult, the question of odor naming is even more complicated. Ordinary prompts used to elicit naming tap into various levels of knowledge, as is indicated by tip-of-the-tongue experiments. For example, a person might have information about the phonological or morphological aspects of a word, or there might be intact semantic knowledge, such as concerning the function or attributes of an object. With odors and their typically weak links with verbal labels, a correct response to prompting in these cases poses a question. Can olfactory stimulation be considered a separate form of prompting, as it taps into an even deeper semantic store, conjoined with affective memory?

This chapter will address that question and others related to the nature of olfactory naming and its relationship to other modalities. It will explore the linkage between olfactory naming and memory, focusing on the all-or-none nature of the latter. It will look at pathological states, particularly misnaming in aphasia and dementia, and the possible parallels in response to olfactory and other-modality stimulation. The chapter will discuss data collected in a population of aphasic and demented subjects with misnaming through the use of a comprehensive assessment battery. Finally, it will suggest that a greater focus on the interrelationship of olfactory and linguistic function be made in the investigation of this much-neglected modality.

Language-Odor Interaction: A Review

Odor Identification and the Tip-of-the Nose State

Despite the relatively inferior status relegated to the human olfactory system by evolution, humans have a profound acuity for odor detection and recognition. It is estimated that trained persons can distinguish among at least 10,000 odors upon presentation out of context, while for some experts (e.g., perfumers) this number can be a high as 100,000 (Dodd, 1989). Further, under carefully controlled conditions of stimulus presentation, odor discrimination approaches levels seen in audition and vision (Wright and Burgess, 1971; Cain, 1977; Mozell, 1983), although odor discrimination depends on successive presentation, whereas simultaneous presentation is possible for the other modalities.

Just as exceptional sensory skill in normal human olfaction has been widely noted, so have there been many descriptions of a profound inability to identify odors. Identification involves detection, discrimination, recognition, and retrieval of a label or name. In normal subjects, misidentification most often occurs at the point of retrieval (Cain, 1977). At least three factors have been shown to impede odor identification: (1) the sluggish formation of associations between odors and names, (2) the failure to retrieve the name in spite of a well-formed association, and (3) the inherent confusability of the stimuli (Cain, 1979). If tested with uncommon stimuli (such as laboratory chemicals), as studied by Engen and Pfaffman (1960), subjects perform poorly because they do not know the appropriate odor-name associations, nor do they have the ability to develop them quickly (Cain, 1979). If tested with common stimuli but required to name them initially with unaided recall, subjects will perform moderately well and will progress as better names come to mind. If tested with common odors and aided in recall, they will perform about as well as inherent confusability will allow (Cain, 1979). But there is also evidence that subjects provided with correct labels do better than those provided only with associations made by others, yet did no better than when given no code (Engen and Ross, 1973). This implies that there is a blocking effect associated with irrelevant associations.

The significant difficulty that most humans have in identifying odors (reported to be as low as 43 percent by Lawless and Engen, 1977) has proved to be frustrating in the clinical setting, as it is not always clear how much of the identification deficit can be attributed to linguistic factors (e.g., absence of lexical labels) and how much to cognitive ones (e.g., memory or perceptual skill), not to mention sensory determinants. With improvements in olfactory testing batteries (Cain and Krause, 1979; Cain et al., 1983; Doty et al., 1984), it is becoming increasingly easier to determine levels of sensory skill (i.e., detection thresholds), therefore allowing greater focus on the linguistic and cognitive mediation necessary for identification.

It is interesting to note that Murphy and Cain (1986) found that blind persons are no better than sighted ones at detecting the presence of an olfactory stimulus, but once it is detected, the blind person is better able to identify it. One interpretation of this is that blind individuals learn exceptional compensation techniques through olfaction to make up for deficiencies in visual skill, particularly in the encoding of a stimulus. An obvious implication of this is that all humans, assuming an intact cognitive profile, have the inherent ability to use olfaction in parallel with other sensory modalities. The pattern is well documented, however, that most humans simply do not identify odors well, even given a normal sense of smell (Cain and Gent, 1986), and this is true even for highly familiar odors such as Juicy Fruit chewing gum or Johnson & Johnson baby powder (Cain and Krause, 1979). Here, a point of interest is that research on absolute judgment of odor quality implies that persons have a better ability to attach names to unfamiliar odors consistently (Engen and Pfaffman, 1960).

Despite the difficulty in actual identification or naming, most individuals are able be provide some information about the stimulus. Statements such as, "I know what it is but I can't come up with the name" are very common, both in everyday life and in the clinical setting. This phenonemon, known as the tip-of-the-nose state (Lawless and Engen, 1977), is analogous to the tip-of-the-tongue state (Brown and McNeill, 1966) found in both normals and aphasics. In a study of aphasic patients, Barton (1971) has shown that a significant number are able to supply the initial letter (62 percent) or syllable length (72 percent) of unevoked nouns, suggesting the presence of the word in some abstract form even though it cannot be articulated. In the tip-of-the-nose state, subjects evidence information about the odor quality but not its name. The similarities seem to end there. While hints can lead to a correct odor label, reading a definition of a word can actually induce the tip-of-the-tongue state. In a series of experiments by Engen and his colleagues (1987), college students were asked two sets of questions following exposure to a variety of odorants: the first set was concerned with the quality of an odor, that is, the name of a similar odor, its general category, an object that might produce it; and the second with information about the odor name, that is, a name with a similar meaning, sound, or number of syllables or letters. Subjects performed well in response to the first set of questions, yet the reverse was true for the second set. This is consistent with the common observation that odor knowledge is often greatly dissociated from odor expression.

Although verbal-odor links are weak and are not spontaneously used in the recall of an odor, verbal labels have been shown to influence performance. Engen and his colleagues (1987) presented subjects with a multiple-choice scratch-and-sniff task that utilized four alternative names per odor, all of which were read to the subject prior to odor presentation. One of each set was correct, while the other three included (1) items greatly different from the target (e.g., for "grape," the foils were pizza, turpentine, and clove) or (2) items very similar to the target (e.g., for "grape," the foils were melon, plum, and strawberry). Performance by the first group was 93.3 percent and by the second, 50 percent. This suggests that odors elicit more generalized semantic categories, the differentiation of which depends on other modalities.

Olfactory Coding via Associations

Categorization of odors is often done in terms of similarity of odors and similarity of context or category of object in which odors may be perceived. Engen (1987) describes the five general ways in which an odorant is named; for example, for "lemon": (1) correct; (2) related odor word ("fruity"); (3) object association ("Pledge"); (4) related sensation ("bitter"); or (5) no response at all. Odors tend to be described in terms of personal references rather than by more universal terms. Idiosyncratic responses are typical. For example, Engen (1982) reported this response: "the smell of a dusty book in the John Hay library," while one of our patients reported, "the smell of a wine cellar I once visited in Normandy."

Recognition performance of odors does not attain the high levels of acquisition found for pictures (Engen et al., 1973) or for sounds, and associations to many odors are acquired more slowly than to visual or verbal stimuli (Davis, 1975; Davis, 1977). However, more rapid acquisitions of associations to familiar odors have been found (Lawless and Engen, 1977, Davis, 1977). Davis (1975) suggests that odors might activate preexisting cognitive mechanisms, the net result of which is a relative improvement of paired associates performance. Lawless and Engen (1977) found that mediational schemes were constructed by their subjects as mnemonics, and that such schemes correlated well with correct choice of response. The schemes were also idiosyncratic, unlike the predictable mediators found by subjects performing nonsense syllable or other linguistic tasks.

Odors are notably less effective than simple geometric figures as stimuli for verbal association tasks (Davis, 1977), but if the odors are highly distinctive, then they compare well with paired associates performance of visual material. In visual confrontation naming, the judgment is dependent on previously learned associations between the visual stimulus and the verbal label, and the same is true for auditory stimulation. In odor stimulation, however, it is difficult to form the same associations. In some odor memory experiments, subjects prefer to work with odor names by forming odor-word associations, but later, when the odor is presented for identification, the odor sensation does not necessarily enhance retrieval of the word from semantic memory (Eich, 1978). Using pairwise discrimination tasks, Engen and Ross (1973) compared performance between olfactory and visual modalities. Using paired items on recognition tasks, subjects are more likely to recognize the odor of an onion if it is paired with Scotch rather than garlic; this was observed especially in Korsakoff's patients (Potter and Butters, 1980). If odors are too similar, subjects' performance approaches chance for all retention intervals (Mair et al., 1980).

To explore subjects' lexicon of names or labels for odors, Engen and Eaton (1975) asked a group of 60 persons to list 20 odors and, for comparison, 10 colors that they could remember as quickly as possible. The average scores were 48 and 12 seconds, respectively, and it is noteworthy that individual differences in labels were especially high for odors. Cain (1980) reports that subjects have difficulty in learning new word associations even when they are used to make odors more identifiable. In one experiment (Engen and Ross, 1973), subjects were presented with a diverse set of 20 odorants, one at a time. They were instructed to identify each odorant by label or by means of a short description of its use. The average number of correct identifications was 54.5 percent.

Engen (1987) points out that the working assumption is that odor stimuli cause sensations of certain qualities that subjects can describe and categorize, and that the problem with this approach is its emphasis on words, which have a tenuous connection to odor perception. It is my view that these connections, however tenuous, may offer a glimpse into the structure of olfactory knowledge and, perhaps more significantly, into meaning systems in general.

Semantic Systems

There have been numerous studies in aphasic patients that support the idea of categorical organization of semantic systems. Distinctions have been observed between abstract and concrete categories of verbal concepts (Goldstein, 1948; Shallice and Warrington, 1975), between animate and inanimate object names (Warrington and Shallice, 1984), between fruits and vegetables (Hart et al., 1985), and between colors and other qualities of objects (Yamadori and Albert, 1973), among others, such as body parts and actions (Goodglass, 1966). Some of the category-specific systems are modality-independent, such as colors in relation to unimodal visual processing, but in most cases, the category specificity has been reported within one modality, such as the abstract:concrete distinction.

There is a suggestion that a pattern of semantic dissociation can be found between visual and verbal modalities (Warrington and Shallice, 1984), with greater consistency of the response within a modality than between modalities. Little work has been done on the distinction of semantic knowledge within the olfactory modality, or the possible dissociation between olfaction and the other modalities. Within olfaction, patterns of performance are relatively consistent. Assuming that modality-specific systems are present, then there is the possibility that an overlap or bias among these systems might give a spurious impression of category specificity. It is interesting to look at how these systems might overlap in creating category specificity. For example, in the organization of knowledge of a category of foods, how much of the semantic system is anchored in olfactory knowledge, how much in visual, in tactile, and so on? Warrington (1975) supports the notion of a hierarchical process of meaning by evidence that in children (Clark, 1975) and adults (Putnam, 1975), there is an incomplete semantic representation for concepts at the limit of the individual's verbal knowledge.

Brown (1988) proposes a convincing argument for the hierarchical arrangement of meaning in his microgenetic description of language disorders. The deepest level in the process of lexical selection concerns the affective and experiential relations of the emerging lexical item. This would correspond to the emotional and experiential associations in olfactory word finding. In Brown's model, the more abstract, categorical relations are achieved through neocortical systems. A strict analogy to olfactory naming, therefore, suggests that affect and experience are entrained through limbic formation, such as the amygdala (Andy et al., 1975), whereas categorical specificity is achieved by way of neocortical systems, such as piriform or orbitofrontal cortex (see Damasio, 1985). Moreover, the limited neocortical representation of olfaction (Davis, 1975) might account for the greater difficulty in olfactory naming, whereas the stronger limbic representation would account for the prominent relation to emotion and memory.

More generally, the vulnerability of subordinate categories (e.g., cat, goat) as compared to superordinate ones (animals) supports a hierarchical arrangement of semantic systems. Disturbances within these systems can also be found. In the study by Hart et al. (1985), which described a specific impairment of superordinate categories (fruits and vegetables), there was a dissociation in categorization ability between performance with lexical instances (which was normal) and with pictorial instances (which was impaired). Hart et al. suggest that lexical categorization could be accomplished on the basis of strictly lexical, as opposed to semantic, information. Although name recognition and name retrieval have often been shown to be dissociable in aphasia, the category-specific dissociation found in Hart et al.'s patient indicates that the output lexicon is addressed by semantically categorized information that can be disrupted highly selectively. These findings are important in the investigation of naming via the olfactory modality. There is evidence that lexical categorization may be performed independent of certain semantic information in the standard language (auditory and visual) modalities. Categorizations made via olfactory stimulation, then, may depend highly (if not solely) on more general semantic categorization, give the natural tendency not to verbalize the olfactory experience. The challenge, then, is to define what "semantic" implies here.

It is true that performance in one language modality might be intact while another is impaired; for example, visual comprehension can be intact in cases of auditory agnosia (Mark and Chobor, 1990), or the reverse could also be true (Hecaen et al., 1974). Warrington and Shallice (1984) point out a number of agnosic phenomena that are difficult to interpret in terms of a single semantic system disconnected from particular input modalities and more easily interpreted in terms of modality-specific semantic systems. They support this with observations by Beauvois et al. (1978) of patients with optic aphasia and tactile aphasia who show separate semantic systems for verbal, visual, and tactile information. The characteristics of olfaction should theoretically make it simple to distinguish such a separate semantic system. For example, the all-or-none quality of odor encoding, the lack of a distinction between a short-and long-term olfactory memory, and the weak links between odors and labels all set olfaction apart from the other modalities. We know that at the basic sensory level, there is considerable bilateral interaction in the olfactory system (shown in bilateral additivity of intensity and in adaptation), making it unique among the senses (Cain, 1977). Patterns of olfactory performance should assist us in delineating a separate olfactory semantic store, further setting this sense apart from the others.

Olfactory Memory

The odor experience has been described as a "unitary perceptual episode" (Engen, 1982), and this is consistent with the observation that olfactory information can be processed more quickly and with less editing than visual and auditory information. (This is true despite the observation that associations to odors are acquired more slowly than to visual or verbal stimuli.) It is obvious that the all-or-none coding for odors might encourage immediate errors, but this same quality makes them quite resistant to subsequent confusions.

Memory for odors is long lasting (Engen and Ross, 1973; Lawless and Cain, 1975; Davis, 1977), as it is less influenced by the passage of time than are auditory and visual memory. Although a poor maximum recognition score of 70 percent for short-term memory of odors has been contrasted with a score of nearly 100 percent for short-term visual recognition (Shepard, 1967), an almost flat retention curve is maintained across all time intervals. In fact, there is a general rise in paired associates performance even when subjects are given different odor-picture pairings on the second day of exposure (Lawless and Engen, 1977). Patterns of interference resemble those for verbal memory (Lawless and Engen, 1977): the second of two lists dominates for a short time after second list learning, but the first recovers, and proactive interference increases with time (Postman et al., 1968).

Rabin and Cain (1984) have shown that the degree to which an odor connects with experience, as indexed by rated familiarity, or label quality (a semantic index) appears to be directly related to subsequent recognition performance. They suggest that a meaningful verbal code is not necessary for recognition, as even imperfectly or vaguely encoded items proved to be recognizable above chance. However, such coding enhances recognizability, as it does in vision (Kurtz and Hovland, 1953) and audition (Bartlett, 1977).

Olfactory information is probably not as much a part of the information input on which cognitive processes operate as it is the determinant that establishes the emotional atmosphere within which cognitive processes are conducted (Davis, 1977). Lawless and Engen (1972) and Engen (1983) have suggested that olfactory memory can be seen as an alerting mechanism that sets the affective tone for subsequent behavior and acts in concert with other sense modalities. Human olfaction is closely related to the limbic system, which includes structures involved in the regulation of emotional and affective states. Odors tend to revive memories from the distant past with emotional impact (Laird, 1935).

It has been suggested that information processed to deeper cognitive levels provides a more discriminable memory "trace" (Jacoby and Craik, 1979). However, Rabin and Cain (1984) point out that the notion of a deeper level is difficult to define apart from the experimental method used. They suggest that experimenters take note of labels produced by their subjects, as information contained within them can assist in determining the level of information processing. For example, their odors with high-information labels (such as "chocolate") fared better in memory retention than those with low-information labels (such as "candy").

Misnaming in Aphasia and Dementia

Aphasia

One approach to anomia involves a (presumed) impairment in the processes through which lexical entry occurs, an impairment that results in defective word finding in all modalities of object presentation, difficulty in naming to description, and, usually, emptiness and circumlocution in conversational speech (Brown, 1972). In the restricted case, speech is fluent without marked disturbance of comprehension, repetition, or reading. In a supposed form of self-cueing, patients will describe the function or structure of an object rather than its name. A common form of anomia involves naming difficulty in restricted perceptual spheres and a heightened threshold for (visual and auditory) modality-dependent naming, although normal word production in conversational speech can be found. Often such patients can name to functional description what they cannot name perceptually, providing evidence for a relatively intact lexical entry mechanism. There is a spectrum from inability to identify objects in one perceptual class to a mildly uneven performance across modalities (Brown, 1972). In modality-specific anomia, the word-finding disturbances and the verbal comprehension disorders observed within the affected modality should be due to a defective transfer of information between a single modality semantic system and the verbal semantics, whereas in aphasic anomia with lexical comprehension disorders, a breakdown of the semantic lexicon should explain the word-finding disorders and the lexical comprehension disturbances observed across various types of sensory modalities (Gainotti, 1987). In patients with a modality-specific naming disorder, naming might be facilitated, in mild cases, by enhancing sensory information within a single modality or by circumventing the modality altogether (Brown, 1972). Historically, this has been investigated via the visual, auditory, and tactile modalities.

Olfaction and Aphasia

Surprisingly little work has been done in the exploration of olfaction as a stimulus modality in the evaluation or treatment of aphasia. This is surprising given the common notion that multimodality stimulation often improves linguistic performance. For example, it is known that aphasics perform better on naming tasks when presented with both visual and auditory information than when presented with either alone.

Vandette (1964) found that some aphasics performed better on naming tasks when confronted with olfactory and olfactory-visual stimuli than when confronted with visual stimuli alone. Mair and Engen (1976) found that both anterior and posterior aphasics were impaired in their ability to match an odor to a directly corresponding visual target. Despite the impairment on an exact matching task, though, performance on a similarity matching task is not significantly disrupted. They also note that aphasics with more massive lesions show more significant difficulty on the exact matching task.

Using visual, tactile, auditory, and olfactory modalities to assess the naming ability of aphasic subjects, Goodglass et al. (1968) reported uniformity of naming ability across modalities, suggesting that a modality-nonspecific process intervenes between stimulus presentation and naming. In this study, subjects were to match items presented in the tactile, auditory, and olfactory modalities to pictures. The experimenter then provided verbal feedback in the form of reinforcement of a correct response or correction of an incorrect response. Subjects were then required to name these same stimuli, and response times were used to analyze the data. It is noteworthy that olfactory stimulation yielded much larger response times for aphasics, nonaphasic brain-injured subjects, and normals. Further, in the screening measure, these latter two groups evidenced difficulty in identifying only olfactory items.

Studies in aphasia provide a basis for the conceptualization of olfactory naming abilities (and deficits). Similarly, studies of dementia increase our understanding of language organization in relation to cognitive processes and, by implication, provide an important population for the investigation of olfaction from the cognitive standpoint. Moreover, studies of the pathological basis of dementia suggest the possibility that substrates involved in olfaction may also be involved in the dementing process.

Dementia

In dementia, particularly dementia of the Alzheimer type, word-finding difficulty is common in the early stages of the disease. This is most obvious in naming, and responses are at least linked conceptually to the object in question. The semantic impairment involves loss of information about specific objects and their names. As the disease progresses, word finding becomes more semantically deviant and may eventually lead to the patient speaking jargon. There is an inability to distinguish among objects that are of the same semantic category. Knowledge of broad categories is retained, but there is difficulty in distinguishing special attributes (Chobor and Brown, 1990). There is often an overextension of attributes (Martin and Fedio, 1983), and impaired ranking of semantic attributes may also be seen (Grober et al., 1985). Generally, early involvement of semantic (word meaning) systems in the context of a fluent aphasia is seen, with progression toward later dysfluency and phonological errors (Chobor and Brown, 1990).

Attempts to define a specific locus for the pathology of word finding—assumed to be somewhere between the "memory store" and the mechanisms involved in the final retrieval of a word—have been hampered by the variability of the disorder and the multiplicity of systems that are involved. The concept of retrieval from a "store" and the implication that this store is disturbed in dementia suggest that olfactory memory may be involved in dementia in a manner similar to contents in other perceptual modalities. There are a number of studies of this problem.

Olfaction and Dementia

It is known that odor identification diminishes with age (Doty et al., 1984) and that olfactory deficits are prominent in dementia of the Alzheimer type. Some authors note intact detection thresholds with impaired identification in the early stages (Rezak, 1987; Koss et al., 1988), whereas others note diminished detection and identification (Doty et al., 1987).

Identification deficits have been described without reference to detection (Serby et al., 1985; Koss et al., 1988; Peabody et al., 1985; Warner et al., 1986). Serby (1991) has pointed out that the discrepancy between some of these findings is due to the differences in the stage of dementia at the time of testing. The pattern that emerges for early-stage patients is that of intact detection despite impaired identification.

Most studies now utilize well-validated odor perception tests such as the UPSIT (University of Pennsylvania Smell Identification Test; Doty et al., 1984) and are sensitive to overall cognitive decline in this population, correlating olfactory deficits with performance on a neuropsychological assessment. However, controls for linguistic ability are often lacking.

Thus, the multiple-choice paradigm of the UPSIT employs foils on the basis of chemosensory strength and related factors rather than semantic relatedness to a target. This does not provide information on deficits in semantic knowledge that may be specific to the olfactory modality (see Bayles and Kaszniak, 1988; Huff et al., 1986, Chertkow and Bub, 1988). In one study, however, Rezak (1987) used a multiple-choice written list to prompt correct responses in Alzheimer's disease cases in a documentation of olfactory identification deficits. Following presentation of a list, the relative difference between healthy and demented subjects remained, yet it was shown that the identification deficit is more complex than an anomia for odors.

Personal Studies in Aphasic and Demented Populations

We have studied patterns of olfactory identification in aphasic, demented, and normal elderly subjects over the past year, compiling pilot data on approximately 10 subjects in each population category. The design of our testing battery (see Appendix for the actual battery and accounts of stimulus selection and subject participation) took into account the basic parameters of age, gender, smoking history, medical history, and basic threshold as discussed by Doty et al. (1984). The focus of its construction was on the potential validity of olfactory scores obtained in reference to the cognitive and linguistic profile of each subject. Olfactory identification tasks included the following: (1) an assessment of the subject's ability to name odors upon first presentation; (2) upon failure of this, an account of the subject's ability to provide some knowledge of the odor (e.g. familiarity, pleasantness, categorization, environmental locus); (3) upon failure of this, a test of odor-object matching; (4) naming of pictures paired with visual or visual/olfactory stimulation; and (5) paired associates recall of an odor with a related or exact picture. General neuropsychological measures were interspersed among these tasks, and included the following: (1) Mini-Mental State (Folstein et al., 1975); (2) paired associates (immediate recall) subtest of the Randt Memory Test (Randt and Brown, 1980); (3) picture recognition (immediate) subtest of the Randt Memory Test; and (4) picture description subtest of the Boston Diagnostic Aphasia Examination (Goodglass and Kaplan, 1972). Further, the above-described visual/odor naming task has inherent comparability.

A preliminary review of our data suggests the following. In normal subjects, a prominent spontaneous odor naming deficit is verified. This is independent of item familiarity or perceived strength. Upon probing, however, these subjects exhibit some knowledge of category membership, as evidenced in response to questions regarding item association and environmental locus. Scores on visual naming and visual/olfactory naming are comparable. Finally, good performance is noted on paired associates recall, for both related and exact pictorial representations. As expected, the aphasic subjects rarely exhibited the ability to spontaneously name odors. Response to probing was similar to that of the normal subjects in that gross category membership was most often successful; however, item associations were not. The most striking performance of the aphasic population was their ability to associate related pictures with odors (e.g., woman-perfume) on an immediate recall task and their inability to associate exact pictures (e.g., garlic-garlic) for this same task. Again, the demented subjects proved incapable of producing spontaneously correct odor labels. There were scarce responses to probes for category membership or item associations and no response to the other probes (such as environmental locus). Interesting performances emerged on the odor/picture paired associates recall subtest; however, no distinguishable patterns have emerged. For example, verbal paraphasias were noted ("toothpick" for "toothpaste"); recurrent perseverations were noted, and categorical substitutions were also noted ("food" for "garlic").

Conclusions

The most striking conclusion that can be formulated based on the studies reviewed here concerns the idea that the linguistic behavior associated with odor identification is a view of the structure of olfactory knowledge. It is not sufficient to make the observation that identification merely is or is not intact given responses to spontaneous naming or matching tasks. When subjects are allowed to expand on odor knowledge through descriptive probes involving the semantic attributes of an item, it is often the case that some intact knowledge is in place. This is particularly important in pathological cases, such as stroke or Alzheimer's disease, where there are known linguistic and cognitive deficits. A focus on intact behavior in these areas can enable the investigator to model olfactory stimulation tasks so that judgments of identification (or even detection) can be made independent of linguistic or perceptual impairment. Mutual benefits are to be gained from an approach such as this one. There is no doubt that more stringent odor presentations and response modes will further our knowledge of olfactory behavior. But at the same time, we will gather information on the structure of semantic systems in language and perception that might be available to us only via the olfactory modality.

References

Andy, O.J., M.F. Jurko, and J.R. Hughes (1975). The amygdala in relation to olfaction *Confinia Neurol. Proc. 6th Symp. Int. Soc. Res. Stereoencephalotomy*, Tokyo, 1973, pt. II, *37*, 215–222.

Bartlett, J.C. (1977). Remembering environmental sounds: The role of verbalization at input. *Memory Cognition*, *5*, 404–414.

Barton, M. (1971). Recall of generic properties of words in aphasic patients. *Cortex*, *7*, 73–82.

Bayles, K., and A. Kaszniak (1987). *Communication and Cognition in Normal Aging and Dementia*. Boston: Little, Brown.

Beauvois, M.F., B. Saillant, V. Meininger, and F. L'Hermitte (1978). Bilateral tactile aphasia: a tacto-verbal dysfunction. *Brain*, *101*, 381–401.

Brown, J.W. (1972). *Aphasia, Apraxia and Agnosia*. Springfield, *IL*: Charles C Thomas.

Brown, J.W. (1988). *The Life of the Mind*. Hillsdale, NJ: Lawrence Erlbaum.

Brown, R., and D. McNeill (1966). The "tip of the tongue" phenomenon. *J. Verbal Learning Verbal Behav.*, *5*, 325–337.

Cain, W.S. (1977). Physical and cognitive limitations on olfactory processing in human beings. In D. Muller-Schwarze and M.M. Mozell (Eds.), *Chemical Signals in Vertebrates*. New York: Plenum, pp. 287–3.

Cain, W.S. (1977). Bilateral interaction in olfaction. *Nature*, *268*, 50–52.

Cain, W.S. (1979). To know with the nose: keys to odor identification. *Science*, *203*, 467–470.

Cain, W.S. (1980). Chemosensation and cognition. In H. van der Starre (Ed.), *Olfaction and Taste*, vol. 7. London: IRL pp. 347–358.

Cain, W.S., and R.J. Krause (1979). Olfactory testing: rules for odor identification. *Neurol. Res.*, *1*, 1–9.

Cain, W.S., J. Gent, F.A. Catalanotto, and R.B. Goodspeed (1983). Clinical evaluation of olfaction. *Am. J. Otolaryngol.*, *4*, 252–256.

Cain, W.S., and J.F. Gent (1986) Using odor identification in clinical testing of olfaction. In H.L. Meiselman and R.S. Rivlin (Eds.), *Clinical Measurement of Taste and Smell*. New York: Macmillan.

Chertkow, H., and D. Bub (1988). Disorders of semantic storage: do they exist in Alzheimer's disease? *Acad. Aphasia*, Montreal.

Chobor, K.L., and J.W. Brown (1990). Semantic deterioration in Alzheimer's: The patterns to expect. *Geriatrics*, *45*(10), 68–75.

Clarke, E.V. (1975). Knowledge, context and strategy in the acquistion of meaning. In D.P. Dato (Ed.), *University Round Table on Languages and Linguistics*. Washington, DC: Georgetown University Press.

Corwin, J., M. Serby, P. Conrad, and J. Rotrosen (1985). Olfactory recognition deficits in Alzheimer's and Parkinsonian dementias. *ICRS Med. Sci.*, *13*, 260.

Damasio, A.R. (1985). The frontal lobes. In K.M. Heilman and E. Valenstein (Eds.), *Clin. Neuropsychol.*, New York: Oxford University Press, 2nd Ed. p. 365.

Davis, R.G. (1975). Acquisition of verbal associations to olfactory stimuli of varying familiarity and to abstract visual stimuli. *J. Exp. Psychol.*, *104*, 134–142.
Davis, R.G. (1977). Acquisition of verbal associations to olfactory and abstract visual stimuli of varying similarity. *J. Exp. Psychol.*, *3*, 37–51.
Doty, R.L., P. Shaman, S.L. Applebaum, R. Gilberson, L. Sikorski, and L. Rosenberg (1984). Smell identification ability: changes with age. *Science*, *226*, 1441–1443.
Doty, R.L., P. Shaman, and M. Dann (1984). Development of the University of Pennsylvania Smell Identification Test: a standard microencapsulated test of olfactory function. *Physiol. Behav.*, *32*, 489–502.
Doty, R.L., P. Reyes, and T. Gregor (1987). Presence of both odor identification and detection deficits in Alzeimer's disease. *Brain Res. Bull.*, *18*, 597–600.
Eich, J.E. (1978). Fragrances as cues for remembering words. *J. Verbal Learning Behav.*, *17*, 103–111.
Engen, T. (1982). *The Perception of Odors*. New York: Academic.
Engen, T. (1983). The human uses of olfaction. *Am. J. Otolaryngol.*, *4*, 250–251.
Engen, T. (1987). Remembering odors and their names. *Am. Sci.*, *75*, 497–503.
Engen, T., and C. Pfaffman (1960). Absolute judgments of odor quality. *J. Exp. Psychol.*, *59*, 214–219.
Engen, T., J.E. Kuisma, and P.D. Eimas (1973). Short-term memory of odors. *J. Exp. Psychol.*, *99*, 222–225.
Engen, T., and B.M. Ross (1973). Long-term memory of odors with and without verbal descriptions. *J. Exp. Psychol.*, *100*, 221–227.
Engen, T., and J. Eaton (1975). Free recall of odor and color names. Unpublished manuscript, Brown University.
Folstein, M.F., S.E. Folstein, and P.R. McHigh (1975). "Mini-Mental State." *J. Psychiatr. Res.*, *12*, 189–198.
Gainotti, G. (1987). The status of the semantic-lexical structures in anomia. *Aphasiology*, *1*(6), 449–461.
Goldstein, K. (1948). *Language and Language Disturbances: Aphasic Symptom Complexes and Their Significance for Medicine and Theory of Language*. New York: Grune and Stratton.
Goodglass, H., B. Klein, P. Carey, and K.J. James (1966). Specific semantic word categories in aphasia. *Cortex*, *2*, 74–89.
Goodglass, H., M. Barton, and E. Kaplan (1968). Sensory modality and object-naming in aphasia. *J. Speech Hearing Res.*, *11*, 488–496.
Goodglass, H., and E. Kaplan (1972). *Assessment of Aphasia and Related Disorders*. Philadelphia: Lea and Febiger.
Grober, E., H. Buschke, C. Kawas, and P. Fuld (1985). Impaired ranking of semantic attributes in dementia. *Brain Lang.*, *26*, 276–286.
Hart, J., Jr., R.S. Berndt, and A. Caramazza (1985). Category-specific naming deficit following cerebral infarction. *Nature*, *316*(6027), 439–440.
Hecaen, H., M.C. Goldblum, M.C. Masure, and A.M. Ramier (1974). Une nouvelle observation d'agnosie d'objet. Deficit de l'association, ou de la categorisation, specifique de la modalite visuelle? *Neuropsychologia*, *12*, 447–464.
Huff, F., S. Corkin, and J. Growdon, (1986) Semantic impairment and anomia in Alzheimer's disease. *Brain Lang.*, *28*, 235–249.
Jacoby, L.L., and F.I.M. Craik (1979). Effects of elaboration of processing at encoding and retrieval: Trace distinctiveness and recovery of initial context. In L.S.

Cermak and F.I.M. Craik (Eds.), *Levels of Processing in Human Memory*. Hillsdale, NJ: Lawrence Erlbaum, pp. 1–22.

Koss, E., J. Weiffenbach, J. Haxby, and R. Friedland (1988). Olfactory detection and identification performance are dissociated in early Alzheimer's disease. *Neurology*, *338*(8), 1228–1232.

Kurtz, K.H., and C.I. Hovlan (1953). The effect of verbalization during observation of stimulus objects upon accuracy of recognition and recall. *J. Exp. Psychol.*, *45*(3), 157–164.

Laird, D.A. (1935). What can you do with your nose? *Sci. Monthly*, *41*, 126–130.

Lawless, H.T., and W.S. Cain, (1975). Recognition memory for odors. *Chem. Senses*, *1*, 331–337.

Lawless, H.T., and T. Engen (1977). Associations to odors: interference, mnemonics, and verbal learning. *J. Exp. Psychol. 3*, 52–59.

Mair, R.G., and T. Engen (1976). Some effects of aphasic lesions on odor perception. *Sensory Processes*, *1*, 33–39.

Mair, R.G., C. Capra, W.J. McEntee, and T. Engen (1980). Odor discrimination and memory in Korsakoff's psychosis. *J. Exp. Psychol.*, *6*, 445–458.

Mark, V.W., and K.L. Chobor (1990). Influences on the speech comprehension deficit in cortical auditory disorder. *Aphasiology*, *4*(5), 587–598.

Martin, A., and P. Fedio (1983). Word production and comprehension in Alzheimer's disease: the breakdown of semantic knowledge. *Brain Lang.*, 19; 124–141.

Mozell, M., D.E. Hornung, D.A. Leopold, and S.L. Youngentob (1983). Initial mechanisms basic to olfactory perception. *Am. J. Otolaryngol.*, *4*, 238–245.

Murphy, C., and W.S. Cain (1986). Odor identification: the blind are better. *Physiol. Behav.*, *37*, 177–180.

Peabody, C., and J. Tinklenberg (1985). Olfactory deficits and primary degenerative dementia. *Am. J. Psychiatry*, *142*, 524–525.

Postman, L., K. Stark, and J. Fraser (1968). Temporal changes in interference. *J. Verbal Learning Behav.*, *7*, 672–694.

Potter, H., and N. Butters (1980). An assessment of olfactory deficits in patients with damage to prefrontal cortex. *Neuropsychologia*, *18*, 621–628.

Putnam, H. (1975). The meaning of "meaning." In K. Gunderson (Ed.). *Language, Mind and Knowledge*, Minnesota Studies in the Philosophy of Science. Vol. 7. Minneapolis: University of Minnesota Press.

Rabin, M.D., and W.S. Cain (1984). Odor recognition: familiarity, identifiability, and encoding consistency. *J. Exp. Psychol.*, *10*(2), 316–325.

Randt, C.T., E.R. Brown, and D.P. Osborne (1980). A memory test for longitudinal measurement of mild to moderate deficiencies. *Clin. Neurosychol.*. *2*, 184–194.

Reisberg, B., S.H. Ferris, M.J. de Leon, and T. Crook (1982). Global deterioration scale (GDS) for age-associated cognitive decline and Alzheimer's disease. *Am. J. Psychiatry*, *139*, 1136–1139.

Rezak, D.L. (1987). Olfactory deficits as a neurologic sign in dementia of the Alzheimer type. *Arch. Neurol.*, *44*, 1030–1032.

Serby, M. (1985). Olfactory dysfunction in Alzheimer's disease and Parkinson's disease. *Am. J. Psychiatry*, *142*, 781–782.

Serby, M., P. Larson, D. Kalkstein (1991). The nature and course of olfactory deficits in Alzheimer's disease. *Am. J. Psychiatry*, *148*, 357–360.

Shallice, T., and E.K. Warrington (1975). Word recognition in a phonemic dyslexic patient. *Q. J. Exp. Psychol.*, *27*, 187–199.
Shepard, R.N. (1967). Recognition memory for words, sentences, and pictures. *J. Verbal Learning Behav.*, *6*, 156–163.
Toglia, M.P., and W.F. Battig (1978). *Handbook of Semantic Norms*. Hillsdale, NJ: Lawrence Erlbaum.
Vandette, J.M. (1964). A clinical study of responses to olfactory stimuli in aphasic adults. Unpublished Master's thesis, University of Oregon.
Warner, M., et al. (1986). Olfactory deficits and Alzheimer's disease. *Biol. Psychiat.*, *21*, 116.
Warrington, E., and T. Shallice (1984). Category specific semantic impairment. *Brain*, *107*, 829.
Wright, R.H. (1964). *The Science of Smell*. New York: Basic Books.
Wright, R.H., and R.E. Burgess (1971). Molecular mechanisms of olfactory discrimination and sensitivity. In G. Ohloff and A.F. Thomas (Eds.), *Gustation and Olfaction*. London: Academic.
Yamadori, A., and M.L. Albert (1973). Word category aphasia. *Cortex*, *9*, 112–125.

Appendix I

Subjects

Ten subjects in each of three categories, Alzheimer's disease, multi-infarct dementia, and normals, were chosen to undergo testing. Patients were identified by clinical neurology, clinical neuropsychology, speech-language, radiology (CT and, in some cases, MRI scanning), electroencephalography, laboratory (blood and urine) studies, and clinical history. None were chosen who had other psychiatric, neurological, or medical diseases. The demented patients included those between levels 3 and 4 of the Global Deterioration Scale (Reisberg et al., 1982). Normals were matched according to age, gender, and ethnicity. All subjects were nonsmokers.

Materials

Stimuli used to determine detection threshold included the following: pine, cheese, coffee, flower (rose), and mineral oil (as a blank). Stimuli used throughout the remainder of the test (on tasks of recognition and identification) included the following: garlic, perfume, cigarette, baby powder, paint, coffee, menthol, cheese, cinnamon, chocolate, pickles, soap, orange, dog food, rubbing alcohol, and toothpaste. Stimuli used as object matching items also included the following: glue, mustard, peanuts, wine, and oregano. Items were rated in terms of subjective strength and familiarity, then selected by familiarity, category, and number of attributes according to the clusters determined by Toglia and Battig (1978). In an attempt to compare trigeminal response, 21 percent of the total list of items were considered to be trigeminal stimulants.

PART ONE

Patient's Name: ______________________________
Date of Session: ______________________________
Date of Birth: ______________ GDS Rating: ______________
General/Smoking History:

a) sex: ________
b) circle highest grade completed: 6 7 8 9 10 11 12 college 1 2 3 4 5 +
c) occupation: ________
d) ethnic background: white black oriental other
e) are you currently under treatment for a medical problem?
if yes, explain ______________________________
f) do you suffer from smell and/or taste problems?
if yes, explain ______________________________
g) are you currently taking medication? yes ________ no ________
list: ______________________________
____
h) do you smoke? ________
if yes: how many cigarettes/day ________; cigars/day ________
other: ________ do you inhale? ________ when did you start?
If you did so before, when did you start? ________
when did you stop? ________
how many cigarettes/day ________; cigars/day________
other: ________ did your smell ability change after stopping? if yes, how? ______________________________
____

1. Olfactory Threshold: The purpose of this section is to determine whether or not the patient has the basic ability to smell odors; standard presentation will be used (birhinically using containers held parallel to and approximately one inch away from the subject's upper lip for 15 seconds, with a 5-second interval between presentations; patient will be blindfolded). The patient will be asked, "do you smell something?," to which he/she will respond "yes" or "no."
(stop at 3 consecutive correct responses)
pine/blank/cheese/coffee/blank/flower/blank/blank/coffee/blank/pine/blank/blank/flower/blank/cheese/
2. Mini-Mental State score: ________
3. Section 2 Naming, Describing, and Matching Odors to Objects
This section is designed to obtain a measure of the subject's ability to name and to provide knowledge about specific odors, and his/her ability to perform olfaction-object matching tasks. Standard presentation will be used, and the subject will be asked to name the odors; if unable to do so, he/she will be asked to provide information about those odors (see probe list). If

the subject is still unable to produce correct names, he/she will be required to perform a multiple-choice object matching task. The presentation of each odor will last 15 seconds; the subject will be given 30 seconds in which to respond; following probes, objects will be presented for 30 seconds and the subject will be required to respond within this time frame.

1. garlic
2. perfume
3. cigarette
4. baby powder
5. paint
6. coffee
7. menthol
8. cheese
9. cinnamon
10. chocolate

Probes (to which patient will answer "yes" or "no" or will rate on a scale of 1–5 (low to high) where possible; record all said by the patient.)

a) Is this odor familiar?

1. ______
2. ______
3. ______
4. ______
5. ______
6. ______
7. ______
8. ______
9. ______
10. ______

b) What comes to mind when you smell this?

1. ______
2. ______
3. ______
4. ______
5. ______
6. ______
7. ______
8. ______
9. ______
10. ______

c) Is this pleasant?
1. ______________________________
2. ______________________________
3. ______________________________
4. ______________________________
5. ______________________________
6. ______________________________
7. ______________________________
8. ______________________________
8. ______________________________
9. ______________________________
10. ______________________________

d) Can you describe this odor (pungent, strong, weak, noxious, etc.)?
1. ______________________________
2. ______________________________
3. ______________________________
4. ______________________________
5. ______________________________
6. ______________________________
7. ______________________________
8. ______________________________
9. ______________________________
10. ______________________________

e) Can you describe anything that might smell like this?
1. ______________________________
2. ______________________________
3. ______________________________
4. ______________________________
5. ______________________________
6. ______________________________
7. ______________________________
8. ______________________________
9. ______________________________
10. ______________________________

f) Does this item belong to the food category?
1. ______________________________
2. ______________________________
3. ______________________________
4. ______________________________
5. ______________________________
6. ______________________________

7. __________________________
8. __________________________
9. __________________________
10. __________________________

g) Can you name an item associated with this one (for example, one from the same category)?

1. __________________________
2. __________________________
3. __________________________
4. __________________________
5. __________________________
6. __________________________
7. __________________________
8. __________________________
9. __________________________
10. __________________________

h) Where do you think this odor can be found? (in a kitchen, bathroom, outdoors, etc.)?

1. __________________________
2. __________________________
3. __________________________
4. __________________________
5. __________________________
6. __________________________
7. __________________________
8. __________________________
9. __________________________
10. __________________________

<u>Multiple-choice section</u> (if failed previous two sections) Present each odor again, and ask the patient to choose from the three object choices listed; one try only.

target	choices		
1. garlic	garlic	glue	mustard
2. perfume	peanuts	perfume	rub alc
3. cigarette	cigarette	orange	menthol
4. baby powder	mustard	glue	baby powder
5. paint	rub alc	wine	paint
6. coffee	coffee	orange	rub alc
7. menthol	pickles	menthol	paint
8. cheese	mustard	cheese	rub alc
9. cinnamon	wine	menthol	cinnamon
10. chocolate	chocolate	oregano	glue

PART TWO

1. Section 1: Naming pictures paired with visual or visual/olfactory stimulation

This section is designed to determine the effect of olfactory stimulation on recall for names of pictured items. Half of these pictured items will be presented with relevant olfactory stimulation, along with a spoken label. The other half will be presented without olfactory stimulation but with a spoken label. Each of the odors will be presented birhinically using containers held parallel to and approximately one inch away from the subject's upper lip. Presentation of each odor or odor/picture pair will last 15 seconds. After a delay (with interposed unrelated tasks), the same sequence of pictures will be presented, this time without olfactory stimulation. The subject will be asked if he/she had seen it earlier and will then be required to provide the correct name for each. Semantic and phonemic cues will be provided if necessary. (* Indicates olfactory presentation.) Again, all are presented with a spoken label, and those with * are presented with olfactory stimulation.

1. pickles (*)
2. pipe
3. chocolate (*)
4. cigarette (*)
5. gasoline
6. peanuts (*)
7. flower
8. wine
9. soap (*)
10. paint
11. orange (*)
12. candle
13. mustard (*)
14. pine
15. coffee (*)

2. Randt Memory Test Paired Associates Learning:
(immediate recall)

key-hole ____________
cake-tent ____________
sheep-wool ____________
job-song ____________
bear-skin ____________
head-hair ____________
total: ____________

3. Section 3: Paired Associates: Odors/Words

The purpose of this section is to determine whether olfactory stimulation paired with relevant pictures will aid recall. Words will be either exact or related to the presented odor.

Each of six odors will be presented for 15 seconds, with simultaneous presentation of a (related or exact) picture. After presentation of these six pairs, the odors will again be presented for 15 seconds each, and the subject will be required to describe the previously associated picture.

Odor	Association	Response
1. perfume	woman	________
2. garlic	garlic	________
3. talcum powder	baby	________
4. dog food	dog food	________
5. rubbing alcohol	hospital	________
6. toothpaste	toothpaste	________

4. Randt Memory Test Picture Recall:
   (immediate)

clock bridge box telephone ear bee horse pen
table shoe window pipe glass sword corn

5. Olfactory/Visual Presentation: part 2

(no concomitant olfactory presentation)

1. pickles
   (used as a relish)
2. cigarette
   (used for smoking)
3. chocolate
   (a snack to eat)
4. pipe
   (used for smoking)
5. gasoline
   (used for fuel)
6. peanuts
   (a snack to eat)
7. flower
   (found in a garden)
8. wine
   (a social drink)
9. soap
   (used for washing)
10. paint
   (used by artists)

11. orange
    (a fruit to eat)
12. candle
    (used for light)
13. mustard
    (used on sandwiches)
14. pine
    (comes from a tree)
15. coffee
    (a hot drink)

6. Picture Description: Cookie Theft (BDAE)

13 Human Olfaction in Infancy and Early Childhood

HILARY J. SCHMIDT and GARY K. BEAUCHAMP

Introduction

The past half century has witnessed a proliferation of interest and resources devoted to the study of the development of sensory and perceptual systems in humans. New, innovative methodologies have revolutionized our understanding of many of the sensory and perceptual capacities of the newborn and the developing preverbal child. The early characterization of the sensory world of the infant as a "blooming, buzzing, confusion," and a "tabula rasa" has been replaced by the view that the neonate possesses many well-structured and coordinated perceptual abilities.

These advances in research, however, have been confined primarily to areas of visual, auditory, and tactile perception, with a notable neglect of the chemical senses. This presents a curious paradox: Despite a formal interest in the development of the chemical senses in humans that predates study of other senses, and despite philosophical perspectives asserting the phylogenetic primacy of the chemical senses, and despite the suggestion that chemical senses may be of particular significance in infancy and early childhood while the other sensory systems continue to mature (Freud, 1962; Darwin, 1872), the chemical senses have received remarkably little attention.

The goal of this chapter is to review and critique the research on the development of olfactory perception in infancy and early childhood and to offer suggestions for future work. Although studies of the development of human olfaction can be traced back to the mid nineteenth century (see the Chapter by Harrington in this Book), the existing body of research is fragmented. There is a preponderance of research on the first two weeks of life and the early school years, with a notable hiatus from about two weeks to four years of age. The fragmented nature of the area best supports a discussion organized around topics within olfaction rather than around developmental periods, since some periods have been neglected altogether. Hence, the approach here will be to review the development of (1) olfactory sensitivity and discrimination and (2) hedonic discrimination, and finally (3) to consider the potential functional importance of the sense of smell for the developing infant. Discussion of developmental animal research will be limited to cases of potential heuristic value. As discussed in the following sections, the vast majority of studies have not separated purely olfactory (cranial nerve I) stimulation from trigeminal (cranial nerve V) effects. Consequently, many of the behavioral responses attributed to olfaction may be due in part or in entirely whole to trigeminal effect. Nevertheless, the terms "odor" and "odorant" will be used here to denote odorous chemicals that have behavioral effects.

Sensitivity and Discrimination

A number of studies have shown that infants can detect and discriminate among a variety of qualitatively distinct odorants (Engen, Lipsitt, and Kaye, 1963; Self, Horowitz, and Paden, 1972; Rieser, Yonas, and Wikner, 1976; Macfarlane, 1975; Russell, 1976; Cernoch and Porter, 1985; Balogh and Porter, 1986; Makin and Porter, 1989). In one of the earliest controlled studies, Engen, Lipsitt, and Kaye (1963) compared newborns' responses to odor pairs the members of which varied in intensity and hedonic valence: 100 percent acetic acid and phenyl-ethyl alcohol (experiment 1) and anise oil and tincture of asafetida (experiment 2). Odors were presented (via Q-tip) in alternation with no-odor control trials over 20 trials; one odor was tested during the first 10 trials, then a switch was made to the second odor. Infants were tested during sleep, and general bodily activity, leg withdrawal, respiration (experiments 1 and 2), and heart rate (experiment 2) were monitored with polygraphic equipment during the alternating odor and no-odor trials.

Infants were sensitive to the odors, as evidenced by responses of higher magnitude on the various odor trials relative to the no-odor controls. They discriminated among the odors, as evidenced by responses of different magnitudes for the various odors: the most reliable elicitor of a response was acetic acid, followed in descending order by asafetida, anise, and phenyl-ethyl alcohol. Differential response patterns over the trial series provided further evidence that infants discriminated among these odors: for both anise and asafetida (but not for acetic acid and phenyl-ethyl alcohol), the likelihood of a response declined over the 10 trials and recovered following a temporal delay. Since the odors were not equated for perceived intensity, it is unclear whether these differences are best accounted for by differences in quality, intensity, hedonicity, or trigeminal component. For example, 100 percent acetic acid probably stimulates not only olfactory nerve I receptors but also receptors sensitive to irritation (trigeminal) in the nose and in the eyes.

Self and colleagues (1972), in an attempt to explore olfactory as opposed to trigeminal reactions, tested newborns' respiratory and behavioral responses to anise, asafoetida, valerian, and lavender. They too observed infants less than 4 days of age to respond to these odors relative to no-odor controls. In general, asafetida produced the most consistent reactions, followed by valerian, lavender, and anise, however, these differences were not reliable. Exactly what motivated the selection of these odors as pure olfactory stimuli is unclear. Subsequent empirical investigations of the relative contributions of olfactory and trigeminal components of various odors (Doty et al., 1978; see Silver, 1987) have not included all of these odors, but anethole—a constituent of anise—and valeric acid—a constituent of valerian—are both potent stimulants of the trigeminal nerve. Therefore, while this study provides additional evidence that newborns are sensitive to a variety of odorants, it is not obvious how the differential responding sheds light on olfactory versus trigeminal responses. In addition, these authors failed to report whether the stimuli were equated for perceived intensity, and therefore the observed trends could be due to intensity as opposed to quality differences.

Together these studies support the credibility of much earlier uncontrolled observational findings concerning neonates' ability to detect odorous chemicals (Kussmaul, 1896; Guzner, 1873; Kroner, 1881, all cited by Peterson and Rainey, 1910; Peterson and Rainey, 1910). Collectively these early studies reported that infants responded with respiratory and/or behavioral changes (e.g., facial expressions, sucking, crying) to asafetida, orange, oil of rose gentian, mother's milk, oil dippelli, oil succ., oil menth. pipp., and oil of amber. In these early studies, experimenters made the observations and did not compare reactions on odor trials to base rate responding in the absence of odorant stimuli. However, since the aforementioned well-controlled studies confirm reactions to asafetida, these early findings suggest that the range of odorants that the newborn can detect includes these additional odors.

The ability to detect odors may develop prenatally. Limited research (Humphrey, 1940; see Bossy, 1980; Brown, 1974; for detailed reviews) on the embryological origins of the olfactory system reveals very early differentiation of the major structural components. The olfactory bulb and receptors have the adult pattern by the 11.5th week of gestation, the maxillary branch of the trigeminal nerve that innervates parts of the nasal cavity has differentiated by about the 10th week (Humphrey, 1940; Brown, 1974), and the vomeronasal structures are formed between the 5th and 18th weeks, but then degenerate so that only a vestigial vomeronasal organ is seen in most older fetuses and neonates (Humphrey, 1940). Recent data suggest that the vomeronasal organ may not regress in all individuals (Johnson, Josephson, and Hawke, 1985); however, there is no evidence to suggest that this structure is functional.

Data concerning the functional onset of these structures are remarkably limited; however, several lines of evidence suggest that around 28 weeks, the olfactory system is capable of detecting chemical stimuli. First, studies of premature newborns have demonstrated reactions to odorants in infants as young as 28 weeks' gestation (Sarnat, 1978; Kussmaul, 1896, cited by Peterson and Rainey, 1910). For example, 8 of 11 sleeping, preterm infants (29 to 32 weeks) initiated sucking or displayed arousal-withdrawal within 30 seconds of the presentation of a peppermint extract odorant (Sarnat, 1978). Such responses were not observed on control trials when no odor was presented. Although Sarnat (1978) observed similar responses in a younger infant of 28 weeks' gestation, this was not characteristic of other infants (5) this age. All infants less than 28 weeks of age failed to respond differentially to the peppermint and control trials, while all infants between 32 and 36 weeks' gestation did so.

Indirect support that olfactory perception may have its inception around the 28th postovulatory week also comes from studies on the ontogeny of olfactory marker protein at various layers of the olfactory system. This protein, apparently present only in mature olfactory tissue (Margolis et al., 1985), is evident at the epithelial layer by 28 weeks but not prior to this. It is not present, however, in the olfactory bulb until 32 weeks or in the glomerular layer until 35 weeks (Chuah and Zheng, 1987). The graded appearance of this protein at these anatomical levels closely parallels the increment in the reliability with which a response to an olfactory stimulus can be elicited in preterm infants of 28 to 35 weeks as described above.

While this parallel appears to support the assumptions (1) that olfactory marker protein is involved in olfactory perception and (2) that olfaction has its inception around 28 gestational weeks, three caveats must be put forth. First, the failure of infants less than 28 weeks' gestation to respond to an odor stimulus in a reliable manner may reflect a response limitation as opposed to an inability to detect the stimulus. In studies of full-term neonates, change in respiratory rate provides a more reliable measure of the ability to detect odorants than motoric responses (Engen, Lipsitt, and Kaye, 1963; Engen and Lipsitt, 1965) and therefore may provide a more sensitive test of preterm infants' ability to detect odorants. Second, it is not clear whether responses in the aforementioned studies were mediated by olfactory, trigeminal, and/or vomeronasal structures. Finally, these studies tested only a small subset of odorants, and it is possible that different odorants might be capable of eliciting responses in younger preterm infants. Animal studies have demonstrated that in utero exposure to odorants can markedly affect postnatal reactions to an odor (e.g., Pederson and Blass, 1982; Smotherman, 1982; Hepper, 1988). Similar effects may occur in humans; not only may prenatal olfaction be possible, but olfactory experience in utero may affect postnatal responses to odors. Schaal (1988) has argued that the natural human intrauterine environment could provide olfactory stimuli and support chemosensory function.

Although the neonate, and possibly the third-trimester fetus, is capable of detecting highly concentrated odorants, Lipsitt and his colleagues (1963) suggest that olfactory sensitivity may increase over the first few days of life.

They tested 10 infants' responsiveness to increasing concentrations of asafetida (from 1.56 percent to 100 percent from a log 2 concentration dilution series) over the first 4 days of life. Odorants were presented on saturated Q-tips just below the infants' nostrils during sleep, and heart rate, respiratory rate, leg withdrawal, and general movement were monitored. The concentration step required to produce a response relative to no-odor control trials decreased by two steps for 7 of the 10 infants, and by one step for the others, over the 4 days. Lipsitt et al. (1963) concluded that these data reflect a decrease in olfactory threshold during the first four days of life, similar to documented threshold changes seen in other sensory modalities.

It is not clear whether this effect is best explained in terms of (1) a stimulus-independent postnatal maturation of peripheral or central chemosensory receptors or neurons, (2) the induction of sensitivity contingent on odor exposure, or merely (3) the development of a more coordinated response system. While alternative (1) or (3) is probably most likely, three findings are consistent with the second alternative. First, Self et al. (1972) observed that newborns who were tested in a longitudinal design over the first three days of life showed more reliable respiratory responses to several suprathreshold (as assessed by adults) odors on the third day of testing than three-day-old infants who had not received any prior exposure to the odorants. If postnatal maturation independent of exposure, or increased ability to coordinate any response to an olfactory stimulus were governing the increased responsiveness to odors over the first few days of life, then the two groups should have been equal in their responsiveness on day three of testing. Second, Wysocki et al. (1989) have demonstrated in adults that repeated exposure to an odorant (androstenone) can induce the ability to detect the odorant. Specifically, a subset of adults who were anosmic to androstenone [as are about 50 percent of the adult population (Wysocki and Beauchamp, 1984)] developed the ability to detect it following regular and repeated exposure to the substance. This effect, however, has not been observed for any other odorants. Finally, there are developmental changes in the olfactory system of rats exposed to odors as neonates (Wilson, Sullivan, and Leon, 1985; Woo and Leon, 1987; Rosselli-Austin and Williams, 1990). These findings are merely suggestive, and clearly human neonatal studies that directly manipulate odor exposure conditions in relation to age and developing olfactory sensitivity are necessary to discriminate among these alternatives.

Although some improvement in olfactory sensitivity may occur during the first few days of life, by the end of the first or second week, neonates appear to be capable of making some very fine discriminations. Several studies have demonstrated that breast-fed infants between 6 and 14 days old can differentiate between the smell of their own mother and the smell of another lactating female. Specifically, breast-fed infants turned their heads towards a breast (Macfarlane, 1975; Russell, 1976; Schaal et al.,

1980) or axillary odor (Cernoch and Porter, 1985) collected from their own mothers when paired with the same odors from an unfamiliar lactating female. In addition, newborns in this age range responded with a suppression of general motoric activity when presented with the odor of their mother's breast or neck relative to activity levels in the presence of comparable odors from another lactating female or a no-odor control (Schaal, 1986). These studies suggest exceptional olfactory acuity in the newborn, since such discriminations are difficult for adults (Schmidt and Beauchamp, 1989; Porter, 1990), if not altogether impossible (Macfarlane, 1975).

Further studies that formally contrast adults' and neonates' thresholds for a variety of olfactory stimuli are clearly necessary before any definitive conclusion can be drawn. However, additional anecdotal evidence suggests that the neonate's exceptional sensitivity may generalize to other nonbiological odors such as the homologous alcohols (Engen, 1965; Rovee, 1972); in testing sensitivity in newborns, both authors comment that observed threshold levels are quite low. The observation that preschoolers (Schmidt and Beauchamp, 1988) and prepubescent school-age children (Dorries et al., 1989) are more likely to be sensitive to another biological odor, androstenone, relative to older children and adults lends further credence to the possibility that newborns' olfactory acuity may exceed that of adults. Finally, comparisons of olfactory neuroepithelium in fetuses and normal adults with no history of exposure to infection or toxic substances reveals degeneration in adults, and a higher ratio of olfactory to respiratory epithelium in fetuses (Nakashima, Kimmelman, and Snow, 1984); these anatomical findings may provide a partial explanation for any sensitivity differences that can be confirmed between newborns and adults.

What underlies the newborn's ability to discriminate the mother's odor from the odor of another lactating female? Recent data suggest both early experience and a genetically determined preference for the breast odor of a lactating female contribute to this ability. The role of experience is suggested by several findings. First, bottle-fed infants do not make similar discriminations (Cernoch and Porter, 1985). Second, breast-feeding infants do not discriminate the axillary odors of their own father from those of another fathers (Cernoch and Porter, 1985). Three, at seven days, breast-fed newborns will turn toward a perfume that has been worn on the mother's breast during feedings in preference to perfumes worn by other mothers (Schleidt and Genzel, 1990). Engen (1988) tested two- and three-day-old newborns' responses to perfumes worn by their mothers, but failed to obtain differential responding between the familiar perfume and a novel perfume using respiratory, heart-rate, sucking, and motoric changes as dependent measures. While it is possible that this discrepancy between Schleidt's and Engen's finding is due to developmental changes and/or different exposure durations, additional data suggest that Engen's measures may be less sensitive indices of discrimination abilities in the neonate than the head-turning measure. Balogh and Porter (1986) demonstrated that, after a 20- to 24-hour period of exposure to an artificial odor (wild cherry or ginger) in the ambient air, female newborns (but not males) less than 2.5 days turned their heads in the direction of the odorant presented on a gauze pad when paired with an equally pleasant but unfamiliar odorant. The fact that this limited exposure, not associated with the mother or feeding, resulted in an olfactory preference in infants less than 2.5 days old suggests that head-turning contingent on simultaneously available stimuli may be a more sensitive test of infants' ability to discriminate highly similar odors than changes in respiration, heart rate, sucking, and general activity level measured with successive presentation of the stimuli.

Using head turning as a measure, an innate component mediating the neonate's ability to discriminate maternal odors has also been suggested (Makin and Porter, 1989). Bottle-fed female neonates (two weeks old) with no prior breast-feeding experience orient toward the breast odor of an unfamiliar lactating female when paired with (1) the breast odor of a nonlactating female or a no-odor control or (2) the axillary odor of the same lactating female. These experiments suggest that some unique component of the breast odor such as milk or sebum, and not intensity differences, mediates this effect, as infants did not orient selectively toward the more intense of two other stimuli: the a lactating female's axillary odor paired with a no-odor control (Makin and Porter, 1989). It is also possible, however, that human milk and formula share volatile components, and that babies in this study are orienting towards a familiar odor component that has been consistently associated with feeding.

While the newborn may be capable of some olfactory discriminations that adults have difficulty making, several additional studies suggest that other olfactory psychophysical functions have the adult pattern at birth or shortly thereafter. Recall the finding of Engen and Lipsitt (1965) that repeated exposure to an odorant may result in a decrement of response to the odor when assessed as a change in respiratory rate. A subsequent study by the same group provided evidence that this phenomenon is best explained in terms of response habituation as opposed to sensory fatigue (Engen and Lipsitt, 1965). Infants who showed a decrement in response to a mixture of two odorants (anise oil and asafetida, or amyl acetate and heptanol) recovered the response when tested with a component of the mixture at the same concentration as in the mixture. However, the magnitude of the recovery response was a function of how perceptually similar (as rated by adults) the component was to the mixture even when the concentrations of the components were equated for perceived intensity. This study is significant, since it suggests that perceived similarity among odors is much the same in the newborn as in the adult. However, it would be premature to endorse this view without demonstrating similar effects with other odors. The possibility that odor categories are not dependent on a language system or much experience with odors is indeed an intriguing possibility and worthy of further study. As yet there is no agreed-upon method for categorizing odors, and the extent to which different schemes of categorization are dependent upon experience and language is unclear. Studies of odor similarity in infants could aid in determining "basic" categories for odorous chemicals.

The relationship between perceived intensity and molecular structure of homologous alcohols appears to be much the same in adults and the newborn. The likelihood of responding (Engen, 1965) (using change in respiratory rate as a dependent measure) to each in a series of full-strength homologous alcohols (ethanol to decanol) decreased as the chain length increased, in much the same way as the perceived intensity of these alcohols for adults decreases as chain length increases (Engen, 1965). It is interesting that while the short-chain alcohols elicited the greatest response when tested at full concentration, the long-chain alcohols elicited responses more systematically when threshold concentrations were evaluated. Adult psychophysical data suggest an explanation, showing that the effect of dilution on perceived intensity of these alcohols varies as a function of the chain length; for example a 50 percent dilution of a long-chain alcohol results in a smaller change in perceived intensity than the same dilution of a short-chain alcohol (Engen, 1965). A subsequent study in which newborns' responses were scaled (using degree of general movement as a measure), as opposed to measured in an all-or-none manner, revealed a similar relationship between intensity functions and alcohol chain length (Rovee, 1969), confirming an essential similarity between olfactory processing in adults and newborns.

The ability to localize an odor source appears to be another perceptual capacity of the newborn. Reiser et al. (1976) observed infants (1 to 5.5 days old) to systematically turn their heads away from a source of ammoniumhydroxide located to either the right or left of midline just below the nostrils. As with many of the previous phenomena, it is unclear whether this ability is mediated by the olfactory nerve and/or the trigeminal system, since ammonium hydroxide is a potent stimulant of the trigeminal nerve. Evaluation of neonates' responses to odorants that appear to selectively stimulate the olfactory nerve, such as phenyl ethyl alcohol or vanillin (Doty et al., 1978), could help discriminate among these possibilities.

Several studies have demonstrated that in adults, exposure to certain olfactory stimuli can inhibit perception of structurally related chemicals. The presumption is that the interaction of the first odor with many of the same receptors responsive to the second results in a decrement in responsiveness to the second odor. Cross-adaptation studies, like similarity judgments, have been taken as ways to explore basic odor categories. The opposite effect, facilitation of one odor by prior exposure to another, is less common but has also been observed (see Corbit and Engen, 1971). Both cross-adaptation and facilitation have been observed in human neonates (Rovee, 1972; Engen and Bosack, 1969), a further indication of the similarity between infants and adults.

Sex differences in responsiveness to odors have been demonstrated repeatedly in adult studies. Women tend to be more sensitive than men to a variety of odorants (Koelega and Koster, 1974; Gilbert, Greenberg, and Beauchamp, 1989) and are better at identifying odorants than men (Doty et al., 1983; Cain, 1982). The predominant theory put forth to explain these differences has posited that sex hormones mediate the documented superiority in females' sensitivity. This theory is consistent with studies that have failed to demonstrate sex differences in sensitivity in prepubescent children (e.g., Koelega and Koster, 1974; Dorries et al., 1989; but see Toulouse and Vaschide, 1899). An alternative to this account, however, is that the tendency for women to engage more frequently in activities that involve olfaction, such as food preparation and child care, leads to the observed superiority. No direct evidence is available in support of either of these possibilities (see Doty, 1986), and some recent findings with infants raise questions about both views. In testing newborns' ability to discriminate a familiar from an unfamiliar odor, Balogh and Porter (1986) observed that only females would selectively orient toward a familiar odor. Male infants, on the other hand, displayed a bias to turn their heads to the right. While it is possible that this sex difference reflects a neurologically less mature motor system in male neonates relative to females (Balogh and Porter, 1986), a recent study of nine-month-olds suggests that this may be a real sex difference in responsiveness to odors. Schmidt and Beauchamp (1989) observed nine-month-old females, but not males, to spend significantly more time exploring an odorized toy than a nonodorized toy when given a simultaneous choice between the two. Male infants, in contrast, spent about equal time exploring the toys. Response patterns suggested that this effect was mediated by the odor and was not due to sex differences in general responsiveness. This result, taken together with the sex difference observed by Balogh and Porter, is provocative, since it suggests that some of the adult gender differences in olfaction may be determined, in part, by innate factors. At the least, it suggests that neither concurrent action of sex hormones nor experience can offer a comprehensive explanation of the adult sex differences.

Hedonic Discriminations

In spite of the spareness of the published research, it appears that the olfactory system in the neonate is organized in much the same way as in the adult: anatomically the olfactory system bears an adultlike pattern; sensitivity appears to be as acute, if not more so than in the adult; perceptual similarity categories appear to be the same between newborns and adults; the ability to localize an odor source is present shortly after birth; even sex differences in responsiveness to olfactory stimuli appear to have their origins in infancy. Given the striking resemblance between the neonate and the adult in all of these olfactory areas, it is somewhat surprising that the origin of hedonic responses to odors has not been unequivocally traced to the neonatal period; in fact, there is substantial controversy on this issue.

This potential difference in olfactory perception between infants and adults presents a puzzle, since hedonic valence is the most salient psychological attribute of an odor for adults (Schiffman, 1974), and anatomical olfactory structures have relatively direct access to brain structures that govern emotion (Kandel and Schwartz, 1985). In addition, from a functional perspective, innately determined hedonic reactions to odors, as to tastes, could play an important role in mediating the selection of foods (Steiner, 1979), since unpleasant odors can signal decay or toxicity, while pleasant odors can signal the safety of a substance for ingestion or inhalation.

Until recently the predominant view has been that all odor preferences and aversions are learned through associational learning processes. Engen (1982, 1988) has argued that even the smell of a skunk would be perceived as pleasant if the conditions under which it was experienced could be controlled. (See the Chapter by Ehiliclman and Boslon in this book) This view is based on several studies that were collectively unable to expose adultlike systematic odor preference patterns in children less than five years of age.

Stein, Ottenberg, and Roulet (1958) observed that in classifying odors as "liked" or "disliked," three- to four-year-old children were as likely to say they liked the smell of amyl acetate (banana/pear-like) as they were to say that they liked the odors of synthetic sweat and feces. In contrast, six-year-olds, in agreement with adults, tended to say they liked the smell of amyl acetate but not that of sweat or feces. In another investigation of 293 children between 1 month and 16 years of age, Peto (1936) observed adultlike disgust reactions (verbal and facial) to odors in only 3 of 92 children under five years of age. Furthermore, in an unpublished experiment by Lipsitt and colleagues (described in Engen, 1982), play behaviors and facial expressions of one- to two-year-olds were not influenced by the surreptitious delivery of odorants judged by adults as unpleasant and pleasant . It has even been suggested that six- to seven-year-old children may not have a completely adultlike hedonic experience of odors. Although adults and six- to seven-year-olds exhibit similar rank-order preferences for odors (Engen, 1974; Kniep et al., 1931; Moncrieff, 1966), they are less consistent than adults in (1) selecting preferred odors from all pairwise comparisons from a set of odorants (Engen, 1974) and (2) classifying odors as "liked" or "disliked" (Kniep et al., 1931).

Recent findings raise the possibility that methodological deficiencies may have contributed to these negative results obtained with young children. Engen and Katz (described in Engen, 1974) observed that children of four years, but not of six years or more, were more likely to respond positively to the odorant butyric acid when asked to "Tell me if this smells pretty" than when asked to "Tell me if this smells ugly." Thus it is possible that some of the prior developmental findings based on opinion polling could reflect a response bias that diminishes with age rather than actual developmental changes in olfactory preferences. Furthermore, since younger children are more prone to attention lapses and have shorter memory spans than older children and adults, any method relying on sustained attention or placing demands on memory could yield spurious age differences.

Indeed, new simplified techniques taking into consideration the attention, memory, and communication limitations of young children have exposed systematic adultlike odor preference patterns in children as young as three years old (Schmidt and Beauchamp, 1988; Strickland, Jessee, and Filsinger, 1988). Essentially adultlike odor preference patterns were obtained for a set of nine odorants (carvone, phenyl ethyl methyl carbinol, methyl salicylate, c-16 aldehyde, eugenol, amyl acetate, androstenone, butyric acid, and pyridine) when three-year-old children were asked to give "good smells" to a likable television puppet (Big Bird) and "bad smells" to a grouchy television character (Oscar the Grouch) (Schmidt and Beauchamp, 1988). Although children and adults in this study did differ in their distributions of hedonic ratings for some odorants, these differences were limited in their ability to explain children's reactions. Using a similar dichotomous forced-choice task in which children pointed to a smiling face or a frowning face to indicate their odor preferences, Strickland and colleagues (1988) also demonstrated adultlike hedonic classifications of two other odorants (benzaldehyde and dimethyl disulfide) in children as young as three years. Together, these studies challenge the empirical foundations of the behavioristic claim that all odor preferences are learned relatively late in the preschool period. At a minimum, these data demand reconsideration of the view that some odors are inherently pleasant or unpleasant.

In fact, some of the neonatal studies discussed previously are consistent with the possibility that affective odor reactions may begin in early infancy. Recall, for example, newborn infants' tendencies to turn toward maternal odors and away from the smell of ammonia. Although behaviors such as these may reflect a tendency to avoid any novel or strong olfactory stimuli, they may also be associated with, or mediated by, hedonic reactions. Arguing against this possibility are the observations that (1) newborns fail to produce reliable and appropriate facial expression in response to adult-rated pleasant and unpleasant odors (cf. Steiner, 1977, 1979; Self et al., 1972; Engen, 1982, 1988), and (2) they do not show differential heart-rate or respiratory-rate changes to odors with different hedonic values (Engen and Lipsitt, 1965; Engen et al., 1963). Steiner (1977, 1979) has reported that a variety of food-related odors elicited adultlike facial expressions of "enjoyment" and "disgust" in newborns with no prior feeding experience. Specifically, he reports that the odors of rotten eggs and strong fish elicited facial expressions of disgust, while the odors of banana, vanilla, butter, and chocolate elicited facial expressions of pleasure. However, other scientists have failed to observe such reliable facial reactions with other odorants of opposite hedonic valence (Self et al., 1972; Engen and Lipsitt, 1965; Peterson and Rainey, 1910). Although Self and colleagues (1972) observed startles and grimaces somewhat more frequently in reaction to two adult-rated unpleasant odors (asafetida and valerian) than to two adult-rated pleasant odors (lavender or anise), these response types represented less than 38 percent of the responses that were analyzed, and lavender and valerian were equally likely to elicit startle reactions (23 percent and 24 percent respectively). Similarly, Engen and his colleagues (Engen and Lipsitt, 1965; Engen, Lipsitt, and Kaye, 1963) failed to obtain evidence that two- to four-day-old newborns discriminated the hedonic valences of odors such as anise, phenyl ethyl alcohol, acetic acid heptanol, amyl acetate, and asafetida on the basis of facial expressions, bodily movements, or respiratory and heart-rate changes. In summarizing his findings, Engen states that "all facial and bodily responses to the odors, whether pleasant or unpleasant to adults, looked like mild startle and escape reactions" (Engen, 1982, p. 131).

However, neither the inconsistency of newborns' facial expressions to pleasant and unpleasant odors nor the absence of differential physiological responses to odors of opposite hedonic valence provides convincing evidence that infants do not experience emotional reactions to odors. First, spontaneous adult facial expressions do not provide a reliable index of the hedonic quality of an odor, and they do not appear to be reflexive in nature (Gilbert, Fridlund, and Sabini, 1987). If adult odor-contingent facial expressions are predominantly social in nature, as suggested by Gilbert and colleagues (1987), then their absence in newborns may more appropriately reflect the absence of social-communicative skills than the absence of hedonic reactions. Second, it has proved extremely difficult to relate systematic heart-rate or respiratory changes to emotional responses of any kind in infants, and therefore it seems tenuous to assume that such responses would be indicative of affective odor reactions in newborns.

A recent study of nine-month-old infants provides the most direct evidence that at least some adultlike hedonic odor discriminations have their origins in infancy. Schmidt (1990) videotaped bodily and facial reactions of nine-month-old infants while they explored in succession each of three rattles that looked alike but had different odors. As judged by adults, one had an unpleasant odor (butyric acid), one had a pleasant smell (methyl salicylate), and the other had no odor. The odors were approximately equated for intensity and for perceived trigeminal component. Eight naive adults who subsequently viewed the videotapes of infants were all significantly better than chance in judging the hedonic valence of each of the rattles. Analysis of specific behaviors failed to reveal any single behavior (e.g., smiles, frowns, fussing, dropping, mouthing, or pushing away the rattle) that was systematically associated with the hedonic quality of the odor: different infants apparently conveyed their hedonic reactions in different manners.

In view of these findings, further investigation of hedonic reactions to other odors throughout infancy should shed light on the origins and modifiability of odor preferences. In particular, new methods for exploring odor hedonios in infancy should be explored. The possible inadequacy of facial expression and the limited behavioral repertoire of the neonate may render head turning toward or away from an odor source the most appropriate, albeit indirect, method of tapping hedonic responses in newborns at this stage.

The Functional Importance of Olfaction

In what way do the apparent olfactory competencies discussed thus far actually serve the infant and young child? The potential superiority of the olfactory system relative to other sensory modalities in the newborn raises the possibility that olfaction is of considerable importance during development; however, the available evidence remains largely speculative. Scientists have hypothesized three potentially important functions for olfaction in infancy: one in modulating feeding behaviors, one in avoidance behaviors, and another recognizing individuals.

It is possible that olfaction guides the newborn's search for food. The observations that newborns will orient toward the odor of the mother's breast (Macfarlane, 1975; Russell, 1976; Schaal et al., 1980; Schaal, 1988; Schleidt and Genzel, 1990) are consistent with this possibility. The most compelling finding in this regard is that bottle-fed neonates oriented toward the breast odor, but not the axillary odor, of an unfamiliar lactating female when this odor was paired with the comparable odor from a nonlactating female (Makin and Porter, 1989). As mentioned earlier, this effect may reflect either a genetic mechanism or a learned association between some shared volatile substrate of breast milk and formula. In the absence of any data to rule out the latter possibility, these "maternal orienting" behaviors may alternatively be characterized as mere avoidance of novel olfactory stimuli. Studies that examine the relationship between prandial state and the reliability of these orienting behaviors could help to discriminate these possibilities.

In addition to guiding the infants' search for a nutritional source, odors may further be involved in feeding by affecting sucking responses. Self and colleagues (1972) reported that anise and lavender elicited sucking more frequently than asafetida, valerian, and a no-odor control stimulus in one- to three-day-old infants. However, these sucking responses represented less than one-third of all the responses observed in reaction to odors, and it is therefore difficult to evaluate the significance of the observation. In addition, prandial state was not considered when testing infants, further confounding the reliability of this effect. Earlier observational studies also reported that specific odors elicited sucking responses in newborns (Peterson and Rainey, 1910; Genzmer, 1873, cited by Peterson and Rainey, 1910), however, base-rate sucking in the absence of odorants was not evaluated, and therefore the significance of the reports cannot be determined. Kroner (1881, cited by Peterson and Rainey, 1910) reported that the application of a novel odorant to a mother's breast inhibited sucking altogether in one neonate, but again the reliability of this effect was not assessed against appropriate base rates. Schmidt and Beauchamp (1989) observed that nine-month-old infants were more likely to put odorized (but tasteless) rattles in the mouth than nonodorized rattles, suggesting that the nine-month-old associates odors and tastes. Collectively these studies suggest that olfaction mediates some aspects of feeding, but additional well-controlled studies are clearly necessary to clarify this possibility. In particular, since food odors are generally experienced retronasally via food actually in the mouth, further studies designed to investigate the effects of odors associated with ingestion of fluids could reveal more profound control of sucking behavior by odor.

Olfaction may also serve a protective function in infancy; newborns may systematically avoid novel or strong odors. The observation that infants turn away from a source of ammonia (Reiser et al., 1976) is the clearest evidence in support of this function. Infants' tendency to turn toward an artificial odor to which they have been exposed for several hours when it is paired with another artificial but unfamiliar odor (Cernoch and Porter, 1985) may also reflect this function. As discussed in the next section, it is also possible to interpret many of the maternal orienting behaviors as avoidance of novel odors as opposed to orienting toward familiar odors.

Another important implication of the studies on infants' discrimination of maternal odors (breast, axillary, neck, perfume) is that odors are an important mechanism in early recognition processes, in particular kin recognition. In this regard, humans may be similar to many other species for whom odors are a primary means by which individuals are recognized. It is intriguing to consider that olfaction is especially important in this regard during the first several weeks of life, before vision is really effective While the data are convincing that newborns of less than two weeks old can recognize the smell of their own mother, the extent to which these social discriminations depend upon the reinforcement provided by social contact is unclear. It is possible that equally fine discriminations and orienting behaviors could emerge with regular and consistent exposure to subtle odors in the absence of social reinforcement.

While olfaction appears to be sufficient for the development of early kin recognition, it is unclear whether it is actually necessary. Schaal (1986) demonstrated that when infants oriented toward a pad impregnated with the odor of the mother's neck or breast, motor activity of the head and arms was significantly less than when newborns oriented toward a no-odor control pad or the odor of an unfamiliar mother. This finding suggests that maternal odors may serve to calm the infant. Studies contrasting the social development of infants with and without olfactory deficits could shed some light on the extent to which olfaction guides social and emotional development. Infants with olfactory deficits might be identified by the methods described earlier for assessing neonatal sensitivity and discrimination capacities, or could be infants with persistent colds or Kallman's syndrome.

Conclusion

Compared with those of the senses of vision, hearing, and taste, developmental studies of human olfaction have not been common until recently. A major reason for this is the absence of an agreed-upon set of methods that can reveal sensitivity to, discrimination of, and preference among odors. Recent developments, however, indicate that with careful analysis of infant behavior, apparent responsiveness to very subtle odors can be revealed. It is now firmly established that the very young infant can detect and respond to a variety of volatile substances, and it is likely that at least some components of these responses are mediated by the olfactory sense. Based upon analogy to recent animal studies, it is possible that lifelong preferences and aversions for odors, as well as other strong emotional associations with odors, are formed during infancy and childhood. It is for future research to critically evaluate these suggestions using current and to-be-developed methods of analysis, some of which could include non-invasive measurement of neurophysiological responses to odors.

Perhaps the major unresolved issue in olfaction is whether there is a small number of primary odors from which all odors are derived by combinatorial rules or whether, as in the immune system, there are many, perhaps thousands or more, independently recognized odors. A problem of psychophysical approaches to this question is that in trying to determine a natural scheme for classifying odors, it is difficult to ignore learned categories and associations. A sophisticated developmental approach to the study of odor similarity and odor adaptation and cross-adaptation could provide insights into this fundamental question of olfactory quality, since the infant and young child would be less burdened by the effects of individual experience. Thus, not only can developmental studies illuminate the origins of early response to odors, but they also will contribute to the general understanding of the nature of olfactory experience.

Acknowledgments. Some of the work described here was supported by a grant from the Fragrance Foundation Research Fund.

References

Balogh, R.D., and R.H. Porter (1986). Olfactory preferences resulting from mere exposure in human neonates. *Infant Behav. Dev.*, *9*, 395–401.

Bossy, J. (1980). Development of olfactory and related structures in staged human embryos. *Anat. Embryol.*, *161*, 223–236.

Brown, J.W. (1974). Prenatal development of the human chief sensory trigeminal nucleus. *J. Comp. Neurol.*, *156*, 307–336.

Cain, W.S. (1982). Odor identification by males and females: predictions vs. performance. *Chem. Senses*, *7*, 129–124.

Cernoch, J.M., and R.H. Porter (1985). Recognition of maternal axillary odors by infants. *Child Dev.*, *56*, 1593–1598.

Chuah, M.I., and D.R. Zheng (1987). Immunohistochemical study of olfactory marker protein in human fetuses. *Neurosci. Lett.*, *Suppl. 28*, S29.

Corbit, T., and T. Engen (1971). Facilitation of olfactory detection. *Perception Psychophys.*, *10*, 433–436.

Darwin, C. (1872). *The Expression of Emotions in Men and Animals.* London: John Murray.

Dorries, K.M., H.J. Schmidt, G.K. Beauchamp, and C.H. Wysocki (1989). Changes in sensitivity to the odor of androstenone during adolescence. *Dev. Psychobiol.*, *22*, 423–435.

Doty, R.L. (1986). Gender and endocrine-related influences on human olfactory perception. In H.L. Meiselman and R.S. Rivlin (Eds.), *Clinical Measurement of Taste and Smell.* New York: Macmillan, pp. 377–413.

Doty, R.L., W.E. Brugger, P.C. Jurs, M.A. Orndorff, P.J. Snyder, and L.D. Lowery (1978). Intranasal trigeminal stimulation from odorous volatiles: Psychometric responses from anosmic and normal humans. *Physiol. Behav.*, *20*, 175–185.

Doty, R.L., S. Applebaum, H. Zusho, and R.G. Settle (1983). Sex differences in odor identification ability: a cross-cultural analysis. *Neuropsychologia*, *23*, 667–672.

Engen, T. (1965). Psychophysical analysis of the odor intensity of homologous alcohols. *J. Exp. Psychol.*, *70*, 611–616.

Engen, T. (1974). Method and theory in the study of odor preferences. In A. Turk, J. Johnson, and D. Moulton (Eds.), *Human Responses to Environmental Odors.* New York: Academic, pp. 121–141.

Engen, T. (1982). *The Perception of Odors.* New York: Academic.

Engen, T. (1988). The acquisition of odour hedonics. In S. Van Toller, and G.H. Dodd (Eds.), *Perfumery: The Psychology and Biology of Fragrance.* London and New York: Chapman & Hill.

Engen, T., L. Lipsitt, and H. Kaye (1963). Olfactory response and adaptation in the human neonate. *J. Comp. Physiol. Psychol.*, *56*, 73–77.

Engen, T., and L. Lipsitt (1965). Decrement and recovery of responses to olfactory stimuli in the human neonate. *J Comp. Physiol. Psychol.*, *59*, 312–316.

Engen, T., and T.N. Bosack (1969). Facilitation in olfactory perception. *J. Comp. Physiol. Psychol.*, *68*, 320–326.

Freud, S. (1962). *Civilization and Its Discontents.* New York: Norton.

Gilbert, A.N., A.J. Fridlund, and J. Sabini (1987). Hedonic and social determinants of facial displays to odors. *Chem. Senses*, *12*, 355–363.

Gilbert, A.N., M.S. Greenberg, and G.K. Beauchamp (1989). Sex, handedness and side of nose modulate human odor perception. *Neuropsychologia*, *27*, 505–511.

Hepper, P.G. (1988). Adaptive fetal learning: Prenatal exposure to garlic affects postnatal preferences. *Animal Behav.*, *36*, 935–936.

Humphrey, T. (1940). The development of the olfactory and the accessory olfactory formations in human embryos and fetuses. *J. Comp. Neurol.*, *73*, 431–468.

Johnson, A., R. Josephson, and M. Hawke (1985). Clinical and histological evidence for the presence of vomeronasal (Jacobson's) organ in adult humans. *J. Otolaryngol.*, *14*, 71–79.

Kandel, E.R., and J.H. Schwartz (1985). *Principles of Neural Science*. New York: Elsevier.

Kniep, E.H., W.L. Morgan, and P.T. Young (1931). Studies in affective psychology. *Am. J. Psychol.*, *43*, 406–421.

Koelega, H.S., and E.P. Koster (1974). Some experiments on sex differences in odor perception. *Ann. NY Acad. Sci.*, *237*, 234–246.

Lipsitt, L.P., T. Engen, and H. Kaye (1963). Developmental changes in the olfactory threshold of the neonate. *Child Dev.*, *34*, 371–376.

Macfarlane, A.J. (1975). Olfaction in the development of social preferences in the human neonate. *CIBA Found. Symp.*, *33*, 103–117.

Makin, J.W., and R.H. Porter (1989). Attractiveness of lactating females' breast odors to neonates. *Child Dev.*, *60*, 803–810.

Margolis, F.L., W. Sydor, Z. Teitlebaum, R. Blacher, M. Grillo, K. Rogers, R. Sun, and U. Gubler (1985). Molecular biological approaches to the olfactory system: Olfactory marker protein. *Chem. Senses*, *10*, 163–174.

Moncrieff, R.W. (1966). *Odour Preferences*. London: Leonard Hill.

Nakashima, T., C.P. Kimmelman, and J. Snow (1984). Structure of human fetal and adult olfactory neuroepithelium. *Arch. Otolaryngol.*, *110*, 641–646.

Pederson, P.A., and E.M. Blass (1982). Prenatal and postnatal determinants of the 1st suckling episode in the albino rat. *Dev. Psychobiol.*, *15*, 349–356.

Peterson, F., and L.H. Rainey (1910). The beginnings of mind in the new born. *Bull. Lying in Hosp. NY*, *7*, 99–122.

Peto, E. (1936). Contribution to the development of smell feeling. *Br. J. Med. Psychiatry*, *15*, 314–320.

Porter, R.H. (1990). Personal communication.

Rieser, J., A. Yonas, and K. Wikner (1976). Radial localization of odors by human newborns. *Child Dev.*, *47*, 856–859.

Rosselli-Austin, L., and J. Williams (1990). Enriched neonatal odor exposure leads to increased numbers of olfactory bulb mitral and granule cells. *Dev. Brain Res.*, *51*, 135–137.

Rovee, C.K. (1969). Psychophysical scaling of olfactory response to the aliphatic alcohols in human neonates. *J. Exp. Child Psychol.*, *7*, 245–254.

Rovee, C.K. (1972). Olfactory cross-adaptation and facilitation in human neonates. *J. Exp. Child Psychol.*, *13*, 368–381.

Russell, M.J. (1976). Human olfactory communication. *Nature*, *260*, 520–522.

Sarnat, H.B. (1978). Olfactory reflexes in the newborn infant. *J. Pediatr.*, *92*, 624–626.

Schaal, B. (1986). Presumed olfactory exchanges between mother and neonate in humans. In J. LeCamus and J. Cosnier (Eds.), *Ethology and Psychology*. Toulouse: Privat, I.E.C., pp. 101–110.

Schaal, B. (1988). Olfaction in infants and children: developmental and functional perspectives. *Chem. Senses*, *13*, 145–190.

Schaal, B., H. Montagner, E. Hertling, D. Bolzoni, A. Moyse, and R. Quichon (1980). Les stimulations olfactives dans les relations entre l'enfant et la mere. *Reprod. Nutr. Dev.*, *20*, 843–858.

Schiffman, S. (1974). Physicochemical correlates of olfactory quality. *Science*, *185*, 112–117.

Schleidt, M., and C. Genzel (1990). The significance of mother's perfume for infants in the first weeks of their life. *Ethol. Sociobiol.*, *11*, 145–154.

Schmidt, H.J. (1990). Adult-like hedonic responses to odors in 9-month-old infants. Achems XII, Sarasota, FL.

Schmidt, H.J., and G.K. Beauchamp (1988). Adult-like odor preferences and aversions in three-year-old children. *Child Dev.*, *59*, 1136–1143.

Schmidt, H.J., and G.K. Beauchamp (1989). Sex differences in responsiveness to odors in 9-month-old infants. Achems XII, Sarasota, FL.

Schmidt, H.J., and G.K. Beauchamp (1989). Unpublished observation.

Self, P.A., F.D. Horowitz, and L.Y. Paden (1972). Olfaction in newborn infants. *Dev. Psychol.*, *7*, 349–363.

Silver, W.L. (1987). The common clinical sense. In T.E. Finger and W.L. Silver (Eds.), *Neurobiology of Taste and Smell.* New York, pp. 65–87.

Smotherman, W.P. (1982). Odor aversion learning by the rat fetus *Physiol. Behav.*, *29*, 769–771.

Stein, M., P. Ottenberg, and N. Roulet (1958). A study of the development of olfactory preferences. *AMA Arch. Neurol. Psychiatry*, *80*, 264–266.

Steiner, J.E. (1977). Facial expressions of the neonate infant indicating the hedonics of food-related chemical stimuli. In J.M. Weiffen (Ed.), *Taste and Development: The Genesis of Sweet Preference.* Bethesda, Md.: NIH-DHEW, pp. 173–188.

Steiner, J.E. (1979). Human facial expressions in response to taste and smell stimulation. *Adv. Child Dev. Behav.*, *13*, 257–295.

Steiner, J.E. (1979). Oral and facial innate motor responses to gustatory and to some olfactory stimuli. In J.H.A. Croese (Ed.), *Preference Behavior and Chemoreception.* London: Human Information Retrieval, pp. 247–262.

Strickland, M., P.O. Jessee, and E.E. Filsinger (1988). A procedure for obtaining young children's reports of olfactory stimuli. *Perception Psychophys.*, *44*, 379–382.

Toulouse, N., and S. Vaschide (1899). Mesure de l'odorat chez l'homme et chez la femme. *Seances Mem. Soc. Biol.*, *51*, 381–383.

Wilson, D.A., R.M. Sullivan, and M. Leon (1985). Odor familiarity alters mitral cell response in the olfactory bulb of neonatal rats. *Dev. Brain Res.*, *22*, 314–317.

Woo, C.C., and M. Leon (1987). Sensitive period for neural and behavioral response development to learned odors. *Dev. Brain Res.*, *36*, 309–313.

Wysocki, C.J., and G.K. Beauchamp (1984). The ability to smell androstenone is genetically determined. Proc. *Nat. Acad. Sci. USA*, *81*, 4899–4902.

Wysocki, C.J., K. Dorries, and G.K. Beauchamp (1989). Ability to smell androstenone can be acquired by ostensibly anosmic people. *Proc. Nat. Acad. Sci. USA*, *86*, 7976–7978.

14
Olfaction and Human Sexuality: A Psychoanalytic Approach

IRVING BIEBER,* TOBY B. BIEBER, and RICHARD C. FRIEDMAN

Introduction

Odor has a decisive role in the sexual development and mating behavior of infrahuman animals. It is reasonable to assume, therefore, that man, too, would have a similar inborn sensate guarantee for heterosexual arousal. Yet, until recently, aside from literateurs, among them Shakespeare, Shelley, Huysman, and Zola, and a few clinicians who wrote about the sensual response to certain human odors, relatively little attention has been given to the relationship between olfaction and sexuality. In an inclusive review of the subject, Gibbon (1986) commented, "Odors reach into all our emotional life. Odors suggest, stimulate associations, evoke, frighten and arouse us, but they seem to lie below conscious thought until someone like the poet, Baudelaire, parts the curtain" (p. 327).

Olfaction and Sexuality: Early Writings

It was not until the latter part of the nineteenth century that psychiatric articles on olfaction began to appear. (See also the chapter by Harrington in this volume.) Havelock Ellis (1929) was the first, with the publication of his volume in 1906, to undertake an extensive and detailed discussion of the effects of odor on human sexuality. Calling attention to the lack of scientific interest in the subject, he stated, "The writer who, more than any other, has in recent years restored the study of the sense of smell from a by-path to its proper position as a highway for investigation, is, without doubt, Professor Zwaardemaker of Utrecht. The invention of his first olfactometer in 1888 and the appearance in 1895 of his great work, *Die Physiologie des Geruchs*, have served to give the physiology of the sense of smell an assured status and to open the way anew for much fruitful inves-

*The editors mourn the loss of Dr. Irving Bieber. His contributions to psychoanalysis and to the field of olfactory sciences were outstanding.

tigation" (p. 52). Ellis praised Zwaardemaker for having shown that odors were powerful nervous system stimulants that play a role in sexual arousal. Hendrick Zwaardemaker, a physiologist, had also tried to categorize odors on a nine-point continuum ranging from very pleasant to nauseating, but his attempt did not prove useful and has been forgotten. Ellis himself described individuals who could identify others by their bodily odors—general skin odor, hair and scalp, breath, armpit, feet, perineum, the odor of preputial smegma in men, and the odor of the mons veneris, the vulvar smegma, and the vaginal mucous and menstrual odor in women. He also reported on individuals who were able to identify others from the odor of their clothing.

Yet, curiously, Ellis, as well as Krafft-Ebing and Freud, the three major sexologists among the psychiatric pioneers of their age, were ambivalent about the role of olfaction. Abounding in contradictions, they tempered their observations by hinting at, alluding to, or frankly ascribing a psychopathological dimension to olfaction in sexuality. Olfaction was largely viewed as animalistic and of relatively little importance in humans, particularly in cultivated people. To be sure, some of the psychiatric patients they described manifested psychopathology through olfactory symptoms that often appeared in a sexual context, but these early writers, who so profoundly influenced those who followed, minimized the role of olfaction or, at best, were ambiguous about normal responses to sexual and human odors. On the one hand, Ellis wrote, "There can be no doubt that . . . the extent to which olfaction influences the sexual sphere in civilized man has been much underestimated" (p. 77). On the other hand, he stated, "Among the apes it [smell] has greatly lost importance and in man it has become almost rudimentary, giving place to the supremacy of vision. . . . If the sense of smell were abolished altogether, the life of mankind would continue as before, with little or no sensible modification, though the pleasure of life and especially of eating and drinking, would be to some extent diminished" (p. 48).

In the same vein, Krafft-Ebing, in a translation of his work (1965), wrote, "I am inclined to doubt that, under normal conditions, olfactory impressions in man, as in animals, play an important role in the excitation of the sexual center. On account of the importance of this sensation for the understanding of *pathological cases* [italics ours] it is necessary here to thoroughly consider the relations existing between the olfactory and sexual senses" (p. 21).

Freud (1961a) speculated that man's assumption of the upright position was associated with the diminished importance of smell in motivation. As people physically "turned away from the earth," they psychologically turned away from the scent of olfactory experience. He observed that young children were attracted to odorous body parts, but that the pleasure in strong-smelling areas was renounced through the socialization process. Psychopathology, however, particularly fetishistic perversions, provided an exception to this normative tendency. "The selection of the fetish depends on a coprophilic smell desire which has been lost by repression." Accordingly, in the perversion, "only the filthy and ill smelling foot is the sexual object."

Freud (1961b) hypothesized that cleanliness and order were essential elements in the ability of people to group together and form intimate social units, from which evolved the family. He suggested that publicly dissipated excrement and excremental odors were a bar to social organization, and therefore the repression of what he referred to as smell desire was a social benefit. He did not refer directly to sexual odors. We can only conjecture that his concept of cleanliness and social order included the control not only of coprophilic odors but of sexual odors as well.

Olfaction and Sexuality: Later Views

Even today, beliefs tend to persist that bodily odors, sexual and other, are related to uncleanliness and unacceptability. People who are preoccupied with neatness and cleanliness, living in a deodorized, sanitized culture, tend to become unusually self-conscious and conflicted about having their body odors detected by others. In a study of patients with abdominal colostomies, Ohrbach et al. (1957) noted that in addition to the fear of odor from feces, flatus, and an unwashed body, a fear of sexual odors emanating from the body was sometimes found, particularly in women. The authors divided their cases into three groups based on the criterion of the importance of cleanliness in the patients' preoperative concepts about social acceptance. The first group was most concerned with cleanliness and had the greatest psychological difficulties postoperatively. The second group was intermediate, having mixed attitudes. The third group had a history of freedom from aversions, taboos, and avoidances in relation to dirt and feces. They were more spontaneous and assertive than the first and second groups; they had more gratifying marriages and had better sexual relationships with their spouses. They adapted best to colostomy and were the least concerned postoperatively with the possibility of emanating odors.

Brill[1] (1932) followed Freud's thinking. He, too, stated that smell played a small part in the life of civilized man, but he then qualified that view: "Olfaction has played a great part in the life of animals, primitives and even in semi-enlightened peoples like the desert Arabs," where, he thought, the sense of smell was very important, adding, "But even among more civilized peoples, Oriental, Southern European and South American, the sense of smell plays an important part" (p. 8). Brill presented

[1]The translator of Freud's early works into English, A.A. Brill, read his paper on olfaction before the American Psychoanalytic Association in New York in 1930. It was published in the first edition of the *Psychoanalytic Quarterly* in 1932.

several case histories in which the sense of smell was intimately associated with sexuality. One case was of a blind man who had an insatiable desire to hug and kiss women so that he could smell them. In boyhood, his objects were his mother and sisters, but in later life, his bizarre fantasy included dead women who could not repulse him. He had fantasies of having access to dead partners so that he could smell them as much as he liked. Another example was the case of a man who had always been sensitive to odors, especially his mother's. As a child, he had the habit of rushing to occupy her vacated chair because the odor she had left there attracted him. One psychotic patient had the delusion that objectionable odors continually emanated from his rectum. Because he believed that others detected this, the patient retreated from all social activity. In agreement with Freud, Brill concluded that his clinical material demonstrated that smell played a part in neurotic and psychotic disturbances.

Daly, a psychoanalyst, and White, an entomologist (1930), collaborated in a study of psychological reactions to olfactory stimuli and published a paper the same year Brill gave his. These writers went beyond the idea that patients' olfactory preoccupations were intrinsic to psychopathology. They, too, considered the reasons for repressing the recognition of sexual odors. "The solution is that the recognition of this hypnotic sex attractive smell was lost to man because of the inhibition of the sexual impulse from causes which were the inevitable result of the severity of the primary law of incest and the associated taboos" (p. 80). The proposal linking olfactory repression with sexual inhibition and the incest taboo approaches our own views, to be explicated later on.

A few major points may be distilled from the reports of Brill and others, as interpreted from the perspective of modern psychiatry. Olfaction does not appear to influence the severity of psychopathology. The determinants of psychoses do not involve olfaction any more than do the determinants of neuroses. One set of variables that occurs in the lives of all people, patients and nonpatients, is the chance experience of early life events in which smell is invested with great meaning. Traumatic events may occur in which smell figures prominently. The association of olfactory sensation with meaningful, stressful early life events may then permanently endow olfaction with unusual and idiosyncratic significance. If an individual develops a mental disorder, olfactory perceptions and misperceptions may be cognitively processed as a function of the etiology, severity, and type of psychopathology; that is to say, patient A may develop delusions, whereas patient B may develop olfactory symptoms without becoming psychotic.

From a historical and, indeed, a contemporary point of view, the question of why the role of olfaction in human sexuality has been relegated to the sidelines of psychoanalytic thought is a tantalizing one. Olfaction has remained within the anecdotal area, the area of occasional comment and observation, without having a solid place within a framework of personality theory. Freud developed a structure of personality development that continues to be the most influential one. His genetic theory relies heavily on the role of sexuality in the various phases of personality growth, but excludes the role of olfaction as a relevant one in normal childhood development. In fact, his comments on odor throughout his prolific work are sparse. One would expect a greater emphasis on odor in the anal phase; however, as stated earlier, he assumed that odor sensitivity underwent repression. Freud conceptualized anality as consisting of an active and a passive component. He asserted that in the active component, the discharge of feces is experienced as pleasurable, and that feces is also a valuable body product that is lost through defecation. In the passive component, receptive pleasure is experienced if an object is inserted in the anus, a concept that provided an underpinning for Freud's theoretical constructs about homosexuality. The active anal phase, though more closely juxtaposed to olfaction than the passive phase, mainly concerns character formation, where, in some individuals, feces become endowed with undue value, resulting in feces retention and in such later traits as penuriousness, aggression, and compulsiveness.

Abraham (1954) was interested in the anal character, and devoted passing attention to the role of olfaction in psychopathology. Yet, the major theory builders around Freud, both disciples and dissidents, generally ignored the subject. Jung made no mention of olfaction in his writings, nor did Adler, despite his close attention to the determinants and consequences of social behavior. Even in the revisions and amendments of the next generation of analysts, the role of olfaction in sexuality was not taken up. The major neo-Freudians—Horney, Sullivan, Thompson, Rado, Kardiner, and other distinguished revisionists—also overlooked the role of olfaction in their clinical reports and in their theories. The subject of olfaction, with some notable exceptions, still has not captured the attention of the present-day analytic community. Yet psychosexual development and heterosexual organization is very importantly influenced by early olfactory impressions, as was first delineated in the clinical-theoretical work by I. Bieber (1959).

Olfaction and Psychosexual Development

The clinical research on olfaction and sexuality by I. Bieber over many years consists of observational data and dream material of patients in analysis. A fair number of people, in and out of analysis, are conscious of the erotic effect of sexual odors and the pleasure experienced from the natural odor of a lover or other beloved person, parent, or sibling. Bieber noted that patients who are unaware of their sexual reactions to olfactory stimuli nevertheless make sexually veiled references to smell in dreams and free associations that come into consciousness when such material is pursued analytically. Among the patients he has studied, such productions are generally associated directly with incestuous objects. In olfactory associations, reference may be made to the aroma of a flower, rather than to its color or other attributes. Dreams with olfactory references are characterized by a central sexual theme involving the parent of the opposite sex or, less frequently, an older sibling of the opposite sex. The following samplings of dreams were reported by patients in analysis with I. Bieber, who has observed that when their opening remarks refer to odor, an oedipal theme usually dominates the session.

From a young woman: "There is a man in my bed who has a very unpleasant odor. The odor is coming from his arm-pit. In each arm-pit there is a large swelling. The man looks like the actor, EX." The patient associated the swelling with testicles and commented that the actor reminded her of her father and older brother in appearance and motility. The dream clearly presents the axilla as an erotogenic zone (Ellis, 1929; Brody, 1975).

Another patient, a physician in his mid thirties, had difficulty breathing through a chronically stuffed nose as far back as he could remember. His problem had been ascribed to a deviated septum. He reported the following dream: "A child of about seven or eight is the central figure. The child appears to be half male, half female, with long, black hair and has a dull, adenoidal expression with the mouth hanging open." The patient associated the child with himself and the open mouth with his own stuffed nose. The hair was associated with an older sister toward whom he consciously had sexual feelings. When reassured of the normalcy of sexual feelings toward a sister, his olfactory inhibition gave way; his nose cleared and remained so, even after a follow-up of many years. In a similar case, the patient complained from time to time throughout her analysis that she had a poor sense of smell. With the resolution of her oedipal problems, her sense of smell became a keen one.

The next patient was a sophisticated European, a gourmet, quite given to sensual experience. Although he did not care for perfumes and would be annoyed by his wife's fondness for them, he was aware of their sexually stimulating potential. His dream opened upon a rivalry situation between himself and the analyst for a woman. The dream then shifted to a scene in which a male dog smells a cauliflower, lifts a hind leg, and urinates on the cauliflower. The patient recognized the dog's sniffing and urinating as part of its sexual pattern. Since childhood, cauliflower had been one of his favorite foods, and he stated that the only evidence he ever had of his mother's affection was her desire to please him by cooking favorite foods for him.

In another case, a young man repetitively dreamed a three-act scenario on successive nights that consistently involved his mother. On the first night, he would find himself in a foreign city with her; not infrequently, it would be Paris, to which he associated sensuality and sexuality. They would dine together in a restaurant pervaded with the aroma of good food and wine. The next night he would dream he was being attacked by men. Sometimes he would be chased by a dangerous, armed gang; sometimes it would be a single man out to get him. The patient would be jolted from sleep feeling very anxious. The third night would find him in a homosexual encounter. He had never acted out his homosexual impulses, but he did have the described repetitive dreams as well as homosexual fantasies that, together with his impotence with women, brought him into analysis. This case not only illustrates a connection between olfaction and the oedipal syndrome, as depicted in the patient's sensual dream adventure with his mother and in his associations to the dream, it also illuminates certain underlying psychodynamics—fears of lethal attack by men for his incestuous wishes, followed by a defensive escape into homosexuality.

A woman in her thirties whose sexual anxieties precluded marriage brought this dream: She is out with a male coworker, and as she draws close to him physically, her nose begins to run. In the next scene she is with her father, whom she is about to kiss when a police car appears and interrupts her. The associations to the dream concerned her father and the ways in which various men she had known resembled and differed from him, associations clearly related to an oedipal constellation.

Rosenbaum (1961) noted that olfactory experiences in some of his analytic patients evoked oedipal reactions, which they expressed in the transference. One male patient's positive reaction to cooking aromas emerging from the analyst's living quarters precipitated oedipal stirrings, defensively manifested at first in hostility toward the analyst's wife. Another patient, a woman, revealed her sexual feelings toward Rosenbaum through dreams and associations of roses. She spoke directly of the sexual effect upon her of the odor of roses. A third case was of a man who befouled the office with his body odor in an attempt to have his odor approved of and accepted.

Olfaction and Oedipus

In 1900, Freud (1961c) described the Oedipus complex, his monumental contribution to the understanding of interpersonal and family dynamics. Although its biological aspect is implicit, his emphasis was psychosocial, stressing the significant conflicts this phase of development engendered, that is, anxiety about sexual feelings toward the opposite-sex parent and siblings, and the sexually competitive relationship with the same-sex parent. According to Bieber's theory, referred to earlier and reviewed here, the Oedipus complex is a psychobiological phenomenon that is characterized by two phases: first, the beginning of a capability for heterosexual arousal, and second, competition with members of the same sex. Conflict on a sexually competitive basis with the same-sex parent does not exist before the oedipal phase unfolds, since heterosexual organization has not as yet coalesced. Emphasis on psychosocial aspects has tended to obscure the biological dimension. In Bieber's view, the capacity to respond sexually to the opposite sex is an innate, physiological given, expressed at roughly similar ages within the species where there are relatively normal nutritional conditions and circumstances. This well-documented phase of observable psychosexual development begins somewhere between the ages of two and three, and by three and a half is usually well established. Further, with reasonably appropriate environmental conditions, the onset of heterosexual reactivity is no more a learned response in humans than it is in dogs. The excitatory response is innate, though the refinements of copulatory patterns may be learned. Freud considered heterosexual reactivity and orientation to be a biological potential, not an inevitable biological development. There are, of course, divergences from this general developmental pattern, but a consideration of deviant cases is not germane to this discussion.

According to the proposed theory, olfaction is linked to human sexual development from its earliest manifestations. From birth on, the child continues to smell his or her own odor and those of caretakers and surroundings, preferring the odors of certain objects over others, but, barring inappropriate or abusive parenting, experiences little disturbance until there is a biological intervention—the maturation of neuroendocrine mechanisms that release the olfactory capability of sensing and reacting to sexual odors.

A child's sexual reaction to the odor of a cross-sex parent or surrogate indicates that that individual is emitting a gender-linked sexual odor. The sexual response to heterosexual objects ushers in the Oedipus complex. It is a response mediated first and primarily through olfaction, with other sensory modalities—sight, touch, hearing—soon coming into play. Olfaction is the primary initiating modality, though it need not and usually does not remain the dominating one. Once heterosexual organization has taken place, any single sensory modality, olfaction included, can be abstracted without obliterating heterosexual reactivity. If heterosexual reactivity begins in early childhood, and the weight of observational evidence argues that it does, then linking olfaction to incestuous objects supports the hypothesis that smell participates in early sexual organization and introduces the Oedipus complex.

Kalagerakis (1963) subsequently supported this thesis in his detailed olfactory observations of a boy in the years between two and a half and five. The child first developed an interest in playing with the feet of his young aunt, who lived with the family. He was aware of the odor of her feet, and his interest in foot odor extended to other young women visitors. He would sit under a table, waiting for one of them to take off her shoes. Some weeks after he had begun to play with his aunt's feet, he became interested in newspaper ads depicting little girls with bare feet. He saved the ads. As previously referred to, olfaction is a central aspect of foot fetishism, and this boy had clearly developed a foot fetish. Particularly significant was the rapid integration of the visual modality, represented in his fascination with pictures of bare feet. When he was three and a half, he became explicitly aversive to the odors of both parents after they had had sexual intercourse. These were invariable responses and occurred even after many hours and even after both parents had showered. The boy's reactions subsided after several months, but were followed by very affectionate behavior toward the mother. He wanted very much to cuddle with her; at the same time, he had a marked aversive reaction to the father's odor, an odor that evoked nausea on several occasions. It should be noted that this child was known to be a normal boy. He was not a behavior problem, had no difficulties at school, and played well with other children, and his later development was unremarkable. Kalagerakis has postulated an olfactory determinant for aversive reactions to the same-sex parent as an important component of the competitive phase of the Oedipus complex.

In this connection, the case of a male academician offers additional evidence. This man, aged 35, came into treatment because he was homosexual and wanted to be heterosexual. He had devised for himself a type of behavior therapy whenever he had homosexual impulses. He would have a smell memory of the odor of a man's genitals, an odor he disliked intensely. His aversive reaction turned off his homosexual desire. He would then think of the smell of a woman, an odor he liked.

Ellis (1929) described an incident involving a man who, while in the presence of a woman and another man, detected an axillary odor that be believed was emanating from the man. Initially, he had an aversive reaction, but when he discovered that the odor had come from the woman, his aversion disappeared.

Sexual attraction to the maternal odor was revealed in the case of a young man whose mother had died when he was one year old. The father remarried a woman whose odor, he was aware, was similar to that of his own mother and his late wife. This had been reported during his analysis. The son like his father, was sexually drawn to women who had this odor, though he had no conscious memory of his mother.

The familiar, loved maternal smell develops into a sexually arousing stimulus *pari passu* with neurophysiological development and sexual organization, leading to a qualitative shift in the parent-child relationship. Freud's brilliant developmental observations, which led to the formulation of the Oedipus complex, correctly apply to boys; however, the psychodynamic explication for that phase of development as it applies to girls has bedeviled psychoanalysis from the time the theory was first proposed. Freud asserted that when the female discovers that the male possesses a penis and she does not, the rage against the mother for having withheld this superior genital turns daughter away from mother and toward the phallic, impregnating father; the girl's fantasy that he will have a child with her is compensation for her perceived genital inferiority. Thus, jealousy of the male, penis envy, was theorized as the motivation driving the female away from the mother and into heterosexuality. The question of how females negotiate an orientation to males from a primary bond to the mother can now be realistically addressed. In accord with the oedipal formulation we have presented, it is not penis envy, but rather sensitivity to the male odor that initiates feminine heterosexual reactivity.

As the Oedipus or Elektra complex makes its appearance, the competitive aspect of this syndrome finds the child at cross purposes with an adult, a situation that plays an important role in the fate of olfactory experience. In adapting to this new interpersonal circumstance, a measure of olfactory inhibition may ensue as a defense against the sexual smell of the opposite-sex parent, an inhibition that may remain fixed into adulthood. The cases of two women are illustrative. Each reported during the analysis that she could not be involved sexually with a man whose odor was reminiscent of her father's odor. Each woman could maintain a friendship with a man who smelled like her father, but an odor like the father's was intolerable in a sexual encounter. Attraction to the parental odor, however, may remain unencumbered, as in the case of the father and son described above.

Olfaction and Sexuality: Physiologic Views

At this point, we turn our attention to a few selected investigations that highlight the relationship between olfaction and human sexuality and are of particular cogency to our subject. Detailed physiological, anatomical, and biochemicial considerations are discussed elsewhere in this volume.

The articulation of olfaction and sexual hormones is illustrated in a report by Schon (1958) of patients with breast malignancy. In an earlier era, hypophysectomy was performed. Following surgery, the women were left with an impaired sense of smell and taste. Since only one olfactory tract would be cut, the deterioration of olfactory acuity was thought to be the result of stretching of the other tract, but even after a reasonable recovery period, olfaction remained impaired. In a personal communication, Schon related that several women in her study were given estrogens and experienced olfactory improvement.

LeMagnen (1950) clearly defined the connection between olfaction and sexual hormones in his well-known investigation of male and female reaction to exaltolide, an aromatic substance found in nature and said to have a sexually stimulating effect. Perhaps that is why exaltolide has been synthesized and used as a fixative in perfume. Only about half the men in the study could detect the odor, whereas all the women could. During menstruation, sensitivity decreased, and during the ovulatory period, it increased. Ovariectomized women were significantly less sensitive to the odor than the others, but when treated with estrogen they attained normal levels of acuity. LeMagnen concluded that progesterone reduced olfactory sensitivity and estrogen reestablished it.

Since the 1950s, the still emerging field of pheromone research has led to an increasing interest in the role of olfaction in human behavior. It is generally agreed among investigators that chemical messengers, *pher-*

omones, deliver odor signals throughout the insect, marine, and infrahuman mammalian world. Animals establish territoriality, identify one another, signal danger, seek out food, advertise sexual receptivity, and arouse one another by pheromonal communication. Yet there is uncertainty about the role of pheromones in human sexuality. Various literatures, popular and scientific, report example after example of the olfactory sensitivity of humankind. Richardson and Zucco (1989) examined the role of the olfactory system in memory and cognition, and found in close-range laboratory experiments that the subjects possessed an excellent ability to detect and discriminate odors, but typically had difficulty identifying particular odorants. In this regard, Dodd (1989) stated: "What human beings lack in acuity . . . they make up in powers of discrimination, which rival those of any other mammal." He reported that though estimates vary considerably, there is consensus among researchers that after some training, a healthy person can distinguish between 10,000 and 40,000 odors—an impressive number even at the lower end. Among professional perfumers and whiskey blenders, the number may approach 100,000.

The olfactory vigor manifested by this odor-distinguishing capability makes it all the more surprising that research that suggests a pheromonal influence on human sexual arousal is frequently viewed with considerable caution, and the importance and heuristic value of the results are deemphasized. A case in point is the much-cited study by McClintock (1971), whose subjects were women sharing a dormitory and whose menstrual cycles synchronized, this synchrony suggesting that pheromonal communication was operant among the group. Dodd (1989) concluded that even if, in some undetected way, there is pheromonal influence on reproduction, the synchronized menstrual rhythm had no important effect on the women's sexual behavior. Interestingly, the converse has been shown to be true: Sexual behavior has an effect on menstrual rhythm. Cutler et al. (1986) demonstrated that women who did not have sexual intercourse would go into erratic phases, either too long or too short a cycle, but if they had intercourse even once a week, they continued to synchronize. In the experiments conducted by Preti et al. (1986) and Cutler et al. (1986), the menstrual cycles of volunteers were manipulated by exposing the women to male and female axillary substances. The resulting phenomena were thought to be analgous to previously noted olfactory/reproductive relationships in nonhuman mammals. The studies of these and other researchers of menstrual synchrony and cycle length are attempts to establish with hard evidence the existence of human sexual pheromones that affect behavior and physiology. Preti (1987) has stated: "There may exist *releaser* pheromones (i.e. substances which release a behavior response in a conspecific). Although specific data to support the existence of a releaser pheromone in humans has not been forthcoming, data from several studies suggest the possibility of another more subtle pheromonal effect, i.e. a *primer* pheromone. Such a substance would cause a slow changing or priming of the endocrine system."

In a series of experiments, Michael and Keverne (1970, 1971) isolated male sex attractants from the vaginal secretions of female rhesus monkeys. The attractants had pheromonal components and were associated with elevated estrogen levels.[2]

Sokolov et al. (1976) fractionated the vaginal secretions collected from several sexually excited human females and found pheromonal substances that were chemically similar to those found in rhesus monkeys. The researchers concluded that "some type of human pheromonal system may eventually be elucidated."

The apocrine glands, especially those in the axillae, where the largest and most numerous are located, have been reported by many research groups to be major contributors to human odor. Under laboratory conditions, Russel (1976), among others, has shown that after puberty, when the apocrine glands are fully functioning, the exposure of men and women to axillary substances enables them to discriminate between male and female odors. In life situations, such discrimination is commonly experienced and has been described in this chapter.

Rogel (1978) has evaluated the possibility of human sexual pheromonal communication, and while she does not rule out pheromonal signaling, she remains uncommitted but offers a balanced and informative review.

The axilla and its role in human olfaction has been examined by Gower et al. (1988). They cite an anonymous writer who commented on the ambivalence with which the axilla is viewed. "Many deem the human armpit a dank, unprepossessing corner of the body, fit only for frequent applications of soap and deodorant. Nevertheless, a dedicated few regard the axilla with mixed feelings, recognizing that this charming grotto is full of intriguing surprises" (p. 49).

Though clinical data document the salience of olfaction in sexual development and adult sexuality, validation from the results of laboratory experimentation broadens psychological perspectives. Recent advances in analytical biochemistry such as high-pressure liquid chromatography (Novotny, 1989) and nuclear magnetic resonance spectroscopy (Bottomley, 1989), coupled with emerging insights into the genetic basis of human odor (Yamazaki et al., 1976; Yamaguchi et al., 1981), could lead to the classification of individuals according to specific odor subtypes analogous to human blood types. Interdisciplinary projects involving both laboratory and behavioral scientists might, for example, include studies of choice of

---

[2]The results reported by Michael and Keverne were not obtained in later experiments by Goldfoot et al. (1976) and have not been repeated since; however, an exchange between the researchers revealed that Goldfoot had carried out the experiments somewhat differently.

mate to determine whether similar odor types are mutually attracted or repelled, or whether parental odor types are selected or avoided. There are rich and varied opportunities for research along such lines.

Conclusions

In recent years, scientific investigations of olfactory processes have been proliferating at a rapid rate; however, as has been emphasized, little attention has been paid to olfaction in a psychoanalytic context. We have reviewed both clinical cases and research studies that point to the importance of olfaction and its connection with human sexual behavior. Finally, we have presented a review of the formulation originally proposed by Bieber more than 30 years ago, namely, that olfaction is the triggering modality that initiates heterosexual reactivity and ushers in the Oedipus complex.

References

Abraham, K. (1954). Remarks on the psychoanalysis of a case of foot and corset fetishism (1910). *Selected Papers.* New York: Basic Books, pp. 125–136.

Bieber, I. (1959). Olfaction in sexual development and adult sexual organization. *Am. J. Psychother.*, *13*, 851–859.

Bieber, I. (1960). Olfaction in sexual development and adult sexual organization. In J.H. Masserman (Ed.), *Psychoanalysis and Human Values*, New York: Grune and Stratton.

Bottomley, P. (1989). Humans in vivo, N.M.R, spectroscopy in diagnostic medicine. *Radiology*, *170*, 1–15.

Brill, A.A. (1932) . The sense of smell in the neuroses and psychoses. *Psychoanalytic Q.*, *1*, 7–42.

Brody, B. (1975). The sexual significance of the axillae. *Psychiatry*, *38*, 278–289.

Cutler, W.W., G. Preti, A. Krieger, R.H. Huggins, C.R. Garcia, and H.J. Lawley (1986). Human axillary secretions influence women's menstrual cycles: The role of donor extract from me. *Hormones Behav.*, *20*, 463–473.

Daly, C.D. and R.S. White (1930). Psychological reactions to olfactory stimuli. *Br. J. Med. Psychol.*, *10*, 70–87.

Dodd, E. (1989). The scents around us. *Sciences*, November–December, p. 46.

Ellis, H. (1929). *Studies in the Psychology of Sex*, vol. II. New York: Random House, pp. 44–112.

Freud, S. (1961a). Fragment of an analysis of a case of hysteria. In J. Strachey (Ed.), *The Standard Edition*, Vol. 7. London: Hogarth Press, p. 155.

Freud, S. (1961b). Civilization and Its Discontents. In J. Strachey (Ed.), *The Standard Edition*, Vol. 21. London: Hogarth Press, p. 100.

Freud, S. (1961c). Identification. In J. Strachey (Ed.), *The Standard Edition*, Vol. 18. London: Hogarth Press, pp. 105–108.

Gibbon, B. (1986). The intimate sense of smell. *National Geographic*, *170*(3), 329.

Goldfoot, D.A., M.A. Kravitz, R.W. Goy, and S.K. Freeman (1976). Lack of effect of vaginal lavages and aliphatic acids on ejaculatory responses in rhesus monkeys—Behavioral and chemical analyses. *Hormones Behav.*, *7*, 1–27.

Gower, D.B., A. Nixon, and A.I. Mallet (1988). The significance of odorous steroids in axillary odour. In S. Van Toller and G.H. Dodd (Eds.), *Perfumery*. London and New York: Chapman and Hall, Ltd., p. 49.

Kalagerakis, M. (1963). The role of olfaction in sexual development. *Psychosomat. Med.*, *25*, 420–432.

Krafft-Ebing, R. (1965). *Psychopathia Sexualis*. New York: Bell Publishing Co., p. 21.

LeMagnen, J. (1950). Physiologie des sensations. Nouvelles donnees sur la phenomene de l'exaltolide. *Compt. rend.*, *230*, 1103.

LeMagnen, J. (1953). L'olfaction: le foncionnement olfactif et son intervention dans les regulations psychophysiologiques. *J. Physiol.*, *45*, 285.

McClintock, M.K. (1971). Menstrual synchrony and suppression. *Nature*, *229*, 244–245.

Novotny, M.V. (1989). Recent developments in analytical chromotography. *Science*, *246*, 51–57.

Ohrbach, C.E., M. Bard, and A.M. Sutherland (1957). Fears and defensive adaptations to the loss of anal sphincter control. *Psychoanalyt. Rev.*, *44*, 121–175.

Preti, G., W.B. Cutler, C.M. Christensen, H.J. Lawley, G.R. Huggins, and C.R. Garcia (1987). Human axillary extracts: Analysis of compounds from samples which influence menstrual timing. *J. Chem. Ecol.*, *13*(4), 717–731.

Richardson, J.T.E., and G.M. Zucco (1989). Cognition and olfartion: review. *Psychol. Bull.*, *105*(3), 352–360.

Rogel, M.J. (1978). A critical evaluation of the possibility of higher primate reproductive and sexual pheromones. *Psychol. Bull.*, *85*(4), 810–830.

Rosenbaum, J. (1961). The significance of smell in the transference. *J. Am. Psychoanalyt. Assoc.*, *4*, 312–324.

Russell, M.J. (1976). Human olfactory communication. *Nature*, *260*, 520–522.

Yamaguchi, M., K. Yamazaki, G.K. Beauchamp, J. Bard, L. Thomas, and E.A. Boyce (1981). Distinctive urinary odors governed by the major histocompatability locus of the mouse. *Proc. Nat. Acad. Sci., USA*, *78*, 5817–5820.

Yamazaki, K., E.A. Boyce, V. Mike, H.T. Thaler, B.J. Mathiesco, D. Abbott, J. Boyce, Z.A. Zayas, and L. Thomas (1976). Control of mating preferences in mice by genes in the major histocompatability complex. *J. Exp. Med.*, *44*, 1324–1335.

15
Olfaction and Emotion

Howard Ehrlichman and Linda Bastone

"Our most elusive sense, smell reaches more directly into memory and emotions than other senses" (Gibbons, 1986, p. 337). "Functionally, smell may be to emotion what sight or hearing is to cognition" (Engen, 1982, p. 3).

Introduction

The idea that olfaction and emotion are closely linked has become commonplace in both popular and scientific discussions of the sense of smell. Odors are said to influence mood, evoke powerful experiences of pleasure or displeasure, produce alertness or relaxation, and evoke long-forgotten emotional memories. These effects are often said to reflect the dependence of olfaction on parts of the brain involved in emotional experience. Some writers have even gone as far as dubbing olfaction "our most emotional sense" (Lieff and Alper, 1988). How much of this is fact and how much fancy? Unfortunately, assertions about olfaction and emotion are often made without sufficient justification from the scientific literature, in part because little relevant research has existed until quite recently. Yet, as research on olfaction and emotion grows, it is important to critically examine ideas that have sometimes been taken as self-evident. In this chapter we seek to clarify the various ways in which the sense of smell could be construed as "emotional" by discussing eight "propositions" connecting olfaction to affect. We also explore the possibility that regardless of whether claims for some unique relationship between olfaction and emotion can be substantiated at present, odors may have a role to play in the study of affective reactions; hence a second aim of this chapter is to suggest ways in which olfaction may be profitably used in the study of affective experience.

Comparing Odors to Other Stimuli

Claims for a relationship between olfaction and emotion often carry the implication that odors are more closely connected to affect than are other sensory experiences. However, few studies have actually compared olfactory stimuli to stimuli in other modalities, and where such comparisons have been made, thorny questions arise as to the appropriateness of the comparison stimuli. Suppose one wanted to demonstrate that smells are more emotional than sights. One version of this hypothesis could be that an odor chosen at random has a higher probability of eliciting an emotional response than does a visual stimulus chosen at random. However, while it may be sensible to think of choosing an odor at random, how does one select a visual stimulus at random? Should one choose a property of visual perception such as color, disembodied from any particular object, or should one choose a specific object? Would a representation of the object (i.e, a picture) be permissible, or would only the object itself suffice? In research on odor memory, comparisons between odors and pictures have led to very different conclusions than have comparisons between odors and nonsense shapes (Lawless, 1978). Such research may be useful in helping to understand the nature of encoding processes in olfaction, but does not lend itself to simple comparative statements about olfaction versus vision.

A more useful comparison in the domain of affect might be between the odor of an object (with visual cues absent) and the sight of the same object (with olfactory cues absent). For example, one could compare emotional reactions to the smell of a rose versus the sight of a physically present rose, or to the smell of feces versus the sight of physically present feces. One problem with this approach is that the stimuli would not be representative of objects in general, since not all visual objects emit odor and not all odors are associated with a particular visual stimulus. Nevertheless, if one were to observe that the smell associated with an object was more emotional than the sight of the same object across a wide range of objects, this would lend some support to the assertion that smells are generally more emotional than sights.

One could also ask subjects to recall stimuli experienced in each sense modality, thereby roughly equating stimuli for cognitive accessibility (Higgins, King, and Maven, 1982), and then to rate them for emotionality. For purposes of this chapter we carried out such a study. Fifty-three college students were asked to generate lists of five things that they had seen, heard, smelled, tasted, and touched. They were told to list the first five things that came to mind, but not to include anything they were experiencing at the moment. The sensory modalities were presented to subjects in one of five random orders. After generating the five lists, subjects rated each sensory experience on an emotionality scale ranging from 0 (*not at all emotional*) to +10 (*extremely emotional*) and a pleasantness-unpleasantness scale ranging from −10 (*extremely unpleasant*) to +10 (*extremely pleasant*). Half the subjects rated emotionality first and half rated pleasantness-unpleasantness first. Overall, visual experiences were rated as the most emotional ($M = 5.22$, s.d. $= 2.50$), significantly higher than all other experiences, including smells ($M = 3.97$, s.d. $= 2.53$), $t(51) = -2.95$, $p < 0.005$. Although smells did receive the highest average rating for hedonic intensity (computed as the absolute distance from the neutral point), the difference between smells and sights was not statistically significant (mean for smells = 6.09, s.d. = 1.99; mean for sights = 5.77, s.d. = 2.42).

The results of this study do not lend much support to the claim that olfaction is more emotional than are other senses; in fact, vision appears more deserving of this status. Although this study relied upon recollections of stimuli rather than responses to physically present stimuli, if olfaction is intrinsically more affective than other senses, one would expect some indication of that even in a study of this type. At the very least, the results underscore our contention that the claim for a special relationship between olfaction and emotion cannot be taken as self-evident in the absence of appropriate research evidence.

Olfaction as an Emotional Sense: Eight Propositions

The eight propositions that follow have been gleaned from both the scientific and popular literature. Although they are not without some overlap, we believe they provide a useful framework for discussing olfaction and emotion. Although many of the propositions can be taken in a comparative sense, it is not necessary to say that odors are more (or less) emotional than other sensory experiences in order to say that odors may sometimes be emotional, or to explore the ways in which odor and emotional experience are related. In this spirit, we have stated each proposition in a way that allows for, but does not limit discussion to, a comparison between olfaction and the other senses.

Proposition 1: The Experience of Odors Is Inextricably Linked to Hedonic Tone; the Most Salient Aspect of Odor Is Its Pleasantness or Unpleasantness

There can be little doubt that the primary reaction to odors is liking or disliking. Numerous studies using various scaling and judgment procedures have attested to the saturation of odor experience with hedonic tone. In the early days of experimental psychology, odors were frequently used in studies of pleasantness and unpleasantness, which were conceived of as fundamental properties of experience (Wundt, 1897). More recently, factor analytic and multidimensional scaling studies involving judgments of odor similarities have consistently uncovered a strong pleasantness–unpleasantness dimension (e.g., Wright and Michaels, 1964; Yoshida, 1979). It seems safe to conclude, as Tassinary (1985) did, that "regardless of how the analysis is done, when odors are involved a strong clear hedonic factor always emerges" (p. 26).

While this finding is fairly robust within the realm of olfaction, it remains to be demonstrated that comparable stimuli in other modalities (e.g., visually present objects) would not also be judged primarily on the basis of pleasantness–unpleasantness or liking–disliking. Therefore, despite the evidence supporting the prevalence of the hedonic dimension in judgments of odors, one cannot assume that pleasantness–unpleasantness is any more closely tied to the experience of odors than to the experience of other stimuli. Further, even if such a result were found, the possibility would remain that odors are judged in terms of pleasantness–unpleasantness because the lack of a linguistic or conceptual structure for odor experience leaves individuals little else to go on (Richardson and Zucco, 1989). Yoshida (1964) reported that perfumers ($n = 2$) were less likely to classify odors in terms of their hedonic quality than were nonexperts. This finding suggests that the perfumers were able to discern (or label) perceptual qualities not available to the untrained individuals. Thus, it may not be that olfactory stimuli are more hedonic than stimuli in other modalities, but rather that other stimuli can be categorized more easily by nonhedonic properties. Such a pattern could lead one to confuse the relative importance of hedonic tone as compared to other factors *within* the domain of olfaction with the assertion that odors are more hedonic than other stimuli.

Another reason odors may be judged as more hedonically saturated than other stimuli is that the pleasantness or unpleasantness of odors seems to be a part of the sensory experience itself. In contrast, the pleasantness or unpleasantness of a visual stimulus is most often experienced as a judgment rendered upon the object. Schachtel (1959) claimed that odors and tastes are experienced in terms of how they feel to the observer (what he called "autocentric" perception); sights and sounds are more often experienced in terms of what they are (what he called "allocentric" perception). Another way of stating this is that olfaction is phenomenologically more proximal than is vision or audition. Schachtel spoke of a fusion in the autocentric senses between the sensory quality and the feelings of pleasure or unpleasure, while in the allocentric senses no such relationship, or a less pronounced one, exists. Thus, whether or not one grants that odors are more hedonic than visual or auditory stimuli, by Schachtel's analysis odors do seem to be hedonic in a different way than sights and sounds (although some sights and sounds, such as painfully bright light or piercing noise, can also be experienced autocentrically).

Schachtel's view suggests that the "meaning" of olfactory stimuli (e.g., actual identification of the odorous substance or a judgment of its implications for well-being) may be less critical in the affective experience of odors than in that of sights or sounds. We wonder if this is in fact the case. Mark Twain's "The Invalid's Story" (Clemens, 1882) describes the increasing distress of a man riding in a freight car with a coffin containing a stinking corpse. Finally, unable to stand the smell any longer, he goes out into a winter storm and contracts a severe illness. He later learns that there was no corpse in the coffin after all, that the smell was "a lot of innocent cheese," but the news comes, as he says, too late to save him. What

Twain's story suggests is that the odor would have been experienced quite differently had the poor fellow known its origin.

One might argue that identifying an odor's source could influence only the emotional response to the odor, not the sensory-hedonic experience of the odor itself. But, in fact, how do we know that the actual odor experience is not influenced by the context in which the odor occurs? Appraisal processes occurring rapidly and perhaps unconsciously may actually alter the sensory experience. Rozin and Fallon (1987) discuss this with regard to disgust, claiming that "sensory-affective" factors change as a function of "ideational" factors:

> The claim is illustrated by a hypothetical situation (supported by some informal observations) of a man sniffing the same decay odor from two opaque vials. The man (who likes cheese) is told that one vial contains feces, the other cheese. . . . If he thinks it is cheese, he likes the smell. If told the odor is in fact from feces, though, he suddenly finds it repellant and unpleasant. It is the subject's conception of the object, rather than the sensory properties of the object, that primarily determine its hedonic value (p. 24).

Although Rozin and Fallon cite a great deal of research showing that people's preferences for stimuli are strongly influenced by such ideational factors (e.g., they are reluctant to eat fudge in the shape of dog feces), they provide no direct evidence for changes in sensory-affective experience. Data reported by Moskowitz (1979) do suggest that context can produce such changes. Panelists rated fragrances with or without a brand name label. Ratings of both liking and sweetness were very different under the two conditions. As Moskowitz put it: "It is remarkable that branding a product so dramatically modifies sensory and hedonic reactions" (p. 137). There is also research showing that context can influence odor experience outside the hedonic arena. Zellner and Kautz (in press) have reported that the perceived intensity of food odors was increased when the odor extracts were presented in colored as compared to colorless liquid. Thus, despite the seemingly "intrinsic" character of odor hedonics, in some circumstances expectancies may play a role in odor experience.

While pleasantness–unpleasantness is clearly an important aspect of the experience of odors, the larger claim that olfaction is *emotional* depends on whether, and to what extent, hedonic tone is considered part of affective experience. As Van Toller (1988) has noted, olfaction is rarely mentioned, let alone treated at length, in discussions of emotion. Perhaps this is because research and theory on emotion have traditionally emphasized strong, discrete emotional reactions. However, more recently other kinds of emotional states, such as moods, have received a great deal of attention (e.g., Isen, 1984). The term "affect" is frequently used to refer generically to a wide range of reactions, from very strong emotions, to background mood states, to feelings of attraction to or repulsion from stimuli. Although emotion theorists differ on the centrality of hedonics in affect, almost all agree that at some level pleasantness–unpleasantness must be involved. For some, affect is virtually definable in terms of pleasantness–unpleasantness (or liking–disliking) (e.g., Arnold, 1960; Young, 1967; Zajonc, 1980). For others, pleasantness–unpleasantness represents the most basic level of affect, strongly connected to approach and avoidance, and from which all specific emotions are elaborated (Frijda, 1986; Plutchik, 1980). Still others (Izard, 1977; Johnson-Laird and Oatley, 1988), consider pleasantness–unpleasantess useful only as a convenient way of classifying discrete emotions as "positive" or "negative." There are also theorists who discuss different "levels" of affect. Hoffman (1986) and Livesey (1986) each posit a level of affect that is relatively noncognitive and experientially direct. Both theorists include pleasant and unpleasant smells as examples of affect at this primary level.

Ortony, Clore, and Collins (1988) have recently presented an emotion theory that allows affective reactions to olfactory stimuli to be placed in the same conceptual framework as affective reactions to any other kind of stimuli. According to these authors, emotions are defined as "valenced reactions." However, the specific nature of the emotion derives from the causes of the valenced reaction, with certain emotions reflecting reactions to events, others to agents, and others to objects. Ortony et al.'s discussion of reactions to objects is relevant to the question of whether odor experience, by virtue of being hedonic, is also emotional. Valenced reactions to objects are labeled "attraction" emotions. They are

> momentary reactions of liking and disliking, and as such, they are among the most salient experiences we have. At the same time, they appear to be more immediate, more spontaneous, and less affected by accessible cognitive processes than almost all of the other emotions . . . [and] seem to be directly experienced as relatively undifferentiated affective reactions (p. 156).

Hoffman (1986), Livesey (1986), and Ortony et al. (1988) agree with Schachtel (1959) that affective experience can occur with minimal cognitive involvement. At first sight, this idea might appear to contradict what is perhaps the central concept in most contemporary theories of affect, that people's emotional reactions largely reflect their interpretations of the events around them and the implications of those events for the individual's well-being (e.g., Frijda, 1986; Plutchik, 1980; Smith and Ellsworth, 1985). Although there has been controversy in the literature as to whether "cognitive appraisal" implies a high level of cognitive involvement or may occur as a direct response to stimulus properties (Lazarus, 1984; Zajonc, 1984), recent treatments accept that appraisals may occur at all levels of cognitive complexity (e.g., Leventhal and Scherer, 1987). In our view, the hedonic experience of odors can be thought of as "sensory appraisal," that is, a way of rendering in experience a "judgment" about the likely positive or negative potential of odor-producing objects. Moreover, we believe that these experiences constitute affective reactions generally occuring with a relatively low level of cognitive involvement (although we have seen that they may be subject to higher-level cognitive input). The claim that olfaction is more emotional than other senses may reflect the belief that a stimulus that produces affect with minimal cognitive involvement is more emotional than one that requires a higher level of cognitive mediation. Yet, precisely the opposite could be argued, namely that the most emotionally compelling stimuli are those that are most embedded in complex cognitive structures and that are communicated through language. Nevertheless, the ability of odors to produce affective responses with little complex cognitive mediation does make them interesting from both a theoretical and methodological perspective.

Proposition 2: Odors Directly Influence Mood; Merely Smelling Certain Substances Brings About Mood Change

In this section we discuss mood change as measured by subject self-report or physiological indicators. In a later section (see Proposition 7) we discuss cognitive and behavioral effects of odors that may be mediated by odor-induced changes in affective states. Studies that involve self-reported mood and behavioral effects are discussed in both sections.

The claim that odors can influence moods takes two forms, specific and nonspecific. The nonspecific claim is based on hedonics: pleasant odors can enhance and unpleasant odors can depress mood. Thus, people manipulate their odorous environment (e.g., through the use of incense, air fresheners, perfume, and fragrance additives) in order to produce more positive moods. The specific claim is that certain odorants have particular effects on mood, such as arousal, relaxation, or sensuality. Practitioners of a form of modern herbal medicine known as aromatherapy contend that such specific effects are routinely observed in their work. Unfortunately, until very recently there has been little systematic research that would enable one to evaluate these claims.

A recent patent by International Flavors and Fragrances (IFF) (Warren et al., 1987) describes research showing that substances found in nutmeg oil (as well as in mace, valerian, and neroli) have stress-reducing effects in humans. In four studies, subjects who were presented with nutmeg either alone or in an apple-scented carrier had significantly reduced systolic blood pressure in response to a mild stressor as compared to subjects presented with the apple scent alone or with no fragrance. In addition, subjects smelling nutmeg reported lower anxiety, tension, anger, and embarrassment and increased calmness, relaxation, and happiness relative to control subjects.

Results consistent with the IFF data have been presented by Lorig and Schwartz (1988), who reported that EEG theta activity (presumably reflecting a relaxed state) differed in response to the nutmeg-apple fragrance (which they refer to as "spiced apple") as compared to lavender or eucalyptus. Since the odors did not differ in their rated pleasantness, the specific effect of spiced apple cannot be attributed to hedonics. However, subsequent research suggests that an important difference among the odors could be that only spiced apple is a food-related odor. Lorig and Schwartz (1988–89) found that merely imagining one's favorite dessert produced self-report and EEG changes similar to those produced by relaxation instructions.

In an effort to obtain evidence for the properties attributed by aromatherapists to various essential oils, Torii et al. (1988) measured subjects' contingent negative variation (CNV) amplitude while subjects were presented with various essential oils. Their results supported claims for the stimulating effects of jasmine and the sedative effects of lavender: jasmine oil increased CNV amplitude, while lavender oil decreased this response. However, neither fragrance appeared to affect heart rate, skin potential level, or reaction time. In two recent abstracts, Kanamura et al. (1989) reported that the mood evoked by various fragrances was related to CNV amplitude, and Sugano (1989) reported that lavender increased EEG alpha activity and decreased CNV amplitude, whereas jasmine increased EEG beta activity and increased CNV amplitude.

Warm et al. (1989) carried out a study to investigate whether odors can have behavioral effects consistent with subjects' descriptions of them. Subjects performed a stressful vigilance task in the (intermittent) presence of peppermint, rated as "stimulating," or muguet (lily of the valley), rated as "relaxing," or with no odor present. The results indicated that both fragrances significantly increased detection accuracy relative to no odor, but did not differ from one another. Further, despite the effects on actual performance, subjects did not report less subjective stress in the presence of either odor relative to the no-odor condition. Although these data suggest that odors may influence behavioral indices of arousal without affecting self-reported mood, the authors note that because both odors were judged to be pleasant, the study did not rule out the possibility that the pleasantness of the odors may have influenced vigilance, that is, that the effect was (in our terms) nonspecific.

Effects on affective reactions and mood have also been reported for the androgen steroids androstenone and androstenol. Filsinger, Braun, and Monte (1985) found that these substances reduced self-ratings of sexiness in men and also seemed to influence ratings of sexiness of photographs of others. However, interactions between subject's sex, target's sex, and other factors, as well as major inconsistencies among previous studies, do not allow for a simple generalization concerning the effects of these odorants. Two recent studies of the effects of androstenol did not rely on subjective responses. In a carefully designed experiment, Gustavson, Dawson, and Bonett (1987) found that men avoided a restroom stall that had been treated with androstenol, whereas women's choices of stalls were not influenced by the presence of this odorant. Burleson and Johnston (1989) found that the P3 component of the visual event–related EEG potential (said to indicate stimulus relevance) was attenuated by androstenol at frontal and central sites for both women and men in response to a variety of slides. It remains to be seen if an understanding of the effects of androstenol can be developed that can account for these diverse findings.

The studies we have discussed support the idea that certain fragrances may have specific influences on mood, physiological responses, and behavior. However, it is unclear whether these effects actually involve the sense of smell. Consider that the term "aromatherapy" itself may be a misnomer (Dodd, 1988). Although aromatherapists utilize essential oils in their practice, it is not at all clear that the mode of (putative) action involves the olfactory system. The mere fact that a substance is inhaled and has a scent does not mean the sense of smell is implicated in its therapeutic effect, especially since much aromatherapy involves massage and thus absorption of molecules through the skin (Tisserand, 1988). Dodd (1988) has drawn attention to the distinction between effects mediated by absorption into the bloodstream and those mediated by the olfactory system, and has suggested the term "osmotherapy" for treatments of the latter type. Warren et al. (1987) are explicit on this point; the effects reported in the IFF patent are described as "pharmacologic," with the "active" substances entering the bloodstream either through inhalation or transdermally. Although the substances can be added to fragrances for "aesthetic" appeal, the mode of action is not intended to be through the olfactory system. However, in more recent research IFF scientists have found that these effects seem to be potentiated when the substances are embedded in certain fragrances as compared to others, suggesting that both pharmacological and psychological effects are occurring (S. Warrenburg, personal communication). This distinction is clearly of great importance with regard to understanding the basis for the emotional effects of specific odors. Research that purports to demonstrate such effects must show that the mode of action of the odorant substance depends on the actual experience of a smell, perhaps by showing that similar effects do not occur in anosmic individuals, a strategy that has been used in olfactory research with animals (e.g., Tanida, Tsuchiya, Uenoyama, and Nakayama, 1989).

Turning now to nonspecific effects, what do we know about the influence of hedonic qualities of odors on mood? If one is exposed to a pleasant or unpleasant odor, is one's mood improved or worsened? Surprisingly, little research directly addresses this question; in fact, we have not found any published study with its primary focus on the effects of odor pleasantness on mood, although some researchers have included mood measures in studies of the behavioral effects of odors. Rotton, Barry, Frey, and Soler (1978) measured self-reported mood in a study of the effects of malodors on interpersonal attraction. Using a cover story that a "physiological study" had been recently carried out in the experimental room, subjects were exposed to the odor of butyric acid or ammonium sulfide. The control group was a no-odor (ambient atmosphere) condition. Subjects in the two malodor conditions reported a more negative mood as indicated by a number of Mood Adjective Checklist (MACL) scales (higher scores on aggression, anxiety, fatigue, sadness; lower scores on elation, concentration, affection) than those in the ambient condition. They also reported feeling less comfortable and pleasant. Although these data strongly support the idea that malodor can produce a negative mood, one cannot rule out the possibility that any odor, unpleasant or pleasant, could have such an effect. Conceivably, the presence of any odor in this situation could have been annoying. It is also difficult to know whether to attribute the mood effects to the odor experience per se or to the fact that the odor was present; that is, the effects on mood may have occurred because subjects resented being exposed to the unpleasant residue of another research project.

Similar comments apply to experiment 1 in Rotton (1983), in which exposure to a malodor was seemingly produced by an accomplice "accidently" knocking over a bottle of colorless liquid. Subjects in the malodor condition reported feeling significantly less pleasant than did subjects in a no-odor condition. However, Rotton's (1983) experiment 2 did not attempt to present the malodor as an accidental occurrence; rather, the subjects were told that the study involved "the effects of chemicals on task performance," and the bottle containing the odorant (or water in the no-odor condition) was presented to the subjects. Self-reports indicated that subjects in the malodor condition felt less pleasure, arousal, and dominance than did those in the control condition. Baron (1990) has recently reported a study in which subjects exposed to pleasant ambient scents (air fresheners) rated their affect as more "pleasant" and more "positive" than did subjects in a no-odor condition; however, the groups did not differ in their ratings of "good" versus "bad" feelings. In this study, no explanation for the presence of the scent was given, presumably because air fresheners were used. Ambient odors were also used by Cann and Ross (1989) in a study of the effectiveness of odors as context cues. Mood was assessed with the MACL and rating scales while subjects sat in a room permeated with either an unpleasant odor, a pleasant odor, or no odor. In contrast to the results of Rotton and Baron, these researchers found no differences among the three odor conditions in self-reported mood.

In an unpublished study carried out as a pilot for research involving odor and memory (Ehrlichman and Halpern, 1988), data were obtained on the effects of sustained exposure (approximately 20 minutes) to pleasant or unpleasant odors on self-reported mood. Forty-five women were assigned to one of three experimental conditions. A pleasant odor (mint, orange, almond, or chocolate), an unpleasant odor (pyridine), or no odor (water) was presented to subjects via a surgical mask . After completing a memory task, subjects responded to 12 items from the Differential Emotions Scale

(DES) (Izard, 1972) and then rated the extent to which their current mood could be described as "good" or "bad" on two six-point scales. An analysis of variance indicated that subjects' ratings of their good mood differed across odor conditions: $F(2, 42) = 4.44$, $p < 0.02$. The means for good mood were pleasant odor = 1.80 (s.d. = 0.94), no odor = 2.40 (s.d. = 1.06), and unpleasant odor = 3.07 (s.d. = 1.44), with lower scores indicating a more positive mood. Ratings of bad mood also differed across odor conditions; $F(2, 42) = 5.38$, $p < 0.01$. The means for bad mood were pleasant odor = 5.27 (s.d. = 1.22), no odor = 5.00 (s,d. = 1.07), and unpleasant odor = 3.87 (s.d. = 1.41), with lower scores indicating a more negative mood. Post hoc tests revealed that while the pleasant and unpleasant odor groups differed significantly from each other in their ratings of both good and bad mood, neither group differed significantly from the no-odor group on these measures. Analyses of ratings of the individual DES items indicated that the only descriptor on which the groups differed significantly was "disgust"; $F(2, 42) = 13.61$, $p < 0.0001$. The mean ratings for the groups were pleasant odor = 0.07 (s.d. = 0.26), no odor = 0.40 (s.d. = 0.74), and unpleasant odor = 2.80 (s.d. = 2.60), with higher scores indicating greater disgust. Post hoc tests revealed that the unpleasant odor group differed significantly from both the pleasant and the no-odor groups, while the latter two did not differ from one another.

Taken together, these studies provide some support for the claim that the hedonic quality of odors can influence mood. Except for the study by Cann and Ross (1989), subjects' self-reported affect differed depending on the valence of the odor to which they were exposed. Nevertheless, many questions remain for future research. Perhaps most important is to carefully distinguish between affective self-reports that reflect sensory pleasure and those that reflect a true change in mood state. The most consistent effects seem to be on general scales such as pleasant–unpleasant or good–bad mood, whereas results for the MACL, which assesses specific mood states, have been less supportive. In addition, one must also confront the possibility that when subjects are asked to report on their mood in the presence of an odor, demand characteristics may be operating. Knasko, Gilbert, and Sabini (1990) told subjects that a scented vapor (pleasant, unpleasant, or neutral) had been sprayed in the room when in fact no odor had been presented. Subjects' ratings of the nonexistent pleasant odor showed a shift toward the hedonic quality suggested by the experimenter. Moreover, ratings of positive mood were enhanced by the suggestion of a pleasant odor relative to the suggestion of an unpleasant or neutral odor. Interestingly, the unpleasant odor suggestion did not depress pleasure scores relative to the neutral suggestion. Although the findings of this study may not be easily generalizable to a situation in which odors are actually present, these results should alert odor researchers to the possible effects of expectation on subjects' responses.

Proposition 3: Odors Form Powerful Connections to Memories, Especially Those from the Distant Past Charged with Emotional Significance

This proposition relates to what has been called the "Proust phenomenon." In fact, the belief that odors have a unique capacity to bring to mind long forgotten but emotionally powerful memories long antedates Proust (Heywood and Vortriede, 1905). This idea has been treated by many authors as self-evident despite the almost complete lack of relevant research. For example, Richardson and Zucco (1989) opened their recent review of research on olfaction and cognition with the following: "Odors have an extraordinary ability to remind us of a sometimes far distant past. These memories are often characterized by a strong emotive connotation" (p. 353). The authors cited Laird (1935), who interviewed 254 men and women "Of eminence" about odors' ability to invoke memories of past experiences. Eighty percent of the men and 90 percent of the women reported having had at least one such experience. In addition, few subjects reported that these memories were hedonically neutral (16 percent of the men and 7 percent of the women). Though rich with anecdotes, Laird's study provides scant empirical evidence for a unique connection between olfaction and emotionally charged memories, since no comparisons were made to other sensory modalities. Moreover, Laird's data may be contaminated by bias; some of the subjects had not noticed this interesting power of odors, but when it was pointed out, they were reportedly "overwhelmed" by it. Unfortunately, Richardson and Zucco provided no other evidence for their claim. Their discussion of memory focused on memory for odors, with little attention to the role of odors as elicitors of emotional memories.

Although many studies have examined long-term memory for odors (Engen, 1982) or have used odors as stimuli in memory studies (e.g., paired-associate learning, Davis 1975, 1977; Heywood and Vortriede, 1905; Ratliff, 1938), these paradigms seem tangential to the question of whether odors have a special power to evoke emotional memories. The one study that directly addressed this question produced equivocal results. Rubin, Groth, and Goldsmith (1984) presented subjects with either smells, words, or pictures of common, easily labeled objects (e.g., popcorn, baby powder). Subjects were then asked to recall a memory brought to mind by the stimulus and to answer a number of questions about these memories. In two experiments, Rubin et al. found that the memories elicited by odors had been thought of and talked about less frequently than were those elicited by the comparison stimuli, although there were no differences in how old the memories were. Ratings of hedonic tone gave mixed results. In the first experiment, odor-evoked memories were rated as less hedonically neutral than word-evoked memories, but this result was not replicated in the second experiment. These results offer weak support for the Proust hypothesis: Although odor-evoked memories were more novel, they were neither older nor consistently more emotional. However, as the authors note, the decision to use words, pictures, and odors all referring to the same set of objects may have biased the results against their hypothesis, since this procedure may have entailed selecting objects both more concrete and more emotional than objects chosen at random.

No doubt if Proust could comment on this study, he would question its relevance to the phenomenon he described. In *Swann's Way*, the author experiences a rush of intense joy upon tasting the tea-soaked madeleine, though he cannot place its source. Only after great effort is he able to bring to mind the scene from his childhood. Moreover, he emphasizes how fragile and infrequent such experiences are (indeed, he says that many people may live their whole lives without that particular, crucial smell or taste ever releasing its imprisoned "souls") (Proust, 1913/1982, p. 48). Nevertheless, anecdotes and testimonials cannot substitute for systematic research. After all, one could undoubtedly find people who have had powerful emotional memories evoked by stimuli other than smells. One is forced to conclude that despite the popularity of this claim, there is insufficient evidence at present that odor has a special ability to evoke emotional memories.

Proposition 4: The Olfactory System Has Major Anatomical Connections with Emotional Circuits in the Brain; Olfaction and Emotion Depend on Similar Neural Systems

Before the discoveries of Papez (1937) linking certain cortical and subcortical nuclei to emotion, the area now called the limbic system was known as the rhinencephalon, the "smell brain." There is no doubt that olfaction and emotion/motivation utilize many of the same nervous system structures, including the amygdala, hypothalamus, hippocampus, prepyriform cortex, and preoptic frontal cortex. The close relationship between olfactory and affective circuits in the brain has led many observers to assert, ipso facto, that olfaction and emotion are closely linked in behavior and experience. Although there is probably some truth in this, we wonder whether anatomical relationships necessarily imply psychological relationships. While it may be true that in evolutionary history olfaction and motivation have been closely linked, the shift away from reliance on olfactory stimuli to reliance on visual and auditory stimuli in humans may have radically altered the interactions between the various sense systems and motivational/affective circuits. For example, although the olfactory bulbs have more direct projections to the amygdala than do other sensory inputs (Aggelton and Mishkin, 1986), this does not mean that olfaction is "more directly involved" in amygdalar function than are the more highly processed signals emanating from the visual receptors. Indeed, the amygdala has multiple afferent and efferent connections to all sensory systems, serving as a "sensory gateway to the emotions" (Aggelton and Mishkin, 1986). Whether or not olfactory experience and emotion are closely linked in humans cannot be decided by anatomy alone, although anatomy may provide valuable insights into the mechanisms underlying the psychological level of analysis. That hedonics seem to be more intrinsic to the experience of odors than to other sensory experiences may have much to do with the nature of the olfactory circuitry; however, in the absence of corroborating psychological research, the anatomical arrangement cannot by itself tell us that such is the case.

Proposition 5: Odors Have Powerful Motivational Effects

An important function of odors is to control approach and avoidance behavior. Chemical substances are powerful attractants throughout the phylogenic series, although there is some question as to whether the transduction system involved is in fact olfaction (the vomeronasal chemoreception system may be particularly important in reptiles and infraprimate mammals) (Cain, 1988). It is unclear whether these chemicals should be considered true pheromones in infrahumans (Beauchamp, Doty, Moulton, and Mugford, 1976); certainly no "human pheromone" has been identified (Cain, 1988). In humans, the motivational effects of odors are closely tied to hedonics. As Levine and McBurney (1986) put it:

> Olfaction is certainly the most hedonically potent of the senses. There is no other sense that elicits such a wide range of emotional reactions to simple stimuli that have not acquired meaning on the basis of conditioning. Thus, it is most implausible to imagine a hue or a tone of moderate intensity that could empty a building of occupants as hydrogen sulfide can or could command the high price of some perfumes (p. 183).

We agree with Levine and McBurney that foul odors are probably among the most potent stimuli to elicit withdrawal. For those who have experienced some of the most unpleasant odors, like pyridine, this statement needs no further elaboration. For those who haven't, consider the effects of the smell of vomit in a crowded environment. Perhaps only extreme pain or a threat to one's life could produce such a powerful desire to flee. What is intriguing about this olfactory experience is that it is *not* painful, nor would it be judged as life-threatening.

The motivational power of pleasant odors is less clear. The most powerfully positive odors for humans are probably those related to food, and their motivational strength undoubtedly is related to the individual's satiety (Cabanac, 1979). Even here, except perhaps in cases of starvation, the approach tendencies would not be as potent as the much more easily elicited avoidance tendencies. In this sense, the most pleasant odors, even the finest perfumes, cannot compare to the most unpleasant odors in their capacity to motivate behavior. This asymmetry in motivational strength may reflect something generic to human motivation or may be specific to the biological functions of odor (Engen, 1982).

At least on the negative side, the motivational power of odors does seem to justify the assertion that olfaction has a particularly close tie to affective systems. However, Levine and McBurney's (1986) claim that odor is more "hedonically potent" than other senses rests on the idea that smells "have not acquired meaning on the basis of conditioning." The question of whether the attractive and repulsive power of odors is learned or innate remains controversial. Engen (1988) has made the case for the role of learning via conditioning and modeling in the subjective experience of smell. In his view, one adopts cultural norms regarding odors, for example, that the smell of a skunk is unpleasant. Moreover, individual responses are also learned. Thus, some individuals may actually like the smell of skunk because of its associations with pleasant childhood memories, but are reluctant to admit this because it would be considered peculiar. Most of Engen's arguments concern the age at which children come to express hedonic reactions to odors, an issue that we discuss later. (See also Schmidt and Beauchamp in this book.) However, with regard to the motivational power of odors, the implication of Engen's claim is that odors have no more potency in motivating approach and avoidance behavior than would any other stimulus conditioned to reward or punishment (except insofar as one's culture specified some odor as a particularly potent stimulus).

We believe that common experience renders this an unlikely hypothesis. Although it may be true that a whiff of skunk odor may be pleasant to some people, it is hard to imagine anyone who would find direct proximate exposure to skunk anything other than extremely unpleasant, unless there was some physiological basis for this indifference (e.g., a specific anosmia). In other words, while there is undoubtedly much individual variation in hedonic responses to most odors, there are also undoubtedly some odors that cause powerful and immediate avoidance behavior in virtually everyone. However, such odors may be exceptional and rarely encountered. Thus, even if one grants that the motivational effects of extremely unpleasant odors are unlearned, the appropriate comparison in Levine and McBurney's example might very well be to blinding light or sounds of very high intensity.

Proposition 6: Odors Produce Facial Expressions That Are Closely Identified with Affect

Facial expression is an important component of many theories of affect (Adelmann and Zajonc, 1989). Although people can dissociate their facial expressions from their emotions, when there is little reason to believe such deception or inhibition is operating, the occurrence of certain facial expressions provides strong evidence that the individual is experiencing a particular emotion. Steiner (1974) has reported that infants show facial reactions that adults interpret as indicative of pleasure or displeasure in response to both tastes and odors. Although facial responses to pleasant and unpleasant tastes seem well supported, the data for odors are far less clear (Engen, 1982). Kraut (1982) carried out a study in which adults smelled various odors while being videotaped. Other subjects were then asked to judge which odor the individual was sniffing based solely on the person's facial expression. Although the judges were able to do this to some extent, examination of the data presented in the article clearly indicates that this ability was largely confined to the most unpleasant odors. Since Kraut did not indicate that the chemicals were diluted, we can assume that the odorants were presented at full strength. If so, it is not surprising that facial expressions to full-strength pyridine, butyric acid, and acetic acid were easily distinguished from those to more benign odors. On the other hand, it appears that judges were not able to tell whether the person was smelling formaldehyde or vanilla, even though the pleasantness ratings for these odors differed by 2.56 units on a 7-point scale. In other words, among odors judged by the individual smelling them as within the range of slightly negative to very positive, facial expression was not particularly revealing. This conclusion is strengthened by the results of a study by Gilbert, Fridlund, and Sabini (1987) in which judges rated facial expressions of seven subjects sniffing pleasant, unpleasant, or no odors. Judges reported seeing facial movement on fewer than half the trials and generally were not very accurate in classifying trials. However, there was again some evidence that unpleasant trials were easier to detect than were pleasant trials. Thirty-nine percent of the unpleasant trials and 21 percent of the no-odor trials were categorized as unpleasant. In contrast, 7 percent of the pleasant trials and 9 percent of the no-odor trials were categorized as pleasant. It is important to note that Gilbert et al. used much more dilute concentrations than did Kraut, perhaps accounting in part for the greater ability of Kraut's judges to correctly identify trials with unpleasant odorants.

Gilbert et al. (1987) also included a condition in which subjects were asked to deliberately express with their face their reaction to each odor as they smelled it. Under these instructions, a strong relationship was observed between subjects' facial expressions and observers' judgments of the hedonic tone of the odorant being smelled. These results show that even though facial expressions to odors may not occur "reflexively," they can be used to communicate one's hedonic experience of an odor, and in this sense are consistent with a characterization of odor as emotional. Whether this reflects merely the ability to conventionally show one's evaluation of a stimulus using facial expression (Gilbert et al., 1987), and hence could be applied equally well to any type of stimulus, or indicates something special about the affective salience of olfactory stimuli is not yet known. It is also the case that the absence of visible changes in facial expression does not mean that covert changes are not occurring (Gilbert et al., 1987). Cacioppo, Martzke, Petty, and Tassinary (1988) have shown that EMG recordings of facial muscles can discriminate among affective stimuli in the absence of observable changes in facial expression. A similar study comparing odors to other types of stimuli could be most informative.

Proposition 7: Odor Experience Has Effects on Cognition and Behavior Similar to Those Produced by Affective States

There has been a great deal of interest in recent years in the effects of affective states on behavior and cognition (Blaney, 1986; Isen, 1984). Demonstrating that odors can have similar effects would be powerful evidence for the affective nature of odor experience.

A number of studies have explored the ability of odors to influence cognition and behavior. Based on the idea that pleasant odors would elicit positive evaluations of other people, Baron (1981) had men provide their first impressions of a woman who either wore perfume or did not. Perfume did enhance liking, as well as rated attractiveness, warmth, and romantic qualities, but only when the woman dressed informally. When the woman was dressed formally, wearing perfume resulted in decreased ratings along these dimensions, perhaps, as some men indicated, because she was perceived as conceited and unattainable. In a study of the effects of fragrance use on evaluations of male and female job applicants, Baron (1983) found that scent had opposite effects on the ratings given by male and female interviewers. In general, the men assigned lower ratings when applicants wore perfume or cologne, while the women rated those who wore scent more favorably than those who did not. Self-reports indicated that men felt more influenced by tactics of impression management (i.e., appearance) than did women. As Baron suggests, the men may therefore have felt resentment toward and subsequently downrated applicants wearing scent. In a similar study, Baron (1986) found that men tended to give lower ratings on job-related dimensions when a woman wore perfume and emitted a high level of positive nonverbal cues (smiling, leaning towards the interviewer) than when only one of these tactics was employed. Women gave significantly higher ratings to the applicant when both tactics were adopted. Taken together, these studies indicate that perfume may affect others' impressions of the wearer; however, these reactions are not uniformly positive and are influenced by other aspects of the social situation as well.

Baron (1990) had male and female subjects perform a variety of work-related tasks in the presence of an ambient pleasant odor (air freshener) or no odor. In order to strengthen the effect of the odor manipulation, subjects in the pleasant odor condition smelled five pleasant fragrances and subjects in the neutral condition smelled five neutral scents prior to performing the tasks. Subjects who worked in the presence of the pleasant odor set higher goals and were more likely to employ an efficient strategy than subjects working in the unscented room. In a negotiation task with a same-sex confederate, subjects exposed to the pleasant fragrance set higher monetary goals, were more likely to make concessions, and were less likely to state a preference for confrontational approaches to future negotiations than were subjects in the unscented room. When self-reported mood was used as a covariate, the difference in stated preferences for confrontational strategies was no longer significant. Thus, as Baron suggests, the behavioral and cognitive effects of odors may be largely due to the intervening effect of pleasant odor on mood.

Rotton et al. (1978) had subjects rate their liking for attitudinally similar and dissimilar others while in a room permeated with the smell of ammonium sulfide or in a room that had no odor. A similar stranger was liked more than a dissimilar one in the presence of odor, and also more than either type of stranger in its absence. The authors suggested that the stranger, who was believed by subjects to be experiencing the malodor in another room, was considered a "fellow sufferer." The experience of this "shared stress" may have intensified subjects' liking/empathy for strangers thought to be similar to themselves. Rotton et al. obtained support for this hypothesis in a follow-up study in which subjects were explicitly told that the stranger was not experiencing the unpleasant odor (butyric acid or ammonium sulfide). Malodors reduced attraction toward both similar and dissimilar others compared to the no-odor condition.

Based on previous findings regarding the effects of temperature on aggression, Rotton, Frey, Barry, Milligan, and Fitzpatrick (1979) hypothesized that a moderately unpleasant odor (ethyl mercaptan) would produce more aggression than an extremely unpleasant odor (ammonium sulfide). Men were given the chance to shock a confederate after they were exposed to either an aggressive model or no model. Support for their hypothesis was obtained only in the no-model condition: shock intensity was highest when subjects were exposed to the moderate malodor. The studies by Rotton and his colleagues suggest that malodors act like other environmental stressors (i.e., temperature, crowding, noise) in affecting both evaluations of others and actions toward them. These effects may be mediated, as suggested by the results of the Rotton et al. (1978) study, by the malodors' negative impact on mood. However, as is the case with pleasant odors, the effects are complex in that under certain circumstances, unpleasant odors may actually enhance liking and do not always lead to aggression.

Rotton (1983) had male and female college students evaluate paintings, people in photographs, and descriptions of people while in the presence of either ethyl mercaptan or no odor. Subjects exposed to the malodor judged paintings as less professional and worthy, and gave significantly lower ratings of the well-being of the photographed individuals than did the control group. While malodor did not affect ratings of highly favorable, highly unfavorable, or moderately unfavorable descriptions of others, moderately favorable descriptions were rated significantly less likable by those exposed to the unpleasant odor. In a second experiment, Rotton found that, like other environmental stressors, malodor impaired college students' performance on a complex (proofreading) but not a simple (arithmetic) task as compared to that of those not exposed to the odor. In addition, after leaving the unpleasant-smelling room, the malodor group demonstrated a lower tolerance for a frustrating task (tracing lines on insoluble puzzles) if they had had no control over the earlier presentation of the malodor. Further, these aftereffects were strongest when the subjects had been exposed to the malodor for longer periods of time (30 versus 15 minutes). Thus, the effects of malodors on performance, especially when exposure is prolonged and unavoidable, may persist even after the unpleasant stimulus has been removed.

In order to determine the effects of unpleasant odors on responsiveness to persuasion, Janis, Kaye, and Kirschner (1965) exposed subjects to butyric acid while they read several persuasive messages. Contrary to their prediction, the malodor had no effect on subjects' opinions relative to those of a control group. The authors suggested that subjects' opinions were not changed because they believed the presence of the malodor was due to an accident for which the experimenter was not responsible.

Cann and Ross (1989) compared men's ratings of attractiveness of slides of women in the presence of a pleasant odor (spray cologne), an unpleasant odor (ammonium sulfide), or no odor. Odor condition did not affect self-reported mood or attractiveness ratings. The authors reasoned that the failure of the odors to induce a change in mood was responsible for the absence of an effect on attractiveness ratings. However, the fact that the stimuli were chosen to represent relatively attractive and unattractive women may have biased the results against their hypothesis. There is considerable evidence that affective state can influence responses to ambiguous or neutral stimuli, but not to clearly positive or negative stimuli (Bower, 1981; Isen and Shalker, 1982; Kuykendall and Keating, 1984). It is unlikely, then, that the experience of an odor (even if accompanied by a change in mood) would significantly affect one's ratings of relatively unambiguous stimuli (cf. Rotton, 1983).

Ludvigson and Rottmann (1989) had subjects complete a series of cognitive tasks in the presence of the odors of lavender or cloves, or with no odor present, in two sessions held one week apart. In the first session, subjects exposed to lavender rated the tasks more favorably than did those in the cloves or no-odor condition. However, there was no difference among conditions in self-reported mood. In the second session, lavender produced the least favorable task ratings, again without any differences in self-reported mood. This study, the only one to test subjects in two separate sessions, suggests, first, that odors can influence judgments without a corresponding change in mood, and second, that subsequent exposures to an odorant may have quite different effects from those of initial exposure.

In a study designed to conceptually replicate verbally induced mood effects on memory, Ehrlichman and Halpern (1988) used pleasant and unpleasant odors to produce positive and negative experiences. College women recalled memories cued by neutral words while exposed to either a pleasant odor (almond), an unpleasant odor (pyridine), or no odor. The subjects then rated the recalled events as to how happy or unhappy they were at the time they occured. Women in the pleasant odor condition recalled significantly more happy memories than those in the unpleasant odor condition. Mean ratings of the memories also differed between the two odor conditions, although neither was significantly different from the no-odor condition. The authors argued that like moods, odors may bias retrieval toward hedonically congruent contents. Since no mood ratings were obtained, it is unclear whether these results reflect the shared hedonic valence of the odor experience and the memories, or were in fact mediated by an odor-induced change in mood.

An interesting difference among these studies concerns the salience of the odors for the subjects. In all except the Ehrlichman and Halpern (1988) and Rotton (1983) studies, odors were presented unobtrusively. Although at first it might appear that it is preferable to present odors in an "ecologically valid" manner, the results of a number of the studies suggest a danger in this approach. For example, although Baron (1981) initially hypothesized that perfume would enhance liking/attractiveness because it produces positive feelings, he discovered that the *meaning* of wearing perfume was more important: Subjects do not simply smell an odor, they draw conclusions about what it is, where it came from, what it signifies. Any attempt to make the odor "natural" could provoke such potentially mood-relevant interpretations. Informing subjects that the odor is part of the experiment could, of course, bias their behavior. Ehrlichman and Halpern used the cover story that their experiment was concerned with speed of memory retrieval in the presence of "environmental stimuli"; Rotton's subjects were told that the study involved effects of "chemicals" on task performance. Ehrlichman and Halpern also had a postexperimental interview to assess subjects' beliefs about the study. Although such procedures do not rule out the presence of demand characteristics, presenting odors unobtrusively may not be appropriate when the object is to study effects of odor experience itself rather than affective reactions that reflect subjects' interpretations of the experimental situation.

The recent studies by Baron (1990), Rotton (1983), and Ehrlichman and Halpern (1988) are all consistent with the idea that odor experience can influence cognition and behavior in ways similar to those produced by affective states. Because it does not rely on subjects' self-reports, this line of research seems particularly promising. An important issue in this research is whether changes in self-reported mood must occur in order for odors to produce affect-related changes in cognition or behavior. Consider the differences between odor experience and mood. Odor is experienced as emanating from a source outside the person; mood is experienced as an internal state (i.e., is self-referenced). If an external source produces a feeling state such as disgust or happiness, one can say that phenomenologically the experience is now internalized and may even be labeled as a mood. Although there can be little doubt that the cognitions accompanying externalized versus internalized experiences of pleasantness or unpleasantness are quite different, in either case the pleasantness or unpleasantness itself still exists "in the person" in the sense that it is a product of the experiencer's nervous system. In fact, to the extent that the hedonic aspect of affect (as opposed to the accompanying cognitions) is responsible for effects on cognition and behavior, it may not matter whether pleasantness or unpleasantness is experienced as internal or external (Ehrlichman and Halpern, 1988). Experiencing any valenced state may be sufficient to activate cognitions and behaviors believed to be activated by moods.

Proposition 8: Odor May Act as a Catalyst, Potentiating Affective Responses to Other Stimuli

Whether or not odors by themselves evoke strong affect, they may, under certain conditions, add substantially to the emotionality of an experience. Van Toller (1988) has made the following analogy:

> the sense of smell can be likened to colour in a television picture. It adds a tremendous amount in terms of texture, dimensionality, pleasure and feeling, but in terms of information content a black and white picture will contain virtually all the relevant elements (p. 184).

A fragrance on a person to whom one is sexually attracted may dramatically increase one's affective response, whereas the same fragrance sniffed from a bottle may smell nice, but elicit little if any affective reaction. The smell of the sea may be powerfully evocative and moving at the seashore, but produce indifference or annoyance in an experimental cubicle. King (1988), who uses such aromas in his therapeutic practice, emphasizes the need to contextualize the experience.

The idea that the affective impact of odors is closely tied to context has yet to be studied seriously despite the fact that it raises important practical as well as theoretical questions. For example, successful attempts to "fragrance" environments (Hinds, 1988) might well require that people come to view such environments as appropriately fragranced (as has been so successfully accomplished for fragrances added to cleaning products). Otherwise, individuals might even react negatively to attempts to manipulate their emotional reactions. The results of the study by Baron (1983) in which men reacted negatively to applicants using fragrance in an interview situation is in keeping with this notion. On the other hand, to the extent that specific mood effects of odorants are pharmacologic, context might be relatively unimportant.

The Uses of Odor in the Study of Affect

We have tried to show that the possible connections between olfaction and emotion are multifaceted. Generally, we have found that where sufficient research exists to begin to evaluate such connections, the results are encouraging. In other cases, there is simply not enough evidence to accept or reject the propositions. Until very recently, olfactory research has largely ignored the affective consequences of odor experience. However, it is apparent that interest in olfaction among researchers of many different stripes is rapidly growing, and that for many of these investigators, issues involving affective responses are central. It is important to emphasize that the specific propositions do not have to be, and should not be, taken as a package. For example, it is quite possible that odors could affect mood without their having any special ability to evoke emotional memories. We have also argued that at present there is no basis for claims that odors are more emotional than are other types of stimuli. Whether such claims could even in principle be upheld is questionable.

Whatever the fate of the specific propositions we have discussed, the abilities of odors to produce pleasant and unpleasant experiences and to do so with minimal cognitive mediation are features of odor experience that could be used profitably in studying affect. One role for odors is in research on populations for whom one would like to manipulate hedonic experience without the mediation of films, mood statements, classical music, and so on, for example, in studying brain-injured people for whom the ability to extract the affective meaning of such stimuli may be problematic. Odor could also be useful for studying the development of hedonic experience. Indeed, for many years it has been claimed that children under the age of six are much more accepting of odors that older children and adults reject. Pető (1936) reported that 89 out of 92 children showed no evidence of disliking a variety of odors of "putrification" or "fumes." Specifically, the children neither cried nor indicated by gestures or words that their reaction to the odors was negative. Engen (1988) has claimed that the lack of negative affective responses to such odorants in young children is well established. However, Schmidt and Beauchamp (1988) have presented data suggesting that the failure of previous research to demonstrate negative affective responses in young children reflects problems in both measurement and odorant selection. In their study, three-year-old children showed clear evidence of disliking the odors of butyric acid and pyridine, and liking the odors of strawberry and cloves. It is noteworthy that extremely low concentrations of the unpleasant stimuli were used. One wonders whether the difference between adults and children in their liking of pyridine (none of the adults liked it, as compared to 25 percent of the children) would have remained if the concentration was more than 0.003 percent!

In addition to their use in special populations, odors may be appropriate for research directly addressing the hedonic component of affective experience. For example, an important question about hedonics is whether pleasantness and unpleasantness are best conceived of as ends of a single continuum or as independent psychological states. This issue has been a continual source of controversy in the mood literature. Traditionally, researchers conceptualized and measured positive and negative mood in terms of a single hedonic dimension, but Bradburn (1969) demonstrated that positive and negative affect vary independently. Is displeasure really the opposite of pleasure? Does less pleasure imply more displeasure and vice versa (Beebe-Center, 1932; Young, 1967)? The important issue here is whether pleasantness and unpleasantness function differently or similarly, have different or similar correlates, and are sensitive to different or similar interventions. Much of the research we have reviewed points to the relative salience of unpleasant as compared to pleasant odor experiences in influencing mood and behavior. In addition, while reactions to malodors can often be classified as "disgust," there does not appear to be an equivalent term to apply to reactions to pleasant odors. Perhaps the term "pleasure" comes closest (although words like "calm," "invigorated," "sexy," or "aroused" may also apply at times). This apparent asymmetry in odor experience may prove useful in delineating the more general relationship between positive and negative affective reactions.

There is also a question as to the role of hedonics in individual mood variation. Are so-called "positive" and "negative" moods (Watson and Tellegen, 1985) fundamentally defined by their hedonic tone, or are these terms misnomers, applying instead to differences in types or levels of arousal (e.g., Thayer, 1989)? Are individuals who are emotionally reactive (Larsen and Diener, 1987) more sensitive to the hedonic value of odors than less reactive individuals? Tassinary (1985) demonstrated that hedonic orderings of odors were similar across subjects; however, the point at which the line was drawn between pleasant and unpleasant varied widely. Warrenburg and Schwartz (1988) found differences in skin temperature and skin resistence between subjects who liked the odor of neroli and those who did not. In addition, the two groups differed with respect to their self-rated boredom during the odor trials. The authors interpreted these results to indicate that the "like-neroli" subjects were predisposed to feel more relaxed and attentive as well as more interested in novel odors. Whether these results reflect underlying traitlike differences between the two groups or state differences, they suggest that affective reactions to "ambiguous" odors (those that are not universally viewed as either pleasant or unpleasant) could prove a promising area for future research.

The ability of odors to produce experiences of pleasantness or unpleasantness with little cognitive mediation could be useful when studying the nervous system processes associated with hedonic experience. For example, Harver, Katkin, Bott, Ehrlichman, and Warrenburg (1989) have presented preliminary data suggesting that unpleasant odors produce heart-rate acceleration with increased skin conductance levels, while pleasant odors produce slight deceleration with no effect on skin conductance. Since these psychophysiological measures can be sensitive to ongoing cognitive activity, minimizing such activity in the study of hedonic reactions can be quite important.

As another example, consider the controversy in the neuropsychological literature concerning the differential involvement of the left and right hemispheres in emotional response. Although there is little doubt that the two sides of the brain differ with regard to their role in affect, how to best characterize this difference remains unclear. Most of the research on this topic has involved clinical studies of brain-damaged patients or studies in which normal subjects respond to emotionally evocative stimuli such as words, faces, and films. Yet clearly the affective meaning of such stimuli involves cognitive processes that themselves are likely to be hemispherically distinct. Thus, differences between the two hemispheres could be reflecting cognitive differences in how meaning is extracted from stimuli. In contrast, the minimal cognitive demands of experiencing an odor as pleasant or unpleasant would seem to make odors well suited for addressing questions of differences in hedonic tone between the two hemispheres. Ehrlichman (1986) has reported a study in which pleasant and unpleasant odors were presented to normal subjects' left or right nostrils. Their only task was to rate the odors on a pleasantness–unpleasantness scale. Unpleasant odors presented to the right nostril were rated as more unpleasant than the same odors presented to the left nostril. In contrast, no such difference was found for pleasant odors. Because of the predominant ipsilateral projections of the olfactory nerves, Ehrlichman (1986) interpreted this result as consistent with a model in which the right hemisphere is specialized for mediating negative affect, with no hemispheric asymmetry existing for positive affect.

These examples suggest that in a number of psychological domains, the use of odors can serve as an important complement to studies using more cognitively mediated affective stimuli. We have little doubt that the number of studies addressing relationships between olfaction and emotion will continue to grow. We suspect that the result will be an increased understanding of both olfaction and emotion.

Acknowledgment. The authors wish to thank Deleri Springer and Stephen Warrenburg for their helpful comments on an earlier version of this chapter, and the Fragrance Research Fund for its support and encouragement.

References

Adelmann, P.K., and R.B. Zajonc (1989). Facial efference and the experience of emotion. *Ann. Rev. Psychol.*, *40*, 249–280.

Aggelton, J.P., and M. Mishkin (1986). The amygdala: Sensory gateway to the emotions. In R. Plutchik and H. Kellerman (Eds.), *Emotion: Theory, research, and Experience*, Vol. 3, *Biological Foundations of Emotion.* Orlando: Academic, pp. 281–299.

Arnold, M.B. (1960). *Emotion and Personality*, Vol. 1. New York: Columbia University Press.

Baron, R.A. (1981). Olfaction and human social behavior: Effects of a pleasant scent on attraction and social perception. *Personality Soc. Psychol. Bull.*, *7*, 611–616.

Baron, R.A. (1983). "Sweet smell of success"? The impact of pleasant artificial scents on evaluations of job applicants. *J. Appl. Pschol.*, *4*, 709–713.

Baron, R.A. (1990). Environmentally-induced positive affect: Its impact on self-efficacy, task performance, negotiation, and conflict. *J. Appl. Soc. Psychol.*, *20*, 368–384.

Beauchamp, G.K., R.L. Doty, D.G. Moulton, and R.A. Mugford (1976). The pheromone concept in mammalian chemical communication: A critique. In R.L. Doty (Ed.), *Mammalian Olfaction, Reproductive Processes, and Behavior.* New York: Academic, pp. 143–160.

Beebe-Center, J.G. (1932). *The Psychology of Pleasantness and Unpleasantness.* New York: Russell and Russell.

Blaney, P.H. (1986). Affect and memory: A review. *Psychol. Bull.*, *99*, 229–246.

Bower, G.H. (1981). Mood and memory. *Am. Psychol.*, *36*, 129–148.

Bradburn, N.M. (1969). *The Structure of Psychological Well-being.* Chicago: Aldine.

Burleson, M.H., and V.S. Johnston (1989). *Androstenol Attenuates P3 Component Amplitude in Visual Event-Related Potentials.* Paper presented at the meeting of the American Psychological Association, New Orleans, LA, August.

Cabanac, M. (1979). Sensory pleasure. *Qu. Rev. Biol.*, *54*, 1–29.

Cacioppo, J.T., J.S. Martzke, R.E. Petty, and L.G. Tassinary (1988). Specific forms of facial EMG response index emotions during an interview: From Darwin to the continuous flow hypothesis of affect-laden information processing. *J. Personality Soc. Psychol.*, *54*, 592–604.

Cain, W.S. (1988). Olfaction. In R.C. Atkinson, R.J. Herrnstein, G. Lindzey, and R.D. Luce (Eds.), *Stevens' Handbook of Experimental Psychology*, Vol. 1, *Perception and Motivation*, rev. ed. New York: Wiley, pp. 409–459.

Cann, A., and D.A. Ross (1989). Olfactory stimuli as context cues in human memory. *Am. J. Psychol.*, *102*, 91–102.

Clemens, S.L. (1882). The invalid's story. In C. Neider (Ed.), *The Complete Short Stories of Mark Twain.* Garden City, N.Y.: International Collectors Library, 1957.

Davis, R.G. (1975). Acquisition of verbal associations to olfactory stimuli of varying familiarity and to abstract visual stimuli. *J. Exp. Psychol.*, *104*, 134–142.

Davis, R.G. (1977). Acquisition and retention of verbal associations to olfactory and abstract visual stimuli of varying similarity. *J. Exp. Psychol.*, *3*, 37–51.

Dodd, G.H. (1988). The molecular dimension in perfumery. In S. Van Toller and G.H. Dodd (Eds.), *Perfumery: The Psychology and Biology of Fragrance.* New York: Chapman and Hall, pp. 19–46.

Ehrlichman, H. (1986). Hemispheric asymmetry and positive-negative affect. In D. Ottoson (Ed.), *Duality and Unity of the Brain: Unified Functioning and Specialization of the Hemispheres.* London: Macmillan, pp. 194–206.

Ehrlichman, H., and J.N. Halpern (1988). Affect and memory: Effects of pleasant and unpleasant odors on retrieval of happy and unhappy memories. *J. Personality Soc. Psychol.*, *55*, 769–779.

Engen, T. (1982). *The Perception of Odors.* New York: Academic.

Engen, T. (1988). The acquisition of odour hedonics. In S. Van Toller and G.H. Dodd (Eds.), *Perfumery: The Psychology and Biology of Fragrance.* New York: Chapman and Hall, pp. 79–90.

Filsinger, E.E., J.J. Braun, and W.C. Monte (1985). An examination of the effects of putative pheromones on human judgments. *Ethol. Sociobiol.*, *6*, 227–236.

Frijda, N.H. (1986). *The Emotions.* New York: Cambridge University Press.

Gibbons, B. (1986). The intimate sense of smell. *National Geographic*, *170*, 324–361.

Gilbert, A.N., A.J. Fridlund, and J. Sabini (1987). Hedonic and social determinants of facial displays to odors. *Chem. Senses*, *12*, 355–363.

Gustavson, A.R., M.E. Dawson, and D.G. Bonett (1987). Androstenol, a putative human pheromone, affects human (Homo sapiens) male choice performance. *J. Comp. Psychol.*, *2*, 210–212.

Harver, A., E.S. Katkin, K. Bott, H. Ehrlichman, and S. Warrenburg (1989). Autonomic and affective responses to odors. Presented at the meeting of the Society for Psychophysiological Research, October.

Heywood, A., and H.A. Vortriede (1905). Some experiments on the associative power of smells. *Am. J. Psychol.*, *16*, 537–541.

Higgins, E.T., G.A. King, and G.H. Maven (1982). Individual construct accessibility and subjective impressions and recall. *J. Personality Soc. Psychol.*, *43*, 35–47

Hinds, M. (1988). When scent sets the mood. *New York Times*, July 23, p. 52.

Hoffman, M.L. (1986). Affect, cognition, and motivation. In R.M. Sorrentino and E.T. Higgins (Eds.), *Handbook of Motivation and Cognition: Foundations of Social Behavior.* New York: Guilford, pp. 244–279.

Isen, A.M. (1984). Toward understanding the role of affect in cognition. In R.S. Wyer, Jr., and T.K. Srull (Eds.), *Handbook of Social Cognition.* Hillsdale, NJ: Lawrence Erlbaum, pp. 179–236.

Isen, A.M., and T.E. Shalker (1982). The effect of feeling state on evaluation of positive, neutral, and negative stimuli: When you "accentuate the positive," do you "eliminate the negative"? *Soc. Psychol. Q.*, *45*, 58–63.

Izard, C.E. (1972). *Patterns of Emotions: A New Analysis of Anxiety and Depression.* New York: Academic.

Izard, C.E. (1977). *Human Emotions.* New York: Plenum.

Janis, I.L., D. Kaye, and P. Kirschner (1965). Facilitating effects of "eating-while-reading" on responsiveness to persuasive communications. *J. Personality Soc. Psychol.*, *1*, 181–186.

Johnson-Laird, P.N., and K. Oatley (1988). Are there only two primitive emotions? A reply to Frijda. *Cognition and Emotion*, *2*, 89–93.

Kanamura, S., S. Kiyota, Y. Takashima, T. Kanisawa, M. Indo, G.V. Loveren, H. Fukuda, and S. Torii (1989). Effects of odors on CNV. *Chem. Senses*, *14*, 303–304.

King, J.R. (1988). Anxiety reduction using fragrances. In S. Van Toller and G.H. Dodd (Eds.), *Perfumery: The Psychology and Biology of Fragrance.* New York: Chapman and Hall, pp. 147–165.

Knasko, S.C., A.N. Gilbert, and J. Sabini (1990). *Emotional State, Physical Well-*

being and Performance in the Presence of Feigned Ambient Odor. Submitted. *J. App. Soc. Psychol.*, *20*, 1345–1357.
Kraut, R.E. (1982). Social presence, facial feedback, and emotion. *J. Personality Soc. Psychol.*, *42*, 853–863.
Kuykendall, D., and J.P. Keating (1984). *Mood and the Evaluation of Positive, Neutral, and Negative Words*. Paper presented at the meeting of the American Psychological Association, Toronto, Canada, August.
Laird, D. (1935). What can you do with your nose? *Sci. Monthly*, *41*, 126–130.
Larsen, R.J., and E. Diener (1987). Affect intensity as an individual difference characteristic. *J. Res. Personality*, *21*, 1–39.
Lawless, H.T. (1978). Recognition of common odors, pictures, and simple shapes. *Perception Psychophys.*, *24*, 493–495.
Lazarus, R.S. (1984). On the primacy of cognition. *Am. Psycholog.*, *39*, 124–129.
Leventhal, H., and K. Scherer (1987). The relationship of emotion to cognition: A functional approach to a semantic controversy. *Cognition and Emotion*, *1*, 3–28.
Levine, J.M., and D.H. McBurney (1986). The role of olfaction in social perception and behavior. In C.P. Herman, M.P. Zanna, and E.T. Higgins (Eds.), *Physical Appearance, Stigma, and Social Behavior: The Ontario Symposium*, Vol. 3. Hillsdale, NJ: Lawrence Erlbaum, pp. 179–217.
Lieff, B., and J. Alper (1988). Aroma driven: On the trail of our most emotional sense. *Health*, *20*, 62–67.
Livesey, P.J. (1986). *Learning and Emotion: A Biological Synthesis*, Vol. I, *Evolutionary Processes*. Hillsdale, NJ: Lawrence Erlbaum.
Lorig, T.S., and G.E. Schwartz (1988). Brain and odor: I. Alteration of human EEG by odor administration. *Psychobiology*, *16*, 281–284.
Lorig, T.S., and G.E. Schwartz (1988–89). EEG activity during relaxation and food imagery. *Imagination, Cognition and Personality*, *8*, 201–208.
Lorig, T.S., K.B. Herman, G.E. Schwartz, and W.S. Cain (1990). EEG activity during administration of low concentration odors. Submitted. *Bull. Psychonomic Soc.*, *28*, 405–408.
Ludvigson, H.W., and T.R. Rottman (1989). Effects of ambient odors of lavender and cloves on cognition, memory, affect and mood. *Chem. Senses*, *14*, 525–536.
Moskowitz, H.R. (1979). Mind, body and pleasure: An analysis of factors which influence sensory hedonics. In J.H.A. Kroeze (Ed.), *Preference Behaviour and Chemoreception*. London: IRL, pp. 131–144.
Ortony, A., G.L. Clore, and A. Collins (1988). *The Cognitive Structure of Emotions*. Cambridge: Cambridge University Press.
Papez, J.W. (1937). A proposed mechanism of emotion. *Arch. Neurol. Psychiatry*, *38*, 725–743.
Petó, E. (1936). Contribution to the development of smell feeling. *Br. J. Med. Psychol.*, *15*, 314–320.
Plutchik, R. (1980). *Emotion: A Psychoevolutionary Synthesis*. New York: Harper and Row.
Proust, M. (1982). *Remembrance of Things Past*, Vol. 1, *Swann's Way*. New York: Vintage. (Original work published 1913.)
Ratliff, M.M. (1938). The varying function of affectively toned olfactory, visual, and auditory cues in recall. *Am. J. Psychol.*, *51*, 695–701.
Richardson, J.T., and G.M. Zucco (1989). Cognition and olfaction: A review. *Psychol. Bull.*, *105*, 352–360.

Rotton, J. (1983). Affective and cognitive consequences of malodorous pollution. *Basic Appl. Soc. Psychol.*, *4*, 171–191.

Rotton, J., T. Barry, J. Frey, and E. Soler (1978). Air pollution and interpersonal attraction. *J. Appl. Soc. Psychol.*, *38*, 213–228.

Rotton, J., J. Frey, T. Barry, M. Milligan, and M. Fitzpatrick (1979). The air pollution experience and physical aggression. *J. Appl. Soc. Psychol.*, *9*, 397–412.

Rozin, P., and A.E. Fallon (1987). A perspective on disgust. *Psychol. Bull.*, *94*, 23–41.

Rubin, D.C., E. Groth, and D.J. Goldsmith (1984). Olfactory cuing of autobiographical memory. *Am. J. Psychol.*, *97*, 493–507.

Schachtel, E.G. (1959). *Metamorphosis: On the Development of Affect, Perception, Attention, and Memory.* New York: Basic Books.

Schmidt, H.J., and G.K. Beauchamp (1988). Adult-like odor preferences and aversions in three-year-old children. *Child Dev.*, *59*, 1136–1143.

Smith, C.A., and P.C. Ellsworth (1985). Patterns of cognitive appraisal in emotion. *J. Personality Soc. Psychol.*, *48*, 813–838.

Steiner, J.E. (1974). Discussion paper: Innate, discriminative human facial expressions to taste and smell stimulation. *Ann. NY Acad. Sci.*, *237*, 229–233.

Sugano, H. (1989). Effects of odors on mental function. *Chem. Senses*, *14*, 303.

Tanida, M., T. Tsuchiya, S. Uenoyama, and Y. Nakayama (1989). Effect of odorant inhalation on the sleeping time induced by pentobarbital administration in mice. *Chem. Senses*, *14*, 304.

Tassinary, L.G. (1985). *Odor Hedonics: Psychophysical, Respiratory, and Facial Measures.* Unpublished doctoral dissertation, Dartmouth College, New Hampshire.

Thayer, R.E. (1989). *The Biopsychology of Mood and Arousal.* New York: Oxford University Press.

Tisserand, R. (1988). Essential oils as psychotherapeutic agents. In S. Van Toller and G.H. Dodd (Eds.), *Perfumery: The Psychology and Biology of Fragrance.* New York: Chapman and Hall, pp. 167–181.

Torii, S., H. Fukuda, H. Kanemoto, R. Miyanchi, Y. Hamauzu, and M. Kawasaki (1988). Contingent negative variation (CNV) and the psychological effects of odour. In S. Van Toller and G.H. Dodd (Eds.), *Perfumery: The Psychology and Biology of Fragrance.* New York: Chapman and Hall, pp. 107–120.

Van Toller, S. (1988). Odours, emotion, and psychophysiology. *Int. J. Cosmetic Sci.*, *10*, 171–197.

Warm, J.S., W.H. Dember, R. Parasuraman, J.B. Gluckman, J.A. Thiemann, M. Munteanu, and C. Warren (1989). *The Enhancement of Sustained Attention by Olfactory Stimulation.* Paper presented at the meeting of the Psychonomic Society, Atlanta, GA.

Warren C.B., M.A. Munteanu, G.E. Schwartz, C. Benaim, H.G. Walter, R.S. Leight, D.A. Withycombe, B.D. Mookberjee, and R.W. Trenkle (1987). *Method of Causing the Reduction of Physiological and/or Subjective Reactivity to Stress in Humans Being Subjected to Stress Conditions.* US Patent Number 4,671,959.

Warrenburg, S., and G. Schwartz (1988). A psychological study of three odorants. *Chem. Senses*, *13*, 744.

Watson, D., and A. Tellegen (1985). Toward a consensual structure of mood. *Psychol. Bull.*, *98*, 219–235.

Wright, R.H., and K.M. Michaels (1964). Evaluation of far infrared relation to odors by a standards similarity method. *Ann. NY Acad. Sci.*, *116*, 535–551.

Wundt, W. (1897). *Outlines of Psychology* (C.H. Judd, Trans.). New York: Gustav E. Stechert. (Original work published 1896.).

Yoshida, M. (1964). Studies of psychometric classification of odors (5). *Jap. Psychol. Res.*, *6*, 145–154.

Yoshida, M. (1979). Descriptive and emotional profiles of odours and their preferences. In J.H.A. Kroeze (Ed.), *Preference Behaviour and Chemoreception*. London: IRL, pp. 83–92.

Young, P.T. (1967). Affective arousal: Some implications. *Am. Psychol.*, *22*, 32–40.

Zajonc, R.B. (1980). Feeling and thinking: Preferences need no inferences. *Am. Psychol.*, *35*, 151–175.

Zajonc, R.B. (1984). On the primacy of affect. *Am. Psychol.*, *39*, 117–123.

Zellner, D.A., and M.A. Kautz Color Affects Perceived Odor Intensity. *J. Exp. Psychol.*, In press.

16
Primary Olfactory Disorders: Anosmia, Hyposmia, and Dysosmia

David V. Smith and Heather J. Duncan

Introduction

It has been estimated that at least 2 million Americans suffer from a disorder of taste or smell,[1] but this is probably an underestimation, particularly if one considers the significant degree of smell impairment in the elderly population. Although there are over 200,000 visits to physicians each year for chemosensory disorders, many more of these problems go unreported; they tend to be dismissed by patients as well as by physicians. It is quite obvious, however, that individuals with a taste or smell loss may suffer a significant impairment to their quality of life, both aesthetically and emotionally. Smell and taste alert us to fires, poisonous fumes, leaking gas, and spoiled foods. For this reason, loss of smell is of particular concern to elderly individuals and to those who live alone. The natural pleasures associated with eating can be severely impaired in a person with a taste or smell loss or distortion. Smell and taste impairment can also lead to depression, especially in persons whose occupations (e.g., fireman, chef) critically depend upon these senses.

Patients with taste or olfactory complaints are difficult to diagnose and treat; often they have been to a number of health professionals in an unsuccessful attempt to get help for their problem. Much of the difficulty the medical profession has in dealing with these patients results from the general lack of knowledge about the senses of smell and taste and their disorders. Although an olfactory or taste dysfunction can sometimes occur as a secondary process in a number of disease states, often a reduced ability to smell or taste, a persistent unpleasant taste, or a distorted olfactory experience is a patient's primary complaint. Although much remains to be learned, there has been a good deal of recent attention to these clinical problems (see Doty, 1979; Doty and Kimmelman, 1986; Doty and Snow, 1987; Feldman et al., 1986; Kimmelman, 1986; Leopold, 1986; Schiffman,

[1] Report of the Panel on Communicative Disorders to the National Advisory Neurological and Communicative Disorders and Stroke Council. Washington, DC: Public Health Service, 1979: 319. (NIH publication no. 79- 1914).

1983a, 1983b; Smith, 1988, 1990). The characteristics and common etiologies of primary olfactory disturbances are discussed in the present chapter.

Evaluation of Smell Complaints

The evaluation of patients with olfactory disturbances must involve a careful medical history, with particular attention to antecedent events that might be related to the onset of olfactory loss, such as an upper respiratory infection or a traumatic injury (Doty, 1979; Goodspeed et al., 1987; Leopold, 1986). In addition, the nature of the smell symptoms can be a clue to the most likely cause of the problem (Gent et al., 1987). For example, if a patient with an olfactory loss reports that smell sensitivity fluctuates, nasal or sinus disease is probably involved. A direct question to a patient about smell loss has high predictive value. That is, nearly all patients with a diminished sense of smell will verbally report a decrease in olfactory function (Gent et al., 1987; Smith et al., 1987). On the other hand, most patients with an olfactory loss also report that they have lost their sense of taste. Patients tend to exaggerate taste problems because of their confusion between taste and flavor. The perception of flavor involves olfactory, tactile, and thermal sensations in addition to taste. An impairment in any of these sensory modalities is usually reported as a taste loss, yielding little predictive value in such a question. Direct questions to the patient about the four qualities of taste (salty, sour, sweet, and bitter) are much more revealing of an actual taste dysfunction (Gent et al., 1987).

Dysfunctions of the sense of smell can take the form of complete or partial loss of sensitivity, increased sensitivity, or a distortion of olfactory perception. Although numerous terms have been used to classify these disorders, there is beginning to be some agreement about their usage (Doty, 1979; Kimmelman, 1986). The terms commonly used for olfactory dysfunction include *general* (or total) *anosmia* (or, more simply, *anosmia*), inability to detect any qualitative odor sensation; *hyposmia*, decreased sensitivity to some or all odorants; *hyperosmia*, increased sensitivity to some or all odorants; *dysosmia*, a distortion in the perception of a smell, such as the perception of an unpleasant odor when there is no odor present (*phantosmia*) or the perception of an atypical odor in response to a particular stimulus (*parosmia*). Although several other terms have been used to describe olfactory dysfunction (e.g., Henkin, 1967a; Estrum and Renner, 1987), these terms have found the most common usage.

Psychophysical Measurement in a Clinical Setting

Much of the difficulty in dealing clinically with smell complaints is in obtaining good quantitative measurements of the patient's sensory function. This difficulty is multifaceted; the sensory assessment must involve well-designed and standardized sensory tests and careful stimulus control, and yet must be done in a clinical setting. No aspect of this problem is simple, and it has attracted a great deal of attention in recent years (Cain and Gent, 1986; Cain et al., 1983, 1988; Doty, 1979; Doty et al., 1984a; Gent et al., 1986; Mozell et al., 1986a; Wright, 1987). As a result, several procedures have been developed and standardized on normal populations for use in the clinical testing of smell (Cain et al., 1983, 1988; Doty et al., 1984a; see also Smith, 1988). See also the chapter by Cain, Cometto-Muñiz, and de Wijk in this volume.

The method of presentation of olfactory stimuli for the assessment of smell dysfunction poses a considerable problem for the clinician. For example, controlling the concentration of an odorant is obviously important, but is difficult to achieve. Several approaches have been taken to this problem, including the use of sniff bottles, blast injection into the nostrils, and air dilution olfactometers (see Doty, 1979, for a thorough discussion of stimulus delivery techniques). One of the major considerations for clinical testing is that stimulus delivery needs to be relatively easy and somewhat portable. Recent tests that have been developed for clinical olfactory testing employ the use of sniff bottles and microencapsulated ("scratch and sniff") odorants. The major goal of any kind of sensory testing for smell is to assess the degree of dysfunction. Thus, many different approaches could be (and have been) employed. However, a proper assessment of chemosensory dysfunction requires the adoption of well-designed and psychophysically sound testing procedures.

Traditionally, neurologists have employed odor identification to test the sense of smell (Bickerstaff, 1968). Typically, only a few odors are used, and the response required of the patient is the identification of the odor, or even a simple acknowledgment of whether the odor is smelled. Such tests are not very quantitative and also suffer from the problem that most people have difficulty identifying odors, even if they have a normal sense of smell (Cain and Gent, 1986; Doty et al., 1984a; Sumner, 1962). The ability to identify odors varies considerably across odorous substances, but even common odors are often difficult to name (Cain, 1982). Consequently, odor identification tests developed for clinical use have employed multiple-choice formats, within which persons with normal smell can show almost perfect performance (Cain et al., 1983, 1988; Doty et al., 1984a). A measurement of the confusion among odorants has also been suggested as a clinical assessment method, particularly for patients suffering from dysosmia (Wright, 1987).

Another method of assessing olfactory ability is to ask questions about a patient's detection or recognition of very weak odors, that is, to determine detection or recognition thresholds (Cain et al., 1983, 1988; Doty et al., 1986; Henkin, 1981). Threshold procedures have seen wide use in clinical studies of taste and smell (Ghorbanian et al., 1983; Henkin, 1967b; Henkin et al., 1968; Henkin and Powell, 1962; Henkin and Smith, 1971; Hertz et al., 1975; Mozell et al., 1986b; Ophir et al., 1986; Weiffenbach and McCarthy, 1984). An assessment of a patient's sensory threshold may provide information different from that obtained using suprathreshold (i.e., magnitude scaling) procedures (Bartoshuk, 1978; Bartoshuk and Marks, 1986). For this reason, threshold procedures alone are not sufficient for chemosensory assessment, which should also include suprathreshold methods.

Three tests of olfactory function that include measurements of threshold and odor identification are in common use for the clinical assessment of olfactory deficits. These tests, developed at the Connecticut Chemosensory Clinical Research Center (Cain et al., 1983, 1988) and the University of Pennsylvania Smell and Taste Center (Doty et al., 1984a), provide a reliable and relatively simple means of quantifying a patient's degree of olfactory functioning. The Connecticut (UCONN) test consists of two components, butanol threshold and odor identification. Butanol threshold (Cain et al., 1983, 1988) is assessed with a forced-choice procedure by presenting the patient with an aqueous concentration of butanol in one sniff bottle and water in another and asking the patient to identify the bottle containing the odorant. The concentration of butanol is increased after each incorrect response until there are five correct responses in a row or until the patient fails to correctly identify a bottle containing 4 percent butanol. Each nostril is tested separately. The concentration step at which the patient correctly identifies the butanol on five consecutive trials is taken as the detection threshold. This step is used as a function score that relates the patient's threshold to a normal subject population.

The odor identification test (Cain and Gent, 1986; Cain et al., 1983, 1988) consists of 10 items, presented separately to each nostril in opaque jars. These items include seven odorants (baby powder, chocolate, cinnamon, coffee, mothballs, peanut butter, and soap) and three trigeminal stimulants. The patient is given a list of 20 items, containing the 10 stimuli and 10 additional distractor names, from which to select a response. Feedback is given to the patient after every trial. If the patient is incorrect, a second chance is given to correctly identify the item. The number of odorants correctly identified is used as a function score that relates the patient's performance to that of a normal control group. The performance of a patient on each of these tests is then combined in a composite function score by averaging the scores for butanol threshold and odor identification. Patients with UCONN composite scores less than 2 are considered anosmic, those between 2 and 5.75 are hyposmic, and those with 6 or greater are normosmic (Cain et al., 1988).

In addition to these olfactory tests, we also routinely administer the University of Pennsylvania Smell Identification Test (UPSIT), which is composed of 40 microencapsulated odors (Doty et al., 1984a). Each item of the UPSIT is presented on a "scratch-and-sniff" pad, accompanied by four response alternatives. The test is essentially self-administered, with the patient instructed to guess if unable to identify the item. Anosmic patients score at or near chance on this test, with a mean of close to 10 correct out

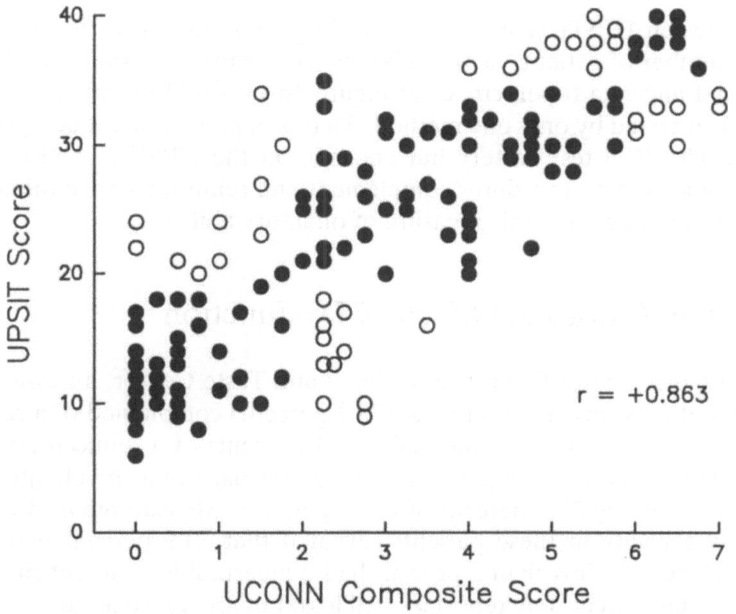

FIGURE 16.1. Scatter plot of the relationship between scores on different olfactory tests ($n = 176$). Scores on the University of Pennsylvania Smell Identification Test (UPSIT) are shown on the ordinate, with 0–19 indicating anosmia, 20–34 hyposmia, and 35 and above normosmia. Composite scores from the butanol threshold and odor identification components of the University of Connecticut (UCONN) test battery are depicted along the abscissa, with 0–1.75 indicative of anosmia, 2–5.75 hyposmia, and scores of 6 or greater indicative of normosmia. Open circles represent patients for whom a diagnosis on one test would be different from that on the other (e.g., normosmic on the UPSIT but hyposmic on the UCONN, or vice versa). Solid circles represent patients for whom the diagnoses agree. The Pearson product moment correlation between these two sets of scores is shown in the lower right of the figure.

of 40. Scores on this test have been obtained for normal individuals across a wide range of ages, and both sex- and age-related norms have been determined. This test has been shown to have excellent short-term ($r = +0.95$) and long-term ($r = +0.92$) test-retest reliability (Doty et al., 1985). Scores below 34 or 35 (depending upon gender) are indicative of hyposmia, and those below 20, of anosmia. Individuals feigning anosmia tend to score much lower on the UPSIT than would be expected by chance, so that extremely low scores are indicative of possible malingering (Doty et al., 1984a).

There is a good correlation between scores on the UPSIT and the composite score of butanol threshold plus odor identification across patients differing widely in olfactory ability (Smith, 1988). A scatter plot of the scores of 176 patients on these tests is shown in Figure 16.1. The correlation between UPSIT scores and UCONN composite scores is +0.863. A small number of patients near the boundaries between normosmia, hyposmia, and anosmia (open circles in Figure 16.1) would be categorized differently if tested by only one method. That is, a patient might be hyposmic on the UCONN test battery but anosmic on the UPSIT, or vice versa. These assessment procedures complement and reinforce one another, and together provide a reliable measure of olfactory ability.

Frequent Causes of Olfactory Dysfunction

At the University of Pennsylvania Smell and Taste Center, an assessment of 750 patients revealed that most (78.1 percent) complained of a reduced ability to smell and more than half (66.4 percent) of a reduced ability to taste (Deems et al., 1990). Dysosmias, either parosmias or phantosmias, were reported by 32.1 percent of these patients. Measurement of chemosensory capacity in these patients revealed that 70.9 percent had olfactory deficits, but less than 3 percent had a measurable taste deficit. Thus, patients quite accurately report a smell loss, but are not so accurate in their reports of taste dysfunction (see also Gent et al., 1987). This discrepancy reflects the common confusion between taste and flavor. Similar results have been reported on a smaller number of patients at the Connecticut Chemosensory Clinical Research Center (Goodspeed et al., 1987) and at the University of Cincinnati Taste and Smell Center (Smith et al., 1987).

Olfactory dysfunction has been associated with a number of systemic diseases and metabolic disorders (Doty, 1979; Doty and Kimmelman, 1986; Doty and Snow, 1987; Estrum and Renner, 1987; Feldman et al., 1986; Henkin, 1981; Kimmelman, 1986; Leopold, 1986; Schiffman, 1983a, 1983b). However, most patients presenting with a primary smell complaint fall into only a few etiologic categories. The distribution of 169 patients with olfactory disturbances seen at the University of Cincinnati Taste and Smell Center is shown in Figure 16.2. This figure indicates the proportion of patients falling into each of several etiologies. Patients with a primary olfactory complaint most likely fall into one of four etiologic categories: (1) nasal and/or sinus disease (i.e., nasal polyposis, chronic sinusitis, allergic rhinitis, etc.), (2) prior upper respiratory infection (i.e., a history of a viral-like upper respiratory illness just prior to the onset of the olfactory loss), (3) head trauma, or (4) idiopathic. These four categories account for about 82 percent of these 169 patients (Figure 16.2). At other centers, these same categories account for 81 percent (Deems et al., 1990), 83 percent (Goodspeed et al., 1987), or 72 percent (Henkin, 1981) of olfactory patients. The terminology in the literature is somewhat different, with Henkin's (1981) postinfluenza hyposmia and hypogeusia (PIHH) and allergic rhinitis categories corresponding to what is referred to here as prior upper respiratory infection and nasal and/or sinus disease, respectively. In addition to these

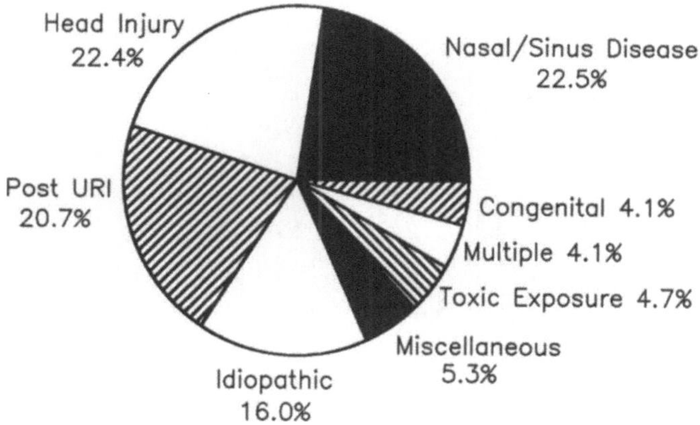

FIGURE 16.2. Most frequent etiologies for olfactory impairment among 169 patients seen at the University of Cincinnati Taste and Smell Center. Four etiologic categories (nasal/sinus disease, head injury, post URI, and idiopathic) account for 81.6 percent of the patients. See text for further details.

four categories, smaller numbers of patients have olfactory impairment due to exposure to toxic chemicals (4.7 percent), a congenital anomaly (4.1 percent), multiple causes (4.1 percent), or several miscellaneous causes (5.3 percent), such as aging, craniofacial surgery, seizure disorders, cerebral vascular accidents, or endocrine disorders (see also Deems et al., 1990; Feldman et al., 1986; Goodspeed et al., 1987; Henkin, 1981). Relatively few patients present with a primary chemosensory complaint resulting from the many widespread systemic or metabolic diseases (e.g., see Schiffman, 1983a, 1983b) that have been associated with taste or smell dysfunction (Goodspeed et al., 1987).

Nasal and/or Sinus Disease

Olfactory disorders can be characterized as (1) conductive disorders, arising from interference with the access of odorants to the olfactory receptors, or (2) sensorineural disorders, resulting from damage to the receptors or olfactory pathways (Doty and Snow, 1987). Unfortunately, olfactory testing alone is not capable at present of distinguishing between these possibilities. Of the four major etiologic categories, only nasal and/or sinus disease typically involves a conductive loss, resulting from the restriction of upper airway patency. This obstruction may arise from several causes, including intranasal polyposis, chronic sinusitis, or allergic rhinitis. A schematic diagram of the paranasal sinuses is shown in Figure 16.3. On the left is an indication of the kind of inflammation that can lead to a conductive olfactory loss, and on the right is shown a patent nasal vault following endoscopic sinus surgery (Seiden and Smith, 1988). The loss of smell resulting from

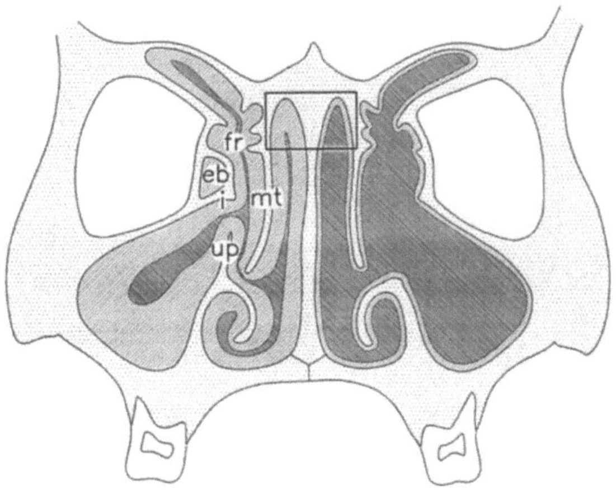

FIGURE 16.3. Schematic drawing (coronal) of the nasal cavity and paranasal sinuses. The left side is typical of chronic rhinitis, with pansinus obstruction. Airflow to the nasal vault (and olfactory epithelium) is obstructed by congestion of the ostiomeatal complex (i, infundibulum; up, uncinate process; eb, ethmoidal bulla; fr, frontal recess; mt, middle turbinate). The right side shows the situation following endoscopic surgery, which opens up the middle meatus, allowing freer airflow to the olfactory mucosa. The superimposed rectangle indicates the general location of the olfactory epithelium, which is actually posterior to this plane of section, on the upper lateral wall of the nasal septum, on the cribriform plate, and on the medial wall of the superior turbinate.

nasal and/or sinus disease is usually quite severe, with most patients characterized as anosmic rather than hyposmic (Goodspeed et al., 1986a, 1987; Smith et al., 1987). Even a moderate degree of ostiomeatal disease without accompanying polyps can result in significant olfactory impairment (Seiden and Smith, 1988). A coronal CT of a patient with hyposmia resulting from a mild amount of allergic disease in the ostiomeatal complex is shown in Figure 16.4. In this case, the airway was patent, but there was significant blockage of airflow to the upper nasal vault, which prevented odorant access to the olfactory epithelium. In a study of patients with allergic rhinitis, about a third of those with IgE-mediated nasal allergy were shown to have a measurable olfactory impairment (Seiden et al., 1989), with the deficit being worse in those patients with nasal polyps.

That a conductive loss is involved in this etiology is demonstrated by the fact that patients with nasal and/or sinus disease are much more likely to verbally report fluctuations in smell sensitivity than those with loss due to a prior upper respiratory illness (Gent et al., 1987). That is, these patients sometimes have a temporary return in their sense of smell, often associated with exercise or with medications. Anosmic patients with allergic rhinitis and/or nasal polyposis have been shown to recover olfactory function following administration of systemic corticosteroids (Fein et al., 1966; Goodspeed et al., 1986b; Hotchkiss, 1956) or after careful intranasal administra-

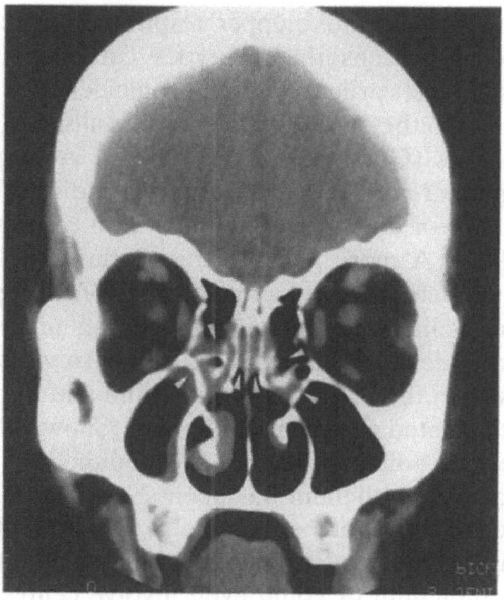

FIGURE 16.4. Coronal CT through the paranasal sinuses of a patient with allergic rhinitis resulting in hyposmia. This patient had undergone a septoplasty, partial inferior turbinectomy, and partial middle turbinectomy three years earlier. However, smell impairment persisted as a result of inflammatory disease of the ostiomeatal complex bilaterally (arrows). Even a mild degree of ostiomeatal disease of the sort shown here, without accompanying nasal polyps, will often produce significant smell loss (Seiden and Smith, 1988). This inflammation is enough to disrupt airflow to the olfactory epithelium even though it does not impair the patient's breathing.

tion of topical corticosteroids (Scott et al., 1988, 1989). Although chronic administration of systemic steroids is not a recommended treatment for anosmia, a short course of steroid therapy may help in diagnosing nasal disease as the cause of an olfactory loss (Leopold, 1986). Surgical procedures that reduce nasal obstruction have also been shown to produce marked improvement in olfactory sensitivity (Ghorbanian et al., 1983; Ophir et al., 1986). Endoscopic ethmoidectomy to alleviate allergic disease in the ostiomeatal complex has been shown to restore olfactory sensitivity (Seiden and Smith, 1988), with some patients maintaining this improvement for as long as two to three years, even without concomitant steroids or immunotherapy. Thus, olfactory losses resulting from an interference with odor access to the receptors, i.e., conductive losses, are often amenable to treatment (Doty and Snow, 1987).

Prior Upper Respiratory Infection

Olfactory losses with sensorineural involvement are presently more difficult to manage than those resulting from conductive disorders. Anosmia or hyposmia following a viral-like upper respiratory infection (Henkin et al., 1975) may be of the sensorineural type. The diagnosis of post-URI anosmia or hyposmia rests primarily on the coincidence of a viral illness or severe cold just prior to the onset of olfactory loss along with an absence of other etiologic factors (Goodspeed et al., 1987). A biopsy of olfactory tissue from one patient with post-URI anosmia indicated extensive scarring and replacement of olfactory epithelium with respiratory epithelium (Douek et al., 1975). A more recent ultrastructural study of the human olfactory epithelium demonstrated a loss of olfactory receptor cells and an absence of olfactory cilia on the remaining receptors in patients with viral-induced smell loss (Jafek and Eller, 1989). Influenza viruses are known to affect ciliary activity in the respiratory epithelium and to produce necrosis of pseudostratified ciliated columnar epithelium (Snow, 1969); thus similar processes may be acting on the olfactory neuroepithelium (Doty, 1979). In addition, the olfactory epithelium has been shown to be a route of viral invasion of the central nervous system (e.g., see Monath et al., 1983). Thus, the long-lasting effects of a viral illness may be the result of damage to the olfactory epithelium or could even involve central olfactory pathways.

In comparison to patients with nasal and/or sinus disease, those with smell dysfunction following an upper respiratory infection are more likely to be women, to be older, and to be hyposmic rather than anosmic (Goodspeed et al., 1987). Patients with these etiologies (nasal disease or post-URI) often report parosmias or phantosmias, almost always of an unpleasant nature (Donnelly et al., 1989; Smith et al., 1987). Not only is the mechanism of viral induced anosmia poorly understood, but the prognosis for this disorder is completely uncertain. It has been suggested that post-URI anosmia may show spontaneous remission (Goodspeed et al., 1986a). It has also been suggested that these patients are unlikely to recover (Marshall and Attia, 1987; Deems et al., 1991). Anecdotal reports suggest that recovery may be possible, and in a recent followup study (Duncan, et al., 1991) we have shown that 17 of 18 patients with post URI-hyposmia have improved olfactory ability after 1 to 5 years. Given the relative frequency of this etiology in the production of olfactory loss, further research on the mechanisms and time course of post-URI anosmia is sorely needed.

Head Trauma

About 5 percent of patients with head injury suffer a resulting olfactory loss (Leigh, 1943; Sumner, 1964; Zusho, 1982; see review by Costanzo and Becker, 1986), and these patients make up about 10 to 22 percent of those presenting with a primary olfactory complaint (Goodspeed et al., 1987; Henkin, 1981; see Figure 16.2; see also the chapter by Costanzo et al. in this volume). The loss of smell sensitivity following head trauma is sometimes accompanied by dysosmia, involving unpleasant smell sensations

(Costanzo and Becker, 1986; Donnelly et al., 1989; Leigh, 1943; Schechter and Henkin, 1974). As in patients with nasal and/or sinus disease, olfactory impairment in head injury patients is more likely to be severe, with a much greater proportion (about 80 percent) showing anosmia than hyposmia (Goodspeed et al., 1986a; Smith et al., 1987). Posttraumatic anosmia is commonly thought to result from shearing of the olfactory nerve filaments or from contusions following a sudden blow to the head (Costanzo and Becker, 1986). The olfactory epithelium in patients with post-traumatic smell loss exhibits marked degenerative changes suggestive of a severed olfactory nerve (Hasegawa et al., 1986; Moran et al., 1985). Intracranial hemmorhage and cerebral ischemia have also been shown in laboratory animals to lead to degeneration of the olfactory epithelium, even though the olfactory nerve is intact (Nakashima et al., 1983, 1984). Thus, a number of mechanisms may be involved in olfactory dysfunction following head trauma.

Since the olfactory system is remarkable in its ability to regenerate (Graziadei, 1973; Graziadei and Monti-Graziadei, 1983; see also the chapter by Morrison and Costanzo in this volume), there is potential for recovery of olfactory function after head injury. Indeed, there is typically recovery of the neuroepithelium in laboratory animals following olfactory axotomy (Monti-Graziadei and Graziadei, 1979; Monti-Graziadei et al., 1980; Simmons and Getchell, 1981). Recordings from cells in the hamster olfactory bulb demonstrate functional recovery within nine months following olfactory nerve transection (Costanzo, 1985). However, the prognosis for the recovery of smell in humans after traumatic injury is generally poor, with various estimates of 15 percent (Zusho, 1982), 33 percent (Costanzo and Becker, 1986), and 39 percent (Sumner, 1964). The time course for recovery is also somewhat questionable, with about 80 percent of recovery occurring within 6 months in one study (Sumner, 1964) but only after about 20 months in another (Costanzo and Becker, 1986). These differences may reflect the way in which olfactory function was measured in these studies and the criteria for recovery. Nonetheless, recovery from posttraumatic anosmia in humans may proceed over a period as long as five years (Costanzo and Becker, 1986; Sumner, 1964). It has been suggested that early recovery may be due to mechanisms such as the disappearance of blood clots or edema and that later recovery may reflect the regeneration of neural elements (Sumner, 1975). Further research is needed to delineate these mechanisms. Some patients may experience parosmia or phantosmia during the recovery process (Goodspeed et al., 1986a; Leigh, 1943), although the mechanisms for these phenomena are unknown. It has been suggested that the occurrence of parosmia may reflect a system that is minimally functional (see Wright, 1987). Consistent with this hypothesis, we have determined that head injury patients reporting parosmia or phantosmia score significantly higher on the UPSIT ($\bar{X} = 18.2$) than those without these symptoms ($\bar{X} = 13.7$; two-tailed t test, $t = 2.19$, df $= 36$; $p = 0.035$).

Idiopathic Smell Dysfunction

A large group of patients with a primary olfactory complaint do not fall easily into any diagnostic category, and the cause of their smell loss remains unknown. These patients have been termed idiopathic and comprise 16 percent (Figure 16.2), 19 percent (Henkin, 1981), and 26 percent (Goodspeed et al., 1987) of patients with olfactory dysfunction in different studies. Although a specific syndrome of idiopathic hypogeusia and hyposmia, with or without dysgeusia and dysosmia, has been described (Henkin et al., 1971), this categorization appears to reflect patients for which no specific etiology can be established (Goodspeed et al., 1987). Many of these patients may have olfactory symptoms as a result of one of a vast number of systemic diseases or neurologic or endocrine disorders (Doty, 1979; Feldman et al., 1986; Jafek, 1982; Kimmelman, 1986; Schiffman, 1983a, 1983b) that are not being detected during the olfactory evaluation. As our ability to provide a differential diagnosis for the cause of olfactory dysfunction improves, the proportion of patients with idiopathic olfactory disturbances will probably decrease (see Feldman et al., 1986; Gent et al., 1987; Liston, 1984).

Other Etiologies

Most patients (about 80 percent) presenting with a primary olfactory complaint will fall into one of the four etiologic categories described above. The remaining patients are diagnosed with one of several other causes. For example, in the study reported by Goodspeed et al. (1987), an additional 4.5 percent of their patients were considered to have multiple causes, such as a patient with a temporally related head trauma who also experienced a viral-like upper respiratory infection at about the same time. For these patients, the establishment of one definitive cause for their olfactory loss was impossible. In several studies, small numbers of patients (5 to 14 percent) have been grouped into a miscellaneous category, which reflects such diverse etiologies as seizure disorders, cerebrovascular accidents, brain surgery, endocrine disorders, and depression (Goodspeed et al., 1987; Henkin, 1981; Smith et al., 1987; Figure 16.2).

A few patients (1 to 5 percent) present with a primary olfactory complaint stemming from exposure to toxic chemicals (Deems et al., 1991; Goodspeed et al., 1987; Henkin, 1981; Smith et al., 1987; Figure 16.2). A number of environmental and industrial agents have been implicated in anosmia or hyposmia, including benzene, benzol, butyl acetate, carbon disulfide, ethyl acetate, formaldehyde, hydrazine, menthol, methyl bromide, paint solvents, oil of peppermint, radiation, trichloroethylene, and a variety of industrial dusts such as cadmium and nickel (see Amoore, 1986; Doty, 1979; and Halpern, 1985, for reviews of this literature). Many of the studies implicating these various chemicals in olfactory dysfunction suffer from methodological problems, so it is hard to draw any general conclusions from these data (Doty, 1979). However, in one well-controlled study, 27.3 percent of workers in an alkaline battery plant who were exposed to high levels of cadmium were found to be hyposmic or anosmic, compared to 4.8 percent of control subjects from a neighboring nonchemical factory (Adams and Crabtree, 1961). Recent investigation of chemical workers exposed to acrylate and methacrylate vapors shows a dose-response relationship between olfactory dysfunction and cumulative exposure (Schwartz et al., 1989). Animal studies have shown degeneration of olfactory epithelium following exposure of rats and mice to methyl methacrylate vapors (Chan et al., 1988). It has also been shown that a 1 percent solution of zinc sulphate will destroy the receptor cells and supporting cells of the olfactory epithelium of mice (Cancalon, 1982). Recovery of the receptor cells and olfactory cilia is seen four to six weeks later. Little is known, however, about recovery following industrial exposure.

Cigarette smoke has been shown to alter the ultrastructure of the olfactory epithelium in certain strains of mice, resulting in a number of changes, including reduced or absent olfactory cilia (Matulionis, 1974). Although there has been relatively little work in this area, there is some suggestion that smoking results in decreased olfactory sensitivity (Ahlström et al., 1987; Hubert et al., 1980; Thumfart et al., 1980; Venstrom and Amoore, 1968). However, other investigators have found no effect of smoking on olfactory function, but have shown a clear reduction in the perception of nasal pungency, mediated through the trigeminal system (Cain, 1981; Cometto-Muniz and Cain, 1982; Dunn et al., 1982). More recent data show that when smoking history is taken into account, there is a significant long-term but reversible effect of smoking on the ability to smell (Frye et al., 1990). The cumulative effects of smoking on odor identification (UPSIT scores) and the recovery after smoking cessation are shown in Figure 16.5. Earlier studies, which simply dichotomized subjects according to current smoking behavior, may have missed seeing the cumulative effects of smoking on olfactory function.

There are many studies suggesting a decline in chemosensory ability in humans with increasing age (Doty et al., 1984b; Doty and Snow, 1988; Leopold et al., 1989; Murphy, 1986; Schiffmann, 1979; Stevens et al., 1984; Van Toller et al., 1985; Weiffenbach et al., 1986; see also the chapter by Schiffman in this volume). Olfactory thresholds have been shown to decrease in old age (Cain et al., 1988; Deems and Doty, 1987; Murphy, 1983; Thumfart et al., 1980; Venstrum and Amoore, 1968). Van Toller and colleagues (1985) have shown significant declines in olfactory thresholds with age across ten different odorants for individuals between the ages of 20 and 80 years. In addition to changes in detectability, the perception of suprathreshold odors also appears to decline with increasing age. This decline is clearly evident in estimates of perceived odor intensity measured with magnitude matching procedures (Stevens and Cain, 1985; Stevens et al.,

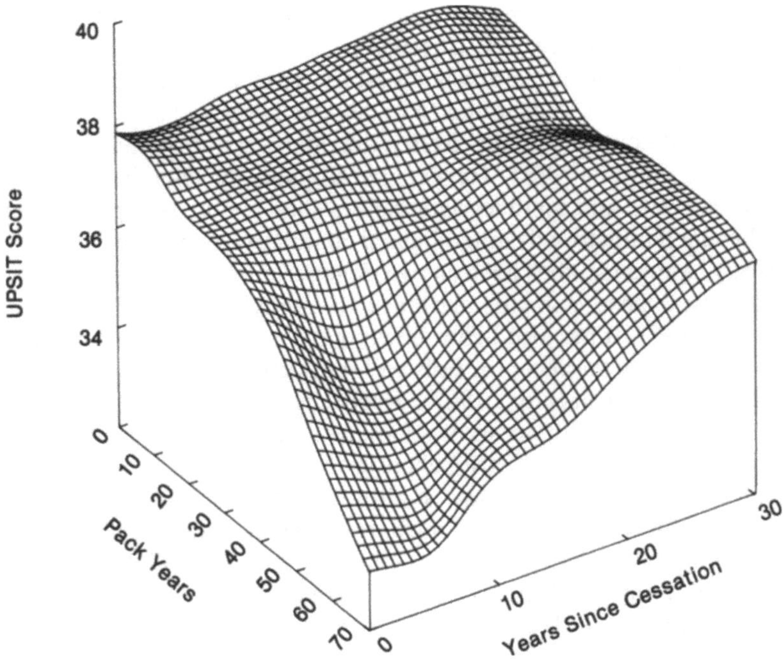

FIGURE 16.5. Effect of cumulative smoking dose (pack years) and years since cessation from smoking on olfactory identification, measured with the University of Pennsylvania Smell Identification Test (UPSIT). The decline in UPSIT scores with increasing smoking exposure is offset by the length of time since smoking ceased. (Reprinted, by permission, from R.E. Frye, B.S. Schwartz, and R.L. Doty Copyright 1990 by the American Medical Association.)

1982, 1984), as well as in the ability to identify odors (Doty et al., 1984b; Murphy, 1986). Although females are slightly better at identifying odors than are males at all ages, there is a sharp decline in the performance of both sexes after the age of about 69 years (Doty et al., 1984b), which is shown in Figure 16.6. Also evident in these data is an increase in the variance of the scores at older ages, which suggests that some older individuals show considerably more impairment than others. However, after the age of 80 years, more than 80 percent of the persons tested showed major olfactory impairment, with nearly 50 percent being anosmic (Doty et al., 1984b). This impairment in odor identification parallels that seen using threshold techniques (Cain et al., 1988; Deems and Doty, 1987). Thus, both the absolute sensitivity to odors and the appreciation of strong suprathreshold odors show a decrement with increasing age. These data help to explain why many elderly persons complain about a lack of flavor in their food (Cohen and Gitman, 1959) and have difficulty detecting life-threatening fires and gas leaks (Cain and Turk, 1985; Doty et al., 1984b).

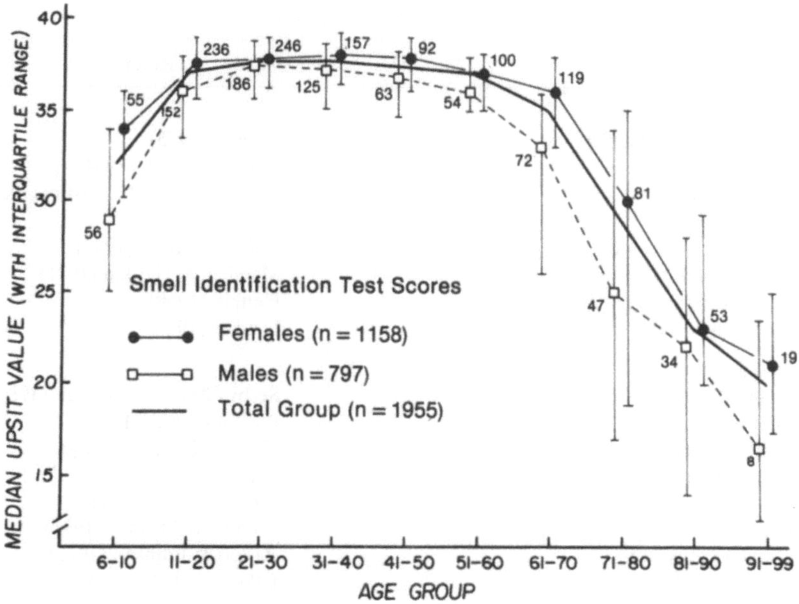

FIGURE 16.6. University of Pennsylvania Smell Identification Test (UPSIT) scores as a function of age and gender. Numbers by the data points indicate sample sizes, and the error bars show the interquartile range. (Reprinted, by permission, from R.L. Doty, P. Shaman, S.L. Applebaum, R. Giberson, L. Siksorski, and L. Rosenberg, Copyright 1984 by the American Association for the Advancement of Science.)

Smell Dysfunction Secondary to Other Diseases

Disorders of smell have been associated with a number of conditions, including endocrine, neurological psychiatric, and nutritional disorders (Doty, 1979; Doty and Kimmelman, 1986; Doty and Snow, 1987; Estrum and Renner, 1987; Feldman et al., 1986; Henkin, 1981; Kimmelman, 1986; Leopold, 1986; Schiffman, 1983a, 1983b; Smith, 1990). Although the occurrence of these disorders in patients presenting with a primary smell complaint is rare, these conditions must always be considered in the evaluation of such a patient. Disorders that have been reported to affect the sense of smell are listed in Table 16.1, which categorizes them into endocrine disorders, neurological disorders, nutritional disorders, local diseases or mechanical obstruction, psychiatric disorders, tumors, and infectious processes. Listed with each of these conditions is a reference, either a research study demonstrating the association of that particular factor with an olfactory disturbance or a review article covering a wide range of studies. The article by Amoore (1986) on the effects of chemical exposure on olfaction is an example of the latter. Some of these conditions were discussed

TABLE 16.1. Disorders associated with olfactory dysfunction

Disorder	Reference
Endocrine	
Adrenal cortical insufficiency	Henkin (1975)
Cushing's syndrome	Kallmann et al. (1944)
Cystic fibrosis	Weiffenbach & McCarthy (1984)
Diabetes mellitus	Settle (1986)
Kallmann's syndrome	Kallmann et al. (1944)
Primary amenorrhea	Marshall & Henkin (1971)
Pseudohypoparathyroidism	Henkin (1968)
Turner's syndrome	Henkin (1967b)
Neurologic	
Alzheimer's disease	Serby (1986), Doty et al. (1987)
Epilepsy	Eskenazi et al. (1986)
Familial dysautonomia	Henkin & Kopin (1964)
Head trauma	Sumner (1964)
Huntington's chorea	Moberg et al. (1984)
Multiple sclerosis	Wender & Szemza (1971)
Parkinson's disease	Ansari & Johnson (1975)
Temporal lobectomy	Eskenazi et al. (1983)
Nutritional	
Chronic renal failure	Schiffman et al. (1978)
Cirrhosis of the liver	Burch et al. (1978)
Cyanocobalamin (B12) deficiency	Rundles (1946)
Korsakoff's psychosis	Jones et al. (1975)
Local diseases/mechanical obstruction	
Adenoid hypertrophy	Ghorbanian et al. (1983)
Allergic rhinitis	Fein et al. (1966)
Atrophic rhinitis (ozena)	Strandbygard (1954)
Bronchial asthma	Fein et al. (1966)
Deformity secondary to trauma	Doty & Kimmelman (1986)
Exposure to toxic chemicals	Amoore (1986)
Laryngectomy	Mozell et al. (1986b)
Leprosy	Barton (1974)
Malignancy of paranasal sinuses with extension	Doty & Kimmelman (1986)
Nasal polyposis	Fein et al. (1966)
Nasal surgery	Champion (1966)
Sinusitis	Ryan & Ryan (1974)
Sjögren's syndrome	Henkin et al. (1972)
Tumors of nasopharynx with extension	Doty & Kimmelman (1986)
Vasomotor rhinitis	Griffith (1976)
Psychiatric	
Depression	Pryse-Phillips (1971)
Olfactory reference syndrome	Pryse-Phillips (1971)
Schizophrenia	Kerekovic (1972)

TABLE 16.1. (*cont.*)

Disorder	Reference
Intracranial tumors	
Aneurysms of the anterior communicating bifurcation	Jefferson (1961)
Frontal lobe glioma	Elsberg (1935)
Olfactory groove meningioma	Bakay & Cares (1972)
Suprasellar meningioma	Elsberg (1935)
Temporal lobe tumors	Furstenberg et al. (1943)
Intranasal tumors	
Adenocarcinoma	Skolnik et al. (1966)
Esthesioneuroepithelioma	Takahashi et al. (1987)
Inverted papilloma	Skolnik et al. (1966)
Melanoma	Skolnik et al. (1966)
Neuroblastoma	Joachims et al. (1975)
Squamous-cell carcinoma	Skolnik et al. (1966)
Viral and infectious	
Acute viral hepatitis	Henkin & Smith (1971)
Herpes simplex	Twomey et al. (1979)
Influenza-like infections	Henkin et al. (1975)

above as common causes of olfactory impairment. Others represent relationships described in the literature but uncommon in patients presenting primarily with smell symptoms (Goodspeed et al., 1987). Because many of these studies have measured only thresholds, which alone do not provide a complete picture of a patient's sensory impairment (Bartoshuk, 1978; Bartoshuk and Marks, 1986), these relationships must be evaluated with caution. Some of the earlier studies involved only a patient's subjective report of smell symptoms and are even less reliable than those reporting the results of threshold testing. In some cases, such as in cystic fibrosis, for example, earlier reports (Henkin and Powell, 1962) have been contradicted by more recent studies (Hertz et al., 1975; Weiffenbach and McCarthy, 1984).

A number of endocrine disorders have been reported to influence smell function. For example, olfactory impairment is seen in several kinds of gonadal dysfunction. Patients with hypogonadotropic hypogonadism (Kallmann's syndrome) often exhibit a congenital anosmia (Kallmann et al., 1944), which has been shown to follow an autosomal dominant mode of inheritance with incomplete expressivity (Santen and Paulsen, 1973). Although Kallmann's syndrome usually occurs in men, some women with hypogonadotropic hypogonadism have also been described as anosmic (Tagatz et al., 1970), as have those suffering from primary amenorrhea (Marshall and Henkin, 1971). Patients with chromatin-negative gonadal dysgenesis (Turner's syndrome) have been reported to have smell impairment (Henkin, 1967b). Changes in smell sensitivity also occur during the human menstrual cycle, with peaks in sensitivity at midcycle, during the midluteal phase, and sometimes in the latter part of the menses (Doty et al., 1981). Since these changes are also seen in women on oral contraceptives, it is not clear whether these fluctuations are modulated by gonadal hormones or are solely under the control of central neural factors (Doty, 1979; Schneider et al., 1958).

Olfactory dysfunctions are sometimes observed in patients with neurologic impairment. Head trauma was discussed above as one of the major etiologic factors in patients with primary olfactory complaints. In addition, when neurological disease involves peripheral or central olfactory pathways, smell function can be compromised. The demyelinating effects of multiple sclerosis have been shown to produce deficits in taste and perhaps in smell (Catalanotto et al., 1986; Cohen, 1965; Doty et al., 1984a; Pinching, 1977; Wender and Szemza, 1971), although there are questions regarding these olfactory deficits. Early work in this area is conflicting, with some studies (measuring odor identification) suggesting olfactory impairment (Wender and Szemza, 1971) and others (measuring detection thresholds) no deficit at all (Ansari, 1976). These differences may reflect the effects of multiple sclerosis on the higher cognitive functions involved in odor identification (Pinching, 1977), although recent work with the UPSIT (Doty et al., 1984a) shows only a small number of multiple sclerosis patients with smell impairment. Thus, smell may be compromised in multiple sclerosis only in some patients, probably depending upon the particular distribution of the disease.

Recent work has shown that a number of central neural disorders seem to involve olfactory deficits, including epilepsy (Eskenazi et al., 1986), Parkinson's disease (Ansari and Johnson, 1975; Corwin et al., 1985; Doty et al., 1988; Quinn et al., 1987; Ward et al., 1983), and Alzheimer's disease (Corwin et al., 1985; Doty et al., 1987; Serby, 1986; see also the chapter by Serby in this volume). Patients with temporal lobe epilepsy are slightly deficient in odor identification and odor recognition memory, and these deficiencies are exacerbated by temporal lobectomy, with greater impairment for the nostril ipsilateral to the lesion (Eskenazi et al., 1986). There does not appear to be any difference between normal controls and epileptic patients in butanol threshold (Eskenazi et al., 1986). Since these patients were also deficient in recognition memory for amorphous shapes, these impairments may have more to do with odor memory processing than with a sensory impairment.

Several studies suggest that patients with Parkinson's disease show deficits in olfactory ability. Early work by Ansari and Johnson (1975) demonstrated that 10 out of 22 Parkinsonian patients had increased thresholds for amyl acetate. This deficit was confirmed by Ward et al. (1983) using an odor identification task, in which 49 percent of 72 patients were unable to identify the odor of coffee and 35 percent the odor of cinnamon. Parkinsonian patients who scored in the normal range on a picture identification test similar to the UPSIT in format were tested for their detection and recognition of odors (Doty et al., 1988). Of these 81 patients, 63 had scores on the UPSIT below the 25th percentile for their age and sex. These patients also showed marked impairment on a phenyl ethyl alcohol detection threshold task. Olfactory thresholds have been shown to be unrelated to pharmacological manipulation of dopaminergic or cholinergic status in Parkinsonian patients (Quinn et al., 1987), suggesting little or no role for these neurotransmitters in the olfactory impairment.

Finally, it has recently been established that Alzheimer's disease is associated with olfactory deficits in both odor identification and threshold (Doty et al., 1987; Serby, 1986). Relative to age-, gender-, and race-matched controls, patients with well-defined Alzheimer's disease were deficient on the UPSIT and in phenyl ethyl alcohol threshold (Doty et al., 1987). All of these patients scored in the normal range on the picture identification test, which demonstrates that they had the cognitive ability to take the UPSIT. Only 2 of the 34 Alzheimer's patients were aware of an olfactory impairment before testing. The olfactory deficits seen in Alzheimer's disease are probably related to the existence of neuritic plaques and neurofibrillary tangles throughout olfactory-related brain structures (Esiri and Wilcock, 1984; Pearson et al., 1985). It has been reported that Alzheimer's disease is accompanied by phosphorylated neurofilament proteins in the olfactory epithelium (Talamo et al., 1989). However, in this study there was a significant difference (two-tailed t test, $t = 3.736$, $df = 21$, $p < 0.01$) in the distributions of ages between the Alzheimer's ($\bar{X} = 83.3$ years) and non-Alzheimer's ($\bar{X} = 60.4$ years) cases. Individuals of the age of the Alzheimer's group would very probably show olfactory deficits as a result of normal aging (see Figure 16.6 above). Thus, the specific role of these pathological changes in the olfactory epithelium requires further study. Other dementia-related disorders that result in lesions in the olfactory pathways are also accompanied by olfactory deficits, including Huntington's chorea (Moberg et al., 1984) and Korsakoff's psychosis (Jones et al., 1975; Mair et al., 1986). It has even been proposed that the olfactory system could be the route of invasion of etiologic factors responsible for some of these diseases (Shipley, 1985).

Several kinds of nutritional deficiencies have been implicated in smell dysfunction. Disturbances of taste or smell in malnutrition or in other systemic conditions, such as pellagra or pernicious anemia, have been attributed to vitamin or trace metal imbalances (Green, 1971; Rundles, 1946). Early work by Henkin and his colleagues (Henkin et al., 1967; Henkin and Bradley, 1969; Schecter et al., 1972) suggested that reduced serum copper levels resulted in depressed taste sensitivity, which could be reversed by $CuSO_4$ or $ZnSO_4$ administration. As a result, it has become relatively standard practice to prescribe zinc for taste and even for smell deficiencies, regardless of their etiologies (see Estrum and Renner, 1987). However, there is little evidence that zinc therapy is an effective treatment for these conditions (Price, 1986). Double-blind clinical trials using a crossover design, in which zinc was compared to a placebo, showed no significant effects of zinc in the treatment of hypogeusia (Henkin et al., 1976). In cases of zinc deficiency due to renal disease (Atkin-Thor et al., 1978; Mahajan et al., 1980) and hepatic cirrhosis (Weisman et al., 1979), double-blind studies have shown that decreased taste acuity can be restored by zinc treatment. Thus, when hypogeusia is accompanied by zinc deficiency, correction of that deficiency often restores taste sensitivity. However, there is little probability that any particular case of taste or smell dysfunction is the result of zinc deficiency or that it would be helped by zinc therapy (Price, 1986).

Many of the remaining disorders listed in Table 16.1 represent factors discussed above as common causes of smell dysfunction, such as allergic rhinitis, influenza-like infections, sinusitis, and exposure to toxic chemicals. A few additional disorders, such as various intranasal or intracranial tumors, are well known as potential causes of olfactory loss or distortion (Doty, 1979; Feldman et al., 1986; Kimmelman, 1986; Leopold, 1986), although their occurrence in patients presenting only with olfactory symptoms is relatively rare (Goodspeed et al., 1986a, 1987). Similarly, although olfactory symptoms may be manifested in psychiatric disturbances (Amsterdam et al., 1987; Kerekovic, 1972; Pryse-Phillips, 1971), these are not typically the kind of patients presenting with primary smell complaints. However, it is clear that disturbances of smell may result from a wide variety of underlying etiologies (Table 16.1), making the diagnosis and treatment of these disorders a difficult and challenging problem. As more specific information becomes available from the various taste and smell research centers, these patients should become easier to manage.

Acknowledgments. This chapter was prepared while the senior author was supported by Claude Pepper Award DC-00353 from the National Institute on Deafness and Other Communication Disorders.

References

Adams R.G., and N. Crabtree (1961). Anosmia in alkaline battery workers. *Brit. J. Indust. Med.*, *18*, 216–222.

Ahlström, R., B. Berglund, U. Berglund, T. Engen, and T. Lindvall (1987). A comparison of odor perception in smokers, nonsmokers, and passive smokers. *Am. J. Otolaryngol.*, *8*, 1–6.

Amoore, J.E. (1986). Effects of chemical exposure on olfaction in humans. In C.S. Barrow (Ed.), *Toxicology of the Nasal Passages*. Washington, DC: Hemisphere Publishing Co., pp. 155–190.

Amsterdam, J.D., R.G. Settle, R.L. Doty, E. Abelman, and A. Winokur (1987). Taste and smell perception in depression. *Biol. Psychiat.*, *22*, 1481–1485.

Ansari, K.A. (1976). Olfaction in multiple sclerosis. *Eur. Neurol.*, *4*, 138–145.

Ansari, K.A., and A. Johnson (1975). Olfactory function in patients with Parkinson's disease. *J. Chronic. Dis.*, *28*, 493–497.

Atkin-Thor, E., B.W. Goddard, J. O'Nion, R.L. Stephen, and W.J. Kolff (1978). Hypogeusia and zinc depletion in chronic dialysis patients. *Am. J. Clin. Nutr.*, *31*, 1948–1951.

Bakay, L., and H.L. Cares (1972). Olfactory meningiomas. Report on a series of twenty-five cases. *Acta Neurochir.*, *26*, 1–12.

Barton, R.P.E. (1974). Olfaction in leprosy. *J. Laryngol. Otol.*, *88*, 355–361.

Bartoshuk, L.M. (1978). The psychophysics of taste. *Am. J. Clin Nutr.*, *31*, 1068–1077.

Bartoshuk, L.M., and L.E. Marks (1986). Ratio scaling. In H.L. Meiselman and R.S. Rivlin (Eds.), *Clinical Measurement of Taste and Smell*. New York: Macmillan, pp. 50–65.

Bickerstaff, E.R. (1968). *Neurological Examination in Clinical Practice*. Oxford: Blackwell Scientific Publications.

Burch, R.E., D.A. Sackin, J.A. Ursick, M.M. Jetton, and J.F. Sullivan (1978). Decreased taste and smell acuity in cirrhosis. *Arch. Intern. Med.*, *138*, 743–746.

Cain, W.S. (1981). Olfaction and the common chemical sense: Similarities, differences, and interaction. In H.R. Moskowitz and C.B. Warren (Eds.), *Odor Quality and Chemical Structure*. Washington, DC: American Chemical Society, pp. 109–121.

Cain, W.S. (1982). Odor identification by males and females: Predictions vs. performance. *Chem. Senses*, *7*, 129–142.

Cain, W.S., J. Gent, F.A. Catalanotto, and R.B. Goodspeed (1983). Clinical evaluation of olfaction. *Am. J. Otolaryngol.*, *4*, 252–256.

Cain, W.S., and A. Turk (1985). Smell of danger: An analysis of LP-gas odorization. *Am. Ind. Hyg. Assoc. J.*, *46*, 115–126.

Cain, W.S., and J.F. Gent (1986). Use of odor identification in clinical testing of olfaction. In H.L. Meiselman and R.S. Rivlin (Eds.), *Clinical Measurement of Taste and Smell*. New York: Macmillan, pp. 170–186.

Cain, W.S., J.F. Gent, R.B. Goodspeed, and G. Leonard (1988). Evaluation of olfactory dysfunction in the Connecticut Chemosensory Clinical Research Center. *Laryngoscope*, *98*, 83–88.

Cancalon, P. (1982). Degeneration and regeneration of olfactory cells induced by $ZnSO_4$ and other chemicals. *Tiss. Cell*, *14*, 717–733.

Catalanotto, F.A., P. Dore-Duffy, and J.O. Donaldson (1986). Taste and smell in multiple sclerosis. In H.L. Meiselman and R.S. Rivlin (Eds.), *Clinical Measurement of Taste and Smell*. New York: Macmillan, pp. 519– 528.

Champion, R. (1966). Anosmia associated with corrective rhinoplasty. *Brit. J. Plast. Surg.*, *19*, 182–185.

Chan, P.O., S.L. Eustis, J.E. Huff, J.K. Haseman and H. Ragan (1988). Two-year inhalation carcinogenesis studies of methyl methacrylate in rats and mice: inflammation and degeneration of nasal epithelium. *Toxicology*, *52*, 237–252.

Cohen, L. (1965). Disturbance of taste as a symptom of multiple sclerosis. *Brit. J. Oral Surg.*, *2*, 184–185.

Cohen, T., and L. Gitman (1959). Oral complaints and taste perception in the aged. *J. Gerontol.*, *14*, 294–298.

Cometto-Muniz, J.E., and W.S. Cain (1982). Perception of nasal pungency in smokers and nonsmokers. *Physiol. Behav.*, *29*, 727–731.

Corwin, J., M. Serby, P. Conrad, and J. Rotrosen (1985). Olfactory recognition deficit in Alzheimer's and Parkinsonian dementias. *IRCS Med. Sci.*, *13*, 260.

Costanzo, R.M. (1985). Neural regeneration and functional reconnection following olfactory nerve transection in hamster. *Brain Res.*, *361*, 258–266.

Costanzo, R.M., and D.P. Becker (1986). Smell and taste disorders in head injury and neurosurgery patients. In H.L. Meiselman and R.S. Rivlin (Eds.), *Clinical Measurement of Taste and Smell.* New York: Macmillan, pp. 565– 578.

Deems, D.A., and R.L Doty (1987). Odor detection ability decreases with age. *Chem. Senses*, *12*, 650.

Deems, D.A., R.L. Doty, R.G. Settle, V. Moore-Gillon, P. Shaman, A.F. Mester, C.P. Kimmelman, V.J. Brightman, and J.B. Snow, Jr. (1991). Smell and taste disorders, a study of 750 patients from the University of Pennsylvania Smell and Taste Center. *Arch. Otolaryngol. Head Neck Surg.* *117*, 519–528.

Donnelly, J.W., W.S. Cain, and A. Scott (1989). Parosmia among patients with olfactory complaints. *Chem. Senses*, *14*, 695.

Doty, R.L (1979). A review of olfactory dysfunctions in man. *Am. J. Otolaryngol.*, *1*, 57–79.

Doty, R.L., P.J. Snyder, G.R. Huggins, and L.D. Lowry (1981). Endocrine, cardiovascular, and psychological correlates of olfactory sensitivity changes during the human menstrual cycle. *J. Comp. Physiol. Psychol*, *95*, 45–60.

Doty, R.L., P. Shaman, and M. Dann (1984a). Development of the University of Pennsylvania Smell Identification Test: A standardized microencapsulated test of olfactory function. *Physiol. Behav.*, *32*, 489–502.

Doty, R.L., P. Shaman, S.L. Applebaum, R. Giberson, L. Siksorski, and L. Rosenberg (1984b). Smell identification ability: changes with age. *Science*, *226*, 1441–1443.

Doty, R.L., M.G. Newhouse, and J.D. Azzaline (1985). Internal consistency and short-term test-retest reliability of the University of Pennsylvania Smell Identification Test. *Chem. Senses*, *10*, 297–300.

Doty, R.L., T.P. Gregor and R.G. Settle (1986). Influence of intertrial interval and sniff-bottle volume on phenyl ethyl alcohol odor detection thresholds. *Chem. Senses*, *11*, 259–264.

Doty, R.L., and C.P. Kimmelman (1986). Smell and taste and their disorders. In A.K. Asbury, G.M. McKhann, and W.I. McDonald (Eds.), *Diseases of the Nervous System: Clinical Neurobiology*, Vol. 1. Philadelphia: W.B. Saunders, pp. 446–478.

Doty, R.L., P.F. Reyes, and T. Gregor (1987). Presence of both odor identification and detection deficits in Alzheimer's disease. *Brain Res. Bull.*, *18*, 597–600.

Doty, R.L., and J.B. Snow, Jr. (1987). Olfaction. In J. Goldman (Ed.), *The Principles and Practice of Rhinology*. New York: Wiley, pp. 761–785.

Doty, R.L., D.A. Deems, and S. Stellar (1988). Olfactory dysfunction in Parkinsonism: A general deficit unrelated to neurologic signs, disease stage, or disease duration. *Neurology*, *38*, 1237–1244.

Doty, R.L., and J.B. Snow, Jr. (1988). Age-related alterations in olfactory structure and function. In F. Margolis and T. Getchell (Eds.), *Molecular Neurobiology of the Olfactory System*. New York: Plenum, pp. 355–374.

Douek, E., L.H. Bannister, and H.C. Dodson (1975). Recent advances in the pathology of olfaction. *Proc. R. Soc. Med.*, *68*, 467–470.

Duncan, H.J., A.M. Seiden, S.-I. Paik, and D.V. Smith (1991). Differences among patients with smell impairment resulting from head trauma, nasal disease, or prior upper respiratory infection. *Chem. Senses*, 16, 517.

Dunn, J.D., J.E. Cometto-Muniz, and W.S. Cain (1982). Nasal reflexes: Reduced sensitivity to CO_2 irritation in cigarette smokers. *J. Appl. Toxicol.*, *2*, 176–178.

Elsberg, C.A. (1935). The sense of smell. XII. The localization of tumors of the frontal lobe of the brain by quantitative olfactory tests. *Bull. Neurol. Inst. NY*, *4*, 535–543.

Esiri, M.M., and G.K. Wilcock (1984). The olfactory bulbs in Alzheimer's disease. *J. Neurol. Neurosurg. Psychiat.*, *47*, 56–60.

Eskenazi, B., W.S. Cain, R.A. Novelly, and K.B. Friend (1983). Olfactory functioning in temporal lobectomy patients. *Neuropsychologia*, *21*, 365–374.

Eskenazi, B., W.S. Cain, R.A. Novelly, and R. Mattson (1986). Odor perception in temporal lobe epilepsy patients with and without temporal lobectomy. *Neuropsychologia*, *24*, 553–562.

Estrum, S.A., and G. Renner (1987). Disorders of taste and smell. *The Otolaryngology Clinics of North America*, Vol. 20, no. 1. Philadelphia: W.B. Saunders, pp. 133–147.

Fein, B.T., P.B. Kamin, and N.N. Fein (1966). The loss of sense of smell in nasal allergy. *Ann. Allergy*, *24*, 278–283.

Feldman, J.I., H.N. Wright, and D.A. Leopold (1986). The initial evaluation of dysosmia. *Am. J. Otolaryngol.*, *4*, 431–444.

Frye, R.E., B.S. Schwartz, and R.L. Doty (1990). Dose-related effects of cigarette smoking on olfactory function. *JAMA*, *263*, 1233–1236.

Furstenberg, A.C., E. Crosby, and B. Farrior (1943). Neurologic lesions which influence the sense of smell. *Arch. Otolaryngol.*, *48*, 529–530.

Gent, J.F., W.S. Cain, and L.M. Bartoshuk (1986). Taste and smell measurement in a clinical setting. In H.L. Meiselman and R.S. Rivlin (Eds.), *Clinical Measurement of Taste and Smell*. New York: Macmillan, pp. 107–116.

Gent, J.F., R.B. Goodspeed, R.T. Zagraniski and F.A. Catalanotto (1987). Taste and smell problems: validation of questions for the clinical history. *Yale J. Biol. Med.*, *60*, 27–35.

Ghorbanian, S.N., J.L. Paradise, and R.L. Doty (1983). Odor perception in children in relation to nasal obstruction. *Pediatrics*, *72*, 510–516.

Goodspeed, R.B., F.A. Catalanotto, J.F. Gent, W.S. Cain, L.M. Bartoshuk, G. Leonard, and J.O. Donaldson (1986a). Clinical characteristics of patients with taste and smell disorders. In H.L Meiselman and R.S. Rivlin (Eds.), *Clinical Measurement of Taste and Smell*. New York: Macmillan, pp. 451–466.

Goodspeed, R.B., J.F. Gent, F.A. Catalanotto, W.S. Cain, and R.T. Zagraniski (1986b). Corticosteroids in olfactory dysfunction. In H.L. Meiselman and R.S. Rivlin (Eds.), *Clinical Measurement of Taste and Smell*. New York: Macmillan, pp. 514–518.

Goodspeed, R.B., J.F. Gent, and F.A. Catalanotto (1987). Chemosensory dysfunction: Clinical evaluation results from a taste and smell clinic. *Postgrad. Med.*, *81*, 251–260.

Graziadei, P.P.C. (1973). Cell dynamics in the olfactory mucosa. *Tiss. Cell*, *5*, 113–131.

Graziadei, P.P.C., and G.A. Monti-Graziadei (1983). Regeneration in the olfactory system of vertebrates. *Am. J. Otolaryngol.*, *4*, 228–233.

Green, R.F. (1971). Subclinical pellagra and idiopathic hypogeusia. *JAMA*, *218*, 1303.

Griffith, I.P. (1976). Abnormalities of smell and taste. *Practitioner*, *217*, 907–913.

Halpern, B.P. (1985). Environmental factors affecting chemoreceptors: An overview. In A.W. Hayes (Ed.), *Toxicology of the Eye, Ear, and Other Special Senses*. New York: Raven Press, pp. 195–211.

Hasegawa, S., M. Yamagishi, and Y. Nakano (1986). Microscopic studies of human olfactory epithelia following traumatic anosmia. *Arch. Otolaryngol., 243*, 112–116.

Henkin, R.I. (1967a). The definition of primary and accessory areas of olfaction as the basis for a classification of decreased olfactory acuity. In T. Hayashi (Ed.), *Olfaction and Taste II*. Oxford: Pergamon, pp. 235–252.

Henkin, R.I. (1967b). Abnormalities of taste and olfaction in patients with chromatin negative gonadal dysgenesis. *J. Clin. Endocrinol. Metab.*, *27*, 1436–1440.

Henkin, R.I. (1968). Impairment of olfaction and the losses of sour and bitter in pseudohypoparathyroidism. *J. Clin. Endocrinol. Metab.*, *28*, 624–628.

Henkin, R.I. (1975). The role of adrenal corticosteroids in sensory processes. In H. Blaschko, A.D. Smith, and G. Sayers (Eds.), *Handbook of Physiology*, Section 7, Vol. 6. Washington, DC: American Physiological Society, pp. 209–230.

Henkin, R.I. (1981). Olfaction in human disease. In G.M. English (Ed.), *Otolaryngology*, vol. 2, Philadelphia: J.B. Lippincott, pp. 1–39.

Henkin, R.I., and G.F. Powell (1962). Increased sensitivity of taste and smell in cystic fibrosis. *Science*, *138*, 1107–1108.

Henkin, R.I., and I.J. Kopin (1964). Abnormalities of taste and smell thresholds in familial dysautonomia: Improvement with methacholine. *Life Sci.*, *3*, 1319–1325.

Henkin, R.I., H.R. Keiser, I.A. Jaffe, I. Sternlieb, and I.H. Scheinberg (1967). Decreased taste sensitivity after D-penicillamine reversed by copper administration. *Lancet*, *2*, 1268–1271.

Henkin, R.I., and D.F. Bradley (1969). Regulation of taste acuity by thiols and metal ions. *Proc. Natl. Acad. Sci. USA*, *62*, 30–37.

Henkin, R.I., P.J. Schechter, R. Hoye, and C.F.T. Mattern (1971). Idiopathic hypogeusia with dysgeusia, hyposmia, and dysosmia: A new syndrome. *JAMA*, *217*, 434–440.

Henkin, R.I., and F.R. Smith (1971). Hyposmia in acute viral hepatitis. *Lancet*, *1*, 823–826.

Henkin, R.I., N. Talal, A.L. Larson, and C.F.T. Mattern (1972). Abnormalities of taste and smell in Sjögren's syndrome. *Ann. Intern. Med.*, *76*, 375–383.

Henkin, R.I., A.L. Larson, and R.D. Powell (1975). Hypogeusia, dysgeusia, hyposmia, and dysosmia following influenza-like infection. *Ann. Otolaryngol.*, *84*, 672–682.

Henkin, R.I., P.J. Schechter, W.T. Friedewald, D.L. Demets, and M. Raff (1976). A double-blind study of the effects of zinc sulfate on taste and smell dysfunction. *Am. J. Med. Sci.*, *272*, 285–299.

Hertz, J., W.S. Cain, L.M. Bartoshuk, and T.F. Dolan (1975). Olfactory and taste sensitivity in children with cystic fibrosis. *Physiol. Behav.*, *14*, 89–94.

Hotchkiss, W.T. (1956). Influence of prednisone on nasal polyposis with anosmia. *Arch. Otolaryngol.*, *64*, 478–479.

Hubert, H.B., R.R. Fabsitz, M. Feinleib and K.S. Brown (1980). Olfactory sensitivity in humans: Genetic versus environmental control. *Science*, *208*, 607–609.

Jafek, B.W. (1982). Anosmia and ageusia. In G.A. Gates (Ed.), *Current Therapy in Otolaryngology—Head and Neck Surgery 1982–1983*. St. Louis, MO: C.V. Mosby Co., pp. 279–282.

Jafek, B.W., and P.M. Eller (1989). Post-viral olfactory dysfunction. *Chem. Senses*, *14*, 713.

Jefferson, M. (1961). Anosmia and parosmia. *Practitioner*, *187*, 715–718.

Joachims, H.Z., M.M. Altman, and S.W. Mayer (1975). Olfactory neuroblastoma. *J. Laryngol. Otol.*, *89*, 335–343.

Jones, B.A., H.R. Moskowitz, and N. Butters (1975). Olfactory discrimination in alcoholic Korsakoff patients. *Neuropsychologia*, *13*, 173–179.

Kallmann, F.J., W.A. Schoenfeld, and S.E. Barrera (1944). The genetic aspects of primary eunuchoidism. *Am. J. Ment. Defic.*, *48*, 203– 236.

Kerekovic, M. (1972). The relationship between objective disorders of smell and olfactory hallucinations. *Acta Otorhinolaryngol. Belg.*, *26*, 518–523.

Kimmelman, C.P. (1986). *Disorders of Taste and Smell.* Washington, DC: American Academy of Otolaryngology—Head and Neck Surgery Foundation.

Leigh, A.D. (1943). Defects of smell after head injury. *Lancet*, *1*, 38–40.

Leopold, D.A. (1986). Physiology of olfaction. In C.W. Cummings, J.M. Fredrickson, L.A. Harker, C.J. Krause, and D.E. Schuller (Eds.), *Otolaryngology—Head and Neck Surgery*, Vol. 1, *General, Face, Nose, Paranasal Sinuses*. St. Louis, MO: C.V. Mosby, pp. 527–545.

Leopold, D.A., L.M. Bartoshuk, R.L. Doty, B. Jafek, D.V. Smith, and J.B. Snow, Jr. (1989). Aging of the upper airway and the senses of taste and smell. *Otolaryngol. Head Neck Surg.*, *100*, 287–289.

Liston, S.L. (1984). Olfactory disorders. In G.R. Holt, D.E. Mattox, and G.A. Gates (Eds.), *Decision Making in Otolaryngology*. St. Louis, MO: C.V. Mosby Co., pp. 28–29.

Mahajan, S.K, A.S. Prasad, J. Lambujon, A.A. Abbasi, W.A. Briggs, and F.D. McDonald (1980). Improvement of uremic hypogeusia by zinc: A double-blind study. *Am. J. Clin. Nutr.*, *33*, 1517–1521.

Mair, R.G., R.L. Doty, K.M. Kelly, C.S. Wilson, P.J. Langlais, W.J. McEntee, and T.A. Vollmecke (1986). Multimodal sensory discrimination deficits in Korsakoff's psychosis. *Neuropsychologia*, *24*, 831–839.

Marshall, J.R., and R.I. Henkin (1971). Olfactory acuity, menstrual abnormalities and oocyte status. *Ann. Intern. Med.*, *75*, 207–211.

Marshall, K.G., and E.L. Attia (1987). Anosmia, hyposmia and dysosmia. In K.G. Marshall and E.L. Attia (Eds.), *Disorders of the Nose and Paranasal Sinuses* Littleton, MA: PSG Publishing Co., pp. 235–249.

Matulionis, D.H. (1974). Ultrastructure of olfactory epithelia in mice after smoke exposure. *Ann. Otolaryngol*, *83*, 192–201.

Moberg, P.J., G.D. Pearlson, L.J. Speedie, J.R. Lipsey, and S.E. Folstein (1984). Deficits in olfactory, but not visual or verbal recognition, in early affected Huntington's disease patients. *Soc. Neurosci. Abstr.*, *10*, 318.

Monath, T.P., C.B. Cropp, and A.K. Harrison (1983). Mode of entry of a neurotropic arbovirus into the central nervous system: Reinvestigation of an old controversy. *Lab. Invest.*, *48*, 399–410.

Monti-Graziadei, G.A., and P.P.C. Graziadei (1979). Neurogenesis and neuron regeneration in the olfactory system of mammals. II. Degeneration and reconstitution of the olfactory sensory neurons after axotomy. *J. Neurocytol.*, *8*, 197–213.

Monti-Graziadei, G.A., M.S. Karlan, J.J. Bernstein, and P.P.C. Graziadei (1980). Reinnervation of the olfactory bulb after section of the olfactory nerve in monkey (*Samiri sciureus*). *Brain Res.*, *189*, 343–354.

Moran, D.T., B.W. Jafek, J.C. Rowley, and P.M. Eller (1985). Electron microscopy of olfactory epithelia in two patients with anosmia. *Arch. Otolaryngol.*, *111*, 122–126.

Mozell, M.M., D.E. Hornung, P.R. Saheehe, and D.B. Kurtz (1986a). What should be controlled in studies of smell? In H.L. Meiselman and R.S. Rivlin (Eds.), *Clinical Measurement of Taste and Smell.* New York: Macmillan, pp. 154–169.

Mozell, M.M., D.N. Schwartz, S.L. Youngentob, D.A. Leopold, D.E. Hornung, and P.R. Sheehe (1986b). Reversal of hyposmia in laryngectomized patients. *Chem. Senses*, *11*, 397–410.

Murphy, C. (1983). Age related effects on the threshold, psychophysical function and pleasantness of menthol. *J. Gerontol.*, *38*, 217–222.

Murphy, C. (1986). Taste and smell in the elderly. In H.L. Meiselman and R.S. Rivlin (Eds.), *Clinical Measurement of Taste and Smell.* New York: Macmillan, pp. 343–371.

Nakashima, T., C.P. Kimmelman, and J.B. Snow, Jr. (1983). Progressive olfactory degeneration due to ischemia. *Surg. Forum*, *34*, 566–568.

Nakashima, T., C.P. Kimmelman, and J.B. Snow, Jr. (1984). Effect of olfactory nerve section and hemorrhage on the olfactory neuroepithelium. *Surg. Forum*, *35*, 562–564.

Ophir, D., R. Gross-Isseroff, D. Lancet, and G. Marshak (1986). Changes in olfactory acuity induced by total inferior turbinectomy. *Arch. Otolaryngol. Head Neck Surg.*, *112*, 195–197.

Pearson, R.C.A., M.M. Esiri, R.W. Hiorns, G.K. Wilcock, and T.P.S. Powell (1985). Anatomic correlates of the distribution of the pathological changes in the neocortex in Alzheimer's disease. *Proc. Natl. Acad. Sci. USA.*, *82*, 4531–4534.

Pinching, A.J. (1977). Clinical testing of olfaction reassessed. *Brain*, *100*, 377–388.

Price, S. (1986). The role of zinc in taste and smell. In H.L. Meiselman and R.S. Rivlin (Eds.), *Clinical Measurement of Taste and Smell.* New York: Macmillan, pp. 443–445.

Pryse-Phillips, W. (1971). An olfactory reference syndrome. *Acta Psychiat. Scand.*, *47*, 484–509.

Quinn, N.P., M.N. Rossor, and C.D. Marsden (1987). Olfactory thresholds in Parkinson's disease. *J. Neurol. Neurosurg. Psychiat.*, *50*, 88–89.

Rundles, R.W. (1946). Prognosis in the neurologic manifestations of pernicious anemia. *Blood*, *1*, 209–219.

Ryan, R.E., Sr., and R.E. Ryan, Jr. (1974). Acute nasal sinusitis. *Postgrad. Med.*, *56*, 159–162.

Santen, R.J., and C.A. Paulsen (1973). Hypogonadotropic eunuchoidism. I. Clinical study of the mode of inheritance. *J. Clin. Endocrinol. Metab.*, *36*, 47–54.

Schechter, P.J., W.T. Friedewald, D.A. Bronzert, M.S. Raff, and R.I. Henkin (1972). Idiopathic hypogeusia: A description of the syndrome and a single blind study with zinc sulfate. *Int. Rev. Neurobiol. Suppl.*, *1*, 125–140.

Schechter, P.J., and R.I. Henkin (1974). Abnormalities of taste and smell after head trauma. *J. Neurol. Neurosurg. Psychiat.*, *37*, 802–810.

Schiffman, S.S. (1979). Changes in taste and smell with age: Psychophysical aspects. In J.M. Ordy and K. Brizzee (Eds.), *Sensory Systems and Communications in the Elderly*, Vol. 10, *Aging*. New York: Raven Press, pp. 227–246.

Schiffman, S.S. (1983a). Taste and smell in disease. I. *N. Engl. J. Med.*, *308*, 1275–1279.

Schiffman, S.S. (1983b). Taste and smell in disease. II. *N. Engl. J. Med.*, *308*, 1337–1343.

Schiffman, S.S., M.L. Nash, and C. Dackis (1978). Reduced olfactory discriminations in patients on chronic renal hemodialysis. *Physiol. Behav.*, *21*, 239–242.

Schneider, R.A., J.P. Castiloe, R.P. Howard, and S. Wolf (1958). Olfactory perception thresholds in hypogonadal women: Changes accompanying administration of androgen and estrogen. *J. Clin. Endocrinol. Metab.*, *18*, 379–390.

Schwartz, B.S., R.L. Doty, C. Monroe, R. Frye, and S. Barker (1989). Olfactory function in chemical workers exposed to acrylate and methacrylate vapors. *Am. J. Pub. Health*, *79*, 613–618.

Scott, A.E., W.S. Cain, and G. Clavet (1988). Topical corticosteroids can alleviate olfactory dysfunction. *Chem. Senses*, *13*, 735.

Scott, A.E., W.S. Cain, and G. Leonard (1989). Nasal/sinus disease and olfactory loss at the Connecticut Chemosensory Clinical Research Center (CCCRC). *Chem. Senses*, *14*, 745.

Seiden, A.M., and D.V. Smith (1988). Endoscopic intranasal surgery as an approach to restoring olfactory function. *Chem. Senses*, *13*, 736.

Seiden, A.M., A. Litwin, and D.V. Smith (1989). Olfactory deficits in allergic rhinitis. *Chem. Senses*, *14*, 746–747.

Serby, M. (1986). Olfaction and Alzheimer's disease. *Prog. Neuro-Psychopharmacol. Biol. Psychiat.*, *10*, 579–586.

Settle, R.G. (1986). Diabetes mellitus and the chemical senses. In H.L. Meiselman and R.S. Rivlin (Eds.), *Clinical Measurement of Taste and Smell*. New York: Macmillan, pp. 487–513.

Shipley, M.T. (1985). Transport of molecules from nose to brain: Transneuronal anterograde and retrograde labeling in the rat olfactory system by wheat germ agglutinin-horseradish peroxidase applied to the nasal epithelium. *Brain Res. Bull.*, *15*, 129–142.

Simmons, P.A., and T.V. Getchell (1981). Physiological activity of newly differentiated olfactory receptor neurons following olfactory nerve section. *J. Comp. Neurol.*, *197*, 237–257.

Skolnik, E.M., F.S. Massari, and L.T. Tenta (1966). Olfactory neuro-epithelioma: Review of the world literature and presentation of two cases. *Arch. Otolaryngol.*, *84*, 649–653.

Smith, D.V. (1988). Assessment of patients with taste and smell disorders. *Acta Otolaryngol. Stockh. Suppl.*, *458*, 129–133.

Smith, D.V. (1990). Taste and smell dysfunction. In M.M. Paparella, D.A. Shumrick, J.L. Gluckman, and W.L. Meyerhoff (Eds.), *Otolaryngology*, 3d Ed., vol. 3, *Head and Neck*, Philadelphia: W.B. Saunders. pp 1911–1934.

Smith, D.V., R.A. Frank, M.L. Pensak, and A.M. Seiden (1987). Characteristics of chemosensory patients and a comparison of olfactory assessment procedures. *Chem. Senses*, *12*, 698.

Snow, J.B., Jr. (1969). The classification of respiratory viruses and their clinical manifestation. *Laryngoscope*, *79*, 1485–1493.

Stevens, J.C., A. Plantinga, and W.S. Cain (1982). Reduction of odor and nasal pungency associated with aging. *Neurobiol. Aging*, *3*, 125–132.

Stevens, J.C., L.M. Bartoshuk, and W.S. Cain (1984). Chemical senses and aging: Taste versus smell. *Chem. Senses*, *9*, 167–179.

Stevens, J.C., and W.S. Cain (1985). Age-related deficiency in the perceived strength of six odorants. *Chem. Senses*, *10*, 517–529.

Strandbygard, E. (1954). Treatment of ozena and rhinopharyngitis chronica sicca with vitamin A. *Arch. Otolaryngol.*, *59*, 485–490.
Sumner, D. (1962). On testing the sense of smell. *Lancet*, *2*, 895–897.
Sumner, D. (1964). Post-traumatic anosmia. *Brain*, *87*, 107–120.
Sumner, D. (1975). Disturbances of the sense of smell and taste after head injuries. Part II. Injuries of the brain and skull. In P.J. Vinken and G.W. Bruyn (Eds.), *Handbook of Clinical Neurology*, Vol. 24. New York: Elsevier, pp. 1–25.
Tagatz, G.T., P.J. Fialkow, D. Smith, and L. Spadoni (1970). Hypogonadotropic hypogonadism associated with anosmia in the female. *N. Engl. J. Med.*, *283*, 1326–1329.
Takahashi, H., S. Ohara, M. Yamada, F. Ikuta, K. Tanimura, and Y. Honda (1987). Esthesioneuroepithelioma: A tumor of true olfactory epithelium origin. *Acta Neuropathol. Berl.*, *75*, 147–155.
Talamo, B.R., R.A. Rudel, K.S. Kosik, V.M.-Y. Lee, S. Neff, L. Adelman, and J.S. Kauer (1989). Pathological changes in olfactory neurons in patients with Alzheimer's disease. *Nature*, *337*, 736–739.
Thumfart, W., K.H. Plattig, and N. Schlicht (1980). Taste and smell sensitivity in the elderly human. *Z. Gerontol.*, *13*, 158–188.
Twomey, J.A., C.M. Barker, G. Robinson, and D.A. Howell (1979). Olfactory mucosa in herpes simplex encephalitis. *J. Neurol. Neurosurg. Psychiat.*, *42*, 983–987.
Van Toller, C., G.H. Dodd, and A. Billing (1985). *Ageing and the Sense of Smell*. Springfield, IL: Charles C Thomas.
Venstrom, D., and J.E. Amoore (1968). Olfactory threshold in age, sex or smoking. *J. Food Sci.*, *33*, 264–265.
Ward, C.D., W.A. Hess, and D.B. Calne (1983). Olfactory impairment in Parkinson's disease. *Neurology*, *33*, 943–946.
Weiffenbach, J.M., and V.P. McCarthy (1984). Olfactory deficits in cystic fibrosis: Distribution and severity. *Chem. Senses*, *9*, 193–199.
Weiffenbach, J.M., B.J. Cowart, and B.J. Baum (1986). Taste intensity perception in aging. *J. Gerontol.*, *41*, 460–468.
Weisman, K., E. Christensen, and V. Dreyer (1979). Zinc supplementation in alcoholic cirrhosis: A double-blind clinical trial. *Acta Med. Scand.*, *205*, 361–366.
Wender, M., and Z. Szemza (1971). Examination of hearing, vestibular system function, taste and olfactory systems in patients with disseminated sclerosis. *Neurol. Neurochir. Pol.*, *64*, 767–772.
Wright, H.N. (1987). Characterization of olfactory dysfunction. *Arch. Otolaryngol. Head Neck Surg.*, *113*, 163–168.
Zusho, H. (1982). Posttraumatic anosmia. *Arch. Otolaryngol.*, *108*, 90–92.

17
Olfactory Hallucinations

Mark S. Greenberg

Introduction

Hallucinations are false perceptions—sensory experiences that are generated in the absence of an adequate stimulus. They occur in all the basic sensory modalities: visual, auditory, tactile, gustatory, and olfactory. The diversity of contexts in which they arise—toxic states, metabolic disorders, schizophrenic episodes, epilepsy, nocturnal dreaming, hypnosis—suggests that their elicitation must involve multiple mechanisms.

Olfactory hallucinations (OHs) entail the discrete perception of an odor that is not objectively there. An OH may be identified as a familiar, real-life smell, or it can be deemed to be novel. If familiar, it can be recognized as a general semantic class of smells (e.g., floral, smoky, pungent, chemical) or as a specific exemplar of such a class (e.g., roses, pipe tobacco, ginger root, benzene). Authors studying visual hallucinations have used the terms "simple" versus "complex" or "elaborate" and "unformed" versus "formed" to differentiate hallucinations along the dimension of complexity (Cogan, 1973). Yet this intuitively reasonable distinction does not necessarily translate readily to the olfactory modality. For example, a simple visual hallucination might consist of flashes of light or a sensation of pure color, while a complex visual hallucination might be of an object. But there is much controversy about the basic components of olfactory experience and the existence of odor "primaries" (Amoore, 1977). One way an OH can be "elaborated upon" is through its linkage to a specific biographical recollection (e.g., "the roses given to me on my fortieth birthday"). Another related dimension is the tendency for the OH to evoke personal associations and reminiscences that are outside the olfactory realm (e.g., "that smell reminds me of the time we visited my Uncle Charlie in Cleveland on Christmas"). Another basic qualitative category concerns the hedonics of the OH—its perceived pleasantness or aversiveness, and its ability to elicit corresponding affect: disgust or pleasure. Yet another dimension is the individual's subjective localization of the source of the odor: Is it believed to emanate from the inside of the nose or within the body cavity, from the surface of the skin, or from the near or distant environment? Is it lateralized to one nostril or one side of external space?

Jaspers (1963) makes the further distinction between hallucination and pseudo-hallucination. The sensory qualities are more or less equivalent, but in the latter case the subject is aware that the experience is a false perception and doesn't actually believe the percept is "out there." Although it possesses vivid perceptual features, the pseudo-hallucination is localized in imaginal as opposed to extracorporeal space. Some authors term this the "insight" dimension. A patient retains insight if he or she has the experience of smelling cigar smoke though knowing there is no smoke there. There is an "as if" quality to such reports—"it smells *as if* someone were smoking a cigar" (Reed, 1972). Among neurological cases, the presence or absence of insight may be a function of the integrity of surrounding brain regions, especially the bilateral frontal and right parietal lobes. Individuals with pseudo-hallucinations are often reticent about acknowledging their experiences because they realize that their perception is nonveridical, and do no wish to appear "crazy." For example, the incidence of visual hallucination among the elderly who have experienced ocular disease is greatly underestimated because patients do not spontaneously complain of such events. When patients are questioned in a nonthreatening manner by a trusted physician, an astonishing number of reports emerge (Damas-Mora, 1981).

Documentation of the above qualitative features of OHs faces the same difficulty as the study of other subjective phenomena: The investigator must rely on the verbal self-report of the patient. As in the case of visual and auditory hallucinations, the unit of analysis is not the private, internal experience itself, but the patient's public account of that experience. Individual differences in ability and willingness to introspect, verbalize, and elaborate such reports can fundamentally distort the end product and obscure true patterns in the experience *per se*. This dilemma has always plagued the study of nocturnal dream hallucinations. This obstacle seems particularly relevant in the case of psychiatric patients, who may have other disturbances of thought, and neurological patients, who exhibit alterations of consciousness and other cognitive deficits (Greenberg and Farah, 1981).

Quantitative parameters of interest include the frequency and duration of the OH. These should seem rather straightforward and less prone to distortion than other descriptive features of the OH, but in many cases the experience of OH is not limited to discrete episodes, and in both seizure patients and psychotics time estimation is unlikely to be reliable.

Hallucinations must be distinguished from two related but distinct phenomena: delusions of smell and distortions of smell. Smell delusions are false beliefs—the mistaken notion that one is smelling something that is not objectively there, *without* the actual false perceptual experience of the odor. The false conviction that toxic fumes are seeping into one's basement from an adjoining property is a typical olfactory delusion. (The fact that toxic environmental exposures do regularly occur may account for the popularity of this delusional idea.) Does the patient actually have the experience of smelling the offending substance? If so, it would qualify as a true OH. While hallucinations of smell and delusions of smell may provoke similar verbal reports, they should be distinguishable from a phenomenological perspective. However, verifying this distinction among patients with neurologically based cognitive deficits and psychiatrically based impairments in reality testing will prove to be quite problematic.

Distortions of smell perception, defined as errors in perception of odors that are actually present, are also common clinical occurrences. Different authors use differing terminologies for categorizing errors in smell perception, including dysosmia, microsmia, hyposmia, cacosmia, and heterosmia (Schiffman, 1983; Zilstorff, 1966; Henkin, 1971). Smell distortions can be further divided into two basic types, negative and positive. Negative symptoms involve loss of function—reduced afferent information. This condition may manifest clinically as diminished sensitivity (higher psychophysical thresholds), inability to discriminate, or perceived faintness of smell. The terms "anosmia," "hyposmia," and "microsmia" refer to this loss of olfactory sensitivity. The subject, however, may have no awareness of this deficit. While negative symptoms are characterized by an absence of function, positive symptoms involve the presence of an abnormal response. Dysosmia (or cacosmia or heterosmia) occurs when a familiar neutral or pleasant odor is perceived as aversive or somehow "different." Illusions of smell, in which an odor is misidentified as another, are another category of positive olfactory symptoms. In all the above examples, the distortions of smell function are manifest in the presence of actual eliciting stimuli.

While the above distinction between true OH and dysosmia seems clear in principle, in clinical practice one is not always confident in making such determinations. For example, the source of a perceived foul odor may be internal (emanating from the oral cavity, sinuses, lungs, or stomach), and therefore its detection may depend on the examiner's sensitivity. It is common practice to search for an occult source for the misperception (Schiffman, 1983), and if none is found to label it an hallucination (or delusion) by exclusion. A recent case report of Leopold, Preti, Mozell, Youngentob, and Wright (1990) illustrates this point clearly. They describe the case of a woman with a 20-year history of complaints of unpleasant odors and tastes occurring during unilateral or bilateral nasal exhalation. Nostril pinching obliterated the percepts. Odor identification ability tested as borderline normal and then well into the hyposmic range at two and four months later. Repeated workups failed to elicit a source of the odors, and various treatment regimens, including psychiatric intervention, failed to alter her complaints. A diagnosis of trimethylaminuria (fish-odor syndrome—one of the inborn errors of metabolism) was finally confirmed when this substance was detected in a urine sample following ingestion of a choline challenge, and her self-report was ultimately validated. Dietary alterations (avoiding choline) had a salutatory effect. This case underscores the often exclusionary nature of the differential diagnosis of OH.

Another subjective phenomenon that can be confounded with true OH is the well-known experience of an actual smell percept triggering a sense of déja vû or an emotionally laden remembrance. This becomes particularly problematic in cases of epileptic OHs, since feelings of familiarity and intrusive memories are commonly experienced in the course of complex partial seizures.

A review of the English-language literature on olfactory hallucinations reveals much fragmentation. Epilepsy and brain stimulation studies are segregated from the clinical literature on smell loss (the "-osmias"), and the neurological literature is separated from the psychiatric literature on hallucinations and delusions of smell. This situation may account for the common clinical impression that olfactory hallucinations herald an underlying organic process. This stance is often expressed in a gustatory metaphor: The presence of an OH imparts an organic "flavor" to a patient. Another clinical truism holds that OHs are diagnostic of (or even pathognomonic of) temporal lobe epilepsy. On the contrary, a wide range of disorders are associated with OHs. Indeed, perhaps the most common settings for OHs are psychiatric and not neurological disorders.

The present chapter will attempt to counter the prevailing pattern of fragmentation by surveying the various organic and functional syndromes in which OHs are manifest. The predominant focus will be on neurological entities in which OHs have been described. Psychiatric conditions posing differential diagnostic dilemmas will then be considered. Cases will be described to illustrate facets of the various syndromes. Where possible, the quantitative and qualitative features of the OHs will be extracted from the case material. Such an attempt may uncover commonalities and distinctions in the experience of OHs among various clinical entities. If discrete patterns emerge across individual patients within specific diagnostic groups, one will be better able to argue that these reflect the actual nature of the OH rather than the mere verbal reports of the hallucinatory experience.

With these basic distinctions and qualifications in mind, we now turn to an analysis of the major neurological and psychiatric syndromes in which OHs have been reported.

Neurological Syndromes

OH in Epilepsy

Complex partial seizures (also termed limbic or psychomotor seizures) are a diverse set of clinical manifestations, with the common characteristic of a sudden alteration in consciousness. This perturbation or awareness is due to an abnormal electrical discharge of a group of neurons released from their usual physiological control. The site of origin of this seizure focus is most often located in the deep nuclei of the limbic system—the hippocampus or amygdala. There are myriad forms of complex partial seizures, spanning virtually the entire range of conscious experience. Perceptual alterations include illusions, distortions, and hallucinations in all modalities. Motor manifestations range from movements such as lip-smacking, grimacing, vocalization, and laughter to more complex, coordinated, automatic behaviors such as running, disrobing, or even doing the laundry. Autonomic discharges such as salivation, flushing, tachycardia, and urinary incontinence are common. Cognitive manifestations range from brief lapses ("absences"), intrusive thoughts, vivid memories, or the sense of déja vû to profound feelings of mystical revelation. Emotional components of complex partial spells include fear, paranoia, and dysphoria, and, much more rarely, elation and even ecstasy (Bear, Freeman, and Greenberg, 1984). Given this broad range of experiences, discussion of the prototypical limbic seizure patient is quite problematic. While in many cases the sequence of ictal events is stereotypic and invariant across episodes, the situation is commonly further complicated by the existence of several distinct seizure types in the same individual and the suppressive effects of anticonvulsant regimens. In many cases the complex partial seizure evolves into a generalized (major motor, grand mal, or tonic-clonic) spell in which posture cannot be maintained and contact with the environment is transiently lost. Upon recovery, the patient may be confused, agitated, and/or amnestic for the episode. Such postictal confusion further confounds the validity of reports of the subtle subjective states that precede loss of consciousness. The reader is referred to Daly (1975), Spiers et al. (1985), and Greenberg and Seidman (in press) for a review of these and other clinical aspects of temporal lobe seizures.

John Hughlings-Jackson is generally recognized as the first neurologist to appreciate the significance of OHs in epilepsy, though credit for the initial case reports is attributed by Daly (1958) to others (see also Gowers, 1910). Certainly Jackson was the first to elaborate upon the subjective experience of OHs, establish clinical associations, and attempt a localization of the responsible "discharging lesion." The report of Jackson and Purves Stewart (1899) describes a case that illustrates the essential clinical features of epileptic OHs. Mr. A.B., a 51-year-old, developed seizures following a systemic infection with fever of 105 degrees. He suffered periods of excitement, a personality change, and memory loss. His seizures were heralded by nausea, followed by "giddiness" and frontal headaches. He reported a strong sensation of smell, "like camphor or ether," followed by a dreamy state with a deep sense of familiarity, in which he "felt as if he were saying, doing, and looking at things which he had experienced before." He also reported visual illusions, including micropsia, in which objects in his visual field looked as if they were "far away." After only four weeks, his seizure pattern changed; the OHs stopped and vestibular phenomena were more prominent, along with ictal fear. His condition worsened over the next four weeks, with increasing headache, mild left hemiparesis of face and leg, optic neuritis of the right eye, and an extensor plantar response on the left. He was able to identify and name two odors (clove and asafetida) bilaterally, and taste sensation was judged to be normal. He expired in his sleep and no autopsy was performed—a circumstance deeply regretted by the authors. They localized the lesion, presumably an abscess or tumor, to the right temporal-sphenoidal region on clinical grounds.

The epileptic OH is typically an intrusive, brief, paroxysmal experience. The perception of smell most often occurs early in the seizure and may be the "aura" or first ictal manifestation. OHs can also precede the ictus proper by minutes, hours, or even days—so-called epileptic prodromes. Such distinctions among prodrome, aura, ictus, and postictal state are always difficult to substantiate (Bear, Freeman, and Greenberg, 1984). A common clinical seizure pattern is, for example, a feeling of irritability for hours (the prodrome), followed by the sudden onset of visceral sensations (the aura), an OH, the feeling of déja vû, and a brief absence, with somnolence when consciousness is restored. The actual ictus itself generally lasts from 10 seconds to a few minutes at most (Adams and Victor, 1981).

While it is commonly held that epileptic OHs are invariably unpleasant, neutral and even pleasant smells are also reported. Common aversive OHs include odors of putrefaction, such as "rotting meat," "garbage," or "decayed animals"; incendiary smells of "burning rubber," "charred wood," or "stale tobacco smoke"; and chemical odors such as "ether," "diesel fuel," or "tar." Oftentimes patients use such descriptors in an attempt to convey the quality of the OH, but when pressed for certainty (or confronted with the actual odor reported), admit they cannot really specify the exact identity of the smell (Gowers, 1910). Other patients spontaneously report the odor as "indescribable." It is relatively common for other perceptual, cognitive, and interoceptive ictal phenomena to defy verbal characterization as well. Moreover, anomia—difficulty with naming ability—is a common finding among patients with temporal lobe epilepsy (Mayuex et al., 1980). (On the other hand, patients afflicted with nonepileptic OHs also report difficulty in capturing the odor in words.) Adjectives such as "horrifying," "sickening" and "disgusting" are not uncommon. It is plausible that some of this aversive quality may derive from the frequently associated visceral and affective seizure components, such as abdominal distress, nausea, and dysphoria. Representative pleasant and neutral odors include fruits, spices, and other foodstuffs ("peaches," "apple pie," "cinnamon," "cooking"), botanical scents ("pines," "perfume," "flowers," "woods"), and environmental odors ("leather," "traffic," "sea breeze.") Some spontaneous descriptions reveal the arbitrary nature of such categorization schemes. For example, a patient of Weil (1955) smelled "stuffed cabbage in a dirty outhouse" during a seizure.

The other ictal features of spells in which OHs are found are difficult to specify with precision, though some clear patterns emerge. Hughlings-Jackson used the phrase "uncinate group of fits" for a category of ictal experiences that include but are not limited to OHs. This term was derived from Ferrier's stimulation experiments to the uncus of monkeys (Daly, 1958). Jackson was well aware of the imprecision of this localization and allowed that "the discharge-lesions in these cases are made up of some cells, not of the uncinate gyrus alone, but of some cells of different parts of a region of which this gyrus is a part—a very vague circumscription I admit—the uncinate region." (Hughlings-Jackson and Stewart, 1899, p. 538). Other uncinate manifestations include abdominal sensations (often a rising sense of pressure emanating from the viscera underneath the sternum—the "epigastric aura"), nausea, hallucinations of taste, movements related to feeding and ingestion (lip smacking, swallowing, licking, chewing, biting), and sensations from within the nasal passageways. It was Jackson's belief that such uncinate spells were all functionally linked to the process of feeding. Furthermore, he believed that such "crude" lower-level phenomena frequently co-accompanied the most elaborate psychical manifestations ("intellectual auras" and "dreamy states") in the same seizure:

> I think it will be found in many, I dare not say in most, cases, the voluminous mental state occurs in patients who have at the onset of their seizures some "digestive" sensation—smell, epigastric sensations, taste, or, in cases where there are movements implying excitation of centers for some such sensations, such movements as those of mastication (Hughlings-Jackson, 1931, p. 303).

Daly (1958) disputes the association with ingestive behavior both on anatomic grounds and empirically, by citing the relatively low co-occurrence of OHs and "oral automatisms" in both his series and that of Magnus, Penfield, and Jasper (1952). Stevens (1957) attempted to document patterns in the sequencing of complex partial seizure phenomena. She selected 40 patients with epileptic but not destructive lesions of the temporal lobe who were able to clearly describe the progression of their seizures. Eleven had OHs. In the eight patients who reported unpleasant odors, the OH was always preceded by one or two other experiences, most often a gastrointestinal sensation followed by fear. In five of these eight, the OHs gave way to dysphoric states, which were then followed by various automatisms such as running, drinking, lip smacking, gagging, chewing, and choking in four of the eight. In the three patients who described pleasant OHs, the OH was the first sign of the seizure. Stevens emphasized the different course of these latter spells, with OHs giving way to illusions and delusions, and an absence of "fight or flight" manifestations, such as fear states or running.

Weil (1955) remarks on the co-occurrence of epileptic OHs and depressive reactions (some of which lasted for weeks into the postictal period) in six patients. Serial EEG recordings in two patients established a clear association between paroxysmal temporal lobe activity and the depressive episodes, and when the depressions resolved (spontaneously or in response to medication), the OHs also remitted.

A patient followed by the present author reveals the tangled interactions between affective disturbance and epileptic OHs. This 58-year-old dextral female with a long history of chronic, intractable depression and suicidality (and a strong family history of the same) suffered the onset of seizures at age 45. Her spells, which occurred up to 10 times per month, consisted of intrusive thoughts ("a thought coming into my head, like something I didn't want to remember . . . like a thought that never came"), followed by a vague visual hallucination ("like a face in color") localized in the right visual field, speech arrest, and an epigastric sensation ("a sinking feeling in my abdomen"). In some but not all spells she experienced an OH ("it reminds me of a flower I smelled in Samoa"). Her elemental neurological exam was entirely normal. Neuropsychological testing documented defects in attention, remote memory, and new learning in both the visual and verbal domains. Head CT and subsequent angiography revealed two small aneurysms, one in the left anterior cerebral artery and one at the bifurcation of the left middle cerebral artery. Repeat EEGs were typically normal, though they sometimes revealed *right* temporal spike activity. She was placed on various anticonvulsants, and excellent seizure control was maintained on 1400 mg of carbemazepine. This patient did not display prominent interictal behavior traits (other than her profound, chronic depression and hyposexuality). She began to report recurrent sustained episodes of OH consisting of "urine" and "cigar smoke" that lasted for days. Convinced of their reality by the salience of the percepts, she kept searching the environment for the source of the smell, even after she accepted that they were related to her epilepsy! We have never been able to establish a clear temporal relation between these sustained OHs and her seizure control or fluctuations in her depression.

A report by Potolicchio, Lossing, O'Doherty, and Henkin (1986) purports to describe a "new and distinct seizure disorder" termed "cyclic phantosmia." Thirty-two patients, 23 of whom were female, complained of intermittent foul odors persisting for minutes to hours. In 25 percent of the cases the smell bilateralized. These OHs had been present for anywhere from 3 months to 28 years. Medical, neurological, and psychiatric histories were reported as benign. A viral syndrome preceded the OH in one-fifth of the cases. The authors report that all patients "discovered that a massive Valsalva relieved the odor, remissions lasting minutes to days." The patients were thought to manifest refractory temporal lobe seizures by their referring doctors, as they did not respond to "major anticonvulsants." Structural lesions were identified on head CT in only three patients. Nonepileptiform EEG abnormalities were noted in one-third. A positive clinical response to either valproic acid and/or butyrophenones was reported for "most patients" (dosage unstated). The authors classify this syndrome as a simple partial sensory seizure to distinguish it from complex partial spells with OH. There are several difficulties with their interpretation as described in this brief abstract. First, as noted above, it is not altogether clear what the criteria for, or parameters of, a simple (or basic or primary) olfactory percept are. In these 32 cases the designation of "simple" cannot be assessed because there were no qualitative descriptions of the OHs (other than their aversiveness) in the report. Moreover, one can challenge the diagnosis of partial epilepsy on the basis of the long duration of the OH, the lack of EEG confirmation, and the absence of plausible etiological conditions (such as head trauma). The further clarification of these issues seems necessary before proposing a new clinical syndrome.

While the above clinical observations raise the fascinating possibility of separate subtypes of olfactory seizures with distinct ictal components and differing clinical correlates, the relative paucity of cases precludes any definitive conclusions. Larger series of seizure patients are required, along with ambulatory EEG monitoring and quantitative assessment of both central and peripheral taste and smell function to better address these issues.

There is much variability in the patient's localization of the OH. Jackson noted that patients typically experience the OH as inside their nasal passages: "it must be added that the rule is that the patient feels the smell 'in his nose'" (Hughlings-Jackson, 1931, p. 303). Yet in the same paragraph he also describes other patients who search the environment or pinch their nostrils shut, convinced that there is an external source of the odor. Daly (1958) discusses intranasal sensations, chiefly tickling and a sense of stimulation, as components of uncinate spells. He terms these "rhinostomal sensations." Gilbert (1990) has suggested that such manifestations may represent seizure discharges from regions subserving trigeminal function. The presence of such interoceptive feedback may be the basis for the patient's belief about odor localization (Case F.N., below, demonstrates this feature). With regard to source attribution, patients with epileptic OHs do not attribute the odors to offensive bodily smells.[1]

The degree of the patient's insight into the nature of the OH—that is, whether or not it is recognized as a false perception—can be a function of the frequency and pattern of the spells. Upon the first seizure, the smell is believed to be real, but with continued, repeated, stereotypic spells, most patients come to realize that the odors are part and parcel of their seizures—no matter how compelling the experience. Whether they retain this insight at the exact moment of the hallucination or only upon *post hoc* reflection is undetermined, and perhaps indeterminate.

The true epidemiology of uncinate seizures is obscure, since clinical

[1] One possible exception may be Toone's (1978) case, discussed below.

series are subject to various sampling biases. Gastaut et al. (1955) found a prevalence of 5.9 percent among 1000 patients, whereas Lennox and Cobb (1933) documented a rate of 2 percent in their series of 1359 epilepsy cases. Mulder and Daly (1952) collected 100 noninstitutionalized patients with various temporal lobe lesions (approximately 25 percent of which were neoplastic), and found 16 cases of OH. Gibbs and Gibbs (1952) observed OHs in 0.3 percent of grand mal patients and 1.5 percent of those with limbic epilepsy. Penfield and Perot (1963) report on 25 years of neurosurgical experience. Out of a total operative series of 1132 patients, 520 had temporal lobe seizures. Only 7 of this group (0.6 percent) had ictal OHs! Howe and Gibson (1982) reviewed 273 patients with complex partial seizures and found 22 (8 percent) with olfactory and 6 (2 percent) with both olfactory and gustatory hallucinations.

The anatomical substrate of epileptic OHs has been the source of some controversy. The traditional characterization of the limbic system as the *rhinencephalon* underscores the acceptance of the functional-anatomical association between olfaction and this region. Yet rhinencephalic structures are imprecisely defined and may include: the olfactory bulb, tract, tubercle, trigone, and striae; the anterior perforate space; the anterior olfactory nucleus; the amygdala; prepyriform cortex; entorhinal cortex; the hippocampus; the septal complex; the supracallosal gyrus; and the gyrus fornicatus, which includes the subcallosal gyrus, cingulum, retrosplenial gyrus, and uncus (Nauta and Freitag, 1986; Carpenter, 1991; Locke, 1966). As noted above, Hughlings-Jackson's designation of the uncus as the primary site of seizure discharge was predicated on Ferrier's stimulation experiments (Hughlings-Jackson, 1931; Daly, 1958). A large body of data on experimental brain lesions and comparative neuroanatomy among nonhuman vertebrate species supports this basic finding (Sarnat and Netsky, 1981; Jenkins, 1978). In humans and other primates, the situation is predictably more complex. While the human olfactory pathways have been well described for years (Cajal, 1909), there is still uncertainty about central projections, especially the connectivity of the so-called medial olfactory projections (Carpenter, 1991), and whether the olfactory bulb projects directly to the amygdala (Nauta and Freitag, 1986). Moreover, in mammals the specialization of the limbic system has evolved to subserve a puzzling set of functions involving not only "drive-related" capabilities such as regulation of aggression, sexuality, and affect, but also complex social behaviors (parenting, attachment, dominance hierarchy), cognitive functions (attention, perception, and memory), and the structure of personality traits (Kluver and Bucy, 1939; Papez, 1937; Milner, 1954; Gloor, 1972; Mishkin, 1972; Bear, 1979). Yet the clinical correlation of discharges from the limbic region and the positive symptoms of epileptic OHs would seem to confirm at least a gross association between this region and smell representation. Converging evidence is also available from destructive lesions. For example, it is well known that patients develop side-specific impairments in smell discrimination and identification following anterior temporal lobectomy (Eskanazi et al., 1983). However, neither finding provides for precise localization, failing to establish the minimal lesion necessary for disrupting smell function. Because focal epileptic discharges by their nature spread to adjacent and even distant neural regions, localization on the basis of seizure foci is notoriously hazardous.

This point is underscored by some of the contradictions in the human temporal lobe stimulation literature. Penfield and his colleagues localized such "experiential" phenomena as auditory and visual hallucinations, reminiscences, and déja vû to the temporal neocortex (chiefly the lateral and superior surfaces of the first temporal convolution) based on the Montreal group's wide experience performing operative mapping of cortical regions (Penfield and Perot, 1963). Gloor and his colleagues (1982), using somewhat different stimulation parameters, simultaneously recorded from both deep limbic and temporal neocortical structures while stimulating and observing spontaneous seizures of same. They reported that activation of the deep limbic structures (amygdala, hippocampus, and parahippocampal gyrus) was *necessary and sufficient* for eliciting experiential responses. While it is tempting, then, to converge on these limbic nuclei as the regions critical for epileptic OHs, several obstacles prevent this conclusion. First, experiential responses are elicited in a highly variable proportion of patients undergoing stimulation [about 40/520 in Penfield's series, 18/29 in Gloor's series, and 31/36 in the Halgren, Walter, Cherlow, and Crandall (1978) series]. Furthermore, even among those who manifested the response, there is large individual variability in the number of such responses, with some patients having rare responses and other having dozens. Halgren et al. (1978) thus speculate that individual patient factors, as opposed to site of stimulation, may be the crucial determinant of when and where an experiential response is elicited. More problematic, however, is the overall paucity of OHs in the stimulation literature. Penfield does not report even a single instance of neocortical stimulation eliciting an olfactory hallucination, even though seven of his cases report OHs in their spontaneous seizure patterns. Gloor reports OHs in only 1 of 29 patients. Patient 5 responded to a left amygdaloid stimulation by a 3-mA current with an olfactory "sensation" that the patient described as "something to do with animals" that he claimed he had smelled before. In Halgren's et al. (1978) series, 3495 stimulation trials failed to produce even one single OH! Weingarten, Cherlow, and Holmgren (1977) stimulated deep temporal regions in 39 epileptics and did not report any OHs, though gastrointestinal and alimentary responses were common.[2] Other studies have noted a

[2]One possible exception may be Case 1, a 24-year-old female who reported a myriad of complex "experiential responses" to stimulation, including "as though she was standing around the chlorinated swimming pool of the YMCA as a child" (p. 203).

somewhat higher incidence of OH upon stimulation (e.g. van Buren, 1961).

Equally puzzling from a localizationist point of view is the report of Nashold and Wilson (1969). These authors describe five patients who underwent unilateral stimulation of various discrete thalamic nuclei (ventrolateral, ventroanterior, intralaminar, zona incerta, medial dorsalis) in preparation for stereotactic thalamic or midbrain ablations for relief of Parkinson's disease, chronic pain, or chorea (three, one, and one patients, respectively). All five reported OHs such as "acrid odor," "burning rubber," "cloves," and "new mown hay." One can conjecture that primary olfactory cortex may have been indirectly activated via its connections to the nucleus medial dorsalis, or that the reciprocal connections between thalamus and amygdala may have provided a path for transmission of impulses. Or else, in the first three patients, the destructive pathology of PD may have lowered the threshold for eliciting OHs (see the section "Degenerative Diseases" below).

An unusual lesion case bolsters the "thalamic connection" to OHs. Faris and Terrence (1989) report the case of a 46-year-old dextral male with a one-year history of paroxysmal, prolonged taste, smell, and visceral intrusions. A sweet, sickening taste for two to three days was followed by a smell "like diesel fuel" for another two to three days, followed by nausea and headaches. He reported other spells consisting of transient dizziness and visual illusions. Neurological exam and bedside mental status were unremarkable. CT visualized a 1-cm nonenhancing lesion of the third ventricle impinging on the fornix. Surgical removal of this colloid cyst led to total resolution of his symptoms. After excluding seizures as the cause of his episodes, the authors posit that the cyst exerted a mass effect on diencephalic structures, specifically the anterior and dorsomedial thalamus. They note that no transient ventricular enlargement was observed in the symptomatic phase, but explain that this can occur regularly with colloid cysts.

A case report by Andy (1967) documents a 38-year-old man who developed complex partial seizures following a closed head injury, with a right subdural hematoma that was surgically evacuated. He developed spells in which he smelled "vomitus," felt light-headed, heard auditory hallucinations, and underwent periods of automatic behavior. EEG revealed a right temporal focus. He was explored surgically and stimulated with bipolar electrodes in the amygdala and hippocampus. Amygdala stimulation re-created his seizures, including the foul smell, but hippocampal stimulation did not. These structures were surgically disconnected (to control for the effect of spreading after-discharges), and the same results obtained. The right temporal tip and amygdala were then excised with the hippocampus left intact. During a six-month follow-up period the patient remained completely seizure-free.

In conclusion, while the gross localization of epileptic OHs to activation of deep limbic nuclei in a proportion of patients seems secure, the exact region(s) necessary and sufficient for subserving the generation of false perceptions of smell, the precise mechanism by which this is achieved, and the factors that determine individual susceptibility all remain obscure.

Destructive Lesions

A destructive lesion might produce OHs by triggering complex partial seizures (as in Jackson and Stewart's case A.B. cited above). Alternatively, some cases of apparent epileptic OHs might represent destructive lesions of the temporal (or frontal) lobe that could interfere with smell function *independently* of the seizure discharge. Some authors argue that cases with temporal lobe neoplasms are preferentially associated with uncinate-type spells (Penfield and Perot, 1963; Feindel, 1974), though this is disputed by others (Howe and Gibson, 1982). Other common destructive etiologies include trauma, anoxia, stroke, and infection. Representative processes are temporal lobe gliomas, contusions and hematomas of the temporal poles and/or orbitofrontal cortex, olfactory groove meningiomas, pituitary adenomas and hamartomas, and herpes simplex encephalitis. With such central destructive insults, one would expect defects in "higher" olfactory function, such as odor matching, odor identification, and/or odor naming. In these instances OHs may be generated not by electrical discharges but by distorted processing of smell information, secondary to central dysfunction. An analogy with the visual system may be instructive. It is well known that destructive lesions leading to field defects—especially homonymous hemianopsia—can lead to complex visual hallucinations in the blind field. Cogan (1973) terms these "release" hallucinations because they are released under conditions of reduced afference. [Testing of the olfactory sense has no direct counterpart to the visual field exam. The closest analog might be left versus right nostril asymmetries (Gilbert, Greenberg, and Beauchamp, 1989).] Therefore some of the cases of epileptic OH may be solely due to central destructive lesion effects or the combined effects of central and peripheral lesions.

Damage to peripheral nasal structures and afferent pathways typically leads to disturbances in basic smell perception, including elevated thresholds, distorted psychophysical scaling, and impaired discrimination. Clinically these defects would be manifest as reduced sensitivity (hyposmia/anosmia) and/or distorted smell perceptions (dysosmia). The cranial nerve I is often inadequately tested, with odorants presented at superthreshold concentrations or with substances that stimulate trigeminal chemoreceptors (Schiffman, 1983). (What separates dysosmia from hallucinations is that the former distortion occurs only upon actual stimulation. In the terminology outlined at the outset of this chapter, these would conform to misperceptions or illusions, not hallucinations, because of the presence of an adequate stimulus.) However, following this framework, it is plausible that reduced or distorted afferent information could trigger true (or pseudo-) hallucinations of smell. Again, the visual system provides a clear analogy. Ocular disease and peripheral nerve damage leading to diminished visual input are common settings for developing visual hallucinations. Cataracts, traumatic eye injuries, and tumors of the optic tracts have all been linked to a lowering of the threshold for experiencing false perceptual experiences. Upper respiratory infections are a common cause of temporary and more rarely permanent anosmia (Schiffman, 1983). This leads to the question as to why more cases of peripheral anosmia (due to infection or trauma) don't produce OHs. The answer might be that many do but go unreported. In Schechter and Henkin's (1974) series of 29 traumatic brain injuries with taste and smell complaints, four were reported to manifest "phantosmia"—their term for OH. The situation is potentially confounded by the increased incidence of developing epilepsy among the head-injured, as the following case demonstrates.

F.N., a 32-year-old single, dextral female, sustained a closed head injury and multiple orthopedic injuries in a motor vehicle accident in which the driver, her boyfriend, was killed. She was briefly unconscious and was in an agitated confusional state for several days. Head CT was unremarkable. She underwent surgery for a fractured tibia and required high doses of opiates for pain control. Six weeks after the event her cognition revealed moderately severe deficits in psychomotor speed, attention and concentration, visual and verbal memory registration and retrieval, planning and executing complex tasks, and abstraction. Three months later her cognition had improved to the mild level of impairment. Three months later her psychometric tests were consistent with her premorbid cognitive capacity. Odor identification, assessed with the UPSIT (Doty, 1984), revealed mild microsmia. She had a difficult emotional recovery, with periods of depression and frustration that were clearly linked to her ongoing medical problems, her disability status, and the insensitivity of her employer and insurance carrier. She manifested elements of post-traumatic stress disorder (PTSD)—with nightmares about the accident and a startle response, especially around the annual anniversary of the accident—but did not meet accepted criteria for the full-blown syndrome. She was medicated with desipramine 50–100 mg. She began reporting paroxysmal episodes of OHs that increased in frequency over time. She noted "a pungent, chemical, foul smell that bristles in my nose . . . it feels like it is inside my face." Curiously, many of the episodes, each lasting 60 or so seconds, took place on her way to work. The odor was so powerful and intrusive that she asked passersby to validate her experience. Multiple sleep-deprived EEGs with NP leads were borderline normal, with rhythmic bursts of theta slowing with temporal predominance bilaterally. There were no typical electroencephalographic manifestations of complex partial seizures. However, she did experience a personality change, adopting many of the features of the interictal personality syndrome of temporal lobe epilepsy (Bear, 1979; Bear,

Freeman, and Greenberg, 1984). She began reacting to trivial events—such as being cut off in traffic—with intense rage. Whereas she once would shrug things off with a smile or a humorous aside, she reacted with moral indignation to perceived slights, petty annoyances, and minor transgressions of her friends and coworkers. Even though there were prominent rhinostomal features and she actually felt the odor in her nasal passages, she suspected that the odors were the result of a toxic leak. She thought of writing an angry diatribe against the medical institutions housed in the buildings she was passing while the OHs occurred. (The tendency to write at length—hypergraphia—is a well-known element of the interictal behavior syndrome.) This case illustrates why head trauma cases yield poor localizing information. One cannot be sure if the site of damage to olfactory regions is peripheral or central or if the experiences are epileptic or "release" phenomena. The topic of PTSD will be addressed in detail later.

A case of Schneider, Crosby, Bagchi, and Clahoun (1961), also points to the difficulty in teasing apart the effects of destructive and epileptogenic lesions and peripheral and central effects. Their case 1 was a 30-year-old dextral female with a five-week history of mild head trauma followed by headaches, discrete episodes of complex visual and auditory hallucinations, and OHs consisting of "the smell of cigar smoke." An EEG revealed focal slowing in the frontal regions bilaterally and in the left anterior and mid-temporal areas, with no spikes or epileptiform activity. Further investigation uncovered a brain abscess at the tip of the left frontal pole abutting the olfactory tract. Repeated left temporal lobe intraoperative stimulation failed to elicit hallucinations in any modalities. Following surgical excision of the abscess and a postoperative reinfection, the spells and OHs resolved.

Kaufman, Lassiter, and Shenoy (1988) describe a 31-year-old woman with an 11-year history of intermittent OHs of the same unfamiliar, indescribable odor that was only perceived in the left nostril. Durations ranged from seconds to hours, and frequencies varied from several daily to once a week. One year prior to the development of the OH, she suffered a minor head trauma with no loss of consciousness and a residual severe headache that lasted for days. Neurological exams and EEGs were normal. Several empirical drug trials were ineffective. Basic taste and smell function were tested and judged to be normal. The patient underwent craniotomy and excision of the entire left olfactory bulb. Pathological examination revealed nonspecific degenerative changes. The operation completely eliminated the OH for at least a three-year follow-up period! Leopold, Youngentob, Schwob, Hornung, Mozell, and Wright (1990) report a case of OH ("a mixture of rotten eggs and vinegar") that responded to unilateral excision of the nasal epithelia, effectively disconnecting the olfactory mucosa from the bulb. Two weeks postoperatively the OH had ceased, and basic olfaction was impaired. At five weeks no OHs were noted and olfactory sensitivity had returned.

Siegel (1978) screened 85 recreational users of cocaine (minimum of 1 gram of intranasal cocaine ingestion per month) for hallucinatory experiences. Six subjects reported OHs, which they attributed to heightened sensitivity to odors actually present. Various odors included "smoke," "natural gas," "feces," "urine," and "garbage." The authors raise the possibility that nasal pathology secondary to the route of drug administration may have produced "abnormal firings (input) to the bulb" (p. 312). Three subjects evidenced obvious rhinitis, and one had lesions of the walls of the nares. An alternative explanation is that chronic cocaine abuse lowered the seizure threshold, and the users were experiencing typical epileptic OHs. Not enough case information was provided to evaluate this latter hypothesis. Yet another hypothesis is that the induction of OHs may have been due to direct pharmacological effects on dopaminergic olfactory receptors. This is lent indirect support by Zilstorff's (1965) finding that application of intranasal 2 to 10 percent cocaine hydrochloride solution transiently diminished olfactory sensitivity ipsilaterally in four subjects. This finding raises the possibility that such a pharmacological effect on peripheral nerve tissue may have interacted synergistically with nasal pathology in chronic cocaine abusers to generate the observed OHs. Herpes simplex encephalitis (HSE) typically presents with focal or generalized seizures and is well known to selectively invade and destroy the temporal and frontal lobes. The route of entry into the CNS may be through retrograde transport of the virus along the olfactory tracts. Lehrich (1979) was the discussant in a clinicopathologic conference on herpes simplex encephalitis in which OHs figured prominently. The patient was a 59-year-old woman who was receiving antibiotic therapy for an ear infection for one month, followed by an upper respiratory infection and bitemporal headaches. She developed a fever and manifested confusion. She reported smelling the odor of fish on her hands. One day later she seized and was admitted. Her mental status deteriorated rapidly, with disorientation, visual hallucinations, and probable hallucinations of "perfume." EEG revealed delta slowing on the right side, with sharp waves more prominent on the right. CT only revealed diffuse swelling. In spite of treatment with adenine arabinoside, she continued to seize, became comatose, and died three weeks post-admission. At postmortem, necrosis of the inferior frontotemporal regions bilaterally was prominent, affecting the orbitofrontal cortex, hippocampus, anterior and inferior temporal lobe, insular cortex, cingulate gyrus, and olfactory bulbs and tracts. Similar cases are reported with other CNS infections—subacute sclerosing panencephalitis (Rosman, 1974) and Powassan virus encephalitis (Embril, Canfield, Artsob, and Chase, 1983). These cases are difficult to localize, since they involve a combination of seizures and structural damage to central and (probably) peripheral olfactory regions.

The author had the opportunity to follow a survivor of HSE who presented with a classic Kluver-Bucy syndrome—agitation and restless exploration of the environment, hyperorality (incessant eating and smoking) with a 50-lb weight gain, loss of maternal interest, but no inappropriate sexuality. Head CT revealed extensive necrotic changes in the orbitofrontal and inferior temporal regions, more prominently on the right side. Consistent with this finding, memory function was moderately impaired (visual worse than verbal), but language function was relatively intact. Smell identification function was tested with the UPSIT and found to be severely compromised. She often mistook nonfood odors for foodstuffs. On a graded series or pleasant (eugenol) and unpleasant (butyric acid) odorants, she was completely insensitive to changes in concentration. Most striking was her rating of all odors as highly pleasant. This was particularly dramatic because most normal volunteers reacted with immediate disgust to the highest concentration of butyric acid. This finding may suggest a clue to one of the most puzzling aspects of the Kluver-Bucy syndrome: increased and inappropriate orality. Perhaps the Kluver-Bucy monkey has lost the primary ability to discriminate food from nonfood on the basis of smell cues. It then indiscriminately mouths objects in the environment and uses intraoral cues to avoid swallowing harmful substances such as metal objects.[3] In spite of the increased mouthing of objects, Kluver-Bucy animals do not typically exhibit true hyperphagia or weight gain. Therefore it is possible that the patient reported here sustained damage to hypothalamic regions regulating appetite and metabolism. At follow-up two years later, her cognition had significantly improved, and her cigarette smoking had diminished along with her weight. She still failed to evidence any maternal interest in her two young children. She remains institutionalized in a chronic rehabilitation facility.

Degenerative Lesions

Defects in smell perception (including elevated odor detection thresholds and impaired discrimination) have been well documented among Parkinsonian patients (Ward, Hess, and Calne, 1983; Quinn, Rossor, and Marsden, 1987). Loss of smell function may precede the development of the motoric signs that are the defining feature of the disorder (Korten and Meulstee, 1980).

Korten and Meulstee (1980) questioned Parkinson's patients about their sensitivity to smell and found 41 who complained of impairment. (They also apparently tested odor discrimination, but the odorants used, methods employed, and results were not specified.) The authors also mention that there were no complaints of OHs ("parosmias"). Yet Sandyk (1981) re-

[3] A similar line of reasoning has been suggested by Bear (1979), but focuses on the Kluver-Bucy monkey's inability to identify foods via the visual channel, necessitating oral exploration.

ports four cases in which OHs occurred prior to or in parallel with the motor stigmata of Parkinson's disease. There was no history of epilepsy or depression, and EEGs and head CTs were negative. They report that L-Dopa therapy (dosages unstated) had a beneficial effect on the OHs. They implicate dopamine depletion in the mesolimbic pathways as a possible mechanism. At first glance these findings are at odds with the previously mentioned findings regarding cocaine usage, with acute cocaine ingestion leading to transiently impaired olfactory perception and chronic abuse lowering the threshold for OHs. Cocaine *increases* postsynaptic activity at dopaminergic synapses by blocking reuptake. If Zilstorff's (1965) observation is reliable and enhanced dopaminergic action does indeed reduce olfactory sensitivity, why should Parkinson's patients—with presumed dopamine depletion—demonstrate similar defects in olfaction? Moreover, the report of the reversibility of the OHs by an exogenously administered dopamine agonist further confuses the matter. This apparent paradox may be ultimately explained by invoking complex receptor kinetics and interactions.

Neuropathological findings in brain regions subserving olfaction are common among Alzheimer's patients, as are defects in smell perception (Greenberg and Gilbert, 1988; cf. Serby, this volume). Yet no case reports of OH were found for this disorder. This is difficult to interpret given the Alzheimer's victim's typical tendency to deny any impairment early in the course of the illness (and neurologists' and other physicians' general tendency to neglect the realm of olfaction). The pronounced confusion and communication deficits that characterize the later stages of that illness might account for the absence of self-reported OHs.

Perhaps a more basic question is: Given the well-documented decline of smell with age (Schiffman et al., 1979), why don't more elderly "normals" experience and complain of OHs? Both vision and audition predictably decline with normal aging, lowering the threshold for visual hallucinations and pseudo-paranoid ideation, respectively. This observation leads to another, perhaps more fundamental issue: Is there a normal analog to the neurologically induced OH? There is no established "base rate" for OHs among healthy, nonpsychiatric samples. The unavailability of such rudimentary descriptive data speaks to the above-noted neglect of olfaction in the medical world.

One place to search for a normal analog of OH would be in the realm of the dream. Are OHs generated in this familiar, endogenous alteration in consciousness, as are hallucinatory impressions in the visual, kinesthetic, auditory, interoceptive, and vestibular modalities? Early introspectionist studies of the phenomenology of dreaming (Thompson, 1914; Weed and Hallam, 1896) revealed a striking paucity of olfactory (and gustatory) images or ideation in dream reports. This observation—which accords well with clinical experience—remains unexplained.

Migraine

Classic migraine presents with simple visual hallucinations (scintillating scotomata), nausea and vomiting, and paresthesias, which typically precede onset of a hemicranial headache by 5 to 20 minutes. However, various neuropsychiatric phenomena, including motoric, perceptual, and cognitive symptoms, have been reported to occur both before and after headache onset, sometimes leaving persistent deficits following headache offset. Isolated case reports of OHs in migraine have appeared, suggesting that the incidence is low but not rare. Ardila and Sanchez (1988) collected 200 vascular headache cases and found two patients who experienced prodromal OHs: "a smell like acid" and "a smell like sewage" (p. 69). These authors suggest that such "neuropsychologic symptoms" appear only after many years of uncomplicated migraines. Wolberg and Ziegler (1982) report a typical case of a 32-year-old dextral female with a 15-plus-year history of mostly right-sided unilateral headaches with nausea and vomiting, which were heralded by moving "dark spots or lines" and transient odors described as "like a decaying animal." Neurological exam and neuroimaging studies failed to isolate a structural etiology. Crosley and Dhamoon (1983) report a case of an eight-year-old girl who developed unilateral, throbbing, predominantly right-sided headaches about two years after a minor head injury. The headaches were preceded by smells of "gas, burning cookies, or wood chips." Interestingly, her mother reported headaches that were preceded by "the smell of the inside of a pencil sharpener," raising the possibility that the child's case was purely functional (Crosley and Dhamoon, 1983, p. 459). Fuller and Guiloff (1987) document three cases of migrainous OHs. A 30-year-old woman had had headaches preceded by a stereotypic "repulsive smell" since the age of 12. The odors lasted from 5 to 120 minutes, followed by dysphoria and auditory hallucinations and then mostly right-sided hemicrania. Her EEG revealed bilateral, generalized paroxysmal slowing with theta and delta waves in the temporal lobes, especially on the right side. There was no response to anticonvulsant medication. Their second case was a 37-year-old female who reported a childhood onset of headaches preceded by "an unusual unidentifiable smell" lasting from 5 minutes to 24 hours, followed most often by a left-sided headache and nausea. Again there was no response to anticonvulsant therapy. Their final case was a 25-year-old woman who was afflicted with right- and left-sided headaches lasting 3 to 4 hours since age 12. They included pain, tinnitus, vomiting, visual loss, and dizziness. On three occasions she reported vivid OHs of 5 minutes duration of "her grandfather's cigars or peanut butter." EEG and CT were normal. The headaches responded to propranolol.

The mechanism for migrainous OH is as obscure as that of the headache and more typical visual hallucinations. Leading candidates are spreading cortical depression or vascular hypoperfusion, presumably affecting limbic regions. Obviously there is much overlap between these cases and uncinate seizures. On clinical grounds, seizures are much briefer, and are usually associated with obvious alterations in consciousness. While ictal headaches are not uncommon, they are also typically brief and lead to loss of consciousness.

To further confuse the picture, strong environmental smells have been reported to trigger migraine attacks. Fuller and Guiloff (1987) remark that some of these cases may in fact be OHs.

Psychiatric Syndromes

The focus of this chapter to this point has been on describing the various neurological entities in which OHs have been reported. We now turn to a brief review of the functional psychiatric conditions that manifest OH and the related phenomenon of olfactory delusions. Knowledge of such syndromes is of practical import, establishing the differential diagnostic possibilities, and of theoretic import, shedding light on the underlying mechanism by which OHs are generated.

Schizophrenia

A growing literature has documented both neuropathological and neuropsychological markers among schizophrenics. Two subgroups have been suggested. One group of patients have so-called negative symptoms, with constricted affect, social withdrawal, attention deficits, a poor premorbid adjustment, and a poor prognosis. Negative schizophrenics manifest many so-called frontal lobe features and often demonstrate enlarged lateral ventricles on head CT or MRI. The second group tends to have more positive symptoms, such as hallucinations and paranoia, suggesting a possible temporal-limbic locus of dysfunction. These latter patients tend to have grossly normal CTs and are more likely to have subtle cognitive deficits that overlap considerably with those of patients with idiopathic temporal lobe epilepsy (Seidman, 1990; Greenberg and Seidman, 1990). Moreover, electrical dysrhythmias, which are not however epileptiform, are commonly found among the EEGs of schizophrenics (Stevens, 1988). Thus, such "organic" aspects of schizophrenia could potentially involve the sense of smell at multiple levels. It is therefore not surprising that OHs have been documented among schizophrenic patients since the last century (Bullen, 1899). Alliez and Noseda (1945) surveyed 5400 patients at the Saint-Pierre Asylum from 1925 to 1938. They found OHs (and gustatory hallucinations) among 95 (about 2 percent) of the patients, 55 of whom carried the diagnosis of chronic hallucinatory psychosis. Bromberg and Schilder (1934) uncovered 40 cases of OH among an unstated number of Bellevue Hospital inpatients—a sample consisting of schizophrenics, alcoholics, and patients with assorted neurological disorders. Rupert, Hollander, and Mehrhof (1961) surveyed all schizophrenic patients known to have manifested hallucinations in any modality at both an acute and a chronic psychiatric facility and inquired about olfactory phenomena ($n = 36$). They found that 83 percent of the patients reported OHs (19 of 24 acute and 11 of 12 chronic cases). Unpleasant odors predominated by a ratio of 3:1. Contrary to accepted clinical lore, among acute patients the presence of OHs did not portend a poor outcome. Goldberg (1985) discussed the issue of positive versus negative symptoms and presented data supporting his contention that the "organic" symptoms (including OHs) generally did not respond to neuroleptic medication, whereas negative symptoms generally did. Ndetei and Singh (1983) surveyed 80 Kenyan schizophrenics and classified them into positive and negative subtypes using the New Haven Schizophrenia Index. Among the 51 positive patients, 13 (25 percent) reported OHs as compared to not one of 29 negative patients. Two studies further cloud the issue of the true incidence of OHs by documenting wide cross-cultural differences in percentages of patients acknowledging the symptom (Teggin, Elk, Ben-Aire, and Gillis, 1985; Ndetei and Vadher, 1984).

Qualitatively, the smell hallucinations of schizophrenics consist of the usual combination of noxious and fetid fumes. Fecal and other bodily odors seems to predominate, though patients usually localize or attribute such human odors to environmental sources and *not* their own bodies (Pryse-Phillips, 1971). Furthermore, patients tend not to draw attention to their OHs, and typically reveal their existence only upon inquiry (Rupert, Hollander, and Mehrhof, 1961; Kerekovic, 1972). Schizophrenics tend to describe their OHs in a matter-of-fact fashion that suggests that these OHs play a peripheral role in their experience. Not surprisingly, there is often an unmistakable thought-disordered quality to the report of the olfactory percept itself. For example, Bromberg and Schilder's case 1 reported, "fumes are administered under my door and through the radiator pipes. I hear them saying that they are going to kill me . . . Some of the smells go right through you, you inhale it, it goes into your bowels" (p. 477). The OHs can be integrated into delusional or classically paranoid schemes. Ideas of passivity and ideas of influence are often intertwined with the perceived source of the smell. Hallucinated odors therefore commonly seep in insidiously, penetrate the patient's environment or body, and exert a malign influence. It can therefore be exceedingly difficult for the examiner to distinguish between actual false perceptions of smell (true OHs) and delusionally based reports of perceived smells.

The mechanism of OH among schizophrenics is unknown and presumed to be of central origin. Kerekovic (1972) approached this population from the vantage point of the otolaryngologist and uncovered evidence of widespread nasal pathology. Thorough nasal examinations were performed on 434 psychiatric patients, 195 of whom were diagnosed as schizophrenic, and 40 (21 percent) of this group reported OHs. Pleasant OHs were found in 12 patients. Nasal pathology was common, observed in 46 (24 percent) of the patients. Of those with nasal pathology, 13 of 46 (28 percent) had OHs. Olfactory sensitivity was also examined. Of patients with unilateral anosmia, 4 of 9 (44 percent) hallucinated, but only 2 of 6 (33 percent) patients with bilateral anosmia did so. While these latter findings are of limited value because of the small number of observations, this study underscores the importance of assessing peripheral olfactory structure and function in psychiatric samples. The mechanism for many of the cases of presumed psychotic OHs may in fact be reduced afference secondary to peripheral causes. Or, as Kerekovic argues, in the schizophrenic patient there may be an admixture of peripheral and central mechanisms: "It may be supposed that in schizophrenics, as a consequence of altered associative processes, the pathologic changes in the nose act as a trigger for the precipitation of corticosensory events" (p. 521).

Most authors marvel at the ubiquity of OHs among schizophrenics. This is in marked contrast to the lack of interest in this phenomenon on the part of most clinicians. While this trend may reflect the aforementioned general neglect of olfaction throughout the medical community, it is also plausible that the patients' lack of concern about experiencing OHs plays an important role. The low levels of anxiety that OHs typically generate for the schizophrenic patient, especially in comparison to hallucinations in the auditory and visual realms, is quite striking. For example, patient 1 of Rupert et al. (1961) was asked: "Did you, when you first came into the hospital, tell the doctors about this—that you smelled things?" He replied, "Not when I first came in . . . I was more frightened of the noise and sights I thought I saw than I was of odors" (p. 316).

Olfactory Reference Syndrome (ORS)

While the OH is usually an insignificant feature of the schizophrenic's mental life, it is the very epicenter of the condition known as the olfactory reference syndrome. In ORS the individual is convinced that there is an offensive odor emanating from his or her body. The source is always attributed to an internal source and is not linked to the environment. The patient may localize the site of origin of the perceived odor to a bodily orifice, commonly the anus, genitalia, or the mouth. The onset is typically insidious, and the course is chronic. Often the odor is constant and cannot be named. It is typically but not universally aversive or even repugnant. Patients are unswerving in their conviction that the odor actually exists. They seek out multiple medical check-ups, often with dentists, ENTs, or gastrointestinal or pulmonary specialists—without ever identifying a plausible source. As these interventions fail to remedy the situation, patients get frustrated and then disgusted and hop from expert to expert in search of a medical explanation for the perceived odor. Patients with ORS develop the preoccupation with the validity of their illness that is characteristic of other hypochondriacal states. They frequently engage in compulsive ritualistic behavior revolving around orderliness and personal hygiene, in ill-fated attempts to control the odors. In some cases characterological obsessive-compulsive trends and preoccupations with contamination predate the onset of the syndrome. Withdrawal from social contact, disabling shame and humiliation, and depression often ensue. The male-to-female ratio is unknown, and the age of onset seems to cut across the lifespan.

Most remarkably, patients with the ORS do not exhibit obvious signs of a generalized thought disorder or psychotic ideation outside the realm of the fixation on their bodily odor (Viderbech, 1961; Pryse-Phillips, 1971). In recognition of this cardinal feature, Beary and Cobb (1981) termed the syndrome a "Solitary Psychosis." ORS patients also do not report other environmental OHs. Auditory hallucinations and first-rank signs of schizophrenia are typically absent. Depressive features are common but are not typically as severe as those seen in major (endogenous) affective disorders.

Viderbach (1967) reviews the literature, presents five cases in which olfactory ideas of reference are prominent, and discusses phenomenological and nosological issues. In particular, he addresses the distinction between hallucination and delusion and concludes that only two of the five patients actually smelled anything. For example, his case 1 reported a 20-year preoccupation with befouling his environment through "flatus expelled from his mouth." Any social situation that made him nervous triggered this reaction. The abrupt onset occurred after his employer on one occasion turned away from him with a groan of disgust! He was convinced of the reality of his foul breath, and consulted a dentist and attended religiously to his oral hygiene, to no avail. Progressive social and occupational constriction and fluctuating depression ensued. He was first treated as an outpatient and finally admitted on three occasions for psychiatric treatments, including chemotherapy, insulin coma therapy, and finally frontal leukotomy, none of which had any apparent impact on his delusional system. When interviewed in follow-up three years after the operation, he acknowledged that he had never really perceived the smell himself but only deduced it from the reactions of others. Even without such first-hand verification of the odor, he remained convinced with unshakable certainty of the *idea* that he was malodorous. This degree of conviction is characteristic of delusionally held ideas. Other patients with the ORS clearly report that they do actually perceive the offending odor and therefore exhibit true OHs in addition to olfactory delusions.

As this case demonstrates, the false belief that one is emitting horrid smells is commonly predicated upon misinterpretations of others' actions. Patients overinterpret or misread meaningless gestures, comments, acts, and facial expressions and react catastrophically, concluding that they have behaved offensively. This elicits feelings of shame and a tendency to withdraw from social interaction. The author recalls a student who telephoned and apologized poignantly for repeatedly disrupting my lectures; he was convinced that he was emanating a vile odor from an unnamed stomach ailment, and equally convinced that all were aware of this "fact." When I remarked that I had no idea what he was talking about, and that I wasn't even sure if I could place a face to his name, he thanked me for trying to spare his feelings. Further protestations on my part had no apparent impact on the certainty with which he clung to his ideas of redolence, and he ultimately dropped out of the class.

Delusions of halitosis were reviewed recently by Goldberg, Buongiorno, and Henkin (1985) with mention of six new cases. As the authors note, this syndrome is actually a specific clinical variant of the ORS. They raise the notion that the raw material for the false conviction that one reeks malodorously can be a true distortion of actual taste or smell input. The authors emphasize that all six of their cases had subjective complaints of dysgeusia that preceded the preoccupation with their odors! Two patients claimed to be able to smell their own odors, and they also apparently reported nonself OHs. Other unexpected findings were a family history of complaints of halitosis and a history of nasal signs and symptoms: chronic intermittent rhinorrhea, congestion, headache, sinus pain without purulent discharge, and seasonal and allergic rhinitis. The uncovering of preexisting peripheral nasal distress and universal dysgeusia is highly suggestive of a significant role for aversive peripheral input as the raw material upon which the ORS is built. This notion receives support from a study of psychological adjustment following colostomy. Orbach, Bard, and Southerland (1952) found that a proportion of postoperative patients developed preoccupations about emanating offensive bodily odors and compulsive hygienic rituals that were strikingly similar to those seen in the ORS. Pryse-Phillips (1971) has replicated this finding.

The case of Toone (1978) is instructive because it contains features of peripheral deafference, complex partial seizures, and the ORS. A 44-year-old man suffered from psychomotor seizures consisting of absence and occasional automatisms for five years, for which he received anticonvulsants. He then developed symptoms of the ORS with the conviction that he gave off a peculiar burning odor "of old leaves or a bonfire" (p. 62). The percept was constant, but intensified in social situations. He misinterpreted actions of others—for example, the switching on of an exhaust fan in a pub—as proof of his malodorousness. He used copious amounts of deodorant to mask the smell. He then began experiencing both hyposmia and hyperosmia and true OHs. These OHs, however, were not associated with his usual seizure pattern. Neurological examination revealed unilateral right-sided anosmia, and imaging studies documented a large AVM of the right medial and inferior frontal lobe. There was no further intervention, but he reportedly gained some insight from the explanation that the odors were the product of a neurological condition. This case may be the exception to the rule that epileptic OHs are always attributed to external sources.

From a taxonomic viewpoint, the ORS is considered to represent an isolated, encapsulated psychosis (or fixed delusional idea) that is distinct from other psychotic disturbances. Although the DSM-III-R (1987) classifies it as one of the delusional (paranoid) disorders, the ORS includes various ingredients of schizophrenia, hypochondriasis, obsessive-compulsive disorder, phobic disturbance, and psychotic-level depression.

Depression

OHs can be an associated symptom of major depression. The typical patient is an elderly individual with negativism, somatic and cognitive complaints, and prominent vegetative signs. In retarded depression, apathy, anergia, anhedonia, and withdrawal are present. In agitated depressions, irritability, agitation, and insomnia are observed, with impaired concentration and ruminations. Lack of appetite with consequent weight loss and complaints of GI malfunction are common in both subtypes. Preoccupation with death, deterioration, and despair may dominate a patient's ideational space. Such ideas are often intimately linked to feelings of guilt and shame. When such preoccupations become rigidly engraved in the psyche and dominate a patient's thinking, they are termed fixations. Bizarre, irrational ideas that one is dead, dying, or decomposing—termed nihilistic delusions—are extreme manifestations of this syndrome. It is in this setting that OHs can arise and be spontaneously reported or readily elicited upon interview. The quality of the smells are congruent with the patient's mood and ideation—that is, they refer to rotting, decay, disintegration, and disease. The hallucinations tend to be more or less constant and not limited to discrete episodes. As with the syndromes described above, there can be great difficulty distinguishing smells that are actually perceived from the *idea* that one is giving off an odor (somatic delusions) (Munro, 1983; Malasi, El-Hilu, Mirza, and Fakhr El-Islam, 1990). To further complicate the matter, such patients often neglect their oral hygiene and grooming, perhaps providing stimuli that comprise the "raw material" upon which OHs and delusions are then elaborated.

Pryse-Phillips (1971), while acknowledging the considerable overlap between the ORS and depressive OHs, presents clinical data that suggest that there are features that distinguish the two entities. He reported that ORS patients either deny subjective dysphoria or experience it only in reaction to their feelings of shame and humiliation and their resulting restricted life style. On the other hand, in depressive OHs, the dysphoria invariably precedes the development of the OH, and vegetative symptoms are prominent. Thus major depression should be entertained in neurologically intact patients presenting with OH accompanied by marked dysphoria, morbid, depressive ideation, and prominent vegetative signs. The OHs—being part and parcel of the depressive symptomology—remit and reoccur in parallel with the depression. They often respond to aggressive anti-depressant therapy—chemotherapy and/or ECT. That seizures induced by ECT can ameliorate OHs via an antidepressant effect seems paradoxical given the etiological association between spontaneously occurring seizures and OHs.

In practice, though, discriminating depression from the ORS may be difficult. Malasi et al. (1990) report a case (3) of a 53-year-old Kuwaiti porter in which the authors diagnosed depression. He complained of sadness and had crying spells because he believed a foul odor emanated from his mouth. He avoided social interactions and lost weight and became suicidal. He hallucinated coworkers talking about his bad odor. After two months of treatment with trimipramine and thioridazine, his symptoms—which had persisted for 20 years—remitted. Alternatively, one could argue that this individual was manifesting the ORS rather than a primary depression. Clearly there is much overlap between the two entities. Lewis, Feldman, Greene, and Martinez-Mustardo (1984) selected 12 cases of bipolar disorder and inquired about the presence of epileptoid manifestations. Six reported olfactory hallucinations, including "funny things," "smoke," "flowers," "cologne," and "coffee." This observation suggests that OH may be more prevalent among manic-depressive patients than is generally appreciated.

Post Traumatic Stress Disorder (PTSD)

PTSD is a well-described cluster of emotional and psychophysiological reactions that can arise in the wake of exposure to an overwhelming life event. PTSD has been documented subsequent to combat experiences, enforced captivity, civilian disasters, rape, fires, transportation accidents, and a multitude of related misfortunes. The syndrome of PTSD is an admixture of two polar clusters of symptoms. One set of symptoms relates to the patient's attempts at suppressing the memories and affects engendered by the trauma. These include denial, psychic numbing, social withdrawal, avoidance, amnesia for all or part of the event, and even frank dissociation. The second cluster arises when the suppressed memories and feelings break through to consciousness in the form of intrusive images, repetitive nightmares, hypervigilent overreactions to innocuous situations, and behavioral reenactments of aspects of the original event (Van der Kolk, 1986). Following exposure to a traumatic experience in which there were strong actual odors present, a patient can reexperience them at later intervals in a brief altered state of consciousness that is similar to but potentially distinguishable from a seizure. This phenomenon has been termed an "olfactory flashback." In some cases (Kline and Rausch, 1985) the flashback is precipitated by smelling a similar odor. Their case 1 was exposed to blood and carnage during an air evacuation of wounded and deceased comrades. Years later, while helping to bandage his neighbor's wounded hand, he experienced a traumatic intrusive memory: "I could smell the blood and I was back in Nam" (p. 383). In other cases there is no obvious olfactory cue that triggers the prior memories. Burstein's (1987) case 2 was of a man who was overcome by smoke inhalation in a fire. Five days after the event, he spontaneously reported OHs in which he experienced the same smell of smoke that he had during the fire. The available case material is quite limited, and the actual prevalence of this reaction is unknown. Given that repression and amnesia are defining characteristics of the syndrome, it would not be at all surprising if the link between a real-life exposure and a subsequent OH was routinely overlooked by both patient and clinician. Malasi et al (1990) reveal that their case 1—a fairly typical presentation of ORS with the OH of "urine and feces emanating from his mouth" (p. 256)—experienced a homosexual assault in childhood. Pryse-Phillips' (1971) case 9 had the onset of the vague OH of "body odor or something like that" (p. 492) immediately after having to wear clothes that were impregnated with smoke from a house fire, subjecting him to ridicule by his classmates.

Of course, in any accident or toxic exposure involving compromise to the CNS, epilepsy and damage to the nasal pathways must be ruled out. A case seen in consultation by the present author demonstrates the difficulty in doing so with confidence. A graduate student in his mid-twenties was transiently exposed to organotin during a laboratory accident. He was hospitalized for treatment of superficial burns and for observation. Three days after the exposure he became confused, with a prominent amnestic disturbance. Serum and urine tin levels documented exposure. His confusion cleared within days to weeks, but he complained of residual cognitive difficulties. On his first day back to the laboratory where the accident took place, he felt a burning sensation on his face, smelled vapors, began choking, and lost consciousness. He was confused when discovered by a friend some moments later. EEGs revealed mild and inconsistent slowing of the temporal lobes bilaterally, with no spikes or seizure discharges. He was medicated with anticonvulsants for a presumed seizure disorder secondary to a toxic encephalopathy. He demonstrated prominent symptoms of withdrawal, apathy, anhedonia, and constriction. He complained of frequent spells that were indescribable, brief lapses in consciousness. He denied other episodes of OH. In this case there were features consistent with both TLE and PTSD, and the OH he experienced may have been either a traumatic reminiscence or a limbic discharge.

The mechanism of OHs in PTSD is not known. A Pavlovian conditioning model has been proposed to account for the perceptual flashbacks in other modalities in PTSD (Pitman, 1988). It seems plausible that a life-threatening environmental situation that primes the limbic system can set the stage for a powerful and perhaps indelible learning trial in which an external smell cue is associated with an internal state of visceral arousal and anxiety. When the individual reencounters contextual cues related to the original exposure, the visceral reaction is reelicited. If this model is correct, such olfactory flashbacks should in principle respond to deconditioning techniques such as flooding and desensitization. Although OHs in PTSD may be an uncommon occurrence, they may teach us something important about the formation of emotional memories and also offer a window through which one can glimpse the inner workings of the intact limbic system.

Conclusion

This chapter has surveyed historical, clinical, epidemiological, and phenomenological aspects of olfactory hallucinations. This curious symptom is associated with a rather astonishing range of medical, neurological, and psychiatric conditions. As such, the OH lies at the interface between psychiatry, neurology, and otolaryngology. The literature consists mostly of compilations of selected samples of patients and small series of case reports. Attempts at integrating this wealth of clinical data are hampered by our limited understanding of the neural code for olfaction and the interactions between peripheral and central structures in odor perception. The future explication of the underlying mechanism(s) of OHs will require the incorporation of sophisticated and standardized tests of basic and higher-order olfaction into the neurological exam, support for longitudinal studies of the natural history of the various syndromes with careful epidemiological and clinicopathological correlations, and advances in our basic knowledge of "normal" olfaction.

Acknowledgment. The author wishes to gratefully acknowledge the assistance of Ronald Calvanio, Ph.D., who read an earlier draft of this manuscript and suggested many changes that were subsequently incorporated into the text.

References

Adams, R.D., and M. Victor (1985). *Principles of Neurology*, 3rd Ed. New York: McGraw Hill.

Alliez, J., and M. Noseda (1945). Considerations statistiques et cliniques sur les hallucinations olfactives et gustatives. *Ann. Med. Psychol.*, *103*, 134–141.

Amoore, J.E. (1977). Specific anosmia and the concept of primary odors. *Chem. Senses*, *2*, 267–281.

Andy, O.J. (1967). The amygdala and hippocampus in olfactory aura. *EEG Clin. Neuophysiol.*, *23*, 291–293.

Ardila, A., and E. Sanchez (1988). Neuropsychologic symptoms in the migraine syndrome. *Cephalgia*, *8*, 67–70.

Bear, D.M. (1979). Temporal-lobe epilepsy—a syndrome of sensory-limbic hyperconnections. *Cortex*, *15*, 357–384.

Bear, D.M., R.L. Freeman, and M.S. Greenberg (1984). Behavioral alterations in patients with temporal lobe epilepsy. In D. Blumer (Ed.), *Psychiatric Aspects of Epilepsy*. Washington, DC: American Psychiatric Association.

Beary, M.D., and J.P. Cobb (1981). Three cases of monosymptomatic delusion of elementary stench treated with behavioural psychotherapy. *Br. J. Psychol.*, *138*, 64–66.

Bromberg, W., and P. Schilder (1934). Olfactory imagination and olfactory hallucinations. *Arch. Neurol. Psychiatry*, *32*, 467–492.

Bullen, F.S.J. (1899). Olfactory hallucinations in the insane. *J. Men. Sci.*, *45*, 513–533.

Burstein, A. (1987). Olfactory hallucinations. *Hosp. Community Psychiatry*, *38*, 80.

Cajal, R.S. (1909). *Histologie du Systeme Nerveux del'Homme et des Vertebres*. Paris: Norbet Maloine.

Carpenter, M.B. (1991). *Core Text Of Neuroanatomy*. 4th Ed. Baltimore: Williams and Wilkins.

Cogan, D.G. (1973). Visual hallucinations as release phenomena. *Albrecht von Graefes Arch. klin. exp. Opthalmol.*, *188*, 139–150.

Crosley, C.J., and S. Dhamoon (1983). Migrainous olfactory aura in a family. *Arch. Neurol.*, *40*, 459.

Daly, D. (1958). Uncinate fits. *Neurology*, *8*, 250–260.

Daly, D. (1975). Complex partial seizures. In J.K Penry and D.D. Daly, *Adv. Neurol.*, Vol. 11. New York: Raven Press.

Damas-Mora, J., M. Skelton-Robinson, and F.A. Jenner (1982). The Charles-Bonnet Syndrome in perspective. *Psychol. Med.*, *12*, 251–261.

Diagnostic and Statistical Manual of Mental Disorders, 3d Ed., Revised. (1987). Washington, DC: American Psychiatric Association.

Doty, R.L., P. Shaman, and M. Dann (1984). Development of the University of Pennsylvania Smell Identification test: a microencapsulated test of olfactory function. *Physiol. Behav.*, *32*, 489–522.

Dyck, P. (1985). Sylvian lipoma causing auditory hallucinations: case report. *Neurosurgery*, *16*(1), 64–67.

Embril, J.A., P. Camfield, H. Artsob, and D.P. Chase (1983). Powassan virus encephalitis resembling herpes simplex encephalitis. *Arch. Intern. Med.*, *143*, 341–343.

Eskenazi, B., W.S. Cain, R.A. Novelly, and R. Mattson (1986). Odor perception in temporal lobe epilepsy patients with and without temporal lobectomy. *Neuropsychologia*, *24*, 553–562.

Faris, A.A., and C.F. Terrence (1989). Limbic system symptomology associated with colloid cyst of the third ventricle. *J. Neurol.*, *236*, 60–61.

Feindel, W. (1974). Temporal lobe epilepsy. In O. Magnes and A.M. Lorentz de Haas, *Handbook of Clinical Neurology*, Vol. 15. Amsterdam: Elsevier.

Fuller; G.N., and R.J. Guiloff (1887). Migrainous olfactory hallucinations. *J. Neurol. Neurosurg. Psychiatry*, *50*, 1688–1690.

Gastaut, H., J. Roger, and C. Giove (1955). Troubles de l'olfaction de la gustation et de l'appetite chez les epileptiques psychomoteurs. *Annales de le Med-psychol*, *113*, 177–206.

Gibbs, F.A., and E.L. Gibbs (1952). *Atlas of Electroencephalography*, Vol. II. Cambridge, Mass. Addison-Wesley.

Gilbert, A.N. (1990). Personal communication.
Gilbert, A.N., M.S. Greenberg, and G. Beauchamp (1989). Handedness, sex and side of nostril modulate human odor perception. *Neuropsychologia*, *27*(4), 505–511.
Gloor, P. (1972). Temporal lobe epilepsy: its possible contribution to the understanding of the functional significance of the amygdala and of its interaction with neocortical-temporal mechanisms. In B.E. Eleftheriou, *The Neurobiology of the Amygdala*. New York: Plenum.
Gloor, P., A. Olivier, L.F. Quesney, F. Andermann, and S. Horowitz (1982). The role of the limbic system in experiential phenomena of temporal lobe epilepsy. *Ann. Neurol.*, *12*, 129–144.
Goldberg, R.L., P.A. Buongiorno, and R.I. Henkin (1985). Delusions of halitosis. *Psychosomatics*, *26*(4), 325–331.
Goldberg, S.C. (1985). Negative and deficit symptoms in schizophrenia do respond to neuroleptics. *Schizophrenia Bull.*, *11*(3), 453–456.
Gowers, W.R. (1910). The Hughlings-Jackson lecture on special sense discharges from organic disease. *Brain*, *32*, 303–326.
Greenberg, M.S., and M.J. Farah (1986). The laterality of dreaming. *Brain and Cognition*, *5*, 307–321.
Greenberg, M.S., and A.N. Gilbert (1988). Le Systeme olfactif dans la maladie d'Alzheimer. *Alzheimer's Actualities*, *28*, 6–9.
Greenberg, M.S., and L.J. Seidman Neuropsychological aspects of temporal lobe epilepsy. In R.F. White (Ed.), *Neuropsychology: A Practitioner's Handbook*. Submitted.
Halgren, E., R.D. Walter, D.G. Cherlow, and P.C. Crandall (1978). Mental phenomena evoked by stimulation of the human hippocampal formation and amygdala. *Brain*, *101*, 83–117.
Henkin, R. (1971). Disorders of taste and smell. *JAMA*, *218*, 1946.
Howe, J.G., and J.D. Gibson (1982). Uncinate seizures and tumors, a myth re-examined. *Ann. Neurol.*, *12*, 227.
Hughlings-Jackson, J. (1931). Lectures on the diagnosis of epilepsy. In J. Taylor, *Selected Writings of John Hughlings-Jackson*. London: Hodder and Stoughton.
Hughlings-Jackson, J., and P. Stewart (1899). Epileptic attacks with a warning of a crude sensation of smell and with the intellectual aura (dreamy state) in a patient who had symptoms pointing to gross organic disease of the right temporal-sphenoidal lobe. *Brain*, *22*, 534–549.
Jaspers, K. (1963). *General Psychopathology*. Manchester, England: J. Hoenig & M. Hamilton.
Jenkins, T.W. (1978). Functional Mammalian Neuroanatomy. 2nd Ed. Philadelphia: Lea & Febiger.
Kaufman, M.D., K.R.L. Lassiter, and B.V. Shenoy, (1988). Paroxsysmal unilateral dysosmia: a cured patient. *Ann. Neurol.*, *24*, 450–451.
Kerekovic, M. (1972). The relationship between objective disorders of smell and olfactory hallucinations. *Acta Oto-Rhino-Laryngol. Belg.*, *26*(5), 518–523.
Kline, N.A., and J.L. Rausch (1985). Olfactory precipitants of flashbacks in posttraumatic stress disorder: case reports. *J. Clin. Psychiatry*, *46*(9), 383–384.
Kluver, H., and P.C. Bucy (1939). Preliminary analysis of functions of the temporal lobes in monkeys. *Arch. Neurol. Psychiatry*, *42*, 979–1000.

Korten, J.J., and J. Meulstee (1980). Olfactory disturbance in Parkinsonism. *Clin. Neurol. Neurosurg.*, *82*(2), 113–118.

Lehrich, J.R. (1979). Case records of the Massachusetts General Hospital—Case #44-1979. *N. Engl. J. Med.*, *301*(18), 987–994.

Lennox, W.G., and S. Cobb (1933). Epilepsy XIII. Aura in Epilepsy: A statistical review of 1,359 cases. *Arch. Neurol. Psychiatry*, *30*, 374–387.

Leopold, D.A., G. Preti, M.M. Mozell, and S.L. Youngentob, and H.N. Wright (1990). Fish-odor syndrome presenting as dysosmia. *Arch. Otolaryngol. Head Neck Surg.*, *116*, 354–355.

Leopold, D.A., S.L. Youngentob, J.E. Schwob, D.E. Hornung, M.M. Mozell, and H.N. Wright (1990). Successful treatment of phantosmia with preservation of olfaction: a case report. *Abstr. 12th Ann. Meeting Assoc. Chemoreception Sci.*, 185.

Lewis, D.O., M. Feldman, M. Greene, and Y. Martinez-Mustardo (1984). Psychomotor epileptic symptoms in six patients with bipolar mood disorders. *Am. J. Psychiatry*, *141*(12), 1583–1586.

Locke, S. (1966). *Neurology*. Boston: Little Brown.

Magnus, O., W. Penfield, and H. Jasper (1952). Mastication and consciousness in epileptic seizures. *Acta Psychiatr. Neurol. Scand.*, *27*, 91–115.

Malasi, T.H., S.R. El-Hilu, I.A. Miraza, and M. Fakhr El-Islam (1990). Olfactory delusional syndrome with various etiologies. *Br. J. Psychiatry*, *156*, 256–260.

Mayeux, R., J. Brandt, J. Rosen, and D.F. Benson (1980). Interictal language and memory impairment in temporal lobe epilepsy. *Neurology*, *30*, 120–125.

Milner, B. (1954). Intellectual functions of the temporal lobes. *Psychol. Bull.*, *51*(1), 42–62.

Mishkin, M. (1972). Cortical visual areas and their interactions. In A. G. Karczmar and J.C. Eccles, *Symp. Brain Human Behav.*, Berlin: Springer-Verlag.

Mulder, D.W., and D. Daly (1952). Psychiatric symptoms associated with lesions of the temporal lobes. *JAMA*, *150*, 173–176.

Munro, A. (1983). Paranoia revisited. *Br. J. Psychiatry*, *141*, 344–349.

Nashold, B.S., and W.P. Wilson (1970). Olfactory hallucinations evoked from stimulation of human thalamus. *Confin. Neurology*, *32*, 298–307.

Nauta, W.J.H., and M. Feirtag (1986). *Fundamental Neuroanatomy*. New York: W.H. Freeman.

Ndetei, D.M., and A. Singh (1983). Hallucinations in Kenyan schizophrenic patients. *Acta Psychiatr. Scand.*, *67*, 144–147.

Ndetei, D.M., and A. Vadher (1984). A comparative cross-cultural study of the frequencies of hallucination in schizophrenia. *Acta Psychiatr. Scand.*, *70*, 545–549.

Orbach, C.E., M. Bard, and A. Southerland (1952). Fears and defensive adaptations to the loss of anal sphincter control. *Psychoanalyt. Rev.*, *44*, 121–175.

Papez, J.W. (1937). A proposed mechanism of emotion. *Arch. Neurol. Psychiatry*, *38*, 725–743.

Penfield, W., and P. Perot (1963). The brain's record of auditory and visual experience. A final summary and discussion. *Brain*, *86*, 595–696.

Pitman, R.K. (1988). Post traumatic stress disorder, conditioning and network theory. *Psychiatr. Ann.*, *18*, 182–189.

Potolicchio, S.J., J.H. Lossong, D.S. O'Dohertt, and R.I. Henkin (1986). Partial

seizures with simple psychosensory symptomology (cyclic phantosmia): a new and distinct seizure disorder. *Clin. Res.*, *34*(2), 635A.
Pryse-Phillips, W. (1971). An olfactory reference syndrome. *Acta Psychiatr. Scand.*, *47*(4), 484–509.
Quinn, N.P., M.N. Rossor, and C.D. Marsden (1987). Olfactory threshold in Parkinson's disease. *J. Neurol. Neurosurg. Psychiatry*, *50*, 88–89.
Reed, G. (1972). *The Psychology of Anomalous Experience.* London: Hutchinson.
Rosman, P. (1974). Case records of the Massachusetts General Hospital—Case #28-1974. *N. Engl. J. Med.*, *291*(3), 141–149.
Rupert, S.L., M.H. Hollender, and E.G. Mehrhof (1961). Olfactory hallucinations. *Arch. Gen. Psychiatry*, *5*, 313–318.
Sandyk, R. (1981). Olfactory hallucinations in Parkinson's disease. *South African Med. J.*, *60*, 950.
Sarnat, H.B., and M.G. Netsky (1981). Evolution of the Nervous System. 2nd Ed. New York: Oxford University Press.
Schechter, P.J., and R.I. Henkin (1974). Abnormalities of taste and smell after head trauma. *J. Neurol. Neurosurg. Psychiatry*, *37*, 802–810.
Schiffman, S.S. (1983). Taste and smell in disease (first of two parts). *N. Engl. J. Med.*, *21*, 1275–1279.
Schiffman, S.S., M. Orlandi, and R. Erickson (1979). Changes in taste and smell with age: biological aspects, pp. 247–268. In J.M. Ordy and K. Brizzee, *Sensory Systems and Communication in the Elderly*. New York: Raven Press.
Schneider, R.C., E.C. Crosby, B.K. Bagchi, and H.D. Calhoun (1961). Temporal or occipital lobe hallucinations triggered from frontal lobe lesions. *Neurology*, *11*, 172–179.
Seidman, L.J. (1991). The neuropsychology of schizophrenia. *J. Neuropsychiatry Clin. Neurosc.*, In press.
Siegel, R.K. (1978). Cocaine hallucinations. *Am. J. Psychiatry*, *135*, 309–314.
Spiers, P.A., D.L. Schomer, H.W. Blume, and M.-M. Mesulam (1985). Temporalimbic epilepsy, pp. 289–326. In M.-M. Mesulam, *Principles of Behavioral Neurology*. Philadephia: F.A. Davis.
Stevens, J.R. (1957). The "march" of temporal lobe epilepsy. *AMA Arch. Neurol. Psychiatry*, *77*, 227–236.
Stevens, J.R. (1988). Epilepsy, psychosis and schizophrenia. *Schizophrenia Res.*, *1*, 79–89.
Teggin, A.F., R. Elk, and O.G. Ben-Aire (1985). A comparison of Catego class 'S' schizophrenia in three ethnic groups: psychiatric manifestations. *Br. J. Psychiatry*, *147*, 683–687.
Thompson, E.R. (1914). An inquiry into some questions connected with imagery in dreams. *Br. J. Psychology*, *1*, 300–318.
Toone, B.K. (1978). Psychomotor seizures, arterio-venous malformation and the olfactory reference syndrome. *Acta Psychiatr. Scand.*, *58*, 61–66.
Van Buren, J.M. (1961). Sensory, motor, and autonomic effects of mesial temporal stimulation in man. *J. Neurosurg.*, *18*, 272–288.
Van der Kolk, B. (1986). The human response to overwhelming life events. In B. Van der Kolk, *Psychological Trauma*. Washington, DC: American Psychiatric Association.
Videbech, T. (1967). Chronic olfactory paranoid syndromes. *Acta Psychiatr. Scand.*, *42*, 182–213.

Ward, C.D., W.A. Hess, and D.B. Calne (1983). Olfactory impairment in Parkinson's disease. *Neurology*, *33*, 943–946.

Weed, S., and F.M. Hallam (1896). A study of the dream consciousness. *Am. J. Psychol.*, *7*, 405–411.

Weil, A.A. (1955). Depressive reactions associated with temporal lobe—uncinate seizures. *J. Nervous Ment. Dis.*, *121*, 505–510.

Weingarten, S.M., D.G. Cherlow, and E. Holmgren (1977). The relationship of hallucinations to the depth structures of the temporal lobe. *Acta Neurochirurgica, Supplement 24*, 199–216.

Wolberg, F.L., and D.K. Ziegler (1982). Olfactory hallucination in migraine. *Arch. Neurol.*, *39*, 382.

Zilstorff, K. (1965). Sense of smell alterations by cocaine and tetracaine. *Arch. Otolaryngol.*, *82*, 53–55.

Zilstorff, K. (1966). Parosmia. *Acta J. Laryngo. Oto.*, *80*, 1102–1104.

18
Olfaction in Aging and Medical Disorders

Susan S. Schiffman

Introduction

Considerable shifts in demographics are taking place on a global scale, and the most significant change is a massive increase in the elderly population. This shift will present challenges to medical practitioners because the elderly have special sensory and nutritional needs that must be taken into account when providing medical care. The sense of smell is especially vulnerable to loss in old age, and this can result in inadequate intake, especially in the sick elderly (Schiffman, 1983, 1987; Schiffman and Warwick, 1988).

As we enter the twenty-first century, the number of persons with olfactory losses will be staggering as a result of the projected growth in the absolute and relative size of the elderly populations. In the United States, 21.8 percent (or 67.4 million persons) will be at least 65 years of age by the year 2050. Globally, the number of persons aged 60 years or more is expected to increase from 376 million in 1980 to 1.121 billion in 2025. On a percentage basis, the elderly will increase from 8.5 percent to 13.6 percent of the world population during that period. In 2025, 72 percent of the world's elderly will reside in developing countries. Losses in olfactory sensitivity may exacerbate the malnutrition that is prevalent and persistent in developing nations where there is chronic poverty.

This chapter will describe the chemosensory losses that occur in older persons. These decrements result from a combination of causes, including normal aging, diseases (especially Alzheimer's disease), and drug therapy. Methods of compensating for these losses will be given. In addition, a model that relates sensory and anatomical losses will be described.

Perceptual Losses in Smell in the Elderly

There is a decline in olfactory perception with increasing age. These losses generally begin in the sixties (although they can begin earlier) and become more severe in persons over 70 years of age. Three classifications of olfactory losses have been described: anosmia (absence of smell), hyposmia (diminished sensitivity of smell), and dysosmia (distortion of normal smell). Subcategories of dysosmia include phantosmia (perception of an unpleasant smell when none is present) and parosmia (perception of an unusual odor in response to a particular stimulus). While dysosmia is difficult to quantify in a psychophysical experiment, anosmia and hyposmia can be quantified by measuring responses to odors at both threshold and suprathreshold concentrations.

Olfactory losses have both disadvantages and advantages (Schiffman, 1979, 1983). Reduced ability to perceive odors increases the risk from hazards such as leaking gas, spoiled food, and smoke. It also decreases the appreciation of food and can even lead to malnutrition. On the other hand, losses in perception for unpleasant odors (e.g., urine) are clearly an advantage. Elderly patients in nursing homes do not complain about unpleasant urinous odors because they do not perceive them.

Threshold Studies

Elevated detection and recognition thresholds for olfactory and trigeminal stimulants are generally found in elderly persons. Elevated thresholds have been reported for *n*-butanol (Kimbrell and Furchtgott, 1963); coal gas (Chalke and Dewhurst, 1957; Chalke et al., 1958); coffee and citral (Megighian, 1958); food odors including cherry, grape, and lemon (Schiffman, 1979; Schiffman et al., 1976); menthol (Murphy, 1983); pyridine and thiophene (Perry et al., 1980); 18 purified odorants (Venstrom and Amoore, 1968); and citralva (Schiffman and Warwick, 1990). A review of these studies indicates that thresholds for elderly persons over 70 years of age are from 2 to 10 times higher than for young persons in their twenties. The sick elderly, especially those taking multiple drugs, tend to have the highest thresholds.

Suprathreshold Studies

Three methodological approaches have been used to investigate suprathreshold functioning in older individuals: magnitude estimation, identification, and discrimination studies. Magnitude estimation experiments, in which numbers are assigned to odors in proportion to their perceived intensities, suggest that the elderly perceive suprathreshold odors to be half as intense as do their younger counterparts. Magnitude estimates reveal losses for pleasant (benzaldehyde and *d*-limonene), foul (pyridine), and neutral (ethyl alcohol and isoamyl alcohol) odorants (Stevens and Cain, 1985). Similar losses using magnitude estimation have been reported for isoamyl butyrate (Stevens et al., 1984; Stevens and Cain, 1985; Stevens et al., 1982); menthol (Murphy, 1983); eight odorants, including citralva, geraniol, citronellal, 2-methoxy-3-isobutyl-pyrazine, benzaldehyde, 2-methoxypyrazine, limonene, and acetic acid (Schiffman and Warwick,

1990); and CO_2, which stimulates the trigeminal nerve (Stevens et al., 1982).

Reduced scores on identification tasks are also found in elderly persons. Losses have been reported for coffee, peppermint, coal tar, and oil of almonds (Anand, 1964), a wide range of foods (Murphy, 1985; Schiffman, 1977), steroids (Schiffman, 1979), 40 common substances (Schemper et al., 1981), and a microencapsulated battery of 40 to 50 odors (Doty et al., 1984a and b). Scores for healthy elderly persons over 70 years of age are generally from 60 to 75 percent of those for young subjects.

Multidimensional scaling experiments in which odors are arranged in a space on the basis of their perceived similarities indicate that the elderly have reduced capacity to discriminate the degree of difference between odors of different qualities. Multidimensional scaling experiments have found losses in discrimination for food odors (Schiffman and Pasternak, 1979), common odors (Stevens and Lawless, 1981), and pyrazines (Schiffman and Leffingwell, 1981).

An Experiment That Reveals the Age of Onset of Olfactory Losses

A cross-sectional study of persons in seven decades of life suggests that a systematic decrement in performance on multiple olfactory tasks begins around 60 years of age and becomes significantly worse in the seventies and eighties (Schiffman and Warwick, 1990). In this study, four tasks were used to assess olfactory and trigeminal functioning. Three tasks assessed the ability to discern differences among suprathreshold odors: a discrimination task, adjective ratings, and identification scores. The fourth task measured detection thresholds.

One hundred forty three persons distributed into seven decades—10–19, 20–29, 30–39, 40–49, 50–59, 60–69, and 70–79—took part in the study. The seven groups were approximately equivalent in intelligence, gender, socioeconomic status, race, cultural eating patterns, and smoking patterns. The subjects were healthy and taking a minimal number of medications. The stimuli for the suprathreshold tasks were nine odorants that range widely in quality, hedonicity, chemical structure, and olfactory quality. These included benzaldehyde, *n*-butanol, caproic acid, citral, citronellal, geraniol, guaiacol, menthol, and methyl salicylate. The odorants were diluted in an odorless grade of diethyl phthalate and were adjusted until they were approximately equal in moderate odor intensity for 40 young subjects (20–25 years old).

Discrimination Task

The discrimination task entailed two steps: confusability ratings and similarity ratings. In the confusability task, subjects sniffed three bottles one at

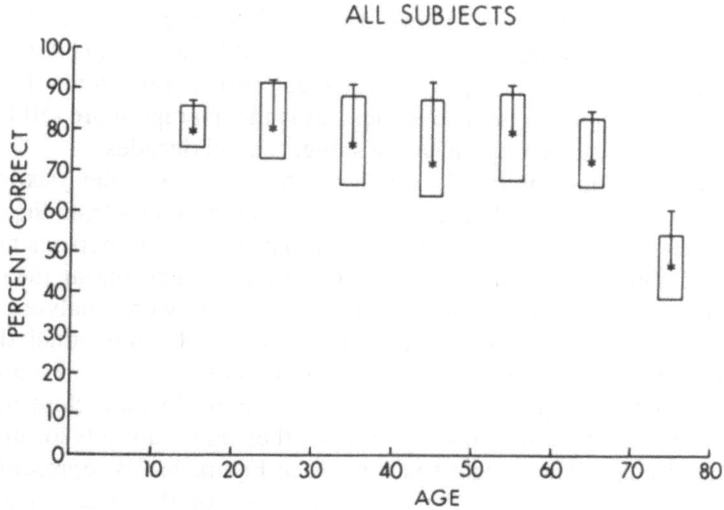

FIGURE 18.1. Mean percentages of 36 triads correctly identified by decade along with standard deviations. Percentages span from the 25th to 75th percentile. Persons in their seventies performed significantly worse than persons in other decades.

a time; two of the bottles contained the same odorant. In a second step, subjects marked the similarity of the two different odorants on a nine-point scale from "identical" to "completely different." The results of the confusability experiment are given in Figure 18.1. The mean percentages of the 36 triads correctly identified by those in each decade are represented as asterisks, and the standard deviations are given by lines. The rectangles span from the 25th to the 75th percentile. Persons in their seventies performed significantly worse than those in previous decades. No gender differences were found.

The percentages of the 36 triads correctly identified in the confusability experiment are represented as percentiles for each decade in Table 18.1.

TABLE 18.1. Percentile in which a person would be classified by decade and over all subjects based on percent correct score

Percentile	*10s*	*20s*	*30s*	*40s*	*50s*	*60s*	*70s*	*Overall*
99	91.7	100.0	97.2	100.0	97.2	91.7	77.8	100.0
95	91.6	99.7	96.9	99.4	96.7	91.7	77.8	94.4
90	88.9	94.4	91.7	93.9	91.1	91.7	72.8	91.7
75	85.4	91.0	88.4	87.4	88.9	83.3	54.2	86.1
50	80.6	79.2	83.3	77.8	86.1	69.4	47.2	77.8
25	75.7	72.9	66.7	63.9	68.1	66.7	38.9	61.1
10	66.9	61.8	53.1	42.2	61.6	58.3	30.3	45.5
5	64.0	61.1	50.1	29.2	58.6	50.0	23.0	39.5
1	63.9	61.1	50.0	27.8	58.3	50.0	22.2	24.7

The scores are interpreted as follows: a score of 54.2 percent correct would place a 75-year-old in the 75th percentile but would relegate an 18-year-old to the 1st percentile. A score of 77.8 percent correct represents the 99th percentile for persons in their seventies but is the average score (50th percentile) for the entire group including subjects in all decades.

Although the confusability data indicate that subjects in their sixties retain the capacity to select that stimulus in a triad that differs from the other two, the ratings on similarity scales reveal that these same persons have a diminished ability to discriminate the *degree* of difference among the odorants. The mean similarity matrices for each decade were analyzed by a multidimensional scaling procedure called ALSCAL (individual differences option). The similarities common to all decades are represented in the two-dimensional space in Figure 18.2A; stimuli located close to one another were rated more similar to one another than odorants located far from one another. The weight space given in Figure 18.2B represents the salience (or importance) of the two dimensions of the space in Figure 18.2A that is idiosyncratic for each decade. The multidimensional spaces for decades with weights far from the origin are similar to the stimulus arrangement in Figure 18.2A. The positions close to the origin of the weights for the sixth and seventh decades indicate that the space in Figure 18.2A does not represent persons in their sixties and seventies. Individual multidimensional spaces based on mean similarity scores indicate that *subjects in their sixties and seventies have difficulty rating the degree of similarity* between two different odor stimuli. The degree of loss in ability to rate the degree of similarity is considerably greater in the seventies than in the sixties.

Adjective Ratings

Ratings of the nine odorants were made on a series of nine-point adjective scales: floral, burnt, sweet, animal, citrus, minty, fuel, almond, and pleasant. These "verbally based" data were collected to compare with the confusability and similarity judgments, which are "nonverbal" in nature. An analysis of the ratings indicated that older subjects (in their sixties and seventies) perceived benzaldehyde as less sweet; *n*-butanol as less burnt; caproic acid as less animal; citral as less sweet, citrus, and pleasant; citronellal and geraniol as less citrus; menthol as less minty; and methyl salicylate as less sweet and minty. No statistically significant gender differences were found.

Odor Identification

Subjects were asked to identify each odorant after they had made ratings on adjective scales. Persons in their sixties and seventies were significantly worse at odor identification than persons from 10 to 59 years of age. The greatest losses were for geraniol, guaiacol, citronellal, and caproic

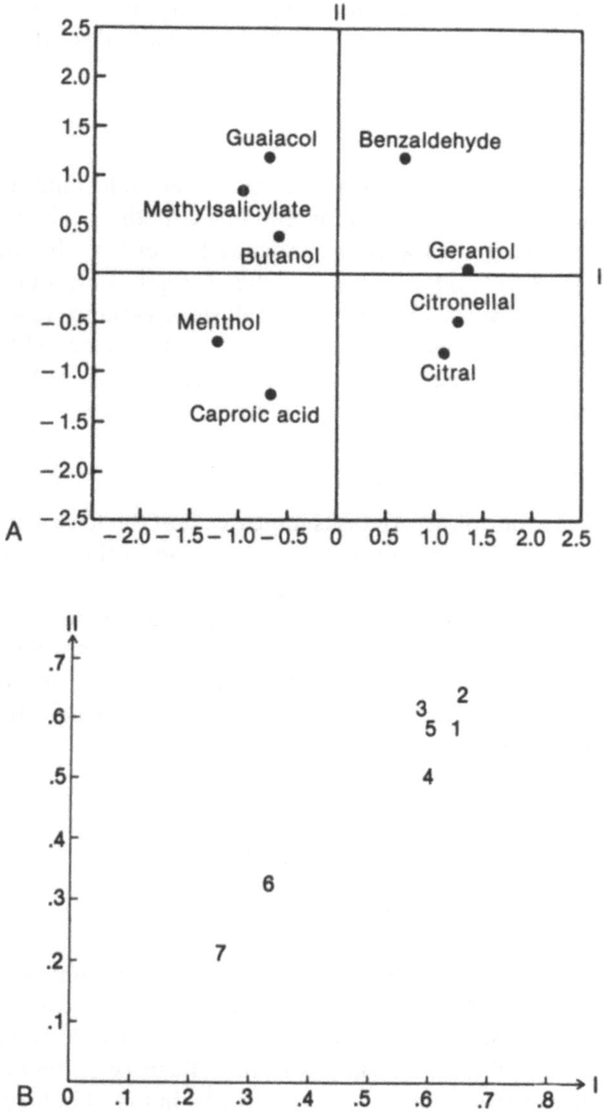

FIGURE 18.2. A. Two-dimensional space representing the similarities among stimuli for all decades. Stimuli arranged close to one another are similar in odor quality for the group as a whole. B. Weight space that represents the salience (or importance) of the two dimensions of the space in A for each decade. Weights for the sixth and seventh decades are close to the origin, which indicates that persons in their sixties and seventies do not perceive the similarities among odors in the same way as younger persons.

acid, and the smallest losses were for citral (lemon) and methyl salicylate (minty) odors. Females were significantly better at odor identification than males.

Detection Thresholds

Detection thresholds were determined for three odorants (geraniol, guaiacol, and benzaldehyde) using a forced-choice method. The thresholds for persons in their seventies were numerically higher than for persons in their teens—1.87, 2.42, and 2.01 times higher for geraniol, guaiacol, and benzaldehyde, respectively. However, an analysis of variance revealed that these differences did not reach the 0.05 level of significance. This is probably due to the fact that the elderly tested in this study were healthy, active, and relatively drug free.

Conclusion Regarding When Olfactory Loss Occurs

An overview of the four olfactory tasks indicates that the losses in olfaction begin in the sixties and become progressively worse in the seventies. The first losses are expressed as a diminished capacity to discriminate the *degree of difference among odorants* on similarity scales. Losses in the ability to rate and identify the qualitative aspects of odors also begin at this time. Females outperform males on identification tasks, but not on discrimination tasts. Thus, females may not have a better sense of smell than males, but rather are better at assigning labels to odors.

Anatomic and Physiologic Losses in the Olfactory System with Age

A range of anatomic and physiologic modifications parallel the sensory losses in the olfactory system during the aging process. These losses have been reported throughout the olfactory system and include reduced protein synthesis and structural alterations in the olfactory epithelium (Dodson and Bannister, 1980; Naessen, 1971), atrophy in the olfactory bulb and nerve (Brizzee et al., 1975; Hinds and McNelly, 1977, 1978, 1979, 1981, 1982; Liss and Gomez, 1958; Smith, 1942), presence of senile plaque and neurofibrillary tangles in the hippocampus and amygdaloid complex (Scheibel and Scheibel, 1975; Tomlinson and Henderson, 1976), hypothalamic degeneration, including disruption of the hypothalamic architecture paralleled by deterioration and loss of dendritic surface (Machado-Salas et al., 1977), altered calcium homeostasis in the hippocampus, leading to elevated intracellular calcium (Landfield and Pitler, 1984), and hippocampal pathology, including an increase in reactive astrocytes associated with elevated plasma adrenocorticoids (Landfield et al., 1978).

Causes of Olfactory Losses with Age

The decrements in olfactory perception with age can result from the anatomical and physiological changes associated with normal aging (just described) as well as disease states (Table 18.2), drugs (Table 18.3), and pollutants (Table 18.4). Diseases, drugs, and pollutants can affect the sense of smell in young and middle-aged persons as well. Most losses prior to old age result from local disorders (nasal and/or sinus disease), upper respiratory infections, and head trauma. Nasal and/or sinus disease can result in blockage of odorants from the olfactory epithelium or in damage to receptors and olfactory pathways (Doty and Snow, 1987). The losses in olfaction after upper respiratory infections probably result from direct damage to the olfactory epithelium and replacement of the olfactory epithelium with respiratory epithelium (Douek et al., 1975); olfactory cilia may be damaged as well (Jafek et al., 1989). Anosmia after head trauma is thought to result from the shearing of olfactory nerve fibers as they pass through the cribriform plate.

Aside from nasal disease, respiratory infections, and head trauma, the mechanisms by which disease, drugs, and pollutants alter olfactory perception are not well understood, and there are many possible candidates for the mechanistic cause. For example, they may interfere with transmembrane signaling mediated by one of the three major second messenger systems: the adenylate cyclase system, the phosphatidyl-inositol system, and the arachidonic acid system. The adenylate cyclase system, which regulates intracellular cAMP levels through activation or inhibition of adenylate cyclase via a regulatory GTP-binding protein (see Gilman, 1989, for review), is frequently altered by medical conditions and drugs. The phosphoinositol system (see Baraban et al., 1989), which involves the hydrolysis of phosphatidyl-inositol 4,5-bisphosphate (PIP_2) by phospholipase C, generating two products, the sugar phosphate inositol-triphosphate IP_3 and diacylglycerol (DAG), may also be altered. The arachidonic acid system, in which phospholipase A_2 is activated, generating arachidonic acid and its metabolites (see Axelrod et al., 1988), may also be affected. Alterations in membrane channels (e.g., sodium, potassium, or calcium channels) and neurotransmitter levels may also play a role. Interference with the olfactory receptor protein may also occur. Further research is necessary to determine the mechanisms by which disease, drugs, and pollutants alter olfactory perception.

Perceptual Losses of Smell in Alzheimer's Disease

The olfactory losses in Alzheimer's disease (AD), a dementing illness characterized by progressive memory loss and deteriorating cognitive functioning, are severe and are substantially worse than in age-matched controls for

TABLE 18.2. Medical conditions affecting smell

Nervous	
Alzheimer's disease	Schiffman et al. (1990); Serby et al. (1985a and b)
Down's syndrome	Warner et al. (1988)
Epilepsy	Currie et al. (1971)
Head trauma	Leigh (1943); Levin et al. (1985); Moran et al. (1985); Schecter and Henkin (1974); Sumner (1964)
Korsakoff's syndrome	Jones et al. (1975, 1978); Mair et al. (1986)
Migraine	Wolberg and Ziegler (1986); Crosley and Dhamoon (1983)
Multiple sclerosis	Pinching (1977)
Parkinson's disease	Ansari and Johnson (1975); Ward et al. (1983); Doty et al. (1988)
Pick's disease	Kretschmar (1984)
Tumors and lesions	Bakay (1984); Furstenberg et al. (1943); Olsen and DeSanto (1983)
Nutritional	
Chronic renal failure	Schiffman et al. (1978)
Liver disease including cirrhosis	Burch et al. (1978); Garrett-Laster et al. (1984)
Vitamin B_{12} deficiency	Rundles (1946)
Endocrine	
Adrenal cortical insufficiency	Henkin (1975)
Cushing's syndrome	Henkin (1975)
Hypothyroidism	McConnel et al. (1975); Schaupp and Seilz (1969)
Diabetes mellitus	Jorgensen and Buch (1961)
Gonadal dysgenesis (Turner's syndrome)	Henkin (1967)
Hypogonadotropic hypogonadism (Kallman's syndrome)	Kallmann et at (1944); Males et al. (1973)
Primary amenorrhea	Marshall and Henkin (1971)
Pseudohypoparathyroidism	Henkin (1968)
X-linked ichthyosis due to steroid sulfatase deficiency	Sunohara et al (1986); Andria et al. (1987)
Local	
Adenoid hypertrophy	Ghorbanian et al. (1978)
Allergic rhinitis, atopy, and bronchial asthma	Church et al. (1978); Fein et al. (1966)
Crouzon's syndrome	Das and Munro (1979)
Leprosy	Barton (1974)
Ozena	Strandbygard (1954)
Paranasal sinus exenteration	Hoye et al. (1970)
Sinusitis and polyposis	Fein et al. (1966); Hotchkiss (1956); Ryan and Ryan (1974)
Sjögren's syndrome	Henkin et al. (1972)

TABLE 18.2. (*cont.*)

Viral and Infectious	
Acute viral hepatitis	Henkin and Smith (1971)
Influenzalike infections	Henkin et al. (1975)
Other	
Amyloidosis and sarcoidosis	Schellinger et al. (1983); Delaney et al. (1977)
Familial (genetic)	Singh et al. (1970)
Laryngectomy	Henkin et al. (1968); Henkin and Larson (1972)
Psychiatric disorders	Meats (1988)

Adapted by permission, from Schiffman (1983, 1987).

TABLE 18.3. Drugs affecting smell

Anesthetics, local
Cocaine hydrochloride and tetracaine hydrochloride (Zilstorff, 1965)
Antianginal drugs
Diltiazem (Berman, 1985); Nifedipine (Levinson and Kennedy, 1985)
Antimicrobial agents
Allicin (Body, 1986); Streptomycin (Zilstorff and Herbild, 1979); tyrothricin (Seydell and McKnight, 1948)
Antithyroid agents
Carbimazole (Erikssen et al., 1975); Methimazole (Erikssen et al., 1975; Hallman and Hurst, 1953); Methylthiouracil (Schneeberg, 1952); Propylthiouracil (Grossman, 1953)
Opiates
Codeine, hydromorphone hydrochloride, and morphine (Macht and Macht, 1940)
Radiation theraphy
Radiation to head (Carmichael et al., 1984)
Sympathomimetic drugs
Amphetamines (Goetzl and Stone, 1948; Schiffman, 1983; Turner, 1965); Phenmetrazine theoclate with fenbutrazate hydrochloride (Turner, 1965)
Vasodilator
Diltiazem (Berman, 1985)
Other
Acetylcholine-like substances (Skouby and Zilstorff-Pedersen, 1954)

Adapted by permission, from Schiffman (1983, 1987).

suprathreshold tasks (see the Chapter by Serby et al.). Losses in recognition and identification of odorants are salient in the earliest phases of the disease. Decrements in the ability to remember odorants have also been found. Altered functioning at the threshold level is more controversial. The sensory decrements in Alzheimer's disease result from histopathological and neurochemical changes in the olfactory system.

TABLE 18.4. Industrial pollutants that can cause hyposmia or anosmia

Acids	Naus, 1967
Aerosols of paints	Rossberg et al., 1966
Benzene	Naus, 1976
Carbon disulfide	Naus, 1976
Formaldehyde	Gorman, 1964, Emmet, 1976
Hydrazine	Rossberg et al., 1966
Hydrogen sulfide	Ahlborg, 1951
Hydrogen selenide	Symanski, 1950
Insecticides	Schiffman, 1983
Menthol	Naus, 1976, Naus, 1968
Niltous gases	Naus, 1976
Oil perppermint	Lehnhardt and Rollin, 1969
Pepper and creosol mixtures	Lehnhardt and Rollin, 1969
Phosphorous oxychloride	MacIntyre, 1971
Solvents	Dragomiretskii et al., 1970; Emmet, 1976
Trichloroethylene	James, 1963
Dusts	Naus, 1976; Adams and Crabtree, 1961
Coke (coke kiln)	
Grain (tractor station)	
SiO_2 (mine)	
Tobacco (cigarette factory)	
Spices (spice mill)	
Flour (flour mill)	
Cotton (knitting factory)	
Paper (paper-packing factory)	
Cement (cement works)	
Cadmium (battery works)	
Ashes (incinerator)	
Lead (lead works)	
Coal (coal bunker)	
Chromium (dye shop, chemistry lab)	
Nickel (nickel refinery)	

Recognition and Identification Tasks in Alzheimer's Patients

All studies that tested the ability to recognize or identify odors in Alzheimer's disease have found losses. Waldton (1974) first reported that persons with AD had difficulty identifying the presence of an odor. In both a cross-sectional and longitudinal study he found diminished ability to identify tinctura asafetida, camphor, citrus fruit, eau de cologne, menthol, and petrol. Peabody and Tinklenberg (1985) found a reduced ability to smell and identify strong odors (lemon, peppermint, coffee, maple, and vinegar) from memory or to select the correct identity of odors from a list. In a 10-pair forced-choice identification task in which the patient was asked to chose which of two vials contained the odor named by the experimenter, Alzheimer's patients also performed significantly worse than age-matched controls (Serby et al.,1985a and b; Serby, 1986, 1987). The ten pairs of odors tested in this experiment were garlic–onion, lemon–orange, water–pineapple, chocolate–maple, bacon–curry, rose–vanilla, coffee–peanuts, vinegar–soap, cinnamon–licorice, and ketchup–mint.

Knupfer and Spiegel (1986) found that Alzheimer's patients had difficulty determining whether the second stimulus of a pair was identical to the first one. Pairs of identical, similar, and different odors were presented, including coconut–cloves, eucalyptus–smoked bacon, aniseed–orange, rum–licorice, asparagus–pine needles, moth balls–camphor, marzipan–bitter almond, grass–hay, lemon–flower, and peppermint–spearmint. Knupfer and Speigel's patients with AD were also asked to choose which of two vials contained an odor named by the experimenter, and again they performed poorly. Performance on a standardized scratch and sniff test containing 36 to 40 odors impregnated in tapes also show reduced recognition and identification of odors in AD (Warner et al., 1986; Koss et al., 1987, 1988; Kesslak et al., 1988; and Doty et al., 1987). Rezek (1987) found that patients with AD performed poorly when they were asked to smell and identify cinnamon oil, lemon oil, peppermint oil, turpentine, and ground coffee and select the name of the odor from a 10-item list. Thus, there is total agreement among these studies that the ability to recognize and identify odors in Alzheimer's disease is impaired.

Odor Memory Tasks in Alzheimer's Patients

Odor memory tasks indicate dramatic losses in Alzheimer's disease. Moberg et al. (1987) found that Alzheimer's patients performed poorly on a task in which 10 target odors, including onion, vanilla, almond, peppermint, lemon, cinnamon, pine, mothballs, smoke, and banana, were first presented. Five minutes later, 20 odorants were presented: the original 10 target odors plus 5 distractants (odors similar to the original 5) and 5 dissimilar odors. The subject was asked to determine if any of the 20 odorants were the same as the original 10. Kesslak et al. (1988) also found poor olfactory memory in Alzheimer's patients on an odor match to sample test. One of 15 uncommon odors (target) was followed by three choice odors. The task was to choose which of the three choice odors matched the target.

Olfactory Threshold Studies in Alzheimer's Patients

The data on losses at threshold concentrations in Alzheimer's disease are equivocal. St. Clair (1985) found no marked impairment in AD compared to aged-matched controls. Rezek (1987) found little or no threshold losses in early AD on a two-bottle forced-choice task in which amyl alcohol and cinnamon oil were dissolved in heavy mineral oil. Koss et al. (1987, 1988) also found no statistically significant losses for pyridine, although performance was poor for 3 of 10 patients with early AD. On the other hand, Knupfer and Spiegel (1986) found threshold losses for eucalyptol, citral, and prunolide (coconut) using a three-bottle forced-choice procedure; one bottle contained the odor plus solvent, and the other two bottles contained only solvent. Doty et al. (1987) reported a threshold loss for phenyl ethyl alcohol in mild to moderate AD using a single staircase forced-choice procedure. Murphy et al. (1987) found a loss for butanol in AD using a two-alternative forced-choice ascending series.

Causes of Olfactory Losses in Alzheimer's Disease

The sensory losses in Alzheimer's disease are in part a consequence of the histopathological changes that occur in olfactory pathways, including the olfactory epithelium (Talamo et al, 1989), olfactory bulb (Ohm and Braak, 1987), anterior olfactory nucleus (Averback, 1983; Esiri and Wilcock, 1984; Ohm and Braak, 1987; and Kahn et al., 1987), olfactory tubercle (Simpson et al., 1984), amygdala (Brun and Gustafson, 1976; Herzog and Kemper, 1980), prepiriform cortex (Reyes et al., 1987), hippocampus (Brun and Gustafson, 1976; Ball, 1977; Hyman et al., 1984; Pearson et al., 1985; Kahn et al., 1987), entorhinal cortex (Brun and Gustafson, 1976; Hyman et al., 1984), uncus (Brun and Gustafson, 1976), and subiculum (Hyman et al., 1984). These histopathological changes are predominantly neuritic plaques, neurofibrillary tangles, and granulovacuolar degeneration. Abnormal masses of neurites (axonal and dendritic processes of neurons) and compositional changes in subunits of neurofilament proteins are found in olfactory epithelium, and reduced choline acetyltransferase (CAT) levels have been reported in olfactory tubercle. Altered neurotransmitter levels also play a role in the olfactory losses associated with AD (Hardy et al., 1985, 1986; Whitehouse et al., 1981).

Treatment of Age-Related Decrements in Odor Perception with Flavor (Odor) Enhancers

The addition of commercial "flavors" to foods both during and after preparation is a practical means of counteracting chemosensory losses in the elderly. Beef flavor can be added to beef gravies, and carrot flavor can be added to carrots. Cheese and bacon flavors can be added to soups and vegetables. Flavors are mixtures of odorous molecules that can be extracted directly from natural products or can be simulated after chromatographic and mass spectrographic analysis of natural products. Flavor amplification of food for the sick elderly has been found to increase appetite, reverse anorexia, and improve nutritional status (Schiffman and Warwick, 1988). For the nonsick elderly, many flavor-enhanced foods are rated as hedonically superior and are eaten in greater amounts than unenhanced foods. Flavor amplification is not effective for persons who have complete anosmia (such as persons with advanced Alzheimer's disease) because even highly concentrated odors cannot be detected.

Flavor Enhancement of Food for the Sick Elderly

Addition of flavors to foods has been used clinically by our group at Duke University Medical Center and in a local North Carolina nursing home for 20 years to promote weight gain and improve nutritional status in elderly persons. For sick persons (especially those who are consuming less than 80 percent of the U.S. Recommended Daily Allowance, have below average weight for their height and age, have clinical manifestations of malnutrition, or have a recent weight loss of at least 6 percent), the addition of flavors to foods increases intake and body weight. Improvements after flavor enhancement have also been found in plasma protein levels, including somatomedin-C/insulinlike growth factor I, albumin, and transferrin. T-lymphocyte counts frequently move from abnormal values to the normal range. Muscle strength improves in some patients as measured by hand grip strength using a Jamar hand dynamometer.

Flavor Enhancement of Food for Persons on Diets Recommended by a Physician

Flavor amplification can also be helpful for a subgroup of elderly persons who must change their food choices to meet the nutritional goals recommended by a physician. The elderly are frequently asked to lower the total fat and saturated fat in the diet. Most elderly find it difficult to put that nutritional advice into practice for several reasons that relate to olfaction. First, most odorants are fat-soluble, and the removal of fat from the diet also reduces the odor of the food. Second, the elderly find it difficult and demanding to change long-standing flavor preferences. Addition of familiar flavors to low-fat food improves the probability that an elderly patient will observe the nutritional advice of his or her physician (personal observation).

Flavor Enhancement of Food in a Retirement Home

Addition of flavors to cafeteria food also increases intake of some foods for independent-living elderly persons in a retirement center. This finding was based on an extensive study of 39 Caucasian elderly persons (mean age = 84.59) who were divided into two matched groups. The participants in the study were selected by the dietitian as persons who had had some eating problems over the last year. The first group received food that was unenhanced by flavor for the first three weeks, and food enhanced by

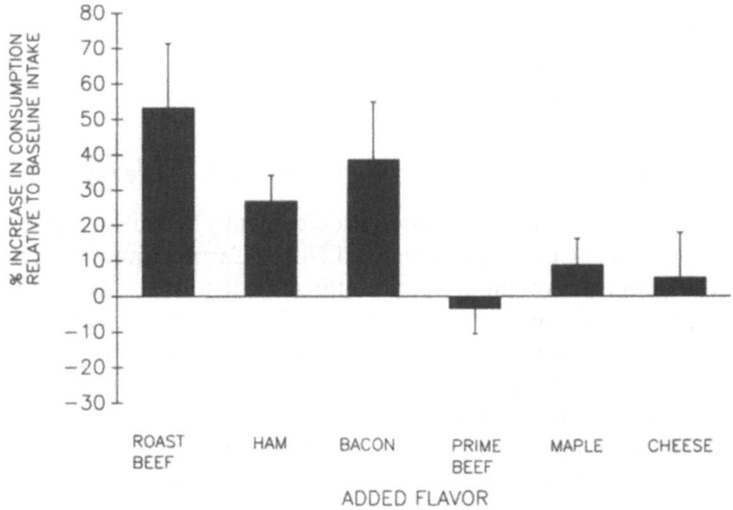

FIGURE 18.3. Percent increase in consumption of flavor-enhanced foods when compared with unenhanced foods. Elderly ate significantly more food that was flavored with roast beef, ham, or bacon flavor.

flavor for the second three-week period. The order was reversed for the second group. Six flavors selected by professional flavorists who were over 60 years of age were utilized throughout the study: roast beef, ham, bacon, prime beef, maple, and cheese. These flavors were added to soups, gravies, eggs and egg substitutes, vegetables, grits, stews, sauces, pancakes, oatmeal, and macaroni.

Six types of measurements were obtained during the six weeks that each person participated in the study. Biochemical measures and anthropometric measurements were obtained at the beginning of the study, at the end of three weeks, and at the end of six weeks. Plate wastage (a measure of food consumption), taste and smell measurements, and functional parameters were also obtained. The biochemical measures were somatomedin-C/insulinlike growth factor I, transferrin, total T and B-lymphocytes, and routine blood chemistries including albumin and creatinine. The anthropometric measures were weight, stature, midarm circumference, and triceps skinfold thickness. Taste thresholds were determined for NaCl and sucrose, and smell thresholds for benzaldehyde, geraniol, and citralva. The functional parameters were hand grip strength and pinch strength.

The major finding was that the elderly ate significantly more food when it was flavor-enhanced with roast beef, ham, or bacon flavor (see Figure 18.3), and the persons with the greatest olfactory losses ate the most flavor-enhanced food. However, these increases in consumption did not shift the overall nutrient balance because the intake was almost identical in the enhanced and unenhanced conditions. The diets consumed in both the unenhanced and enhanced conditions were fairly well balanced, although they tended to be low in dietary fiber, pantothenic acid, calcium, copper, magnesium, potassium, selenium, and zinc. All biochemical measures, including somatomedin-C/insulinlike growth factor I, transferrin, total T and B-lymphocytes, albumin, creatinine, and routine blood chemistries, were within the normal range both before and after the study, suggesting that there may be less malnutrition in the elderly than is often presumed. No changes were found in anthropometric measurements, taste and smell assessments, or functional parameters throughout the study. The only pertinent statistical change was a slight improvement in lymphocyte counts in the enhanced condition, which suggests that addition of flavors may improve immune status, perhaps by stimulating the limbic system. This study indicates that the independent-living elderly in retirement homes are not particularly malnourished but will eat more flavor-enhanced food, which appears to improve their immune status.

Legal Aspects and Safety of Flavors

The study just described, as well as a review of the literature, suggests that addition of flavors to foods is a safe procedure (see Evenhuis, 1986). There are four multinational groups that monitor the safety of flavors: Council of Europe, European Community (EC), Codex Alimentarium Commission established by FAO/WHO, and the International Organization of the Flavor Industry (IOFI). In the United States, an expert panel of the Flavor and Extract Manufacturers' Association (FEMA) confers GRAS ("generally regarded as safe") status on a flavor based on chemical identity, purity, structural analogy, natural occurrence in food, concentration of the chemical in food and in the total diet, toxicity and metabolism in animals, and, where possible, metabolism in man. These safety evaluations are carried out by a non-industry-related expert panel of scientists, and the lists of GRAS substances are published by FEMA and in the Federal Register.

Flavors present far fewer potential hazards than microbial contaminants (e.g., pathogenic microorganisms), malnutrition (e.g., excess fat or alcohol), environmental contaminants (e.g., DDT, lead, or cadmium), natural toxicants (e.g., glycoalkaloids or nitrosamines), or pesticide residues (e.g., halogenated hydrocarbons). The risk is also negligible because the quantity of flavor presently consumed by the average person per day is quite small. In Germany, for example, 15 percent of the food consumed per person is flavored. The amount of flavoring consumed by an individual is 8.5 g per year, or approximately 25 mg per day. These 25 mg consist of hundreds of different types of molecules, so that the risk to metabolism from any one molecule is insignificant. When flavor levels are elevated to improve food intake in the elderly, the exposure is still minimal.

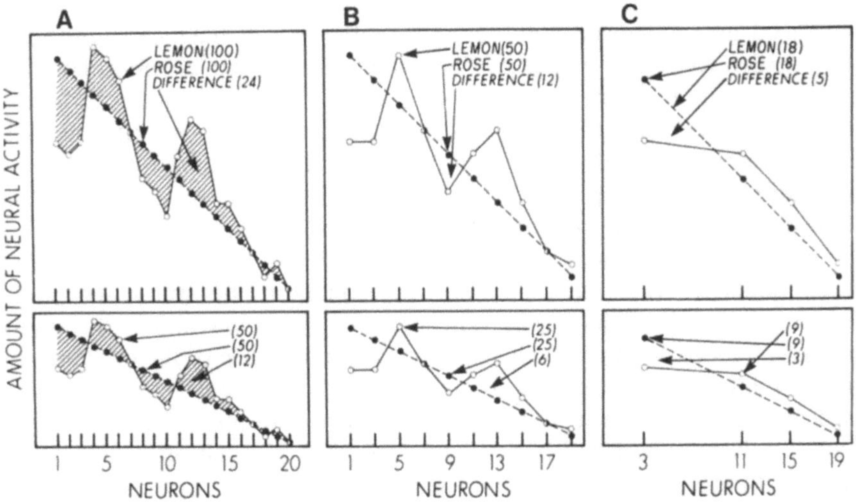

FIGURE 18.4. A. Neural codes across 20 neurons for two different odor stimuli, rose and lemon, at two different intensities (upper and lower panel). The amount of activity in each neuron is depicted by each point. B. Neural codes across 10 neurons (one-half the neurons have been lost with age) for rose and lemon at two different intensities (upper and lower panel). C. Neural codes across 4 neurons (representing severe neural loss, as in Alzheimer's disease) for two different odors at two different intensities (upper and lower panel).

A Model of Neural Coding That Accounts for Odor Losses in Normal Aging and Alzheimer's Disease

A theoretical model that can account for the observations on olfactory losses in the elderly has been proposed by Schiffman et al. (1979). This model is based on the "across-fiber pattern" theory of Erickson (1963), which proposes that the pattern of neural activity across many neurons codes the sensation of a particular stimulus. Two different odorants would be represented by two different "across-fiber patterns." Figure 18.4A, B, and C gives a schematic representation of the neural codes for two different odorants, each at two different intensities. Each point depicts the amount of neural activity in a given neuron. The neural code for "rose" is represented in Figure 18.4A by the pattern of progressively lesser amounts of activity from left to right in these olfactory neurons; a weaker "rose" odor utilizes the same pattern at a lower level in the lower portion of Figure 18.4A. The irregular patterns in Figure 18.4A represent the code for "lemon" at two different intensities. The perceived strength of each odor is directly related to the amount of neural activity evoked (termed here "neural mass"). The neural mass for the stronger rose sensation in the upper panel is 100, and that for the weaker rose sensation in the lower panel is 50. The neural masses for stronger and weaker lemon sensations are also 100 and 50, respectively. Figure 18.4B and C depicts the patterns of Figure 18.4A with progressive losses in the number of olfactory neurons. Thus for both rose and lemon in Figure 18.4B, the stronger and weaker patterns have neural masses of 50 and 25. With even greater loss, in Figure 18.4C, the neural masses for stronger and weaker stimuli are 18 and 9.

Relation of Neural Loss to Olfactory Detection Thresholds

A certain neural mass (e.g., 10) is presumed necessary to detect a stimulus. Thus, with the neural losses in Figure 18.4B, both rose and lemon odors will still be perceived, but as weaker than in Figure 18.4A. With the neural losses in Figure 18.4C, the weaker stimuli in the lower panel would not be perceived because their neural masses are less than 10. Hence, the neural loss results in an increase in threshold value.

Relation of Neural Loss to Olfactory Quality Discriminations

The perceptual differences between rose and lemon result from the differences in their neural patterns, represented schematically in Figure 18.4A. The shaded region in Figure 18.4A represents the difference between rose and lemon in neural mass units, that is, 24. The value 24 is a measure of the ease with which these two stimuli can be discriminated from one another. The difference between the two weaker stimuli in the lower panel of Figure 18.4A is only 12 neural mass units; thus it is more difficult to discern the differences between these two stimuli than between the stronger ones. With progressive neural loss, the differences between the patterns become smaller, and the distinctions between rose and lemon become less clear.

If the difference between the patterns must be at least 10 neural mass units to distinguish the differences between two different odors, then in Figure 18.4A discriminations for both strong and weak stimuli are possible. With some neural losses, discriminations for the stronger stimuli in Figure 18.4B are still possible, although lemon and rose will smell more alike than before. However, the weaker odorants in the lower panel of Figure 18.4B cannot be distinguished from each other, although their presence can still be detected. With further neural loss in Figure 18.4C, the weaker stimuli can neither be distinguished from one another nor detected because there is not enough neural mass for either judgment.

Thus, this model proposes that the amount of neural mass determines olfactory perception in the elderly. The consequence of reduced amounts of evoked neural mass (resulting from neural loss or histopathological changes) are (1) elevated thresholds, (2) reduced intensity of sensations, and (3) smaller distinctions between stimuli.

References

Adams, R.G., and N. Crabtree (1961). Anosmia in alkaline battery workers. *Br. J. Indust. Med.*, *18*, 216–221.

Ahlborg, G. (1951). Hydrogen sulfide poisoning in shale oil industry. *Arch. Indust. Hyg.*, *3*, 247–266.

Anand, M.P. (1964). Accidents in the home. In W.F. Anderson and B. Isaacs (Eds.), *Current Achievements in Geriatrics.* London: Cassell, pp. 239–245.

Andria, G., A. Ballabio, and G. Parenti (1987). X-linked ichthyosis due to steroid sulfatase deficiency associated with hypogonadism and anosmia [letter]. *Ann. Neurol.*, *22*, 98–99.

Ansari, K.A., and A. Johnson (1975). Olfactory function in patients with Parkinson's disease. *J. Chronic Dis.*, *28*, 493–497.

Axelrod, J., R.M. Burch, and C.L. Jelsema (1988). Receptor-mediated activation of phospholipase A_2 via GTP-binding proteins: arachidonic acid and its metabolites as second messengers. *Trends Neurol. Sci.*, *11*, 117–122.

Averback, P. (1983). Two new lesions in Alzheimer's disease. *Lancet*, *2*, 1203.

Bakay, L. (1984). Olfactory meningiomas: The missed diagnosis. *JAMA*, *251*, 53–55.

Ball, M.J. (1977). Neuronal loss, neurofibrillary tangles and granulovacuolar degeneration in the hippocampus with ageing and dementia: A quantitative study. *Acta Neuropathol. (Berl.)*, *37*, 111–118.

Baraban, J.M., P.F. Worley, and S.H. Snyder (1989). Second messenger systems and psychoactive drug action: focus on the phosphoinositide system and lithium. *Am. J. Psychiatry*, *146*, 1251–1260.

Barton, R.P.E. (1974). Olfaction in leprosy. *J. Laryngol. Otol.*, *88*, 355–361.

Berman, J.L. (1985). Dysosmia, dysgeusia, and diltiazem [letter]. *Ann. Intern. Med.*, *102*, 717.

Body, S.C. (1986). A taste of allicin? *Anaesth. Intensive Care*, *14*, 94.

Brizzee, K.R., P. Klara, and J.E. Johnson (1975). Changes in microanatomy, neurocytology, and fine structure with aging. In J.M. Ordy and K.R. Brizzee (Eds.), *Advances in Behavioral Biology*, Vol. 16. New York: Plenum.

Brun, A., and L. Gustafson (1976). Distribution of cerebral degeneration in Alzheimer's disease. A clinico-pathological study. *Arch. Psychiat. Nervenkr.*, *223*, 15–33.

Burch, R.E., D.A. Sackin, J.A. Ursick, M.M. Jetton, and J.F. Sullivan (1978). Decreased taste and smell acuity in cirrhosis. *Arch. Intern. Med.*, *138*, 743–746.

Carmichael, K.A., A.S. Jennings, and R.L. Doty (1984). Reversible anosmia after pituitary irradiation. *Ann. Intern. Med.*, *100*, 532–533.

Chalke, H.D., and J.R. Dewhurst (1957). Accidental coal-gas poisoning: Loss of sense of smell as a possible contributory factory with old people. *Br. Med. J.*, *2*, 915–917.

Chalke, H.D., J.R. Dewhurst, and C.W. Ward (1958). Loss of sense of smell in old people. *Public Health (Lond.)*, *72*, 223–230.

Church, J.A., H. Bauer, J.A. Bellanti, R.A. Satterly, and R.I. Henkin (1978). Hyposmia associated with atopy. *Ann. Allergy*, *40*, 105–109.

Crosley, C.J., and S. Dhamoon (1983). Migrainous olfactory aura in a family [letter]. *Arch. Neurol.*, *40*, 459.

Currie, S., K.W.G. Heathfield, R.A. Henson, and D.F. Scott (1971). Clinical course and prognosis of temporal lobe epilepsy. A survey of 666 patients. *Brain*, *94*, 173–190.

Das, S.K., and I.R. Munro (1979). Anosmia in Crouzon's syndrome and its recovery following cranio-facial reconstruction. *Br. J. Plast. Sur.*, *32*, 55–56.

Delaney, P., R.I. Henkin, H. Manz, R.A. Satterly, and H. Bauer (1977). Olfactory sarcoidosis. *Arch. Otolaryngol.*, *103*, 717–724.

Dodson, H.C., and L.H. Bannister (1980). Structural aspects of ageing in the olfactory and vomeronasal epithelia in mice. In H. van der Starre (Ed.), *Olfaction and Taste VII*. London: IRL.

Doty, R.L., P. Shaman, S.L. Applebaum, R. Gilberson, L. Siksorski, and L. Rosenberg (1984a). Smell identification ability: Changes with age. *Science*, *226*, 1441–1443.

Doty, R.L., P. Shaman, and M. Dann (1984b). Development of the University of Pennsylvania Smell Identification Test: A standardized microencapsulated test of olfactory function. *Physiol. Behav.*, *32*, 489–502.

Doty, R.L., and J.B. Snow, Jr. (1987). Olfaction. In J. Goldman (Ed), *The Principles and Practice of Rhinology*. New York: Willey, pp. 761–785.

Doty, R.L., P.F. Reyes, and T. Gregor (1987). Presence of both odor identification and detection deficits in Alzheimer's disease. *Brain Res. Bull.*, *18*, 597–600.

Doty, R.L., D.A. Deems, and S. Stellar (1988). Olfactory dysfunction in Parkinsonism: A general deficit unrelated to neurologic signs, disease stage, or disease duration. *Neurology*, *38*, 1237–1244.

Douek, E., L.H. Bannister, and H.C. Dodson (1975). Recent advances in the pathology of olfaction. *Proc. R. Soc. Med.*, *68*, 467–470.

Dragomiretskii, V.D., T.V. Rasskazova, and E.I. Berezner (1970). The effect of small concentrations of solvents on the state of the mucous membrane of the upper respiratory tract. *Zh. Ushn. Nos. Gorl. Bolezn.*, *30*, 16–18.

Emmet, E.A. (1976). Parosmia and hyposmia induced by solvent exposure. *Br. J. Indust. Med.*, *33*, 196–198.

Erickson, R.P. (1963). Sensory neural patterns and gustation. In Y. Zotterman (Ed.), *Olfaction and Taste*. Oxford: Pergamon.

Erikssen, J., E. Seegaard, and K. Naess (1975). Side-effect of thiocarbamides. *Lancet*, *1*, 231–232.

Esiri, M.M., and A.H. Tomlinson (1984). Herpes simplex encephalitis: Immunohistochemical demonstration of spread of virus via olfactory and trigeminal pathways after infection of facial skin in mice. *J. Neurol. Sci.*, *64*, 213–217.

Esiri, M.M., and G.K. Wilcock (1984). The olfactory bulbs in Alzheimer's disease. *J. Neurol. Neurosurg. Psychiatr.*, *47*, 56–60.

Evenhuis, B. (1986). General, legal and safety aspects of the use of flavourings. *Perfumer and Flavorist*, *11*, 17–28.

Fein, B.T., P.B. Kamin, and N.N. Fein (1966). The loss of sense of smell in nasal allergy. *Ann. Allergy*, *24*, 278–283.

Furstenberg, A.C., E. Crosby, and B. Farrior (1943). Neurologic lesions which influence the sense of smell. *Arch. Otolaryngol.*, *38*, 529–530.

Garrett-Laster, M., R.M. Russell, and P.F. Jacques (1984). Impairment of taste and olfaction in patients with cirrhosis: The role of vitamin A. *Hum. Nutr.: Clin. Nutr.*, *38C*, 203–214.

Ghorbanian, S.N., J.L. Paradise, and R.L. Doty (1978). Odor perception in children in relation to nasal obstruction. *Pediatr. Res.*, *12*, 371.

Gilman, A.G. (1989). G proteins and regulation of adenylyl cyclase. The Albert Lasker Medical Awards. *JAMA*, *262*, 1819–1825.
Goetzl, F.R., and F. Stone (1948). The influence of amphetamine sulfate upon olfactory acuity and appetite. *Gastroenterology*, *10*, 708–713.
Gorman, W. (1964). The sense of smell. A clinical review. *Eye Ear Nose Throat Mon.*, *43*, 54–58.
Grossman, S. (1953). Loss of taste and smell due to propylthiouracil therapy. *NY State J. Med.*, *53*, 1236.
Hallman, B.L., and J.W. Hurst (1953). Loss of taste as toxic effect of methimazole (Tapazole) therapy: Report of three cases. *JAMA*, *152*, 322.
Hardy, J.A., R. Adolfsson, I. Alafuzoff, G. Bucht, J. Marcusson, P. Nyberg, E. Perdahl, P. Wester, and B. Winblad (1985). Transmitter deficits in Alzheimer's disease. *Neurochem. Int.*, *7*, 545–563.
Hardy, J.A., D.M. Mann, P. Wester, and B. Winblad (1986). An integrative hypothesis concerning the pathogenesis and progression of Alzheimer's disease. *Neurobiol. Aging*, *7*, 489–502.
Henkin, R.I. (1967). Abnormalities of taste and olfaction in patients with chromatin negative gonadal dysgenesis. *J. Clin. Endocrinol. Metab.*, *27*, 1436–1440.
Henkin, R.I. (1968). Impairment of olfaction and of the tastes of sour and bitter in pseudohypoparathyroidism. *J. Clin. Endocrinol. Metab.*, *28*, 624–628.
Henkin, R.I. (1975). The role of adrenal corticosteroids in sensory processes. In H. Blaschko, A.D. Smith, and G. Sayers (Eds.), *Handbook of Physiology*, Section 7, Vol. 6. Washington, DC: American Physiological Society, pp. 209–230.
Henkin, R.I., R.C. Hoye, A.S. Ketcham, and W.J. Gould (1968). Hyposmia following laryngectomy. *Lancet*, *2*, 479–481.
Henkin, R.I., and F.R. Smith (1971). Hyposmia in acute viral hepatitis. *Lancet*, *1*, 823–826.
Henkin, R.I., and A.L. Larson (1972). On the mechanism of hyposmia following laryngectomy in man. *Laryngoscope*, *82*, 836–843.
Henkin, R.I., N. Talal, A.L. Larson, and C.F.T. Mattern (1972). Abnormalities of taste and smell in Sjögren's syndrome. *Ann. Intern. Med.*, *76*, 375–383.
Henkin, R.I., A.L. Larson, and R.D. Powell (1975). Hypogeusia, dysgeusia, hyposmia, and dysosmia following influenza-like infection. *Ann Otol. Rhinol. Laryngol.*, *84*, 672–682.
Herzog, A.G., and T.L. Kemper (1980). Amygdaloid changes in aging and dementia. *Arch. Neurol.*, *37*, 625–629.
Hinds, J. W., and N.A. McNelly (1977). Aging of the rat olfactory bulb: Growth and atrophy of constituent layers and changes in size and number of mitral cells. *J. Comp. Neurol.*, *171*, 345–368.
Hinds, J.W., and N.A. McNelly (1978). Dispersion of cisternae of rough endoplasmic reticulum in aging CNS neurons: A strictly linear trend. *Am. J. Anat.*, *152*, 433–439.
Hinds, J.W., and N.A. McNelly (1979). Aging in the rat olfactory bulb: Quantitative changes in mitral cell organelles and somatodendritic synapses. *J. Comp. Neurol.*, *184*, 811–820.
Hinds, J.W., and N.A. McNelly (1981). Aging in the rat olfactory system: Correlation of changes in the olfactory epithelium and olfactory bulb. *J. Comp. Neurol.*, *203*, 441–453.

Hinds, J.W., and N.A. McNelly (1982). Capillaries in aging rat olfactory bulb: A quantitative light and electron microscopic analysis. *Neurobiol. Aging*, *3*, 197–207.

Hotchkiss, W.T. (1956). Influence of prednisone on nasal polyposis with anosmia. *Arch. Otolaryngol.*, *64*, 478–479.

Hoye, R.C., A.S. Ketcham, and R.I. Henkin (1970). Hyposmia after paranasal sinus exenteration or laryngectomy. *Am. J. Surg.*, *120*, 485–491.

Hyman, B.T., G.W. Van Hoesen, A.R. Damasio, and C.L. Barnes (1984). Alzheimer's disease: Cell-specific pathology isolates the hippocampal formation. *Science*, *225*, 1168–1170.

Jafek, B.W., P.M. Eller, B.A. Esses, and D.T. Horan (1989). Post-traumatic anosmia. Ultrastructural correlates. *Arch. Neurol.*, *46*, 300–304.

James, W.R.L. (1963). Fatal addiction to trichloroethylene. *Brit. J. Indust. Med.*, *20*, 47–49.

Jones, B.P., H.R. Moskowitz, and N. Butters (1975). Olfactory discrimination in alcoholic Korsakoff's patients. *Neuropsychologia*, *13*, 173–179.

Jones, B.P., N. Butters, H.R. Moskowitz, and K. Montgomery (1978). Olfactory and gustatory capacities of alcoholic Korsakoff patients. *Neuropsychologia*, *16*, 323–337.

Jorgensen, M.B., and N.H. Buch (1961). Studies on the sense of smell and taste in diabetics. *Acta Otolaryngol.*, *53*, 539–545.

Kahn, J., B.H. Anderton, C.C. Miller, J.N. Wood, and M.M. Esiri (1987). Staining with monoclonal antibodies to neurofilaments distinguishes between subpopulations of neurofibrillary tangles, between groups of axons and between groups of dendrites. *J. Neurol.*, *234*, 241–246.

Kallman, F.J., W.A. Schoenfeld, and S.E. Barrera (1944). The genetic aspects of primary eunuchoidism. *Am. J. Ment. Defic.*, *48*, 203–236.

Kesslak, J.P., C.W. Cotman, H.C. Chui, S. van den Noort, H. Fang, R. Pfeffer, and G. Lynch (1988). Olfactory tests as possible probes for detecting and monitoring Alzheimer's disease. *Neurobiol. Aging*, *9*, 399–403.

Kimbrell, G.M., and E. Furchtgott (1963). The effect of aging on olfactory threshold. *J. Gerontol.*, *18*, 364–365.

Knupfer, L., and R. Spiegel (1986). Differences in olfactory test performance between normal aged, Alzheimer and vascular type dementia individuals. *Int. J. Geriatr. Psychiatry*, *1*, 3–14.

Koss, E., J.M. Weiffenbach, J.V. Haxby, and R.P. Friedland (1987). Olfactory detection and recognition in Alzheimer's disease. *Lancet*, *1*, 622.

Koss, E., J.M. Weiffenbach, J.V. Haxby, and R.P. Friedland (1988). Olfactory detection and identification performance are dissociated in early Alzheimer's disease. *Neurology*, *38*, 1228–1232.

Kretschmar, J.H. (1984) Anosmia as a symptom of incipient Pick's atrophy? *Z. Gerontol.*, *17*, 126–127.

Landfield, P.W., J.C. Waymire, and G. Lynch (1978). Hippocampal aging and adrenocorticoids: Quantitative correlations. *Science*, *202*, 1098–1102.

Landfield, P.W., and T.A. Pitler (1984). Prolonged Ca^{2+}-dependent after hyperpolarizations in hippocampal neurons in aged rats. *Science*, *226*, 1089–1092.

Lehnhardt, E., and H. Rollin (1969). Berufsbedingte Riechstorungen. *HNO*, *17*, 104–106.

Leigh, A.D. (1943). Defects of smell after head injury. *Lancet*, *1*, 38–40.
Levin, H.S., W.M. High, and H.M. Eisenberg (1985). Impairment of olfactory recognition after closed head injury. *Brain*, *108*, 579–591.
Levinson, J.L., and K. Kennedy (1985). Dysosmia, dysgeusia, and nifedipine [letter]. *Ann. Intern. Med.*, *102*, 135–136.
Liss, L., and F. Gomez (1958). The nature of senile changes of the human olfactory bulb and tract. *Arch. Otolaryngol.*, *67*, 167–171.
Machado-Salas, J., M.E. Scheibel, and A.B. Scheibel (1977). Morphologic changes in the hypothalamus of the old mouse. *Exp. Neurol.*, *57*, 102–111.
Macht, D.I., and M.B. Macht (1940). Comparison of effect of cobra venom and opiates on olfactory sense. *Am. J. Physiol.*, *129*, P411–412.
MacIntyre, I. (1971). Prolonged anosmia. *Br. Med. J.*, *2*, 709.
Mair, R.G., R.L. Doty, K.M. Kelly, C.S. Wilson, P.J. Langlais, W.J. McEntee, and T.A. Vollmecke (1986). Multimodal sensory discrimination deficits in Korsakoff's psychosis. *Neuropsychologia*, *24*, 831–839.
Males, J.L., J.L. Townsend, and R.A. Schneider (1973). Hypogonadotrophic hypogonadism with anosmia—Kallman's syndrome. A disorder of olfactory and hypothalamic function. *Arch. Intern. Med.*, *131*, 501–507.
Marshall, J.R., and R.I. Henkin (1971). Olfactory acuity, menstrual abnormalities, and oocyte status. *Ann. Intern. Med.*, *75*, 207–211.
McConnell, R.J., C.E. Menendez, F.R. Smith, R.I. Henkin, and R.S. Rivlin (1975). Defects of taste and smell in patients with hypothyroidism. *Am. J. Med.*, *59*, 354–364.
Meats, P. (1988). Olfactory hallucinations [letter]. *Br. Med. J.*, *296*, 645.
Megighian, D. (1958). Variazioni della soglia olfattiva nell'ets senile. *Minerva Otorinolaringol.*, *9*, 331–337.
Moberg, P.J., G.D. Pearlson, L.J. Speedie, J.R. Lipsey, M.E. Strauss, and S.E. Folstein (1987). Olfactory recognition: Differential impairments in early and late Huntington's and Alzheimer's diseases. *J. Clin. Exp. Neuropsychol.*, *9*, 650–664.
Moran, D.T., B.W. Jafek, J.C. Rowley, and P.M. Eller (1985). Electron microscopy of olfactory epithelia in two patients with anosmia. *Arch. Otolaryngol.*, *111*, 122–126.
Murphy, C. (1983). Age-related effects on the threshold, psychophysical function, and pleasantness of menthol. *J. Gerontol.*, *38*, 217–222.
Murphy, C. (1985). Cognitive and chemosensory influences on age-related changes in the ability to identify blended foods. *J. Gerontol.*, *40*, 47–52.
Murphy, C., B.R. Lasker, and D.P. Salmon (1987). Olfactory dysfunction and odor memory in Alzheimer's disease, Huntington's disease and normal aging. *Soc. Neurosci. Abstr.*, Vol. 13, Part 1, 17th Annual Meeting, New Orleans, LA, p. 1403.
Naessen, R. (1971). An inquiry on the morphological characteristics and possible changes with age in the olfactory region of man. *Acta Otolaryngol.*, *71*, 49–62.
Naus, A. (1968). Alterations of the smell acuity caused by menthol. *J. Laryngol. Otol.*, *82*, 1009–1011.
Naus, A. (1976). *Olphactoric Properties of Industrial Matters*. Prague: Charles University, pp. 55–65.
Ohm, T.G., and H. Braak (1987). Olfactory bulb changes in Alzheimer's disease. *Acta Neuropathol. (Berl.)*, *73*, 365–369.

Olsen, K.D., and L.W. DeSanto (1983). Olfactory neuroblastoma. *Arch. Otolaryngol.*, *109*, 797–803.
Peabody, C.A., and J.R. Tinklenberg (1985). Olfactory deficits and primary degenerative dementia. *Am. J. Psychiatry*, *142*, 524–525.
Pearson, R.C., M.M. Esiri, R.W. Hiorns, G.K. Wilcock, and T.P. Powell (1985). Anatomical correlates of the distribution of the pathological changes in the neocortex in Alzheimer disease. *Proc. Natl. Acad. Sci. USA*, *82*, 4531–4534.
Perry, J.D., S. Frisch, B. Jafek, and M. Jafek (1980). Olfactory detection thresholds using pyridine, thiophene, and phenethyl alcohol. *Otolaryngol. Head Neck Surg.*, *88*, 778–782.
Pinching, A.J. (1977). Clinical testing of olfaction reassessed. *Brain*, *100*, 377–388.
Reyes, P.F., G.T. Golden, P.L. Fagel, R.G. Fariello, L. Katz, and E. Carner (1987). The prepiriform cortex in dementia of the Alzheimer type. *Arch. Neurol.*, *44*, 644–645.
Rezek, D.L. (1987). Olfactory deficits as a neurologic sign in dementia of the Alzheimer type. *Arch. Neurol.*, *44*, 1030–1032.
Rossberg, G. von, H. Schaupp, and W. Schmidt (1966). Geruchs- und geschmack vergogen bei Arbeitern der chemischen und metallverarbeitenden Industrie. *Z. Laryngol. Rhinol.*, *45*, 571–590.
Rundles, R.W. (1946). Prognosis in the neurologic manifestations of pernicious anemia. *Blood*, *1*, 209–219.
Ryan, R.E., Sr. and R.E. Ryan, Jr. (1974). Acute nasal sinusitis. *Postgrad. Med.*, *56*, 159–162.
Schaupp, H., and J. Seilz (1969). Geruch und Geschmack bei endokrinen Erkrankungen. *Arch. Klin. Exp. Ohr. Nas. u. Kehlk. Heilk.*, *195*, 179–191.
Schechter, P.J., and R.I. Henkin (1974). Abnormalities of taste and smell after head trauma. *J. Neurol. Neurosurg. Psychiatry*, *37*, 802–810.
Scheibel, M.E., and A.B. Scheibel (1975). Structural changes in the aging brain. In H. Brody, D. Harman, and J.M. Ordy (Eds.), *Aging*, Vol. 1, *Clinical, Morphological, and Neurochemical Aspects in the Aging Central Nervous System.* New York: Raven Press, pp. 11–37.
Schellinger, D., R.T. Henkin, and J.G. Smirniotopoulos (1983). CT of the brain in taste and smell dysfunction. *Am. J. Neuroradiog. (AJNR)*, *41*, 752–754.
Schemper, T., S. Voss, and W.S. Cain (1981). Odor identification in young and elderly persons: Sensory and cognitive limitations. *J. Gerontol.*, *36*, 446–452.
Schiffman, S.S. (1977). Food recognition by the elderly. *J. Gerontol.*, *36*, 446–452.
Schiffman, S.S. (1979). Changes in taste and smell with age: Psychophysical aspects. In J.M. Ordy and K. Brizzee (Eds.), *Sensory Systems and Communication in the Elderly*, Vol. 10. *Aging*. New York: Raven Press, pp. 227–246.
Schiffman, S.S. (1983). Taste and smell in disease. *N. Engl. J. Med.*, *308*, 1275–1279, 1337–1343.
Schiffman, S.S. (1987). Smell. In G.L. Maddox (Ed.), *Encyclopedia of Aging*. New York: Springer Verlag, pp. 618–619.
Schiffman, S.S., J. Moss, and R.P. Erickson (1976). Thresholds of food odors in the elderly. *Exp. Aging Res.*, *2*, 389–398.
Schiffman, S.S., M.L. Nash, and C. Dackis (1978). Reduced olfactory discrimination in patients on chronic hemodialysis. *Physiol. Behav.*, *21*, 239–242.
Schiffman, S.S., M. Orlandi, and R.P. Erickson (1979). Changes in taste and smell with age: Biological aspects. In J.M. Ordy and K. Brizzee (Eds.), *Sensory Systems and Communication in the Elderly*, Vol. 10, *Aging*. New York: Raven

Press, pp. 247–268.

Schiffman, S.S., and M. Pasternak (1979). Decreased discrimination of food odors in the elderly. *J. Gerontol.*, *34*, 73–79.

Schiffman, S.S., and J.C. Leffingwell (1981). Perception of odors of simple pyrazines by young and elderly subjects: A multidimensional analysis. *Pharmacol. Biochem. Behav.*, *14*, 787–798.

Schiffman, S.S., and Z.S. Warwick (1988). Flavor enhancement of foods for the elderly can reverse anorexia. *Neurobiol. Aging*, *9*, 24–26.

Schiffman, S.S., C.M. Clark, and Z.S. Warwick (1990). Gustatory and olfactory dysfunction in dementia: Not specific to Alzheimer's disease. *Neurobiol. Aging.*, *11*, 597–600.

Schiffman, S.S., and Z.S. Warwick (1990). Changes in taste and smell over the life span: Effects on appetite and nutrition in the elderly. In M.I. Friedman, M.E. Tordoff, and M.R. Kare 1991. Chemical Senses. Vol. 4. Appetite and Nutrition. New York, Marcel Dekker, pp. 341–365.

Schneeberg, N.G. (1952). Loss of sense of taste due to methylthiouracil therapy. *JAMA*, *149*, 1091–1093.

Serby, M. (1986). Olfaction and Alzheimer's disease. *Prog. Neuropsychopharmacol. Biol. Psychiatry*, *10*, 579–586.

Serby, M. (1987). Olfactory deficits in Alzheimer's disease. *J. Neural Transm.*, *[Suppl] 24*, 69–77.

Serby, M., J. Corwin, P. Conrad, and J. Rotrosen (1985a). Olfactory dysfunction in Alzheimer's disease and Parkinson's disease. *Am. J. Psychiatry*, *142*, 781–782.

Serby, M., J. Corwin, A. Novatt, P. Conrad, and J. Rotrosen (1985b). Olfaction in dementia. *J. Neurol. Neurosurg. Psychiatry*, *48*, 848–849.

Seydell, E.M., and W.P. McKnight (1948). Disturbances of olfaction resulting from intranasal use of tyrothricin: A clinical report of seven cases. *Arch. Otolaryngol.*, *47*, 465–470.

Simpson, J., C.M. Yates, A. Gordon and D.M. St. Clair (1984). Olfactory tubercle choline acetyltransferase activity in Alzheimer-type dementia, Down's syndrome and Huntington's chorea. *J. Neurol. Neurosurg. Psychiatry*, *47*, 1138–1139.

Singh, N., M.S. Grewal, and J.H. Austin (1970). Familial anosmia. *Arch. Neurol.*, *22*, 40–44.

Skouby, A.P., and K. Zilstorff-Pedersen (1954). The influence of acetylcholine-like substances, menthol and strychnine on olfactory receptors in man. *Acta Physiol. Scand.*, *32*, 252–258.

Smith, C.G. (1942). Age incidence of atrophy of olfactory nerves in man. *J. Comp. Neurol.*, *77*, 589–595.

St. Clair, D.M., J. Simpson, C.M. Yates, and A. Gordon (1985). Olfaction in dementia: a response. *J. Neurol. Neurosurg. Psychiatry*, *48*, 849.

Stevens, D.A., and H.T. Lawless (1981). Age related changes in flavor perception. *Appetite*, *2*, 127–136.

Stevens, J.C., A. Plantinga, and W.S. Cain (1982). Reduction of odor and nasal pungency associated with aging. *Neurobiol. Aging*, *3*, 125–132.

Stevens, J.C., L.M. Bartoshuk, and W.S. Cain (1984). Chemical senses and aging: Taste versus smell. *Chem. Senses*, *9*, 167–179.

Stevens, J.C., and W.S. Cain (1985). Age-related deficiency in the perceived strength of six odorants. *Chem. Senses*, *10*, 517–529.

Strandbygard, E. (1954). Treatment of ozena and rhinopharyngitis chronica sicca with vitamin A. *Arch. Otolaryngol.*, *59*, 485–490.

Sumner, D. (1964). Post-traumatic anosmia. *Brain*, *87*, 107–120.

Sunohara, N., N. Sakuragawa, E. Satoyoshi, A. Tanae, and L.J. Shapiro (1986). A new syndrome of anosmia, ichthyosis, hypogonadism, and various neurological manifestations with deficiency of steroid sulfatase and arylsulfatase C. *Ann. Neurol.*, *191*, 98–99.

Symanski, H. (1950). Ein Fall von Selenwasserstoffvergiftung. *Deutsche Med. Wochenschr.*, *75*, 1730.

Talamo, B.R., J.S.R. Rudel, K.S. Kosik, V.M. Lee, S. Neff, L. Adelman, and J.S. Kauer (1989). Pathological changes in olfactory neurons in patients with Alzheimer's disease. *Nature*, *337*, 736–739.

Tomlinson, B.E., and G. Henderson (1976). Some quantitative cerebral findings in normal and demented old people. In R.D. Terry and S. Gershon (Eds.), *Neurobiology of Aging* Vol. 3. New York: Raven Press, pp. 183–204.

Turner, P. (1965). Some observations on centrally-acting drugs in man. *Proc. R. Soc. Med.*, *58*, 913–914.

Venstrom, D., and J.E. Amoore (1968). Olfactory threshold in relation to age, sex, or smoking. *J. Food Sci.*, *33*, 264–265.

Waldton, S. (1974). Clinical observations of impaired cranial nerve function in senile dementia. *Acta Psychiatr. Scand.*, *50*, 539–547.

Ward, C.D., W.A. Hess, and D.B. Caine (1983). Olfactory impairment in Parkinson's disease. *Neurology*, *33*, 943–946.

Warner, M.D., C.A. Peabody, J.J. Flattery, and J.R. Tinklenberg (1986). Olfactory deficits and Alzheimer's disease. *Biol. Psychiatry*, *21*, 116–118.

Warner, M.D., C.A. Peabody, and P.A. Berger (1988). Olfactory deficits and Down's syndrome. *Biol. Psychiat.*, *23*, 836–839.

Whitehouse, P.J., D.L. Price, A.W. Clark, J.T. Coylee, and M.R. DeLong (1981). Alzheimer disease: evidence for selective loss of cholinergic neurons in the nucleus basalis. *Ann. Neurol.*, *10*, 122–126.

Wolberg, F.L., and D.K. Ziegler (1986). Olfactory hallucination in migraine. *Arch. Neurol.*, *39*, 382.

Zilstorff, K. (1965). Sense of smell alterations by cocaine and tetracaine. *Archiv. Otolaryngol.*, *82*, 53–55.

Zilstorff, K., and O. Herbild (1979). Parosmia. *Acta Otolaryngol. [Suppl.]*, *360*, 40–41.

19
Olfactory Impairment in Korsakoff's Syndrome

R.G. Mair and D.L. Flint

Introduction

Odor information is transmitted from receptors in the nose to the olfactory bulb and then through several layers of processing within the brain. Beyond the bulb, analyses of olfactory function have proved to be as much a matter of conjecture as they are of scientific inference. Until the second half of this century, it was commonly taught that the rhinencephalon, or olfactory brain, included the hippocampus and much of the related limbic system (cf. Krieg, 1942). This belief has been ascribed to the influence of comparative neruroanatomists, who emphasized the role of olfactory structures in the evolution of the vertebrate forebrain, observations of olfactory auras during uncinate seizures (Jackson and Beevor, 1890), and Ferrier's (1876) demonstration of lip and nostril movements during electrical stimulation of the hippocampal lobe in monkeys. Opinions changed shortly after the second world war with the demonstration, by degeneration methods, that olfactory tracts end in primitive cortical areas on the ventral surface of the forebrain (cf. Brodal, 1947; Allison, 1953). The dominant view at this time can be summarized by the suggestion of Pribram and Kruger (1954) that the capacity to perform simple olfactory discriminations requires only an intact olfactory bulb and its direct projections to a "primary olfactory system" consisting of the olfactory tubercle, the area of the diagonal band, pyriform cortex, and the corticomedial nuclei of the amygdala. Other areas previously included in the "rhinencephalon" were now regarded as being more importantly involved in emotional, reproductive, or homeostatic functions.

A more modern era in the neurology of olfaction began with the discovery by Powell et al. (1963, 1965) of a projection from pyriform cortex to the mediodorsal nucleus (MDn) of thalamus in rats. Prior to this, central olfactory pathways were thought to consist of projections from the amygdala through the stria terminalis to the hypothalamus, presumably important for the chemosensory regulation of reproductive and feeding behaviors (cf. Scott and Pfaffmann, 1966; Leonard and Scott, 1971). The identification of projections to MDn was important in its implication that there might be a neocortical projection zone within orbitofrontal cortex (cf. Powell et al., 1963), presumably involved in other aspects of olfactory perception. Subsequent studies in rodents have shown a direct projection to orbitofrontal cortex as well as to MDn from the endopyriform nucleus, a structure deep in pyriform cortex that appears to function as a projection zone for primary olfactory cortex (cf. Price, 1985).

Lesion studies have provided evidence that these thalamo-neocortical pathways are critical for olfactory discrimination. In rats (Eichenbaum et al., 1980; Slotnick and Kaneko, 1981; Staubli et al., 1987) and hamsters (Sapolsky and Eichenbaum, 1980), lesions of the thalamus that include MDn have been reported to impair performance on a variety of tasks measuring olfactory discrimination learning. On the other hand, lesions of the amygdala have no such effect (Slotnick, 1985). Other evidence indicates that although damage to the hippocampus can impair relational responding to olfactory stimuli, it does not interfere with simpler olfactory discrimination tasks thought to be sensitive to thalamic lesions (see Otto and Eichenbaum, this volume).

There is controversy over how MDn lesions disrupt performance on odor-guided tasks. Eichenbaum et al. (1980) showed that the deficit is more severe for similar than for dissimilar odorants, a finding consistent with a diminished capacity for sensory discrimination. Slotnick and Kaneko (1981) argued that the deficit involves higher-order aspects of olfactory learning based on their observation that MDn lesions do not affect go/no-go dicrimination between propyl and ethyl acetate, but impair subsequent serial reversal learning with these same stimuli. Their argument is supported by a body of evidence that reversal learning in other sensory modalities is disrupted by lesions or cyrogenic blockade of thalamocortical fibers from MDn (cf. Skinner and Lindsley, 1973) or by lesions of its cortical targets (Mishkin, 1964; Gross and Weiskrantz, 1964; Milner, 1974). More recently, Staubli et al. (1987) have provided evidence that thalamic lesions can impair olfactory learning without disrupting sensory discrimination in an odor-guided maze learning task. After exhibiting impairments on the first several discrimination problems in the maze task, rats with MDn lesions were able to learn subsequent problems at a normal rate. The normal performance in learning the later problems is inconsistent with an impairment of olfactory discrimination. Staubli et al. (1987) argued that lesions of MDn impair the learning of procedural aspects of the task.

Olfactory Deficits in Korsakoff's Disease

The analysis of sensory deficits in animals is limited by the inherent difficulty of learning about internal perceptual events through studies of overt conditioned behaviors. Although animal studies have shown that lesions of

MDn can disrupt behaviors that require olfactory discrimination, they have provided scant information about the precise nature of the impairment that produces this deficit. From the results of animal studies, it is by no means certain whether MDn lesions disrupt (1) olfactory discrimination; (2) the consolidation, storage, or retrieval of odor memories; (3) the association of odors with other sensory stimuli; or (4) a more general process related to attention, learning, memory, or successive stimulus comparison that cannot reasonably be construed as a specific olfactory function.

The goal of this chapter is to consider the nature of olfactory impairments produced by brain disease. In doing this, we will place particular emphasis on studies of human neuropsychology that utilize the methods of psychophysics and cognitive psychology to study introspective aspects of cognition and perception. Our focus will be on Korsakoff's disease, a disorder associated with consistent lesions of MDn and perhaps the most thoroughly studied example of a central olfactory deficit.

Korsakoff's disease, the chronic phase of the Wernicke-Korsakoff syndrome, is the most common cause of diencephalic amnesia. This disease is characterized by a bilateral pattern of lesions located along the midline of the brainstem and periventricular region. Pathophysiologic studies of the memory impairment have emphasized lesions of the mediodorsal thalamic nucleus and the medial mammillary bodies, structures reported to be lesioned consistently among patients exhibiting this form of amnesia (Victor et al., 1971; Mair et al., 1979). Others have presented evidence that lesions involving noradrenergic projections from the locus coeruleus may also contribute to the amnesic symptoms of this disease (Mair and McEntee, 1983; Mayes et al., 1988).

Early clinical reports provided the first evidence of olfactory deficits in Korsakoff's amnesia (cf. Talland, 1965). Later, Jones et al. (1975a) reported that Korsakoff patients were impaired in both identification of common kitchen odorants and short-term odor recognition memory. Unfortunately, performance of the Korsakoff patients was not above chance at even the shortest memory delays, and thus it was not possible to determine whether the deficit was related to a rapid decay of memory prior to the shortest delay interval or to an inability to discriminate or recognize different odorants. In two subsequent studies, Jones et al. (1975b, 1978) used psychophysical methods to compare the scaling of olfactory, gustatory, auditory, and visual intensity by Korsakoff and control subjects. The results showed differential impairment for olfactory stimuli using several scaling procedures designed to compensate for memory impairment. They argued that Korsakoff patients have a fundamental inability to detect odorants, based on elevated estimates of thresholds obtained by extrapolation of the suprathreshold scaling results.

The results of Jones et al. (1975a and b, 1978) provided the first experimental evidence of severe olfactory impairments in Korsakoff's disease. Their results, however, raised a number of questions about the nature of the impairment. First, the chance-level performance in recognition memory prevented analysis of the nature of the apparent discrimination deficit. Second, fundamentally different patterns of impairment were observed in each of the scaling studies. In other sensory modalities, impairments in sensitivity are associated with recruitment functions, in which subjects underestimate the strength of weak but not strong stimuli. This typically results in scaling functions with elevated slope constants over the higher ranges of stimulus intensity (cf. Stevens, 1975). While Jones et al. (1975b) report magnitude estimation results consistent with a recruitment deficit, Jones et al. (1978) describe Korsakoff patients as making normal responses for weak stimuli and underestimating the strength of stronger ones. The differences in these results suggest an incosistency in rating odorants, more indicative of an impairment in memory or discrimination than in sensitivity. Third, Jones et al. (1975b, 1978) based their sensitivity hypothesis on the observation that Korsakoff patients applied intensity ratings of 0 to stronger stimuli than did controls. Such an extrapolation from scaling data must be questioned on several counts. It is based on a series of stimuli that were all well above threshold. Thus subjects were not given stimuli in the range of normal absolute sensitivity, and they were subjected to uncontrolled effects of adaptation by intermittent presentations of strong odorous stimuli. Further, the rating procedure did not instruct subjects to report the occurrence of weak stimuli, and "thresholds" were ultimately based on responses to three or four stimulus presentations. Thus there is no indication that Korsakoff and control subjects knowingly made detection judgments, and there are no data to indicate whether they applied comparable criteria in making their ratings.

Mair et al. (1980) tested the sensitivity hypothesis directly by comparing the performances of Korsakoff ($N = 10$) and nonalcoholic control ($N = 8$) subjects on a detection task using weak concentrations of *n*-butanol. Subjects were first given 10 practice trials in which they were instructed to attend to differences between a weak odorous stimulus (the vapor over an aqueous solution of 2.05 mM *n*-butanol in an equilibrium sniffer) and a blank (the vapor over distilled water in a similar sniffer). Following this, they were given 60 yes/no detection trials, 30 with the stimulus and 30 with the blank. There was a 30-second interval between each stimulus, and feedback was given after each trial. The results (Table 19.1) showed that the Korsakoff subjects adopted more conservative criteria, making fewer affirmative responses on both stimulus (hit) and blank (false alarm) trials. Computation of a bias-resistant measure of sensitivity (d') indicated that the detection performance of the two groups was virtually identical. Thus the sensitivity hypothesis was rejected.

These same groups of Korsakoff and control subjects were also tested for discrimination using a short-term recognition memory procedure. There were two stimulus conditions: one with pairs of odorants rated as highly similar (hard to discriminate), and the other with pairs rated as dissimilar

TABLE 19.1. Number of hits (H), false alarms (FA), and corresponding d' values for Korsakoff and control subjects

Korsakoff patients			Nonalcoholic controls		
H	FA	d'	H	FA	d'
22	16	0.54	15	18	0.26
8	1	1.27	23	4	1.87
18	6	1.11	15	13	0.18
2	0	—	17	17	0
16	11	0.42	7	5	0.22
9	4	0.60	17	8	0.78
28	8	2.10	20	2	1.92
17	13	0.36	23	11	1.08
10	8	0.18	—	—	—
11	14	0.26	—	—	—
M 14.1	8.1	0.70	17.1	9.7	0.72
SD 7.6	5.5	0.70	5.2	6.0	0.84

Data from Mair et al. (1980).

(easy to discriminate). On a typical trial, subjects sampled an odorant, performed a distractor task during a delay interval (5, 15, or 30 seconds), sampled a second odorant, and judged whether the two odors were the same or different.

For both sets of stimuli, Korsakoff patients performed significantly worse than controls, although they showed no evidence of a decay in performance over the intervals tested (Figure 19.1). Since Korsakoff patients did not perform above chance for similar odorant pairs, their lack of temporal decay in this condition can be ascribed to a floor effect. However, since they scored significantly above chance in the easy discrimination task, this performance can be taken as evidence that their impaired ability to recognize odorants does not result from an abnormally rapid decay in odor memory. The apparent lack of temporal decay is consistent with studies of odor memory in normal subjects (Engen et al., 1973) and is in marked contrast to the performance of these same Korsakoff subjects on short-term recognition of pictures of human faces and recall of consonant trigrams (Figure 19.1). The observation of impaired odor recognition mem-

FIGURE 19.1. Short-term memory for Korsakoff and control patients as measured for: (A) recall of consonant trigrams, (B) recognition memory for odors, and (C) recognition memory for pictures of unfamiliar human faces. All subjects were engaged in a backwards counting distractor task during delay intervals for all three tests. Korsakoff patients show more rapid decay of short-term memory in the consonant trigrams and face recognition tasks, but not for odors. Initial performance for face and odor recognition memory was impaired in Korsakoff patients. Data are taken from Mair et al. (1980). See text for details.

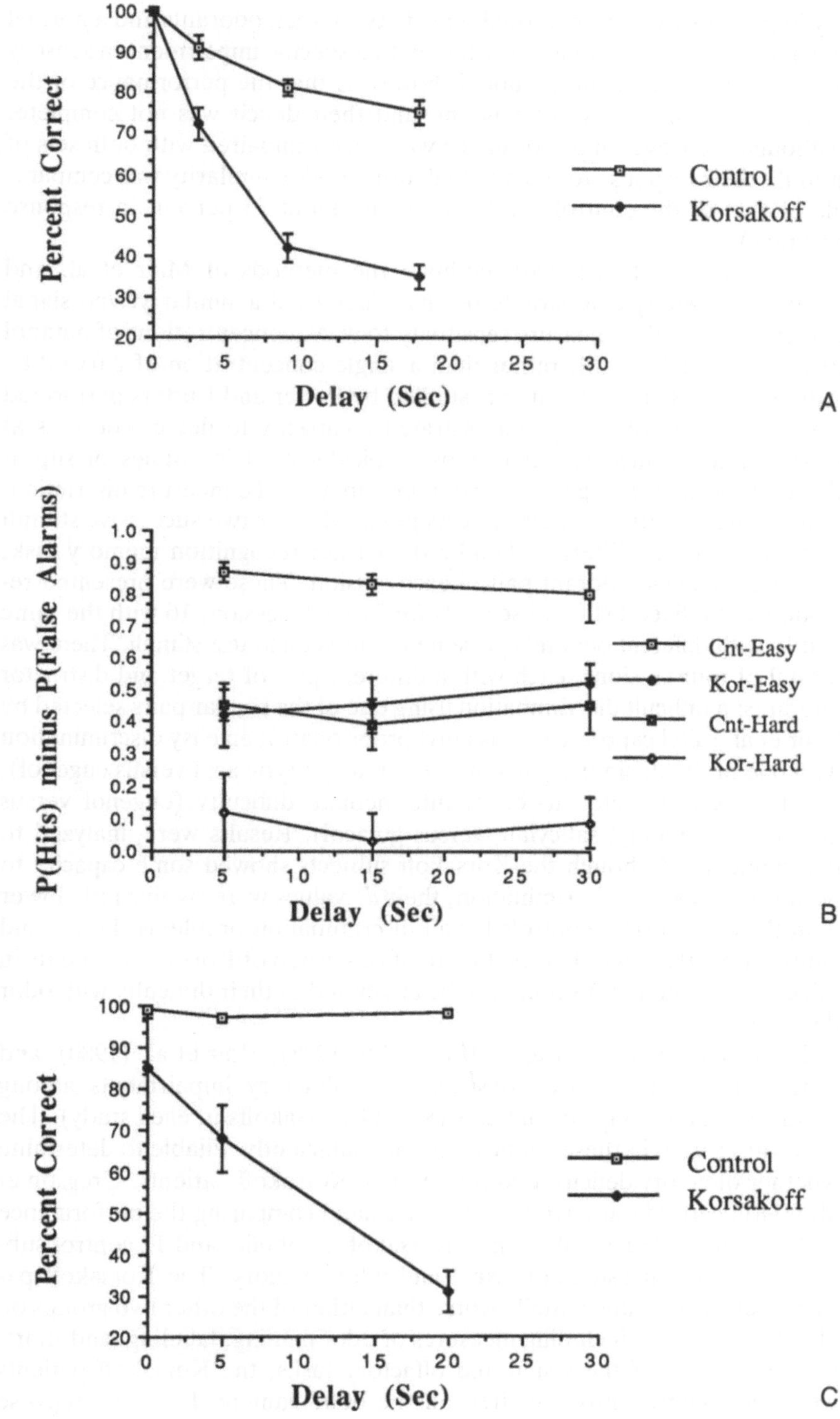

Percent Correct
100
90
80
70
60
50
40
30
20
0 5 10 15 20 25 30
Delay (Sec)
Control
Korsakoff
A
P(Hits) minus P(False Alarms)
1.0
0.9
0.8
0.7
0.6
0.5
0.4
0.3
0.2
0.1
0.0
0 5 10 15 20 25 30
Delay (Sec)
Cnt-Easy
Kor-Easy
Cnt-Hard
Kor-Hard
B
Percent Correct
100
90
80
70
60
50
40
30
20
0 5 10 15 20 25 30
Delay (Sec)
Control
Korsakoff
C

ory in patients exhibiting normal sensitivity to weak odorants and a normal (flat) forgetting function is consistent with a specific impairment in sensory discrimination. It should be noted, however, that the performance of the Korsakoff patients was not random, and their deficit was not complete. Although their overall performance was clearly impaired with both sets of stimuli, their response to the manipulation of odor similarity was comparable to that of the controls (a difference of about 35 percent in response accuracy).

Potter and Butters (1980) modified the methods of Mair et al. and obtained generally comparable results. They used a similar yes/no signal detection method to measure sensitivity to weak concentrations of butanol (0.13 mM and 125 mM, rather than a single concentration of 2.05 mM). Although the Korsakoff patients studied by Potter and Butters performed worse than controls, they demonstrated a capacity to detect odorants at concentrations much less than those typically used in studies of suprathreshold scaling, recognition, and discrimination. To measure discrimination, Potter and Butters had subjects judge whether two successive stimuli were the same or different. Unlike the earlier recognition memory task, they used a single odorant pair in each session. These were presented repeatedly at a fixed ISI of 15 seconds for 32 trials/session, 16 with the same and 16 with different odorants presented as target and test stimuli. There was a total of four sessions, each with a different pair of target and distractor odorants: a difficult discrimination using one of the similar pairs selected by Mair et al. (allyl caproate versus amyl proprionate); an easy discrimination with one of the dissimilar pairs of Mair et al. (butyric acid versus eugenol); and two pairs selected to be of intermediate difficulty (eugenol versus guaiacol and methyl salicylate versus guiacol). Results were analyzed to determine *d'*. Although the Korsakoff subjects showed some capacity to perform the easiest discrimination, their *d'* values were significantly lower than those of normal controls for all discrimination problems. Potter and Butters (1980) argued that the failure of this group of Korsakoff patients in olfactory discrimination could not be attributed to their difficulty with odor detection.

The results of Jones et al. (1975a and b, 1978), Mair et al. (1980), and Potter and Butters show consistency in olfactory impairments among relatively small groups of subjects (8 to 14 Korsakoffs in each study). The procedures used in these studies were not sufficiently reliable to determine whether olfactory deficits are present in all Korsakoff patients. Gregson et al. (1981) carried out a study in New Zealand comparing the performance of 12 alcoholic Korsakoff, 12 non-Korsakoff alcoholic, and 12 control subjects on seven measures of visual and odor memory. The Korsakoff patients performed substantially worse than either of the other two groups on all olfactory tasks, including measures of odor naming, labeling, and short-term memory. Of the visual and olfactory tasks, the Korsakoff patients were most consistently impaired on the odor naming. Further, stepwise discriminant analysis showed that the combination of the three olfactory measures provided a strong separation of the three patient groups. Unlike the scaling and recognition memory tasks described above, the naming and labeling procedures of Gregson et al. do not require a comparison of successively presented odorants. Thus the deficits observed on these tasks provide further evidence that the olfactory impairments of Korsakoff patients do not result from an inability to hold one odor in memory while comparing it to a second subsequent smell. Based on their results, Gregson et al. (1981) argured that olfactory deficits are sufficiently prevalent in Korsakoff's disease to provide a diagnostic test for this disorder.

The clinical assessment of olfactory function was advanced considerably by R.L. Doty and his colleagues with the development of the UPSIT, a standardized multiple-choice test of odor identification (Doty et al., 1984). Each of the 40 items on the UPSIT consists of a microencapsulated odorant and four verbal labels, one of which matches the odorant. The incorrect distractor labels used in the UPSIT tend to be associated with odors that are quite distinct from the stimulus substance. For instance, a cherry odor is presented with the labels whiskey, honey, lime, and cherry. The use of diverse labels does not require a particularly precise olfactory judgment. Engen (1986) has shown that normals perform much more poorly at identifying UPSIT stimuli when distractor labels are selected that are associated with smells similar to the target. Not surprisingly, normals tend to do well on the UPSIT, generally scoring above 90 percent correct. Although it is not particularly sensitive to precise aspects of discrimination, the UPSIT has proven to be valuable in the clinic as a measure of olfactory function that normals can perform with great reliability. Just as the Snellen chart is useful for detecting impairments in visual acuity, the UPSIT provides a metric that can identify persons with subnormal capacity to identify odors.

Mair et al. (1986) compared the perfomance of alcoholic Korsakoff and non-Korsakoff alcoholic control subjects on the UPSIT. To make certain that they could smell the stimuli, all subjects were screened clinically for anosmia and required to detect, without error, the presence or absence of UPSIT stimuli in a series of pretest trials. The Korsakoff subjects ($N = 26$) averaged 41 percent correct on the UPSIT, with only two (8 percent) of the Korsakoff patients scoring above 65 percent correct, a level reached by virtually every subject in the normative sample ($N = 1256$) and seven of eight alcoholic controls matched with the Korsakoff patients for age, education, and performance on intelligence tests.

To determine whether nonolfactory cognitive demands of the task affected the performance of the Korsakoff patients on the UPSIT, seven of the Korsakoff and eight matched alcoholic control subjects were also tested on the Picture Identification Test (PIT). The PIT is identical to the UPSIT, using the same sets of target and distractor labels, except that subjects are presented with a picture rather than an odor of the target label. The Korsakoff patients scored an average of 90 percent correct, a level significantly better than their UPSIT performance (40 percent correct) and not significantly different from the PIT performance of the alcoholic controls. The preserved capacity to perform the PIT demonstrates that the failure of the Korsakoff patients on the UPSIT did not result from difficulty understanding the meaning of the verbal labels or matching them to a sensory stimulus. This finding is consistent with other evidence that Korsakoff patients have a preserved ability to make use of information in semantic memory (cf. Butters and Cermak, 1980).

The results reviewed above provide a strong case for arguing that impaired olfactory discrimination is as consistent a sign of Korsakoff's disease as the more apparent, and better known, symptoms of amnesia. Deficits have been demonstrated in different laboratories for a number of tasks that require an ability to discriminate differences between odorous stimuli, including (1) the identification of odorants by either naming or matching to labels, (2) judging the similarity or difference of two odorants presented successively in recognition memory or two-choice detection tasks, and (3) magnitude estimation or category scaling of the intensity of superthreshold odorants. Other results argue against alternative explanations for these deficits.

Simple anosmia can be ruled out as a cause for the impairment by evidence of (1) recognition memory deficits in patients with normal sensitivity to weak odorants (Mair et al., 1980), (2) impaired capacity to detect differences between odorants at concentrations substantially above the threshold of patients showing signs of hyposmia (Potter and Butters, 1980), (3) severe and consistent deficits on the UPSIT among patients free from clinical signs of anosmia and able to detect the presence of absence of UPSIT stimuli without error (Mair et al., 1986), and (4) impairments in scaling that are not always consistent with recruitment functions (Jones et al., 1978). The argument that these deficits result from rapid decay of odor memory fails to explain (1) the lack of decay in recognition memory for dissimilar odorants (Mair et al., 1980), (2) the failure of Korsakoffs to apply labels to odorants when they can successfully apply these same labels to pictures (Mair et al., 1986), and (3) the observation of magnitude estimation deficits when the memory load was minimized by presentation of the modulus before each test stimulus (Jones et al., 1975b).

Is the Discrimination Deficit in Korsakoff's Disease Specifically Olfactory?

It has long been recognized that patients with Korsakoff's disease exhibit signs of perceptual impairment (see Wechsler, 1917, for an early discussion). In the recent past, deficits have been described for a variety of tasks, including dichotic listening (Glosser et al., 1976; Parkinson, 1979), discriminating short musical passages (Talland, 1965), locating simple figures in complex visual designs (Talland, 1965; Glosser et al., 1977), matching photographs of unfamiliar human faces (Dricker et al., 1978), and utilizing multiple stimulus dimensions of complex visual stimuli (Oscar-Berman and Samuels, 1977). In fact, it has been argued that failure to encode semantic and perceptual features of stimuli may contribute to be memory impairments that characterize this disease (cf. Butters and Cermak, 1980).

What seems to separate these findings from the olfactory deficits described above is the complexity of stimulus information required to produce failure among Korsakoff subjects. Olfactory deficits have been reported for relatively simple recognition or identification tasks. In contrast, Glosser et al. (1976) report that although Korsakoff patients are impaired in dichotic listening tasks such as monitoring inputs to both ears for a critical binaural stimulus, they perform normally under less demanding conditions in which subjects report the occurrence of a simple monoaural stimulus only. Further, even the dichotic deficit can be eliminated by increasing the interval between successive stimuli from 1.2 to 2.0 seconds.

Other evidence has suggested that the visuo-perceptual deficits of this disease do not result from an inability to discriminate visual information, but rather from a failure to attend to or encode multiple dimensions of complex visual stimuli. For instance, Dricker et al. (1978) report that Korsakoff patients are more easily misled than are controls in recognizing the identity of faces confused by superficial cues such as articles of clothing or facial expressions. These authors argued that the poor performance of Korsakoff patients in face recognition results from a tendency to rely on superficial cues rather than configurational information. Huppert and Piercy (1977) showed that the poor initial performance of Korsakoff patients in picture recognition can be eliminated by allowing increased time to study stimulus items. These data have been interpreted as evidence that although Korsakoff patients retain a capacity to fully encode pictures, they will not unless given extra time to do so. Similar arguments have been applied to the encoding of verbal information by Korsakoff patients (cf. Butters and Cermak, 1980):

> [I]t appears that alcoholic Korsakoff patients' recognition memory for verbal information can be affected by the level to which they process that information, at least under some conditions. This, in turn, lends support to previous contentions that alcoholic Korsakoff patients' anterograde amnesia stems, at least in part, from a preference not to perform semantic alalyses of to-be-retained verbal information (p. 91).

It seems, then, that the argument for odor perception being selectively impaired in Korsakoff's disease is predicated on the idea that the perceptual deficits in olfaction are at a more fundamental level of discrimination than those in other sensory modalities. Is this so? Certainly the striking disparity between the performance of Korsakoff patients on picture (PIT) and odor (UPSIT) identification suggests that something is going on in olfaction that is not happening in vision. Early psychophysical scaling studies also provided evidence of specificity. Jones et al. (1975b) used three scaling methods to compare psychophysical functions for olfactory (*n*-butanol), visual (shades of gray), and auditory (white noise) stimuli. While the Korsakoff subjects were impaired with all three methods for scaling the intensity of butanol, they produced psychophysical functions resembling those of normals in all three visual and two of three auditory tasks. Later, Jones et al. (1978) reported significant deficits in scaling the taste of sodium chloride as well as the smell of *n*-butanol among Korsakoff patients who showed "mild impairments" scaling visual stimuli (shades of gray, size of circles, and complexity of grids) and normal performance scaling the loudness of white noise.

Potter and Butters (1980) have provided perhaps the strongest evidence that olfactory discrimination is disrupted selectively in Korsakoff's disease. They compared odor and hue discrimination by measuring the detection of differences between four odorant pairs and four hue pairs selected to represent four levels of discrimination difficulty. Stimuli were selected to provide roughly comparable levels of difficulty between modalities, with d' scores of controls ranging from about 2 to 4 for odorant pairs and 0.4 to 4 for hues. While Korsakoff patients were significantly impaired for all odorant pairs, they were not significantly impaired for any of the hue pairs. It should be noted, however, that there was a trend for the Korsakoff patients to perform more poorly than controls and other patient groups for all but the hardest hue discrimination, which was so difficult that all groups fell to near chance-level performance.

Other studies have provided evidence against the specificity argument. The 10 Korsakoff patients reported to be impaired in odor recognition memory by Mair et al. (1980) were also found to be deficient in their recognition memory for pictures of human faces. Since the deficit was significant even when target and test pictures were presented simultaneously, it seems clear that perceptual factors must have contributed to this deficit. It could be argued, however, that this impairment reflected the informational complexity of the pictures (cf. Dricker et al., 1978).

Mair et al. (1986) approached the issue more directly by comparing the performance of Korsakoffs and alcoholic control subjects on measures of sensory discrimination for olfactory (the UPSIT), color hue (the Farnsworth-Munsell 100 Hue test), and auditory (the Seashore Measures of Musical Talent) stimuli. Significant deficits were observed for all three modalities. The deficits observed on the UPSIT are described above. Because of differences in testing procedure, it was not possible to compare the severity of impairment in the different modalities. However, it should be noted that the consistency of impairment was at least roughly comparable. On the UPSIT, one of eight controls scored below the 75th percentile of

Korsakoff performance. By comparison, two of eight controls scored below the 75th percentile of Korsakoffs on the Farnsworth-Munsell and one of eight on the overall Seashore score. The Farnsworth-Munsell deficit is of particular note, as it involves the arrangement of color chips within the field of vision and thus does not require comparisons based on memory images.

We have recently (Flint and Mair, 1990) compared Korsakoff and abstinent alcoholic controls for recognition memory of color hues. As in the previous study of odor recognition memory (Mair et al., 1980), we used several sets of stimuli that varied in the similarity of target and distractor items. Our results verified the existence of a hue discrimination deficit among a group of Korsakoff patients distinct from those studied in our earlier experiment (Mair et al., 1986). Corrected for guessing, the severity of the hue recognition memory deficit was comparable to that described earlier for odors (Mair et al., 1980). Further, like the odor recognition memory deficit, there was a substantial and significant hue recognition deficit even when there was no delay between the presentations of the target and test items. Thus there was no basis for arguing that the hue discrimination impairment resulted from a rapid decay of short-term memory. This conclusion is, of course, supported by the impairment described above on the simultaneous ordering procedure of the Farnsworth-Munsell 100 Hue test.

It can be argued that the deficits observed in olfaction reflect pathologic mechanisms distinct from those that impair perceptual processes in other sensory modalities. Such an argument is consistent with anatomical evidence that olfactory information can reach neocortex without thalamic relay and without interacting with any other major sensory pathway. It would be difficult to conceive of any focal lesion that could disrupt the primary neocortical projections of the visual, auditory, and olfactory systems. Parsimony, however, is against this argument. Any theory explaining the pathophysiology of Korsakoff's disease must account for the coincidence of disorders affecting (1) anterograde and retrograde memory, (2) higher-order aspects of stimulus perception and encoding, and (3) sensory discrimination deficits affecting olfaction and other sensory modalities. In one manner of thinking, it is more direct and concrete to consider these symptoms separately as impairments of distinct mental faculties that happen to co-occur in Korsakoff's disease. Such a theory, however, requires one to postulate the existence of numerous distinct faculties and to assume that these are disrupted separately by the disease process.

By the criteria of Occam's razor, it is more economical to hypothesize that Korsakoff's disease disrupts a single, more general cognitive process that results in the diverse pattern of deficits described above. In his classic paper "In search of the engram", Lashley (1950) argued in favor of such a view of amnessia:

I believe that the evidence strongly favors the view that amnesia from brain injury rarely, if ever, is due to the destruction of specific memory traces. Rather, the amnesias represent a lowered level of vigilance, a greater difficulty in activating the organized patterns of traces, or a disturbance of some broader system of organized functions (p. 471).

There is now convergent evidence to support lashley's view. Talland (1965) was among the first to emphasize the importance of deficits in attention and perception. He argued that the amnesic symptoms of Korsakoff's disease result from a "premature closure of activation" or an inability to sustain internal cognitive processes. Since the publication of his monograph, evidence has accumulated to support his contention that patients with Korsakoff's disease are impaired on tasks requiring selected or divided attention (cf. Oscar-Berman, 1980; Mair and McEntee, 1983). Other evidence of general cognitive dysfunction has come from studies of reaction time. Talland (1965) reported that although Korsakoff patients exhibit normal simple reaction times, they are abnormally slow when the cognitive demands of the task are increased:

Motor performance showed defects in the Korsakoff patients when the task demanded that they divide their attention between two dissimilar operations or that they process information presented to them successively and for a limited time (p. 129).

Still others have shown that Korsakoff patients do better on a variety of perceptual and memory tasks when allowed more time to process or respond to stimuli (above; see also Cermak and Butters, 1980; Mair and McEntee, 1983).

More recently, we have found evidence of a general processing deficit in studies of cued reaction time (Mair et al., 1988). As in previous studies, Korsakoffs and controls were observed to respond at virtually identical speeds in a simple reaction time situation, executing a single finger tap in response to simple visual and auditory cues. Finger tap reaction times were slowed under two conditions: (1) when the finger tap occurred as the first element of a two- or three-digit tapping sequence, and (2) when a visual cue was preceded by a visual alerting stimulus that required the execution of an endogenously controlled covert shift in attention. In the first circumstance, sensory processing and motor performance requirements were held constant, and thus the impairment seemed to depend on whether the tap occurred as a part of a sequential motor program. In the second circumstance, the motor response and sensory cue were held constant, and the deficit depended on the relative locations of the alerting stimulus and the cue. There was no deficit when their spatial locations coincided and subjects needed only to fixate on the location of the alerting stimulus to focus on the subsequent cue. In the terminology of Posner (1978), this is a form of exogenous or stimulus-driven attention. Korsakoff subjects were impaired when the cue and the alerting stimulus occurred in spatially distinct locations, requiring subjects to disengage attention from the location of the alerting stimulus, shift it to the predicted location of the subsequent cue, and refocus visual attention in the new location. In the terminology of Posner (1978), this is a form of endogenous or subject-driven attention. The response is considered covert, as it does not involve any overt orienting behaviors (including eye movements).

The pattern of impairment on reaction-time tasks is consistent with a general deficit in endogenous cognitive processing (cf. Mair et al., 1988). Given a simple motor response, Korsakoffs are as fast as normals. Their responses are slow when they require the execution of an internal (or endogenous) cognitive process, whether that process is related to the execution of an overt motor program, as in the finger-tapping task, or a covert shift in attention.

From the view of olfaction, the question can be reduced to one of competence versus performance. Does the apparent discrimination deficit reflect an absolute inability (incompetence) to distinguish between odorants, or is discrimination performance impaired by some more general nonolfactory cognitive dysfunction? At this point, there are no data that allow either of these two possibilities to be discarded.

What Is the Pathologic Basis of Olfactory Deficits in Korsakoff's Disease?

Neuropsychological studies have established that olfactory dysfunction is a common sign of Korsakoff's disease. There is uncertainty, however, over the precise nature of the impairment. An analysis of the olfactory deficits (above) supports the argument that this disease spares absolute sensitivity while affecting the fundamental capacity to discriminate differences between odorants. Other evidence (also above) has indicated that discrimination deficits may occur in other sensory modalities and that Korsakoff's disease may be associated with global deficits in endogenous cognitive processes that might account for these multimodal sensory deficits. Whether odor perception is impaired selectively or as a result of a more global cognitive disorder, the consistency of olfactory impairment indicates that Korsakoff's disease must somehow disrupt processes necessary for normal odor perception. In the remainder of this chapter, we will focus on evidence bearing on the pathophysiologic basis of the olfactory deficits in Korsakoff's disease.

Since its discovery over a century ago, there have been numerous reports describing lesions observed at post mortem in patients with Korsakoff's disease. Analyses of large series of cases (Malamud and Skillicorn, 1956; Victor et al., 1971) have described a characteristic pattern of small hemorrhagic and/or necrotic lesions found in a bilateral distribution in periventricular regions of the brainstem and diencephalon. Pathophysiologic analyses have emphasized the consistent involvement of the mediodorsal nucleus (MDn) of the thalamus and the mammillary bpdies (MB) in cases of Korsakoff's amnesia (Victor et al., 1971; Mair et al., 1979). Correlational studies have suggested that lesions of MDn are the only pathologic sign that consistently predicts the occurrence of amnesia (Victor et al., 1971).

Other evidence has suggested that lesions affecting norepinephrine-(NE) containing neurons within the locus coeruleus (LC) may contribute to the amnesic symptoms of this disease (McEntee and Mair, 1978). Pathologic studies have shown lesions involving LC (Victor et al., 1971; Mayes et al., 1988). Korsakoff patients have been reported to show abnormally low indices of brain NE activity, namely, reduced concentrations of the NE metabolite MHPG in cerebrospinal fluid (McEntee et al., 1984). The concentration of CSF MHPG correlates negatively with psychometric measures of memory function in Korsakoff patients (Mair et al., 1985). Finally, treatment with catecholamine agonist medications has been found to relieve some of the psychometric deficits associated with this disease (McEntee and Mair, 1980; Mair and McEntee, 1986; Langlais et al., 1988).

An argument could be made that either lesions of MDn or impairments of brain NE activity might also affect olfactory function. Of the two possibilities, diminished NE activity has a weaker empirical basis. NE-containing axons originating in the brainstem innervate the olfactory bulb and its projection sites in the basal forebrain (cf. Lindvall and Bjorklund, 1983). Although the functional significance of these pathways is not certain, a number of authors have argued that NE serves a modulatory role within the brain, and it may have an important influence on sensory perception or attentional processes (cf. Mair and McEntee, 1983). Multiple regression analyses have shown significant correlations between the concentration CSF MHPG, an index of brain NE activity, and the performance of Korsakoff patients on odor recognition memory (Mair et al., 1985) and on odor identification on the UPSIT (Mair et al., 1986). However, since this same index also correlates with measures of memory and other cognitive processes, it is by no means certain that the relationship between CSF MHPG and olfactory function reflects anything more than variations among patients in disease severity. The possibility that NE is critical for odor discrimination is further weakened by evidence that rats depleted of NE by the neurotoxin 6-OHDA exhibit a normal capacity to discriminate between different odorants (cf. Mair and Harrison, 1990).

The MDn receives inputs from primary olfactory cortex and is reciprocally connected to neocortical areas thought to be involved in higher-order olfactory processes (cf. Price, 1985). In fact, MDn is the only structure consistently affected by this disease that is also known to receive a specific olfactory input (albeit in infrahuman species). Further, animal studies (see introduction) have provided direct evidence that experimental lesions of the thalamus affecting MDn can disrupt performance of odor-guided behaviors.

There are several questions about the role of the MDn. First, it is not certain that the thalamic lesions of Korsakoff's disease affect specific olfactory projections to the MDn. There are at least two reports of cases in which thalamic lesions are limited to the medial extent of the thalamus, outside the region of the MDn (Mair et al., 1979; Mayes et al., 1988). In other cases, thalamic lesions are described as affecting limited portions of medial MDn (Malamud and Skillicorn, 1956; Victor et al., 1971). Second, one must be cautious in extending results of MDn lesion studies from rodents to man given the immense phylogenetic variation in this structure. The MDn in man is much larger and projects to frontal cortical zones that are not only larger but more differentiated and characterized by well-developed granular layers. Third, the MDn receives robust inputs from multimodal sensory areas, including the amygdata, temporal cortex, and all areas of frontogranular cortex, and thus seems a questionable site for a specific olfactory deficit (cf. Nauta, 1964; Krettek and Price, 1977a and b).

The importance of the MDn in human olfactory perception has been supported by studies of patients with lesions of frontal cortex. Potter and Butters (1980) compared olfactory functions in alcoholic Korsakoffs ($N = 8$), patients who underwent unilateral craniotomy for removal of prefrontal tumors or intracerebral hematomas, and appropriate controls. For the side ipsilateral to surgery, prefrontal patients exhibited impairments in odor discrimination comparable to those of Korsakoff patients. These same surgical patients exhibited normal detection of weak odorants, supporting the contention that quality discrimination can be disrupted in patients who have normal sensivity to weak odorants. This same pattern of results has been described in rodents with thalamic lesions that include the MDn (see introduction). More recently, Jones-Gotman and Zatorre (1988) measured butanol sensitivity and UPSIT performance among 120 patients with focal surgical lesions of cerebral cortex. None of the patients exhibited impairments in absolute sensitivity. Of patients with frontal removals, those in whom the lesions included orbital cortex ($N = 16$) were significantly impaired on the UPSIT, and those with orbital sparing ($N = 13$) were not. Taken with the results of Potter and Butters (1980), these data are consistent with the hypothesis that lesions affecting MDn target areas in orbitofrontal cortex disrupt the ability to discriminate and identify odors without impairing absolute olfactory sensitivity. The close parallels with deficits observed among Korsakoff patients provides strong support for the argument that MDn lesions account for the olfactory deficits associated with this disease.

Conclusions

1. Patients with Korsakoff's disease perform poorly on a variety of tasks that require them to discriminate differences between odorants. These

deficits in odor perception are as consistent a sign of this disease as are the better known impairments of memory.
2. Although groups of Korsakoff patients have been reported to show signs of diminished sensitivity to weak odorants, there are reports that show a clear dissociation between signs of anosmia or hyposmia and deficits in odorant discrimination.
3. Korsakoff's disease is also associated with perceptual impairments affecting stimuli in other sensory modalities. Unlike those in olfaction, many deficits in other modalities require manipulations to increase information processing demands to produce failure. Color perception may provide an exception. There are reports (although not in total agreement) that deficits occur in simultaneous ranking of and in recognition of differences between hue samples.
4. There is also evidence that Korsakoff patients have a global mental impairment that affects attentional and endogenous cognitive processes. Taken with evidence of multimodal perceptual deficits, these results raise the possibility that the olfactory discrimination deficits of this disease may result from a more general, nonolfactory cognitive disorder. It is not established that these patients have a selective impairment of odor perception, although this remains a possibility.
5. Pathophysiologic studies of Korsakoff patients suggest two processes that are disrupted by this disease that might also contribute to deficits in memory and olfaction. These include the mediodorsal nucleus of the thalamus and noradrenergic pathways that ascend from the brainstem to innervate widespread areas of the forebrain.
6. The MDn is more strongly implicated by available empirical evidence. It is (a) consistently lesioned in cases of Korsakoff's amnesia, (b) an important part of central olfactory projections, (c) the site of lesions known to produce olfactory (and other sensory) deficits in experimental animals, and (d) the source of thalamic fibers to regions of frontal cortex that when surgically damaged produce signs of olfactory discrimination deficits without accompanying hyposmia.

Acknowledgment. This work was supported by Veterans Administration Medical Research Funds and by grant RO-1-NS26855 from NINDS to RGM.

References

Allison, A.C. (1953). The morphology of the olfactory system in the vertebrates. *Biol. Rev.*, *28*, 195–244.

Brodal, A. (1947). The hippocampus and the sense of smell. A review. *Brain*, *70*, 179–222.

Butters, N., and L.S. Cermak (1980). *Alcoholic Korsakoff's Syndrome: An Information-Processing Approach to Amnesia*. New York: Academic.

Doty, R.L., P. Shaman, and M. Dann, (1984). Development of the University of

Pennsylvania Smell Identification Test: a standardized microencapsulated test of olfactory function. *Physiol. Behav.*, *32*, 489–502.

Dricker, J., N. Butters, G. Berman, I. Samuels, and S. Carey (1978). Recognition and encoding of faces by alcoholic Korsakoff and right hemisphere patients. *Neuropsychologia*, *16*, 683–695.

Eichenbaum, H., K.J. Shedlack, and K.W. Eckmann, (1980). Thalamocortical mechanisms in odor-guided behavior. I. Effects of lesions of the mediodorsal thalamic nucleus and frontal cortex on olfactory discrimination in the rat. *Brain Behav. Evol.*, *17*, 255–275.

Engen, T. (1986). Remembering odors and their names. *Am. Scient.*, *75*, 497–504.

Engen, T., J.E. Kuisma, and P.D. Eimas (1973). Short-term memory of odors. *J. Exp. Psychol.*, *99*, 222–225.

Ferrier, D. (1876). *Functions of the Brain* (Reprinted, 1966). London, Dawsons of Pall Mall.

Flint, D., and R.G. Mair (1990). *Recognition memory in Korsakoff's disease: enhancement by increased frequency of exposure but not by discriminability of visual stimuli.* In preparation.

Glosser, G., N. Butters, and I. Samuels (1976). Failures in information processing in patients with Korsakoff's syndrome. *Neuropsychologia*, *14*, 327–334.

Glosser, G., N. Butters, and E. Kaplan (1977). Visuoperceptual processes in brain-damaged patients on the digit symbol substitution tests. *Int. J. Neurosci.*, *7*, 59–66.

Gregson, R.A.M., M.I. Free, and M.W. Abbott (1981). Olfaction in Korsakoff's, alcoholics and normals. *Brit. J. Clin. Psych*, *20*, 3–10.

Gross, C.G., and L. Weiskrantz (1964). Some changes in behavior produced by lateral frontal lesions in the Macaque. In J.M. Warren and K. Akert (Eds.), *The Frontal Granular Cortex and Behavior*. New York: McGraw-Hill, pp. 74–98.

Huppert, F., and M. Piercy (1977). Recognition memory in amnesia patients: a defect of acquisition? *Neuropsychologia*, *15*, 643–652.

Jackson, J.H., and C.E. Beevor (1890). Case of tumour of the right temporosphenoidal lobe bearing on the localization of the sense of smell and on the interpretation of a particular variety of epilepsy. *Brain*, *12*, 346–357.

Jones, B.P., H. Moskowitz, and N. Butters (1975a). Olfactory discrimination in alcoholic Korsakoff patients. *Neuropsychologia*, *13*, 173–179.

Jones, B.P., H. Moskowitz, N. Butters, and G. Glosser (1975b). Psychophysical scaling of olfactory, visual, and auditory stimuli by alcoholic Korsakoo patients. *Neuropsychologia*, *13*, 387–393.

Jones, B.P., N. Butters, H. Moskowitz, and K. Montgomery (1978). Olfactory and gustatory capacities of alcoholic Korsakoff patients. *Neuropsychologia*, *16*, 332–337.

Jones-Gotman, M., and R.J. Zatorre (1988). Olfactory identification deficits in patients with focal cerebral excision. *Neuropsychologia*, *26*, 387–400.

Kreig, W.J.S. (1942). *Functional Neuroanatomy*. Philadelphia: Blakiston Co., pp. 289–306.

Krettek, J.E., and J.L. Price (1977a). The cortical projection of the mediodorsal nucleus and adjacent thalamic nuclei in the rat. *J. Comp. Neural.*, *171*, 157–192.

Krettek, J.E., and J.L. Price (1977b). Projections from the amygdaloid complex to the cerebral cortex and thalamus in the rat and cat. *J. Comp. Neurol.*, *172*, 687–772.

Langlais, P.J., R.G. Mair, P.J. Whalen, W. McCourt, and W.J. McEntee (1988). Memory effect of dl-threo-3,4-dihyroxyphenylserine (DOPS) in human Korsakoff's disease. *Psychopharmacology*, *95*, 250–254.

Lashley, K.S. (1950). In search of the Engram. *Symp. Soc. Exp. Biol.*, *4*, 454–482.

Leonard, C.M., and J.W. Scott (1971). Origin and distribution of amygdalofugal pathways in the rat: an experimental neuroanatomical study. *J. Comp. Neurol.*, *141*, 313–330.

Lindvall, O., and A. Bjorklund (1983). Dopamine- and Norepinephrine-containing neuron systems: their anatomy in the rat brain. In P.C. Emson (Ed.), *Chemical Neuroanatomy*. New York: Raven Press, pp. 229–255.

Mair, R.G., C. Capra, W.J. McEntee, and T. Engen (1980). Odor discrimination and memory in Korsakoff's psychosis. *J. Exp. Psych.: Hum. Perc. Perf.*, *6*, 445–458.

Mair, R.G. and W.J. McEntee (1983). Korsakoff's amnesia: noradrenergic systems and cognitive impairment. *Behav. Brain Res*, *9*, 1–32.

Mair, R.G., R.J. Zatorre, and W.J. McEntee (1985). Monoamine activity correlates with psychometric deficits in Korsakoff's disease. *Behav. Brain Res*, *15*, 247–254.

Mair, R.G., R.L. Doty, K.M. Kelly, C.S. Wilson, P.J. Langlais, W.J. McEntee, and T.A. Vollmecke (1986). Multimodal sensory discrimination deficits in Korsakoff's psychosis. *Neuropsychologia*, *24*, 831–839.

Mair, R.G., and W.J. McEntee (1986). Cognitive enhancement in Korsakoff's disease by clonidine: a comparison will 1-dopa and ephedrine. *Psychopharmacology*, *88*, 374–380.

Mair, R.G., D. Flint, A. Inhoff, and P.J. Langlais (1988). Voluntary control of attention is impaired in Korsakoff's syndrome. *Neurosci. Abstr.*, *14*, 1044.

Mair, R.G., and L.M. Harrison (1990). Influence of drugs on smell function and memory. In D.G. Laing, R.L. Doty, and W. Breipol (Eds.), *The Human Sense of Smell*. Berlin: Springer-Verlag.

Mair, W.P.G., E.K. Warrington, and L. Weiskrantz (1979). Memory disorder in Korsakoff's psychosis: a neuropathological and neuropsychological investigation of two cases. *Brain*, *102*, 749–785.

Mallamud, N., and S.A. Skillicorn (1956). Relationship between the Wernicke and Korsakoff syndrome: a clinicopathologic study of seventy cases. *AMA Arch. Neurol. Psychiatr.*, *76*, 585–596.

Mayes, A.R., P.R. Meudell, D. Mann, and A. Pickering (1988). Location of lesions in Korsakoff's syndrome: neuropsychological and neuropathological data on two patients. *Cortex*, *24*, 367–388.

McEntee, W.J., and R.G. Mair (1978). Memory impairment in Korsakoff's psychosis: a correlation with brain noradrenergic activity. *Science*, *202*, 905– 907.

McEntee, W.J., and R.G. Mair (1980). Memory enhancement in Korsakoff's disease by clonidine: further evidence of a noradrenergic deficit. *Ann. Neurol.*, *7*, 466–470.

McEntee, W.J., R.G. Mair, and P.J. Langlais (1984). Neurochemical pathology in Korsakoff's psychosis: implications for other cognitive disorders. *Neurology*, *34*, 648–652.

Milner, B. (1974). Hemispheric specialization: scopes and limits. In F.O. Schmitt and F.G. Worden (Eds.), *The Neurosciences Third Study Program*. Cambridge, MA: MIT Press, pp. 75–89.

Mishkin, M. (1964). Perseveration of central sets after frontal lesions in monkeys. In J.M. Warren and K. Akert (Eds.), *The Frontal Granular Cortex and Behavior*. New York: McGraw-Hill, pp. 219–237.

Nauta, W. (1964). Some efferent connections of the prefrontal cortex in the monkey. In J.M. Warren and K. Akert (Eds.), *The Frontal Granular Cortex and Behavior*. New York: McGraw-Hill, pp. 397–407.

Oscar-Berman, M. (1980). Neuropsychological consequences of long-term alcoholism. *Am. Scient.*, *68*, 410–419.

Oscar-Berman, M., and I. Samuels (1977). Stimulus preference and memory factors in Korsakoff's syndrome. *Neuropsychologia*, *15*, 99–106.

Otto, T., and H. Eichenbaum (1990). Olfactory learning and memory in the rat: a "model system" for studies of the neurobiology of memory. This volume.

Parkinson, S.R. (1979). The amnesic Korsakoff syndrome: a study of selected and divided attention. *Neuropsychologia*, *17*, 67–75.

Posner, M. (1978). *Chronometric Explorations of the Mind* Hillsdale, NJ: Lawrence Erlbaum.

Potter, H., and N. Butters (1980). An assessment of olfactory deficits in patients with damage to prefrontal cortex. *Neuropsychologia*, *18*, 621–628.

Powell, T.P.S., W.M. Cowan, and G. Raisman (1963). Olfactory relationships of the diencephalon. *Nature*, *199*, 710–712.

Powell, T.P.S., W.M. Cowan, and G. Raisman (1965). The central olfactory connections. *J. Anat.*, *99*, 791–813.

Pribram, K.H., and L. Kruger (1954). Functions of the "olfactory brain". *Ann. NY Acad. Sci.*, *32*, 109–139.

Price, J.L. (1985). Beyond the primary olfactory cortex: olfactory-related areas in the neocortex, thalamus, and hypothalamus. *Chem. Senses*, *10*, 239–258.

Sapolsky, R.M., and H. Eichenbaum (1980). Thalamo-cortical mechanisms in odor-guided behavior II. Effects of lesions of the mediodorsal thalamic nucleus and frontal cortex on odor preferences and sexual behavior in the hamster. *Brain Behav. Evol.*, *17*, 276–290.

Scott, J., and C. Pfaffmann (1967). Olfactory input to the hypothalamus: electrophysiological evidence. *Science*, *158*(1), 592–594.

Skinner, J., and D. Lindsley (1973). The non-specific mediothalamic-frontocortical system: its influence on electrocortical activity and behavior. In K.H. Pribram and A.R. Luria (Eds.), *Pathophysiology of the Frontal Lobes*. New York: Academic, pp. 185–234.

Slotnick, B.M. (1985). Olfactory discrimination in rats with anterior amygdala lesions. *Behav. Neurosci.*, *99*, 956–963.

Slotnick, B.M., and N. Kaneko (1981). Role of mediodorsal thalamic nucleus in olfactory discrimination learning in rats. *Science*, *214*, 91–92.

Staubli, U., F. Schottler, and D. Nejat-Bina (1987). Role of dorsomedial thalamic nucleus and piriform cortex in processing olfactory information. *Behav. Brain Res.*, *25*, 117–129.

Stevens, S.S. (1975). *Psychophysics*. New York: Wiley.

Talland, G.A. (1965). *Deranged Memory*. New York: Academic.

Victor, M., G.H. Collins, and R.D. Adams (1971). *The Wernicke-Korsakoff Syndrome*. Philadelphia: F.A. Davis.

Wechsler, D. (1917). A study of retention in Korsakoff psychosis. *Psychiatr. Bull. NY State Hosp.*, *2*, 403–451.

20
Olfaction and Head Injury

RICHARD M. COSTANZO, JOHN D. WARD, and HAROLD F. YOUNG

Introduction

An estimated 500,000 Americans each year suffer from head injuries. These individuals develop a variety of medical problems resulting from injuries to the brain and often require extensive treatment and rehabilitation. One of the many well-documented problems in head injury is anosmia, or the loss of smell sensations. This loss can involve a complete anosmia or a partial loss of or disturbance in olfactory function. Although the loss of olfactory function is often overshadowed by life-threatening issues at the time of injury, it can become more important during the recovery and rehabilitation period. A number of different mechanisms that can lead to posttraumatic anosmia have been identified. The type and severity of the injury is an important factor in predicting the nature and extent of the olfactory loss. Patients with posttraumatic anosmia are often unaware of their olfactory loss. They sometimes report a loss of "taste" sensation. However, this is usually due to their inability to sense the "flavor" or aromas of food. The recent development of clinical olfactory function tests has proven useful in providing quantitative assessments of sensory losses in head injury patients. These tests have also been of some help in localizing lesions to olfactory regions of the brain. Such tests should prove helpful in studying the impact of olfactory loss on long-term outcome and rehabilitation of the head-injured patient. This chapter reviews studies of posttraumatic olfactory losses following head injury and discusses some new areas of clinical interest in this field.

Reports of Posttraumatic Anosmia

It is well established that anosmia is likely to occur in some cases following injuries to the head and face (Cairns, 1937; Hagan, 1967; Rauh, 1967; Roseburg and Fikentscher, 1976; Sumner, 1964; Zusho, 1982). The first report of posttraumatic anosmia appeared in volume 1 of the London

Hospital Reports (Jackson, 1864). In his paper on diseases of the nervous system, Hughlings Jackson included a description of a 50-year-old man who suffered a concussion after apparently being knocked off his horse by a highwayman. He reports that one of the consequences of this head injury was that the patient suffered a complete loss of his sense of smell.

Shortly after this report, additional articles appeared in the literature describing olfactory disturbances following head injury. Legg (1873) reported a single case of a 36-year-old man, Notta (1870) reported six cases of posttraumatic anosmia, and Ferrier (1876) reported a case in which there was posttraumatic anosmia and ageusia (loss of taste). Ogle (1870) was the first to describe taste loss in patients with posttraumatic anosmia. He suggested that in many cases patients confuse a taste loss with a loss of the "flavor" of foods. This loss in flavor is typically associated with smell loss, or the inability to detect the aromas of foods. In a review of the literature on taste and smell disorders following head injury, Costanzo and Becker (1986) estimated that the actual occurrence of taste loss was only about 0.4 to 0.5 percent. In addition, they noted that in almost all cases of posttraumatic taste loss, the sense of smell was also impaired or completely abolished.

Estimates of Posttraumatic Anosmia

In several large-scale studies of head injury, attempts were made to estimate the incidence of posttraumatic anosmia (Leigh, 1943; Mifka, 1964; Sumner, 1975; Sumner, 1964; Zusho, 1982). These studies all evaluated large populations of head injury patients (from 1000 to 5000). The combined results from these studies suggest that posttraumatic anosmia occurs in about 4 to 8 percent of all head-injured patients.

More recent studies (Costanzo et al., 1987; Costanzo and Becker, 1986; Levin et al., 1985; Rauh, 1967; Rutherford, 1977) suggest that the incidence of olfactory disorders among head-injured patients may be considerably higher. For example, Costanzo and Becker (1986) found olfactory impairment to be present in 22.1 percent of head injury patients. In their study of 592 moderate and severe head injuries, they noted that they were able to assess olfactory function in only 348 (58.8 percent). They point out that in many studies of head injury, patients are often not evaluated or tested for olfactory function. This may be due to the fact that some patients are untestable (unconscious or in a coma), or, in the case of retrospective studies, there may be no information on smell function available in the patient's record. Such sampling problems may contribute to the variability in published estimates of the frequency of posttraumatic anosmia. Of the 348 patients that were evaluated for smell in the Costanzo and Becker study, a total of 77 (22.1 percent) had some type of olfactory disturbance. Complete anosmia was found in 13.5 percent of the patients, and a partial olfactory impairment in 8.6 percent. Levin et al. (1985) examined a

TABLE 20.1. Incidence of anosmia as a function of posttraumatic amnesia

	All patients	Patients with anosmia	
Duration of amnesia	Number	Number	Percent
Less than 1 hour	945	44	4.8
1–24 hours	104	15	14.5
1–7 days	46	14	31.5
More than 7 days	72	14	19.5
Totals	1167	87	7.5

Data from Sumner (1964).

population of patients with closed head injury who had no evidence of anosmia. Although these patients could detect odors, 33 percent of them had impairment of olfactory naming and recognition. Although some studies suggest that the incidence of total anosmia among head injuries may be relatively low, the number of patients with some degree of olfactory disturbance has probably been frequently underestimated.

Rutherford (1977) reviewed the symptoms of 145 patients who had concussion from minor head injuries and found that the status of these patients can change considerably. He reported that anosmia was present in 23 (16 percent) of these patients when evaluated 24 hours post-injury. After a period of 6 weeks, however, this number decreased to 4 (2.8 percent). These data suggest that some olfactory disturbances, at least in minor head injuries, may be of a temporary nature and in many cases may be resolved with time.

Severity of Injury

A number of studies demonstrate that the incidence of anosmia increases with severity of injury (Rutherford, 1977; Levin et al., 1985; Caruso et al., 1969; Rauh, 1967; Sumner, 1975; Sumner, 1964; Costanzo and Becker, 1986). For example, Rutherford (1977) reports 16 percent anosmia in minor head injury, while in moderate and severe head injuries (Costanzo and Becker, 1986) the incidence is 19.4 and 24.5 percent, respectively. Rauh (1967) assigned head injury patients to categories according to increasing severity of injury (grades I, II, and III) and found the incidence of anosmia to be 0, 15, and 30 percent, respectively. Sumner (1964) used the duration of amnesia as a way of assessing severity of injury, and his findings show that the percent of patients with anosmia increases with the duration of posttraumatic amnesia (Table 20.1). The more severe cases, with amnesia lasting for periods of one to several days, had the highest percentage of anosmia.

In a sample of 25 patients who had severe head injuries and received follow-up evaluations at least two months following injury, olfactory testing revealed that over 50 percent were classified as anosmic at the time of follow-up (Costanzo et al., 1987). These data suggest that the more severely injured patients (those who require follow-up and rehabilitation) are most likely to have a loss of olfactory function.

Mechanisms of Posttraumatic Anosmia

The cause most often associated with posttraumatic anosmia is a tearing of the olfactory nerves (Bickerstaff, 1973; Costanzo and Becker, 1986; Ogle, 1870). Although this is probably one of the more important mechanisms of anosmia following head injury, it is not the only mechanism that could result in olfactory dysfunction. Cortical contusions and hemorrhage are often important factors in head injury and can lead to impairment of olfactory function. In addition, mechanical injury to the normal nasal passages can limit access of odors to the olfactory regions and result in decreased or total loss of smell.

Injury to the Nasal Cavity

Head injuries, especially those involving the face and nose, can result in a displacement of the nasal bones, causing a blockage of air passages. Such injuries can occur with or without damage to the olfactory epithelium lining the nasal cavity. Surgical correction of nasal structures would be likely to restore olfactory function disrupted by mechanical blockage of air flow to the olfactory region.

Tearing of the Olfactory Nerves

The olfactory nerves are fine processes that project from the nasal cavity, pass through small openings at the base of the skull (cribriform plate), and connect to cortical cells in the olfactory bulb (Figure 20.1). These small axon fibers are easily torn by shearing forces that can occur when the brain is suddenly displaced within the cranium. This often occurs when there is a blow to the head and coup, contra-coup forces are of sufficient magnitude to cause tearing of the olfactory nerves. This would interrupt the sensory pathway by which olfactory signals reach the brain. Following damage to the olfactory nerves, the sensory neurons in the nasal cavity undergo retrograde degeneration. Biopsies taken from patients who had anosmia following head injury show degenerative changes among olfactory vesicles and the disappearance of sensory cilia (Hasegawa et al., 1986; Jafek et al., 1989; Moran et al., 1985; Yamagishi et al., 1988).

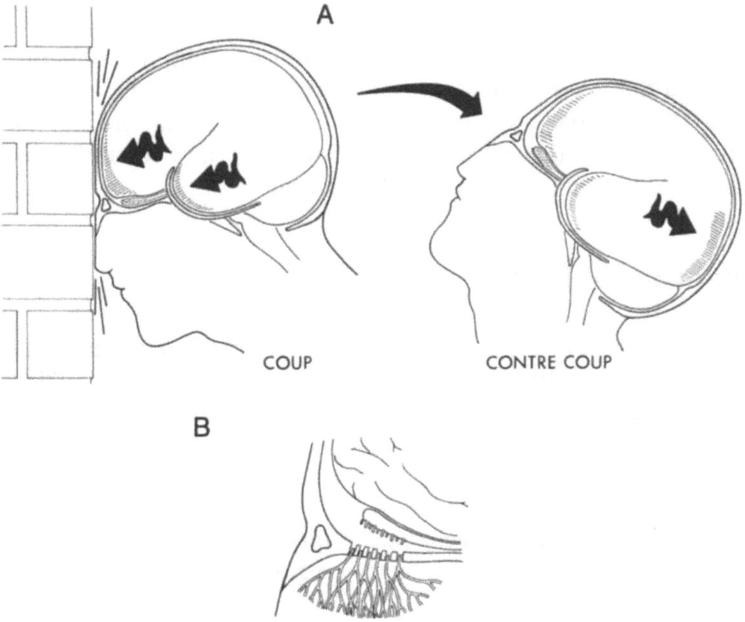

FIGURE 20.1. Mechanisms of posttraumatic anosmia. (A) Brain contusions and hemorrhage resulting from coup contre-coup forces; (B) tearing of olfactory nerves as they pass through the cribriform plate at the base of the skull. (Reprinted, by permission, from Costanzo and Becker 1986.)

Contusions and Hemorrhage

Contusions of the frontal and temporal lobes are another common occurrence in head injury that can result in olfactory impairment. In such cases, complete anosmia may not be present, and the ability to detect odors may remain intact. However, odor identification and recognition may be severely impaired because of the damage to underlying cortical regions that process olfactory information (Eslinger et al., 1982; Levin et al., 1985). Increases in intracranial pressure, blood clots, and hemorrhage can also interfere with cortical regions involved in olfactory function and therefore can lead to a partial or complete loss of olfactory function. Newly developed methods for assessing olfactory function have proved to be helpful in determining the nature of the damage (Cain et al., 1983; Cain et al., 1988; Levin et al., 1985).

Site of Injury

Sumner (1975) studied the incidence of posttraumatic anosmia in 584 patients with relatively mild head injuries (posttraumatic amnesia < 1 hour or no amnesia) to determine the relationship between site of injury and incidence of anosmia. He limited his study to mild head injuries because he thought that in more severe injuries the intracranial forces might be more complex. Although he found that the most common injuries were to the front of the head (459), anosmia occurred in only 18 (3.9 percent) of these subjects. In contrast to frontal injuries, anosmia was present in 8 of the 38 head injuries where there was a blow to the occipital region. This represents a 21 percent incidence of anosmia, a significantly higher occurrence than for frontal injuries. The remaining 87 subjects who received vertical head injuries had a 10.4 percent incidence of anosmia. These data suggest that anosmia is most likely to occur with blows to the back of the head (occipital) rather than frontal or vertical injuries.

Accident Type

Costanzo and Becker (1986) examined the incidence of olfactory disorders by accident type and found that the highest percentage of anosmia (26.6 percent) occurred with domestic falls. Motor vehicle (23.6 percent) and cycle (22.5 percent) accidents also showed high percentages. Although most of these injuries would be expected to be frontal in nature, it is possible that a higher percentage of domestic falls might be to the back of the head. This could explain the slightly higher incidence of anosmia in this group.

Clinical Evaluation and Testing

Neurological Exam (CN I and V)

Test of cranial nerve I should be included in the neurological exam of all head injury patients. Often this is accomplished by presenting the patient with one or two familiar odor stimuli (e.g., perfume, soap, coffee, peanut butter) and asking the patient if he or she can detect and identify each odor. Trigeminal nasal sensations should also be evaluated. Chemicals such as ammonia, rubbing alcohol, and Vicks® are typically used to test trigeminal nasal function. A common problem in evaluating posttraumatic anosmia is that a trigeminal sensation is generated in response to some odors. These sensations are often confused with olfactory sensations and are sometimes incorrectly interpreted as intact olfactory function. Careful selection of test odors and a separate assessment of olfactory and trigeminal nasal sensations is useful following head injury. This is especially important in medical-legal cases (Labayle and Hadengue, 1973), where an objective assessment of olfactory function is needed, and suspected malingerers need to be identified (Cabral, 1972). In such cases, quantitative olfactory function tests are clearly indicated. Miller and Cartlidge (1972) noted the importance of testing trigeminal nasal sensations as a way of identifying malingerers. As they point out, "the patient who claims to be unable to smell strong ammonia despite lacrimation is bogus."

Olfactory Function Tests

Clinical olfactory function tests are now available (Cain et al., 1988; Doty et al., 1984), and results can be compared with published norms (Cain et al., 1983; Doty et al., 1984; Heywood and Costanzo, 1986). For example, the Connecticut Chemosensory Clinical Research Center (CCCRC) test includes a measurement of both odor detection threshold and odor identification (Cain et al., 1988). Odor detection threshold is determined using an ascending series of butanol concentrations in a two-bottle forced-choice paradigm. Odor identification testing involves sampling a series of jars containing common odorants (baby powder, chocolate, cinnamon, coffee, mothballs, peanut butter, and soap). Awareness of olfactory disturbances following head injury has increased with the development of diagnostic tests to measure olfactory function. Olfactory function tests such as the CCRC test not only provide objective quantification of the degree of sensory loss, but may also help localize lesions in regions of the brain used in odor identification. Although screening tests can often alert the physician to olfactory disturbances, there are cases where simple odor detection is normal, yet an olfactory dysfunction of central origin is present. In some cases, comprehensive olfactory function testing is recommended. Levin et al. (1985) reported such patients in a study where olfactory discrimination was impaired in the presence of normal detection. He also tested general sensory discrimination function using a visual discrimination task; and found this to be intact. His results indicated that there was a selective deficit in olfactory discrimination. Such findings are not uncommon among head injury patients. Olfactory function tests that include a discrimination or identification task in addition to odor detection play an important role in the assessment of olfactory function in head injury patients. These tests often provide an excellent opportunity to measure the degree of impairment and monitor the patient's progress during recovery periods.

Central versus Peripheral

Cortical regions responsible for processing olfactory information have been reported to include the orbital frontal pole in addition to the olfactory bulb and pyriform cortex (Potter and Butters, 1980; Tanabe et al., 1975; Yarita et al., 1980). In head injury, contusions or hemorrhage in these areas could account for olfactory impairment observed with olfactory function testing. Varney (1988) recently reported psychosocial deficits in patients with posttraumatic anosmia of a type typically associated with damage to the orbital frontal pole. This report suggests that posttraumatic anosmia is often a sign of orbital frontal damage. CTs and MRIs of brain-injured heads have in some cases confirmed the correlation of cortical lesions with selective deficits in olfactory recognition and discrimination

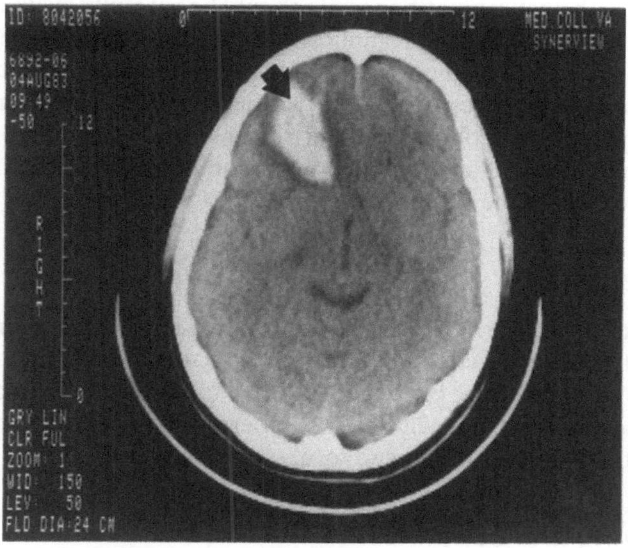

FIGURE 20.2. CAT scan of patient following severe head injury. Patient was able to detect odors but could not discriminate or identify odors. Large mass in right frontal lobe (arrow) suggests that olfactory impairment is due to a central (intracranial) lesion. (Reprinted, by permission, from Costanzo et al. 1987.)

(Costanzo et al., 1987). One such case that illustrates the diagnostic utility of olfactory function testing is that of a 26-year-old male who sustained a blow to the head in a motorcycle accident. Olfactory function testing revealed that the patient could detect odors but could not discriminate or identify the odor stimuli. A central lesion was confirmed by the presence of a large mass (hematoma) in the right frontal lobe (Figure 20.2). The inability of patients to identify odors in the presence of normal detection thresholds has been reported by several investigators (Costanzo et al., 1987; Jones-Gotman and Zatorre, 1988; Varney, 1988). These studies confirm the relationship between central lesions and selective deficits in olfactory recognition and discrimination.

Patient Management and Treatment

Related Medical Problems

Anosmia following head injury may alert the physician to other complications that may need attention. CSF leaks and subsequent infections can result if the olfactory nerves and dural barrier are torn (Cairns, 1937; Raskind and Doria, 1966). This could leave an open access for viruses and other vectors from the nasal cavity to enter the normally protected brain environment (Gibson and Kurukchy, 1974; Leech, 1974; Lundh et al., 1987).

Patients with posttraumatic anosmia should be carefully evaluated to determine if there are medical complications associated with their anosmia. CSF leaks should be ruled out, if they are present, medications to prevent infection and surgical repair may be warranted. Intracranial pressure or hemorrhage involving olfactory brain regions (e.g., orbital frontal pole, prepyriform cortex, olfactory bulbs) may require medical attention. Fractures to the base of the skull may be present.

Treatments

Treatments to restore olfactory function in cases of head injury are not feasible at this time. Attempts to repair olfactory nerves that may have been torn or severed by coup, contra-coup forces are not practical; and even if they were technically feasible, there is no diagnostic method to verify that nerve fibers have been torn.

Counseling and Social Issues

If and when recovery of olfactory function occurs, it will most likely take place on its own. Patients with anosmia should, however, be counseled about their sensory loss and advised of the risks and hazards associated with loss of olfactory function. These patients are often at risk of exposure to gas leaks, are unable to detect smoke from fires, and occasionally accidentally ingest toxic or harmful substances or spoiled foods.

Social issues related to loss of smell may be less important from a medical management point of view, but may have psychological and emotional impact. Anosmic patients often get depressed, have difficulty enjoying foods, and miss the pleasant smells associated with experiences, such as the smell of fresh flowers, the small of grass in the springtime, and the scent of a familiar perfume. The inability to smell has been associated with depression and other psychological problems involving mood and behavior.

Prognosis and Recovery

Severity of Injury

The prognosis for recovery of olfactory function following head injury depends on the severity and type of injury. Although it is thought that olfactory neurons can be replaced within the nasal cavity, reconnection of newly growing nerve fibers is unlikely, as they are most probably blocked from regaining access to the olfactory bulb. Recovery of olfactory function is more likely to occur following contusions or reversible disruption of blood supply to cortical olfactory regions. Sumner reported 50 percent recovery in patients with amnesia lasting less than 24 hours, whereas recovery was

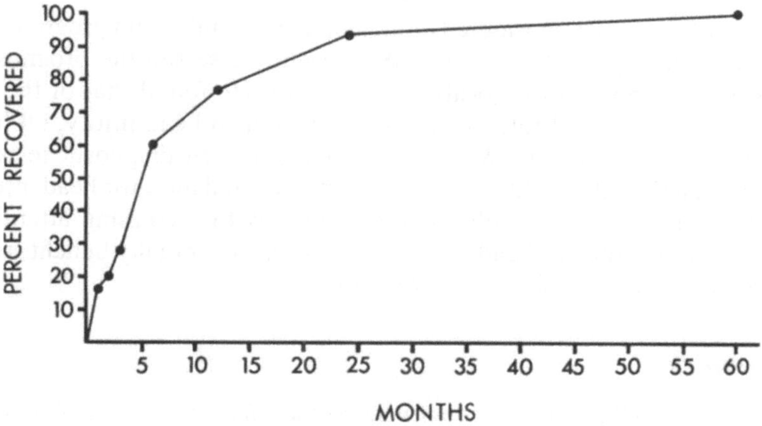

FIGURE 20.3. Graph of data from 25 patients who recovered olfactory function, showing percent recovered at various times after head injury. (Modified, by permission, from Costanzo and Becker 1986.)

only about 20 percent in the more severe cases. Recovery has been estimated to occur on the average in about one-third of head injury patients (Costanzo and Becker, 1986; Sumner, 1964).

Likelihood versus Time

The likelihood of recovery is much higher in the first six months to a year; after that the chances of recovery are quite low (Figure 20.3). Sumner (1975) suggested that there was more than one mechanism involved in recovery. He suggested that early recovery (in the first few weeks) could be explained by resolution of blood clots and edema, whereas slower recovery, observed at 8 to 12 weeks, might be related to some neural recovery or regeneration process. Although regeneration of olfactory neurons following head injury has not been demonstrated in humans, animal experiments (Cancalon, 1982; Graziadei and Monti Graziadei, 1980; Monti Graziadei et al., 1980) have shown that damaged olfactory cells can replace themselves, and that functional connections can be reestablished with the CNS (Costanzo, 1985; Simmons and Getchell, 1981). It may be that such a process could also occur in humans, though this would be difficult to confirm.

Rehabilitation

One issue that is just beginning to emerge is that of the impact of olfactory loss on recovery and rehabilitation. What are the implications of this sensory impairment in returning to the workplace and home environments? How does it affect preparation and enjoyment of food, or risk and hazards from smoke and potentially dangerous substances? These are issues that until recently were not considered or included in rehabilitation programs of the head-injured patient. Varney (1988) recently assessed the prognostic significance of anosmia for vocational disability. He found that of the 40 patients who developed total anosmia subsequent to head injury, all had major vocational problems. With the availability of new diagnostic tests of olfactory function, it should be possible to identify and monitor head injury patients who suffer loss of olfactory function. With increasing attention devoted to this problem, head injury patients may eventually benefit from new treatments and management protocols.

References

Bickerstaff, E.R. (1973). The first cranial nerve, the olfactory nerve. In *Neurological Examination in Clinical Practice.* Oxford, England: Blackwell Scientific Publications, pp. 33–35.

Cabral, R.J. (1972). Simulation and malingering after injuries to the brain and spinal cord. *Lancet, 1*, 953–954.

Cain, W.S., J. Gent, F.A. Catalanotto, and R.B. Goodspeed (1983). Clinical evaluation of olfaction. *Am. J. Otolaryngol.*, *4*, 252–256.

Cain, W.S., J.F. Gent, R.B. Goodspeed, and G. Leonard (1988). Evaluation of olfactory dysfunction in the Connecticut Chemosensory Clinical Research Center. *Laryngoscope*, *98*, 83–88.

Cairns, H. (1937). Injuries of the frontal and ethmoidal sinuses with special reference to cerebrospinal rhinorrhoea and aeroceles. *J. Laryngol. Otol.*, *52*, 589–623.

Cancalon, P. (1982). Degeneration and regeneration of olfactory cells induced by ZnSO4 and other chemicals. *Tiss. Cell*, *14*, 717–733.

Caruso, V., J. Hagan, and H. Manning (1969). Quantitative olfactometry in the measurement of posttraumatic hyposmia. A simple method. *Arch. Otolaryngol.*, *90*, 500–503.

Costanzo, R.M. (1985). Neural regeneration and functional reconnection following olfactory nerve transection in hamster. *Brain. Res.*, *361*, 258–266.

Costanzo, R.M., and D.P. Becker (1986). Smell and taste disorders in head injury and neurosurgery patients. In H.L. Meiselman and R.S. Rivlin (Eds.), *Clinical Measurements of Taste and Smell.* New York: MacMillian, pp. 565–578.

Costanzo, R.M., P.G. Heywood, J.D. Ward, and H.F. Young (1987). Neurosurgical applications of clinical olfactory assessment. *NY Acad. Sci.*, *510*, 242–244.

Doty, R.L., P. Shaman, and M. Dann (1984). Development of the University of Pennsylvania Smell Identification Test: A standardized microencapsulated test of olfactory function. *Physiol. Behav.*, *32*, 489–502.

Eslinger, P.J., A.R. Damasio, and G.W. Van Hoesen (1982). Olfactory dysfunction in man: anatomical and behavioral aspects. *Brain. Cogn.*, *1*, 259–285.

Ferrier, D. (1876). The hemispheres considered physiologically. In *The Functions of the Brain.* New York: G.P. Putnam's Sons, pp. 181–211.

Gibson, R.M., and T. Kurukchy (1974). Neurosurgical aspects of recurrent meningitis. *Proc. R. Soc. Med.*, *67*, 1150–1154.

Graziadei, P.P.C., and G.A. Monti Graziadei (1980). Neurogenesis and neuron regeneration in the olfactory system of mammals. III. Deafferentation and reinnervation of the olfactory bulb following section of the fila olfactoria in rat. *J. Neurocytol.*, *9*, 145–162.

Hagan, P.J. (1967). Posttraumatic anosmia. *Arch. Otolaryngol.*, *85*, 85–89.

Hasegawa, S., M. Yamagishi, and Y. Nakano (1986). Microscopic studies of human olfactory epithelia following traumatic anosmia. *Arch. Otorhinolaryngol.*, *243*, 112–116.

Heywood, P.G., and R.M. Costanzo (1986). Identifying normosmics: a comparison of two populations. *Am. J. Otolaryngol.*, *7*, 194–199.

Jackson, J.H. (1864). Illustrations of diseases of the nervous system. *Lond. Hosp. Rep.*, *1*, 470–471.

Jafek, B.W., P.M. Eller, B.A. Esses, and D.T. Moran (1989). Post-traumatic anosmia. Ultrastructural correlates. *Arch. Neurol.*, *46*, 300–304.

Jones-Gotman, M., and R.J. Zatorre (1988). Olfactory identification deficits in patients with focal cerebral excision. *Neuropsychologia*, *26*, 387–400.

Labayle, J., and A. Hadengue (1973). [Post-traumatic anosmia and its medico-legal implications: occupational consequences]. *Med. Leg. Dommage. Corpor.*, *6*, 343–355.

Leech, P. (1974). Cerebrospinal fluid leakage, dural fistulae and meningitis after basal skull fractures. *Injury*, *6*, 141–149.

Legg, J.W. (1873). A case of anosmia following a blow. *Lancet*, *2*, 659–660.

Leigh, A.D. (1943). Defects of smell after head injury. *Lancet*, *1*, 38–40.

Levin, H.S., W.M. High, and H.M. Eisenberg (1985). Impairment of olfactory recognition after closed head injury. *Brain*, *108*, 579–591.

Lundh, B., K. Kristensson, and E. Norrby (1987). Selective infections of olfactory and respiratory epithelium by vesicular stomatitis and Sendai viruses. *Neuropathol. Appl. Neurobiol.*, *13*, 111–122.

Mifka, P. (1964). Post-traumatische anosmie. *Wien. Med. Wschr.*, *114*, 793.

Miller, H., and N. Cartlidge (1972). Simulation and malingering after injuries to the brain and spinal cord. *Lancet*, March 11, 580–585.

Monti Graziadei, G.A., M.S. Karlan, J.J. Bernstein, and P.P.C. Graziadei (1980). Reinnervation of the olfactory bulb after section of the olfactory nerve in monkey (*Saimiri sciureus*). *Brain Res.*, *189*, 343–354.

Moran, D.T., B.W. Jafek, J. Rowley, III, and P.M. Eller (1985). Electron microscopy of olfactory epithelia in two patients with anosmia. *Arch. Otolaryngol.*, *111*, 122–126.

Notta, A. (1870) Recherches sur la perte de l'odorat. *Arch. Gen. Med.*, *15*, 385–407.

Ogle, W. (1870). Anosmia or cases illustrating the physiology and pathology of the sense of smell. *Med. Chir. Trans.*, *53*, 263.

Potter, H., and N. Butters (1980). An assessment of olfactory deficits in patients with damage to prefrontal cortex. *Neuropsychologia*, *18*, 621–628.

Raskind, R., and A. Doria (1966). Cerebrospinal fluid rhinorrhea and otorrhea of traumatic origin. *Int. Surg.*, *46*, 223–227.

Rauh, C. (1967). [Anosmia following skull injuries]. *HNO*, *15*, 271–273.

Roseburg, B., and R. Fikentscher (1976). [Traumatic olfactory and taste disorders]. *Z. Arztl. Fortbild. (Jena)*. *70*, 132–135.

Rutherford, W.H. (1977). Sequelae of concussion caused by minor head injuries. *Lancet*, *1*, 1–4.

Simmons, P.A., and T.V. Getchell (1981). Physiological activity of newly differentiated olfactory receptor neurons correlated with morphological recovery from olfactory nerve section in the salamander. *J. Neurophysiol.*, *45*, 529–549.

Sumner, D. (1964). Post-traumatic anosmia. *Brain*, *87*, 107–120.

Sumner, D. (1975). Disturbance of the senses of smell and taste after head injuries. In P.J. Vinken and G.W. Bruyn (Eds.), *Handbook of Clinical Neurology*. Amsterdam: North-Holland Publishing Co., pp. 1–25.

Tanabe, T., H. Yarita, M. Iino, Y. Ooshima, and S.F. Takagi (1975). An olfactory projection area in orbitofrontal cortex of the monkey. *J. Neurophysiol.*, *38*, 1269–1283.

Varney, N.R. (1988). Prognostic significance of anosmia in patients with closed-head trauma. *J. Clin. Exp. Neuropsychol.*, *10*, 250–254.

Yamagishi, M., S. Hasegawa, and Y. Nakano (1988). Examination and classification of human olfactory mucosa in patients with clinical olfactory disturbances. *Arch. Otorhinolaryngol.*, *245*, 316–320.

Yarita, H., M. Iino, T. Tanabe, S. Kogure, and S.F. Takagi (1980). A transthalamic olfactory pathway to orbitofrontal cortex in the monkey. *J. Neurophysiol.*, *43*, 69–85.

Zusho, H. (1982). Posttraumatic anosmia. *Arch. Otolaryngol.*, *108*, 90–92.

21 Olfaction and Neuropsychiatry

MICHAEL J. SERBY, PAMELA M. LARSON, and DAVINA KALKSTEIN

Introduction

Olfactory performance is impaired in a variety of neurological and psychiatric disorders. This chapter will review work accomplished in some of these disorders, most notably Alzheimer's disease (AD), depression, schizophrenia, and the dementia of the acquired immune deficiency syndrome (AIDS). There is a major focus on AD, reflecting the thrust of our own research and the recent emphasis on olfactory changes in this dementia. Olfactory testing has proved to be a safe and productive method of gaining information about the central nervous system (CNS). We have used tests of odor detection and identification to investigate a number of clinical and basic science questions pertinent to AD. The findings in major depression, schizophrenia, and AIDS are more restricted and preliminary in nature. It is hoped that these studies will engender further understanding of CNS diseases.

Olfaction and Alzheimer's Disease

Background

Senile plaques and neurofibrillary tangles, the pathological hallmarks of AD, are present in the olfactory structures of affected individuals, including the olfactory bulb and anterior olfactory nucleus (Averback, 1986; Esiri and Wilcock, 1984). In addition to an anatomical link between AD and the sense of smell, there are neurochemical ties as well. Acetylcholine (ACh), which is clearly deficient in AD cortex and hippocampus (Bartus, Dean, Beer, and Lippa, 1982), is a transmitter in the olfactory system (Halasz and Shepherd, 1983). The activity of the important enzyme choline acetyltransferase (ChAT) is reduced in olfactory tubercles from AD brains (Simpson, Yates, Gordon, and St.-Clair, 1984). Studies of olfactory performance in AD are grounded in these observations.

Serby et al. noted impaired odor identification in AD patients compared to age-matched controls and demented alcoholics (Serby, Corwin, Conrad, and Rotrosen, 1985). This study utilized a simple forced-choice procedure with common odorants contained in bottles. Both AD patients and those with Parkinsonian dementia scored barely above chance. A second group of investigators found a deficit in odor identification in a small sample of AD patients (Peabody and Tinklenberg, 1985) and later extended their observations (Warner, Peabody, Flattery, and Tinklenberg, 1986) by employing a standardized methodology, the University of Pennsylvania Smell Identification Test (UPSIT) (Doty, Shaman, and Dann, 1984a). There have subsequently been a number of studies of olfactory identification in AD, all reporting deficits in this disorder (Knupfer and Spiegel, 1986; Koss, Weiffenbach, Haxby, and Friedland, 1987; Rezek, 1987; Doty, Reyes, and Gregor, 1987).

A few groups have also examined the status of odor detection capabilities in AD. Detecting the presence of an odor is different from identifying one in that the process of detection is less cognitively dependent (Corwin, Serby, and Rotrosen, 1986). Olfactory detection and identification also seem to be mediated via different neuroanatomical substrates (Eichenbaum, Morton, Potter, and Corkin, 1983). The existence of odor detection deficits in AD was first reported by Knupfer and Speigel (1986). The AD group in this latter report was several years older than controls and patients with vascular dementia, but there was no correlation between age and performance in the total sample studied, suggesting that diagnosis, not age, may have been the determining factor. Subsequent investigations have yielded conflicting data regarding the presence or absence of impaired detection. Doty and associates (1987) supported the finding of altered detection of phenyl ethyl alcohol. On the other hand, Koss et al. (1987) and Rezek (1987) reported normal thresholds in AD patients. It is unclear why the results are conflicting. Differences may result from diagnostic criteria or experimental methodology, including the evaluation of thresholds for different substances. The clarification of this question is an obvious imperative. Later in this chapter we will present data on odor detection in a large sample of AD patients.

The nature of the pattern of deficit is just one of many issues raised by the olfactory studies. There is as yet no definitive answer as to the specificity of the hyposmia. How do demented groups without AD perform? Will the sense of smell prove to be a valuable diagnostic tool? Are there parallel disturbances in disorders that are related to AD? Is the impairment truly sensory in nature, is it cognitive, or is it a combination of both? How early does the hyposmia appear, and how does it progress? What is the relationship of the olfactory deficits to demonstrable anatomical changes associated with AD? What are the neurochemical bases of impaired performance in odor detection and identification? Over the past few years we have been investigating these questions. We will discuss our findings in this chapter.

We will also consider the theory that the nose is a portal of entry for toxins that lead to AD. This hypothesis was expounded by Roberts (1986), who suggested that aluminosilicates may bypass a natural barrier in some predisposed individuals and produce a cascade of events, culminating in AD. A recent study of autopsy tissue in AD and control subjects demonstrated the existence of pathological cytoskeletal proteins in nasal mucosa (Talamo et al., 1989). It may be that peripheral involvement of olfactory structures precedes more central destruction. This possibility establishes the need for a new direction in olfactory research in AD.

Methodology

We have administered one or more olfactory tests to 281 subjects of various ages and in different diagnostic categories. Our database includes 155 controls, 71 of them above age 60 (old controls). We have obtained data from 62 patients with AD. Subjects were patients, families, and staff recruited from the New York Veterans Administration Medical Center, the Aging and Dementia Research Program of New York University, and various senior citizen centers. All subjects were screened for cognitive, medical, and psychiatric disorders that might interfere with olfactory scores. The preliminary interviews included the Mini-Mental State Exam (MMS) (Folstein, Folstein, and McHugh, 1975), the Hamilton Depression Scale (Ham-D) (Hamilton, 1967), and questions about specific exclusion criteria (Appendix I). AD patients were rated on the Global Deterioration Scale (GDS), a seven-stage system designed to reflect the characteristic presentation and course of age-associated cognitive impairment and AD (Reisberg, Ferris, deLeon, and Crook, 1982). GDS stage 3 describes a group of patients who have early cognitive and behavioral changes but may not decline further into AD (GDS 4 through 7). Patients who are at GDS 4 level have moderate cognitive decline but may be oriented and may recognize familiar people and places. At GDS 5 the patient can no longer survive without assistance and is disoriented; GDS 6 reflects severe deterioration in personality and cognition.

Most subjects were tested once. Some AD patients were tested either twice (designated as one interval for each patient) or three times (two intervals per patient). The intervals were at least six months in duration.

To evaluate odor identification, we administered the UPSIT. This is a standardized 40-odorant task utilizing microencapsulated suprathreshold concentrations of common odorants. The subject is instructed to scratch and sniff the strip that contains the odorant and then to identify the correct answer from four supplied possibilities. The choices have some internal consistency on various scales, including pleasantness–unpleasantness, "hot–cold," and others.

We have pursued an alternative method of examining subjects' abilities in identifying odors, the Yes-No Identification task. Twenty microencapsulated odors were arranged in 10 pairs that were similar in quality (e.g., orange and lemon). Each odorant was presented twice; on one trial the odorant matched a verbal label supplied (e.g., lemon was presented and "Is this lemon?" was asked), and in the other trial the question was asked with the remaining item in the pair (e.g., lemon was presented and "Is this orange?" was asked). The order (first or second presentation) of true match of odorant to label was randomized. The procedure yielded 40 trials, 20 signal and 20 noise. Hit and false alarm rates were converted to measures of discrimination (Pr) and response bias (Br) (Snodgrass and Corwin, 1988).

To test odor detection, we used a simple method designed to eliminate cognitive factors as much as possible, considering the difficulties inherent in research with elderly and demented subjects. A forced-choice ascending method of limits procedure was administered. Three sniff bottles were presented in each of nine trials; two bottles contained odorless fluid, while a third bottle in each set contained diluted geraniol, an olfactory stimulant used in the perfume industry. The concentration of geraniol increased in each successive triad of bottles to minimize adaptation effects. The position of the geraniol bottle in each set was randomized. A threshold value was designated as the lowest of three consecutive concentrations that were correctly identified. The threshold score was designated from 1 (lowest concentration) through 9 (highest); a score of 10 signifies no perception of odorant. Microencapsulated strips of the various geraniol concentrations (and odorless mineral water as a control) have been developed (3M Co.) and tested, and they will henceforth replace the bottles in testing.

Aging and Olfactory Function

Aging reportedly affects various aspects of olfactory performance, although not all studies have demonstrated such changes. Doty and associates measured odor identification in 1955 subjects and found a decline in the later decades of life (Doty et al., 1984b). Age did not affect identification ability in a much smaller group of subjects who were tested with a forced-choice procedure (Serby, Corwin, Conrad, and Rotrosen, 1985). Although Engen found no increase in olfactory thresholds with age (Engen, 1977), most studies document such changes in odor detection (Schiffman, Moss, and Erickson, 1976; Murphy, 1983; Kimbrell and Furchgott, 1963).

Deficits in smell perception in the elderly are both clinically and theoretically important. From the standpoint of dementia research, it is necessary to partial out effects of aging per se and to control for such effects. We have, therefore, tested scores of healthy individuals and categorized them by decade.

We found that in the UPSIT there was a significant ANOVA [$F(6,147) = 12.99$; $p < 0.0001$] demonstrating a decline in acuity in the later decades. This result was verified by the Scheffe method. A regression analysis demonstrated a strong linear trend [$F(1,141) = 35.84$;

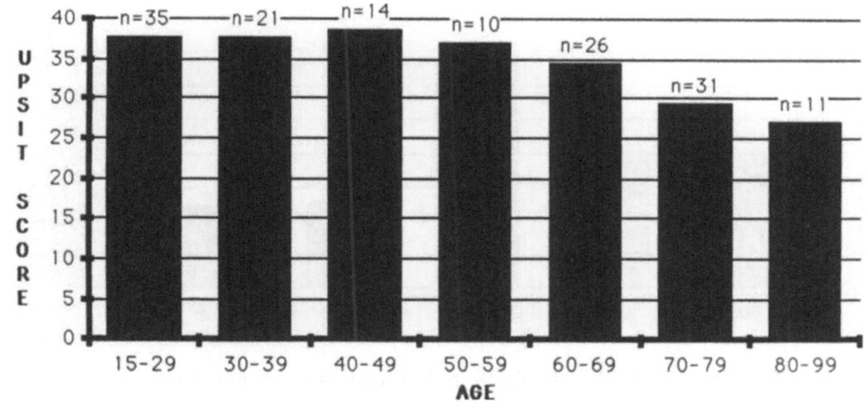

FIGURE 21.1. UPSIT by decade in normal controls.

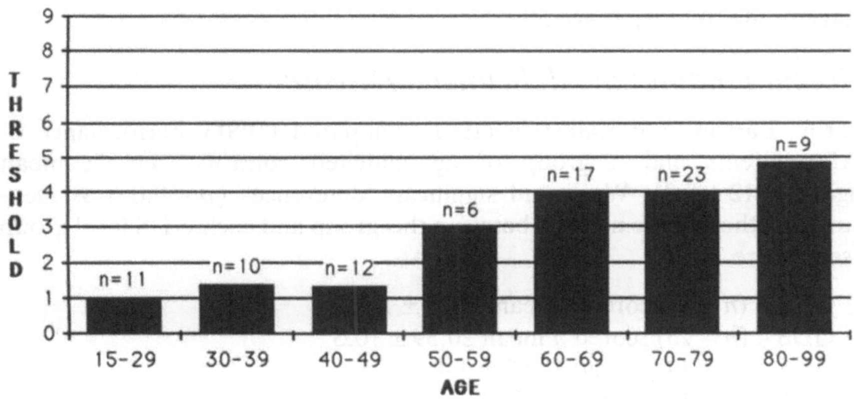

FIGURE 21.2. Olfactory threshold by decade in normal controls.

$p < 0.0001$). Our findings paralleled those of Doty et al. (1984b) in their study of people tested with the UPSIT. Our data show no differences between decades below 60, but significant decrements with each decade thereafter (see Figure 21.1).

There was also an effect of age across decades in normals from 20 to 89 years old in the Pr component of the Yes-No Identification Task. The effect [$F(6,119) = 3.34$; $p < 0.005$] was smaller than that seen in the UPSIT. There were no significant between-group (by decade) differences as measured by the Scheffe post hoc method. The bias measure (Br) was not significant [$F(6,119) = 1.15$; $p < 0.337$].

There was a strong age effect apparent in the geraniol threshold test [$F(6,87) = 5.74$; $p < 0.0001$]. The change in threshold across decades can be seen in Figure 21.2.

In summary, there is evidence of multimodal olfactory dysfunction with age. Identification as measured by the UPSIT does not change until rel-

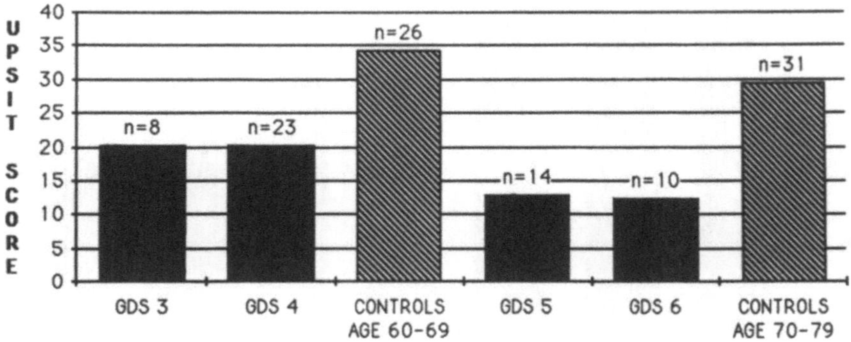

FIGURE 21.3. UPSIT by GDS in AD with age-matched controls.

atively late in life. The geraniol threshold begins to decline at the same stage of life, that is, at age 50.

Recent Findings in Alzheimer's Disease

Serby, Larson, and Kalkstein (1991) compared UPSIT performance of AD patients and a group of age-matched controls ($n = 57$; mean age $= 70.12 \pm 5.3$). We found significant differences ($p < 0.05$) as measured by the Scheffe method between this group and each GDS level group (see Figure 21.3):

GDS 3 ($n = 8$) scored a mean 20.25 ± 7.1
GDS 4 ($n = 23$) scored a mean 20.39 ± 10.3
GDS 5 ($n = 14$) scored 12.93 ± 3.4
GDS 6 ($n = 10$) scored 12.5 ± 4.7

There were no significant differences between any of the GDS levels individually. When GDS 3 and GDS 4 were combined (20.36 ± 9.5), they scored significantly better than the combined GDS 5 and GDS 6 groups (12.5 ± 4.7).

Particular attention should be paid to the data on GDS 3 subjects because of the uncertainty of the diagnosis of dementia. Not all GDS 3 patients decline further into AD; in fact, in one study only 25 percent of 36 GDS 3 subjects reached the GDS 4 level after 3.5 years (Barry Reisberg, M.D., personal communication). Some of these GDS 3 patients may be hyposmic, which might predict decline. As noted above, the GDS 3 group did show impaired performance on the UPSIT; the scores were not different from those of GDS 4 patients. It is unclear why the degree of olfactory impairment is as severe in the GDS 3 group. It is possible that most if not all GDS 3 patients will deteriorate into dementia after many years and that the hyposmia is an early indicator. It does appear that olfactory loss should be evaluated further as an early marker for AD.

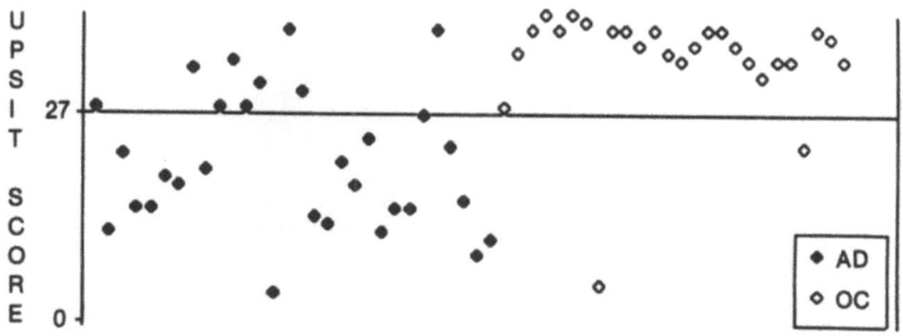

FIGURE 21.4. UPSIT by scores, GDS 3–GDS 4 AD subjects versus age 60–69 normal controls.

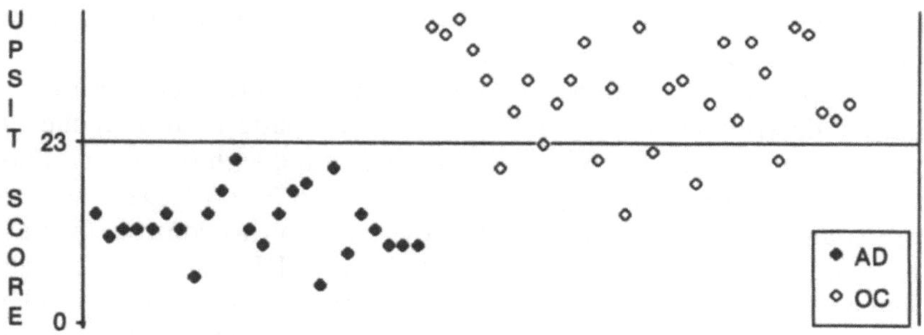

FIGURE 21.5. UPSIT scores, GDS 5–GDS 6 AD subjects versus age 70–79 normal controls.

A number of studies have found identification differences between AD patients and old controls. We decided to evaluate the potential utility of the UPSIT as a discriminant between this dementia and normal aging. We chose a cutoff score of 27 to reflect a presumptive diagnosis of AD in GDS 3 and GDS 4 patients. This point is one standard deviation below the mean for the seventh-decade controls. We also chose a cutoff UPSIT of 23 for the GDS 5 and GDS 6 subjects. This is one standard deviation below the mean for the eighth decade controls. We found that 9 of the 31 GDS 3 and GDS 4 patients (29 percent) scored above 27, while only 2 of 26 age-matched controls (7.7 percent) scored below 28. None of the 24 GDS 5 and GDS 6 patients scored above 23 on the UPSIT, and 7 out of 31 (22.6 percent) of the age-matched controls scored 23 and below (Figures 21.4 and 21.5).

The overlap between patients and normals limits the utility of this test as a diagnostic tool. This is especially true because the greatest overlap occurs in the earliest phases of the disorder, when diagnosis may be obscure. The greatest potential for such clinical tests rests in the possibility that olfactory dysfunction is an early marker of pathological changes in otherwise "nor-

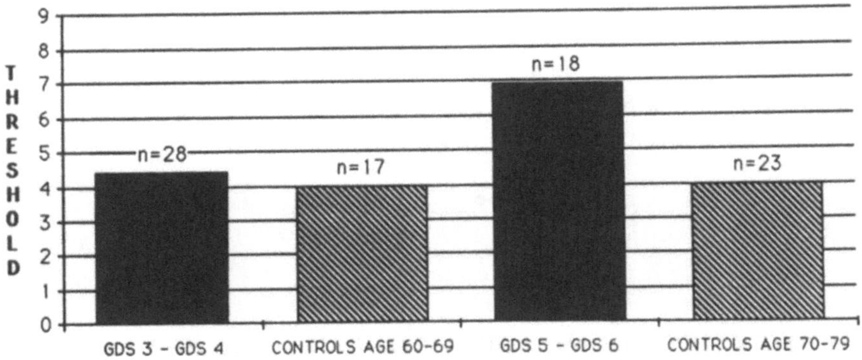

FIGURE 21.6. Olfactory threshold by GDS in AD with age-matched controls.

mal" geriatric subjects. Testing may prove to be more useful in combination with other diagnostic procedures.

It is apparent that there is progressive change in odor identification as patients deteriorate across GDS levels. This pattern also applies to threshold values. Because fewer people were tested with the geraniol task than with the identification paradigms, the GDS 3 and GDS 4 subjects were counted as one group ($n=28$) in comparison to controls ($n=17$). The more advanced AD cases, at GDS 5 and GDS 6 stages, were also combined ($n=18$) and compared to the appropriate decade group of controls ($n=23$). The GDS 3–4 patients had a mean S.E.M. score of 4.46 ± 0.64, not significantly different from normals in the 60–69 age group (4.0 ± 0.75). While the 70–79-year-old controls had a threshold of 4.0 ± 0.51, the more advanced AD patients demonstrated impaired odor detection, with a threshold of 6.94 ± 0.81 [S.E.M.]. This advanced dementia sample performed worse than their controls (Scheffę F test $= 3.02$; $p<0.05$) and worse than the GDS 3–4 group [$F(1,45)=5.86$; $p<0.02$). These results are depicted in Figure 21.6.

We have attempted to track the course of olfactory deficits in AD as the disorder progresses. We have verified changes by repeated testing of AD subjects. A number of AD patients were retested at intervals of 6 to 36 months after their baseline measures. The UPSIT was administered on repeated occasions 15 times: twice to seven patients and three times to four patients. The threshold test was repeated 10 times: twice to six people and on three occasions to two. Paired t tests demonstrated significant decline in UPSIT performance ($t=3.17$; $p<0.007$) and geraniol detection ($t=2.98$; $p<0.015$). The change was greatest in those subjects who also deteriorated on the GDS scale. Those AD patients who changed by one GDS level between tests had a mean loss of 6.4 points on the UPSIT ($n=8$ intervals) and increased their threshold by 4.4 concentrations ($n=5$ intervals). The comparable scores in AD patients without a change in GDS were 3.6 points for the UPSIT ($n=7$ intervals) and 2.0 levels for threshold

($n = 5$ intervals). The number of subjects is small, and GDS change is a gross measure. However, there is a suggestion of increasing olfactory impairment as the course of AD proceeds.

Related Disorders

The data presented so far confirm the existence of olfactory deficits in AD. The specificity of this hyposmia is an unresolved issue. Our original study demonstrated no dysfunction in odor identification in demented alcoholics (Serby, Corwin, Conrad, and Rotrosen, 1985). Patients with dementia secondary to vascular disease performed significantly better than AD subjects but worse than normals in one study (Knupfer and Spiegel, 1986). Recent work suggests that olfactory deficits are not limited to AD but are common in a variety of dementias (Dr. Susan Schiffman, personal communication). We explored this question by testing a group of patients with multi-infarct dementia (MID).

Six patients were diagnosed as having MID by a consulting neurologist. Each MID patient was compared with the average score of a group of AD patients matched for age and MMS score. On the UPSIT, the performance of the MID patients [19.17 (mean) ± 3.74 (S.E.M.)] did not differ from that of the matched AD group (18.72 ± 2.11). On the other hand, threshold scores were significantly higher in five MID patients (9.25 ± 0.2) than in the matched AD sample (5.98 ± 0.88) ($t = 3.58$, df = 8, $p \pm 0.05$). These preliminary findings conflict with the data of Knupfer and Spiegel, who noted greater hyposmia in AD than in vascular dementia. In part, this may be explained by the fact that their vascular subjects were six years younger than their AD patients. However, the data from our small sample of subjects must be considered inconclusive. Future studies using larger samples should focus on the relationships between specific anatomical areas of infarct and olfactory deficits.

Down's syndrome (DS) is another disease with particular relevance to AD. Patients with Down's syndrome who live to 35 to 40 years of age develop the pathological brain changes of AD (Burger and Vogel, 1973). Warner and associates found that DS subjects had significantly worse performance on the UPSIT than age-matched normal controls (Warner, Peabody, and Berger, 1989). We have completed a study of olfaction in DS using more appropriate controls, a larger sample, and more extensive testing (Hemdal, Corwin, Oster, and Serby, 1989). Twenty DS subjects were matched for age and IQ with 15 people who had idiopathic mental retardation (IMR) and for age alone with 21 normal controls (NC). All subjects were tested with a modified UPSIT (the 20 most familiar odors), the Yes/No Task, and the geraniol threshold test (see above) as well as three different tactile tasks designed to duplicate the procedures of the olfactory tests.

Table 21.1 shows summary statistics for demographic information and

TABLE 21.1. Demographic information and olfactory task performance.

	N		Age	IQ	Threshold	Hit rate	FA rate	Pr	Br	Mult. choice
DS	20	Mean	23.10	45.85	6.42	0.77	0.63	0.14	0.75	10.00
		SD	2.30	1.34	0.65	0.04	0.04	0.03	0.04	0.78
IMR	15	Mean	20.60	49.87	6.07	0.87	0.48	0.39	0.75	13.67
		SD	1.83	2.15	0.69	0.02	0.06	0.05	0.05	0.63
NC	21	Mean	23.81	NA	2.17	0.85	0.14	0.71	0.50	19.00
		SD	1.89	NA	0.39	0.02	0.02	0.02	0.05	0.26

olfactory task performance. (FA = false alarms; Pr is a discrimination index, specifically hits minus FAs; Br is a measure of response bias; multi choice-modified UPSIT score.)

The DS and IMR groups had similar odor threshold scores, both significantly higher than those of NCs. However, the pattern of results for the tactile task was identical, suggesting that olfactory threshold differences could have been related to cognitive demand.

DS performed worse than the other groups regarding both olfactory identification tasks. This finding cannot be accounted for by general cognitive factors, as the DS and IMR groups performed near ceiling on the tactile task.

Results specific to the DS group show that olfactory identification ability declines as age increases (Figure 21.7). It is unlikely that age alone could explain the correlation in subjects who are this young. This suggests that the development of AD changes in DS patients further compromises olfactory performance.

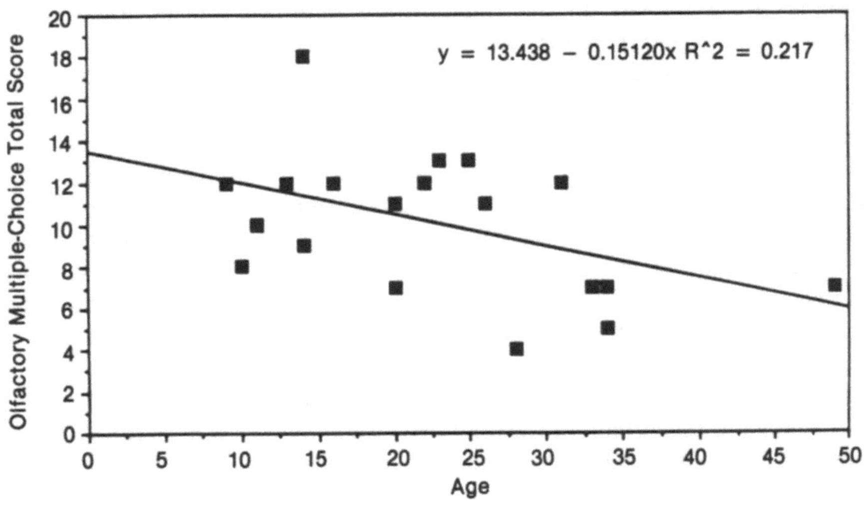

FIGURE 21.7. Age and olfactory multiple-choice, Down's syndrome only.

Pharmacologic Studies

Acetylcholine (ACh) and bethanechol enhance odor detection when applied to the nasal mucosa in man (Skouby and Zilstorff-Pedersen, 1954). Given the many links between olfaction and the cholinergic system, it is possible that the olfactory deficits in AD are related to reduced ACh neurotransmission.

One means of learning more about the neurochemistry of olfaction and of AD is to observe the consequences of pharmacologic manipulations. Serby et al. studied the effects of the cholinergic antagonist scopolamine on both odor detection and odor identification in young normals (Serby, Flicker, Rypma, Weber, Rotrosen, and Ferris, 1989). Thirty volunteers (18 to 30 years old) were seen on two occasions, one week apart. In a randomized, double-blind fashion, each subject received a subcutaneous injection of scopolamine hydrobromide [either 0.22 ($n = 10$), 0.43 ($n = 10$), or 0.65 mg/70 kg ($n = 10$)] on one visit and an equal volume of saline on the other visit.

To evaluate odor identification, the UPSIT was administered to all 30 subjects 20 minutes after injection. Odor detection ability was determined by means of a signal detection method requiring the subject to specify the presence or absence of odor produced by geraniol. Twenty-six of the volunteers were tested 90 minutes postinjection (nine in each of the low- and high-dose scopolamine groups, eight in the mid-dose group). There were 60 quasi-randomized trials: five trials with each of nine concentrations of geraniol (6.3, 9.97, 15.8, 25, 39.6, 62.8, 99.6, 157.9, and 250 cm^3 dissolved in odorless diethyl phthalate to produce one liter of each concentration) interspersed with 15 blank trials. A measure of sensitivity (d') was derived from the hit and false alarm rates in keeping with signal detection theory (Green and Swets, 1974). Subjects were also tested on a cognitive test battery consisting of measures of memory, language, psychomotor speed, and visuospatial praxis.

In this study, there were no significant differences between UPSIT scores (mean $\pm$ S.E.M) following scopolamine [37.5 $\pm$ 0.4 (S.E.M.)] and PBO (38.1 $\pm$ 0.4 (S.E.M.)] [($F(1,27) = 1.50$, ns) nor was there a significant drug $\times$ dosage group interaction [$F(2,23) = 0.13$, ns]. For the detection measure, scopolamine caused a significant impairment compared to placebo [$F(1.23) = 16.18, p < 0.0005$). The mean d' was 1.62 $\pm$ 0.20 (S.E.M.) for the 26 scopolamine trials and 2.60 $\pm$ 0.21 (S.E.M.) for the 26 PBO trials (Figure 21.8). The effect of scopolamine on detection was not dose-dependent, as indicated by the lack of a drug $\times$ dosage group interaction [$F(2,23) = 0.17$, ns]. The impairment in odor detection was noted even at the low dose of scopolamine [$F(1,8) = 8.09, p < 0.03$] (Figure 21.9) despite normal cognitive performance.

The results demonstrated a significant impairment in odor detection but not in odor identification as a consequence of scopolamine administration. These results suggest that ACh may play a role in the physiology of odor

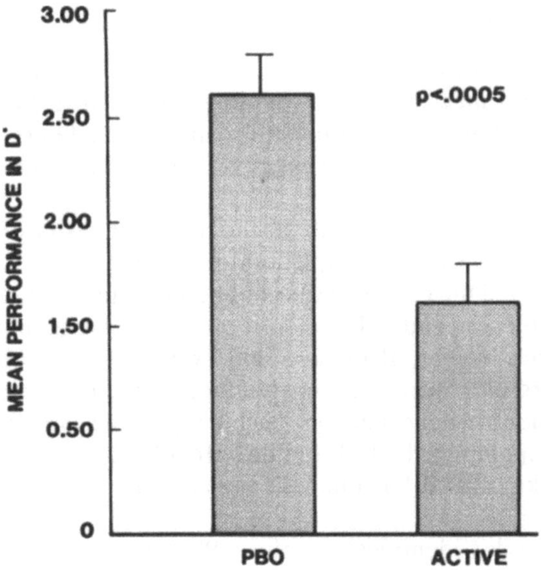

FIGURE 21.8. Active scopolamine produced a significant decrease in *d'* scores compared to placebo (PBO). (Reprinted, by permission, from Serby, Flicker, Rypma, et al. Copyright 1990 by the Society of Biological Psychiatry.)

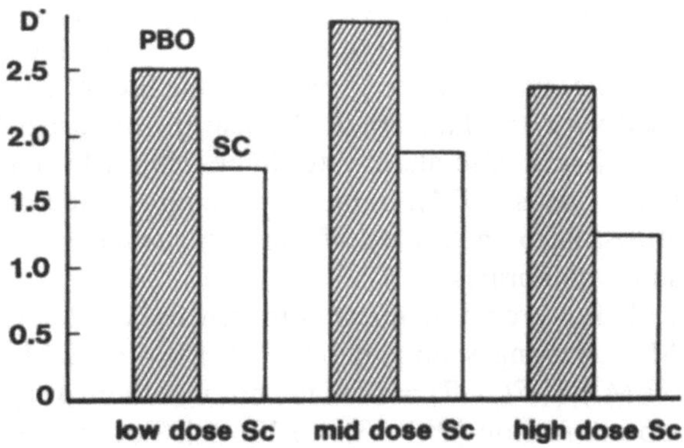

FIGURE 21.9. All doses of scopolamine (Sc) yielded significantly lower *d'* scores. (Reprinted, by permission, from Serby, Flicker, Rypma, et al. Copyright 1990 by the Society of Biological Psychiatry.)

detection but not identification. A cholinergic influence on detection was previously suggested by the lowering of thresholds after direct nasal application of ACh or bethanechol (Skouby and Zilstorff-Pedersen, 1954). Thus, cholinergic blockade impairs olfactory detection, and cholinergic stimulation enhances it.

The impact of scopolamine on odor detection seemed to be independent of drying of the nasal mucosa, cognitive changes, or dampening of general arousal mechanisms. While detection deficits in AD may reflect ACh deficiency, it seems likely that some of the olfactory losses in AD patients, especially alterations in odor recognition, depend upon additional physiological, neurochemical, or cognitive changes.

Although ACh blockade and perhaps ACh deficits may produce olfactory impairment, we have been unable to improve the sense of smell through enhancement of the cholinergic system. We gave RS-86, a cholinergic agonist, to 18 AD subjects and measured their olfactory performance at various points in the protocol. There was no change noted in this sample (unpublished data). It is not surprising that the olfactory effects of ACh blockade are more potent than the effects of ACh stimulation in AD patients; the same pattern holds true for cognition. Additional studies in normals should help determine the extent of such effects.

We have also studied the olfactory effects of a drug that is a putative treatment for dementia. AD patients received either placebo or Hydergine (ergoloid mesylates), either 4.5 or 9.0 mg per day, for 24 weeks (unpublished data). Of the six patients who received 9.0 mg per day, only one completed olfactory testing; this group's performance cannot be considered. Comparing the 4.5-mg group to those patients on placebo, the following pattern emerges. Treatment with the drug increased UPSIT scores in four out of five cases by a mean increase of 4.5 points; one of the five patients had a decline in UPSIT score. In the placebo group, only one of four subjects improved (by only one point) and the other three worsened, perhaps as a consequence of deteriorating disease over 24 weeks. In the geraniol threshold task, two of five in the treated group improved, by three and four concentrations, respectively; two others maintained the same scores; and one got worse by two dilutions. In the placebo group, one of four improved by only one concentration, while three patients worsened. These results are preliminary but intriguing. This ergoloid mesylate compound may increase ChAT levels and act as a dopamine agonist (Krassner, 1984). Either of these neurochemical effects could conceivably modify olfactory performance. Further studies should include a much larger number of subjects to confirm significant results.

Alzheimer's and Olfaction: New Directions

It has been suggested that the study of olfactory performance in the cognitively impaired could lead to the development of an ante-mortem diagnostic marker, which is sorely needed in AD. The search for valid markers has led researchers in many directions. There have been many efforts to identify constituents of cerebrospinal fluid, including peptides, neurotransmitters, and enzymes, as indicators of the presence of AD. Abnormal fingerprint patterns have also been investigated. We have been exploring the utility of the identification and quantification of ACh receptors in eccrine sweat glands in human skin. The standard EEG has also been explored as a diagnostic aid. The EEG is not a reliable discriminant between AD

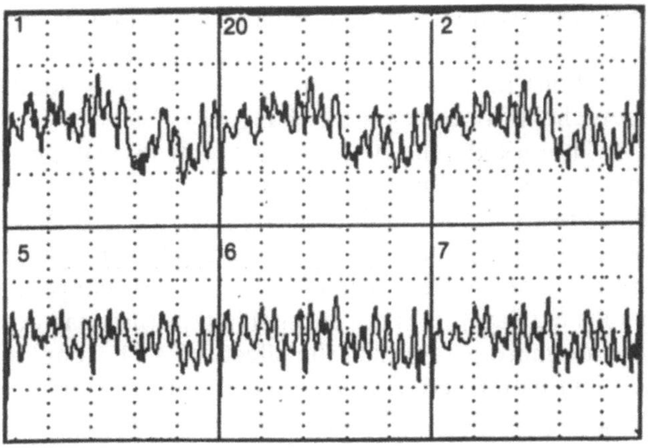

FIGURE 21.10. Electrophysiologic response to blank odorant presentation.

patients and old controls, even though a normal EEG occurs four times more often in normals and certain features are noted exclusively in AD (Rae-Grant et al., 1987).

Another way of potentially developing a test for AD is to measure electrophysiological responses to olfactory stimuli. A similar strategy has been reported using visual EPs in AD (Pollock et al., 1989). The authors found a significant prolongation of the P100 potential evoked in response to pattern reversal of a checkerboard (PRVEP) (unidirectional *t* test), but not in response to a flash (FVEP). The difference in PRVEP was small and probably not detectable in all AD patients, although this is not specifically reported. Therefore it is unlikely that PRVEPs can serve as reliable diagnostic discriminants. Given the impoverished olfactory performance in AD, the use of odorant-generated EPs might prove more useful. A study published in the German literature (Gerulli, Mielke, and Mrowinski, 1981) reported the existence of EPs following olfactory stimulation (OEPs). One of the stimuli used was geraniol. Recently olfactory EPs were again elicited in normals by eight different odorants, with characteristic responses at 300 to 500 ms (Kobal and Hummel, 1988). The authors concluded that such a procedure might be useful in AD.

We therefore tested two young normal subjects by delivering odorant to a face mask via an olfactometer and simultaneously measuring the EP. There was no evidence of an olfactory EP when the stimulus was a blank (Figure 21.10). However, an EP appeared in tracings from both subjects at 1400 ms during geraniol inhalations (Figure 21.11A and B). An EP was also noted after butanol inhalation (Figure 21.12).

Further development of odor-induced EPs may prove useful as an aid in diagnosis, a yardstick of response in drug trials, a predictive test vis-a-vis course of disease, and a means of exploring physiologic questions in AD.

A

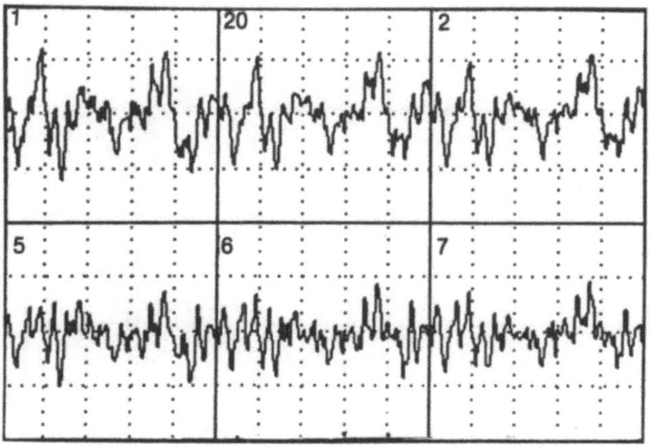

B

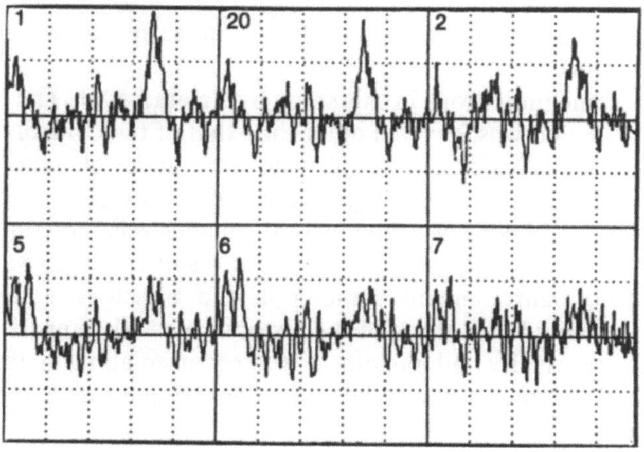

FIGURE 21.11. Electrophysiologic responses of two subjects to geraniol inhalations.

Another potentially productive area of study involves the exploration of pathological changes in the olfactory mucosal surface. Investigators have used antibodies to specific cytoskeletal proteins to examine autopsy olfactory epithelium in AD and have detected cellular changes (Talamo et al., 1989). The major alterations were a disruption and possible hyperplasia of the cell body layer of the neuroepithelia and the expression of phosphorylated forms of all size species of neurofilament proteins extending from this layer. In addition, an abnormal accumulation of neurites (which may include dendrites) was observed. These neurites were positive for phosphorylated neurofilaments and for the Tau protein, which is found in the paired helical filaments that form a major part of the neurofibrillary tangles in AD (Selkoe, 1986). This preliminary study, then, demonstrates the

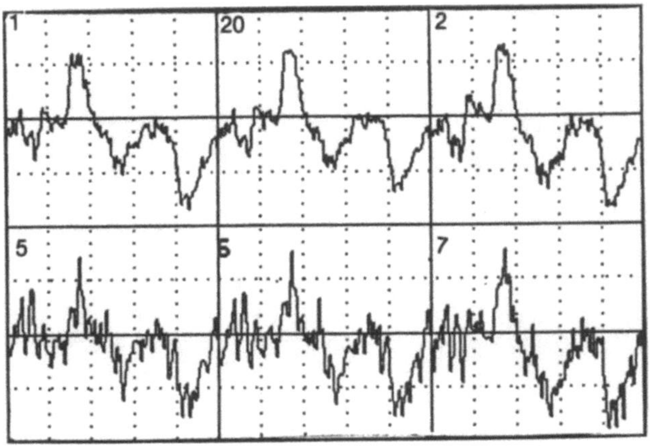

FIGURE 21.12. Electrophysiologic response to butanol inhalation.

involvement of the most peripheral olfactory anatomical area in the AD pathological process.

There are some limitations in this study by Talamo et al. The AD group had a mean age that was 23 years older than that of the controls. This was partly due to the inclusion of controls who were under 60 years old when they died (6 of 14). In fact, the effect of age was not considered. The one AD subject who did not have the mucosal changes was the youngest (and the only one below age 72). Of the five controls over 72 years, three had phosphorylated neurofilament staining in the olfactory epithelium. It should also be noted that three of the six controls with cancers had these changes. The specificity and meaning of the reported abnormalities should be established. Further study of olfactory sensory neuron changes in AD seems warranted.

A third domain of study deals with the relationships between results of olfactory testing and specific brain areas. A recent report documented a significant correlation between odor identification deficits and loss of hippocampal metabolic activity in AD as measured by PET scanning (Buchsbaum et al., 1989). We have investigated the relationship between UPSIT scores and hippocampal lesions apparent on the CT scans of AD brains. Nineteen subjects had scans and olfaction tests within 12 months of each other. Group 1 patients had either questionable lesions or none ($n = 6$); group 2 had definite hippocampal lesions on at least one side ($n = 13$). We found that UPSIT scores were significantly lower in group 2 [$F(1,18) = 4.5$; $p < 0.05$]. Thresholds, however, did not differ [$F(1,18) = 1.12$; ns) (unpublished data). These results not only confirm an anatomical relationship between odor identification and the temporal lobe area, they also suggest a dichotomy in which olfactory detection is independent of temporal lobe function. Bilateral temporal lobe ablation has previously been shown to cause a loss of the ability to assess qualitative aspects of odorants (e.g., identification) while leaving quantitative capacities (e.g., detection) intact (Eichenbaum, Morton, Potter, and Corkin, 1983). However, it should be noted that this pattern is not universally accepted. One group found detection impairment in patients with temporal lobe lesions (Rausch and Serafetinides, 1975), but another set of investigators did not (Eskenazi, Cain, Novelly, and Mattson, 1986). The differences in conclusions may derive from the techniques employed. The bulk of data at this point does suggest different anatomical loci for various olfactory functions. The clarification of this matter may contribute to our understanding of AD.

Alzheimer's Disease and Olfaction: Conclusions

There are deficits in both odor detection and identification in AD. The loss of identification ability begins early, perhaps as a harbinger of disease, but the change in detection acuity is a later phenomenon that is reserved for advanced stages of the disease. This pattern of hyposmia explains the contradictory findings in earlier studies. The two reports that found no impaired detection recruited "mild" patients (Koss et al., 1987; Rezek, 1987). On the other hand, increased thresholds were present in a predominantly moderate group of AD subjects (Doty et al., 1987) and in a sample that contained patients who had "rather advanced dementia" (Knupfer and Spiegel, 1986).

The differences in time course of detection and identification changes in AD may call into question the theory that more peripheral structures are the earliest to be involved. Roberts (1986) hypothesized a process whereby aluminosilicates enter the nasal mucosa and, via retrograde absorption into the CNS, lead to a cascade of events culminating in AD. One might expect odor detection to be impaired early in the disease and identification later if Roberts is correct, but the pattern of olfactory deficit is the reverse of this hypothetical picture. It could be that the physiologic processes involved in odor detection are more redundant and resistent to functional change than those that are involved in identification, but there is as yet no evidence that this is the case.

The utility of olfactory testing for differential diagnosis of dementia is limited at this time. The overlap with other disorders, such as MID, restricts the feasibility of this approach. Hyposmia is an inherent part of the AD process. For example, the deterioration of odor identification in DS temporally parallels the onset of AD in these people. However, a refinement of olfactory testing is an essential prelude to any development of a diagnostic tool.

Pharmacologic protocols have yielded some interesting preliminary facts about the neurochemistry of the olfactory system. Future approaches should focus on the cholinergic and dopaminergic systems.

Finally, there is reason to hope that new information regarding AD will emerge from studies of olfactory EPs, mucosal protein changes, and neuroimaging correlates to hyposmia.

Olfaction and Depression

There are a number of links between affective disorders and olfactory issues. For example, psychotically depressed individuals may be delusionally concerned with their body odors, and manics may have olfactory hallucinations. Both detection and identification of odors have been reported to be abnormal in major depression (Suffin and Gitlin, 1986; Moberg, Pearlson, Speedie, Lipsey, and DePaulo, 1986), a condition frequently confused with dementia in the clinical setting. However, Amsterdam and associates found normal UPSIT performance in depressives (Amsterdam, Settle, Doty, Abelman, and Winokur, 1987). Recently normal odor detection was reported in a small group of melancholic subjects (Ackerman, Mattis, Sechzer, Russ, Burton, and Henriques, 1988).

We have administered olfactory tests to nine men with major depression (DSM-III R criteria) (Serby, Larson, and Kalkstein, 1990). These subjects had a mean Hamilton-D score of 19.9 ± 1.6 (S.E.M.). All subjects were free of antidepressant medications for at least ten days. Depressed subjects performed significiantly worse than 10 controls on olfactory measures (Figure 21.13) . On the UPSIT, depressed patients scored a mean of 29.8 ± 2.1 (S.E.M.), and a group of controls matched to the depressives' mean age (50–59) scored a mean of 37.0 ± 1.3 (S.E.M.); this difference is significant [$F(1,18) = 8.58$, $p < 0.01$).

Performance on the Yes/No identification task did not show a difference in the Pr means: 0.582 for depressed and 0.707 for controls. The difference between UPSIT and Yes/No performance in depressives may be a function of task-specific difficulty.

There was no significant difference in the olfactory threshold task between depressed patients and age-matched controls [$F(1,12) = 3.10$, $p < 0.10$).

Patients were tested a second time on antidepressant medication for all olfactory tasks. The results indicated no difference between trial 1 (off medication) and trial 2 (on medication) for all olfactory tasks despite a significant decline in the Hamilton-D scores [$F(1,15) = 7.26$, $p < 0.02$). The average Hamilton-D on medication was 12, a decrease of 8 points from baseline.

We must consider these data preliminary. There are more caveats than conclusions from these results. The fact that the only dysfunction occurred in the UPSIT might suggest a strongly cognitive basis. However, it is conceivable that a larger sample would yield significant data in all the tests employed. The lack of change in olfactory scores during affective improvement on drugs implies that the UPSIT deficit may be trait-dependent. This can only be addressed by longer follow-up testing.

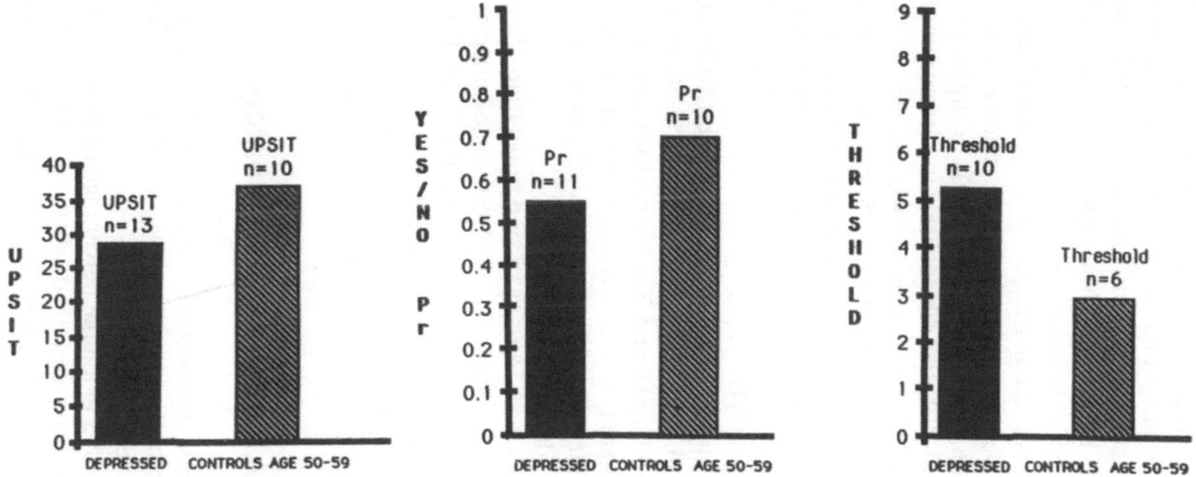

FIGURE 21.13. Olfactory performance in depressed patients and age-matched controls.

Olfaction and Schizophrenia

Although aspects of olfactory performance have been evaluated in schizophrenia (Connolly and Gittleson, 1971; Bradley, 1984; Moberg, Pearlson, Kim, and Tune, 1985), the results are preliminary in nature. The impact of neuroleptics is also of interest because of the established olfactory deficits in Parkinson's disease (Ansari and Johnson, 1975; Ward, Hess, and Calne, 1983; Quinn, Rossor, and Marsden, 1987), a disorder that can be mimicked by neuroleptic dopamine blockade.

Olfactory tests were administered to 14 schizophrenic (DSM-III R criteria) males who were free of antipsychotic medication for a period of 10 days (Serby, Larson, and Kalkstein, 1990). Compared to their age-matched controls (40–49), the schizophrenics' olfactory performance was abnormal on the UPSIT and threshold tasks (Figure 21.14). The UPSIT exhibited a significant impairment [$F(1,25) = 22.8$, $p \pm 0.0004$). The UPSIT mean for schizophrenics was 27.8, and that for age-matched controls was 39.0. Similarly, the olfactory threshold scores were significantly different between schizophrenics and controls, indicating a main effect of $F(1,23) = 15.47, p < 0.0007$.

The data did not suggest a difference in the Yes/No Pr task at $p < 0.05$. The Pr means were 0.676 for schizophrenics and 0.709 for controls. Again the difference in the UPSIT and Yes/No performances in schizophrenics, as in depressed patients, may be related to olfactory task difficulty. It is important to note that, in contrast to the depressives, the schizophrenics have higher thresholds than normals.

The schizophrenics were tested a second time (10 to 15 days later) while on an antipsychotic medication. The schizophrenics' olfactory performance did not change while on medication. Since the Brief Psychiatric Rating Scale (BPRS) did not indicate a change in psychiatric status while these patients were on medication, it is unknown whether clinical improvement would be coupled with enhanced olfactory function.

Olfaction and AIDS Spectrum Disorders

We recently reported impaired olfactory identification performance in patients infected with human immunodeficiency virus (HIV) (Brody, Serby, Etienne, and Kalkstein, 1989).

HIV infects the CNS in a large percentage of cases. Cerebrospinal fluid abnormalities have been described in 60 percent of asymptomatic HIV seropositive individuals (Marshall, Bry, Cahill, et al., 1988). It is unclear whether this early infection produces any change in brain function. AIDS Dementia Complex (ADC) occurs in approximately 40 percent of AIDS patients (Price et al., 1988).

As a measure of CNS function, we tested odor identification ability by administering the UPSIT to various groups of infected people: 10 asymp-

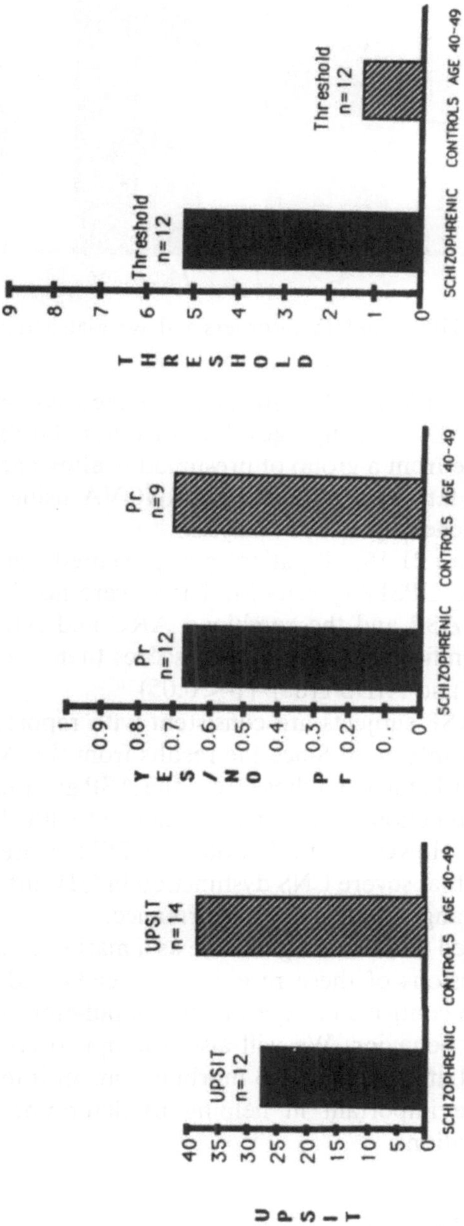

FIGURE 21.14. Olfactory performance in schizophrenics and age-matched controls.

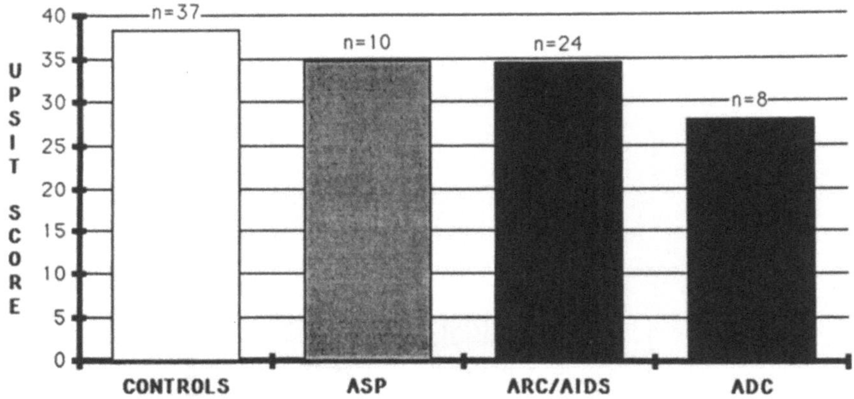

FIGURE 21.15. UPSIT in HIV disorders and age-matched controls.

tomatic seropositives (ASP), 24 with AIDS-related complex (ARC) or AIDS, and 8 with ADC. Group ages did not differ. UPSIT scores were compared with those from a group of presumed healthy age-matched controls ($n = 37$). The data were analyzed by ANOVA using arc-sine transformation and the Scheffe post hoc analysis.

As shown in Figure 21.15, all patient groups scored significantly lower than controls on the UPSIT ($p < 0.05$). There were no significant differences between the ASP and the combined ARC and AIDS group. The ADC group had significantly lower UPSIT scores than both the ASP and the combined ARC and AIDS group ($p < 0.05$)

The findings in ASP subjects are consistent with reports of early CNS involvement in HIV infection. Since the results from the ARC and AIDS group are indistinguishable from those from the ASP group, it appears that worsened immune function need not be associated with deterioration in odor identification. However, the fall-off in UPSIT scores in the ADC group may indicate that severe CNS dysfunction in HIV infection is associated with further change in olfactory performance.

Clinically, impaired olfaction might serve as a marker of early CNS HIV involvement. Extensions of these results are already underway. We are attempting to match control and experimental populations for socioeconomic status and risk behavior. We will also attempt to ascertain whether interventions aimed at reducing CNS morbidity might improve olfaction. Such work may be important in helping to determine the timing of appropriate interventions.

Olfaction and Neuropsychiatry

The olfactory dysfunction in AD has been appreciated for several years and is beginning to yield information about many facets of this dementia. The presence of alterations in odor detection and identification in such diverse entities as Down's syndrome, major depression, schizophrenia, and AIDS dementia complex might lead to the conclusion that such changes are nonspecific and not particularly enlightening. It is important to recognize the fact that olfactory testing can be used to explore a variety of areas in neuropsychiatric disorders. It is hoped that hyposmia will continue to shed light on the course of AD and on the nature and course of other CNS disturbances.

Acknowledgments. The authors gratefully acknowledge the cooperation and participation of the staff and patrons of the New York Veterans Administration Medical Center; the New York University Aging and Dementia Research Program; the Caring Community, Inc.; and the Council Center for Senior Citizens. We also thank Dr. James Schmeidler for his statistical suggestions and Dr. Patricia Warne for her manuscript review. This work was supported by the Veterans Administration.

References

Ackerman, S.H., S. Mattis, J.A. Sechzer, M. Russ, L. Burton, and J.B. Henriques (1986). Gustatory and olfactory changes associated with depression. *Neurosci. Abstr.*, *12*, 1359.

Amsterdam, J.D., R.G. Settle, R.L. Doty, E. Abelman, and A. Winokur (1987). Taste and smell perception in depression. *Biol. Psychiat.*, *22*, 1481–1485.

Ansari, K.A., and A. Johnson (1975). Olfactory function in patients with Parkinson's disease. *J. Chronic. Dis.*, *28*, 497.

Averbach, P. (1983). Two new lesions in Alzheimer's disease. *Lancet*, *2*, 1203.

Bartus, R.T., R.C. Dean, III, B. Beer, and A. Lippa (1982). The cholinergic hypothesis of geriatric memory dysfunction: A critical review. *Science*, *217*, 408–417.

Bradley, E.A. (1984). Olfactory acuity to a pheromonal substance and psychotic illness. *Biol. Psychiat.*, *19*, 899–905.

Brody, D., M. Serby, N. Etienne, and D. Kalkstein (1989). Olfactory function and HIV-related illness. International Conference on AIDS: Neurological and Neuropsychological Complications of HIV infection. Quebec City, May 31–June 3.

Buchsbaum, M.S., P. Kesslak, C. Cotman, G. Lynch, H. Chui, J. Wu, N. Sicotte, and F. Hazlett (1989). PET studies of hippocampal metabolism in patients with Alzheimer's disease during an olfactory memory task. *J. Neural Trans.* [P–D Sect.], *1*, 30.

Burger, P.C., and F.S. Vogel (1973). The development of the pathologic changes of Alzheimer's disease and senile dementia in patients with Down's syndrome. *Am. J. Pathol.*, *73*, 457–476.

Connolly, F.H., and N.L. Gittleson (1971). The relationship between delusions of sexual change and olfactory and gustatory hallucinations in schizophrenia. *Br. J. Psychiatry*, *119*, 443–444.

Corwin, J., M. Serby, and J. Rotrosen (1986). Olfactory deficits in AD: What we know about the nose. *Neurobiol. Aging*, *7*, 580–582.

Doty, R.L., P. Shaman, and M. Dann (1984a). Development of the University of

Pennsylvania Smell Identification Test. A standard micro-encapsulated test of olfactory function. *Physiol. Behav,. 32*, 489–502.

Doty, R.L., P. Shaman, S.L. Applebaum, R. Giberson, L. Siksorski, and L. Rosenberg (1984b). Smell identification ability: changes with age. *Science, 226*, 1441–1443.

Doty, R.L., P.F. Reyes, and T. Gregor (1987). Presence of both odor identification and detection deficits in Alzheimer's disease. *Brain Res. Bull., 18*, 597–600.

Eichenbaum, H., T.H. Morton, H. Potter, H. and S. Corkin (1983). Selective olfactory deficits in Case H.M. *Brain, 106*, 459, 472.

Engen, T. (1977). Taste and smell. In J.E. Birren and K.W. Schaie, (Eds.), *Handbook of the Psychology of Aging*. New York: Van Nostrand Reinhold, pp. 554–561.

Esiri, M.M., and G.K. Wilcock (1984). The olfactory bulbs in Alzheimer's disease. *J. Neurol. Neurosurg. Psychiatry, 47*, 56–60.

Eskenazi, B., W.S. Cain, R.A. Novelly, and R. Mattson (1986). Odor perception in temporal lobe epilepsy patients with and without temporal lobectomy. *Neuropsychologia 24*, 553–562.

Folstein, M.J., S.E. Folstein, and P.R. McHugh (1975). Mini-mental state: A practical method for grading the cognitive state of patients for the clinician. *J. Psychiat. Res., 12*, 189–198.

Gerulli, G., G. Mielke, and D. Mrowinski (1981). Contingent negative variation bei olfaktorischer Reizung. *Z. EEG-EMG., 12*, 125–127.

Green, D.L. M., and J.A. Swet (1974). *Signal Detection Theory and Psychophysics*. Huntington, NY: Krieger.

Halasz, N., and G.M. Shepherd (1983). Neurochemistry of the verterbrate olfactory bulb. *Neuroscience, 10*, 579–619.

Hamilton, M. (1967). Development of a rating scale for primary depressive illness. *Br. J Clin. Psychol., 6*, 278–296.

Hemdal, P., J. Corwin, H. Oster, and M. Serby (1989). Olfactory identification deficits in Down's Syndrome and Idiopathic Mental Retardation. Presented at the *International Neuropsychological Association* North American Annual Meeting, February.

Kimbrell, G., and E. Furchgott (1963). The effect of aging on olfactory threshold. *J. Gerontol., 18*, 364–365.

Knupfer, L., and R. Speigel (1986). Distinction between normal aging, Alzheimer-type and vascular dementia by means of olfactory test. *Int. J. Geriatr. Psychiatry, 1*, 3–14.

Kobal, G., and C. Hummel (1988). Cerebral chemosensory evoked potentials elicited by chemical stimulaton of the human olfactory and respiratory nasal mucosa. *Electroencephalography and Clinical Neurophys., 71*, 241–250.

Koss, E., J.M. Weiffenbach, J.V. Haxby, and R.P. Freidland (1987). Olfactory detection and recognition in Alzheimer's disease. *Lancet, 1*, 622.

Krassner, M.B. (1984). Mechanism of action of Hydergine (ergoloid mesylates) in relation to organic brain disorders. *Advances in Therapy, 1*(3), 172–189.

Marshall, D.W., Brey, R.L., Cahill. W.T., Houk, R.W., Zajack, R.A., and R.N. Boswell (1988). Spectrum of cerebrospinal fluid findings in various stages of human immunodeficiency virus infection. *Arch. Neurol., 45*, 954–958.

Moberg, P.J., G.D. Pearlson, W.S. Kim, and L.E. Tune (1985) Olfactory recall deficits and VBR in schizophrenia. Presented at American Psychiatric Associa-

tion. Annual Meeting, Dallas, May.
Moberg, P.J., G.D. Pearlson, L.J. Speedie, J.R. Lipsey, and J.R. DePaulo (1986). Olfactory recognition and mood in major depression. Presented at American Psychiatric Association Annual Meeting, Washington, DC, May.
Murphy, C. (1983). Age-related effects on the threshold, psychophysical function, and pleasantness of menthol. *J. Gerontol.*, *38*, 217–222.
Peabody, C.A., and J.R. Tinklenberg (1985). Olfactory deficits and primary degenerative dementia. *Am. J. Psychiatry*, *142*, 524–525.
Pollock, V.E., L.S. Schneider, H.C. Chui, V. Henderson, M. Zemansky, and R.B. Sloane (1989). Visual Evoked Potentials in dementia: Meta-analysis and empirical study of Alzheimer's disease patients. *Biol. Psychiat.*, *25*, 1003–1013.
Price, R.W., B. Brew, J. Sidtis, M. Rosenblum, A.C. Scheck, and P. Cleary (1988). The AIDS dementia complex. The brain in AIDS: Central Nervous System HIV-1 infection and AIDS. *Science*, *239*, 586–592.
Quinn, N.P., M.N. Rossor, and C.D. Marsden (1987). Olfactory threshold in Parkinson's disease. *J. Neurol. Neurosurg. Psychiatry.*, *50*, 88–89.
Rae-Grant, A., W. Blume, D. Lau, V.C. Hachinski, M. Fisman, and H. Merskey (1987). The EEG in Alzheimer-type dementia. *Arch. Neurol.*, *44*, 50–54.
Rausch, R., and E.A. Serafetinides (1975). Specific alteration of olfaction function in humans with temporal lobe lesions. *Nature*, *225*, 557–558.
Reisberg, B., S.H. Ferris, J.J. de Leon, and T. Crook (1982). The global deterioration scale for assessment of primary degenerative dementia. *Am. J. Psychiatry*, *139*, 1136–1139.
Reisberg, B., S.H. Ferris, E. Shulman, G. Steinberg, C. Buttinger, E. Sinaiko, J. Borenstein, M.J. deLeon, and J. Cohen (1986). Longitudinal course of normal aging and progressive dementia of the Alzheimer's type: A prospective study of 106 subjects over a 3.6 year mean interval. *Prog. Neuro-Psychopharmacol. Biol. Psychiat*, *10* (3–5), 571–578.
Rezek, D.L. (1987). Olfactory deficits as a neurologic sign in dementia of the Alzheimer type. *Arch. Neurol.*, *44*, 1030–1032.
Roberts, E. (1986). Alzheimer's disease may begin in the nose and may be caused by aluminosilicates. *Neurobiol. Aging*, *7*, 561–567.
Schiffman, S., J. Moss, and R.P. Erickson (1976). Thresholds of food odors in the elderly. *Exp. Aging Res.*, *2*, 389–398.
Selkoe. D.J. (1986). Altered structural proteins in plaques and tangles: What do they tell us about the biology of Alzheimer's disease? *Neurobiol. Aging*, *7*, 425–432.
Serby, M., J. Corwin, P. Conrad and J. Rotrosen (1985). Olfactory dysfunction in Alzheimer's disease and Parkinson's disease. *Am. J. Psychiatry*, *142*, 781–782.
Serby, M., C. Flicker, B. Rypma, S. Weber, J.P. Rotrosen, and S.H. Ferris (1990). Scopolamine and olfactory function. *Biol. Psychiat. 28*, 79–82.
Serby, M., P. Larson, and D. Kalkstein (1990). Olfactory sense in psychoses. *Biol. Psychiat.*, *28*, 830.
Serby, M., P. Larson, and D. Kalkstein (1991). The nature and course of olfactory deficits in Alzheimer's disease. *Am. J. Psychiatry. 148*, 357–360.
Simpson, J., C.M. Yates, A. Gordon, and D.M. St. Clair (1984). Olfactory tubercle choline acetyltransferase activity in Alzheimer-type dementia, Down's syndrome and Huntington's disease. *J. Neurol. Neurosurg. Psychiatry*, *47*, 1138–1139.

Skouby, A.P., and K. Zilstorff-Pedersen (1954). The influence of acetylcholine-like substances, menthol and strychnine on olfactory receptors in man. *Acta Physiologica Scand.*, *32*, 252–258.

Snodgrass, J.G and J. Corwin (1988). Pragmatics of measuring recognition memory: Applications to dementia and amnesia. *J. of Exp. Psychology: General*, *117*, 34–50

Suffin, S.C., and M. Giltin (1986). Olfaction in depression and recovery: A new marker. Presented at American Psychiatric Association. Annual Meeting, Washington, DC, May.

Talamo, B.R., R.A. Rudel, K.S. Kosik, V.M.-Y Lee, S. Neff, L. Adelman, and J.S. Kauer (1989). Pathological changes in olfactory neurons in patients with Alzheimer's disease. *Nature*, *337*, 736–739.

Ward, D.D., W.A. Hess, and D.B. Calne (1983). Olfactory impairment in Parkinson's disease. *Neurology (Cleveland)*, *33*, 943–946.

Warner, M.D., C.A. Peabody, J.J. Flatter, and J.R. Tinklenberg (1986). Olfactory deficits and Alzheimer's disease. *Biol. Psychiat*, *21*, 116–118.

Warner, M.D., C.A. Peabody, and P.A. Berger. Olfactory deficits and Down's syndrome. *Biol. Psychiat.* In press.

Index

MIX
Papier aus verantwortungsvollen Quellen
Paper from responsible sources
FSC® C105338

If you have any concerns about our products,
you can contact us on
ProductSafety@springernature.com

In case Publisher is established outside the EU,
the EU authorized representative is:
Springer Nature Customer Service Center GmbH
Europaplatz 3, 69115 Heidelberg, Germany

Printed by Libri Plureos GmbH
in Hamburg, Germany